Graduate Texts in Mathematics 228

More information about this series at http://www.springer.com/series/136

Graduate Texts in Mathematics

(continued after references)

Fred Diamond
Jerry Shurman

A First Course
in Modular Forms

Springer

Fred Diamond
Department of Mathematics
King's College London
Strand, London WC2R 2LS
United Kingdom
fred.diamond@kcl.ac.uk

Jerry Shurman
Department of Mathematics
Reed College
Portland, OR 97202
USA
jerry@reed.edu

Mathematics Subject Classification (2000): 25001, 11019

Library of Congress Cataloging-in-Publication Data
Diamond, Fred.
    A first course in modular forms / Fred Diamond and Jerry Shurman.
       p.   cm. — (Graduate texts in mathematics ; 228)
    Includes bibliographical references and index.
    ISBN 0-387-23229-X
    1. Forms, Modular.   I. Shurman, Jerry Michael.   II. Title.   III. Series.
    QA243.D47   2005
    512.7′3—dc22   2004058971

ISBN 0-387-23229-X   Printed on acid-free paper.
ISBN 978-0-387-23229-4 (hardcover)   ISBN 978-0-387-27226-9   (eBook)
ISBN 978-1-4419-2005-8 (softcover)
DOI 10.1007/978-0-387-27226-9

Springer is part of Springer Science+Business Media (www.springer.com)

For our parents

Contents

Preface

This book explains a result called the Modularity Theorem:

> *All rational elliptic curves arise from modular forms.*

Taniyama first suggested in the 1950's that a statement along these lines might be true, and a precise conjecture was formulated by Shimura. A paper of Weil [Wei67] provided strong theoretical evidence for the conjecture. The theorem was proved for a large class of elliptic curves by Wiles [Wil95] with a key ingredient supplied by joint work with Taylor [TW95], completing the proof of Fermat's Last Theorem after some 350 years. The Modularity Theorem was proved completely by Breuil, Conrad, Taylor, and the first author of this book [BCDT01]. Different forms of it are stated here in Chapters 2, 6, 7, 8, and 9.

To describe the theorem very simply for now, first consider a situation from elementary number theory. Take a quadratic equation

$$Q : x^2 = d, \qquad d \in \mathbb{Z}, \; d \neq 0,$$

and for each prime number p define an integer $a_p(Q)$,

$$a_p(Q) = \left(\begin{array}{c} \text{the number of solutions } x \text{ of equation } Q \\ \text{working modulo } p \end{array} \right) - 1.$$

The values $a_p(Q)$ extend multiplicatively to values $a_n(Q)$ for all positive integers n, meaning that $a_{mn}(Q) = a_m(Q)a_n(Q)$ for all m and n.

Since by definition $a_p(Q)$ is the Legendre symbol (d/p) for all $p > 2$, one statement of the Quadratic Reciprocity Theorem is that $a_p(Q)$ depends only on the value of p modulo $4|d|$. This can be reinterpreted as a statement that the sequence of solution-counts $\{a_2(Q), a_3(Q), a_5(Q), \dots\}$ arises as a system of eigenvalues on a finite-dimensional complex vector space associated to the equation Q. Let $N = 4|d|$, let $G = (\mathbb{Z}/N\mathbb{Z})^*$ be the multiplicative group of integer residue classes modulo N, and let V_N be the vector space of complex-valued functions on G,

$$V_N = \{f : G \longrightarrow \mathbb{C}\}.$$

For each prime p define a linear operator T_p on V_N,

$$T_p : V_N \longrightarrow V_N, \qquad (T_p f)(n) = \begin{cases} f(pn) & \text{if } p \nmid N, \\ 0 & \text{if } p \mid N, \end{cases}$$

where the product $pn \in G$ uses the reduction of p modulo N. Consider a particular function $f = f_Q$ in V_N,

$$f : G \longrightarrow \mathbb{C}, \qquad f(n) = a_n(Q) \text{ for } n \in G.$$

This is well defined by Quadratic Reciprocity as stated above. It is immediate that f is an eigenvector for the operators T_p,

$$(T_p f)(n) = \begin{cases} f(pn) = a_{pn}(Q) = a_p(Q)a_n(Q) & \text{if } p \nmid N, \\ 0 & \text{if } p \mid N \end{cases}$$

$$= a_p(Q)f(n) \quad \text{in all cases.}$$

That is,

$$T_p f = a_p(Q)f \quad \text{for all primes } p.$$

This shows that the sequence $\{a_p(Q)\}$ is a system of eigenvalues as claimed.

The Modularity Theorem can be viewed as giving an analogous result. Consider a cubic equation

$$E : y^2 = 4x^3 - g_2 x - g_3, \qquad g_2, g_3 \in \mathbb{Z}, \ g_2^3 - 27g_3^2 \neq 0.$$

Such equations define *elliptic curves*, objects central to this book. For each prime number p define a number $a_p(E)$ akin to $a_p(Q)$ from before,

$$a_p(E) = p - \left(\begin{array}{c} \text{the number of solutions } (x, y) \text{ of equation } E \\ \text{working modulo } p \end{array} \right).$$

One statement of Modularity is that again the sequence of solution-counts $\{a_p(E)\}$ arises as a system of eigenvalues. Understanding this requires some vocabulary.

A *modular form* is a function on the complex upper half plane that satisfies certain transformation conditions and holomorphy conditions. Let τ be a variable in the upper half plane. Then a modular form necessarily has a Fourier expansion,

$$f(\tau) = \sum_{n=0}^{\infty} a_n(f)e^{2\pi i n \tau}, \quad a_n(f) \in \mathbb{C} \text{ for all } n.$$

Each nonzero modular form has two associated integers k and N called its *weight* and its *level*. The modular forms of any given weight and level form a vector space. Linear operators called the *Hecke operators*, including an operator T_p for each prime p, act on these vector spaces. An *eigenform* is a modular form that is a simultaneous eigenvector for all the Hecke operators. By analogy to the situation from elementary number theory, the Modularity Theorem associates to the equation E an eigenform $f = f_E$ in a vector space V_N of weight 2 modular forms at a level N called the *conductor* of E. The eigenvalues of f are its Fourier coefficients,

$$T_p f = a_p(f)f \quad \text{for all primes } p,$$

and a version of Modularity is that *the Fourier coefficients give the solution-counts*,

$$a_p(f) = a_p(E) \quad \text{for all primes } p. \tag{0.1}$$

That is, the solution-counts of equation E are a system of eigenvalues, like the solution-counts of equation Q, but this time they arise from modular forms,

$$T_p f = a_p(E)f \quad \text{for all primes } p.$$

This version of the Modularity Theorem will be stated in Chapter 8.

Chapter 1 gives the basic definitions and some first examples of modular forms. It introduces elliptic curves in the context of the complex numbers, where they are defined as tori and then related to equations like E but with $g_2, g_3 \in \mathbb{C}$. And it introduces *modular curves*, quotients of the upper half plane that are in some sense more natural domains of modular forms than the upper half plane itself. Complex elliptic curves are compact Riemann surfaces, meaning they are indistinguishable in the small from the complex plane. Chapter 2 shows that modular curves can be made into compact Riemann surfaces as well. It ends with the book's first statement of the Modularity Theorem, relating elliptic curves and modular curves as Riemann surfaces: *If the complex number $j = 1728g_2^3/(g_2^3 - 27g_3^2)$ is rational then the elliptic curve is the holomorphic image of a modular curve.* This is notated

$$X_0(N) \longrightarrow E.$$

Much of what follows over the next six chapters is carried out with an eye to going from this complex analytic version of Modularity to the arithmetic version (0.1). Thus this book's aim is not to prove Modularity but to state its different versions, showing some of the relations among them and how they connect to different areas of mathematics.

Modular forms make up finite-dimensional vector spaces. To compute their dimensions Chapter 3 further studies modular curves as Riemann surfaces. Two complementary types of modular forms are *Eisenstein series* and *cusp forms*. Chapter 4 discusses Eisenstein series and computes their Fourier expansions. In the process it introduces ideas that will be used later in the book, especially the idea of an *L-function*,

$$L(s) = \sum_{n=1}^{\infty} \frac{a_n}{n^s}.$$

Here s is a complex variable restricted to some right half plane to make the series converge, and the coefficients a_n can arise from different contexts. For instance, they can be the Fourier coefficients $a_n(f)$ of a modular form. Chapter 5 shows that if f is a Hecke eigenform of weight 2 and level N then its L-function has an *Euler factorization*

$$L(s, f) = \prod_{p} (1 - a_p(f)p^{-s} + \mathbf{1}_N(p)p^{1-2s})^{-1}.$$

The product is taken over primes p, and $\mathbf{1}_N(p)$ is 1 when $p \nmid N$ (true for all but finitely many p) but is 0 when $p \mid N$.

Chapter 6 introduces the *Jacobian* of a modular curve, a complex torus like a complex elliptic curve but possibly of higher dimension. The Jacobian thus has Abelian group structure. Another version of the Modularity Theorem says that every complex elliptic curve with a rational j-value is the holomorphic homomorphic image of a Jacobian,

$$J_0(N) \longrightarrow E.$$

Modularity refines to say that the elliptic curve is in fact the image of a quotient of a Jacobian, the *Abelian variety* associated to a weight 2 eigenform,

$$A_f \longrightarrow E.$$

This version of Modularity associates a cusp form f to the elliptic curve E.

Chapter 7 brings algebraic geometry into the picture and moves toward number theory by shifting the environment from the complex numbers to the rational numbers. Every complex elliptic curve with rational j-invariant can be associated to the solution set of an equation E with $g_2, g_3 \in \mathbb{Q}$. Modular curves, Jacobians, and Abelian varieties are similarly associated to solution sets of systems of polynomial equations over $\mathbb{Q}$, algebraic objects in contrast to the earlier complex analytic ones. The formulations of Modularity already in play rephrase algebraically to statements about objects and maps defined by polynomials over $\mathbb{Q}$,

$$X_0(N)_{\mathrm{alg}} \longrightarrow E, \qquad J_0(N)_{\mathrm{alg}} \longrightarrow E, \qquad A_{f,\mathrm{alg}} \longrightarrow E.$$

We discuss only the first of these in detail since $X_0(N)_{\mathrm{alg}}$ is a curve while $J_0(N)_{\mathrm{alg}}$ and $A_{f,\mathrm{alg}}$ are higher-dimensional objects beyond the scope of this book. These algebraic versions of Modularity have applications to number theory, for example constructing rational points on elliptic curves using points called Heegner points on modular curves.

Chapter 8 develops the *Eichler–Shimura relation*, describing the Hecke operator T_p in characteristic p. This relation and the versions of Modularity already stated help to establish two more versions of the Modularity Theorem. One is the arithmetic version that $a_p(f) = a_p(E)$ for all p, as above. For the other, define the *Hasse–Weil L-function* of an elliptic curve E in terms of the solution-counts $a_p(E)$ and the conductor N of E,

$$L(s, E) = \prod_p (1 - a_p(E)p^{-s} + \mathbf{1}_N(p)p^{1-2s})^{-1}.$$

Comparing this to the Euler product form of $L(s, f)$ above gives a version of Modularity equivalent to the arithmetic one: *The L-function of the modular form is the L-function of the elliptic curve,*

$$L(s, f) = L(s, E).$$

As a function of the complex variable s, both L-functions are initially defined only on a right half plane, but Chapter 5 shows that $L(s, f)$ extends analytically to all of $\mathbb{C}$. By Modularity the same now holds for $L(s, E)$. This is important because we want to understand E as an Abelian group, and the conjecture of Birch and Swinnerton-Dyer is that the analytically continued $L(s, E)$ contains information about the group's structure.

Chapter 9 introduces ℓ-adic Galois representations, certain homomorphisms of Galois groups into matrix groups. The simplest nontrivial such representation arises from the Quadratic Reciprocity example at the beginning of this preface. Galois representations are also associated to elliptic curves and to modular forms, incorporating the ideas from Chapters 6 through 8 into a framework rich in additional algebraic structure. The corresponding version of the Modularity Theorem is: *Every Galois representation associated to an elliptic curve over $\mathbb{Q}$ arises from a Galois representation associated to a modular form,*

$$\rho_{f,\ell} \sim \rho_{E,\ell}.$$

This is the version of Modularity that was proved. The book ends by discussing the broader relation between Galois representations and modular forms.

Many good books on modular forms already exist, but they can be daunting for a beginner. Although some of the difficulty lies in the material itself, the authors believe that a more expansive narrative with exercises will help students into the subject. We also believe that algebraic aspects of modular forms, necessary to understand their role in number theory, can be made accessible to students without previous background in algebraic number theory and algebraic geometry. In the last four chapters we have tried to do so by introducing elements of these subjects as necessary but not letting them take over the text. We gratefully acknowledge our debt to the other books, especially to Shimura [Shi73].

The minimal prerequisites are undergraduate semester courses in linear algebra, modern algebra, real analysis, complex analysis, and elementary number theory. Topics such as holomorphic and meromorphic functions, congruences, Euler's totient function, the Chinese Remainder Theorem, basics of general point set topology, and the structure theorem for modules over a principal ideal domain are used freely from the beginning, and the Spectral Theorem of linear algebra is cited in Chapter 5. A few facts about representations and tensor products are also cited in Chapter 5, and Galois theory is used extensively in the later chapters. Chapter 3 quotes formulas from Riemann surface theory, and later in the book Chapters 6 through 9 cite steadily more results from Riemann surface theory, algebraic geometry, and algebraic number theory. Seeing these presented in context should help the reader absorb the new language necessary en route to the arithmetic and representation theoretic versions of Modularity.

We thank our colleagues Joe Buhler, David Cox, Paul Garrett, Cris Poor, Richard Taylor, and David Yuen, Reed College students Asher Auel, Rachel Epstein, Harold Gabel, Michael Lieberman, Peter McMahan, and John Saller, and Brandeis University student Makis Dousmanis for looking at drafts. We thank Joe Lipman and Shaul Zemel for their helpful comments on this book's early printings, and the many other readers who have since alerted us to errors. Comments and corrections should be sent to the second author at jerry@reed.edu. Errata lists for this book's successive printings can be found at http://www.reed.edu/~jerry/. However, the fourth printing contains added subject matter in chapter 5 (mostly Section 5.9) and revisions and updates to chapter 9, creating too many changes from the third printing to detail in a list. The first author received support from NSF Grant 0300434 while working on this book.

July 2004, May 2016

Fred Diamond
King's College London
London, UK

Jerry Shurman
Reed College
Portland, OR

1

Modular Forms, Elliptic Curves, and Modular Curves

This chapter introduces three central objects of the book.

Modular forms are functions on the complex upper half plane. A matrix group called the modular group acts on the upper half plane, and modular forms are the functions that transform in a nearly invariant way under the action and satisfy a holomorphy condition. Restricting the action to subgroups of the modular group called congruence subgroups gives rise to more modular forms.

A *complex elliptic curve* is a quotient of the complex plane by a lattice. As such it is an Abelian group, a compact Riemann surface, a torus, and—nonobviously—in bijective correspondence with the set of ordered pairs of complex numbers satisfying a cubic equation of the form E in the preface.

A *modular curve* is a quotient of the upper half plane by the action of a congruence subgroup. That is, two points are considered the same if the group takes one to the other.

These three kinds of object are closely related. Modular curves are mapped to by *moduli spaces*, equivalence classes of complex elliptic curves enhanced by associated torsion data. Thus the points of modular curves represent enhanced elliptic curves. Consequently, functions on the moduli spaces satisfying a homogeneity condition are essentially the same thing as modular forms.

Related reading: Gunning [Gun62], Koblitz [Kob93], Schoeneberg [Sch74], and Chapter 7 of Serre [Ser73] are standard first texts on this subject. For modern expositions of classical modular forms in action see [Cox84] (reprinted in [BBB00]) and [Cox97].

1.1 First definitions and examples

The *modular group* is the group of 2-by-2 matrices with integer entries and determinant 1,

$$\mathrm{SL}_2(\mathbb{Z}) = \left\{ \begin{bmatrix} a & b \\ c & d \end{bmatrix} : a, b, c, d \in \mathbb{Z}, ad - bc = 1 \right\}.$$

© Springer Science+Business Media New York 2005
F. Diamond, J. Shurman, *A First Course in Modular Forms*,
Graduate Texts in Mathematics 228, DOI 10.1007/978-0-387-27226-9_1

The modular group is generated by the two matrices

$$\begin{bmatrix} 1 & 1 \\ 0 & 1 \end{bmatrix} \quad \text{and} \quad \begin{bmatrix} 0 & -1 \\ 1 & 0 \end{bmatrix}$$

(Exercise 1.1.1). Each element of the modular group is also viewed as an automorphism (invertible self-map) of the Riemann sphere $\widehat{\mathbb{C}} = \mathbb{C} \cup \{\infty\}$, the fractional linear transformation

$$\begin{bmatrix} a & b \\ c & d \end{bmatrix} (\tau) = \frac{a\tau + b}{c\tau + d}, \quad \tau \in \widehat{\mathbb{C}}.$$

This is understood to mean that if $c \neq 0$ then $-d/c$ maps to ∞ and ∞ maps to a/c, and if $c = 0$ then ∞ maps to ∞. The identity matrix I and its negative $-I$ both give the identity transformation, and more generally each pair $\pm\gamma$ of matrices in $\mathrm{SL}_2(\mathbb{Z})$ gives a single transformation. The group of transformations defined by the modular group is generated by the maps described by the two matrix generators,

$$\tau \mapsto \tau + 1 \quad \text{and} \quad \tau \mapsto -1/\tau.$$

The *upper half plane* is

$$\mathcal{H} = \{\tau \in \mathbb{C} : \mathrm{Im}(\tau) > 0\}.$$

Readers with some background in Riemann surface theory—which is not necessary to read this book—may recognize $\mathcal{H}$ as one of the three simply connected Riemann surfaces, the other two being the plane $\mathbb{C}$ and the sphere $\widehat{\mathbb{C}}$. The formula

$$\mathrm{Im}(\gamma(\tau)) = \frac{\mathrm{Im}(\tau)}{|c\tau + d|^2}, \quad \gamma = \begin{bmatrix} a & b \\ c & d \end{bmatrix} \in \mathrm{SL}_2(\mathbb{Z})$$

(Exercise 1.1.2(a)) shows that if $\gamma \in \mathrm{SL}_2(\mathbb{Z})$ and $\tau \in \mathcal{H}$ then also $\gamma(\tau) \in \mathcal{H}$, i.e., the modular group maps the upper half plane back to itself. In fact the modular group acts on the upper half plane, meaning that $I(\tau) = \tau$ where I is the identity matrix (as was already noted) and $(\gamma\gamma')(\tau) = \gamma(\gamma'(\tau))$ for all $\gamma, \gamma' \in \mathrm{SL}_2(\mathbb{Z})$ and $\tau \in \mathcal{H}$. This last formula is easy to check (Exercise 1.1.2(b)).

Definition 1.1.1. *Let k be an integer. A meromorphic function $f : \mathcal{H} \longrightarrow \mathbb{C}$ is **weakly modular of weight k** if*

$$f(\gamma(\tau)) = (c\tau + d)^k f(\tau) \quad \text{for } \gamma = \begin{bmatrix} a & b \\ c & d \end{bmatrix} \in \mathrm{SL}_2(\mathbb{Z}) \text{ and } \tau \in \mathcal{H}.$$

Section 1.2 will show that if this transformation law holds when γ is each of the generators $\begin{bmatrix} 1 & 1 \\ 0 & 1 \end{bmatrix}$ and $\begin{bmatrix} 0 & -1 \\ 1 & 0 \end{bmatrix}$ then it holds for all $\gamma \in \mathrm{SL}_2(\mathbb{Z})$. In other words, f is weakly modular of weight k if

$$f(\tau + 1) = f(\tau) \quad \text{and} \quad f(-1/\tau) = \tau^k f(\tau).$$

Weak modularity of weight 0 is simply $\mathrm{SL}_2(\mathbb{Z})$-invariance, $f \circ \gamma = f$ for all $\gamma \in \mathrm{SL}_2(\mathbb{Z})$. Weak modularity of weight 2 is also natural: complex analysis relies on path integrals of differentials $f(\tau)d\tau$, and $\mathrm{SL}_2(\mathbb{Z})$-invariant path integration on the upper half plane requires such differentials to be invariant when τ is replaced by any $\gamma(\tau)$. But (Exercise 1.1.2(c))

$$d\gamma(\tau) = (c\tau + d)^{-2}d\tau,$$

and so the relation $f(\gamma(\tau))d(\gamma(\tau)) = f(\tau)d\tau$ is

$$f(\gamma(\tau)) = (c\tau + d)^2 f(\tau),$$

giving Definition 1.1.1 with weight $k = 2$. Weight 2 will play an especially important role later in this book since it is the weight of the modular form in the Modularity Theorem. The weight 2 case also leads inexorably to higher even weights—multiplying two weakly modular functions of weight 2 gives a weakly modular function of weight 4, and so on. Letting $\gamma = -I$ in Definition 1.1.1 gives $f = (-1)^k f$, showing that the only weakly modular function of any odd weight k is the zero function, but nonzero odd weight examples exist in more general contexts to be developed soon. Another motivating idea for weak modularity is that while it does not make a function f fully $\mathrm{SL}_2(\mathbb{Z})$-invariant, at least $f(\tau)$ and $f(\gamma(\tau))$ always have the same zeros and poles since the factor $c\tau + d$ on $\mathcal{H}$ has neither.

Modular forms are weakly modular functions that are also holomorphic on the upper half plane and holomorphic at ∞. To define this last notion, recall that $\mathrm{SL}_2(\mathbb{Z})$ contains the translation matrix

$$\begin{bmatrix} 1 & 1 \\ 0 & 1 \end{bmatrix} : \tau \mapsto \tau + 1,$$

for which the factor $c\tau + d$ is simply 1, so that $f(\tau + 1) = f(\tau)$ for every weakly modular function $f : \mathcal{H} \longrightarrow \mathbb{C}$. That is, weakly modular functions are $\mathbb{Z}$-periodic. Let $D = \{q \in \mathbb{C} : |q| < 1\}$ be the open complex unit disk, let $D' = D - \{0\}$, and recall from complex analysis that the $\mathbb{Z}$-periodic holomorphic map $\tau \mapsto e^{2\pi i\tau} = q$ takes $\mathcal{H}$ to D'. Thus, corresponding to f, the function $g : D' \longrightarrow \mathbb{C}$ where $g(q) = f(\log(q)/(2\pi i))$ is well defined even though the logarithm is determined only up to $2\pi i\mathbb{Z}$, and $f(\tau) = g(e^{2\pi i\tau})$. If f is holomorphic on the upper half plane then the composition g is holomorphic on the punctured disk since the logarithm can be defined holomorphically about each point, and so g has a Laurent expansion $g(q) = \sum_{n \in \mathbb{Z}} a_n q^n$ for $q \in D'$. The relation $|q| = e^{-2\pi \mathrm{Im}(\tau)}$ shows that $q \to 0$ as $\mathrm{Im}(\tau) \to \infty$. So, thinking of ∞ as lying far in the imaginary direction, define f to be *holomorphic at* ∞ if g extends holomorphically to the puncture point $q = 0$, i.e., the Laurent series sums over $n \in \mathbb{N}$. This means that f has a *Fourier expansion*

$$f(\tau) = \sum_{n=0}^{\infty} a_n(f) q^n, \quad q = e^{2\pi i \tau}.$$

Since $q \to 0$ if and only if $\mathrm{Im}(\tau) \to \infty$, showing that a weakly modular holomorphic function $f : \mathcal{H} \longrightarrow \mathbb{C}$ is holomorphic at ∞ doesn't require computing its Fourier expansion, only showing that $\lim_{\mathrm{Im}(\tau) \to \infty} f(\tau)$ exists or even just that $f(\tau)$ is bounded as $\mathrm{Im}(\tau) \to \infty$.

Definition 1.1.2. *Let k be an integer. A function $f : \mathcal{H} \longrightarrow \mathbb{C}$ is a **modular form of weight** k if*

(1) f is holomorphic on $\mathcal{H}$,
(2) f is weakly modular of weight k,
(3) f is holomorphic at ∞.

The set of modular forms of weight k is denoted $\mathcal{M}_k(\mathrm{SL}_2(\mathbb{Z}))$.

It is easy to check that $\mathcal{M}_k(\mathrm{SL}_2(\mathbb{Z}))$ forms a vector space over $\mathbb{C}$ (Exercise 1.1.3(a)). Holomorphy at ∞ will make the dimension of this space, and of more spaces of modular forms to be defined in the next section, finite. We will compute many dimension formulas in Chapter 3. When f is holomorphic at ∞ it is tempting to define $f(\infty) = g(0) = a_0$, but the next section will show that this doesn't work in a more general context.

The product of a modular form of weight k with a modular form of weight l is a modular form of weight $k + l$ (Exercise 1.1.3(b)). Thus the sum

$$\mathcal{M}(\mathrm{SL}_2(\mathbb{Z})) = \bigoplus_{k \in \mathbb{Z}} \mathcal{M}_k(\mathrm{SL}_2(\mathbb{Z}))$$

forms a ring, a so-called graded ring because of its structure as a sum.

The zero function on $\mathcal{H}$ is a modular form of every weight, and every constant function on $\mathcal{H}$ is a modular form of weight 0. For nontrivial examples of modular forms, let $k > 2$ be an even integer and define the *Eisenstein series of weight* k to be a 2-dimensional analog of the Riemann zeta function $\zeta(k) = \sum_{d=1}^{\infty} 1/d^k$,

$$G_k(\tau) = \sum_{(c,d)}' \frac{1}{(c\tau + d)^k}, \quad \tau \in \mathcal{H},$$

where the primed summation sign means to sum over nonzero integer pairs $(c, d) \in \mathbb{Z}^2 - \{(0,0)\}$. The sum is absolutely convergent and converges uniformly on compact subsets of $\mathcal{H}$ (Exercise 1.1.4(c)), so G_k is holomorphic on $\mathcal{H}$ and its terms may be rearranged. For any $\gamma = \begin{bmatrix} a & b \\ c & d \end{bmatrix} \in \mathrm{SL}_2(\mathbb{Z})$, compute that

$$G_k(\gamma(\tau)) = \sideset{}{'}\sum_{(c',d')} \frac{1}{\left(c'\left(\frac{a\tau+b}{c\tau+d}\right)+d'\right)^k}.$$

$$= (c\tau+d)^k \sideset{}{'}\sum_{(c',d')} \frac{1}{((c'a+d'c)\tau+(c'b+d'd))^k}.$$

But as (c',d') runs through $\mathbb{Z}^2 - \{(0,0)\}$, so does $(c'a + d'c, c'b + d'd) = (c',d')\left[\begin{smallmatrix} a & b \\ c & d \end{smallmatrix}\right]$ (Exercise 1.1.4(d)), and so the right side is $(c\tau + d)^k G_k(\tau)$, showing that G_k is weakly modular of weight k. Finally, G_k is bounded as $\mathrm{Im}(\tau) \to \infty$ (Exercise 1.1.4(e)), so it is a modular form.

To compute the Fourier series for G_k, continue to let $\tau \in \mathcal{H}$ and begin with the identities

$$\frac{1}{\tau} + \sum_{d=1}^{\infty}\left(\frac{1}{\tau-d}+\frac{1}{\tau+d}\right) = \pi \cot \pi\tau = \pi i - 2\pi i \sum_{m=0}^{\infty} q^m, \quad q = e^{2\pi i\tau} \quad (1.1)$$

(Exercise 1.1.5—the reader who is unhappy with this unmotivated invocation of unfamiliar expressions for a trigonometric function should be reassured that it is a standard rite of passage into modular forms; but also, Exercise 1.1.6 provides other proofs, perhaps more natural, of the following formula (1.2)). Differentiating (1.1) $k-1$ times with respect to τ gives for $\tau \in \mathcal{H}$ and $q = e^{2\pi i\tau}$,

$$\sum_{d \in \mathbb{Z}} \frac{1}{(\tau+d)^k} = \frac{(-2\pi i)^k}{(k-1)!} \sum_{m=1}^{\infty} m^{k-1}q^m, \quad k \geq 2. \quad (1.2)$$

For even $k > 2$,

$$\sideset{}{'}\sum_{(c,d)} \frac{1}{(c\tau+d)^k} = \sum_{d \neq 0} \frac{1}{d^k} + 2\sum_{c=1}^{\infty}\left(\sum_{d \in \mathbb{Z}} \frac{1}{(c\tau+d)^k}\right),$$

so again letting ζ denote the Riemann zeta function and using (1.2) gives

$$\sideset{}{'}\sum_{(c,d)} \frac{1}{(c\tau+d)^k} = 2\zeta(k) + 2\frac{(2\pi i)^k}{(k-1)!} \sum_{c=1}^{\infty} \sum_{m=1}^{\infty} m^{k-1}q^{cm}.$$

Rearranging the last expression gives the Fourier expansion

$$G_k(\tau) = 2\zeta(k) + 2\frac{(2\pi i)^k}{(k-1)!} \sum_{n=1}^{\infty} \sigma_{k-1}(n)q^n, \quad k > 2, \ k \text{ even}$$

where the coefficient $\sigma_{k-1}(n)$ is the arithmetic function

$$\sigma_{k-1}(n) = \sum_{\substack{m|n \\ m>0}} m^{k-1}.$$

Exercise 1.1.7(b) shows that dividing by the leading coefficient gives a series having rational coefficients with a common denominator. This *normalized Eisenstein series* $G_k(\tau)/(2\zeta(k))$ is denoted $E_k(\tau)$. The Riemann zeta function will be discussed further in Chapter 4.

Since the set of modular forms is a graded ring, we can make modular forms out of various sums of products of the Eisenstein series. For example, $\mathcal{M}_8(\mathrm{SL}_2(\mathbb{Z}))$ turns out to be 1-dimensional. The functions $E_4(\tau)^2$ and $E_8(\tau)$ both belong to this space, making them equal up to a scalar multiple and therefore equal since both have leading term 1. Expanding out the relation $E_4^2 = E_8$ gives a relation between the divisor-sum functions σ_3 and σ_7 (Exercise 1.1.7(c)),

$$\sigma_7(n) = \sigma_3(n) + 120 \sum_{i=1}^{n-1} \sigma_3(i)\sigma_3(n-i), \quad n \geq 1. \tag{1.3}$$

The modular forms that, unlike Eisenstein series, have constant term equal to 0 play an important role in the subject.

Definition 1.1.3. *A **cusp form** of weight k is a modular form of weight k whose Fourier expansion has leading coefficient $a_0 = 0$, i.e.,*

$$f(\tau) = \sum_{n=1}^{\infty} a_n q^n, \quad q = e^{2\pi i \tau}.$$

The set of cusp forms is denoted $\mathcal{S}_k(\mathrm{SL}_2(\mathbb{Z}))$.

So a modular form is a cusp form when $\lim_{\mathrm{Im}(\tau)\to\infty} f(\tau) = 0$. The limit point ∞ of $\mathcal{H}$ is called the *cusp of* $\mathrm{SL}_2(\mathbb{Z})$ for geometric reasons to be explained in Chapter 2, and a cusp form can be viewed as vanishing at the cusp. The cusp forms $\mathcal{S}_k(\mathrm{SL}_2(\mathbb{Z}))$ form a vector subspace of the modular forms $\mathcal{M}_k(\mathrm{SL}_2(\mathbb{Z}))$, and the graded Abelian group

$$\mathcal{S}(\mathrm{SL}_2(\mathbb{Z})) = \bigoplus_{k \in \mathbb{Z}} \mathcal{S}_k(\mathrm{SL}_2(\mathbb{Z}))$$

is an ideal in $\mathcal{M}(\mathrm{SL}_2(\mathbb{Z}))$ (Exercise 1.1.3(c)).

For an example of a cusp form, let

$$g_2(\tau) = 60 G_4(\tau), \qquad g_3(\tau) = 140 G_6(\tau),$$

and define the *discriminant function*

$$\Delta : \mathcal{H} \longrightarrow \mathbb{C}, \quad \Delta(\tau) = (g_2(\tau))^3 - 27(g_3(\tau))^2.$$

Then Δ is weakly modular of weight 12 and holomorphic on $\mathcal{H}$, and $a_0 = 0$, $a_1 = (2\pi)^{12}$ in the Fourier expansion of Δ (Exercise 1.1.7(d)). So indeed

$\Delta \in \mathcal{S}_{12}(\mathrm{SL}_2(\mathbb{Z}))$, and Δ is not the zero function. Section 1.4 will show that in fact $\Delta(\tau) \neq 0$ for all $\tau \in \mathcal{H}$ so that the only zero of Δ is at ∞.

It follows that the *modular function*

$$j : \mathcal{H} \longrightarrow \mathbb{C}, \qquad j(\tau) = 1728 \frac{(g_2(\tau))^3}{\Delta(\tau)}$$

is holomorphic on $\mathcal{H}$. Since the numerator and denominator of j have the same weight, j is $\mathrm{SL}_2(\mathbb{Z})$-invariant,

$$j(\gamma(\tau)) = j(\tau), \qquad \gamma \in \mathrm{SL}_2(\mathbb{Z}), \ \tau \in \mathcal{H},$$

and in fact it is also called *the modular invariant*. The expansion

$$j(\tau) = \frac{(2\pi)^{12} + \cdots}{(2\pi)^{12} q + \cdots} = \frac{1}{q} + \cdots$$

shows that j has a simple pole at ∞ (and is normalized to have residue 1 at the pole), so it is not quite a modular form. Let μ_3 denote the complex cube root of unity $e^{2\pi i/3}$. Easy calculations (Exercise 1.1.8) show that $g_3(i) = 0$ so that $g_2(i) \neq 0$ and $j(i) = 1728$, and $g_2(\mu_3) = 0$ so that $g_3(\mu_3) \neq 0$ and $j(\mu_3) = 0$. One can further show (see [Ros81], [CS05]) that

$$g_2(i) = 4\varpi_4^4, \qquad \varpi_4 = 2\int_0^1 \frac{dt}{\sqrt{1 - t^4}} = 2\sqrt{\pi} \frac{\Gamma(5/4)}{\Gamma(3/4)}$$

and

$$g_3(\mu_3) = (27/16)\varpi_3^6, \qquad \varpi_3 = 2\int_0^1 \frac{dt}{\sqrt{1 - t^3}} = 2\sqrt{\pi} \frac{\Gamma(4/3)}{\Gamma(5/6)}.$$

Here the integrals are *elliptic integrals*, and Γ is Euler's *gamma function*, to be defined in Chapter 4. Finally, Exercise 1.1.9 shows that the j-function surjects from $\mathcal{H}$ to $\mathbb{C}$.

Exercises

1.1.1. Let Γ be the subgroup of $\mathrm{SL}_2(\mathbb{Z})$ generated by the two matrices $\begin{bmatrix} 1 & 1 \\ 0 & 1 \end{bmatrix}$ and $\begin{bmatrix} 0 & -1 \\ 1 & 0 \end{bmatrix}$. Note that $\begin{bmatrix} 1 & n \\ 0 & 1 \end{bmatrix} = \begin{bmatrix} 1 & 1 \\ 0 & 1 \end{bmatrix}^n \in \Gamma$ for all $n \in \mathbb{Z}$. Let $\alpha = \begin{bmatrix} a & b \\ c & d \end{bmatrix}$ be a matrix in $\mathrm{SL}_2(\mathbb{Z})$. Use the identity

$$\begin{bmatrix} a & b \\ c & d \end{bmatrix} \begin{bmatrix} 1 & n \\ 0 & 1 \end{bmatrix} = \begin{bmatrix} a & b' \\ c & nc + d \end{bmatrix}$$

to show that unless $c = 0$, some matrix $\alpha\gamma$ with $\gamma \in \Gamma$ has bottom row (c, d') with $|d'| \leq |c|/2$. Use the identity

$$\begin{bmatrix} a & b \\ c & d \end{bmatrix} \begin{bmatrix} 0 & -1 \\ 1 & 0 \end{bmatrix} = \begin{bmatrix} b & -a \\ d & -c \end{bmatrix}$$

to show that this process can be iterated until some matrix $\alpha\gamma$ with $\gamma \in \Gamma$ has bottom row $(0, *)$. Show that in fact the bottom row is $(0, \pm 1)$, and since $\begin{bmatrix} 0 & -1 \\ 1 & 0 \end{bmatrix}^2 = -I$ it can be taken to be $(0, 1)$. Show that therefore $\alpha\gamma \in \Gamma$ and so $\alpha \in \Gamma$. Thus Γ is all of $\mathrm{SL}_2(\mathbb{Z})$.

1.1.2. (a) Show that $\mathrm{Im}(\gamma(\tau)) = \mathrm{Im}(\tau)/|c\tau + d|^2$ for all $\gamma = \begin{bmatrix} a & b \\ c & d \end{bmatrix} \in \mathrm{SL}_2(\mathbb{Z})$.
  (b) Show that $(\gamma\gamma')(\tau) = \gamma(\gamma'(\tau))$ for all $\gamma, \gamma' \in \mathrm{SL}_2(\mathbb{Z})$ and $\tau \in \mathcal{H}$.
  (c) Show that $d\gamma(\tau)/d\tau = 1/(c\tau + d)^2$ for $\gamma = \begin{bmatrix} a & b \\ c & d \end{bmatrix} \in \mathrm{SL}_2(\mathbb{Z})$.

1.1.3. (a) Show that the set $\mathcal{M}_k(\mathrm{SL}_2(\mathbb{Z}))$ of modular forms of weight k forms a vector space over $\mathbb{C}$.
  (b) If f is a modular form of weight k and g is a modular form of weight l, show that fg is a modular form of weight $k + l$.
  (c) Show that $\mathcal{S}_k(\mathrm{SL}_2(\mathbb{Z}))$ is a vector subspace of $\mathcal{M}_k(\mathrm{SL}_2(\mathbb{Z}))$ and that $\mathcal{S}(\mathrm{SL}_2(\mathbb{Z}))$ is an ideal in $\mathcal{M}(\mathrm{SL}_2(\mathbb{Z}))$.

1.1.4. Let $k \geq 3$ be an integer and let $L' = \mathbb{Z}^2 - \{(0,0)\}$.
  (a) Show that the series $\sum_{(c,d)\in L'}(\sup\{|c|, |d|\})^{-k}$ converges by considering the partial sums over expanding squares.
  (b) Fix positive numbers A and B and let

$$\Omega = \{\tau \in \mathcal{H} : |\mathrm{Re}(\tau)| \leq A, \mathrm{Im}(\tau) \geq B\}.$$

Prove that there is a constant $C > 0$ such that $|\tau + \delta| > C\sup\{1, |\delta|\}$ for all $\tau \in \Omega$ and $\delta \in \mathbb{R}$. (Hints for this exercise are at the end of the book.)
  (c) Use parts (a) and (b) to prove that the series defining $G_k(\tau)$ converges absolutely and uniformly for $\tau \in \Omega$. Conclude that G_k is holomorphic on $\mathcal{H}$.
  (d) Show that for $\gamma \in \mathrm{SL}_2(\mathbb{Z})$, right multiplication by γ defines a bijection from L' to L'.
  (e) Use the calculation from (c) to show that G_k is bounded on Ω. From the text and part (d), G_k is weakly modular so in particular $G_k(\tau + 1) = G_k(\tau)$. Show that therefore $G_k(\tau)$ is bounded as $\mathrm{Im}(\tau) \to \infty$.

1.1.5. Establish the two formulas for $\pi \cot \pi\tau$ in (1.1). (A hint for this exercise is at the end of the book.)

1.1.6. This exercise obtains formula (1.2) without using the cotangent. Let $f(\tau) = \sum_{d\in\mathbb{Z}} 1/(\tau + d)^k$ for $k \geq 2$ and $\tau \in \mathcal{H}$. Since f is holomorphic (by the method of Exercise 1.1.4) and $\mathbb{Z}$-periodic and since $\lim_{\mathrm{Im}(\tau)\to\infty} f(\tau) = 0$, there is a Fourier expansion $f(\tau) = \sum_{m=1}^{\infty} a_m q^m = g(q)$ as in the section, where $q = e^{2\pi i\tau}$ and

$$a_m = \frac{1}{2\pi i} \int_\gamma \frac{g(q)}{q^{m+1}} dq$$

is a path integral once counterclockwise over a circle about 0 in the punctured disk D'.
  (a) Show that

$$a_m = \int_{\tau=0+iy}^{1+iy} f(\tau)e^{-2\pi im\tau}\,d\tau = \int_{\tau=-\infty+iy}^{+\infty+iy} \tau^{-k}e^{-2\pi im\tau}\,d\tau \quad \text{for any } y > 0.$$

(b) Let $g_m(\tau) = \tau^{-k}e^{-2\pi im\tau}$, a meromorphic function on $\mathbb{C}$ with its only singularity at the origin. Show that

$$-2\pi i\,\mathrm{Res}_{\tau=0}g_m(\tau) = \frac{(-2\pi i)^k}{(k-1)!}m^{k-1}.$$

(c) Establish (1.2) by integrating $g_m(\tau)$ clockwise about a large rectangular path and applying the Residue Theorem. Argue that the integral along the top side goes to a_m and the integrals along the other three sides go to 0.

(d) Let $h : \mathbb{R} \longrightarrow \mathbb{C}$ be a function such that the integral $\int_{-\infty}^{\infty} |h(x)|\,dx$ is finite and the sum $\sum_{d\in\mathbb{Z}} h(x+d)$ converges absolutely and uniformly on compact subsets and is infinitely differentiable. Then the *Poisson summation formula* says that

$$\sum_{d\in\mathbb{Z}} h(x+d) = \sum_{m\in\mathbb{Z}} \hat{h}(m)e^{2\pi imx}$$

where $\hat{h}$ is the *Fourier transform* of h,

$$\hat{h}(x) = \int_{t=-\infty}^{\infty} h(t)e^{-2\pi ixt}\,dt.$$

We will not prove this, but the idea is that the left side sum symmetrizes h to a function of period 1 and the right side sum is the Fourier series of the left side since the mth Fourier coefficient is $\int_{t=0}^{1} \sum_{d\in\mathbb{Z}} h(t+d)e^{-2\pi imt}\,dt = \hat{h}(m)$. Letting $h(x) = 1/\tau^k$ where $\tau = x+iy$ with $y > 0$, show that h meets the conditions for Poisson summation. Show that $\hat{h}(m) = e^{-2\pi my}a_m$ with a_m from above for $m > 0$, and that $\hat{h}(m) = 0$ for $m \leq 0$. Establish formula (1.2) again, this time as a special case of Poisson summation. We will see more Poisson summation and Fourier analysis in connection with Eisenstein series in Chapter 4. (A hint for this exercise is at the end of the book.)

1.1.7. The *Bernoulli numbers* B_k are defined by the formal power series expansion

$$\frac{t}{e^t - 1} = \sum_{k=0}^{\infty} B_k \frac{t^k}{k!}.$$

Thus they are calculable in succession by matching coefficients in the power series identity

$$t = (e^t - 1)\sum_{k=0}^{\infty} B_k \frac{t^k}{k!} = \sum_{n=1}^{\infty}\left(\sum_{k=0}^{n-1}\binom{n}{k}B_k\right)\frac{t^n}{n!}$$

(i.e., the nth parenthesized sum is 1 if $n = 1$ and 0 otherwise) and they are rational. Since the expression

$$\frac{t}{e^t - 1} + \frac{t}{2} = \frac{t}{2} \cdot \frac{e^t + 1}{e^t - 1}$$

is even, it follows that $B_1 = -1/2$ and $B_k = 0$ for all other odd k. The Bernoulli numbers will be motivated, discussed, and generalized in Chapter 4.

(a) Show that $B_2 = 1/6$, $B_4 = -1/30$, and $B_6 = 1/42$.

(b) Use the expressions for $\pi \cot \pi\tau$ from the section to show

$$1 - 2\sum_{k=1}^{\infty} \zeta(2k)\tau^{2k} = \pi\tau \cot \pi\tau = \pi i\tau + \sum_{k=0}^{\infty} B_k \frac{(2\pi i\tau)^k}{k!}.$$

Use these to show that for $k \geq 2$ even, the Riemann zeta function satisfies

$$2\zeta(k) = -\frac{(2\pi i)^k}{k!} B_k,$$

so in particular $\zeta(2) = \pi^2/6$, $\zeta(4) = \pi^4/90$, and $\zeta(6) = \pi^6/945$. Also, this shows that the normalized Eisenstein series of weight k

$$E_k(\tau) = \frac{G_k(\tau)}{2\zeta(k)} = 1 - \frac{2k}{B_k} \sum_{n=1}^{\infty} \sigma_{k-1}(n)q^n$$

has rational coefficients with a common denominator.

(c) Equate coefficients in the relation $E_8(\tau) = E_4(\tau)^2$ to establish formula (1.3).

(d) Show that $a_0 = 0$ and $a_1 = (2\pi)^{12}$ in the Fourier expansion of the discriminant function Δ from the text.

1.1.8. Recall that μ_3 denotes the complex cube root of unity $e^{2\pi i/3}$. Show that $\left[\begin{smallmatrix} 1 & -1 \\ 1 & 0 \end{smallmatrix}\right](\mu_3) = \mu_3 + 1$ so that by periodicity $g_2(\left[\begin{smallmatrix} 0 & -1 \\ 1 & 0 \end{smallmatrix}\right](\mu_3)) = g_2(\mu_3)$. Show that by modularity also $g_2(\left[\begin{smallmatrix} 0 & -1 \\ 1 & 0 \end{smallmatrix}\right](\mu_3)) = \mu_3^4 g_2(\mu_3)$ and therefore $g_2(\mu_3) = 0$. Conclude that $g_3(\mu_3) \neq 0$ and $j(\mu_3) = 0$. Argue similarly to show that $g_3(i) = 0$, $g_2(i) \neq 0$, and $j(i) = 1728$.

1.1.9. This exercise shows that the modular invariant $j : \mathcal{H} \longrightarrow \mathbb{C}$ is a surjection. Suppose that $c \in \mathbb{C}$ and $j(\tau) \neq c$ for all $\tau \in \mathcal{H}$. Consider the integral

$$\frac{1}{2\pi i} \int_\gamma \frac{j'(\tau)d\tau}{j(\tau) - c}$$

where γ is the contour shown in Figure 1.1 containing an arc of the unit circle from $(-1 + i\sqrt{3})/2$ to $(1 + i\sqrt{3})/2$, two vertical segments up to any height greater than 1, and a horizontal segment. By the Argument Principle the integral is 0. Use the fact that j is invariant under $\left[\begin{smallmatrix} 1 & 1 \\ 0 & 1 \end{smallmatrix}\right]$ to show that the integrals over the two vertical segments cancel. Use the fact that j is invariant under $\left[\begin{smallmatrix} 0 & -1 \\ 1 & 0 \end{smallmatrix}\right]$ to show that the integrals over the two halves of the circular arc cancel. For the integral over the remaining piece of γ make the change of coordinates $q = e^{2\pi i\tau}$, remembering that $j'(\tau)$ denotes derivative with respect to τ and that $j(\tau) = 1/q + \cdots$, and compute that it equals 1. This contradiction shows that $j(\tau) = c$ for some $\tau \in \mathcal{H}$ and j surjects.

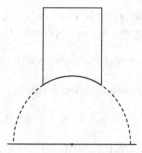

Figure 1.1. A contour

1.2 Congruence subgroups

Section 1.1 stated that if a meromorphic function $f : \mathcal{H} \longrightarrow \mathbb{C}$ satisfies

$$f(\gamma(\tau)) = (c\tau + d)^k f(\tau) \quad \text{for } \gamma = \begin{bmatrix} 1 & 1 \\ 0 & 1 \end{bmatrix} \text{ and } \gamma = \begin{bmatrix} 0 & -1 \\ 1 & 0 \end{bmatrix}$$

then f is weakly modular, i.e,

$$f(\gamma(\tau)) = (c\tau + d)^k f(\tau) \quad \text{for all } \gamma = \begin{bmatrix} a & b \\ c & d \end{bmatrix} \in \mathrm{SL}_2(\mathbb{Z}).$$

Replacing the modular group $\mathrm{SL}_2(\mathbb{Z})$ in this last condition by a subgroup Γ generalizes the notion of weak modularity, allowing more examples of weakly modular functions.

For example, a subgroup arises from the *four squares problem* in number theory, to find the number of ways (if any) that a given nonnegative integer n can be expressed as the sum of four integer squares. To address this, define more generally for nonnegative integers n and k the *representation number of n by k squares*,

$$r(n, k) = \#\{v \in \mathbb{Z}^k : n = v_1^2 + \cdots + v_k^2\}.$$

Note that if $i + j = k$ then $r(n, k) = \sum_{l+m=n} r(l, i) r(m, j)$, summing over nonnegative values of l and m that add to n (Exercise 1.2.1). This looks like the rule $c_n = \sum_{l+m=n} a_l b_m$ relating the coefficients in the formal product of two power series,

$$\left(\sum_{l=0}^{\infty} a_l q^l \right) \left(\sum_{m=0}^{\infty} b_m q^m \right) = \sum_{n=0}^{\infty} c_n q^n.$$

So consider the *generating function* of the representation numbers, meaning the power series with nth coefficient $r(n, k)$,

$$\theta(\tau, k) = \sum_{n=0}^{\infty} r(n, k) q^n, \quad q = e^{2\pi i \tau}, \ \tau \in \mathcal{H}.$$

Chapter 4 will show that the series is absolutely convergent, and so the formal product relation between power series applies to the θ series,

$$\theta(\tau, k_1)\theta(\tau, k_2) = \theta(\tau, k_1 + k_2), \quad \tau \in \mathcal{H}.$$

Clearly θ satisfies the transformation formula

$$\theta(\tau + 1, k) = \theta(\tau, k).$$

To obtain another such law for θ, write $\theta(\tau)$ for $\theta(\tau, 1)$ and note that the definition of the representation numbers combines with absolute convergence to give the rearrangement

$$\theta(\tau) = \sum_{d \in \mathbb{Z}} e^{2\pi i d^2 \tau}.$$

This looks like the left side of the Poisson summation formula from Exercise 1.1.6(d) with $x = 0$ and $h(d) = e^{2\pi i d^2 \tau}$. Chapter 4 will carry out the summation, obtaining the transformation law

$$\theta(-1/(4\tau)) = \sqrt{-2i\tau}\,\theta(\tau).$$

(This formula calls for the principal branch of the square root, making sense since $-2i\tau$ lies in the right half plane.) The matrix $\begin{bmatrix} 0 & -1 \\ 4 & 0 \end{bmatrix}$ taking τ to $-1/(4\tau)$ does not have determinant 1, but the product $\begin{bmatrix} 0 & 1/4 \\ -1 & 0 \end{bmatrix} \begin{bmatrix} 1 & -1 \\ 0 & 1 \end{bmatrix} \begin{bmatrix} 0 & -1 \\ 4 & 0 \end{bmatrix} = \begin{bmatrix} 1 & 0 \\ 4 & 1 \end{bmatrix}$, taking τ to $\tau/(4\tau + 1)$, does. Applying the corresponding succession of transformations and the transformation laws for θ gives

$$\theta\left(\frac{\tau}{4\tau + 1}\right) = \theta\left(-\frac{1}{4(-1/(4\tau) - 1)}\right) = \sqrt{2i\left(\frac{1}{4\tau} + 1\right)}\,\theta\left(-\frac{1}{4\tau} - 1\right)$$

$$= \sqrt{2i\left(\frac{1}{4\tau} + 1\right)}\,\theta\left(-\frac{1}{4\tau}\right) = \sqrt{2i\left(\frac{1}{4\tau} + 1\right)(-2i\tau)}\,\theta(\tau)$$

$$= \sqrt{4\tau + 1}\,\theta(\tau).$$

(Note here that $\sqrt{-i\tau_1}\sqrt{-i\tau_2} = \sqrt{(-i\tau_1)(-i\tau_2)}$ for $\tau_1, \tau_2 \in \mathcal{H}$ and the principal square root.) Now the product relation $\theta(\tau, 4) = \theta(\tau)^4$ gives a second transformation formula for the four squares problem,

$$\theta\left(\frac{\tau}{4\tau + 1}, 4\right) = (4\tau + 1)^2 \theta(\tau, 4).$$

That is, analogously to the first formula of this section,

$$\theta(\gamma(\tau), 4) = (c\tau + d)^2 \theta(\tau, 4) \quad \text{for } \gamma = \pm \begin{bmatrix} 1 & 1 \\ 0 & 1 \end{bmatrix} \text{ and } \gamma = \pm \begin{bmatrix} 1 & 0 \\ 4 & 1 \end{bmatrix}.$$

The subgroup of $\mathrm{SL}_2(\mathbb{Z})$ associated to the four squares problem is the group Γ_θ generated by $\pm \left[\begin{smallmatrix} 1 & 1 \\ 0 & 1 \end{smallmatrix}\right]$ and $\pm \left[\begin{smallmatrix} 1 & 0 \\ 4 & 1 \end{smallmatrix}\right]$. We will soon see that this is a special instance of the subgroups about to be introduced more generally. The four squares problem will be solved at the end of the section once more ideas are in place.

Let N be a positive integer. The *principal congruence subgroup of level N* is

$$\Gamma(N) = \left\{ \begin{bmatrix} a & b \\ c & d \end{bmatrix} \in \mathrm{SL}_2(\mathbb{Z}) : \begin{bmatrix} a & b \\ c & d \end{bmatrix} \equiv \begin{bmatrix} 1 & 0 \\ 0 & 1 \end{bmatrix} \ (\mathrm{mod} \ N) \right\}.$$

(The matrix congruence is interpreted entrywise, i.e., $a \equiv 1 \ (\mathrm{mod} \ N)$ and so on.) In particular $\Gamma(1) = \mathrm{SL}_2(\mathbb{Z})$. Being the kernel of the natural homomorphism $\mathrm{SL}_2(\mathbb{Z}) \longrightarrow \mathrm{SL}_2(\mathbb{Z}/N\mathbb{Z})$, the subgroup $\Gamma(N)$ is normal in $\mathrm{SL}_2(\mathbb{Z})$. In fact the map is a surjection (Exercise 1.2.2(b)), inducing an isomorphism

$$\mathrm{SL}_2(\mathbb{Z})/\Gamma(N) \overset{\sim}{\longrightarrow} \mathrm{SL}_2(\mathbb{Z}/N\mathbb{Z}).$$

This shows that $[\mathrm{SL}_2(\mathbb{Z}) : \Gamma(N)]$ is finite for all N. Specifically, the index is

$$[\mathrm{SL}_2(\mathbb{Z}) : \Gamma(N)] = N^3 \prod_{p|N} \left(1 - \frac{1}{p^2} \right),$$

where the product is taken over all prime divisors of N (Exercise 1.2.3(b)).

Definition 1.2.1. *A subgroup Γ of $\mathrm{SL}_2(\mathbb{Z})$ is a **congruence subgroup** if $\Gamma(N) \subset \Gamma$ for some $N \in \mathbb{Z}^+$, in which case Γ is a congruence subgroup of level N.*

Thus every congruence subgroup Γ has finite index in $\mathrm{SL}_2(\mathbb{Z})$. Besides the principal congruence subgroups, the most important congruence subgroups are

$$\Gamma_0(N) = \left\{ \begin{bmatrix} a & b \\ c & d \end{bmatrix} \in \mathrm{SL}_2(\mathbb{Z}) : \begin{bmatrix} a & b \\ c & d \end{bmatrix} \equiv \begin{bmatrix} * & * \\ 0 & * \end{bmatrix} \ (\mathrm{mod} \ N) \right\}$$

(where "$*$" means "unspecified") and

$$\Gamma_1(N) = \left\{ \begin{bmatrix} a & b \\ c & d \end{bmatrix} \in \mathrm{SL}_2(\mathbb{Z}) : \begin{bmatrix} a & b \\ c & d \end{bmatrix} \equiv \begin{bmatrix} 1 & * \\ 0 & 1 \end{bmatrix} \ (\mathrm{mod} \ N) \right\},$$

satisfying

$$\Gamma(N) \subset \Gamma_1(N) \subset \Gamma_0(N) \subset \mathrm{SL}_2(\mathbb{Z}).$$

The map

$$\Gamma_1(N) \longrightarrow \mathbb{Z}/N\mathbb{Z}, \qquad \begin{bmatrix} a & b \\ c & d \end{bmatrix} \mapsto b \ (\mathrm{mod} \ N)$$

is a surjection with kernel $\Gamma(N)$ (Exercise 1.2.3(c)). Therefore $\Gamma(N) \lhd \Gamma_1(N)$ and

$$\Gamma_1(N)/\Gamma(N) \xrightarrow{\sim} \mathbb{Z}/N\mathbb{Z}, \quad [\Gamma_1(N) : \Gamma(N)] = N.$$

Similarly the map

$$\Gamma_0(N) \longrightarrow (\mathbb{Z}/N\mathbb{Z})^*, \qquad \begin{bmatrix} a & b \\ c & d \end{bmatrix} \mapsto d \pmod{N}$$

is a surjection with kernel $\Gamma_1(N)$ (Exercise 1.2.3(d)), so that $\Gamma_1(N) \triangleleft \Gamma_0(N)$ and

$$\Gamma_0(N)/\Gamma_1(N) \xrightarrow{\sim} (\mathbb{Z}/N\mathbb{Z})^*, \quad [\Gamma_0(N) : \Gamma_1(N)] = \phi(N),$$

where ϕ is the Euler totient function from number theory. (Recall that $\phi(N)$ counts the elements of $\{0, \dots, N-1\}$ relatively prime to N, so that $\phi(1) = 1$ and $\phi(N) = |(\mathbb{Z}/N\mathbb{Z})^*|$ for $N > 1$; see, for example, [IR92] Chapter 2.) It follows that $[\mathrm{SL}_2(\mathbb{Z}) : \Gamma_0(N)] = N \prod_{p|N}(1 + 1/p)$, the product taken over all primes dividing N (Exercise 1.2.3(e)).

Returning briefly to the example of the four squares problem, its associated group Γ_θ generated by the matrices $\pm \begin{bmatrix} 1 & 1 \\ 0 & 1 \end{bmatrix}$ and $\pm \begin{bmatrix} 1 & 0 \\ 4 & 1 \end{bmatrix}$ is $\Gamma_0(4)$ (Exercise 1.2.4).

Two pieces of notation are essential before we continue. For any matrix $\gamma = \begin{bmatrix} a & b \\ c & d \end{bmatrix} \in \mathrm{SL}_2(\mathbb{Z})$ define the *factor of automorphy* $j(\gamma, \tau) \in \mathbb{C}$ for $\tau \in \mathcal{H}$ to be

$$j(\gamma, \tau) = c\tau + d,$$

and for $\gamma \in \mathrm{SL}_2(\mathbb{Z})$ and any integer k define the *weight-k operator* $[\gamma]_k$ on functions $f : \mathcal{H} \longrightarrow \mathbb{C}$ by

$$(f[\gamma]_k)(\tau) = j(\gamma, \tau)^{-k} f(\gamma(\tau)), \quad \tau \in \mathcal{H}.$$

Since the factor of automorphy is never zero or infinity, if f is meromorphic then $f[\gamma]_k$ is also meromorphic and has the same zeros and poles as f. Now we can define *weakly modular of weight k with respect to Γ* to mean meromorphic and weight-k invariant under Γ, that is, a meromorphic function f on $\mathcal{H}$ is weakly modular of weight k if

$$f[\gamma]_k = f \quad \text{for all } \gamma \in \Gamma.$$

If f is weakly modular of weight k with respect to Γ then its zeros and poles are Γ-invariant as sets. The factor of automorphy and the weight-k operator will be ubiquitous from now on. The next lemma lists some of their basic properties.

Lemma 1.2.2. *For all $\gamma, \gamma' \in \mathrm{SL}_2(\mathbb{Z})$ and $\tau \in \mathcal{H}$,*

(a) $j(\gamma\gamma', \tau) = j(\gamma, \gamma'(\tau))j(\gamma', \tau)$,
(b) $(\gamma\gamma')(\tau) = \gamma(\gamma'(\tau))$,
(c) $[\gamma\gamma']_k = [\gamma]_k[\gamma']_k$ (*this is an equality of operators*),
(d) $\mathrm{Im}(\gamma(\tau)) = \dfrac{\mathrm{Im}(\tau)}{|j(\gamma, \tau)|^2}$,

(e) $\dfrac{d\gamma(\tau)}{d\tau} = \dfrac{1}{j(\gamma,\tau)^2}$.

Proof. Some of these have already been established in Exercise 1.1.2 by computing with the matrix entries, and the rest can be shown similarly. But the results are now phrased intrinsically, i.e., with no direct reference to the matrix entries, and so a proof in the right spirit should be intrinsic as well. Each element γ of $\mathrm{SL}_2(\mathbb{Z})$ acts on column vectors via multiplication and acts on points as a fractional linear transformation, and computing with the matrix entries only once shows that the relation between these actions is

$$\gamma \begin{bmatrix} \tau \\ 1 \end{bmatrix} = \begin{bmatrix} \gamma(\tau) \\ 1 \end{bmatrix} j(\gamma,\tau).$$

Applying this identity repeatedly gives

$$\gamma\gamma' \cdot \begin{bmatrix} \tau \\ 1 \end{bmatrix} = \begin{bmatrix} (\gamma\gamma')(\tau) \\ 1 \end{bmatrix} j(\gamma\gamma',\tau),$$

$$\gamma \cdot \gamma' \begin{bmatrix} \tau \\ 1 \end{bmatrix} = \gamma \begin{bmatrix} \gamma'(\tau) \\ 1 \end{bmatrix} j(\gamma',\tau) = \begin{bmatrix} \gamma(\gamma'(\tau)) \\ 1 \end{bmatrix} j(\gamma,\gamma'(\tau))j(\gamma',\tau).$$

The left sides are equal, hence so are the right sides. Equating the lower entries of the right sides proves (a), and then equating the upper entries proves (b). Next, for any $f : \mathcal{H} \longrightarrow \mathbb{C}$ compute that

$$(f[\gamma\gamma']_k)(\tau) = j(\gamma\gamma',\tau)^{-k} f((\gamma\gamma')(\tau)),$$

$$((f[\gamma]_k)[\gamma']_k)(\tau) = j(\gamma',\tau)^{-k}(f[\gamma]_k)(\gamma'(\tau))$$

$$= j(\gamma',\tau)^{-k} j(\gamma,\gamma'(\tau))^{-k} f(\gamma(\gamma'(\tau))).$$

The right sides are equal by parts (a) and (b), hence so are the left sides, proving (c). For parts (d) and (e), juxtapose two copies of the relation between the actions of γ side by side to get

$$\gamma \begin{bmatrix} \tau & \tau' \\ 1 & 1 \end{bmatrix} = \begin{bmatrix} \gamma(\tau) & \gamma(\tau') \\ 1 & 1 \end{bmatrix} \begin{bmatrix} j(\gamma,\tau) & 0 \\ 0 & j(\gamma,\tau') \end{bmatrix}.$$

Taking determinants and letting $\tau' \to \tau$ gives (e). Setting $\tau' = \overline{\tau}$ (complex conjugate), noting $\gamma(\overline{\tau}) = \overline{\gamma(\tau)}$ and $j(\gamma,\overline{\tau}) = \overline{j(\gamma,\tau)}$, and taking determinants gives (d). $\qquad\square$

One consequence of the lemma is that if a function $f : \mathcal{H} \longrightarrow \mathbb{C}$ is weakly modular of weight k with respect to some set of matrices then f is weakly modular of weight k with respect to the group of matrices the set generates. Thus the fact that the matrices $\left[\begin{smallmatrix} 1 & 1 \\ 0 & 1 \end{smallmatrix}\right]$ and $\left[\begin{smallmatrix} 0 & -1 \\ 1 & 0 \end{smallmatrix}\right]$ generate $\mathrm{SL}_2(\mathbb{Z})$ justifies the claim that weak modularity needs to be checked only for these matrices in order that it hold for the full group. Similarly, the function $\theta(\tau,4)$ from the four squares problem is weakly modular of weight 2 with respect to the group $\Gamma_\theta = \Gamma_0(4)$ generated by $\pm\begin{bmatrix} 1 & 1 \\ 0 & 1 \end{bmatrix}$ and $\pm\begin{bmatrix} 1 & 0 \\ 4 & 1 \end{bmatrix}$.

We now develop the definition of a modular form with respect to a congruence subgroup. Let k be an integer and let Γ be a congruence subgroup of $SL_2(\mathbb{Z})$. A function $f : \mathcal{H} \longrightarrow \mathbb{C}$ is a modular form of weight k with respect to Γ if it is weakly modular of weight k with respect to Γ and satisfies a holomorphy condition to be described below. When $-I \notin \Gamma$, nonzero modular forms of odd weight may well exist, in contrast to the case of $SL_2(\mathbb{Z})$ in Section 1.1.

Each congruence subgroup Γ of $SL_2(\mathbb{Z})$ contains a translation matrix of the form

$$\begin{bmatrix} 1 & h \\ 0 & 1 \end{bmatrix} : \tau \mapsto \tau + h$$

for some minimal $h \in \mathbb{Z}^+$. This is because Γ contains $\Gamma(N)$ for some N, but h may properly divide N—for example, the group $\Gamma_1(N)$ contains the translation matrix $\begin{bmatrix} 1 & 1 \\ 0 & 1 \end{bmatrix}$. Every function $f : \mathcal{H} \longrightarrow \mathbb{C}$ that is weakly modular with respect to Γ therefore is $h\mathbb{Z}$-periodic and thus has a corresponding function $g : D' \longrightarrow \mathbb{C}$ where again D' is the punctured disk but now $f(\tau) = g(q_h)$ where $q_h = e^{2\pi i \tau / h}$. As before, if f is also holomorphic on the upper half plane then g is holomorphic on the punctured disk and so it has a Laurent expansion. Define such f to be *holomorphic at* ∞ if g extends holomorphically to $q = 0$. Thus f has a Fourier expansion

$$f(\tau) = \sum_{n=0}^{\infty} a_n q_h^n, \quad q_h = e^{2\pi i \tau / h}.$$

To keep the vector spaces of modular forms finite-dimensional, modular forms need to be holomorphic not only on $\mathcal{H}$ but at limit points. For a congruence subgroup Γ the idea is to adjoin not only ∞ but also the rational numbers $\mathbb{Q}$ to $\mathcal{H}$, and then to identify adjoined points under Γ-equivalence. A Γ-equivalence class of points in $\mathbb{Q} \cup \{\infty\}$ is called a *cusp of* Γ. The geometry motivating the term "cusp" will be explained in Chapter 2. When $\Gamma = SL_2(\mathbb{Z})$ all rational numbers are Γ-equivalent to ∞ and so $SL_2(\mathbb{Z})$ has only one cusp, represented by ∞. But when Γ is a proper subgroup of $SL_2(\mathbb{Z})$ fewer points are Γ-equivalent and so Γ will have other cusps as well, represented by rational numbers. Since each $s \in \mathbb{Q}$ takes the form $s = \alpha(\infty)$ for some $\alpha \in SL_2(\mathbb{Z})$, the number of cusps is at most the number of cosets $\Gamma\alpha$ in $SL_2(\mathbb{Z})$ but possibly fewer, a finite number since the index $[SL_2(\mathbb{Z}) : \Gamma]$ is finite. We will count the cusps in Chapter 3.

A modular form with respect to a congruence subgroup Γ should be holomorphic at the cusps. Writing any $s \in \mathbb{Q} \cup \{\infty\}$ as $s = \alpha(\infty)$, holomorphy at s is naturally defined in terms of holomorphy at ∞ via the $[\alpha]_k$ operator. Since $f[\alpha]_k$ is holomorphic on $\mathcal{H}$ and weakly modular with respect to $\alpha^{-1}\Gamma\alpha$, again a congruence subgroup of $SL_2(\mathbb{Z})$ (Exercise 1.2.5), the notion of its holomorphy at ∞ makes sense.

Definition 1.2.3. *Let Γ be a congruence subgroup of $\mathrm{SL}_2(\mathbb{Z})$ and let k be an integer. A function $f : \mathcal{H} \longrightarrow \mathbb{C}$ is a* **modular form of weight k with respect to Γ** *if*

(1) *f is holomorphic,*
(2) *f is weight-k invariant under Γ,*
(3) *$f[\alpha]_k$ is holomorphic at ∞ for all $\alpha \in \mathrm{SL}_2(\mathbb{Z})$.*

If in addition,

(4) *$a_0 = 0$ in the Fourier expansion of $f[\alpha]_k$ for all $\alpha \in \mathrm{SL}_2(\mathbb{Z})$,*

then f is a **cusp form** *of weight k with respect to Γ. The modular forms of weight k with respect to Γ are denoted $\mathcal{M}_k(\Gamma)$, the cusp forms $\mathcal{S}_k(\Gamma)$.*

The cusp conditions (3) and (4) are phrased independently of the congruence subgroup Γ. But in fact $f[\alpha]_k$ needs to be checked in (3) and (4) only for the finitely many coset representatives α_j in any decomposition $\mathrm{SL}_2(\mathbb{Z}) = \bigcup_j \Gamma \alpha_j$ since $f[\gamma \alpha_j]_k = f[\alpha_j]_k$ for all $\gamma \in \Gamma$ by the second condition, and these representatives depend on Γ. In connection with the third condition, we will need the following result (Exercise 1.2.6) in Chapter 4.

Proposition 1.2.4. *Let Γ be a congruence subgroup of $\mathrm{SL}_2(\mathbb{Z})$ of level N, and let $q_N = e^{2\pi i \tau / N}$ for $\tau \in \mathcal{H}$. Suppose that the function $f : \mathcal{H} \longrightarrow \mathbb{C}$ satisfies conditions (1) and (2) in Definition 1.2.3 and satisfies*

(3′) *f is holomorphic at ∞, and in the Fourier expansion $f(\tau) = \sum_{n=0}^{\infty} a_n q_N^n$ the coefficients for $n > 0$ satisfy the condition*

$$|a_n| \le C n^r \quad \text{for some positive constants } C \text{ and } r.$$

Then f also satisfies condition (3) of Definition 1.2.3, and so $f \in \mathcal{M}_k(\Gamma)$.

For instance, the proposition easily shows that $\theta(\tau, 4) \in \mathcal{M}_2(\Gamma_0(4))$. Condition (3) also implies condition (3′), as we will finally see in Section 5.9.

Definition 1.2.3 reduces to previous ones when $\Gamma = \mathrm{SL}_2(\mathbb{Z})$. As before, the modular forms and the cusp forms are vector spaces and subspaces, and the sums

$$\mathcal{M}(\Gamma) = \bigoplus_{k \in \mathbb{Z}} \mathcal{M}_k(\Gamma) \quad \text{and} \quad \mathcal{S}(\Gamma) = \bigoplus_{k \in \mathbb{Z}} \mathcal{S}_k(\Gamma)$$

form a graded ring and a graded ideal.

The third condition of Definition 1.2.3 does not associate a unique Fourier series expansion of f to a rational number s or even to $s = \infty$. If the matrix $\alpha \in \mathrm{SL}_2(\mathbb{Z})$ takes ∞ to s then so does any matrix $\pm \alpha \beta$ where $\beta = \left[\begin{smallmatrix} 1 & j \\ 0 & 1 \end{smallmatrix}\right]$ with $j \in \mathbb{Z}$. Note that $(f[\pm \alpha \beta]_k)(\tau) = (\pm 1)^k (f[\alpha]_k)(\tau + j)$ (Exercise 1.2.7). Let h' be the smallest positive integer such that $\left[\begin{smallmatrix} 1 & h' \\ 0 & 1 \end{smallmatrix}\right] \in \alpha^{-1} \Gamma \alpha$. (Since $\Gamma(N) \subset \Gamma$ and $\Gamma(N)$ is normal in $\mathrm{SL}_2(\mathbb{Z})$, it follows that $\Gamma(N) \subset \alpha^{-1} \Gamma \alpha$ and so $h' \mid N$.) If

$$(f[\alpha]_k)(\tau) = \sum_{n=0}^{\infty} a_n q_{h'}^n, \quad q_{h'} = e^{2\pi i\tau/h'}$$

then since $e^{2\pi i(\tau+j)/h'} = \mu_{h'}^j q_{h'}$ where $\mu_{h'} = e^{2\pi i/h'}$ is the complex h'th root of unity, it follows that

$$(f[\pm\alpha\beta]_k)(\tau) = (\pm 1)^k \sum_{n=0}^{\infty} a_n \mu_{h'}^{nj} q_{h'}^n, \quad q_{h'} = e^{2\pi i\tau/h'}$$

and all such expansions are equally plausible Fourier series of f at s. In particular, when k is odd the leading coefficient is determined only up to sign, and thinking of $f(s)$ as a_0 does not give it a well defined value. What is well defined is whether a_0 is 0, and so the intuition that a cusp form vanishes at all the cusps makes sense.

For more examples of modular forms with respect to congruence subgroups, start from the weight 2 Eisenstein series

$$G_2(\tau) = \sum_{c\in\mathbb{Z}} \sum_{d\in\mathbb{Z}'_c} \frac{1}{(c\tau + d)^2}$$

where $\mathbb{Z}'_c = \mathbb{Z} - \{0\}$ if $c = 0$ and $\mathbb{Z}'_c = \mathbb{Z}$ otherwise. This series converges only conditionally, but the terms are arranged so that in specializing equation (1.2) to $k = 2$, the ensuing calculation remains valid to give

$$G_2(\tau) = 2\zeta(2) - 8\pi^2 \sum_{n=1}^{\infty} \sigma(n)q^n, \quad q = e^{2\pi i\tau}, \ \sigma(n) = \sum_{\substack{d\mid n \\ d>0}} d$$

(Exercise 1.2.8(a)). Conditional convergence keeps G_2 from being weakly modular. Instead, a calculation that should leave the reader deeply appreciative of absolute convergence in the future shows that

$$(G_2[\gamma]_2)(\tau) = G_2(\tau) - \frac{2\pi i c}{c\tau + d} \quad \text{for } \gamma = \begin{bmatrix} a & b \\ c & d \end{bmatrix} \in \mathrm{SL}_2(\mathbb{Z}) \qquad (1.4)$$

(Exercise 1.2.8(b–c)). The corrected function $G_2(\tau) - \pi/\mathrm{Im}(\tau)$ is weight-2 invariant under $\mathrm{SL}_2(\mathbb{Z})$ (Exercise 1.2.8(d)), but it is not holomorphic. However, for any positive integer N, if

$$G_{2,N}(\tau) = G_2(\tau) - NG_2(N\tau)$$

then $G_{2,N} \in \mathcal{M}_2(\Gamma_0(N))$ (Exercise 1.2.8(e)). We will see many more Eisenstein series in Chapter 4.

Weight 2 Eisenstein series solve the four squares problem from the beginning of the section. The modular forms $G_{2,2}$ and $G_{2,4}$ work out to (Exercise 1.2.9)

$$G_{2,2}(\tau) = -\frac{\pi^2}{3}\left(1 + 24\sum_{n=1}^{\infty}\left(\sum_{\substack{0<d|n \\ d \text{ odd}}} d\right)q^n\right)$$

and

$$G_{2,4}(\tau) = -\pi^2\left(1 + 8\sum_{n=1}^{\infty}\left(\sum_{\substack{0<d|n \\ 4\nmid d}} d\right)q^n\right).$$

Now, $G_{2,2} \in \mathcal{M}_2(\Gamma_0(2)) \subset \mathcal{M}_2(\Gamma_0(4))$ (the smaller group allows more weight-2 invariant functions and the other conditions in Definition 1.2.3 make no reference to a congruence subgroup, so the smaller group has more modular forms) and $G_{2,4} \in \mathcal{M}_2(\Gamma_0(4))$. Exercise 3.9.3 will show that $\dim(\mathcal{M}_2(\Gamma_0(4))) = 2$, so $G_{2,2}$ and $G_{2,4}$, which visibly are linearly independent, are a basis. Recall that the function $\theta(\tau, 4)$ also lies in the space $\mathcal{M}_2(\Gamma_0(4))$. Thus $\theta = aG_{2,2} + bG_{2,4}$ for some $a, b \in \mathbb{C}$, and the expansions

$$\theta(\tau, 4) = 1 + 8q + \cdots,$$

$$-\frac{3}{\pi^2}G_{2,2}(\tau) = 1 + 24q + \cdots,$$

$$-\frac{1}{\pi^2}G_{2,4}(\tau) = 1 + 8q + \cdots,$$

show that $\theta(\tau, 4) = -(1/\pi^2)G_{2,4}(\tau)$. Equating the Fourier coefficients gives the representation number of n as a sum of four squares,

$$r(n, 4) = 8\sum_{\substack{0<d|n \\ 4\nmid d}} d, \quad n \geq 1.$$

In particular, if $4 \nmid n$ then $r(n, 4) = 8\sigma_1(n)$. The two squares problem, the six squares problem, and the eight squares problem are solved similarly once additional machinery is in place. For any even $s \geq 10$ the same methods give an asymptotic solution $\tilde{r}(n, s)$ to the s squares problem, meaning that $\lim_{n\to\infty}\tilde{r}(n, s)/r(n, s) = 1$. Exercise 4.8.7 will discuss all of this.

For another application of weight 2 Eisenstein series, first normalize G_2 to

$$E_2(\tau) = \frac{G_2(\tau)}{2\zeta(2)} = 1 - 24\sum_{n=1}^{\infty}\sigma(n)q^n.$$

Then equation (1.4) with $\gamma = \begin{bmatrix} 0 & -1 \\ 1 & 0 \end{bmatrix}$ specializes to

$$\tau^{-2}E_2(-1/\tau) = E_2(\tau) + \frac{12}{2\pi i\tau}. \tag{1.5}$$

The *Dedekind eta function* is the infinite product

$$\eta(\tau) = q_{24}\prod_{n=1}^{\infty}(1 - q^n), \quad q_{24} = e^{2\pi i\tau/24}, \quad q = e^{2\pi i\tau}.$$

Since the series $S(\tau) = \sum_{n=1}^{\infty} \log(1 - q^n)$ converges absolutely and uniformly on compact subsets of $\mathcal{H}$ (Exercise 1.2.10(a)), a theorem from complex analysis (see for example [JS87]) says that η is holomorphic on $\mathcal{H}$ and satisfies the logarithmic differentiation formula for products, $(\log \prod_n f_n)' = \sum_n f_n'/f_n$ (Exercise 1.2.10(b)). The eta function satisfies a relation similar to that satisfied by the theta function.

Proposition 1.2.5. *The Dedekind eta function satisfies the transformation law*

$$\eta(-1/\tau) = \sqrt{-i\tau}\,\eta(\tau), \quad \tau \in \mathcal{H}.$$

Proof. Compute the logarithmic derivative

$$\frac{d}{d\tau}\log(\eta(\tau)) = \frac{\pi i}{12} - 2\pi i \sum_{d=1}^{\infty} \frac{dq^d}{1-q^d} = \frac{\pi i}{12} - 2\pi i \sum_{d=1}^{\infty} d \sum_{m=1}^{\infty} q^{dm}$$

$$= \frac{\pi i}{12} - 2\pi i \sum_{m=1}^{\infty} \sum_{d=1}^{\infty} dq^{dm} = \frac{\pi i}{12} - 2\pi i \sum_{n=1}^{\infty} \left(\sum_{0<d|n} d \right) q^n$$

$$= \frac{\pi i}{12} E_2(\tau).$$

It follows that

$$\frac{d}{d\tau}\log(\eta(-1/\tau)) = \frac{\pi i}{12}\tau^{-2}E_2(-1/\tau)$$

and

$$\frac{d}{d\tau}\log(\sqrt{-i\tau}\,\eta(\tau)) = \frac{1}{2\tau} + \frac{\pi i}{12}E_2(\tau) = \frac{\pi i}{12}\left(E_2(\tau) + \frac{12}{2\pi i\tau} \right).$$

These are equal by (1.5), so the desired relation holds up to a multiplicative constant. Setting $\tau = i$ shows that the constant is 1. $\qquad\square$

The function

$$\eta^{24}(\tau) = q \prod_{n=1}^{\infty}(1 - q^n)^{24}, \quad q = e^{2\pi i\tau}$$

is invariant under $\tau \mapsto \tau + 1$ and satisfies $\eta^{24}(-1/\tau) = \tau^{12}\eta^{24}(\tau)$. Also, $\lim_{\mathrm{Im}(\tau)\to\infty} \eta^{24}(\tau) = 0$, so $\eta^{24} \in \mathcal{S}_{12}(\mathrm{SL}_2(\mathbb{Z}))$. We will see that this space, which also contains the discriminant function Δ from Section 1.1, is 1-dimensional. Comparing the leading terms of the Fourier expansions $\eta^{24}(\tau) = q + \cdots$ and $\Delta(\tau) = (2\pi)^{12}q + \cdots$ thus gives the remarkable identity

$$\Delta = (2\pi)^{12}\eta^{24},$$

or

$$\left(60 \sum_{(c,d)}' \frac{1}{(c\tau + d)^4} \right)^3 - 27\left(140 \sum_{(c,d)}' \frac{1}{(c\tau + d)^6} \right)^2 = (2\pi)^{12}q \prod_{n=1}^{\infty}(1 - q^n)^{24}.$$

More on the eta function, especially its connection with theta functions per the beginning of this section, is in Chapter 1 of [Bum97].

For a more general example, to make more modular forms with respect to congruence subgroups, start from any modular form f. Let d be a positive integer and let $g(\tau) = f(d\tau)$. Then g is a modular form with respect to $\Gamma_0(d)$. The same process takes modular forms with respect to $\Gamma_0(N)$ to modular forms with respect to $\Gamma_0(dN)$ and takes modular forms with respect to $\Gamma_1(N)$ to modular forms with respect to $\Gamma_1(dN)$ (Exercise 1.2.11).

Exercises

1.2.1. Show that $r(n, k) = \sum_{l+m=n} r(l, i) r(m, j)$ when $i + j = k$.

1.2.2. Let N be a positive integer. Let $\gamma \in \mathrm{SL}_2(\mathbb{Z}/N\mathbb{Z})$ be given. Lift γ to a matrix $\left[\begin{smallmatrix} a & b \\ c & d \end{smallmatrix}\right] \in \mathrm{M}_2(\mathbb{Z})$.

(a) Show that $\gcd(c, d, N) = 1$. Show that $\gcd(c', d') = 1$ for some $c' = c + sN$ and $d' = d + tN$ where $s, t \in \mathbb{Z}$. (A hint for this exercise is at the end of the book.)

(b) Show that some lift $\left[\begin{smallmatrix} a+kN & b+lN \\ c' & d' \end{smallmatrix}\right]$ of γ lies in $\mathrm{SL}_2(\mathbb{Z})$. Thus the map $\mathrm{SL}_2(\mathbb{Z}) \longrightarrow \mathrm{SL}_2(\mathbb{Z}/N\mathbb{Z})$ surjects.

1.2.3. (a) Let p be a prime and let e be a positive integer. Show that $|\mathrm{SL}_2(\mathbb{Z}/p^e\mathbb{Z})| = p^{3e}(1 - 1/p^2)$. (Hints for this exercise are at the end of the book.)

(b) Show that $|\mathrm{SL}_2(\mathbb{Z}/N\mathbb{Z})| = N^3 \prod_{p|N}(1 - 1/p^2)$, so this is also the index $[\mathrm{SL}_2(\mathbb{Z}) : \Gamma(N)]$.

(c) Show that the map $\Gamma_1(N) \longrightarrow \mathbb{Z}/N\mathbb{Z}$ given by $\left[\begin{smallmatrix} a & b \\ c & d \end{smallmatrix}\right] \mapsto b \pmod{N}$ surjects and has kernel $\Gamma(N)$.

(d) Show that the map $\Gamma_0(N) \longrightarrow (\mathbb{Z}/N\mathbb{Z})^*$ given by $\left[\begin{smallmatrix} a & b \\ c & d \end{smallmatrix}\right] \mapsto d \pmod{N}$ surjects and has kernel $\Gamma_1(N)$.

(e) Show that $[\mathrm{SL}_2(\mathbb{Z}) : \Gamma_0(N)] = N \prod_{p|N}(1 + 1/p)$.

1.2.4. Let Γ_θ be the group generated by the matrices $\pm \left[\begin{smallmatrix} 1 & 1 \\ 0 & 1 \end{smallmatrix}\right]$ and $\pm \left[\begin{smallmatrix} 1 & 0 \\ 4 & 1 \end{smallmatrix}\right]$. This exercise shows that $\Gamma_\theta = \Gamma_0(4)$. The containment "⊂" is clear. For the other containment, let $\alpha = \left[\begin{smallmatrix} a & b \\ c & d \end{smallmatrix}\right]$ be a matrix in $\Gamma_0(4)$. Similarly to Exercise 1.1.1, the identity

$$\begin{bmatrix} a & b \\ c & d \end{bmatrix} \begin{bmatrix} 1 & n \\ 0 & 1 \end{bmatrix} = \begin{bmatrix} a & b' \\ c & nc+d \end{bmatrix}$$

shows that unless $c = 0$, some matrix $\alpha\gamma$ with $\gamma \in \Gamma_\theta$ has bottom row (c', d') with $|d'| < |c'|/2$. (The inequality is strict because $c' \equiv 0 \pmod{4}$ and d' is odd.) Use the identity

$$\begin{bmatrix} a & b \\ c & d \end{bmatrix} \begin{bmatrix} 1 & 0 \\ 4n & 1 \end{bmatrix} = \begin{bmatrix} a' & b \\ c+4nd & d \end{bmatrix}$$

to show that because $d \neq 0$ (no $\Gamma_0(4)$ element can have $d = 0$), some matrix $\alpha\gamma$ with $\gamma \in \Gamma_\theta$ has bottom row (c', d') with $|c'| < 2|d'|$. (The inequality is strict because of properties of c' and d' modulo 4.) Each multiplication reduces the positive integer quantity $\min\{|c|, 2|d|\}$, so the process must stop with $c = 0$. Show that in fact this means $\alpha\gamma \in \Gamma_\theta$ for some $\gamma \in \Gamma_\theta$ and so $\alpha \in \Gamma_\theta$.

1.2.5. If Γ is a congruence subgroup of $\mathrm{SL}_2(\mathbb{Z})$ and $\gamma \in \mathrm{SL}_2(\mathbb{Z})$, show that $\gamma^{-1}\Gamma\gamma$ is again a congruence subgroup of $\mathrm{SL}_2(\mathbb{Z})$.

1.2.6. This exercise proves Proposition 1.2.4. Let Γ be a congruence subgroup of $\mathrm{SL}_2(\mathbb{Z})$, thus containing $\Gamma(N)$ for some N, and suppose that the function $f : \mathcal{H} \longrightarrow \mathbb{C}$ is holomorphic and weight-k invariant under Γ. Suppose also that in the Fourier expansion $f(\tau) = \sum_{n=0}^\infty a_n q_N^n$, the coefficients for $n > 0$ satisfy $|a_n| \leq Cn^r$ for some positive constants C and r.

(a) Show that for any $\tau = x + iy \in \mathcal{H}$,

$$|f(\tau)| \leq |a_0| + C\sum_{n=1}^\infty n^r e^{-2\pi ny/N}. \tag{1.6}$$

Changing n to a nonnegative real variable t, show that the continuous version $g(t) = t^r e^{-2\pi ty/N}$ of the summand increases monotonically on the interval $[0, \frac{rN}{2\pi y}]$ and then decreases monotonically on $[\frac{rN}{2\pi y}, \infty)$. Using this, represent all but two terms of the sum in (1.6) as unit-wide boxes under the graph of g and consider the missing terms individually to establish the estimate (where in this exercise C_0 and C can denote different constants in different places)

$$|f(\tau)| \leq C_0 + C\left(\int_{t=0}^\infty g(t)dt + 1/y^r\right).$$

After a change of variable the integral takes the form $C/y^{r+1} \cdot \int_{t=0}^\infty t^r e^{-t}dt$. This last integral is a *gamma function integral*, to be introduced formally in Section 4.4, but at any rate it converges at both ends and is independent of y. In sum,

$$|f(\tau)| \leq C_0 + C/y^r \quad \text{as } y \to \infty.$$

(b) For every $\alpha \in \mathrm{SL}_2(\mathbb{Z})$, the transformed function $(f[\alpha]_k)(\tau)$ is holomorphic and weight-k invariant under $\alpha^{-1}\Gamma\alpha$ and therefore has a Laurent expansion

$$(f[\alpha]_k)(\tau) = \sum_{n\in\mathbb{Z}} a_n' q_N^n, \quad q_N = e^{2\pi i\tau/N}.$$

To show that the Laurent series truncates from the left to a power series it suffices to show that

$$\lim_{q_N \to 0}((f[\alpha]_k)(\tau) \cdot q_N) = 0.$$

If α fixes ∞ then this is immediate from the Fourier series of f itself. Otherwise show that the transformed function $(f[\alpha]_k)(\tau) = (c\tau + d)^{-k}f(\alpha(\tau))$ satisfies

$$\lim_{q_N \to 0} |(f[\alpha]_k)(\tau) \cdot q_N| \le C \lim_{q_N \to 0} (y^{r-k}|q_N|).$$

Recalling that $q_N = e^{2\pi i(x+iy)/N}$, show that $y = C\log(1/|q_N|)$, and use the fact that polynomials dominate logarithms to complete the proof. (A hint for this exercise is at the end of the book.)

1.2.7. Show that $(f[\pm\alpha\beta]_k)(\tau) = (\pm 1)^k (f[\alpha]_k)(\tau + j)$ in the comment after Proposition 1.2.4.

1.2.8. (a) Verify the Fourier expansion of G_2.

(b) Suppose that G_2 satisfies the identity in (1.4) for two particular matrices $\gamma_1, \gamma_2 \in \mathrm{SL}_2(\mathbb{Z})$. Show that G_2 then satisfies the identity for the product $\gamma_1\gamma_2$ and the inverse γ_1^{-1} as well. Thus to establish (1.4) it suffices to prove the identity for a set of generators for $\mathrm{SL}_2(\mathbb{Z})$. (Hints for this exercise are at the end of the book.)

(c) Recall that $\mathrm{SL}_2(\mathbb{Z})$ is generated by $\begin{bmatrix} 1 & 1 \\ 0 & 1 \end{bmatrix}$ and $\begin{bmatrix} 0 & -1 \\ 1 & 0 \end{bmatrix}$. The Fourier expansion for G_2 shows that it satisfies (1.4) for $\gamma = \begin{bmatrix} 1 & 1 \\ 0 & 1 \end{bmatrix}$. For the other generator, show that

$$\left(G_2\left[\begin{bmatrix} 0 & -1 \\ 1 & 0 \end{bmatrix}\right]_2\right)(\tau) = \tau^{-2}G_2(-1/\tau) = \sum_{d \in \mathbb{Z}} \sum_{c \in \mathbb{Z}'_d} \frac{1}{(c\tau + d)^2}$$

$$= 2\zeta(2) + \sum_{d \in \mathbb{Z}} \sum_{c \neq 0} \frac{1}{(c\tau + d)^2},$$

which differs from $G_2(\tau) = 2\zeta(2) + \sum_{c \neq 0} \sum_{d \in \mathbb{Z}} (c\tau + d)^{-2}$ in the reversed order of summation. Next use partial fractions to show that

$$\sum_{c \neq 0} \sum_{d \in \mathbb{Z}} \frac{1}{(c\tau + d)(c\tau + d + 1)} = 0.$$

Subtract this from G_2 to get

$$G_2(\tau) = 2\zeta(2) + \sum_{\substack{c \neq 0 \\ d \in \mathbb{Z}}} \frac{1}{(c\tau + d)^2(c\tau + d + 1)}$$

where now the sum is absolutely convergent by Exercise 1.1.4 and the limit comparison test, and thus it can have its order of summation reversed. Doing so and separating the convergence terms back out shows that

$$G_2(\tau) = \tau^{-2}G_2(-1/\tau) - \sum_{d \in \mathbb{Z}} \sum_{c \neq 0} \frac{1}{(c\tau + d)(c\tau + d + 1)}.$$

The error term is $-\lim_{N \to \infty} \sum_{d=-N}^{N-1} \sum_{c \neq 0} \left(\frac{1}{c\tau + d} - \frac{1}{c\tau + d + 1} \right)$. Reverse the order and the inner sum telescopes. Manipulate the result into an expression including a sum for $\pi\cot(\pi N/\tau)$ per the first equality of (1.1). Finally, use the other side of (1.1) to take the limit.

(d) Show that

$$\frac{\pi}{j(\gamma,\tau)^2 \mathrm{Im}(\gamma(\tau))} = \frac{\pi}{\mathrm{Im}(\tau)} - \frac{2\pi i c}{c\tau+d} \quad \text{for } \gamma = \begin{bmatrix} a & b \\ c & d \end{bmatrix} \in \mathrm{SL}_2(\mathbb{Z}),$$

and therefore the function $G_2(\tau) - \pi/\mathrm{Im}(\tau)$ is weight-2 invariant under $\mathrm{SL}_2(\mathbb{Z})$.

(e) Show that $G_{2,N} \in \mathcal{M}_2(\Gamma_0(N))$.

(f) Compute $G_2(i)$ and $G_2(\mu_3)$ where $\mu_3 = e^{2\pi i/3}$.

1.2.9. Verify the Fourier expansions of $G_{2,2}$ and $G_{2,4}$ given in the section. (A hint for this exercise is at the end of the book.)

1.2.10. (a) Show that the series $S(\tau) = \sum_{n=1}^{\infty} \log(1-q^n)$ converges absolutely and uniformly on compact subsets of $\mathcal{H}$. (A hint for this exercise is at the end of the book.)

(b) Show the logarithmic differentiation formula $(\log(\prod f_n))' = \sum f_n'/f_n$ for finite products. The analogous result for infinite products applies to the Dedekind eta function.

1.2.11. Let $\mathrm{GL}_2^+(\mathbb{Q})$ be the group of 2-by-2 matrices with positive determinant and rational entries. The elements of this group act as fractional linear transformations on the upper half plane since $\mathrm{Im}(\gamma(\tau)) = \det\gamma \cdot \mathrm{Im}(\tau)/|c\tau+d|^2$ for $\gamma \in \mathrm{GL}_2^+(\mathbb{Q})$. Let k be a positive integer. For $\tau \in \mathcal{H}$, extend the formula $j(\gamma,\tau) = c\tau+d$ to $\gamma \in \mathrm{GL}_2^+(\mathbb{Q})$, and extend the weight-$k$ operator to $\mathrm{GL}_2^+(\mathbb{Q})$ by the rule

$$(f[\gamma]_k)(\tau) = (\det\gamma)^{k-1} j(\gamma,\tau)^{-k} f(\gamma(\tau)) \quad \text{for } f : \mathcal{H} \longrightarrow \mathbb{C}.$$

(a) Show that $[\gamma\gamma']_k = [\gamma]_k[\gamma']_k$ for all $\gamma, \gamma' \in \mathrm{GL}_2^+(\mathbb{Q})$, generalizing Lemma 1.2.2(b).

(b) Show that every $\gamma \in \mathrm{GL}_2^+(\mathbb{Q})$ satisfies $\gamma = \alpha\gamma'$ where $\alpha \in \mathrm{SL}_2(\mathbb{Z})$ and $\gamma' = r\begin{bmatrix} a & b \\ 0 & d \end{bmatrix}$ with $r \in \mathbb{Q}^+$ and $a, b, d \in \mathbb{Z}$ relatively prime. Use this to show that given $f \in \mathcal{M}_k(\Gamma)$ for some congruence subgroup Γ and given such $\gamma = \alpha\gamma'$, since $f[\alpha]_k$ has a Fourier expansion, so does $f[\gamma]_k$. Show that if the Fourier expansion for $f[\alpha]_k$ has constant term 0 then so does the Fourier expansion for $f[\gamma]_k$. (A hint for this exercise is at the end of the book.)

(c) Suppose that Γ_1 and Γ_2 are congruence subgroups of $\mathrm{SL}_2(\mathbb{Z})$ and suppose that $\Gamma_1 \supset \gamma\Gamma_2\gamma^{-1}$ for some $\gamma \in \mathrm{GL}_2^+(\mathbb{Q})$. Prove that if $f \in \mathcal{M}_k(\Gamma_1)$ then $f[\gamma]_k \in \mathcal{M}_k(\Gamma_2)$. Show that the same result holds for cusp forms.

(d) Let N and d be positive integers. Suppose $f \in \mathcal{M}_k(\Gamma_0(N))$. Show that if $g(\tau) = f(d\tau)$ then $g \in \mathcal{M}_k(\Gamma_0(dN))$, and the same result holds for cusp forms. Show that if $f \in \mathcal{M}_k(\Gamma_1(N))$ then $g \in \mathcal{M}_k(\Gamma_1(dN))$, and the same result holds for cusp forms.

1.3 Complex tori

This section gives a sketch of results about complex tori, also known as complex elliptic curves for reasons to be explained in the next section. Since the material is covered in many texts (for example, [JS87] is a very nice second complex analysis book), some facts will be stated without proof. Complex tori will be related to modular forms in Section 1.5.

A *lattice in* $\mathbb{C}$ is a set $\Lambda = \omega_1\mathbb{Z} \oplus \omega_2\mathbb{Z}$ with $\{\omega_1, \omega_2\}$ a basis for $\mathbb{C}$ over $\mathbb{R}$. We make the normalizing convention $\omega_1/\omega_2 \in \mathcal{H}$, but this still does not specify a basis given a lattice. Instead,

Lemma 1.3.1. *Consider two lattices* $\Lambda = \omega_1\mathbb{Z} \oplus \omega_2\mathbb{Z}$ *and* $\Lambda' = \omega_1'\mathbb{Z} \oplus \omega_2'\mathbb{Z}$ *with* $\omega_1/\omega_2 \in \mathcal{H}$ *and* $\omega_1'/\omega_2' \in \mathcal{H}$. *Then* $\Lambda' = \Lambda$ *if and only if*

$$\begin{bmatrix} \omega_1' \\ \omega_2' \end{bmatrix} = \begin{bmatrix} a & b \\ c & d \end{bmatrix} \begin{bmatrix} \omega_1 \\ \omega_2 \end{bmatrix} \quad \text{for some} \quad \begin{bmatrix} a & b \\ c & d \end{bmatrix} \in \mathrm{SL}_2(\mathbb{Z}).$$

Proof. Exercise 1.3.1.   □

A *complex torus* is a quotient of the complex plane by a lattice,

$$\mathbb{C}/\Lambda = \{z + \Lambda : z \in \mathbb{C}\}.$$

Algebraically a complex torus is an Abelian group under the addition it inherits from $\mathbb{C}$. Geometrically a complex torus is a parallelogram spanned by $\{\omega_1, \omega_2\}$ with its sides identified in opposing pairs. Identifying one pair of sides rolls the parallelogram into a tube, and then identifying the other pair bends the tube into a torus. But the flat model of the complex torus with neighborhoods extending across the sides (see Figure 1.2) better illustrates that every complex torus is a *Riemann surface*, roughly meaning a connected set that looks like the complex plane $\mathbb{C}$ in the small. The precise definition of a Riemann surface will be demonstrated by example in Chapter 2.

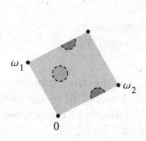

Figure 1.2. A complex torus

The notion of a holomorphic map makes sense for Riemann surfaces since it is local. Any holomorphic map between compact Riemann surfaces is either a surjection or a map to one point. To see this, suppose X and Y are compact Riemann surfaces and $f : X \longrightarrow Y$ is holomorphic. Since f is continuous and X is compact and connected, so is the image $f(X)$, making $f(X)$ closed. Unless f is constant f is open by the Open Mapping Theorem of complex analysis, applicable to Riemann surfaces since it is a local result, making $f(X)$ open as well. So $f(X)$ is either a single point or a connected, open, closed subset of the connected set Y, i.e., all of Y. As a special case of this, any nonconstant holomorphic map from one complex torus to another is a surjection.

Proposition 1.3.2. *Suppose $\varphi : \mathbb{C}/\Lambda \longrightarrow \mathbb{C}/\Lambda'$ is a holomorphic map between complex tori. Then there exist complex numbers m, b with $m\Lambda \subset \Lambda'$ such that $\varphi(z + \Lambda) = mz + b + \Lambda'$. The map is invertible if and only if $m\Lambda = \Lambda'$.*

Proof. (Sketch.) The key is to lift φ to a holomorphic map $\tilde{\varphi} : \mathbb{C} \longrightarrow \mathbb{C}$ by using topology. (The plane is the so-called *universal covering space* of the torus—see a topology text such as [Mun00] for the definition and the relevant lifting theorem.) With the map lifted, consider for any $\lambda \in \Lambda$ the function $f_\lambda(z) = \tilde{\varphi}(z + \lambda) - \tilde{\varphi}(z)$. Since $\tilde{\varphi}$ lifts a map between the quotients, the continuous function f_λ maps to the discrete set Λ' and is therefore constant. Differentiating gives $\tilde{\varphi}'(z+\lambda) = \tilde{\varphi}'(z)$. Thus $\tilde{\varphi}'$ is holomorphic and Λ-periodic, making it bounded and therefore constant by Liouville's Theorem. Now $\tilde{\varphi}$ is a first degree polynomial $\tilde{\varphi}(z) = mz + b$, and again since this lifts a map between quotients, necessarily $m\Lambda \subset \Lambda'$. The original map thus has the form asserted in the proposition. If the containment $m\Lambda \subset \Lambda'$ is proper then φ is not injective: some $z \in \Lambda'$ satisfies $z/m \notin \Lambda$ but $\varphi(z/m + \Lambda) = b + \Lambda' = \varphi(\Lambda)$. If $m\Lambda = \Lambda'$ then $(1/m)\Lambda' = \Lambda$ and the map $\psi : \mathbb{C}/\Lambda' \longrightarrow \mathbb{C}/\Lambda$ given by $\psi(w + \Lambda') = (w - b)/m + \Lambda$ inverts φ. $\qquad\square$

Corollary 1.3.3. *Suppose $\varphi : \mathbb{C}/\Lambda \longrightarrow \mathbb{C}/\Lambda'$ is a holomorphic map between complex tori, $\varphi(z + \Lambda) = mz + b + \Lambda'$ with $m\Lambda \subset \Lambda'$. Then the following are equivalent:*

(1) φ is a group homomorphism,
(2) $b \in \Lambda'$, so $\varphi(z + \Lambda) = mz + \Lambda'$,
(3) $\varphi(0) = 0$.

In particular, there exists a nonzero holomorphic group homomorphism between the complex tori $\mathbb{C}/\Lambda$ and $\mathbb{C}/\Lambda'$ if and only if there exists some nonzero $m \in \mathbb{C}$ such that $m\Lambda \subset \Lambda'$, and there exists a holomorphic group isomorphism between the complex tori $\mathbb{C}/\Lambda$ and $\mathbb{C}/\Lambda'$ if and only if there exists some $m \in \mathbb{C}$ such that $m\Lambda = \Lambda'$.

Proof. Exercise 1.3.2. $\qquad\square$

For one isomorphism of particular interest, start from an arbitrary lattice $\Lambda = \omega_1\mathbb{Z} \oplus \omega_2\mathbb{Z}$ with $\omega_1/\omega_2 \in \mathcal{H}$. Let $\tau = \omega_1/\omega_2$ and let $\Lambda_\tau = \tau\mathbb{Z} \oplus \mathbb{Z}$. Then since $(1/\omega_2)\Lambda = \Lambda_\tau$, Corollary 1.3.3 shows that the map $\varphi_\tau : \mathbb{C}/\Lambda \longrightarrow \mathbb{C}/\Lambda_\tau$ given by $\varphi(z + \Lambda) = z/\omega_2 + \Lambda_\tau$ is an isomorphism. This shows that every complex torus is isomorphic to a complex torus whose lattice is generated by a complex number $\tau \in \mathcal{H}$ and by 1. This τ is not unique, but if $\tau' \in \mathcal{H}$ is another such number then $\tau' = \omega_1'/\omega_2'$ where $\Lambda = \omega_1'\mathbb{Z} \oplus \omega_2'\mathbb{Z}$, and so by Lemma 1.3.1 $\tau' = \gamma(\tau)$ for some $\gamma \in \mathrm{SL}_2(\mathbb{Z})$. Thus each complex torus determines a point $\tau \in \mathcal{H}$ up to the action of $\mathrm{SL}_2(\mathbb{Z})$. Section 1.5 will show as part of a more general argument that in fact the isomorphism classes of complex tori biject to the orbits $\mathrm{SL}_2(\mathbb{Z})\tau$ in $\mathcal{H}$.

Definition 1.3.4. *A nonzero holomorphic homomorphism between complex tori is called an* **isogeny**.

In particular, every holomorphic isomorphism is an isogeny. Every isogeny surjects and has finite kernel—the kernel is finite because it is discrete (otherwise complex analysis shows that the map is zero) and complex tori are compact.

Multiply-by-integer maps are isogenies but not isomorphisms. For any positive integer N and lattice Λ consider the map

$$[N] : \mathbb{C}/\Lambda \longrightarrow \mathbb{C}/\Lambda, \qquad z + \Lambda \mapsto Nz + \Lambda.$$

This is an isogeny since $N\Lambda \subset \Lambda$. Its kernel, the points $z + \Lambda \in \mathbb{C}/\Lambda$ such that $[N](z + \Lambda) = 0$, is the set of N-*torsion points of* $\mathbb{C}/\Lambda$, a subgroup isomorphic to $\mathbb{Z}/N\mathbb{Z} \times \mathbb{Z}/N\mathbb{Z}$. (See Figure 1.3.) Letting E denote the torus $\mathbb{C}/\Lambda$ (for reasons to be explained soon), this subgroup is denoted $E[N]$.

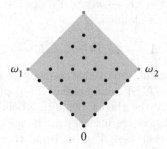

Figure 1.3. $E[5]$: the 5-torsion points of a torus

Cyclic quotient maps are also isogenies but not isomorphisms. Let $\mathbb{C}/\Lambda$ be a complex torus, let N be a positive integer, and let C be a cyclic subgroup of $E[N]$ isomorphic to $\mathbb{Z}/N\mathbb{Z}$. The elements of C are cosets $\{c + \Lambda\}$ and so as a set C forms a superlattice of Λ. Slightly abusing notation we use the same symbol for the subgroup and the superlattice. Then the cyclic quotient map

$$\pi : \mathbb{C}/\Lambda \longrightarrow \mathbb{C}/C, \qquad z + \Lambda \mapsto z + C$$

is an isogeny with kernel C. (See Figure 1.4, where Λ is of the form Λ_τ and the kernel is $C = (1/5)\mathbb{Z} + \Lambda$.)

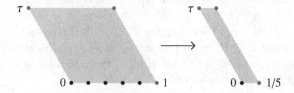

Figure 1.4. A quotient isogeny with cyclic kernel of order 5

In fact every isogeny is a composition of the examples already given. To see this, consider an arbitrary isogeny

$$\varphi : \mathbb{C}/\Lambda \longrightarrow \mathbb{C}/\Lambda', \qquad z + \Lambda \mapsto mz + \Lambda'$$

and let K denote its kernel, the finite subgroup $K = m^{-1}\Lambda'/\Lambda$ of $\mathbb{C}/\Lambda$ also viewed as the superlattice $K = m^{-1}\Lambda'$ of Λ. If N is the order of K as a subgroup then $K \subset E[N] \cong \mathbb{Z}/N\mathbb{Z} \times \mathbb{Z}/N\mathbb{Z}$, and so by the theory of finite Abelian groups $K \cong \mathbb{Z}/n\mathbb{Z} \times \mathbb{Z}/nn'\mathbb{Z}$ for some positive integers n and n'. The multiply-by-n isogeny $[n]$ of $\mathbb{C}/\Lambda$ takes K to a cyclic subgroup nK isomorphic to $\mathbb{Z}/n'\mathbb{Z}$, and then the quotient isogeny π from $\mathbb{C}/\Lambda$ to $\mathbb{C}/nK$ has kernel nK. Follow this by the map $\mathbb{C}/nK \longrightarrow \mathbb{C}/\Lambda'$ given by $z + nK \mapsto (m/n)z + (m/n)nK$, now viewing nK as a lattice in $\mathbb{C}$. This map makes sense and is an isomorphism since $(m/n)nK = mK = \Lambda'$. The composition of the three maps is $z + \Lambda \mapsto nz + \Lambda \mapsto nz + nK \mapsto mz + \Lambda' = \varphi(z + \Lambda)$. That is, the general isogeny is a composition as claimed,

$$\varphi : \mathbb{C}/\Lambda \xrightarrow{[n]} \mathbb{C}/\Lambda \xrightarrow{\pi} \mathbb{C}/nK \xrightarrow{\sim} \mathbb{C}/\Lambda'.$$

A very similar argument shows that isogeny is an equivalence relation. Suppose that $\varphi : \mathbb{C}/\Lambda \longrightarrow \mathbb{C}/\Lambda'$ is an isogeny. Thus $\varphi(z+\Lambda) = mz + \Lambda'$ where $m \neq 0$ and $m\Lambda \subset \Lambda'$. By the theory of finite Abelian groups there exists a basis $\{\omega_1, \omega_2\}$ of Λ' and positive integers n_1, n_2 such that $\{n_1\omega_1, n_2\omega_2\}$ is a basis of $m\Lambda$. It follows that $n_1 n_2 \Lambda' \subset m\Lambda$ and therefore $(n_1 n_2/m)\Lambda' \subset \Lambda$. Thus there is a *dual isogeny* $\hat{\varphi} : \mathbb{C}/\Lambda' \longrightarrow \mathbb{C}/\Lambda$ back in the other direction, $\hat{\varphi}(z + \Lambda') = (n_1 n_2/m)z + \Lambda$. Note that $(\hat{\varphi} \circ \varphi)(z + \Lambda) = n_1 n_2 z + \Lambda$, i.e., the isogeny followed by its dual is multiplication by a positive integer. The integer $n_1 n_2$ in question is the degree of the original isogeny since $\{\omega_1/m, \omega_2/m\}$ is a basis of $\ker(\varphi)$ and $\{n_1\omega_1/m, n_2\omega_2/m\}$ is a basis of Λ, making $\ker(\varphi) \cong \mathbb{Z}/n_1\mathbb{Z} \times \mathbb{Z}/n_2\mathbb{Z}$ and showing that φ is $n_1 n_2$-to-1. That is,

$$\hat{\varphi} \circ \varphi = [\deg(\varphi)].$$

This condition specifies $\hat{\varphi}$ uniquely since φ surjects. Since the map $[\deg(\varphi)]$ has degree $(\deg(\varphi))^2$ and the degree of a composition is the product of the degrees, the dual isogeny has degree $\deg(\hat{\varphi}) = \deg(\varphi)$. The dual isogeny of a multiply-by-integer map is itself. The dual isogeny of a cyclic quotient isogeny quotients the torus in a second direction by a cyclic group of the same order to restore its shape and then expands it back to full size. The dual isogeny of an isomorphism is its inverse. The dual of a composition of isogenies is the composition of the duals in the reverse order. If φ is an isogeny and $\hat{\varphi}$ is its dual then the formulas $\varphi(z + \Lambda) = mz + \Lambda'$, $\hat{\varphi}(z' + \Lambda') = (\deg(\varphi)/m)z' + \Lambda$ show that also

$$\varphi \circ \hat{\varphi} = [\deg(\varphi)] = [\deg(\hat{\varphi})],$$

so that φ is in turn the dual isogeny of its dual $\hat{\varphi}$. Isogeny of complex tori, rather than isomorphism, will turn out to be the appropriate equivalence relation in the context of modular forms.

If $\varphi_1, \varphi_2 : \mathbb{C}/\Lambda \longrightarrow \mathbb{C}/\Lambda'$ are isogenies, and $\varphi_1 + \varphi_2 \neq 0$ so that their sum is again an isogeny, then the dual of the sum is the sum of their duals. To see this, let $\varphi : \mathbb{C}/\Lambda \longrightarrow \mathbb{C}/\Lambda'$ be an isogeny. Thus $\varphi(z + \Lambda) = mz + \Lambda'$ where $m \neq 0$ and $m\Lambda \subset \Lambda'$. Let $\Lambda = \omega_1\mathbb{Z} \oplus \omega_2\mathbb{Z}$ with $\omega_1/\omega_2 \in \mathcal{H}$, and similarly for Λ'. Consequently

$$\begin{bmatrix} m\omega_1 \\ m\omega_2 \end{bmatrix} = \alpha \begin{bmatrix} \omega_1' \\ \omega_2' \end{bmatrix} \quad \text{for some } \alpha = \begin{bmatrix} a & b \\ c & d \end{bmatrix} \in \mathrm{M}_2(\mathbb{Z}).$$

Homogenizing this equality gives $\omega_1/\omega_2 = \alpha(\omega_1'/\omega_2')$ where now α acts as a fractional linear transformation, showing that $\det\alpha \neq 0$ and hence $\det\alpha > 0$ because in general $\mathrm{Im}(\alpha(\tau)) = \det\alpha \cdot \mathrm{Im}(\tau)/|j(\alpha, \tau)|^2$ for $\alpha \in \mathrm{GL}_2(\mathbb{R})$. This justifies the last step of the calculation

$$\deg(\varphi) = |\ker(\varphi)| = [m^{-1}\Lambda' : \Lambda] = [\Lambda' : m\Lambda] = \det\alpha.$$

Since $\hat{\varphi} \circ \varphi = [\deg(\varphi)]$ and the matrix of a composition is the right-to-left product of the matrices, the dual isogeny must induce the matrix

$$\hat{\alpha} = \det\alpha \cdot \alpha^{-1} = \begin{bmatrix} d & -b \\ -c & a \end{bmatrix},$$

and conversely this matrix determines the dual isogeny. Now let φ_1 and φ_2 be isogenies from $\mathbb{C}/\Lambda$ to $\mathbb{C}/\Lambda'$ with $\varphi_1 + \varphi_2 \neq 0$. Their sum $(\varphi_1 + \varphi_2)(z + \Lambda) = (m_1 + m_2)z + \Lambda'$ gives rise to the matrix

$$\alpha_1 + \alpha_2 = \begin{bmatrix} a_1 + a_2 & b_1 + b_2 \\ c_1 + c_2 & d_1 + d_2 \end{bmatrix},$$

and so correspondingly the dual isogeny of the sum is determined by the matrix

$$\begin{bmatrix} d_1 + d_2 & -b_1 - b_2 \\ -c_1 - c_2 & a_1 + a_2 \end{bmatrix} = \hat{\alpha}_1 + \hat{\alpha}_2,$$

the sum of the matrices determining the dual isogenies. This proves the claim at the beginning of the paragraph,

$$\widehat{\varphi_1 + \varphi_2} = \hat{\varphi}_1 + \hat{\varphi}_2 \quad \text{if } \varphi_1 + \varphi_2 \neq 0. \tag{1.7}$$

We will cite this fact in Chapter 6.

For one more example, some complex tori have endomorphisms other than the multiply-by-N maps $[N]$, in which case they have *complex multiplication*. Let $\tau = \sqrt{d}$ for some squarefree $d \in \mathbb{Z}^-$ such that $d \equiv 2, 3 \pmod{4}$, or let $\tau = (-1 + \sqrt{d})/2$ for squarefree $d \in \mathbb{Z}^-$, $d \equiv 1 \pmod 4$. Then the set $\mathcal{O} = \tau\mathbb{Z} \oplus \mathbb{Z}$ is a ring. (Readers with background in number theory will recognize it as the ring of integers in the imaginary quadratic number field $\mathbb{Q}(\sqrt{d})$.) Let Λ be any ideal of $\mathcal{O}$ and let m be any element of $\mathcal{O}$. Then $m\Lambda \subset \Lambda$, so multiplying by m gives an endomorphism of $\mathbb{C}/\Lambda$. In particular, the ring of endomorphisms of $\mathbb{C}/\Lambda_i$ is isomorphic to $\Lambda_i = i\mathbb{Z} \oplus \mathbb{Z}$ rather than to $\mathbb{Z}$, and similarly for the ring of endomorphisms of $\mathbb{C}/\Lambda_{\mu_3}$ where $\mu_3 = e^{2\pi i/3}$.

Let Λ be a lattice. The N-torsion subgroup of the additive torus group $\mathbb{C}/\Lambda$,

$$E[N] = \{P \in \mathbb{C}/\Lambda : [N]P = 0\} = \langle \omega_1/N + \Lambda \rangle \times \langle \omega_2/N + \Lambda \rangle,$$

is analogous to the N-torsion subgroup of the multiplicative circle group $\mathbb{C}^*/\mathbb{R}^+ \cong \{z \in \mathbb{C} : |z| = 1\} \cong \mathbb{R}/\mathbb{Z}$, the complex Nth roots of unity

$$\boldsymbol{\mu}_N = \{z \in \mathbb{C} : z^N = 1\} = \langle e^{2\pi i/N} \rangle.$$

A sort of inner product exists on $E[N]$ with values in $\boldsymbol{\mu}_N$, the *Weil pairing*

$$e_N : E[N] \times E[N] \longrightarrow \boldsymbol{\mu}_N.$$

To define this, let P and Q be points in $E[N]$, possibly equal. If $\Lambda = \omega_1\mathbb{Z} \oplus \omega_2\mathbb{Z}$ with $\omega_1/\omega_2 \in \mathcal{H}$ then

$$\begin{bmatrix} P \\ Q \end{bmatrix} = \gamma \begin{bmatrix} \omega_1/N + \Lambda \\ \omega_2/N + \Lambda \end{bmatrix} \quad \text{for some } \gamma \in \mathrm{M}_2(\mathbb{Z}/N\mathbb{Z})$$

since $\omega_1/N + \Lambda$ and $\omega_2/N + \Lambda$ generate $E[N]$. The Weil pairing of P and Q is

$$e_N(P, Q) = e^{2\pi i \det \gamma/N}.$$

This makes sense even though $\det \gamma$ is defined only modulo N. It is independent of how the basis $\{\omega_1, \omega_2\}$ is chosen (and once the basis is chosen the matrix γ is uniquely determined since its entries are reduced modulo N), remembering the normalization $\omega_1/\omega_2 \in \mathcal{H}$ (Exercise 1.3.3(a)). If P and Q generate $E[N]$ then the matrix γ lies in the group $\mathrm{GL}_2(\mathbb{Z}/N\mathbb{Z})$ of invertible 2-by-2 matrices with entries in $\mathbb{Z}/N\mathbb{Z}$, making $\det \gamma$ invertible modulo N and $e_N(P, Q)$ therefore a primitive complex Nth root of unity. See Exercise 1.3.3(b–d) for more properties of the Weil pairing, in particular that the Weil pairing is preserved under isomorphisms of complex tori. We will use the Weil pairing in the Section 1.5.

Exercises

1.3.1. Prove Lemma 1.3.1.

1.3.2. Prove Corollary 1.3.3. (A hint for this exercise is at the end of the book.)

1.3.3. (a) Show that the Weil pairing is independent of which basis $\{\omega_1, \omega_2\}$ is used, provided $\omega_1/\omega_2 \in \mathcal{H}$.

(b) Show that the Weil pairing is bilinear, alternating, and nondegenerate. (Remember that the group $\boldsymbol{\mu}_N$ is multiplicative.)

(c) Show that the Weil pairing is compatible with N. This means that for positive integers N and d, the diagram

$$
\begin{array}{ccc}
E[dN] \times E[dN] & \xrightarrow{\ e_{dN}(\cdot,\cdot)\ } & \boldsymbol{\mu}_{dN} \\
{\scriptstyle d(\cdot,\cdot)}\Big\downarrow & & \Big\downarrow{\scriptstyle .d} \\
E[N] \times E[N] & \xrightarrow{\ e_{N}(\cdot,\cdot)\ } & \boldsymbol{\mu}_{N}
\end{array}
$$

commutes, where the vertical maps are suitable multiplications by d. (A hint for this exercise is at the end of the book.)

(d) Let Λ and Λ' be lattices with $m\Lambda = \Lambda'$ for some $m \in \mathbb{C}$. Show that the isomorphism of complex elliptic curves $\mathbb{C}/\Lambda \xrightarrow{\sim} \mathbb{C}/\Lambda'$ given by $z + \Lambda \mapsto mz + \Lambda'$ preserves the Weil pairing.

1.4 Complex tori as elliptic curves

This section shows how complex tori $\mathbb{C}/\Lambda$ can also be viewed as cubic curves of the sort mentioned back in the preface. These cubic curves are called *elliptic* despite not being ellipses, due to a connection between them and the arc length of an actual ellipse. The presentation here is terse, so the reader may want to consult a relevant complex analysis text such as [JS87].

The meromorphic functions on a complex torus are what relate it to a cubic curve. Given a lattice Λ, the meromorphic functions $f : \mathbb{C}/\Lambda \longrightarrow \widehat{\mathbb{C}}$ on the torus are naturally identified with the Λ-periodic meromorphic functions $f : \mathbb{C} \longrightarrow \widehat{\mathbb{C}}$ on the plane. Exercise 1.4.1 derives some basic properties of these functions in general. The most important specific example is the *Weierstrass $\wp$-function*

$$
\wp(z) = \frac{1}{z^2} + {\sum_{\omega \in \Lambda}}' \left(\frac{1}{(z-\omega)^2} - \frac{1}{\omega^2} \right), \quad z \in \mathbb{C}, \ z \notin \Lambda.
$$

(The primed summation means to omit $\omega = 0$.) Subtracting $1/\omega^2$ from $1/(z-\omega)^2$ makes the summand roughly z/ω^3, cf. the sketched proof of Proposition 1.4.1 to follow, so the sum converges absolutely and uniformly on compact subsets of $\mathbb{C}$ away from Λ. Correcting the summand this way prevents the terms of the sum from being permuted when z is translated by a lattice element, so $\wp$ doesn't obviously have periods Λ. But the derivative

$$\wp'(z) = -2 \sum_{w \in \Lambda} \frac{1}{(z-w)^3}.$$

clearly does have periods Λ, and combining this with the fact that $\wp$ is an even function quickly shows that in fact $\wp$ has periods Λ as well (Exercise 1.4.2). It turns out that $\wp$ and $\wp'$ are the only basic examples we need since the field of meromorphic functions on $\mathbb{C}/\Lambda$ is $\mathbb{C}(\wp, \wp')$, the rational expressions in these two functions. Since the Weierstrass $\wp$-function depends on the lattice Λ as well as the variable z we will sometimes write $\wp_\Lambda(z)$ and $\wp'_\Lambda(z)$; in particular for lattices Λ_τ we will write $\wp_\tau(z)$ and $\wp'_\tau(z)$.

Eisenstein series generalize to functions of a variable lattice,

$$G_k(\Lambda) = {\sum_{w \in \Lambda}}' \frac{1}{w^k}, \quad k > 2 \text{ even},$$

so that $G_k(\tau)$ from before is now $G_k(\Lambda_\tau)$. As lattice functions Eisenstein series satisfy the *homogeneity condition* $G_k(m\Lambda) = m^{-k}G_k(\Lambda)$ for all nonzero $m \in \mathbb{C}$. Part (a) of the next result shows that these lattice Eisenstein series appear in the Laurent expansion of the Weierstrass $\wp$-function for Λ. Part (b) relates the functions $\wp(z)$ and $\wp'(z)$ in a cubic equation whose coefficients are also lattice Eisenstein series.

Proposition 1.4.1. *Let $\wp$ be the Weierstrass function with respect to a lattice Λ. Then*

(a) *The Laurent expansion of $\wp$ is*

$$\wp(z) = \frac{1}{z^2} + \sum_{\substack{n=2 \\ n \text{ even}}}^{\infty} (n+1)G_{n+2}(\Lambda)z^n$$

for all z such that $0 < |z| < \inf\{|\omega| : \omega \in \Lambda - \{0\}\}$.
(b) *The functions $\wp$ and $\wp'$ satisfy the relation*

$$(\wp'(z))^2 = 4(\wp(z))^3 - g_2(\Lambda)\wp(z) - g_3(\Lambda)$$

where $g_2(\Lambda) = 60G_4(\Lambda)$ and $g_3(\Lambda) = 140G_6(\Lambda)$.
(c) *Let $\Lambda = \omega_1\mathbb{Z} \oplus \omega_2\mathbb{Z}$ and let $\omega_3 = \omega_1 + \omega_2$. Then the cubic equation satisfied by $\wp$ and $\wp'$, $y^2 = 4x^3 - g_2(\Lambda)x - g_3(\Lambda)$, is*

$$y^2 = 4(x - e_1)(x - e_2)(x - e_3), \qquad e_i = \wp(\omega_i/2) \text{ for } i = 1, 2, 3.$$

This equation is nonsingular, meaning its right side has distinct roots.

Proof. (Sketch.) For (a), if $|z| < |\omega|$ then

$$\frac{1}{(z-\omega)^2} - \frac{1}{\omega^2} = \frac{1}{\omega^2}\left(\frac{1}{(1-z/\omega)^2} - 1\right)$$

and the geometric series squares to $\sum_{n=0}^{\infty}(n+1)z^n/\omega^n$, making the summand $2z/\omega^3 + 3z^2/\omega^4 + \cdots$. Convergence results from Exercise 1.1.4 allow the resulting double sum to be rearranged, and then the inner sum cancels when n is odd.

For (b), one uses part (a) to show that the nonpositive terms of the Laurent series of both sides are equal. Specifically, since

$$\wp(z) = \frac{1}{z^2} + 3G_4(\Lambda)z^2 + 5G_6(\Lambda)z^4 + \mathcal{O}(z^6)$$

(where "$\mathcal{O}$" means "a quantity on the order of") and

$$\wp'(z) = -\frac{2}{z^3} + 6G_4(\Lambda)z + 20G_6(\Lambda)z^3 + \mathcal{O}(z^5),$$

a little algebra shows that both $(\wp'(z))^2$ and $4(\wp(z))^3 - g_2(\Lambda)\wp(z) - g_3(\Lambda)$ work out to $4/z^6 - 24G_4(\Lambda)/z^2 - 80G_6(\Lambda) + \mathcal{O}(z^2)$. So their difference is holomorphic and Λ-periodic, therefore bounded, therefore constant, therefore zero since it is $\mathcal{O}(z^2)$ as $z \to 0$.

For (c), since $\wp'$ is odd, it has zeros at the order 2 points of $\mathbb{C}/\Lambda$: if $z \equiv -z \pmod{\Lambda}$ then $\wp'(z) = \wp'(-z) = -\wp'(z)$ and thus $\wp'(z) = 0$. Letting $\Lambda = \omega_1\mathbb{Z} \oplus \omega_2\mathbb{Z}$, the order 2 points are $z_i = \omega_i/2$ with $\wp'(z_i) = 0$ for $i = 1, 2, 3$. The relation between $\wp$ and $\wp'$ from (b) shows that the corresponding values $x_i = \wp(z_i)$ for $i = 1, 2, 3$ are roots of the cubic polynomial $p_\Lambda(x) = 4x^3 - g_2(\Lambda)x - g_3(\Lambda)$, so it factors as claimed. Each x_i is a double value of $\wp$ since $\wp'(z_i) = 0$, and since $\wp$ has degree 2, meaning it takes each value twice counting multiplicity (see Exercise 1.4.1(b)), this makes the three x_i distinct. That is, the cubic polynomial p_Λ has distinct roots. $\qquad\square$

Part (b) of the proposition shows that the map $z \mapsto (\wp_\Lambda(z), \wp'_\Lambda(z))$ takes nonlattice points of $\mathbb{C}$ to points $(x, y) \in \mathbb{C}^2$ satisfying the nonsingular cubic equation of part (c), $y^2 = 4x^3 - g_2(\Lambda)x - g_3(\Lambda)$. The map bijects since generally a value $x \in \mathbb{C}$ is taken by $\wp_\Lambda$ twice on $\mathbb{C}/\Lambda$, that is, $x = \wp_\Lambda(\pm z + \Lambda)$, and then the two y-values satisfying the cubic equation are $\wp'(\pm z + \Lambda) = \pm\wp'(z + \Lambda)$. The exceptional x-values where $y = 0$ occur at the order-2 points of $\mathbb{C}/\Lambda$, so they are taken once by $\wp_\Lambda$ as necessary. The map extends to all $z \in \mathbb{C}$ by mapping lattice points to a suitably defined point at infinity. In sum, for every lattice the associated Weierstrass $\wp$-function and its derivative give a bijection

$$(\wp, \wp') : \text{complex torus} \longrightarrow \text{elliptic curve.}$$

For example, the value $g_3(i) = 0$ from Section 1.1 shows that the complex torus $\mathbb{C}/\Lambda_i$ bijects to the elliptic curve with equation $y^2 = 4x^3 - g_2(i)x$.

Similarly the complex torus $\mathbb{C}/\Lambda_{\mu_3}$ (where again $\mu_3 = e^{2\pi i/3}$) bijects to the elliptic curve with equation $y^2 = 4x^3 - g_3(\mu_3)$. See Exercise 1.4.3 for some values of the functions $\wp$ and $\wp'$ in connection with these two lattices.

The map $(\wp, \wp')$ transfers the group law from the complex torus to the elliptic curve. To understand addition on the curve, let $z_1 + \Lambda$ and $z_2 + \Lambda$ be nonzero points of the torus. The image points $(\wp(z_1), \wp'(z_1))$ and $(\wp(z_2), \wp'(z_2))$ on the curve determine a secant or tangent line of the curve in $\mathbb{C}^2$, $ax + by + c = 0$. Consider the function

$$f(z) = a\wp(z) + b\wp'(z) + c.$$

This is meromorphic on $\mathbb{C}/\Lambda$. When $b \neq 0$ it has a triple pole at $0 + \Lambda$ and zeros at $z_1 + \Lambda$ and $z_2 + \Lambda$, and Exercise 1.4.1(c) shows that its third zero is at the point $z_3 + \Lambda$ such that $z_1 + z_2 + z_3 + \Lambda = 0 + \Lambda$ in $\mathbb{C}/\Lambda$. When $b = 0$, f has a double pole at $0 + \Lambda$ and zeros at $z_1 + \Lambda$ and $z_2 + \Lambda$, and Exercise 1.4.1(c) shows that $z_1 + z_2 + \Lambda = 0 + \Lambda$ in $\mathbb{C}/\Lambda$. In this case let $z_3 = 0 + \Lambda$ so that again $z_1 + z_2 + z_3 + \Lambda = 0 + \Lambda$, and since the line is vertical view it as containing the infinite point $(\wp(0), \wp'(0))$ whose second coordinate arises from a pole of higher order than the first. Thus for any value of b the elliptic curve points on the line $ax + by + c = 0$ are the points $(x_i, y_i) = (\wp(z_i), \wp'(z_i))$ for $i = 1, 2, 3$. Since $z_1 + z_2 + z_3 + \Lambda = 0 + \Lambda$ on the torus in all cases, the resulting group law on the curve is that *collinear triples sum to zero*.

Recall that a holomorphic isomorphism of complex tori takes the form $z + \Lambda \mapsto mz + \Lambda'$ where $\Lambda' = m\Lambda$. Since $\wp_{m\Lambda}(mz) = m^{-2}\wp_\Lambda(z)$ and $\wp'_{m\Lambda}(mz) = m^{-3}\wp'_\Lambda(z)$, the corresponding isomorphism of elliptic curves is $(x, y) \mapsto (m^{-2}x, m^{-3}y)$ or equivalently the substitution $(x, y) = (m^2 x', m^3 y')$, changing the cubic equation $y^2 = 4x^3 - g_2 x - g_3$ associated to Λ to the equation $y^2 = 4x^3 - m^{-4}g_2 x - m^{-6}g_3$ associated to Λ'. Suitable choices of m (Exercise 1.4.3 again) normalize the elliptic curves associated to $\mathbb{C}/m\Lambda_i$ and $\mathbb{C}/m\Lambda_{\mu_3}$ to have equations

$$y^2 = 4x(x - 1)(x + 1), \qquad y^2 = 4(x - 1)(x - \mu_3)(x - \mu_3^2).$$

We will return to these ideas in Chapter 7.

The appearance of Eisenstein series as coefficients of a nonsingular curve lets us prove as a corollary to Proposition 1.4.1 that the discriminant function from Section 1.1 has no zeros in the upper half plane.

Corollary 1.4.2. *The function Δ is nonvanishing on $\mathcal{H}$. That is, $\Delta(\tau) \neq 0$ for all $\tau \in \mathcal{H}$.*

Proof. For any $\tau \in \mathcal{H}$, specialize the lattice Λ in the proposition to Λ_τ. By part (c) of the proposition, the cubic polynomial $p_\tau(x) = 4x^3 - g_2(\tau)x - g_3(\tau)$ has distinct roots. Exercise 1.4.4 shows that $\Delta(\tau)$ is the discriminant of p_τ up to constant multiple (hence its name), so $\Delta(\tau) \neq 0$. $\square$

Not only does every complex torus $\mathbb{C}/\Lambda$ lead via the Weierstrass $\wp$-function to an elliptic curve

$$y^2 = 4x^3 - a_2 x - a_3, \quad a_2^3 - 27a_3^2 \neq 0 \tag{1.8}$$

with $a_2 = g_2(\Lambda)$ and $a_3 = g_3(\Lambda)$, but the converse holds as well.

Proposition 1.4.3. *Given an elliptic curve* (1.8), *there exists a lattice* Λ *such that* $a_2 = g_2(\Lambda)$ *and* $a_3 = g_3(\Lambda)$.

Proof. The case $a_2 = 0$ and the case $a_3 = 0$ are Exercise 1.4.5. For the case $a_2 \neq 0$ and $a_3 \neq 0$, since $j : \mathcal{H} \longrightarrow \mathbb{C}$ surjects there exists $\tau \in \mathcal{H}$ such that $j(\tau) = 1728a_2^3/(a_2^3 - 27a_3^2)$. This gives

$$\frac{g_2(\tau)^3}{g_2(\tau)^3 - 27g_3(\tau)^2} = \frac{a_2^3}{a_2^3 - 27a_3^2},$$

or, after taking reciprocals and doing a little algebra,

$$\frac{a_2^3}{g_2(\tau)^3} = \frac{a_3^2}{g_3(\tau)^2}. \tag{1.9}$$

For any nonzero complex number ω_2, let $\omega_1 = \tau\omega_2$ and $\Lambda = \omega_1\mathbb{Z} \oplus \omega_2\mathbb{Z}$. Then

$$g_2(\Lambda) = \omega_2^{-4} g_2(\tau) \quad \text{and} \quad g_3(\Lambda) = \omega_2^{-6} g_3(\tau).$$

Thus we are done if we can choose ω_2 such that

$$\omega_2^{-4} = a_2/g_2(\tau) \quad \text{and} \quad \omega_2^{-6} = a_3/g_3(\tau).$$

Choose ω_2 to satisfy the first condition, so that $\omega_2^{-12} = a_2^3/g_2(\tau)^3$. Then by (1.9) $\omega_2^{-6} = \pm a_3/g_3(\tau)$, and replacing ω_2 by $i\omega_2$ if necessary completes the proof. $\qquad\square$

It follows that any map of elliptic curves $(x, y) \mapsto (m^{-2}x, m^{-3}y)$, changing a cubic equation $y^2 = 4x^3 - a_2 x - a_3$ to the equation $y^2 = 4x^3 - m^{-4}a_2 x - m^{-6}a_3$, comes from the holomorphic isomorphism of complex tori $z + \Lambda \mapsto mz + \Lambda'$ where $a_2 = g_2(\Lambda)$, $a_3 = g_3(\Lambda)$, and $\Lambda' = m\Lambda$. This makes the map of elliptic curves an isomorphism as well.

Thus complex tori (Riemann surfaces, complex analytic objects) and elliptic curves (solution sets of cubic polynomials, algebraic objects) are interchangeable. With the connection between them in hand, let the term *complex elliptic curve* be a synonym for "complex torus" and call meromorphic functions with periods Λ *elliptic functions with respect to* Λ.

The lattice Eisenstein series $G_k(\Lambda)$ will appear occasionally from now on but generally we will work with $G_k(\tau)$.

Exercises

1.4.1. Let $E = \mathbb{C}/\Lambda$ be a complex elliptic curve where $\Lambda = \omega_1\mathbb{Z} \oplus \omega_2\mathbb{Z}$, and let f be a nonconstant elliptic function with respect to Λ, viewed either as a meromorphic function on $\mathbb{C}$ with periods Λ or as a meromorphic function on E. Let $P = \{x_1\omega_1 + x_2\omega_2 : x_1, x_2 \in [0, 1]\}$ be the parallelogram representing E when its opposing boundary edges are suitably identified, and let ∂P be the counterclockwise boundary of P. Since f has only finitely many zeros and poles in E, some translation $t + \partial P$ misses them all. This exercise establishes some necessary properties of f. Showing that these properties are sufficient for an appropriate f to exist requires more work, see for example [JS87].

(a) Compute that $1/(2\pi i) \int_{t+\partial P} f(z)dz = 0$. It follows by the Residue Theorem that the sum of the residues of f on E is 0. In particular there is no meromorphic function on E with one simple pole and so the Weierstrass $\wp$-function, with its double pole at Λ, is the simplest nonconstant elliptic function with respect to Λ.

(b) Compute that $1/(2\pi i) \int_{t+\partial P} f'(z)dz/f(z) = 0$. It follows by the Argument Principle that f has as many zeros as poles, counting multiplicity. Replacing f by $f - w$ for any $w \in \mathbb{C}$ shows that f takes every value the same number of times, counting multiplicity. In particular the Weierstrass $\wp$-function on E takes every value twice.

(c) Compute that $1/(2\pi i) \int_{t+\partial P} zf'(z)dz/f(z) \in \Lambda$. Show that this integral is also $\sum_{x \in E} \nu_x(f)x$ where $\nu_x(f)$ is the order of f at x, meaning that $f(z) = (z - x)^{\nu_x(f)}g(z)$ with $g(x) \neq 0$. Note that $\nu_x(f) = 0$ except at zeros and poles of f, so the sum is finite. Thus parts (b) and (c) combine to show that for any nonconstant meromorphic function f on E,

$$\sum_{x \in E} \nu_x(f) = 0 \text{ in } \mathbb{Z} \quad \text{and} \quad \sum_{x \in E} \nu_x(f)x = 0 \text{ in } E.$$

(A hint for this exercise is at the end of the book.)

1.4.2. Let $\Lambda = \omega_1\mathbb{Z} \oplus \omega_2\mathbb{Z}$ be a lattice and let $\wp$ be its associated Weierstrass $\wp$-function.

(a) Show that $\wp$ is even and that $\wp'$ is Λ-periodic.

(b) For $i = 1, 2$ show that the function $\wp(z+\omega_i) - \wp(z)$ is some constant c_i by taking its derivative. Substitute $z = -\omega_i/2$ to show that $c_i = 0$. Conclude that $\wp$ is Λ-periodic.

1.4.3. Let $\Lambda = \Lambda_i$. The derivative $\wp'$ of the corresponding Weierstrass function has a triple pole at 0 and simple zeros at $1/2$, $i/2$, $(1+i)/2$. Since $i\Lambda = \Lambda$ it follows that

$$\wp(iz) = \wp_{i\Lambda}(iz) = i^{-2}\wp(z) = -\wp(z).$$

Show that in particular $\wp((1 + i)/2) = 0$. This is a double zero of $\wp$ since also $\wp'((1 + i)/2) = 0$, making it the only zero of $\wp$ as a function on $\mathbb{C}/\Lambda$. Since $\overline{\Lambda} = \Lambda$ (complex conjugation) it also follows that $\wp(\overline{z}) = \overline{\wp(z)}$, so that

$\wp(1/2)$ is real, as is $\wp(i/2) = -\wp(1/2)$. Compute some dominant terms of $\wp(1/2)$ and $\wp(i/2)$ to show that $\wp(1/2)$ is the positive value. For what m does the complex torus $\mathbb{C}/m\Lambda$ correspond to the elliptic curve with equation $y^2 = 4x(x-1)(x+1)$?

Reason similarly with $\Lambda = \Lambda_{\mu_3}$ to find the zeros of the corresponding Weierstrass function $\wp$ and to show that $\wp(1/2)$ is real. For what m does the complex torus $\mathbb{C}/m\Lambda$ correspond to the elliptic curve with equation $y^2 = 4(x-1)(x-\mu_3)(x-\mu_3^2)$? (A hint for this exercise is at the end of the book.)

1.4.4. For $\tau \in \mathcal{H}$ let $p_\tau(x) = 4x^3 - g_2(\tau)x - g_3(\tau)$. Show that the discriminant of p_τ equals $\Delta(\tau)$ up to constant multiple, where Δ is the cusp form from Section 1.1.

1.4.5. Show that when $a_2 = 0$ in Proposition 1.4.3 the desired lattice is $\Lambda = m\Lambda_{\mu_3}$ for a suitably chosen m. Prove the case $a_3 = 0$ in Proposition 1.4.3 similarly.

1.5 Modular curves and moduli spaces

Recall from Corollary 1.3.3 that two complex elliptic curves $\mathbb{C}/\Lambda$ and $\mathbb{C}/\Lambda'$ are holomorphically group-isomorphic if and only if $m\Lambda = \Lambda'$ for some $m \in \mathbb{C}$. Viewing two such curves as equivalent gives a quotient set of equivalence classes of complex elliptic curves. Similarly, view two points τ and τ' of the upper half plane as equivalent if and only if $\gamma(\tau) = \tau'$ for some $\gamma \in \mathrm{SL}_2(\mathbb{Z})$, and consider the resulting quotient set as well. This section shows that there is a bijection from the first quotient set to the second. That is, the equivalence classes of points in the upper half plane under the action of the modular group are described by the isomorphism classes of complex elliptic curves. Theorem 1.5.1 to follow shows considerably more, that the quotients of the upper half plane by the various congruence subgroups from Section 1.2 are described by the sets of equivalence classes of elliptic curves enhanced by corresponding torsion data.

We begin by describing the relevant torsion data for the congruence subgroups. Let N be a positive integer. An *enhanced elliptic curve for* $\Gamma_0(N)$ is an ordered pair (E, C) where E is a complex elliptic curve and C is a cyclic subgroup of E of order N. Two such pairs (E, C) and (E', C') are *equivalent*, written $(E, C) \sim (E', C')$, if some isomorphism $E \xrightarrow{\sim} E'$ takes C to C'. The set of equivalence classes is denoted

$$S_0(N) = \{\text{enhanced elliptic curves for } \Gamma_0(N)\}/\sim.$$

An element of $S_0(N)$ is denoted $[E, C]$, the square brackets connoting equivalence class.

An *enhanced elliptic curve for* $\Gamma_1(N)$ is a pair (E, Q) where E is a complex elliptic curve and Q is a point of E of order N. (Thus $NQ = 0$ but $nQ \neq 0$

for $0 < n < N$.) Two such pairs (E, Q) and (E', Q') are equivalent if some isomorphism $E \xrightarrow{\sim} E'$ takes Q to Q'. The set of equivalence classes is denoted

$$S_1(N) = \{\text{enhanced elliptic curves for } \Gamma_1(N)\}/ \sim .$$

An element of $S_1(N)$ is denoted $[E, Q]$.

An *enhanced elliptic curve for* $\Gamma(N)$ is a pair $(E, (P, Q))$ where E is a complex elliptic curve and (P, Q) is a pair of points of E that generates the N-torsion subgroup $E[N]$ with Weil pairing $e_N(P, Q) = e^{2\pi i/N}$. From Section 1.3 $e_N(P, Q)$ is some primitive complex Nth root of unity, but this condition is more specific. Two such pairs $(E, (P, Q))$ and $(E', (P', Q'))$ are equivalent if some isomorphism $E \xrightarrow{\sim} E'$ takes P to P' and Q to Q'. The set of equivalence classes is denoted

$$S(N) = \{\text{enhanced elliptic curves for } \Gamma(N)\}/ \sim .$$

An element of $S(N)$ is denoted $[E, (P, Q)]$.

Each of $S_0(N)$, $S_1(N)$, and $S(N)$ is a *space of moduli* or *moduli space* of isomorphism classes of complex elliptic curves and N-torsion data. When $N = 1$ all three moduli spaces reduce to the isomorphism classes of complex elliptic curves as described at the beginning of the section.

For any congruence subgroup Γ of $SL_2(\mathbb{Z})$, acting on the upper half plane $\mathcal{H}$ from the left, the *modular curve* $Y(\Gamma)$ is defined as the quotient space of orbits under Γ,

$$Y(\Gamma) = \Gamma \backslash \mathcal{H} = \{\Gamma\tau : \tau \in \mathcal{H}\}.$$

The modular curves for $\Gamma_0(N)$, $\Gamma_1(N)$, and $\Gamma(N)$ are denoted

$$Y_0(N) = \Gamma_0(N)\backslash\mathcal{H}, \qquad Y_1(N) = \Gamma_1(N)\backslash\mathcal{H}, \qquad Y(N) = \Gamma(N)\backslash\mathcal{H}.$$

Chapter 2 will show that modular curves are Riemann surfaces and they can be compactified. Compact Riemann surfaces are described by polynomial equations. Thus modular curves have complex analytic and algebraic characterizations like complex elliptic curves. Chapter 7 will show how the moduli spaces arise from a single elliptic curve, and it will further show that the polynomials describing $Y_0(N)$ and $Y_1(N)$ have rational coefficients. For now we continue to work complex analytically and show that the moduli spaces map bijectively to noncompactified modular curves. Recall the lattice $\Lambda_\tau = \tau\mathbb{Z} \oplus \mathbb{Z}$ for $\tau \in \mathcal{H}$. Proving the following theorem amounts to checking that the torsion data defining the moduli spaces match the conditions defining the congruence subgroups.

Theorem 1.5.1. *Let N be a positive integer.*

(a) *The moduli space for $\Gamma_0(N)$ is*

$$S_0(N) = \{[E_\tau, \langle 1/N + \Lambda_\tau \rangle] : \tau \in \mathcal{H}\}.$$

Two points $[E_\tau, \langle 1/N + \Lambda_\tau \rangle]$ *and* $[E_{\tau'}, \langle 1/N + \Lambda_{\tau'} \rangle]$ *are equal if and only if* $\Gamma_0(N)\tau = \Gamma_0(N)\tau'$. *Thus there is a bijection*

$$\psi_0 : S_0(N) \xrightarrow{\sim} Y_0(N), \qquad [\mathbb{C}/\Lambda_\tau, \langle 1/N + \Lambda_\tau \rangle] \mapsto \Gamma_0(N)\tau.$$

(b) *The moduli space for* $\Gamma_1(N)$ *is*

$$S_1(N) = \{[E_\tau, 1/N + \Lambda_\tau] : \tau \in \mathcal{H}\}.$$

Two points $[E_\tau, 1/N + \Lambda_\tau]$ *and* $[E_{\tau'}, 1/N + \Lambda_{\tau'}]$ *are equal if and only if* $\Gamma_1(N)\tau = \Gamma_1(N)\tau'$. *Thus there is a bijection*

$$\psi_1 : S_1(N) \xrightarrow{\sim} Y_1(N), \qquad [\mathbb{C}/\Lambda_\tau, 1/N + \Lambda_\tau] \mapsto \Gamma_1(N)\tau.$$

(c) *The moduli space for* $\Gamma(N)$ *is*

$$S(N) = \{[\mathbb{C}/\Lambda_\tau, (\tau/N + \Lambda_\tau, 1/N + \Lambda_\tau)] : \tau \in \mathcal{H}\}.$$

Two points $[\mathbb{C}/\Lambda_\tau, (\tau/N + \Lambda_\tau, 1/N + \Lambda_\tau)]$, $[\mathbb{C}/\Lambda_{\tau'}, (\tau'/N + \Lambda_{\tau'}, 1/N + \Lambda_{\tau'})]$ *are equal if and only if* $\Gamma(N)\tau = \Gamma(N)\tau'$. *Thus there is a bijection*

$$\psi : S(N) \xrightarrow{\sim} Y(N), \qquad [\mathbb{C}/\Lambda_\tau, (\tau/N + \Lambda_\tau, 1/N + \Lambda_\tau)] \mapsto \Gamma(N)\tau.$$

Proof. Parts (a) and (c) are left as Exercise 1.5.1. For (b), take any point $[E, Q]$ of $S_1(N)$. Since E is isomorphic to $\mathbb{C}/\Lambda_{\tau'}$ for some $\tau' \in \mathcal{H}$ as discussed in Section 1.3, we may take $E = \mathbb{C}/\Lambda_{\tau'}$. Thus $Q = (c\tau' + d)/N + \Lambda_{\tau'}$ for some $c, d \in \mathbb{Z}$. Then $\gcd(c, d, N) = 1$ because the order of Q is exactly N, i.e., $ad - bc - kN = 1$ for some a, b, and k, and the matrix $\gamma = \begin{bmatrix} a & b \\ c & d \end{bmatrix} \in M_2(\mathbb{Z})$ reduces modulo N into $SL_2(\mathbb{Z}/N\mathbb{Z})$. Modifying the entries of γ modulo N doesn't affect Q, so since $SL_2(\mathbb{Z})$ surjects to $SL_2(\mathbb{Z}/N\mathbb{Z})$ we may take $\gamma = \begin{bmatrix} a & b \\ c & d \end{bmatrix} \in SL_2(\mathbb{Z})$. Let $\tau = \gamma(\tau')$ and let $m = c\tau' + d$. Then $m\tau = a\tau' + b$, so

$$m\Lambda_\tau = m(\tau\mathbb{Z} \oplus \mathbb{Z}) = (a\tau' + b)\mathbb{Z} \oplus (c\tau' + d)\mathbb{Z} = \tau'\mathbb{Z} \oplus \mathbb{Z} = \Lambda_{\tau'}$$

(using Lemma 1.3.1 for the third equality), and

$$m\left(\frac{1}{N} + \Lambda_\tau\right) = \frac{c\tau' + d}{N} + \Lambda_{\tau'} = Q.$$

This shows that $[E, Q] = [\mathbb{C}/\Lambda_\tau, 1/N + \Lambda_\tau]$ where $\tau \in \mathcal{H}$.

Suppose two points $\tau, \tau' \in \mathcal{H}$ satisfy $\Gamma_1(N)\tau = \Gamma_1(N)\tau'$. Thus $\tau = \gamma(\tau')$ where $\gamma = \begin{bmatrix} a & b \\ c & d \end{bmatrix} \in \Gamma_1(N)$. Again let $m = c\tau' + d$. Then as just shown,

$$m\Lambda_\tau = \Lambda_{\tau'}, \qquad m\left(\frac{1}{N} + \Lambda_\tau\right) = \frac{c\tau' + d}{N} + \Lambda_{\tau'}.$$

But since $(c, d) \equiv (0, 1) \pmod{N}$ the second equality is now $m(1/N + \Lambda_\tau) = 1/N + \Lambda_{\tau'}$. Thus $[\mathbb{C}/\Lambda_\tau, 1/N + \Lambda_\tau] = [\mathbb{C}/\Lambda_{\tau'}, 1/N + \Lambda_{\tau'}]$.

Conversely, suppose $[\mathbb{C}/\Lambda_\tau, 1/N + \Lambda_\tau] = [\mathbb{C}/\Lambda_{\tau'}, 1/N + \Lambda_{\tau'}]$ with $\tau, \tau' \in \mathcal{H}$. Then for some $m \in \mathbb{C}$, $m\Lambda_\tau = \Lambda_{\tau'}$ (by Corollary 1.3.3) and $m(1/N + \Lambda_\tau) = 1/N + \Lambda_{\tau'}$. By Lemma 1.3.1 the first of these conditions means that

$$\begin{bmatrix} m\tau \\ m \end{bmatrix} = \gamma \begin{bmatrix} \tau' \\ 1 \end{bmatrix} \quad \text{for some } \gamma = \begin{bmatrix} a & b \\ c & d \end{bmatrix} \in \mathrm{SL}_2(\mathbb{Z}), \tag{1.10}$$

so in particular $m = c\tau' + d$. Now the second condition becomes

$$\frac{c\tau' + d}{N} + \Lambda_{\tau'} = \frac{1}{N} + \Lambda_{\tau'},$$

showing that $(c, d) \equiv (0, 1) \pmod{N}$ and $\gamma \in \Gamma_1(N)$. Since $\tau = \gamma(\tau')$ by (1.10), it follows that $\Gamma_1(N)\tau = \Gamma_1(N)\tau'$.   □

Specializing to $N = 1$, the theorem shows that the space of isomorphism classes of complex elliptic curves parametrizes the modular curve $Y_0(1) = Y_1(1) = Y(1) = \mathrm{SL}_2(\mathbb{Z})\backslash\mathcal{H}$ as mentioned in Section 1.3 and at the beginning of this section. This lets us associate a complex number to each isomorphism class. Recall the modular invariant j from Section 1.1, an $\mathrm{SL}_2(\mathbb{Z})$-invariant function on $\mathcal{H}$. Each isomorphism class of complex elliptic curves has an associated orbit $\mathrm{SL}_2(\mathbb{Z})\tau \in \mathrm{SL}_2(\mathbb{Z})\backslash\mathcal{H}$ and thus has a well defined invariant $j(\mathrm{SL}_2(\mathbb{Z})\tau)$. This value is also associated to any complex elliptic curve E in the isomorphism class and correspondingly denoted $j(E)$. The Modularity Theorem states that the elliptic curves with rational j-values arise from modular forms, as discussed back in the preface.

It follows from Theorem 1.5.1 that maps of the modular curves $Y_0(N)$, $Y_1(N)$, and $Y(N)$ give rise to maps of moduli spaces. For example, the natural map from $Y_1(N)$ to $Y_0(N)$ taking orbits $\Gamma_1(N)\tau$ to $\Gamma_0(N)\tau$ becomes the map from $S_1(N)$ to $S_0(N)$ taking equivalence classes $[E, Q]$ to $[E, \langle Q\rangle]$, forgetting the generator but keeping the group it generates. For another example, since $\Gamma_1(N)$ is a normal subgroup of $\Gamma_0(N)$ the quotient group acts on $Y_1(N)$ and therefore on $S_1(N)$. The action works out to

$$\Gamma_1(N)\gamma : [E, Q] \mapsto [E, dQ] \quad \text{where } \gamma \equiv \begin{bmatrix} a & b \\ c & d \end{bmatrix} \pmod{N}.$$

This self-map will recur in Chapter 5 and thereafter as a Hecke operator of the sort mentioned in the preface. For details of these examples and others see Exercises 1.5.2 through 1.5.6. The hybrid maps from Exercise 1.5.6,

$$[E, C, Q] \mapsto [E, Q], \qquad [E, C, Q] \mapsto [E/C, Q + C], \tag{1.11}$$

where C is a cyclic subgroup of prime order p and Q is a point of order N and $C \cap \langle Q\rangle = \{0_E\}$, combine to describe the other Hecke operator to be introduced in Chapter 5,

$$[E, Q] \mapsto \sum_C [E/C, Q + C].$$

The bijections between moduli spaces and modular curves give more examples of modular forms. The idea is that a class of functions of enhanced elliptic curves corresponds to the weight-k invariant functions on the upper half plane. Let k be an integer and let Γ be one of $\Gamma_0(N)$, $\Gamma_1(N)$, or $\Gamma(N)$. A complex-valued function F of the enhanced elliptic curves for Γ is *degree-k homogeneous with respect to Γ* if for every nonzero complex number m,

$$\left.\begin{array}{l} F(\mathbb{C}/m\Lambda, mC) \\ F(\mathbb{C}/m\Lambda, mQ) \\ F(\mathbb{C}/m\Lambda, (mP, mQ)) \end{array}\right\} = \begin{cases} m^{-k}F(\mathbb{C}/\Lambda, C) & \text{if } \Gamma = \Gamma_0(N), \\ m^{-k}F(\mathbb{C}/\Lambda, Q) & \text{if } \Gamma = \Gamma_1(N), \\ m^{-k}F(\mathbb{C}/\Lambda, (P, Q)) & \text{if } \Gamma = \Gamma(N). \end{cases} \quad (1.12)$$

Given such a function F, define the corresponding *dehomogenized* function $f : \mathcal{H} \longrightarrow \mathbb{C}$ by the rule

$$f(\tau) = \begin{cases} F(\mathbb{C}/\Lambda_\tau, \langle 1/N + \Lambda_\tau \rangle) & \text{if } \Gamma = \Gamma_0(N), \\ F(\mathbb{C}/\Lambda_\tau, 1/N + \Lambda_\tau) & \text{if } \Gamma = \Gamma_1(N), \\ F(\mathbb{C}/\Lambda_\tau, (\tau/N + \Lambda_\tau, 1/N + \Lambda_\tau)) & \text{if } \Gamma = \Gamma(N). \end{cases} \quad (1.13)$$

Then f is weight-k invariant with respect to Γ. To see this, let $\gamma = \left[\begin{smallmatrix} a & b \\ c & d \end{smallmatrix}\right] \in \Gamma$ and for any $\tau \in \mathcal{H}$ let $m = (c\tau + d)^{-1}$. Then, e.g., for $\Gamma = \Gamma_1(N)$, using the condition $(c, d) \equiv (0, 1) \pmod N$ at the third step,

$$f(\gamma(\tau)) = F(\mathbb{C}/\Lambda_{\gamma(\tau)}, 1/N + \Lambda_{\gamma(\tau)}) = F(\mathbb{C}/m\Lambda_\tau, m(c\tau + d)/N + m\Lambda_\tau)$$
$$= m^{-k}F(\mathbb{C}/\Lambda_\tau, 1/N + \Lambda_\tau) = (c\tau + d)^k f(\tau).$$

For instance, the lattice Eisenstein series from Section 1.4 are degree-k homogeneous with respect to $\mathrm{SL}_2(\mathbb{Z})$, dehomogenizing to Eisenstein series on the upper half plane.

Conversely, let f be weight-k invariant with respect to Γ where Γ is one of $\Gamma_0(N)$, $\Gamma_1(N)$, or $\Gamma(N)$. Then formula (1.13) turns around to define a function F on enhanced elliptic curves of the special type $(\mathbb{C}/\Lambda_\tau, (\text{torsion data}))$ in terms of f (Exercise 1.5.7). If two such enhanced elliptic curves are equivalent, e.g., $(\mathbb{C}/\Lambda_{\tau'}, 1/N + \Lambda_{\tau'}) = (\mathbb{C}/m\Lambda_\tau, m/N + m\Lambda_\tau)$, then $\tau = \gamma(\tau')$ and $m = c\tau' + d$ for some $\gamma = \left[\begin{smallmatrix} a & b \\ c & d \end{smallmatrix}\right] \in \Gamma$ as in the proof of Theorem 1.5.1. Thus $f(\tau) = m^k f(\tau')$ and so F obeys formula (1.12) for those two points,

$$F(\mathbb{C}/\Lambda_{\tau'}, 1/N + \Lambda_{\tau'}) = f(\tau') = m^{-k}f(\tau) = m^{-k}F(\mathbb{C}/\Lambda_\tau, 1/N + \Lambda_\tau).$$

Since every enhanced elliptic curve is equivalent to an enhanced elliptic curve of the special type, formula (1.12) in its full generality extends F to a degree-k homogeneous function of enhanced elliptic curves for Γ.

As an example of the correspondence, let $N > 1$ and let $v = (c_v, d_v) \in \mathbb{Z}^2$ be a vector whose reduction $\bar{v}$ modulo N is nonzero. Define a function of enhanced elliptic curves

$$F_2^{\bar{v}}(\mathbb{C}/\Lambda, (P, Q)) = \frac{1}{N^2}\, \wp_\Lambda(c_v P + d_v Q).$$

(The superscript $\bar{v}$ is just a label, not an exponent.) Then $F_2^{\bar{v}}$ is degree-2 homogeneous with respect to $\Gamma(N)$ (Exercise 1.5.8(a)). Chapter 4 will use the corresponding function

$$f_2^{\bar{v}}(\tau) = \frac{1}{N^2} \, \wp_\tau\!\left(\frac{c_v \tau + d_v}{N}\right)$$

to construct Eisenstein series of weight 2. Similarly, recall the lattice constants $g_2(\Lambda) = 60 G_4(\Lambda)$ and $g_3(\Lambda) = 140 G_6(\Lambda)$, the integer multiples of lattice Eisenstein series from Section 1.4, and define functions of enhanced elliptic curves for $\Gamma(N)$, $\Gamma_1(N)$, and $\Gamma_0(N)$,

$$F_0^{\bar{v}}(\mathbb{C}/\Lambda, (P, Q)) = \frac{g_2(\Lambda)}{g_3(\Lambda)} \, \wp_\Lambda(c_v P + d_v Q),$$

$$F_0^{\bar{d}}(\mathbb{C}/\Lambda, Q) = \frac{g_2(\Lambda)}{g_3(\Lambda)} \, \wp_\Lambda(dQ), \qquad\qquad d \in \mathbb{Z},\, d \not\equiv 0 \ (\mathrm{mod}\ N)$$

$$= F_0^{\overline{(0,d)}}(\mathbb{C}/\Lambda, (P, Q)) \qquad \text{for suitable } P,$$

$$F_0(\mathbb{C}/\Lambda, C) = \frac{g_2(\Lambda)}{g_3(\Lambda)} \sum_{Q \in C - \{0\}} \wp_\Lambda(Q)$$

$$= \sum_{d=1}^{N-1} F_0^{\bar{d}}(\mathbb{C}/\Lambda, dQ) \qquad \text{for any generator } Q \text{ of } C.$$

Each of these functions is degree-0 homogeneous with respect to its group (Exercise 1.5.8(b)). The corresponding weight-0 invariant functions are

$$f_0^{\bar{v}}(\tau) = \frac{g_2(\tau)}{g_3(\tau)} \, \wp_\tau\!\left(\frac{c_v \tau + d_v}{N}\right),$$

$$f_0^{\bar{d}}(\tau) = \frac{g_2(\tau)}{g_3(\tau)} \, \wp_\tau\!\left(\frac{d}{N}\right) = f_0^{\overline{(0,d)}}(\tau),$$

$$f_0(\tau) = \frac{g_2(\tau)}{g_3(\tau)} \sum_{d=1}^{N-1} \wp_\tau\!\left(\frac{d}{N}\right) = \sum_{d=1}^{N-1} f_0^{\bar{d}}(\tau).$$

Chapter 7 will use these functions to show how polynomial equations describe modular curves. We could have defined the functions with $g_2(\tau) g_3(\tau)/\Delta(\tau)$ where we have $g_2(\tau)/g_3(\tau)$, to avoid poles in $\mathcal{H}$, but the meromorphic functions here are normalized more suitably for our later purposes.

Exercises

1.5.1. Prove the other two parts of Theorem 1.5.1. (A hint for this exercise is at the end of the book.)

1.5.2. The containments $\Gamma(N) \subset \Gamma_1(N) \subset \Gamma_0(N)$ give rise to surjections $Y(N) \longrightarrow Y_1(N)$ and $Y_1(N) \longrightarrow Y_0(N)$ given by $\Gamma(N)\tau \mapsto \Gamma_1(N)\tau$ and $\Gamma_1(N)\tau \mapsto \Gamma_0(N)\tau$. Describe the corresponding maps between moduli spaces. (A hint for this exercise is at the end of the book.)

1.5.3. Since the group containments mentioned in the previous exercises are normal, it follows that the quotient $\Gamma_1(N)/\Gamma(N) \cong \mathbb{Z}/N\mathbb{Z}$ (cf. Section 1.2) acts on the modular curve $Y(N)$ by multiplication from the left, and similarly for $\Gamma_0(N)/\Gamma_1(N) \cong (\mathbb{Z}/N\mathbb{Z})^*$ and $Y_1(N)$. Describe the corresponding action of $\mathbb{Z}/N\mathbb{Z}$ on the moduli space $S(N)$, and similarly for $(\mathbb{Z}/N\mathbb{Z})^*$ and $S_1(N)$. (A hint for this exercise is at the end of the book.)

1.5.4. Let $w_N = \left[\begin{smallmatrix} 0 & -1 \\ N & 0 \end{smallmatrix}\right] \in \mathrm{GL}_2^+(\mathbb{Q})$. Show that w_N normalizes the group $\Gamma_0(N)$ and so gives an automorphism $\Gamma_0(N)\tau \mapsto \Gamma_0(N)w_N(\tau)$ of the modular curve $Y_0(N)$. Show that this automorphism is an involution (meaning it has order 2) and describe the corresponding automorphism of the moduli space $S_0(N)$. (A hint for this exercise is at the end of the book.)

1.5.5. Let N be a positive integer and let p be prime. Define maps π_1, π_2 : $Y_0(Np) \longrightarrow Y_0(N)$ to be $\pi_1(\Gamma_0(Np)\tau) = \Gamma_0(N)\tau$ and $\pi_2(\Gamma_0(Np)\tau) = \Gamma_0(N)(p\tau)$. How do the corresponding maps $\hat{\pi}_1, \hat{\pi}_2 : S_0(Np) \longrightarrow S_0(N)$ act on equivalence classes $[E, C]$? Same question, but for $\Gamma_1(Np)$ and $\Gamma_1(N)$ and $S_1(Np)$ and $S_1(N)$, and then again for $\Gamma(Np)$, etc.

1.5.6. Let N be a positive integer and let p be prime. Define congruence subgroups of $\mathrm{SL}_2(\mathbb{Z})$,

$$\Gamma^0(p) = \left\{ \begin{bmatrix} a & b \\ c & d \end{bmatrix} \in \mathrm{SL}_2(\mathbb{Z}) : \begin{bmatrix} a & b \\ c & d \end{bmatrix} \equiv \begin{bmatrix} * & 0 \\ * & * \end{bmatrix} \pmod{p} \right\}$$

and

$$\Gamma_1^0(N, p) = \Gamma_1(N) \cap \Gamma^0(p),$$

and define a corresponding modular curve,

$$Y_1^0(N, p) = Y(\Gamma_1^0(N, p)).$$

An enhanced elliptic curve for $\Gamma_1^0(N, p)$ is an ordered triple (E, C, Q) where E is a complex elliptic curve, C is a cyclic subgroup of E of order p, and Q is a point of E of order N such that $C \cap \langle Q \rangle = \{0_E\}$. Two such triples (E, C, Q) and (E', C', Q') are equivalent if some isomorphism $E \xrightarrow{\sim} E'$ takes C to C' and Q to Q'. The moduli space for $\Gamma_1^0(N, p)$ is the set of equivalence classes,

$$S_1^0(N, p) = \{\text{enhanced elliptic curves for } \Gamma_1^0(N, p)\}/ \sim .$$

An element of $S_1^0(N, p)$ is denoted $[E, C, Q]$.

(a) By the proof of Theorem 1.5.1(b), every element of $S_1^0(N, p)$ takes the form $[\mathbb{C}/\Lambda_{\tau'}, C, 1/N + \Lambda_{\tau'}]$. Show that C has a generator of the form

$(\tau' + j)/p + \Lambda_{\tau'}$ or $1/p + \Lambda_{\tau'}$, this last possibility arising only if $p \nmid N$. (This idea will be elaborated in Section 5.2.) Show that therefore C and Q can be assumed to take the form

$$C = \left\langle \frac{a\tau' + b}{p} + \Lambda_{\tau'} \right\rangle, \qquad Q = \frac{c\tau' + d}{N} + \Lambda_{\tau'}, \qquad \begin{bmatrix} a & b \\ c & d \end{bmatrix} \in \mathrm{SL}_2(\mathbb{Z}).$$

Let $\gamma = \begin{bmatrix} a & b \\ c & d \end{bmatrix}$ and let $\tau = \gamma(\tau')$. Show that $[\mathbb{C}/\Lambda_{\tau'}, C, 1/N + \Lambda_{\tau'}] = [\mathbb{C}/\Lambda_\tau, \langle \tau/p + \Lambda_\tau \rangle, 1/N + \Lambda_\tau]$. (A hint for this exercise is at the end of the book.)

(b) Part (a) shows that the moduli space for $\Gamma_1^0(N, p)$ is

$$S_1^0(N, p) = \{ [\mathbb{C}/\Lambda_\tau, \langle \tau/p + \Lambda_\tau \rangle, 1/N + \Lambda_\tau] : \tau \in \mathcal{H} \}.$$

Show that two points of $S_1^0(N, p)$ corresponding to $\tau, \tau' \in \mathcal{H}$ are equal if and only if $\Gamma_1^0(N, p)\tau = \Gamma_1^0(N, p)\tau'$, and thus there is a bijection

$$\psi_1^0 : S_1^0(N, p) \longrightarrow Y_1^0(N, p), \qquad [\mathbb{C}/\Lambda_\tau, \langle \tau/p + \Lambda_\tau \rangle, 1/N + \Lambda_\tau] \mapsto \Gamma_1^0(N, p)\tau.$$

(c) Show that the maps $\Gamma_1^0(N, p)\tau \mapsto \Gamma_1(N)\tau$ and $\Gamma_1^0(N, p)\tau \mapsto \Gamma_1(N)\tau/p$ from $Y_1^0(N, p)$ to $Y_1(N)$ are well defined. Show that the corresponding maps from $S_1^0(N, p)$ to $S_1(N)$ are as described in (1.11).

1.5.7. Show that if two enhanced elliptic curves for Γ of the special type $(\mathbb{C}/\Lambda_\tau, (\text{torsion data}))$ are equal then $\tau' = \gamma(\tau)$ for some $\gamma = \begin{bmatrix} 1 & b \\ 0 & 1 \end{bmatrix} \in \Gamma$. Show that consequently if $f : \mathcal{H} \longrightarrow \mathbb{C}$ is weight-k invariant with respect to Γ then formula (1.13) gives a well defined function F on enhanced elliptic curves of the special type.

1.5.8. (a) Show that the function $F_2^{\bar{v}}$ from the section is degree-2 homogeneous with respect to $\Gamma(N)$.

(b) Similarly, show that the functions $F_0^{\bar{v}}$, $F_0^{\bar{d}}$, and F_0 are degree-0 homogeneous with respect to $\Gamma(N)$, $\Gamma_1(N)$, and $\Gamma_0(N)$ respectively.

2

Modular Curves as Riemann Surfaces

For any congruence subgroup Γ of $SL_2(\mathbb{Z})$ the corresponding modular curve has been defined as the quotient space $\Gamma \backslash \mathcal{H}$, the set of orbits

$$Y(\Gamma) = \{\Gamma \tau : \tau \in \mathcal{H}\}.$$

This chapter shows that $Y(\Gamma)$ can be made into a Riemann surface that can be compactified. The resulting compact Riemann surface is denoted $X(\Gamma)$.

Related reading: [FK80] is one of many good books about Riemann surfaces.

2.1 Topology

The upper half plane $\mathcal{H}$ inherits the Euclidean topology as a subspace of $\mathbb{R}^2$. The natural surjection

$$\pi : \mathcal{H} \longrightarrow Y(\Gamma), \qquad \pi(\tau) = \Gamma \tau$$

gives $Y(\Gamma)$ the quotient topology, meaning a subset of $Y(\Gamma)$ is open if its inverse image under π in $\mathcal{H}$ is open. This makes π an open mapping (Exercise 2.1.1). And the following equivalence holds (Exercise 2.1.2):

$$\pi(U_1) \cap \pi(U_2) = \emptyset \text{ in } Y(\Gamma) \quad \Longleftrightarrow \quad \Gamma(U_1) \cap U_2 = \emptyset \text{ in } \mathcal{H}. \qquad (2.1)$$

Since $\mathcal{H}$ is connected and π is continuous, the quotient $Y(\Gamma)$ is also connected.

This section will show that $Y(\Gamma)$ is Hausdorff, meaning distinct points have disjoint neighborhoods. The key to this, and to putting coordinate charts on $Y(\Gamma)$ in the next section, is the idea that any two points in $\mathcal{H}$ have neighborhoods small enough that every $SL_2(\mathbb{Z})$ transformation taking one point away from the other also takes its neighborhood away from the other's. In the parlance of topology, the action of $SL_2(\mathbb{Z})$ on $\mathcal{H}$ is *properly discontinuous*. The precise statement is

© Springer Science+Business Media New York 2005
F. Diamond, J. Shurman, *A First Course in Modular Forms*,
Graduate Texts in Mathematics 228, DOI 10.1007/978-0-387-27226-9_2

Proposition 2.1.1. *Let $\tau_1, \tau_2 \in \mathcal{H}$ be given. Then there exist neighborhoods U_1 of τ_1 and U_2 of τ_2 in $\mathcal{H}$ with the property*

for all $\gamma \in \mathrm{SL}_2(\mathbb{Z})$, if $\gamma(U_1) \cap U_2 \neq \emptyset$ then $\gamma(\tau_1) = \tau_2$.

Note that τ_1 and τ_2 in the proposition can be equal.

With the transformations $\mathrm{SL}_2(\mathbb{Z})$ viewed as congruent motions, the upper half plane $\mathcal{H}$ has a geometry concentrated down near the real axis, essentially the same as how the hyperbolic geometry of the disk (perhaps familiar to the reader from M. C. Escher's artwork) is concentrated out near the rim. This idea, which will become clearer in Section 2.3, motivates condition (2.2) in the following proof.

Proof. Let U_1' be any neighborhood of τ_1 with compact closure in $\mathcal{H}$ and similarly for U_2'. Consider the intersection $\gamma(U_1') \cap U_2'$ for $\gamma = \left[\begin{smallmatrix} a & b \\ c & d \end{smallmatrix}\right] \in \mathrm{SL}_2(\mathbb{Z})$. For all but finitely many integer pairs (c,d) with $\gcd(c,d) = 1$ the condition

$$\sup\{\mathrm{Im}(\gamma(\tau)) : \gamma \in \mathrm{SL}_2(\mathbb{Z}) \text{ has bottom row } (c,d), \ \tau \in U_1'\}$$
$$< \inf\{\mathrm{Im}(\tau) : \tau \in U_2'\} \quad (2.2)$$

holds (Exercise 2.1.3(a)), making $\gamma(U_1') \cap U_2'$ empty. Also, for every integer pair (c,d) with $\gcd(c,d) = 1$, the matrices $\gamma \in \mathrm{SL}_2(\mathbb{Z})$ with bottom row (c,d) are

$$\left\{ \begin{bmatrix} 1 & k \\ 0 & 1 \end{bmatrix} \begin{bmatrix} a & b \\ c & d \end{bmatrix} : k \in \mathbb{Z} \right\}$$

where (a,b) is any particular pair such that $ad - bc = 1$ (Exercise 2.1.3(b)). Thus $\gamma(U_1') \cap U_2' = \left(\left[\begin{smallmatrix} a & b \\ c & d \end{smallmatrix}\right] (U_1') + k\right) \cap U_2'$ is empty for all but finitely many γ with bottom row (c,d). Combining these two results shows that $\gamma(U_1')$ intersects U_2' for only finitely many $\gamma \in \mathrm{SL}_2(\mathbb{Z})$. (See Exercise 2.1.3(c) for a less elementary way to prove this. The exercise establishes some facts that will make Proposition 2.2.2, to follow, clearer as well.)

Let $F = \{\gamma \in \mathrm{SL}_2(\mathbb{Z}) : \gamma(U_1') \cap U_2' \neq \emptyset, \ \gamma(\tau_1) \neq \tau_2\}$, a finite set. For each $\gamma \in F$ there exist disjoint neighborhoods $U_{1,\gamma}$ of $\gamma(\tau_1)$ and $U_{2,\gamma}$ of τ_2 in $\mathcal{H}$. Define

$$U_1 = U_1' \cap \left(\bigcap_{\gamma \in F} \gamma^{-1}(U_{1,\gamma}) \right), \quad \text{a neighborhood of } \tau_1 \text{ in } \mathcal{H},$$

$$U_2 = U_2' \cap \left(\bigcap_{\gamma \in F} U_{2,\gamma} \right), \quad \text{a neighborhood of } \tau_2 \text{ in } \mathcal{H}.$$

Take any $\gamma \in \mathrm{SL}_2(\mathbb{Z})$ such that $\gamma(U_1) \cap U_2 \neq \emptyset$. To show $\gamma(\tau_1) = \tau_2$ it suffices to show $\gamma \notin F$. But if $\gamma \in F$, then $\gamma^{-1}(U_{1,\gamma}) \supset U_1$ and $U_{2,\gamma} \supset U_2$, so $U_{1,\gamma} \cap U_{2,\gamma} \supset \gamma(U_1) \cap U_2 \neq \emptyset$, a contradiction since $U_{1,\gamma}$ and $U_{2,\gamma}$ are disjoint. This completes the proof. $\qquad \square$

Corollary 2.1.2. *For any congruence subgroup Γ of $SL_2(\mathbb{Z})$, the modular curve $Y(\Gamma)$ is Hausdorff.*

Proof. Let $\pi(\tau_1)$ and $\pi(\tau_2)$ be distinct points in $Y(\Gamma)$. Take neighborhoods U_1 of τ_1 and U_2 of τ_2 as in Proposition 2.1.1. Since $\gamma(\tau_1) \neq \tau_2$ for all $\gamma \in \Gamma$, the proposition says that $\Gamma(U_1) \cap U_2 = \emptyset$ in $\mathcal{H}$, and so equivalence (2.1) shows that $\pi(U_1)$ and $\pi(U_2)$ are disjoint supersets of $\pi(\tau_1)$ and $\pi(\tau_2)$ in $Y(\Gamma)$. They are neighborhoods since π is an open mapping. $\square$

It needs noting that $Y(\Gamma)$ is small enough to be a Riemann surface. A Riemann surface is a 1-dimensional connected complex manifold, and by definition the set underlying a manifold must have a *second countable* topology, meaning the topology has a countable basis. This is certainly true for Euclidean space and therefore for the quotient $Y(\Gamma)$. For more on this terminology, see any point set topology text, e.g., [Mun00].

Exercises

2.1.1. Show that giving $Y(\Gamma)$ the quotient topology makes the natural surjection π an open mapping.

2.1.2. Establish equivalence (2.1).

2.1.3. (a) Establish inequality (2.2). (A hint for this exercise is at the end of the book.)

(b) Show that the set of matrices in $SL_2(\mathbb{Z})$ with bottom row (c, d) is $\left\{ \begin{bmatrix} 1 & k \\ 0 & 1 \end{bmatrix} \begin{bmatrix} a & b \\ c & d \end{bmatrix} : k \in \mathbb{Z} \right\}$ where (a, b) is any particular pair such that $ad - bc = 1$.

(c) The first part of the proof of Proposition 2.1.1 can also be argued by studying the particular point i. Show that in the group $SL_2(\mathbb{R})$, the subgroup of elements fixing i is the special orthogonal group $SO_2(\mathbb{R})$. Note that the group $SL_2(\mathbb{R})$ takes i to any point $\tau = x + iy \in \mathcal{H}$ since the function $s : \mathcal{H} \longrightarrow SL_2(\mathbb{R})$ given by

$$s(\tau) = \frac{1}{\sqrt{y}} \begin{bmatrix} y & x \\ 0 & 1 \end{bmatrix}$$

satisfies $s(\tau)i = \tau$. (So now the upper half plane has a purely group theoretic description, $\mathcal{H} \cong SL_2(\mathbb{R})/SO_2(\mathbb{R})$.) Use these facts to show that for any $\gamma \in SL_2(\mathbb{R})$ and $e_1, e_2 \in \mathcal{H}$, the conditions $\gamma(e_1) = e_2$ and $\gamma \in s(e_2)SO_2(\mathbb{R})s(e_1)^{-1}$ are equivalent. Let e_1 range over the compact closure of the set U_1' in the proof, and similarly for e_2, to show that $\gamma(U_1')$ intersects U_2' only for γ in a certain compact subset of $SL_2(\mathbb{R})$. Since γ must also lie in the discrete subgroup $SL_2(\mathbb{Z})$, conclude that only finitely many such γ exist.

2.2 Charts

The next task is to put local coordinates on the modular curve $Y(\Gamma)$. This means finding for each point $\pi(\tau) \in Y(\Gamma)$ a neighborhood $\widetilde{U}$ (the symbol U

is reserved for neighborhoods in $\mathcal{H}$) and a homeomorphism $\varphi : \tilde{U} \longrightarrow V \subset \mathbb{C}$ such that the transition maps between the local coordinate systems are holomorphic.

At a point $\pi(\tau)$ where $\tau \in \mathcal{H}$ is fixed only by the identity transformation in Γ, i.e., only by the matrices $\Gamma \cap \{\pm I\}$, this is easy: a small enough neighborhood U of τ in $\mathcal{H}$ is homeomorphic under π to its image $\pi(U)$ in $Y(\Gamma)$, as Proposition 2.1.1 guarantees such a neighborhood with no Γ-equivalent points (Exercise 2.2.1). So a local inverse $\varphi : \pi(U) \longrightarrow U$ could serve as the local coordinate map.

But a point $\pi(\tau)$ where τ has a nontrivial group of fixing transformations in Γ (its *isotropy subgroup*) poses more of a problem. For example, let $\Gamma = \mathrm{SL}_2(\mathbb{Z})$ and let $\tau = i$, a fixed point under $\gamma = \left[\begin{smallmatrix} 0 & -1 \\ 1 & 0 \end{smallmatrix}\right]$. In the small γ acts as 180-degree rotation about i (Exercise 2.2.2), so any neighborhood U of i in $\mathcal{H}$ contains pairs of γ-equivalent points (see Figure 2.1) and thus can't biject to a neighborhood of $\pi(i)$ in $\mathrm{SL}_2(\mathbb{Z})\backslash\mathcal{H}$—projecting to the modular curve identifies the two halves of the neighborhood. This illustrates that the points to worry about satisfy

Definition 2.2.1. *Let Γ be a congruence subgroup of $\mathrm{SL}_2(\mathbb{Z})$. For each point $\tau \in \mathcal{H}$ let Γ_τ denote the isotropy subgroup of τ, i.e., the τ-fixing subgroup of Γ,*

$$\Gamma_\tau = \{\gamma \in \Gamma : \gamma(\tau) = \tau\}.$$

*A point $\tau \in \mathcal{H}$ is an **elliptic point for** Γ (or **of** Γ) if Γ_τ is nontrivial as a group of transformations, that is, if the containment $\{\pm I\}\Gamma_\tau \supset \{\pm I\}$ of matrix groups is proper. The corresponding point $\pi(\tau) \in Y(\Gamma)$ is also called elliptic.*

Note that Γ_τ also fixes the point $\bar{\tau} \in -\mathcal{H}$ since $\Gamma \subset \mathrm{SL}_2(\mathbb{R})$. ("Elliptic point" is unrelated to "elliptic curve" from Chapter 1 other than each having a distant connection to actual ellipses.)

The next result will be established in the following section.

Proposition 2.2.2. *Let Γ be a congruence subgroup of $\mathrm{SL}_2(\mathbb{Z})$. For each elliptic point τ of Γ the isotropy subgroup Γ_τ is finite cyclic.*

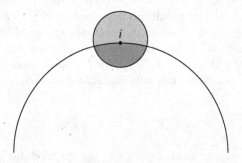

Figure 2.1. Pairwise equivalence about i

Thus each point $\tau \in \mathcal{H}$ has an associated positive integer,

$$h_\tau = |\{\pm I\}\Gamma_\tau/\{\pm I\}| = \begin{cases} |\Gamma_\tau|/2 & \text{if } -I \in \Gamma_\tau, \\ |\Gamma_\tau| & \text{if } -I \notin \Gamma_\tau. \end{cases}$$

This h_τ is called the *period* of τ, with $h_\tau > 1$ only for the elliptic points. If $\tau \in \mathcal{H}$ and $\gamma \in \mathrm{SL}_2(\mathbb{Z})$ then the period of $\gamma(\tau)$ under $\gamma\Gamma\gamma^{-1}$ is the same as the period of τ under Γ (Exercise 2.2.3). In particular, h_τ depends only on $\Gamma\tau$, making the period well defined on $Y(\Gamma)$, and if Γ is normal in $\mathrm{SL}_2(\mathbb{Z})$ then all points of $Y(\Gamma)$ over a point of $Y(\mathrm{SL}_2(\mathbb{Z}))$ have the same period. The space $Y(\Gamma)$ depends on Γ as a group of transformations acting on $\mathcal{H}$, and $-I$ acts trivially, so defining the period as we did rather than simply taking $h_\tau = |\Gamma_\tau|$ is natural. The definition correctly counts the τ-fixing transformations.

To put coordinates on $Y(\Gamma)$ about a point $\pi(\tau)$, first use the map $\delta_\tau = \begin{bmatrix} 1 & -\tau \\ 1 & -\bar{\tau} \end{bmatrix} \in \mathrm{GL}_2(\mathbb{C})$ (the group of invertible 2-by-2 matrices with complex entries) to take τ to 0 and $\bar{\tau}$ to ∞. The isotropy subgroup of 0 in the conjugated transformation group, $(\delta_\tau\{\pm I\}\Gamma\delta_\tau^{-1})_0/\{\pm I\}$, is the conjugate of the isotropy subgroup of τ, $\delta_\tau(\{\pm I\}\Gamma_\tau/\{\pm I\})\delta_\tau^{-1}$, and therefore is cyclic of order h_τ as a group of transformations by Proposition 2.2.2. Since this group of fractional linear transformations fixes 0 and ∞, it consists of maps of the form $z \mapsto az$, and since the group is finite cyclic these must be the rotations through angular multiples of $2\pi/h_\tau$ about the origin. Thus δ_τ is "straightening" neighborhoods of τ to neighborhoods of the origin in the sense that after the map, equivalent points are spaced apart by fixed angles. This suggests (as Figure 2.1 has illustrated with $h_\tau = 2$) that a coordinate neighborhood of $\pi(\tau)$ in $Y(\Gamma)$ should be, roughly, the π-image of a circular sector through angle $2\pi/h_\tau$ about τ in $\mathcal{H}$, and that the identifying action of π is essentially the wrapping action of the h_τth power map, taking the sector to a disk.

To write this down precisely and check that it works, start with another consequence of Proposition 2.1.1 (Exercise 2.2.4),

Corollary 2.2.3. *Let Γ be a congruence subgroup of $\mathrm{SL}_2(\mathbb{Z})$. Each point $\tau \in \mathcal{H}$ has a neighborhood U in $\mathcal{H}$ such that*

$$\text{for all } \gamma \in \Gamma, \text{ if } \gamma(U) \cap U \neq \emptyset \text{ then } \gamma \in \Gamma_\tau.$$

Such a neighborhood has no elliptic points except possibly τ.

Now given any point $\pi(\tau) \in Y(\Gamma)$, take a neighborhood U as in the corollary. Define $\psi : U \longrightarrow \mathbb{C}$ to be $\psi = \rho \circ \delta$ where $\delta = \delta_\tau$ and ρ is the power function $\rho(z) = z^h$ with $h = h_\tau$ as above. Thus $\psi(\tau') = (\delta(\tau'))^h$ acts as the straightening map δ followed by the h-fold wrapping ρ. (See Figure 2.2.) Let $V = \psi(U)$, an open subset of $\mathbb{C}$ by the Open Mapping Theorem from complex analysis. Since the projection π and the wrapping ψ identify the same points of U, there should exist an equivalence between the images of U under the two mappings. To confirm this, consider the situation

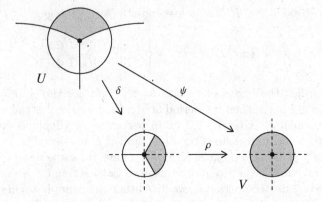

Figure 2.2. Local coordinates at an elliptic point

$$U \xrightarrow{\pi} \pi(U) \subset Y(\Gamma), \qquad U \xrightarrow{\psi} V \subset \mathbb{C}$$

and note that for any points $\tau_1, \tau_2 \in U$,

$$\pi(\tau_1) = \pi(\tau_2) \iff \tau_1 \in \Gamma \tau_2 \iff \tau_1 \in \Gamma_\tau \tau_2 \quad \text{by Corollary 2.2.3}$$
$$\iff \delta(\tau_1) \in (\delta \Gamma_\tau \delta^{-1})(\delta(\tau_2)) \iff \delta(\tau_1) = \mu_h^d(\delta(\tau_2)) \text{ for some } d,$$

where $\mu_h = e^{2\pi i/h}$, since $\delta \Gamma_\tau \delta^{-1}$ is a cyclic transformation group of h rotations. So

$$\pi(\tau_1) = \pi(\tau_2) \iff (\delta(\tau_1))^h = (\delta(\tau_2))^h \iff \psi(\tau_1) = \psi(\tau_2),$$

as desired. Thus there exists an injection $\varphi : \pi(U) \longrightarrow V$ such that the diagram

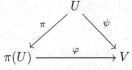

commutes. Also, φ surjects because ψ surjects by definition of V. The map φ is the local coordinate. The coordinate neighborhood about $\pi(\tau)$ in $Y(\Gamma)$ is $\pi(U)$ and the map $\varphi : \pi(U) \longrightarrow V$ is a homeomorphism (Exercise 2.2.5).

In sum, U is a neighborhood with no elliptic point except possibly τ; $\psi : U \longrightarrow V$ is the composition $\psi = \rho \circ \delta$ that matches the identifying action of π by wrapping U onto itself as prescribed by the period of τ; and the local coordinate $\varphi : \pi(U) \longrightarrow V$, defined by the condition $\varphi \circ \pi = \psi$, transfers the identified image of U in the modular curve back to the wrapped image of U in the complex plane.

We need to check that the transition maps between coordinate charts are holomorphic. That is, given overlapping $\pi(U_1)$ and $\pi(U_2)$, we need to check

$\varphi_{2,1}$, the restriction of $\varphi_2 \circ \varphi_1^{-1}$ to $\varphi_1(\pi(U_1) \cap \pi(U_2))$. Let $V_{1,2} = \varphi_1(\pi(U_1) \cap \pi(U_2))$ and $V_{2,1} = \varphi_2(\pi(U_1) \cap \pi(U_2))$ and write the commutative diagram

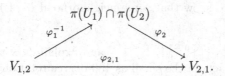

For each $x \in \pi(U_1) \cap \pi(U_2)$ it suffices to check holomorphy in some neighborhood of $\varphi_1(x)$ in $V_{1,2}$. Write $x = \pi(\tau_1) = \pi(\tau_2)$ with $\tau_1 \in U_1$, $\tau_2 \in U_2$, and $\tau_2 = \gamma(\tau_1)$ for some $\gamma \in \Gamma$. Let $U_{1,2} = U_1 \cap \gamma^{-1}(U_2)$. Then the projection $\pi(U_{1,2})$ is a neighborhood of x in $\pi(U_1) \cap \pi(U_2)$ and so $\varphi_1(\pi(U_{1,2}))$ is a corresponding neighborhood of $\varphi_1(x)$ in $V_{1,2}$.

Assume first that $\varphi_1(x) = 0$, i.e., the first straightening map is $\delta_1 = \delta_{\tau_1}$. Then an input point $q = \varphi_1(x')$ to $\varphi_{2,1}$ in this neighborhood takes the form

$$q = \varphi_1(\pi(\tau')) = \psi_1(\tau') = (\delta_1(\tau'))^{h_1} \quad \text{for some } \tau' \in U_{1,2}$$

where h_1 is the period of τ_1. Letting $\tilde{\tau}_2 \in U_2$ be the point such that $\psi_2(\tilde{\tau}_2) = 0$ and letting h_2 be its period, the corresponding output is

$$\begin{aligned}
\varphi_2(x') &= \varphi_2(\pi(\gamma(\tau'))) = \psi_2(\gamma(\tau')) \quad \text{which is defined since } \gamma(\tau') \in U_2 \\
&= (\delta_2(\gamma(\tau')))^{h_2} = ((\delta_2\gamma\delta_1^{-1})(\delta_1(\tau')))^{h_2} \\
&= ((\delta_2\gamma\delta_1^{-1})(q^{1/h_1}))^{h_2}.
\end{aligned}$$

This calculation shows that the only case where the transition map might not be holomorphic is when $h_1 > 1$, meaning τ_1 is elliptic and hence so is $\tau_2 = \gamma(\tau_1)$ with the same period. Recall that U_2 contains at most one elliptic point by construction and then the local coordinate takes it to 0. So when $h_1 > 1$ the point τ_2 is the point $\tilde{\tau}_2 \in U_2$ mentioned above, the second straightening map is $\delta_2 = \delta_{\tau_2}$, and $h_2 = h_1$. Thus

$$0 \xmapsto{\delta_1^{-1}} \tau_1 \xmapsto{\gamma} \tau_2 \xmapsto{\delta_2} 0, \qquad \infty \xmapsto{\delta_1^{-1}} \tau_1 \xmapsto{\gamma} \tilde{\tau}_2 \xmapsto{\delta_2} \infty,$$

showing that $\delta_2\gamma\delta_1^{-1} = \left[\begin{smallmatrix} \alpha & 0 \\ 0 & \beta \end{smallmatrix}\right]$ for some nonzero $\alpha, \beta \in \mathbb{C}$. The formula for $\varphi_{2,1}$ becomes

$$q \mapsto \left(\left[\begin{smallmatrix} \alpha & 0 \\ 0 & \beta \end{smallmatrix}\right] (q^{1/h})\right)^h = (\alpha/\beta)^h q$$

and the map is clearly holomorphic.

So far the argument assumes that $\varphi_1(x) = 0$. But it also covers the case $\varphi_2(x) = 0$ since the inverse of a holomorphic bijection is again holomorphic. And in general, $\varphi_{2,1}$ is a composite $\varphi_{2,3} \circ \varphi_{3,1}$ where $\varphi_3 : \pi(U_3) \longrightarrow V_3$ takes x to 0, so in fact the argument suffices for all cases.

Exercises

2.2.1. Let $\tau \in \mathcal{H}$ be fixed only by the identity transformation in Γ. Use Proposition 2.1.1 to show that some neighborhood U of τ is homeomorphic to its image in $Y(\Gamma)$.

2.2.2. Compute $\begin{bmatrix} 1 & -i \\ 1 & i \end{bmatrix} \begin{bmatrix} 0 & -1 \\ 1 & 0 \end{bmatrix} \begin{bmatrix} 1 & -i \\ 1 & i \end{bmatrix}^{-1}$ and explain how this shows that the map $\gamma(\tau) = -1/\tau$ acts in the small as 180-degree rotation about i.

2.2.3. Show that if $\tau \in \mathcal{H}$ and $\gamma \in \mathrm{SL}_2(\mathbb{Z})$ then the period of $\gamma(\tau)$ under $\gamma\Gamma\gamma^{-1}$ is the same as the period of τ under Γ.

2.2.4. Prove Corollary 2.2.3.

2.2.5. Explain why $\varphi : \pi(U) \longrightarrow V$ as defined in the section is a homeomorphism. (A hint for this exercise is at the end of the book.)

2.3 Elliptic points

It remains to prove Proposition 2.2.2, that for any congruence subgroup Γ of $\mathrm{SL}_2(\mathbb{Z})$ each elliptic point τ of Γ has finite cyclic isotropy subgroup Γ_τ. The process of doing so will develop a much clearer picture of $Y(\Gamma)$ and show that it has only finitely many elliptic points.

The simplest case is $Y(1) = \mathrm{SL}_2(\mathbb{Z})\backslash\mathcal{H}$. The next two lemmas will show that $Y(1)$ can essentially be identified with the set (Figure 2.3)

$$\mathcal{D} = \{\tau \in \mathcal{H} : |\mathrm{Re}(\tau)| \le 1/2, \ |\tau| \ge 1\}.$$

By Section 1.5, this set $\mathcal{D}$ also essentially represents the equivalence classes of complex elliptic curves under isomorphism, the point $\tau \in \mathcal{D}$ representing the class of $\mathbb{C}/\Lambda_\tau$ where $\Lambda_\tau = \tau\mathbb{Z} \oplus \mathbb{Z}$.

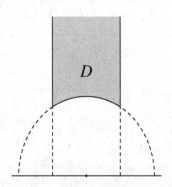

Figure 2.3. The fundamental domain for $\mathrm{SL}_2(\mathbb{Z})$

Lemma 2.3.1. *The map* $\pi : \mathcal{D} \longrightarrow Y(1)$ *surjects, where* π *is the natural projection* $\pi(\tau) = \mathrm{SL}_2(\mathbb{Z})\tau$.

Proof. Given $\tau \in \mathcal{H}$ it suffices to show that τ is $\mathrm{SL}_2(\mathbb{Z})$-equivalent to some point in $\mathcal{D}$. Repeatedly apply one of $\left[\begin{smallmatrix} 1 & \pm 1 \\ 0 & 1 \end{smallmatrix}\right] : \tau \mapsto \tau \pm 1$ to translate τ into the vertical strip $\{|\mathrm{Re}(\tau)| \le 1/2\}$, and replace τ by this transform. Now if $\tau \notin \mathcal{D}$ then $|\tau| < 1$ and so $\mathrm{Im}(-1/\tau) = \mathrm{Im}(-\bar{\tau}/|\tau|^2) = \mathrm{Im}(\tau/|\tau|^2) > \mathrm{Im}(\tau)$; replace τ by $\left[\begin{smallmatrix} 0 & -1 \\ 1 & 0 \end{smallmatrix}\right](\tau) = -1/\tau$ and repeat the process. Since there are only finitely many integer pairs (c, d) such that $|c\tau + d| < 1$ (because there are only finitely many lattice points inside a disk), the formula

$$\mathrm{Im}(\gamma\tau) = \frac{\mathrm{Im}(\tau)}{|c\tau + d|^2} \quad \text{for } \gamma = \begin{bmatrix} a & b \\ c & d \end{bmatrix} \in \mathrm{SL}_2(\mathbb{Z})$$

shows that only finitely many transforms of τ have larger imaginary part. Therefore the algorithm terminates with some $\tau \in \mathcal{D}$. $\qquad\square$

The surjection $\pi : \mathcal{D} \longrightarrow Y(1)$ isn't injective. The translation $\left[\begin{smallmatrix} 1 & 1 \\ 0 & 1 \end{smallmatrix}\right] :$ $\tau \mapsto \tau + 1$ identifies the two boundary half-lines, and the inversion $\left[\begin{smallmatrix} 0 & -1 \\ 1 & 0 \end{smallmatrix}\right] :$ $\tau \mapsto -1/\tau$ identifies the two halves of the boundary arc. But these boundary identifications are the only ones:

Lemma 2.3.2. *Suppose* τ_1 *and* τ_2 *are distinct points in* $\mathcal{D}$ *and that* $\tau_2 = \gamma\tau_1$ *for some* $\gamma \in \mathrm{SL}_2(\mathbb{Z})$. *Then either*

(1) $\mathrm{Re}(\tau_1) = \pm 1/2$ *and* $\tau_2 = \tau_1 \mp 1$, *or*
(2) $|\tau_1| = 1$ *and* $\tau_2 = -1/\tau_1$.

Proof. (Sketch—Exercise 2.3.1 asks for details.) Assume $\mathrm{Im}(\tau_2) \ge \mathrm{Im}(\tau_1)$ by symmetry. Let $\gamma = \left[\begin{smallmatrix} a & b \\ c & d \end{smallmatrix}\right]$. Then $|c\tau_1 + d|^2 \le 1$ since $\mathrm{Im}(\tau_2) \ge \mathrm{Im}(\tau_1)$, and $\mathrm{Im}(\tau_1) \ge \sqrt{3}/2$ since $\tau_1 \in \mathcal{D}$. So

$$|c|\sqrt{3}/2 \le |c|\mathrm{Im}(\tau_1) = |\mathrm{Im}(c\tau_1 + d)| \le |c\tau_1 + d| \le 1,$$

and since $c \in \mathbb{Z}$ this shows that $|c| \in \{0, 1\}$.

If $c = 0$ then $\gamma = \pm\left[\begin{smallmatrix} 1 & b \\ 0 & 1 \end{smallmatrix}\right]$ and $\mathrm{Re}(\tau_2) = \mathrm{Re}(\tau_1) + b$, forcing $|b| = 1$ and (1) holds.

If $|c| = 1$ then the condition $|c\tau_1 + d|^2 \le 1$ becomes $|\tau_1 \pm d|^2 \le 1$, or $(\mathrm{Re}(\tau_1) \pm d)^2 + (\mathrm{Im}(\tau_1))^2 \le 1$, implying $(\mathrm{Re}(\tau_1) \pm d)^2 \le 1 - (\mathrm{Im}(\tau_1))^2 \le 1 - 3/4 = 1/4$, so $|\mathrm{Re}(\tau_1) \pm d| \le 1/2$, forcing $|d| \le 1$.

If $|c| = 1$ and $|d| = 1$ then in the preceding calculation all inequalities must be equalities. It follows that $\mathrm{Im}(\tau_1) = \sqrt{3}/2$ and $|\mathrm{Re}(\tau_1) \pm 1| = 1/2$, so $\mathrm{Re}(\tau_1) = \pm 1/2$ and both (1) and (2) hold.

If $|c| = 1$ and $d = 0$ then the condition $|c\tau_1 + d| \le 1$ becomes $|\tau_1| \le 1$, so in fact $|\tau_1| = 1$ (since $\tau_1 \in \mathcal{D}$) and $\mathrm{Im}(\tau_2) = \mathrm{Im}(\tau_1)$. So also $|\tau_2| = 1$ by symmetry since now τ_1 and τ_2 have the same conditions on their imaginary parts and on the c-entries of the matrices transforming each to the other. Thus τ_1 and τ_2 have the same absolute value and the same imaginary part but are distinct, forcing their real parts to be opposites and (2) holds. $\qquad\square$

So with suitable boundary identification the set $\mathcal{D}$ is a model for $Y(1) = \mathrm{SL}_2(\mathbb{Z})\backslash\mathcal{H}$, also called a *fundamental domain* for $\mathrm{SL}_2(\mathbb{Z})$. Topologically, $\mathcal{D}$ modulo the identification is a plane. Figure 2.4 shows some $\mathrm{SL}_2(\mathbb{Z})$-translates of $\mathcal{D}$. The whole configuration repeats horizontally with period 1 under iterations of $\left[\begin{smallmatrix} 1 & \pm1 \\ 0 & 1 \end{smallmatrix}\right] : \tau \mapsto \tau \pm 1$. The figure shows how the $\mathrm{SL}_2(\mathbb{Z})$-translates of a point $\tau \in \mathcal{H}$ can cluster only down toward the real axis, perhaps giving a more intuitive understanding of Proposition 2.1.1.

Figure 2.4. Some $\mathrm{SL}_2(\mathbb{Z})$-translates of $\mathcal{D}$

Returning to elliptic points, suppose $\tau \in \mathcal{H}$ is fixed by a nontrivial transformation $\gamma = \left[\begin{smallmatrix} a & b \\ c & d \end{smallmatrix}\right] \in \mathrm{SL}_2(\mathbb{Z})$. Then $a\tau + b = c\tau^2 + d\tau$; solving for τ with the quadratic equation ($c = 0$ is impossible since $\tau \notin \mathbb{Q}$) and remembering that $\tau \in \mathcal{H}$ shows that $|a + d| < 2$ (Exercise 2.3.2). Thus the characteristic polynomial of γ is $x^2 + 1$ or $x^2 \pm x + 1$. Since γ satisfies its characteristic polynomial, one of $\gamma^4 = I$, $\gamma^3 = I$, $\gamma^6 = I$ holds, and γ has order 1, 2, 3, 4, or 6 as a matrix. Orders 1 and 2 give the identity transformation (Exercise 2.3.3). So the following proposition describes all nontrivial fixing transformations.

Proposition 2.3.3. *Let $\gamma \in \mathrm{SL}_2(\mathbb{Z})$.*

(a) *If γ has order 3 then γ is conjugate to $\left[\begin{smallmatrix} 0 & 1 \\ -1 & -1 \end{smallmatrix}\right]^{\pm1}$ in $\mathrm{SL}_2(\mathbb{Z})$.*

(b) *If γ has order 4 then γ is conjugate to $\left[\begin{smallmatrix} 0 & -1 \\ 1 & 0 \end{smallmatrix}\right]^{\pm1}$ in $\mathrm{SL}_2(\mathbb{Z})$.*

(c) *If γ has order 6 then γ is conjugate to $\left[\begin{smallmatrix} 0 & -1 \\ 1 & 1 \end{smallmatrix}\right]^{\pm 1}$ in $\mathrm{SL}_2(\mathbb{Z})$.*

Proof. (c) Since $\gamma^6 = I$ the lattice $L = \mathbb{Z}^2$ of integral column vectors is a module over the ring $\mathbb{Z}[\mu_6]$ where $\mu_6 = e^{2\pi i/6}$, defining the scalar-by-vector product $(a + b\mu_6) \cdot v$ for $a, b \in \mathbb{Z}$ and $v \in L$ as the matrix-by-vector product $(aI + b\gamma)v$.

The ring $\mathbb{Z}[\mu_6]$ is known to be a principal ideal domain and L is finitely generated over it. The structure theorem for modules over a principal ideal domain therefore says that L is $\mathbb{Z}[\mu_6]$-isomorphic to a sum $\bigoplus_k \mathbb{Z}[\mu_6]/I_k$ where the I_k are ideals. As an Abelian group, L is free of rank 2. Every nonzero ideal I_k of $\mathbb{Z}[\mu_6]$ has rank 2 as an Abelian group, making the quotient $\mathbb{Z}[\mu_6]/I_k$ a torsion group, so no such terms appear in the sum. Only one free summand appears, for otherwise the sum would be too big as an Abelian group. Thus there is a $\mathbb{Z}[\mu_6]$-module isomorphism $\phi_\gamma : \mathbb{Z}[\mu_6] \longrightarrow L$.

Let $u = \phi_\gamma(1)$ and $v = \phi_\gamma(\mu_6)$ and let $[u\ v]$ denote the matrix with columns u and v. Then $L = \mathbb{Z}u + \mathbb{Z}v$ so $\det[u\ v] \in \{\pm 1\}$ (Exercise 2.3.4(a)). Compute that $\gamma u = \mu_6 \cdot \phi_\gamma(1) = \phi_\gamma(\mu_6) = v$, and similarly $\gamma v = \mu_6 \cdot \phi_\gamma(\mu_6) = \phi_\gamma(\mu_6^2) = \phi_\gamma(-1 + \mu_6) = -u + v$. Thus

$$\gamma[u\ v] = [v\ -u+v] = [u\ v]\left[\begin{smallmatrix} 0 & -1 \\ 1 & 1 \end{smallmatrix}\right], \text{ so } \gamma = [u\ v]\left[\begin{smallmatrix} 0 & -1 \\ 1 & 1 \end{smallmatrix}\right][u\ v]^{-1},$$

and

$$\gamma[v\ u] = [-u+v\ v] = [v\ u]\left[\begin{smallmatrix} 1 & 1 \\ -1 & 0 \end{smallmatrix}\right], \text{ so } \gamma = [v\ u]\left[\begin{smallmatrix} 0 & -1 \\ 1 & 1 \end{smallmatrix}\right]^{-1}[v\ u]^{-1}.$$

One of $[u\ v]$, $[v\ u]$ is in $\mathrm{SL}_2(\mathbb{Z})$, proving (c).

Parts (a) and (b) are Exercise 2.3.4(b). $\square$

Now we can understand the elliptic points and their isotropy subgroups.

Corollary 2.3.4. *The elliptic points for $\mathrm{SL}_2(\mathbb{Z})$ are $\mathrm{SL}_2(\mathbb{Z})i$ and $\mathrm{SL}_2(\mathbb{Z})\mu_3$ where $\mu_3 = e^{2\pi i/3}$. The modular curve $Y(1) = \mathrm{SL}_2(\mathbb{Z})\backslash\mathcal{H}$ has two elliptic points. The isotropy subgroups of i and μ_3 are*

$$\mathrm{SL}_2(\mathbb{Z})_i = \langle\left[\begin{smallmatrix} 0 & -1 \\ 1 & 0 \end{smallmatrix}\right]\rangle \quad and \quad \mathrm{SL}_2(\mathbb{Z})_{\mu_3} = \langle\left[\begin{smallmatrix} 0 & -1 \\ 1 & 1 \end{smallmatrix}\right]\rangle.$$

For each elliptic point τ of $\mathrm{SL}_2(\mathbb{Z})$ the isotropy subgroup $\mathrm{SL}_2(\mathbb{Z})_\tau$ is finite cyclic.

Proof. The fixed points in $\mathcal{H}$ of the matrices in Proposition 2.3.3 are i and μ_3. The first statement follows (Exercise 2.3.5(a)). The second statement follows since i and μ_3 are not equivalent under $\mathrm{SL}_2(\mathbb{Z})$. The third statement can be verified directly (Exercise 2.3.5(a) again), and the fourth statement follows since all other isotropy subgroups of order greater than 2 are conjugates of $\mathrm{SL}_2(\mathbb{Z})_i$ and $\mathrm{SL}_2(\mathbb{Z})_{\mu_3}$. See Exercise 2.3.5(b) for a more conceptual proof of the fourth statement. $\square$

Proposition 2.2.2 and a bit more now follow.

Corollary 2.3.5. *Let Γ be a congruence subgroup of $SL_2(\mathbb{Z})$. The modular curve $Y(\Gamma)$ has finitely many elliptic points. For each elliptic point τ of Γ the isotropy subgroup Γ_τ is finite cyclic.*

Proof. If $SL_2(\mathbb{Z}) = \bigcup_{j=1}^{d} \Gamma\gamma_j$ then the elliptic points of $Y(\Gamma)$ are a subset of $E_\Gamma = \{\Gamma\gamma_j(i), \Gamma\gamma_j(\mu_3) : 1 \leq j \leq d\}$. The second statement is clear since Γ_τ is a subgroup of $SL_2(\mathbb{Z})_\tau$ for all $\tau \in \mathcal{H}$. $\qquad\square$

For a congruence subgroup Γ, if $SL_2(\mathbb{Z}) = \bigcup_j \{\pm I\}\Gamma\gamma_j$ then the set $\bigcup_j \gamma_j\mathcal{D}$ surjects to $Y(\Gamma)$ (Exercise 2.3.8), and with suitable boundary identification this is a bijection as before. The set need not be a fundamental domain, however, since a fundamental domain is also required to be connected.

Exercises

2.3.1. Fill in any details as necessary in the proof of Lemma 2.3.2.

2.3.2. If the nontrivial transformation $\left[\begin{smallmatrix} a & b \\ c & d \end{smallmatrix}\right] \in SL_2(\mathbb{Z})$ fixes $\tau \in \mathcal{H}$, show that $|a + d| < 2$.

2.3.3. Show that if $\gamma \in SL_2(\mathbb{Z})$ has order 2 then $\gamma = -I$.

2.3.4. (a) In the proof of Proposition 2.3.3, why does the condition $L = \mathbb{Z}u + \mathbb{Z}v$ imply $\det[u\ v] = \pm 1$?

(b) Prove the other two parts of Proposition 2.3.3. (Hints for this exercise are at the end of the book.)

2.3.5. (a) Complete the proof of Corollary 2.3.4. (A hint for this exercise is at the end of the book.)

(b) Give a more conceptual proof of the fourth statement in Corollary 2.3.4 as follows: The results from Exercise 2.1.3(c), that the isotropy subgroup of i in $SL_2(\mathbb{R})$ is $SO_2(\mathbb{R})$ and that $\mathcal{H} \cong SL_2(\mathbb{R})/SO_2(\mathbb{R})$, show that an element $s(\tau)$ of $SL_2(\mathbb{R})$ moves τ to i and the isotropy subgroup of τ correspondingly conjugates to a discrete subgroup of $SO_2(\mathbb{R})$. But since $SO_2(\mathbb{R})$ is the rotations of the circle, any such subgroup is cyclic.

2.3.6. In the proof of Corollary 2.3.5, need the $2d$ points in $Y(\Gamma)$ listed in the description of E_Γ be distinct? Need the points of E_Γ all be elliptic points of Γ?

2.3.7. Prove that there are no elliptic points for the following groups: (a) $\Gamma(N)$ for $N > 1$, (b) $\Gamma_1(N)$ for $N > 3$ (also, find the elliptic points for $\Gamma_1(2)$ and $\Gamma_1(3)$ given that each group has one), (c) $\Gamma_0(N)$ for N divisible by any prime $p \equiv -1 \pmod{12}$. (A hint for this exercise is at the end of the book.) In the next chapter we will extend the technique of proving Proposition 2.3.3 to count the elliptic points of $\Gamma_0(N)$ for all N.

Also, show that if the congruence subgroup Γ does not contain the negative identity matrix $-I$ then Γ has no elliptic points of period 2.

2.3.8. Suppose $SL_2(\mathbb{Z}) = \bigcup_j \{\pm I\} \Gamma \gamma_j$ where Γ is a congruence subgroup. Show that $\bigcup_j \gamma_j \mathcal{D}$ surjects to $Y(\Gamma)$. (A hint for this exercise is at the end of the book.)

2.4 Cusps

Transferring the fundamental domain $\mathcal{D}$ (as pictured in Figure 2.3) to the Riemann sphere via stereographic projection gives a triangle with one vertex missing. (The reader should sketch this before continuing.) Clearly the triangle should be compactified by adjoining the point at infinity, to be thought of as infinitely far up the imaginary axis in the plane or as the north pole on the sphere.

In general, let Γ be a congruence subgroup of $SL_2(\mathbb{Z})$. Chapter 1 defined the cusps of Γ as the Γ-equivalence classes of $\mathbb{Q} \cup \{\infty\}$. This section will adjoin cusps with appropriate local coordinate charts to the modular curve $Y(\Gamma)$, completing $Y(\Gamma)$ to a compact Riemann surface denoted $X(\Gamma)$. The coordinates to be used have already been built into the cusp condition in the definition of modular forms. The compactified $X(\Gamma)$ is also called a modular curve.

The group $GL_2^+(\mathbb{Q})$ of 2-by-2 matrices with positive determinant and rational entries acts on the set $\mathbb{Q} \cup \{\infty\}$ by the rule

$$\begin{bmatrix} a & b \\ c & d \end{bmatrix} \left(\frac{m}{n} \right) = \frac{am + bn}{cm + dn}$$

where this means to take ∞ to a/c and $-d/c$ to ∞ if $c \neq 0$ and to take ∞ to ∞ if $c = 0$. As the matrices are nonsingular no $0/0$ expressions will arise. The subgroup $SL_2(\mathbb{Z})$ acts transitively since any rational number takes the form $s = a/c$ with $a, c \in \mathbb{Z}$ and $\gcd(a,c) = 1$, meaning that $ad - bc = 1$ for some $b, d \in \mathbb{Z}$, and $\begin{bmatrix} a & b \\ c & d \end{bmatrix}(\infty) = s$. (Note how this geometrically motivates part of Exercise 1.2.11(b), that every $\gamma \in GL_2^+(\mathbb{Q})$ factors as $\gamma = \alpha\gamma'$, with $\alpha \in SL_2(\mathbb{Z})$ and $\gamma' = \begin{bmatrix} * & * \\ 0 & * \end{bmatrix}$: if the matrix $\gamma \in GL_2^+(\mathbb{Q})$ doesn't already have lower left entry 0 then it takes ∞ to some rational number s. Left multiplying by a suitable $\alpha^{-1} \in SL_2(\mathbb{Z})$ takes s back to ∞, meaning the product $\gamma' = \alpha^{-1}\gamma$ fixes ∞ and therefore has lower left entry 0 as desired.) If $\alpha \in SL_2(\mathbb{Z})$ takes ∞ to a rational number s then α transforms the fundamental domain $\mathcal{D}$ to a region that tapers to a cusp at s (cf. Figure 2.4), just as $\mathcal{D}$ itself tapers to a cusp at ∞ on the Riemann sphere. The isotropy subgroup of ∞ in $SL_2(\mathbb{Z})$ is the translations,

$$SL_2(\mathbb{Z})_\infty = \left\{ \pm \begin{bmatrix} 1 & m \\ 0 & 1 \end{bmatrix} : m \in \mathbb{Z} \right\}.$$

Let Γ be a congruence subgroup of $SL_2(\mathbb{Z})$. To compactify the modular curve $Y(\Gamma) = \Gamma \backslash \mathcal{H}$, define $\mathcal{H}^* = \mathcal{H} \cup \mathbb{Q} \cup \{\infty\}$ and take the extended quotient

$$X(\Gamma) = \Gamma \backslash \mathcal{H}^* = Y(\Gamma) \cup \Gamma \backslash (\mathbb{Q} \cup \{\infty\}).$$

The points Γs in $\Gamma \backslash (\mathbb{Q} \cup \{\infty\})$ are also called the *cusps of* $X(\Gamma)$. For the congruence subgroups $\Gamma_0(N)$, $\Gamma_1(N)$, and $\Gamma(N)$ we write $X_0(N)$, $X_1(N)$, and $X(N)$.

Lemma 2.4.1. *The modular curve* $X(1) = \mathrm{SL}_2(\mathbb{Z}) \backslash \mathcal{H}^*$ *has one cusp. For any congruence subgroup* Γ *of* $\mathrm{SL}_2(\mathbb{Z})$ *the modular curve* $X(\Gamma)$ *has finitely many cusps.*

Proof. Exercise 2.4.1.   □

The topology on $\mathcal{H}^*$ consisting of its intersections with open complex disks (including disks $\{z : |z| > r\} \cup \{\infty\}$) contains too many points of $\mathbb{Q} \cup \{\infty\}$ in each neighborhood to make the quotient $X(\Gamma)$ Hausdorff. Instead, to put an appropriate topology on $X(\Gamma)$ start by defining for any $M > 0$ a neighborhood

$$\mathcal{N}_M = \{\tau \in \mathcal{H} : \mathrm{Im}(\tau) > M\}.$$

Adjoin to the usual open sets in $\mathcal{H}$ more sets in $\mathcal{H}^*$ to serve as a base of neighborhoods of the cusps, the sets

$$\alpha(\mathcal{N}_M \cup \{\infty\}) \ : \ M > 0, \ \alpha \in \mathrm{SL}_2(\mathbb{Z}),$$

and take the resulting topology on $\mathcal{H}^*$. Since fractional linear transformations are conformal and take circles to circles, if $\alpha(\infty) \in \mathbb{Q}$ then $\alpha(\mathcal{N}_M \cup \{\infty\})$ is a disk tangent to the real axis. (Figure 2.5 shows $\mathcal{N}_1 \cup \{\infty\}$ and some of its $\mathrm{SL}_2(\mathbb{Z})$-translates; note how this quantifies the discussion leading up to Definition 1.2.3.) Under this topology each $\gamma \in \mathrm{SL}_2(\mathbb{Z})$ is a homeomorphism of $\mathcal{H}^*$. Finally, give $X(\Gamma)$ the quotient topology and extend natural projection to $\pi : \mathcal{H}^* \longrightarrow X(\Gamma)$.

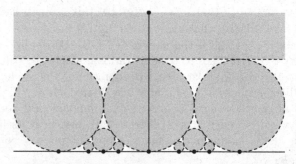

Figure 2.5. Neighborhoods of ∞ and of some rational points

Proposition 2.4.2. *The modular curve* $X(\Gamma)$ *is Hausdorff, connected, and compact.*

Proof. The first statement requires distinct points $x_1, x_2 \in X(\Gamma)$ to have disjoint neighborhoods. The case $x_1 = \Gamma \tau_1$, $x_2 = \Gamma \tau_2$ with $\tau_1, \tau_2 \in \mathcal{H}$ is already established as Corollary 2.1.2.

Suppose $x_1 = \Gamma s_1$, $x_2 = \Gamma \tau_2$ with $s_1 \in \mathbb{Q} \cup \{\infty\}$ and $\tau_2 \in \mathcal{H}$. Then $s_1 = \alpha(\infty)$ for some $\alpha \in \mathrm{SL}_2(\mathbb{Z})$. Let U_2 be any neighborhood of τ_2 with compact closure K. Then the formula

$$\mathrm{Im}(\gamma(\tau)) \leq \max\{\mathrm{Im}(\tau), 1/\mathrm{Im}(\tau)\} \quad \text{for } \tau \in \mathcal{H} \text{ and } \gamma \in \mathrm{SL}_2(\mathbb{Z})$$

(Exercise 2.4.2(a)) shows that for M large enough, $\mathrm{SL}_2(\mathbb{Z}) K \cap \mathcal{N}_M = \emptyset$. Let $U_1 = \alpha(\mathcal{N}_M \cup \{\infty\})$. Then $\pi(U_1)$ and $\pi(U_2)$ are disjoint (Exercise 2.4.2(b)).

Suppose $x_1 = \Gamma s_1$, $x_2 = \Gamma s_2$ with $s_1, s_2 \in \mathbb{Q} \cup \{\infty\}$. Then $s_1 = \alpha_1(\infty)$ and $s_2 = \alpha_2(\infty)$ for some $\alpha_1, \alpha_2 \in \mathrm{SL}_2(\mathbb{Z})$. Let $U_1 = \alpha_1(\mathcal{N}_2 \cup \{\infty\})$ and $U_2 = \alpha_2(\mathcal{N}_2 \cup \{\infty\})$. Then $\pi(U_1)$ and $\pi(U_2)$ are disjoint, for if $\gamma \alpha_1(\tau_1) = \alpha_2(\tau_2)$ for some $\gamma \in \Gamma$ and $\tau_1, \tau_2 \in \mathcal{N}_2$ then $\alpha_2^{-1} \gamma \alpha_1$ takes τ_1 to τ_2 and (since $\mathcal{N}_2$ is tessellated by the integer translates of $\mathcal{D}$ and contains no elliptic points) therefore must be $\pm \left[\begin{smallmatrix} 1 & m \\ 0 & 1 \end{smallmatrix}\right]$ for some $m \in \mathbb{Z}$. Thus $\alpha_2^{-1} \gamma \alpha_1$ fixes ∞ and consequently $\gamma(s_1) = s_2$, contradicting that x_1 and x_2 are distinct. This completes the proof that $X(\Gamma)$ is Hausdorff.

Suppose $\mathcal{H}^* = O_1 \cup O_2$ is a disjoint union of open subsets. Intersect with the connected set $\mathcal{H}$ to conclude that $O_1 \supset \mathcal{H}$ and so $O_2 \subset \mathbb{Q} \cup \{\infty\}$. But then O_2 is not open after all unless it is empty. Thus $\mathcal{H}^*$ is connected and hence so is its continuous image $X(\Gamma)$.

For compactness, first note that the set $\mathcal{D}^* = \mathcal{D} \cup \{\infty\}$ is compact in the $\mathcal{H}^*$ topology (Exercise 2.4.3). Since $\mathcal{H}^* = \mathrm{SL}_2(\mathbb{Z}) \mathcal{D}^* = \bigcup_j \Gamma \gamma_j(\mathcal{D}^*)$ where the γ_j are coset representatives, $X(\Gamma) = \bigcup_j \pi(\gamma_j(\mathcal{D}^*))$. Since each γ_j is continuous and π is continuous and $[\mathrm{SL}_2(\mathbb{Z}) : \Gamma]$ is finite, the result follows. $\square$

Making $X(\Gamma)$ a compact Riemann surface requires giving it charts. Retain the coordinate patches $\pi(U)$ and maps $\varphi : \pi(U) \longrightarrow V$ from Section 2.2 for neighborhoods $U \subset \mathcal{H}$. For each cusp $s \in \mathbb{Q} \cup \{\infty\}$ some $\delta = \delta_s \in \mathrm{SL}_2(\mathbb{Z})$ takes s to ∞. Define the *width* of s to be

$$h_s = h_{s,\Gamma} = |\mathrm{SL}_2(\mathbb{Z})_\infty / (\delta \{\pm I\} \Gamma \delta^{-1})_\infty|.$$

(The Γ subscript is suppressed when context makes the group clear.) This notion is dual to the period of an elliptic point, being inversely proportional to the size of an isotropy subgroup. Recall that the period of an elliptic point is the number of sectors of the disk at the point that are identified under isotropy. At a cusp, infinitely many sectors come together, sectors most easily seen after translating to ∞ as unit vertical strips, and the width of the cusp is the number of such strips are that are distinct under isotropy (see Figure 2.6). The group $\mathrm{SL}_2(\mathbb{Z})_\infty = \{\pm I\} \langle \left[\begin{smallmatrix} 1 & 1 \\ 0 & 1 \end{smallmatrix}\right] \rangle$ is infinite cyclic as a group of transformations, and so the width is characterized by the conditions $\{\pm I\} (\delta \Gamma \delta^{-1})_\infty = \{\pm I\} \langle \left[\begin{smallmatrix} 1 & h \\ 0 & 1 \end{smallmatrix}\right] \rangle$, $h > 0$. The width is finite (Exercise 2.4.4(a)) and independent of δ since in fact $h_s = |\mathrm{SL}_2(\mathbb{Z})_s / \{\pm I\} \Gamma_s|$ (Exercise 2.4.4(b)). If $s \in \mathbb{Q} \cup \{\infty\}$ and $\gamma \in \mathrm{SL}_2(\mathbb{Z})$

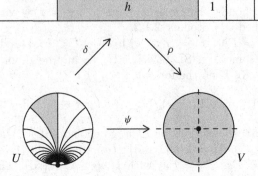

Figure 2.6. Local coordinates at a cusp

then the width of $\gamma(s)$ under $\gamma\Gamma\gamma^{-1}$ is the same as the width of s under Γ (Exercise 2.4.4(c)). In particular, h_s depends only on Γs, making the width well defined on $X(\Gamma)$, and if Γ is normal in $\mathrm{SL}_2(\mathbb{Z})$ then all cusps of $X(\Gamma)$ have the same width. Now define $U = U_s = \delta^{-1}(\mathcal{N}_2 \cup \{\infty\})$ and as before define ψ as a composite $\psi = \rho \circ \delta$ where this time ρ is the h-periodic wrapping map $\rho(z) = e^{2\pi i z/h}$. Note that ψ is exactly the change of variable embedded in condition (3) of Definition 1.2.3. Also as before, let $V = \mathrm{im}(\psi)$, an open subset of $\mathbb{C}$, to get

$$\psi : U \longrightarrow V, \qquad \psi(\tau) = e^{2\pi i \delta(\tau)/h}.$$

As with elliptic points, ψ mimics the identifying action of π. This time the fractional linear transformation δ straightens neighborhoods of s by making identified points differ by a horizontal offset, and then the exponential map ρ wraps the upper half plane into a cylinder which, held to one's eye like a telescope, becomes in perspective a disk with ∞ at its center. (Again see Figure 2.6, noting how the shape of the shaded sector in U motivates the term "cusp.")

To confirm that ψ carries out the same identification as π about s, compute that for $\tau_1, \tau_2 \in U$,

$$\pi(\tau_1) = \pi(\tau_2) \iff \tau_1 = \gamma(\tau_2) \iff \delta(\tau_1) = (\delta\gamma\delta^{-1})(\delta(\tau_2))$$

for some $\gamma \in \Gamma$. But this makes $\delta\gamma\delta^{-1}$ a translation since $\delta(\tau_1)$ and $\delta(\tau_2)$ both lie in $\mathcal{N}_2 \cup \{\infty\}$. So $\delta\gamma\delta^{-1} \in \delta\Gamma\delta^{-1} \cap \mathrm{SL}_2(\mathbb{Z})_\infty = (\delta\Gamma\delta^{-1})_\infty \subset \pm\langle[\begin{smallmatrix}1 & h \\ 0 & 1\end{smallmatrix}]\rangle$ and

$$\pi(\tau_1) = \pi(\tau_2) \iff \delta(\tau_1) = \delta(\tau_2) + mh \quad \text{for some } m \in \mathbb{Z}$$
$$\iff \psi(\tau_1) = \psi(\tau_2).$$

As in Section 2.2 there exists a bijection $\varphi : \pi(U) \longrightarrow V$ such that the diagram

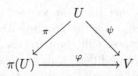

commutes. As before, the coordinate neighborhood about $\pi(\tau)$ in $Y(\Gamma)$ is $\pi(U)$ and the coordinate map is $\varphi : \pi(U) \longrightarrow V$, a homeomorphism (Exercise 2.4.5). Again $\psi : U \longrightarrow V$ is the composition $\psi = \rho \circ \delta$, and the local coordinate $\varphi : \pi(U) \longrightarrow V$ is defined by the condition $\varphi \circ \pi = \psi$.

Checking that the transition maps are holomorphic is similar to the process carried out for coordinate patches $U_1, U_2 \subset \mathcal{H}$ in Section 2.2, but now at least one patch is a cusp neighborhood.

Suppose $U_1 \subset \mathcal{H}$ has corresponding $\delta_1 = \delta_{\tau_1} \in GL_2(\mathbb{C})$ where τ_1 has period h_1, and suppose $U_2 = \delta_2^{-1}(\mathcal{N}_2 \cup \{\infty\})$. As before, for each $x \in \pi(U_1) \cap \pi(U_2)$ write $x = \pi(\tilde{\tau}_1) = \pi(\tau_2)$ with $\tilde{\tau}_1 \in U_1$, $\tau_2 \in U_2$, and $\tau_2 = \gamma(\tilde{\tau}_1)$ for some $\gamma \in \Gamma$. Let $U_{1,2} = U_1 \cap \gamma^{-1}(U_2)$, a neighborhood of $\tilde{\tau}_1$ in $\mathcal{H}$. Then $\varphi_1(\pi(U_{1,2}))$ is a neighborhood of $\varphi_1(x)$ in $V_{1,2} = \varphi_1(\pi(U_1) \cap \pi(U_2))$. Note that if $h_1 > 1$ then $\tau_1 \notin U_{1,2}$, else the point $\delta_2(\gamma(\tau_1)) \in \mathcal{N}_2$ is an elliptic point for Γ, but $\mathcal{N}_2$ contains no elliptic points for $SL_2(\mathbb{Z})$ (Exercise 2.4.6). So if $h_1 > 1$ then $0 \notin \varphi_1(\pi(U_{1,2}))$. As before, an input point $\varphi_1(x')$ to $\varphi_{2,1}$ in $V_{1,2}$ takes the form $q = (\delta_1(\tau'))^{h_1}$. This time the corresponding output is

$$\varphi_2(x') = \varphi_2(\pi(\gamma(\tau'))) = \psi_2(\gamma(\tau')) = \exp(2\pi i \delta_2 \gamma(\tau')/h_2)$$
$$= \exp(2\pi i \delta_2 \gamma \delta_1^{-1}(q^{1/h_1})/h_2).$$

So the only case where the transition map might not be holomorphic is when $h_1 > 1$ and $0 \in \varphi_1(\pi(U_{1,2}))$, but we have seen that this cannot happen. The discussion here also covers the case with the roles of U_1 and U_2 exchanged since the inverse of a holomorphic bijection is again holomorphic.

Suppose $U_1 = \delta_1^{-1}(\mathcal{N}_2 \cup \{\infty\})$ with $\delta_1 : s_1 \mapsto \infty$ and $U_2 = \delta_2^{-1}(\mathcal{N}_2 \cup \{\infty\})$ with $\delta_2 : s_2 \mapsto \infty$. If $\pi(U_1) \cap \pi(U_2) \neq \emptyset$ then $\gamma \delta_1^{-1}(\mathcal{N}_2 \cup \{\infty\})$ meets $\delta_2^{-1}(\mathcal{N}_2 \cup \{\infty\})$ for some $\gamma \in \Gamma$, i.e., $\delta_2 \gamma \delta_1^{-1}$ moves some point in $\mathcal{N}_2 \cup \{\infty\}$ to another and therefore must be a translation $\pm [\begin{smallmatrix} 1 & m \\ 0 & 1 \end{smallmatrix}]$. So $\gamma(s_1) = \gamma \delta_1^{-1}(\infty) = \pm \delta_2^{-1} [\begin{smallmatrix} 1 & m \\ 0 & 1 \end{smallmatrix}] (\infty) = s_2$. It follows that $h_1 = h_2$ and the transition map takes an input point in $\varphi_1(\pi(U_{1,2}))$,

$$q = \psi_1(\tau) = \exp(2\pi i \delta_1(\tau)/h),$$

to the output point

$$\psi_2(\gamma(\tau)) = \exp(2\pi i \delta_2 \gamma \delta_1^{-1}(\delta_1(\tau))/h) = \exp(2\pi i(\delta_1(\tau) + m)/h)$$
$$= e^{2\pi i m/h} q.$$

This is clearly holomorphic.

For any congruence subgroup Γ of $SL_2(\mathbb{Z})$ the extended quotient $X(\Gamma)$ is now a compact Riemann surface. Figure 2.7 summarizes the local coordinate structure for future reference.

Topologically every compact Riemann surface is a g-holed torus for some nonnegative integer g called its *genus*. In particular, complex elliptic curves have genus 1. The next chapter will compute the genus of $X(\Gamma)$ and study the meromorphic functions and differentials on $X(\Gamma)$. This will give counting results about modular forms as well as being interesting in its own right.

$\pi : \mathcal{H}^* \longrightarrow X(\Gamma)$ is natural projection. $U \subset \mathcal{H}^*$ is a neighborhood containing at most one elliptic point or cusp. The local coordinate $\varphi : \pi(U) \xrightarrow{\sim} V$ satisfies $\varphi \circ \pi = \psi$ where $\psi : U \longrightarrow V$ is a composition $\psi = \rho \circ \delta$.	
About $\tau_0 \in \mathcal{H}$:	About $s \in \mathbb{Q} \cup \{\infty\}$:
The straightening map is $z = \delta(\tau)$ where $\delta = \begin{bmatrix} 1 & -\tau_0 \\ 1 & -\overline{\tau_0} \end{bmatrix}$, $\delta(\tau_0) = 0$. $\delta(U)$ is a neighborhood of 0 in $\mathbb{C}$.	The straightening map is $z = \delta(\tau)$ where $\delta \in \mathrm{SL}_2(\mathbb{Z})$, $\delta(s) = \infty$. $\delta(U)$ is a neighborhood of ∞ in $\mathcal{H}^*$.
The wrapping map is $q = \rho(z)$ where $\rho(z) = z^h$, $\rho(0) = 0$ with period $h = |\{\pm I\}\Gamma_{\tau_0}/\{\pm I\}|$. $V = \rho(\delta(U))$ is a neighborhood of 0.	The wrapping map is $q = \rho(z)$ where $\rho(z) = e^{2\pi i z/h}$, $\rho(\infty) = 0$ with width $h = |\mathrm{SL}_2(\mathbb{Z})_s/\{\pm I\}\Gamma_s|$. $V = \rho(\delta(U))$ is a neighborhood of 0.

Figure 2.7. Local coordinates on $X(\Gamma)$

Exercises

2.4.1. Prove Lemma 2.4.1.

2.4.2. (a) Justify the formula $\mathrm{Im}(\gamma(\tau)) \leq \max\{\mathrm{Im}(\tau), 1/\mathrm{Im}(\tau)\}$ for $\tau \in \mathcal{H}$ and $\gamma \in \mathrm{SL}_2(\mathbb{Z})$ used in the proof of Proposition 2.4.2.
  (b) In the same proof, verify that $\pi(U_1)$ and $\pi(U_2)$ are disjoint.

2.4.3. Show that the set $\mathcal{D}^* = \mathcal{D} \cup \{\infty\}$ is compact in the $\mathcal{H}^*$ topology.

2.4.4. (a) Show that the width h_s is finite. (Hints for this exercise are at the end of the book.)
  (b) Show that $h_s = |\mathrm{SL}_2(\mathbb{Z})_s/\{\pm I\}\Gamma_s|$.
  (c) Show that if $s \in \mathbb{Q} \cup \{\infty\}$ and $\gamma \in \mathrm{SL}_2(\mathbb{Z})$ then the width of $\gamma(s)$ under $\gamma\Gamma\gamma^{-1}$ is the same as the width of s under Γ.

2.4.5. Explain why $\varphi : \pi(U) \longrightarrow V$ as defined in the section is a homeomorphism.

2.4.6. Show that $\mathcal{N}_2$ contains no elliptic points for $\mathrm{SL}_2(\mathbb{Z})$. (A hint for this exercise is at the end of the book.)

2.4.7. Suppose that Γ_1 and Γ_2 are congruence subgroups of $\mathrm{SL}_2(\mathbb{Z})$ and that γ is an element of $\mathrm{GL}_2^+(\mathbb{Q})$ such that $\gamma\Gamma_1\gamma^{-1} \subset \Gamma_2$. Show that the formula $\Gamma_1\tau \mapsto \Gamma_2\gamma(\tau)$ defines a holomorphic map $X(\Gamma_1) \longrightarrow X(\Gamma_2)$.

2.5 Modular curves and Modularity

Recall from Section 1.5 that every complex elliptic curve E has a well defined modular invariant $j(E)$. Recall the congruence subgroup $\Gamma_0(N)$ of $\mathrm{SL}_2(\mathbb{Z})$ from Section 1.2 and its associated modular curve $X_0(N)$. Complex analytically the Modularity Theorem says that all elliptic curves with rational invariants come from such modular curves via holomorphic maps, viewing both kinds of curve as compact Riemann surfaces,

Theorem 2.5.1 (Modularity Theorem, Version $X_\mathbb{C}$). *Let E be a complex elliptic curve with $j(E) \in \mathbb{Q}$. Then for some positive integer N there exists a surjective holomorphic function of compact Riemann surfaces from the modular curve $X_0(N)$ to the elliptic curve E,*

$$X_0(N) \longrightarrow E.$$

The function in the theorem is called a *modular parametrization* of E. This first incarnation of Modularity gives little indication of number theory. The statement will be refined in later chapters, with other objects replacing the modular curve $X_0(N)$ and with the field $\mathbb{Q}$ of rational numbers replacing $\mathbb{C}$, as more ideas enter into the book.

3

Dimension Formulas

For any congruence subgroup Γ of $\mathrm{SL}_2(\mathbb{Z})$, the compactified modular curve $X(\Gamma)$ is now a Riemann surface. The genus of $X(\Gamma)$, its number of elliptic points, its number of its cusps, and the meromorphic functions and meromorphic differentials on $X(\Gamma)$ are all used by Riemann surface theory to determine dimension formulas for the vector spaces $\mathcal{M}_k(\Gamma)$ and $\mathcal{S}_k(\Gamma)$.

Related reading: This chapter is based on Chapters 1 and 2 of [Shi73]. The material is also in [Miy89] and (less completely) [Sch74].

3.1 The genus

Let Γ be a congruence subgroup of $\mathrm{SL}_2(\mathbb{Z})$. Topologically the compact Riemann surface $X(\Gamma) = \Gamma \backslash \mathcal{H}^*$ is a sphere with g handles for some $g \in \mathbb{N}$. This g, the *genus* of $X(\Gamma)$, will figure in the pending dimension formulas for vector spaces of modular forms.

Let $f : X \longrightarrow Y$ be a nonconstant holomorphic map between compact Riemann surfaces. As explained in Section 1.3, f surjects. For any $y \in Y$ the inverse image $f^{-1}(y)$ is discrete in X since f is nonconstant and X is closed, and therefore $f^{-1}(y)$ is finite since X is compact. In fact the map f has a well defined *degree* $d \in \mathbb{Z}^+$ such that $|f^{-1}(y)| = d$ for all but finitely many $y \in Y$. More precisely, for each point $x \in X$ let $e_x \in \mathbb{Z}^+$ be the *ramification degree* of f at x, meaning the multiplicity with which f takes 0 to 0 as a map in local coordinates, making f an e_x-to-1 map about x, cf. the Local Mapping Theorem of complex analysis. Then there exists a positive integer d such that

$$\sum_{x \in f^{-1}(y)} e_x = d \quad \text{for all } y \in Y.$$

To see this, let $\mathcal{E} = \{x \in X : e_x > 1\}$ denote the *exceptional points* in X, the finite set of points where f is ramified. Let $X' = X \backslash \mathcal{E}$ and $Y' = Y \backslash f(\mathcal{E})$ be the Riemann surfaces obtained by removing the exceptional points from X

© Springer Science+Business Media New York 2005   65
F. Diamond, J. Shurman, *A First Course in Modular Forms*,
Graduate Texts in Mathematics 228, DOI 10.1007/978-0-387-27226-9_3

and their images from Y. So X' is all but finitely many points of X, making it connected, and similarly for Y'. Let y be a point of Y'. Each $x \in f^{-1}(y)$ has a neighborhood U_x where f is locally bijective. Shrinking these neighborhoods enough, they can be assumed disjoint and assumed to map to the same neighborhood of y. Thus y has a neighborhood V in Y' whose inverse image is a finite disjoint union of neighborhoods U_x in X' with each restriction $f_x : U_x \longrightarrow V$ invertible. This makes the integer-valued function $y \mapsto |f^{-1}(y)|$ on Y' continuous and therefore constant since Y' is connected. Let d be its value; so $\sum_{x \in f^{-1}(y)} e_x = d$ for all $y \in Y'$. The sum extends continuously to all of Y by definition of ramification degree: if $y = f(x)$ then every y' in some neighborhood of y is the image of e_x points x' near x, each with $e_{x'} = 1$. Thus $\sum_{x \in f^{-1}(y)} e_x = |f^{-1}(y')| = d$ as claimed, provided that y' is in a small enough neighborhood of y to avoid ramification except possibly over y itself.

Again let $f : X \longrightarrow Y$ be a nonconstant holomorphic map between compact Riemann surfaces, of degree d. Let g_X and g_Y be the genera of X and Y. To compute the genus of the modular curve $X(\Gamma)$ we will use the *Riemann–Hurwitz formula*,

$$2g_X - 2 = d(2g_Y - 2) + \sum_{x \in X}(e_x - 1).$$

This shows, for instance, that there is no nonconstant map to a surface of higher genus, that a nonconstant map between surfaces of equal genus $g \geq 1$ is unramified, and that there is no meromorphic function of degree $d = 1$ on any compact Riemann surface of positive genus. We sketch a proof of the Riemann–Hurwitz formula for the reader with some background in topology: Triangulate Y using V_Y vertices including all points of $f(\mathcal{E})$, E_Y edges, and F_Y faces. This lifts under f^{-1} to a triangulation of X with $E_X = dE_Y$ edges and $F_X = dF_Y$ faces but only $V_X = dV_Y - \sum_{x \in X}(e_x - 1)$ vertices because of ramification. The Riemann–Hurwitz formula follows because $2 - 2g_X = F_X - E_X + V_X$ and similarly for Y.

As a special case of all this, if Γ_1 and Γ_2 are congruence subgroups of $\mathrm{SL}_2(\mathbb{Z})$ with $\Gamma_1 \subset \Gamma_2$, then the natural projection of corresponding modular curves,

$$f : X(\Gamma_1) \longrightarrow X(\Gamma_2), \quad \Gamma_1\tau \mapsto \Gamma_2\tau,$$

is a nonconstant holomorphic map between compact Riemann surfaces, cf. Exercise 2.4.7. Its degree is (Exercise 3.1.1)

$$\deg(f) = [\{\pm I\}\Gamma_2 : \{\pm I\}\Gamma_1] = \begin{cases} [\Gamma_2 : \Gamma_1]/2 & \text{if } -I \in \Gamma_2 \text{ and } -I \notin \Gamma_1, \\ [\Gamma_2 : \Gamma_1] & \text{otherwise.} \end{cases}$$

Recall from Figure 2.7 the local structure on the Riemann surfaces $X(\Gamma_j)$ for $j = 1, 2$. The commutative diagram

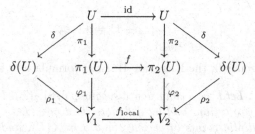

shows that $f_{\text{local}} \circ \psi_1 = \psi_2$, where ψ_j is the composition $\rho_j \circ \delta$. If U is a neighborhood of $\tau \in \mathcal{H}$ then $\rho_1(z) = z^{h_1}$ and $\rho_2(z) = z^{h_2}$ so the local map is $q \mapsto q^{h_2/h_1}$. Let $\Gamma_{j,\tau}$ denote the fixing subgroup of τ in Γ_j for $j = 1, 2$. Then the periods $h_j = |\{\pm I\}\Gamma_{j,\tau}|/2$ lie in $\{1, 2, 3\}$ and since their ratio h_2/h_1 is integral this implies $h_1 = 1$ or $h_1 = h_2$. So the ramification degree is

$$e_{\pi_1(\tau)} = h_2/h_1 = \begin{cases} h_2 & \text{if } \tau \text{ is an elliptic point for } \Gamma_2 \text{ but not for } \Gamma_1, \\ 1 & \text{otherwise} \end{cases}$$

$$= [\{\pm I\}\Gamma_{2,\tau} : \{\pm I\}\Gamma_{1,\tau}].$$

Since the periods h_j are defined on the curves X_j, the ramification degree is indeed defined on X_1. If U is a neighborhood of $s \in \mathbb{Q} \cup \{\infty\}$ then $\rho_1(z) = e^{2\pi i z/h_1}$ and $\rho_2(z) = e^{2\pi i z/h_2}$ so the local map is $q \mapsto q^{h_1/h_2}$. Here the widths are $h_j = [\text{SL}_2(\mathbb{Z})_\infty : \{\pm I\}\Gamma_{j,s}]$ for $j = 1, 2$ and the ramification degree is the reciprocal of their ratio,

$$e_{\pi_1(s)} = h_1/h_2 = [\{\pm I\}\Gamma_{2,s} : \{\pm I\}\Gamma_{1,s}].$$

Again this is well defined on X_1. If Γ_1 is normal in Γ_2 then all points of $X(\Gamma_1)$ lying over a given point of $X(\Gamma_2)$ have the same ramification degree (Exercise 3.1.2).

To compute the genus of $X(\Gamma)$, further specialize to $\Gamma_1 = \Gamma$ and $\Gamma_2 = \text{SL}_2(\mathbb{Z})$. Let $y_2 = \text{SL}_2(\mathbb{Z})i$, $y_3 = \text{SL}_2(\mathbb{Z})\mu_3$, and $y_\infty = \text{SL}_2(\mathbb{Z})\infty$ be the elliptic point of period 2, the elliptic point of period 3, and the cusp of $X(1) = \text{SL}_2(\mathbb{Z})\backslash\mathcal{H}^*$. Let ε_2 and ε_3 be the number of elliptic points of Γ in $f^{-1}(y_2)$ and $f^{-1}(y_3)$, i.e., the number of elliptic points of period 2 and 3 in $X(\Gamma)$, and let ε_∞ be the number of cusps of $X(\Gamma)$. Then recalling that $d = \deg(f)$ and letting $h = 2$ or $h = 3$, the formula for d at the beginning of the section and then the formula for $e_{\pi_1(\tau)}$ at the nonelliptic points and the elliptic points over $\text{SL}_2(\mathbb{Z})y_h$ show that (Exercise 3.1.3(a))

$$d = \sum_{x \in f^{-1}(y_h)} e_x = h \cdot (|f^{-1}(y_h)| - \varepsilon_h) + 1 \cdot \varepsilon_h,$$

and using these equalities twice gives

$$\sum_{x \in f^{-1}(y_h)} (e_x - 1) = (h - 1)(|f^{-1}(y_h)| - \varepsilon_h) = \frac{h-1}{h}(d - \varepsilon_h).$$

Also,

$$\sum_{x \in f^{-1}(y_\infty)} (e_x - 1) = d - \varepsilon_\infty.$$

Since $X(1)$ has genus 0, the Riemann–Hurwitz formula now shows

Theorem 3.1.1. *Let Γ be a congruence subgroup of $\mathrm{SL}_2(\mathbb{Z})$. Let $f : X(\Gamma) \longrightarrow X(1)$ be natural projection, and let d denote its degree. Let ε_2 and ε_3 denote the number of elliptic points of period 2 and 3 in $X(\Gamma)$, and ε_∞ the number of cusps of $X(\Gamma)$. Then the genus of $X(\Gamma)$ is*

$$g = 1 + \frac{d}{12} - \frac{\varepsilon_2}{4} - \frac{\varepsilon_3}{3} - \frac{\varepsilon_\infty}{2}.$$

Proof. Exercise 3.1.3(b). □

The next result will be used in Sections 3.5 and 3.6.

Corollary 3.1.2. *Let Γ, g, ε_2, ε_3, and ε_∞ be as above. Then*

$$2g - 2 + \frac{\varepsilon_2}{2} + \frac{2\varepsilon_3}{3} + \varepsilon_\infty > 0.$$

Proof. Exercise 3.1.3(c). □

Exercises 3.1.4, 3.1.5, and 3.1.6 compute the genus of $X(\Gamma)$ for $\Gamma = \Gamma_0(p)$, $\Gamma = \Gamma_1(p)$, and $\Gamma = \Gamma(p)$ where p is prime. The results for general N will take considerably more work.

A fundamental domain for $\Gamma_0(13)$ is shown in Figure 3.1, including representatives of the two elliptic points of order 2 (the light dots) and the two of order 3 (the dark ones—Exercise 3.1.4(f) explains how the figure was generated and justifies the description to follow in this paragraph and the next). Each boundary arc with an order-2 elliptic point folds together at the elliptic point, identifying its two halves with each other, and each pair of boundary arcs joined at an order-3 elliptic point folds together at the elliptic point, identifying with each other. The remaining two curved boundary arcs in the left half that don't include the cusp 0 are identified with each other, and similarly in the right half. The vertical boundary segments in the left half identify with those in the right half, and so the remaining two arcs are identified with each other as well. Gluing boundary arcs together under these identifications gives a sphere per the formula in Exercise 3.1.4(e).

The 13 points of $\mathrm{SL}_2(\mathbb{Z})(i)$ in the figure identify to eight points of $X_0(13)$, and the 14 points of $\mathrm{SL}_2(\mathbb{Z})(\mu_3)$ in the figure identify to six points of $X_0(13)$. The boundary arc for each order-2 elliptic point x folds together like the local coordinate about i in $X(1)$ so that the ramification degree of the projection $f : X_0(13) \longrightarrow X(1)$ at these points is $e_x = 1$, while $e_x = 2$ at the other six points x that project to $\mathrm{SL}_2(\mathbb{Z})(i)$ and the ramification degrees sum to $14 = [\mathrm{SL}_2(\mathbb{Z}) : \Gamma_0(13)]$ as they should. Each pair of boundary arcs joined at an

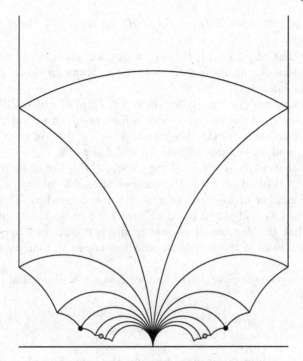

Figure 3.1. Fundamental domain for $\Gamma_0(13)$

order-3 elliptic point x folds together so that $e_x = 1$, while $e_x = 3$ at the other four points x that project to $\mathrm{SL}_2(\mathbb{Z})(\mu_3)$ and again the sum is 14. The ramification degrees at the cusps $\Gamma_0(p)(\infty)$ and $\Gamma_0(13)(0)$ are visibly 1 and 13.

To generate pictures of fundamental domains, see Helena Verrill's web page [Ver].

Exercises

3.1.1. Show that natural projection $f : X(\Gamma_1) \longrightarrow X(\Gamma_2)$ has the degree claimed in the text. (A hint for this exercise is at the end of the book.)

3.1.2. Show that if Γ_1 is normal in Γ_2 then all points of $X(\Gamma_1)$ lying over a given point of $X(\Gamma_2)$ have the same ramification degree. (A hint for this exercise is at the end of the book.)

3.1.3. (a) Fill in details as necessary in the computations preceding Theorem 3.1.1.
  (b) Prove Theorem 3.1.1.
  (c) Prove Corollary 3.1.2.

3.1.4. Let p be prime, let $X_0(p) = \Gamma_0(p)\backslash\mathcal{H}^*$, let $\alpha_j = \left[\begin{smallmatrix} 1 & 0 \\ j & 1 \end{smallmatrix}\right]$ for $j = 0, \ldots, p-1$, and let $\alpha_\infty = \left[\begin{smallmatrix} 1 & -1 \\ 1 & 0 \end{smallmatrix}\right]$.

(a) Show that $SL_2(\mathbb{Z}) = \bigcup_j \Gamma_0(p)\alpha_j$, a disjoint union.

(b) Show that $X_0(p)$ has exactly two cusps. (Hints for this exercise are at the end of the book.)

(c) Show that $\gamma\alpha_j(i) = \alpha_j(i)$ for some $\gamma \in \Gamma_0(p)$ of order 4 if and only if $j^2 + 1 \equiv 0 \pmod{p}$. Thus the number of elliptic points of period 2 in $X_0(p)$ is the number of solutions of the congruence $x^2 + 1 \equiv 0 \pmod{p}$. This number is 2 if $p \equiv 1 \pmod 4$, 0 if $p \equiv 3 \pmod 4$, and 1 if $p = 2$.

(d) Show that $\gamma\alpha_j(\mu_3) = \alpha_j(\mu_3)$ for some $\gamma \in \Gamma_0(p)$ of order 6 if and only if $j^2 - j + 1 \equiv 0 \pmod{p}$. Thus the number of elliptic points of period 3 in $X_0(p)$ is the number of solutions of $x^2 - x + 1 \equiv 0 \pmod{p}$. This number is 2 if $p \equiv 1 \pmod 3$, 0 if $p \equiv 2 \pmod 3$, and 1 if $p = 3$. Along with part (c), this shows that the number of elliptic points is determined by $p \pmod{12}$. The example $p = 13$ in the text is the smallest where all four possible elliptic points exist.

(e) Let g be the genus of $X_0(p)$ and let $k = p + 1$. Show that

$$
g = \begin{cases} \lfloor \frac{k}{12} \rfloor - 1 & \text{if } k \equiv 2 \pmod{12}, \\ \lfloor \frac{k}{12} \rfloor & \text{otherwise.} \end{cases}
$$

(The symbol k is introduced here because the same formula with k as the weight will recur later in the chapter.)

(f) Figure 3.1 was generated using the coset representatives

$$
\beta_j = \left[\begin{smallmatrix} 1 & 0 \\ j & 1 \end{smallmatrix}\right]\left[\begin{smallmatrix} 0 & -1 \\ 1 & 0 \end{smallmatrix}\right], \quad j = -6, \ldots, 6,
$$

and $\beta_\infty = \left[\begin{smallmatrix} 1 & -1 \\ 0 & 1 \end{smallmatrix}\right]\alpha_\infty\left[\begin{smallmatrix} 0 & -1 \\ 1 & 0 \end{smallmatrix}\right]$. Find each translate $\beta_j(\mathcal{D})$ in the figure, where $\mathcal{D}$ is the fundamental domain from Chapter 2. Show that the 13 points of $SL_2(\mathbb{Z})(i)$ in the figure are $\beta_j(i)$ for $j = -6, \ldots, 6$. Since $\left[\begin{smallmatrix} 0 & -1 \\ 1 & 0 \end{smallmatrix}\right]$ fixes i, the elliptic points of order 2 are $\beta_j(i)$ when $j^2 + 1 \equiv 0 \pmod{13}$. Show that $\gamma\beta_j(i) = \beta_{j'}(i)$ for some $\gamma \in \Gamma_0(13)$ with $j' \neq j$ if and only if $jj' + 1 \equiv 0 \pmod{13}$. Use this to partition the 13 points of $SL_2(\mathbb{Z})(i)$ in the figure into eight equivalence classes under $\Gamma_0(13)$, five with two points each where the angle is π, giving a total of 2π; one with one point where the angle is 2π; and two with one point where the angle is π as it is at i in $\mathcal{D}$, representing the unramified points. Identify the boundary arcs pairwise except for two arcs that fold in on themselves. Note that $SL_2(\mathbb{Z})(\mu_3) = SL_2(\mathbb{Z})(\mu_6)$. Since $\left[\begin{smallmatrix} 0 & -1 \\ 1 & 0 \end{smallmatrix}\right]$ takes μ_6 to μ_3, the elliptic points of order 3 are $\beta_j(\mu_6)$ when $j^2 - j + 1 \equiv 0 \pmod{13}$. Show that the 14 points of $SL_2(\mathbb{Z})(\mu_3)$ in the figure are $\beta_j(\mu_6)$ for $j = -6, \ldots, 6, \infty$. Show that $\gamma\beta_j(\mu_6) = \beta_{j'}(\mu_6)$ for some $\gamma \in \Gamma_0(13)$ with $j' \neq j$ if and only if $jj' - j + 1 \equiv 0 \pmod{13}$ or $jj' - j' + 1 \equiv 0 \pmod{13}$. Use this to partition the 14 points of $SL_2(\mathbb{Z})(\mu_3)$ in the figure into six equivalence classes under $\Gamma_0(13)$, one with the four points where two of the angles are $2\pi/3$ and two of the angles are $\pi/3$, giving a total of 2π; two with three points where the angle is $2\pi/3$, again giving 2π; one with the two points where the angle is π, giving 2π; and two classes with one point each where the angle is $2\pi/3$ as it is at μ_3 in $\mathcal{D}$, representing the unramified points. Determine whether each pair of boundary arcs is identified with orientation preserved or reversed. Show that under identification the figure is topologically a sphere.

(g) Carry out the same process for $X_0(11)$ and $X_0(17)$, convincing yourself that they are tori in accordance with part (e).

3.1.5. Let p be prime and let $X_1(p) = \Gamma_1(p)\backslash\mathcal{H}^*$.

(a) Show that if $\gamma \in \mathrm{SL}_2(\mathbb{Z})_s$ where $s \in \mathbb{Q}\cup\{\infty\}$ (and as in Chapter 2 the group is its isotropy subgroup) then γ has trace ± 2. Show that if p is a prime then $\Gamma_0(p)_s = \{\pm I\}\Gamma_1(p)_s$. Deduce that natural projection $f : X_1(p) \longrightarrow X_0(p)$ is unramified at the cusps. Combine this with part (b) of the previous problem to show that $X_1(p)$ has $p-1$ cusps if $p > 2$ and 2 cusps if $p = 2$.

(b) Exercise 2.3.7 showed that $X_1(p)$ has no elliptic points if $p > 3$, that $X_1(3)$ has no elliptic points of period 2 and one elliptic point of period 3, and that $X_1(2)$ has one elliptic point of period 2 and no elliptic points of period 3. Combine this information with part (a) to compute the genus of $X_1(p)$. (A hint for this exercise is at the end of the book.)

3.1.6. Let p be prime and let $X(p) = \Gamma(p)\backslash\mathcal{H}^*$. Compute the genus of $X(p)$. (A hint for this exercise is at the end of the book.)

3.2 Automorphic forms

This section defines automorphic forms, functions on the upper half plane $\mathcal{H}$ similar to modular forms but meromorphic rather than holomorphic. Automorphic forms of positive even weight k with respect to a congruence subgroup Γ are closely related to meromorphic differentials of degree $k/2$ on the quotient Riemann surface $X(\Gamma)$, to be described in the next section. Sections 3.5 and 3.6 will exploit the relation to compute dimension formulas for spaces of modular forms.

Let $\widehat{\mathbb{C}}$ denote the Riemann sphere $\mathbb{C}\cup\{\infty\}$. Recall that for an open subset $V \subset \mathbb{C}$, a function $f : V \longrightarrow \widehat{\mathbb{C}}$ is *meromorphic* if it is the zero function or it has a Laurent series expansion truncated from the left about each point $\tau \in V$,

$$f(t) = \sum_{n=m}^{\infty} a_n(t - \tau)^n \quad \text{for all } t \text{ in some disk about } \tau,$$

with coefficients $a_n \in \mathbb{C}$ and $a_m \neq 0$. The starting index m is the *order of f at τ*, short for "order of vanishing" and denoted $\nu_\tau(f)$; the zero function is defined to have order $\nu_\tau(f) = \infty$. The function f is holomorphic at τ when $\nu_\tau(f) \geq 0$, it vanishes at τ when $\nu_\tau(f) > 0$, and it has a pole at τ when $\nu_\tau(f) < 0$. The set of meromorphic functions on V forms a field.

Let Γ be a congruence subgroup of $\mathrm{SL}_2(\mathbb{Z})$. Recall the weight-$k$ operator,

$$(f[\gamma]_k)(\tau) = j(\gamma, \tau)^{-k} f(\gamma(\tau)), \quad j(\gamma, \tau) = c\tau + d \text{ for } \gamma = \begin{bmatrix} a & b \\ c & d \end{bmatrix}.$$

As in Chapter 1, a meromorphic function $f : \mathcal{H} \longrightarrow \widehat{\mathbb{C}}$ is called *weakly modular of weight k with respect to Γ* if $f[\gamma]_k = f$ for all $\gamma \in \Gamma$. To discuss meromorphy of f at ∞, let $\hbar$ be the smallest positive integer such that $\begin{bmatrix} 1 & \hbar \\ 0 & 1 \end{bmatrix} \in \Gamma$. Thus f has period $\hbar$, although $\hbar$ need not be the smallest period of f. If also f has no poles in some region $\{\tau \in \mathcal{H} : \mathrm{Im}(\tau) > c\}$ then f has a Laurent series on the corresponding punctured disk about 0,

$$f(\tau) = \sum_{n=-\infty}^{\infty} a_n q_\hbar^n \quad \text{if } \mathrm{Im}(\tau) > c, \text{ where } q_\hbar = e^{2\pi i \tau / \hbar}.$$

Then f is *meromorphic at ∞* if this series truncates from the left, starting at some $m \in \mathbb{Z}$ where $a_m \neq 0$, or if $f = 0$. The *order of f at ∞*, denoted $\nu_\infty(f)$, is again defined as the starting index m except when $f = 0$, in which case $\nu_\infty(f) = \infty$. Now automorphic forms are defined the same way as modular forms except with meromorphy in place of holomorphy.

Definition 3.2.1. *Let Γ be a congruence subgroup of $\mathrm{SL}_2(\mathbb{Z})$ and let k be an integer. A function $f : \mathcal{H} \longrightarrow \widehat{\mathbb{C}}$ is an* **automorphic form of weight k with respect to Γ** *if*

(1) *f is meromorphic,*
(2) *f is weight-k invariant under Γ,*
(3) *$f[\alpha]_k$ is meromorphic at ∞ for all $\alpha \in \mathrm{SL}_2(\mathbb{Z})$.*

The set of automorphic forms of weight k with respect to Γ is denoted $\mathcal{A}_k(\Gamma)$.

Condition (3) makes sense since $f[\alpha]_k$ is meromorphic on $\mathcal{H}$ and weakly modular with respect to $\alpha^{-1}\Gamma\alpha$, a congruence subgroup of $\mathrm{SL}_2(\mathbb{Z})$. For any cusp $s \in \mathbb{Q} \cup \{\infty\}$, the order of f at s is defined as

$$\nu_s(f) = \nu_\infty(f[\alpha]_k) \quad \text{where } \alpha(\infty) = s, \ \alpha \in \mathrm{SL}_2(\mathbb{Z}).$$

This is independent of the choice of α and well defined on the quotient $X(\Gamma)$ (Exercise 3.2.1). Condition (3) implies beyond Condition (1) that the poles of f in $\mathcal{H}$ cannot cluster at any $s \in \mathbb{Q} \cup \{\infty\}$.

Setting $k = 0$ in the definition shows that $\mathcal{A}_0(\Gamma)$ is the field of meromorphic functions on $X(\Gamma)$, denoted $\mathbb{C}(X(\Gamma))$. At elliptic points and cusps, where the local coordinate is many-to-one, functions $f \in \mathcal{A}_0(\Gamma)$ are correspondingly periodic and thus meromorphic in the local variable of the Riemann surface structure. This statement will be elaborated later in the section.

For example, recall the modular invariant, defined as

$$j = 1728 \frac{g_2^3}{\Delta} : \mathcal{H} \longrightarrow \widehat{\mathbb{C}}$$

where g_2 and Δ are the weight 4 Eisenstein series and the weight 12 cusp form from Chapter 1. Since the numerator and denominator are both weight-12 invariants under $SL_2(\mathbb{Z})$ and holomorphic at ∞, the quotient is $SL_2(\mathbb{Z})$-invariant and meromorphic at ∞. Thus j is an automorphic form of weight 0 with respect to the full modular group $SL_2(\mathbb{Z})$ and is naturally viewed as a meromorphic function on the quotient, $j : X(1) \longrightarrow \widehat{\mathbb{C}}$. Since Δ is a cusp form, j has a pole at ∞. By Corollary 1.4.2, stating that Δ is nonzero at every point of $\mathcal{H}$, this is its only pole. The leading 1728 normalizes its Laurent series to (Exercise 3.2.2)

$$j(\tau) = \frac{1}{q} + \sum_{n=0}^{\infty} a_n q^n, \quad a_n \in \mathbb{Z}, \ q = e^{2\pi i \tau}. \tag{3.1}$$

Since its one pole is simple, j has degree 1 as a map $j : X(1) \longrightarrow \widehat{\mathbb{C}}$ of compact Riemann surfaces, cf. the beginning of Section 3.1. It follows that j is a homeomorphism and conformal since it is locally analytic with nonvanishing derivative, and it takes $Y(1)$ to $\mathbb{C}$ and ∞ to ∞. In fact

$$\mathcal{A}_0(SL_2(\mathbb{Z})) = \mathbb{C}(j),$$

that is, the field of meromorphic functions on $X(1)$ is the set of rational functions of j. The containment $\mathbb{C}(j) \subset \mathcal{A}_0(SL_2(\mathbb{Z}))$ is clear. For the other containment, let f be meromorphic and nonconstant on $X(1)$ with finite zeros $z_1, \ldots, z_m$ and finite poles $p_1, \ldots, p_n$, listed with multiplicity. Consider a function $g \in \mathbb{C}(j)$,

$$g(\tau) = \frac{\prod_{i=1}^{m}(j(\tau) - j(z_i))}{\prod_{i=1}^{n}(j(\tau) - j(p_i))}.$$

Then g has the same zeros and poles as f away from $SL_2(\mathbb{Z})\infty$, so g must vanish to the same order as f at $SL_2(\mathbb{Z})\infty$ as well since for both functions the total number of zeros minus poles is 0. Thus $f/g : X(1) \longrightarrow \widehat{\mathbb{C}}$ has no zeros or poles, making it constant and showing that $f \in \mathbb{C}(j)$.

Similarly, Chapter 7 will show that

$$\mathcal{A}_0(\Gamma_0(N)) = \mathbb{C}(j, j_N), \quad \text{where } j_N(\tau) = j(N\tau).$$

Chapter 7 will also describe $\mathcal{A}_0(\Gamma_0(N))$, $\mathcal{A}_0(\Gamma_1(N))$, and $\mathcal{A}_0(\Gamma(N))$ using the functions f_0, $f_0^{\vec{d}}$, and $f_0^{\vec{v}}$ from the end of Section 1.5.

Returning to the general discussion, the sum $\mathcal{A}(\Gamma) = \bigoplus_{k \in \mathbb{Z}} \mathcal{A}_k(\Gamma)$ forms a ring. The derivative j' lies in $\mathcal{A}_2(\Gamma)$, and it follows that $\mathcal{A}_k(\Gamma)$ contains nonzero elements for all even positive integers k (Exercise 3.2.3). For any $k \in \mathbb{Z}$, if f is a nonzero element of $\mathcal{A}_k(\Gamma)$ then $\mathcal{A}_k(\Gamma) = \mathbb{C}(X(\Gamma))f = \{f_0 f : f_0 \in \mathbb{C}(X(\Gamma))\}$ (Exercise 3.2.4). These facts will be used to compute dimension formulas later in the chapter.

The transformation law $f(\gamma(\tau)) = j(\gamma, \tau)^k f(\tau)$ shows that an automorphic form f with respect to a congruence subgroup Γ is generally not well defined on the quotient $X(\Gamma)$. (The exceptional cases are when $k = 0$ or $f = 0$.) However, the order of vanishing of any automorphic form f does turn out to make sense on the quotient. Defining this notion appropriately requires taking the Riemann surface structure of $X(\Gamma)$ into account.

Let $\pi(\tau) \in X(\Gamma)$ be a noncusp. Since the factor of automorphy $j(\gamma, \tau)$ has no zeros or poles on $\mathcal{H}$, the order of f is the same at all points of $\Gamma(\tau)$. The local coordinate at $\pi(\tau)$ is essentially $q = (t - \tau)^h$ where τ has period h as an elliptic point of Γ. Excluding the zero function, since $f(t) = a_m(t - \tau)^m + \cdots = a_m q^{m/h} + \cdots$ with $a_m \neq 0$ the natural definition of the order of f at $\pi(\tau) \in X(\Gamma)$ in terms of the order m of f at $\tau \in \mathcal{H}$ is m/h, that is

$$\nu_{\pi(\tau)}(f) = \frac{\nu_\tau(f)}{h} \quad \text{where } \tau \text{ has period } h. \tag{3.2}$$

Thus $\nu_{\pi(\tau)}(f) = \nu_\tau(f)$ except at the elliptic points, where $\nu_{\pi(\tau)}(f)$ can be half-integral or third-integral. As remarked earlier, if $k = 0$ then h divides m and $\nu_{\pi(\tau)}(f)$ is integral.

Defining the order of f at cusps $\pi(s) \in X(\Gamma)$ is trickier. First consider the cusp $\pi(\infty)$. Its local coordinate is $q_h = e^{2\pi i \tau / h}$ where h is the width, characterized by the conditions

$$\{\pm I\}\Gamma_\infty = \{\pm I\}\left\langle \left[\begin{smallmatrix} 1 & h \\ 0 & 1 \end{smallmatrix}\right]\right\rangle, \quad h \in \mathbb{Z}^+.$$

But since the negative identity matrix $-I$ might not lie in Γ this implies only

$$\Gamma_\infty = \{\pm I\}\left\langle \left[\begin{smallmatrix} 1 & h \\ 0 & 1 \end{smallmatrix}\right]\right\rangle \quad \text{or} \quad \Gamma_\infty = \left\langle \left[\begin{smallmatrix} 1 & h \\ 0 & 1 \end{smallmatrix}\right]\right\rangle \quad \text{or} \quad \Gamma_\infty = \left\langle -\left[\begin{smallmatrix} 1 & h \\ 0 & 1 \end{smallmatrix}\right]\right\rangle.$$

Thus the width h here is not necessarily the same as the period $\hbar$ from the discussion before Definition 3.2.1. In all three cases the orbits $\Gamma\tau$ and $\Gamma(\tau+h)$ for $\tau \in \mathcal{H}$ are equal. In the first two cases, f has period $\hbar = h$ compatibly with this. But in the third case, which can arise only when $-I \notin \Gamma$ (see Exercise 3.2.5 for an example), the period of f is $\hbar = 2h$. (The factor of automorphy $j(-\left[\begin{smallmatrix} 1 & h \\ 0 & 1 \end{smallmatrix}\right], \tau) = -1$ shows that if k is even then $f(\tau + h) = f(\tau)$ while if k is odd then $f(\tau + h) = -f(\tau)$ so that $f(\tau + 2h) = f(\tau)$, but nonetheless the period $\hbar$ in the discussion before Definition 3.2.1 is $2h$ independently of k.) In all cases set $h' = 2h$, so that h' is twice the width and usually twice the period, and set $q_{h'} = e^{2\pi i \tau / h'}$. Then $f(\tau) = g(q_{h'})$ for $\text{Im}(\tau) > c$ where c is large enough to avoid the finite poles of f, and g has Laurent expansion

$$g(q_{h'}) = \sum_{n=m}^{\infty} a_n q_{h'}^n, \quad a_m \neq 0.$$

Since $q_{h'} = q_h^{1/2}$ is the square root of the local variable, the order of f at $\pi(\infty)$ is naturally defined as $m/2$, half the lowest power of $q_{h'}$, except when $f = 0$ in which case $\nu_{\pi(\infty)}(f) = \infty$. When h' is twice the period, the process here simply recovers the integer $\nu_\infty(f)$ from before, but in the exceptional case where $\Gamma_\infty = \langle -[\begin{smallmatrix} 1 & h \\ 0 & 1 \end{smallmatrix}] \rangle$, when h' is the period, $\nu_{\pi(\infty)}(f)$ is $\nu_\infty(f)/2$, possibly a half-integer.

For a general cusp $\pi(s) \in X(\Gamma)$ where $s \in \mathbb{Q} \cup \{\infty\}$, take any matrix $\alpha \in \mathrm{SL}_2(\mathbb{Z})$ such that $\alpha(\infty) = s$. Using α to conjugate the discussion above from ∞ to s gives the order of f at $\pi(s) \in X(\Gamma)$ in terms of the order of f at $s \in \mathbb{Q} \cup \{\infty\}$,

$$\nu_{\pi(s)}(f) = \begin{cases} \nu_s(f)/2 & \text{if } (\alpha^{-1}\Gamma\alpha)_\infty = \langle -[\begin{smallmatrix} 1 & h \\ 0 & 1 \end{smallmatrix}] \rangle, \\ \nu_s(f) & \text{otherwise.} \end{cases} \tag{3.3}$$

This is half-integral if $(\alpha^{-1}\Gamma\alpha)_\infty = \langle -[\begin{smallmatrix} 1 & h \\ 0 & 1 \end{smallmatrix}] \rangle$ (when $\pi(s)$ or s itself is called an *irregular cusp* of Γ) and k is odd. For example, $1/2$ is an irregular cusp of $\Gamma_1(4)$ by Exercise 3.2.5. Section 3.8 will show that this is the only example of an irregular cusp for the groups $\Gamma_0(N)$, $\Gamma_1(N)$, and $\Gamma(N)$. Again, if $k = 0$ then $\nu_{\pi(s)}(f)$ is integral at all cusps, regular or irregular.

The results so far in this chapter lead to more examples of modular forms. Recall the Dedekind function $\eta(\tau) = q_{24} \prod_{n=1}^\infty (1 - q^n)$ and the discriminant function $\Delta(\tau) = (2\pi)^{12}\eta(\tau)^{24} = (2\pi)^{12}q\prod_{n=1}^\infty (1 - q^n)^{24}$ from Chapter 1, where as always $\tau \in \mathcal{H}$ and $q = e^{2\pi i \tau}$, and $q_{24} = e^{2\pi i \tau/24}$.

Proposition 3.2.2. *Let k and N be positive integers such that $k(N+1) = 24$. Thus (k, N) is one of $(1, 23)$, $(2, 11)$, $(3, 7)$, $(4, 5)$, $(6, 3)$, $(8, 2)$, or $(12, 1)$. Define a function*

$$\varphi_k(\tau) = \eta(\tau)^k \eta(N\tau)^k.$$

(a) *If $\mathcal{S}_k(\Gamma_1(N))$ is nonzero then $\mathcal{S}_k(\Gamma_1(N)) = \mathbb{C}\varphi_k$.*
(b) *If $\mathcal{S}_k(\Gamma_0(N))$ is nonzero then $\mathcal{S}_k(\Gamma_0(N)) = \mathcal{S}_k(\Gamma_1(N)) = \mathbb{C}\varphi_k$.*

Proof. Consider the function $g = \varphi_k^{N+1}$, i.e.,

$$g(\tau) = (2\pi)^{-24}\Delta(\tau)\Delta(N\tau) = q^{N+1}\prod_{n=1}^\infty (1 - q^n)^{24}(1 - q^{Nn})^{24}.$$

Since $\Delta(\tau) \in \mathcal{S}_{12}(\mathrm{SL}_2(\mathbb{Z}))$ it follows that $\Delta(N\tau) \in \mathcal{S}_{12}(\Gamma_0(N))$ and so $g \in \mathcal{S}_{24}(\Gamma_0(N)) \subset \mathcal{S}_{24}(\Gamma_1(N))$. Let $S = [\begin{smallmatrix} 0 & -1 \\ 1 & 0 \end{smallmatrix}]$ and $q_N = e^{2\pi i \tau/N}$, and compute (Exercise 3.2.6)

$$(g[S]_{24})(\tau) = N^{-12}g(\tau/N) = N^{-12}q_N^{N+1}\prod_{n=1}^\infty (1 - q_N^n)^{24}(1 - q_N^{Nn})^{24}.$$

Let $\pi_0 : \mathcal{H}^* \longrightarrow X_0(N)$ and $\pi_1 : \mathcal{H}^* \longrightarrow X_1(N)$ be the natural projections, e.g., $\pi_0(\tau) = \Gamma_0(N)(\tau)$ for $\tau \in \mathcal{H}^*$. All cusps of $\Gamma_0(N)$ and $\Gamma_1(N)$ are regular since $N \neq 4$. If $N = 1$ then the only cusp of $X_0(N)$ is $\pi_0(\infty)$, of width 1. Otherwise N is prime and Exercise 3.1.4 shows that the cusps of $X_0(N)$ are $\pi_0(\infty)$, of width 1, and $\pi_0(0)$, of width N. Exercise 3.1.5 shows that the natural projection $\pi : X_1(N) \longrightarrow X_0(N)$ is unramified at the cusps when N is prime, and the same assertion holds trivially when $N = 1$. Thus if $\pi_1(s)$ is a cusp of $X_1(N)$ over $\pi_0(\infty)$ then it has width 1, $s = \alpha(\infty)$ for some $\alpha \in \Gamma_0(N)$, and the identity $g[\alpha]_{24} = g$ combines with the displayed product form of g to show that $\nu_{\pi_1(s)}(g) = N + 1$ by (3.3) since the cusp is regular, i.e., g has a zero of order $N + 1$ at $\pi_1(s)$. Similarly, if $\pi_1(s)$ is a cusp over $\pi_0(0)$ then it has width $N + 1$, $s = \alpha(0) = \alpha S(\infty)$ for some $\alpha \in \Gamma_0(N)$, and the identity $g[\alpha S]_{24} = g[S]_{24}$ combines with the displayed product form of $g[S]_{24}$ to show that also $\nu_{\pi_1(s)}(g) = N + 1$ and again g has a zero of order $N + 1$ at $\pi_1(s)$.

Suppose $f \in S_k(\Gamma_1(N))$. Then $f^{N+1} \in S_{24}(\Gamma_1(N))$. The quotient f^{N+1}/g lies in $\mathcal{A}_0(\Gamma_1(N)) = \mathbb{C}(X_1(N))$, it is holomorphic on $Y_1(N)$ because Δ is nonzero on $\mathcal{H}$, and it is holomorphic at the cusps of $X_1(N)$ since f^{N+1} has zeros of order at least $N + 1$ at the cusps. The only holomorphic functions on the compact Riemann surface $X_1(N)$ are the constants, so $f^{N+1} = cg$ for some $c \in \mathbb{C}$. Taking $(N + 1)$st roots gives $f(\tau) = c\mu_N^{e(\tau)}\varphi_k(\tau)$ for each $\tau \in \mathcal{H}$ where $\mu_N = e^{2\pi i/N}$ and $e(\tau)$ is a continuous integer-valued function of τ, making it constant, and c denotes different constants in different places. That is, $f = c\varphi_k$, proving part (a). Part (b) follows immediately since $S_k(\Gamma_0(N)) \subset S_k(\Gamma_1(N))$. $\qquad\square$

The proposition does not show that φ_k is a modular form at all until we know by some other means that a nonzero modular form of the desired type exists; the only possible f in the proof could be $f = 0$, in which case the relation $f = c\varphi_k$ does not imply $\varphi_k = cf$. When $(k, N) = (12, 1)$ the proposition shows that $S_{12}(\mathrm{SL}_2(\mathbb{Z})) = \mathbb{C}\Delta$ since Δ is a nonzero cusp form of weight 12. The dimension-counting formulas to be obtained in this chapter will show that $S_k(\Gamma_0(N))$ is 1-dimensional when (k, N) is any of $(2, 11)$, $(4, 5)$, $(6, 3)$, $(8, 2)$, so part (b) of the proposition applies in these cases. Although $S_k(\Gamma_0(N)) = \{0\}$ when k is odd since $-I \in \Gamma_0(N)$, the dimension formulas will also show that $S_3(\Gamma_1(7))$ is 1-dimensional, and part (a) applies. The formulas will not help with the remaining case $S_1(\Gamma_1(23))$, but here one can still verify directly that φ_1 lies in the space (though we omit this) and so the proposition says that it spans the space. Thus φ_k spans $S_k(\Gamma_1(N))$ whenever $k(N + 1) = 24$, and it also spans $S_k(\Gamma_0(N))$ when k is even.

Exercises

3.2.1. Let $f \in \mathcal{A}_k(\Gamma)$ be an automorphic form and let $s \in \mathbb{Q} \cup \{\infty\}$ be a cusp. Show that the definition of $\nu_s(f)$ given in the text is independent of the choice of α and well defined on the quotient $X(\Gamma)$.

3.2.2. Show that the modular invariant j has Laurent series (3.1). (A hint for this exercise is at the end of the book.)

3.2.3. Let Γ be a congruence subgroup of $\mathrm{SL}_2(\mathbb{Z})$. Show that the derivative of the modular invariant j lies in $\mathcal{A}_2(\Gamma)$. Show that consequently $\mathcal{A}_k(\Gamma)$ contains nonzero elements for all even positive integers k. (A hint for this exercise is at the end of the book.)

3.2.4. Show that for any $k \in \mathbb{Z}$, if f is a nonzero element of $\mathcal{A}_k(\Gamma)$ then $\mathcal{A}_k(\Gamma) = \mathbb{C}(X(\Gamma))f$.

3.2.5. Let $\alpha = \left[\begin{smallmatrix} 1 & -\frac{1}{2} \\ 2 & -1 \end{smallmatrix}\right]$ and let $\Gamma = \alpha^{-1}\Gamma_1(4)\alpha$. Show that $\Gamma_\infty = \langle -\left[\begin{smallmatrix} 1 & 1 \\ 0 & 1 \end{smallmatrix}\right]\rangle$. (A hint for this exercise is at the end of the book.)

3.2.6. Verify the expression for $g[S]_{24}$ in the proof of Proposition 3.2.2.

3.3 Meromorphic differentials

Let Γ be a congruence subgroup of $\mathrm{SL}_2(\mathbb{Z})$. The transformation rule for automorphic forms of weight k with respect to Γ,

$$f(\gamma(\tau)) = j(\gamma, \tau)^k f(\tau), \quad \gamma \in \Gamma,$$

does not make such forms Γ-invariant. On the other hand, Section 1.1 observed that $d\gamma(\tau) = j(\gamma, \tau)^{-2}d\tau$ for $\gamma \in \Gamma$ so that at least formally the differential

$$f(\tau)(d\tau)^{k/2}$$

is truly Γ-invariant. This section makes sense of such differentials on the Riemann surface $X(\Gamma)$. Doing so requires care at the elliptic points and the cusps.

The first step is to define differentials locally. Let V be any open subset of $\mathbb{C}$ and let $n \in \mathbb{N}$ be any natural number. The *meromorphic differentials on V of degree n* are

$$\Omega^{\otimes n}(V) = \{f(q)(dq)^n : f \text{ is meromorphic on } V\}$$

where q is the variable on V. These form a vector space over $\mathbb{C}$ under the natural definitions of addition and scalar multiplication, $f(q)(dq)^n + g(q)(dq)^n = (f + g)(q)(dq)^n$ and $c(f(q)(dq)^n) = (cf)(q)(dq)^n$. The sum over all degrees,

$$\Omega(V) = \bigoplus_{n \in \mathbb{N}} \Omega^{\otimes n}(V),$$

naturally forms a ring under the definition $(dq)^n(dq)^m = (dq)^{n+m}$.

Since a Riemann surface involves not only local patches but also transition maps between them, we next need to study maps between local differentials. Any holomorphic map $\varphi : V_1 \longrightarrow V_2$ between open sets in $\mathbb{C}$ induces the *pullback* map in the other direction between meromorphic differentials,

$$\varphi^* : \Omega^{\otimes n}(V_2) \longrightarrow \Omega^{\otimes n}(V_1),$$
$$\varphi^*(f(q_2)(dq_2)^n) = f(\varphi(q_1))(\varphi'(q_1))^n(dq_1)^n.$$

In other words, taking the pullback means changing variables in the meromorphic differential, doing so compatibly with the change of variable formula for integrals. The pullback is obviously linear. The pullback is also *contravariant*, meaning that if $\varphi_1 = \varphi$ is as before and $\varphi_2 : V_2 \longrightarrow V_3$ is also holomorphic then $(\varphi_2 \circ \varphi_1)^* = \varphi_1^* \circ \varphi_2^*$ (Exercise 3.3.1(a)). If $V_1 \subset V_2$ and $\iota : V_1 \longrightarrow V_2$ is inclusion then its pullback is restriction, $\iota^*(\omega) = \omega|_{V_1}$ for $\omega \in \Omega^{\otimes n}(V_2)$ (Exercise 3.3.1(b)). It follows that if φ is a bijection, making its inverse holomorphic as well by complex analysis, then $(\varphi^*)^{-1} = (\varphi^{-1})^*$ (Exercise 3.3.1(c)). If $\pi : V_1 \longrightarrow V_2$ is a holomorphic surjection of open sets in $\mathbb{C}$ then π^* is an injection (Exercise 3.3.1(d)).

Now we can piece together local differentials on a Riemann surface X. Let X have coordinate charts $\varphi_j : U_j \longrightarrow V_j$, where j runs through some index set J, each U_j is a neighborhood in X, and each V_j is an open set in $\mathbb{C}$. A *meromorphic differential on X of degree n* is a collection of local meromorphic differentials of degree n,

$$(\omega_j)_{j \in J} \in \prod_{j \in J} \Omega^{\otimes n}(V_j),$$

that is *compatible*. To define this, let $V_{j,k} = \varphi_j(U_j \cap U_k)$ and $V_{k,j} = \varphi_k(U_j \cap U_k)$ for $j, k \in J$. Then the compatibility criterion is that pulling back any transition map

$$V_{j,k} \xrightarrow{\varphi_{k,j}} V_{k,j}$$

to get the corresponding map of local meromorphic differentials

$$\Omega^{\otimes n}(V_{j,k}) \xleftarrow{\varphi_{k,j}^*} \Omega^{\otimes n}(V_{k,j})$$

gives

$$\varphi_{k,j}^*(\omega_k|_{V_{k,j}}) = \omega_j|_{V_{j,k}}.$$

The meromorphic differentials of degree n on X are denoted $\Omega^{\otimes n}(X)$. Again this set forms a complex vector space and the sum $\Omega(X) = \bigoplus_{n \in \mathbb{N}} \Omega^{\otimes n}(X)$ forms a ring.

For example, when X is a complex elliptic curve $\mathbb{C}/\Lambda$ the coordinate maps $\varphi_j : U_j \longrightarrow V_j$ are homeomorphic local inverses to natural projection $\pi : \mathbb{C} \longrightarrow X$. Let $\omega = (dz_j)_{j \in J}$. Then ω satisfies the compatibility criterion since the transition maps between coordinate patches take the form $z \mapsto z + \lambda$ for $\lambda \in \Lambda$, and these pull dz_k back to dz_j. Thus the differential dz makes sense globally on a complex torus even though a holomorphic variable z does not.

Returning to the congruence subgroup Γ of $\mathrm{SL}_2(\mathbb{Z})$, we now map each meromorphic differential ω on $X(\Gamma)$ to a meromorphic differential $f(\tau)(d\tau)^n$ on $\mathcal{H}$, the Γ-invariant object on $\mathcal{H}$ mentioned at the beginning of the section. The map is the pullback of the natural projection $\pi : \mathcal{H} \longrightarrow X(\Gamma)$,

$$\pi^* : \Omega^{\otimes n}(X(\Gamma)) \longrightarrow \Omega^{\otimes n}(\mathcal{H}).$$

To define this, recall from Chapter 2 that a collection of coordinate neighborhoods on $X(\Gamma)$ is $\{\pi(U_j) : j \in J\}$ where each $U_j \subset \mathcal{H}^*$ is a neighborhood of a point $\tau_j \in \mathcal{H}$ or of a cusp $s_j \in \mathbb{Q} \cup \{\infty\}$; the local coordinate map $\varphi_j : \pi(U_j) \longrightarrow V_j$ is characterized by the relation $\psi_j = \varphi_j \circ \pi$ where $\psi_j : U_j \longrightarrow V_j$ mimics the identifying action of π but maps into $\mathbb{C}$. Now let $\omega = (\omega_j)_{j \in J}$ be a meromorphic differential on $X(\Gamma)$. For each $j \in J$ set $U'_j = U_j \cap \mathcal{H}$ and $V'_j = \psi_j(U'_j)$ and $\omega'_j = \omega_j|_{V'_j}$. Then the pullback $\pi^*(\omega)$ is defined locally on $\mathcal{H}$ as

$$\pi^*(\omega)|_{U'_j} = \psi_j^*(\omega'_j) \quad \text{for all } j. \tag{3.4}$$

To see that (3.4) gives a well defined global meromorphic differential $\pi^*(\omega) = f(\tau)(d\tau)^n$ on $\mathcal{H}$, consider the commutative diagram

$$
\begin{array}{ccc}
& U_j \cap U_k & \\
{\scriptstyle\psi_j}\swarrow & {\scriptstyle\pi}\downarrow & \searrow{\scriptstyle\psi_k} \\
V_{j,k} \xrightarrow[\varphi_j^{-1}]{} & \pi(U_j \cap U_k) \xrightarrow[\varphi_k]{} & V_{k,j}.
\end{array}
\tag{3.5}
$$

The transition map $\varphi_{k,j} = \varphi_k \varphi_j^{-1}|_{V_{j,k}}$ satisfies $\varphi_{k,j} \circ \psi_j = \psi_k$ on $U_j \cap U_k$, and pulling back gives $\psi_k^* = \psi_j^* \circ \varphi_{k,j}^*$ on $V_{k,j}$. Therefore, letting $V'_{j,k} = \psi_j(U'_j \cap U'_k)$ and $V'_{k,j} = \psi_k(U'_j \cap U'_k)$,

$$\psi_k^*(\omega_k|_{V'_{k,j}}) = \psi_j^*(\varphi_{k,j}^*(\omega_k|_{V'_{k,j}})) = \psi_j^*(\omega_j|_{V'_{j,k}}) \quad \text{by compatibility,}$$

so the overlapping local pullbacks have a common value $f(\tau)(d\tau)^n|_{U'_j \cap U'_k}$ and the global pullback $\pi^*(\omega) = f(\tau)(d\tau)^n$ is well defined.

Since the pullback comes from an object on the quotient $X(\Gamma)$ it must be Γ-invariant. That is, for any $\gamma \in \Gamma$,

$$
\begin{aligned}
f(\tau)(d\tau)^n = \gamma^*(f(\tau)(d\tau)^n) &= (f(\gamma(\tau)))(\gamma'(\tau))^n (d\tau)^n \\
&= j(\gamma, \tau)^{-2n} f(\gamma(\tau))(d\tau)^n = (f[\gamma]_{2n})(\tau)(d\tau)^n.
\end{aligned}
$$

Thus the meromorphic function f defining the pullback is weakly modular of weight $2n$. This is the motivation promised back in Chapter 1 for the definition of weak modularity, at least for even weights.

The function f also satisfies condition (3) in Definition 3.2.1, that $f[\alpha]_{2n}$ is meromorphic at ∞ for any $\alpha \in \mathrm{SL}_2(\mathbb{Z})$. To see this, let $s = \alpha(\infty)$. The local map $\psi : U \longrightarrow V$ about s takes the form $\psi = \rho \circ \delta : \tau \mapsto z \mapsto q$ with $\delta = \alpha^{-1}$ and $\rho(z) = e^{2\pi i z/h}$ where h is the width of s. Since ω is meromorphic at the cusps of $X(\Gamma)$, the local differential $\omega|_V$ takes the form $g(q)(dq)^n$ where g is meromorphic at 0. The function f on $U - \{s\}$ comes from pulling back $\omega|_{V-\{0\}}$ under ψ to $f(\tau)(d\tau)^n$. Computing this gives $f = \tilde{f}[\delta]_{2n}$ where (Exercise 3.3.2(a))

$$\tilde{f}(z) = (2\pi i/h)^n q^n g(q), \quad q = e^{2\pi i z/h}.$$

Thus $(f[\alpha]_{2n})(z) = \tilde{f}(z)$ is meromorphic at ∞ as claimed. In sum, every meromorphic differential ω of degree n on $X(\Gamma)$ pulls back to a meromorphic differential $\pi^*(\omega) = f(\tau)(d\tau)^n$ on $\mathcal{H}$ where f is an automorphic form of weight $2n$ with respect to Γ.

The calculation that $\pi^*(\omega)$ is well defined can be turned around (Exercise 3.3.2(b))—given a collection $(\omega_j) \in \prod_{j\in J} \Omega^{\otimes n}(V_j)$ of local meromorphic differentials, exclude the cusps by setting $U_j' = U_j \cap \mathcal{H}$ and $V_j' = \psi_j(U_j')$ and $\omega_j' = \omega_j|_{V_j'}$ for all $j \in J$. If the ω_j' pull back under ψ_j^* to restrictions of some meromorphic differential $f(\tau)(d\tau)^n$ on $\mathcal{H}$ then the original local differentials ω_j are compatible and so $(\omega_j) \in \Omega^{\otimes n}(X(\Gamma))$. Thus, a given collection (ω_j) of local meromorphic differentials is compatible, giving a meromorphic differential on $X(\Gamma)$, if and only if its local elements pull back to restrictions of some $f(\tau)(d\tau)^n \in \Omega^{\otimes n}(\mathcal{H})$ with $f \in \mathcal{A}_{2n}(\Gamma)$.

Conversely, given an automorphic form $f \in \mathcal{A}_{2n}(\Gamma)$ we will construct a meromorphic differential $\omega(f) \in \Omega^{\otimes n}(X(\Gamma))$ such that $\pi^*(\omega) = f(\tau)(d\tau)^n$. By the previous paragraph it suffices to construct local differentials that pull back to restrictions of $f(\tau)(d\tau)^n$. Thus the idea is to express $f(\tau)(d\tau)^n$ in local coordinates.

Each local map $\psi_j : U_j \longrightarrow V_j$ is a composite $\psi_j = \rho_j \circ \delta_j : \tau \mapsto z \mapsto q$ with $\delta_j \in \mathrm{GL}_2(\mathbb{C})$. Since δ_j is invertible it is easy to transform $f(\tau)(d\tau)^n$ into z-coordinates locally. First extend the weight-k operator $[\gamma]_k$ to matrices $\gamma \in \mathrm{GL}_2(\mathbb{C})$ by defining $j(\gamma,\tau) = c\tau + d$ as before and

$$(f[\gamma]_k)(\tau) = (\det\gamma)^{k/2} j(\gamma,\tau)^{-k} f(\gamma(\tau))$$

when this makes sense. The exponent $k/2$ of $\det\gamma$ is a normalization convenient for the next calculation because of the formula $\gamma'(\tau) = \det\gamma/j(\gamma,\tau)^2$ (Exercise 3.3.3); in other contexts the factor $(\det\gamma)^{k-1}$ is more convenient. Now let $U_j' = U_j \cap \mathcal{H}$. Then $f(\tau)(d\tau)^n|_{U_j'}$ is a pullback $\delta_j^*(\lambda_j)$ where λ_j is the differential obtained by pulling $f(\tau)(d\tau)^n|_{U_j'}$ forward to z-space under δ_j^{-1}. Letting $\alpha = \delta_j^{-1}$,

$$\lambda_j = \alpha^*(f(\tau)(d\tau)^n|_{U_j'}) = f(\alpha(z))(d(\alpha(z)))^n$$
$$= (\det\alpha)^n j(\alpha,z)^{-2n} f(\alpha(z))(dz)^n = (f[\alpha]_{2n})(z)(dz)^n.$$

Thus λ_j is $\delta_j \Gamma \delta_j^{-1}$-invariant since $f(\tau)(d\tau)^n$ is Γ-invariant. Further pushing λ_j forward from z-space to q-space isn't quite as easy since ρ_j is not invertible in general, but it isn't hard to find local forms ω_j that pull back under ρ_j to λ_j as desired.

Specifically, if $U_j \subset \mathcal{H}$, i.e., if we are not working at a cusp, then δ_j takes τ_j to 0 and the quotient $\{\pm I\}(\delta_j \Gamma \delta_j^{-1})_0 / \{\pm I\}$ is cyclic of order h, generated by the rotation $r_h : z \mapsto \mu_h z$ where $\mu_h = e^{2\pi i/h}$. By $\delta_j \Gamma \delta_j^{-1}$-invariance the pullback $r_h^*(\lambda_j) = (f[\alpha]_{2n})(\mu_h z)\mu_h^n (dz)^n$ must equal λ_j, or equivalently $\mu_h^n(f[\alpha]_{2n})(\mu_h z) = (f[\alpha]_{2n})(z)$, or $(\mu_h z)^n(f[\alpha]_{2n})(\mu_h z) = z^n(f[\alpha]_{2n})(z)$. In other words, the function $z^n(f[\alpha]_{2n})(z)$ takes the form $g_j(z^h)$ for some meromorphic function g_j. Note for later reference in proving formula (3.8) that $h\nu_0(g_j) = n + \nu_{\tau_j}(f)$ and thus $\nu_0(g_j) = \nu_{\pi(\tau_j)}(f) + n/h$. Define a local meromorphic differential in q-coordinates,

$$\omega_j = \frac{g_j(q)}{(hq)^n}(dq)^n \quad \text{on } V_j. \tag{3.6}$$

Since $\rho_j(z) = z^h$, this ω_j pulls back under ρ_j to λ_j (Exercise 3.3.4(a)), which in turn pulls back under δ_j to the original $f(\tau)(d\tau)^n|_{U_j}$. Thus, for each $U_j \subset \mathcal{H}$ the differential ω_j on V_j pulls back under ψ_j to a suitable restriction of the global differential $f(\tau)(d\tau)^n$ on $\mathcal{H}$.

On the other hand, if U_j contains a cusp s_j then δ_j takes s_j to ∞ and the function $(f[\alpha]_{2n})(z)$ takes the form $g_j(q_h)$ where h is the width of s and $q_h = e^{2\pi i z/h}$; here g_j is meromorphic in q_h at 0 if the cusp is regular and g_j is meromorphic in $q_h^{1/2} = e^{\pi i z/h}$ at 0 if the cusp is irregular, but we think of g_j as a series in powers of q_h (half-integral powers in the irregular case) so that the order is the index of the leading coefficient. The relevant local differential is now

$$\omega_j = \frac{g_j(q)}{(2\pi i q/h)^n}(dq)^n \quad \text{on } V_j, \tag{3.7}$$

which is meromorphic at $q = 0$. Since $q = \rho_j(z) = e^{2\pi i z/h}$, again ω_j pulls back under ρ_j to λ_j (Exercise 3.3.4(b)), and as before it follows that $\psi_j^*(\omega_j) = f(\tau)(d\tau)^n|_{U_j'}$. Putting all of this together gives

Theorem 3.3.1. *Let $k \in \mathbb{N}$ be even and let Γ be a congruence subgroup of* $\mathrm{SL}_2(\mathbb{Z})$. *The map*

$$\omega : \mathcal{A}_k(\Gamma) \longrightarrow \Omega^{\otimes k/2}(X(\Gamma))$$

$$f \mapsto (\omega_j)_{j \in J} \text{ where } (\omega_j) \text{ pulls back to } f(\tau)(d\tau)^{k/2} \in \Omega^{\otimes k/2}(\mathcal{H})$$

is an isomorphism of complex vector spaces.

Proof. The map ω is defined since we have just constructed $\omega(f)$. Clearly ω is $\mathbb{C}$-linear and injective. And ω is surjective because every $(\omega_j) \in \Omega^{\otimes k/2}(X(\Gamma))$ pulls back to some $f(\tau)(d\tau)^{k/2} \in \Omega^{\otimes k/2}(\mathcal{H})$ with $f \in \mathcal{A}_k(\Gamma)$. $\qquad\qquad \square$

Exercises 3.2.3 and 3.2.4 showed that for k positive and even, $\mathcal{A}_k(\Gamma)$ takes the form $\mathbb{C}(X(\Gamma))f$ where $\mathbb{C}(X(\Gamma))$ is the field of meromorphic functions on $X(\Gamma)$ and f is any nonzero element of $\mathcal{A}_k(\Gamma)$. Thus, Theorem 3.3.1 shows that $\Omega^{\otimes k/2}(X(\Gamma)) = \mathbb{C}(X(\Gamma))\omega(f)$ for such k.

The aim of this chapter is to compute the dimensions of the subspaces $\mathcal{M}_k(\Gamma)$ and $\mathcal{S}_k(\Gamma)$ of $\mathcal{A}_k(\Gamma)$. Now that we know that $\mathcal{A}_k(\Gamma)$ and $\Omega^{\otimes k/2}(X(\Gamma))$ are isomorphic, the final business of this section is to describe the images $\omega(\mathcal{M}_k(\Gamma))$ and $\omega(\mathcal{S}_k(\Gamma))$ in $\Omega^{\otimes k/2}(X(\Gamma))$. Some Riemann surface theory to be presented in the next section will then find the desired dimensions by computing the dimensions of these image subspaces instead in Sections 3.5 and 3.6.

So take any automorphic form $f \in \mathcal{A}_k(\Gamma)$ and let $\omega(f) = (\omega_j)_{j \in J}$. For a point $\tau_j \in \mathcal{H}$, the local differential (3.6) with $n = k/2$ vanishes at $q = 0$ to (integral) order (Exercise 3.3.5)

$$\nu_0(\omega_j) \overset{\text{def}}{=} \nu_0\left(\frac{g_j(q)}{(hq)^{k/2}}\right) = \nu_{\pi(\tau_j)}(f) - \frac{k}{2}\left(1 - \frac{1}{h}\right). \tag{3.8}$$

In particular, at a nonelliptic point, when $h = 1$, the order of vanishing is $\nu_0(\omega_j) = \nu_{\pi(\tau_j)}(f)$, the order of the original function. For a cusp s_j the local differential (3.7) with $n = k/2$ vanishes at $q = 0$ to order (Exercise 3.3.5 again)

$$\nu_0(\omega_j) \overset{\text{def}}{=} \nu_0\left(\frac{g_j(q)}{(2\pi i q/h)^{k/2}}\right) = \nu_{\pi(s_j)}(f) - \frac{k}{2}. \tag{3.9}$$

When $k \in \mathbb{N}$ is even, formulas (3.8) and (3.9) translate the conditions $\nu_{\pi(\tau_j)}(f) \geq 0$ and $\nu_{\pi(s_j)}(f) \geq 0$ characterizing $\mathcal{M}_k(\Gamma)$ as a subspace of $\mathcal{A}_k(\Gamma)$ into conditions characterizing $\omega(\mathcal{M}_k(\Gamma))$ as a subspace of $\Omega^{\otimes k/2}(X(\Gamma))$, and similarly for $\mathcal{S}_k(\Gamma)$ and $\omega(\mathcal{S}_k(\Gamma))$. In particular, the weight 2 cusp forms $\mathcal{S}_2(\Gamma)$ are isomorphic as a complex vector space to the degree 1 *holomorphic* differentials on $X(\Gamma)$, denoted $\Omega^1_{\text{hol}}(X(\Gamma))$ (Exercise 3.3.6). This special case will figure prominently in the later chapters of the book.

Exercises

3.3.1. (a) Show that the pullback is contravariant.

(b) Show that if $\iota : V_1 \longrightarrow V_2$ is inclusion then its pullback is the restriction $\iota^*(\omega) = \omega|_{V_1}$ for $\omega \in \Omega^{\otimes n}(V_2)$.

(c) Show that if φ is a holomorphic bijection of open sets in $\mathbb{C}$ then $(\varphi^{-1})^* = (\varphi^*)^{-1}$.

(d) Show that if $\pi : V_1 \longrightarrow V_2$ is a holomorphic surjection of open sets in $\mathbb{C}$ then π^* is an injection. If $i : V_1 \longrightarrow V_2$ is an injection of open sets in $\mathbb{C}$, need i^* be a surjection?

3.3.2. (a) Let $\psi = \rho \circ \delta$ where $\delta \in \mathrm{SL}_2(\mathbb{Z})$ and $\rho(z) = e^{2\pi i z/h}$. Let $\omega = g(q)(dq)^n$. Show that $\psi^*(\omega) = f(\tau)(d\tau)^n$ with $f = \tilde{f}[\delta]_{2n}$ and $\tilde{f}(z) = (2\pi i/h)^n q^n g(q)$ where $q = e^{2\pi i z/h}$.

(b) Consider the situation $\pi : \mathcal{H} \longrightarrow X(\Gamma)$ as described in the small in diagram (3.5). Show that if the local meromorphic differentials ω_j pull back under ψ_j^* to restrictions of some common global meromorphic differential $f(\tau)d\tau$ on $\mathcal{H}$ then they are compatible. Exercise 3.3.1(d) is relevant.

3.3.3. For $\gamma \in \mathrm{GL}_2(\mathbb{C})$, show that $\gamma'(\tau) = \det \gamma / j(\gamma, \tau)^2$. (This generalizes Lemma 1.2.2(e).)

3.3.4. (a) For a neighborhood $U_j \subset \mathcal{H}$, show that the local differential ω_j in (3.6) pulls back under ρ_j to λ_j. How was ω_j found in the first place?
  (b) Same question for ω_j in (3.7) when U_j contains a cusp.

3.3.5. Prove formulas (3.8) and (3.9).

3.3.6. Prove that $\mathcal{S}_2(\Gamma)$ and $\Omega^1_{\mathrm{hol}}(X(\Gamma))$ are isomorphic as complex vector spaces. Your proof will incidentally show that the elements of $\mathcal{S}_2(\Gamma)$ vanish at the elliptic points of $X(\Gamma)$. Argue this directly as well by examining the text leading up to (3.6).

3.4 Divisors and the Riemann–Roch Theorem

Let X be any compact Riemann surface, though for our purposes X is nearly always $X(\Gamma)$. A *divisor* on X is a finite formal sum of integer multiples of points of X,

$$D = \sum_{x \in X} n_x x, \quad n_x \in \mathbb{Z} \text{ for all } x, \ n_x = 0 \text{ for almost all } x.$$

The set $\mathrm{Div}(X)$ of divisors on X forms an Abelian group, the *free Abelian group on the points of X*, under the natural definition of addition,

$$\sum n_x x + \sum n'_x x = \sum (n_x + n'_x) x.$$

(Since every divisor sums over all points $x \in X$, there is no need to continue notating the set of summation.) For divisors $D = \sum n_x x$ and $D' = \sum n'_x x$, write $D \geq D'$ when $n_x \geq n'_x$ for all x. The *degree* of a divisor is the integer

$$\deg(D) = \sum n_x.$$

The map $\deg : \mathrm{Div}(X) \longrightarrow \mathbb{Z}$ is a homomorphism of Abelian groups.
  Every nonzero meromorphic function $f : X \longrightarrow \widehat{\mathbb{C}}$ has a divisor

$$\mathrm{div}(f) = \sum \nu_x(f) x.$$

Let $\mathbb{C}(X)$ denote the field of meromorphic functions on X and let $\mathbb{C}(X)^*$ be its multiplicative group of nonzero elements. Then the map

$$\mathrm{div} : \mathbb{C}(X)^* \longrightarrow \mathrm{Div}(X)$$

is a homomorphism, meaning $\mathrm{div}(f_1 f_2) = \mathrm{div}(f_1) + \mathrm{div}(f_2)$ for all $f_1, f_2 \in \mathbb{C}(X)^*$. Every $f \in \mathbb{C}(X)^*$ has a divisor of degree 0, so the group $\mathrm{Div}^\ell(X)$ of divisors of nonzero meromorphic functions is a subgroup of the group $\mathrm{Div}^0(X)$ of divisors of degree 0. *Abel's Theorem* (see [FK80]) states to what extent the converse fails: the quotient $\mathrm{Div}^0(X)/\mathrm{Div}^\ell(X)$ is isomorphic to a complex torus $\mathbb{C}^g/\Lambda_g$ where g is the genus of X and Λ_g is a lattice spanning $\mathbb{C}^g$. We will discuss Abel's Theorem further in Chapter 6.

For example, let Λ_i denote the lattice $i\mathbb{Z} \oplus \mathbb{Z}$ in $\mathbb{C}$ as usual and let X be the complex elliptic curve $\mathbb{C}/\Lambda_i$. Exercise 1.4.3 shows that (Exercise 3.4.1)

$$\mathrm{div}(\wp_i'/\wp_i) = -(0 + \Lambda_i) + (1/2 + \Lambda_i) + (i/2 + \Lambda_i) - ((1+i)/2 + \Lambda_i),$$

and one similarly finds $\mathrm{div}(\wp_{\mu_3}'/\wp_{\mu_3})$ on $\mathbb{C}/\Lambda_{\mu_3}$ (Exercise 3.4.1 again).

Define the *linear space* of any divisor D on X to be

$$L(D) = \{f \in \mathbb{C}(X) : f = 0 \text{ or } \mathrm{div}(f) + D \geq 0\}.$$

The idea here is that when $D = \mathrm{div}(\tilde{f})$ for some $\tilde{f} \in \mathbb{G}(X)^*$, the linear space $L(D)$ is those functions f on X vanishing to high enough order to make the product $f\tilde{f}$ holomorphic. The formula

$$\nu_x(f_1 + f_2) \geq \min\{\nu_x(f_1), \nu_x(f_2)\}, \quad f_1, f_2 \in \mathbb{C}(X)$$

(Exercise 3.4.2) shows that $L(D)$ forms a vector space over $\mathbb{C}$. This space turns out to be finite-dimensional and its dimension is denoted $\ell(D)$.

Let $\omega \in \Omega^{\otimes n}(X)$ (where $n \in \mathbb{N}$) be a nonzero meromorphic differential on X. Then ω has a local representation $\omega_x = f_x(q)(dq)^n$ about each point $x \in X$, where q is the local coordinate about x. Define $\nu_x(\omega) = \nu_0(f_x)$ and then define

$$\mathrm{div}(\omega) = \sum \nu_x(\omega)x.$$

So $\mathrm{div}(\omega_1\omega_2) = \mathrm{div}(\omega_1) + \mathrm{div}(\omega_2)$ for any nonzero $\omega_1 \in \Omega^{\otimes n}(X)$ and $\omega_2 \in \Omega^{\otimes m}(X)$. A *canonical divisor on X* is a divisor of the form $\mathrm{div}(\lambda)$ where λ is a nonzero element of $\Omega^1(X)$.

Theorem 3.4.1 (Riemann–Roch). *Let X be a compact Riemann surface of genus g. Let $\mathrm{div}(\lambda)$ be a canonical divisor on X. Then for any divisor $D \in \mathrm{Div}(X)$,*

$$\ell(D) = \deg(D) - g + 1 + \ell(\mathrm{div}(\lambda) - D).$$

The next result gives information about the canonical divisors and a simpler version of the Riemann–Roch Theorem for divisors of large enough degree.

Corollary 3.4.2. *Let X, g, $\mathrm{div}(\lambda)$, and D be as above. Then*

(a) $\ell(\mathrm{div}(\lambda)) = g$.
(b) $\deg(\mathrm{div}(\lambda)) = 2g - 2$.
(c) *If* $\deg(D) < 0$ *then* $\ell(D) = 0$.
(d) *If* $\deg(D) > 2g - 2$ *then* $\ell(D) = \deg(D) - g + 1$.

Proof. Note $\ell(0) = 1$ since $\mathrm{div}(f) \geq 0$ only for the constant functions in $\mathbb{C}(X)^*$, and so letting $D = 0$ in the Riemann–Roch Theorem gives $1 = -g + 1 + \ell(\mathrm{div}(\lambda))$, proving (a). Next set $D = \mathrm{div}(\lambda)$, combine the Riemann–Roch Theorem with (a) to get $g = \deg(\mathrm{div}(\lambda)) - g + 1 + 1$, and (b) follows. For (c), suppose $\ell(D) > 0$. Then some nonzero $f \in L(D)$ exists, so $\mathrm{div}(f) \geq -D$ and taking degrees shows that $\deg(D) \geq 0$, proving (c) by contraposition. Finally, (b) and (c) imply (d) since if $\deg(D) > 2g - 2$ then $\deg(\mathrm{div}(\lambda) - D) < 0$. □

We have seen in Section 3.2 that a nonzero element $f \in \mathcal{A}_2(\Gamma)$ exists for any congruence subgroup Γ of $\mathrm{SL}_2(\mathbb{Z})$. Let $\lambda = \omega(f) \in \Omega^1(X(\Gamma))$ be the corresponding meromorphic differential, so that $\mathrm{div}(\lambda)$ is canonical and its degree is $2g - 2$ by Corollary 3.4.2(b). For any positive even integer k, the meromorphic differential $\lambda^{k/2}$ belongs to $\Omega^{\otimes k/2}(X(\Gamma))$ and its divisor has degree $k(g - 1)$. Since $\Omega^{\otimes k/2}(X(\Gamma)) = \mathbb{C}(X(\Gamma))\lambda^{k/2}$ and $\deg(\mathrm{div}(f)) = 0$ for all nonzero $f \in \mathbb{C}(X(\Gamma))$, it follows that every nonzero differential $\omega \in \Omega^{\otimes k/2}(X(\Gamma))$ has a divisor of degree $k(g - 1)$ as well.

Exercise 3.4.3 shows that if Γ is a congruence subgroup of $\mathrm{SL}_2(\mathbb{Z})$ and g is the genus of $X(\Gamma)$ then the vector space $\Omega^1_{\mathrm{hol}}(X(\Gamma))$ of holomorphic differentials of degree 1 on $X(\Gamma)$ is isomorphic to the linear space $L(\lambda)$ for any canonical divisor λ, a space of dimension g by Corollary 3.4.2(a). Thus Exercise 3.3.6 shows that $\dim(\mathcal{S}_2(\Gamma)) = g$. The next two sections will broaden the ideas of this argument to obtain general dimension formulas for $\mathcal{M}_k(\Gamma)$ and $\mathcal{S}_k(\Gamma)$.

Exercises

3.4.1. Confirm the section's formula for $\mathrm{div}(\wp'_i/\wp_i)$. What is $\mathrm{div}(\wp'_{\mu_3}/\wp_{\mu_3})$?

3.4.2. Show that $\nu_x(f_1 + f_2) \geq \min\{\nu_x(f_1), \nu_x(f_2)\}$ for $f_1, f_2 \in \mathbb{C}(X)$ where x is any point of the compact Riemann surface X.

3.4.3. Recalling from the discussion immediately after Theorem 3.3.1 that $\Omega^1(X(\Gamma)) = \mathbb{C}(X(\Gamma))\lambda$ for any nonzero $\lambda \in \Omega^1(X(\Gamma))$, show that the map $\Omega^1(X(\Gamma)) \longrightarrow \mathbb{C}(X(\Gamma))$ taking each $f_0\lambda$ to f_0 is a vector space isomorphism mapping the holomorphic subspace $\Omega^1_{\mathrm{hol}}(X(\Gamma))$ to the linear space $L(\lambda)$.

3.5 Dimension formulas for even k

We can now express the results from Section 3.3 in the language of divisors. Let k be any even integer. Then using formulas (3.2) and (3.3), every nonzero automorphic form $f \in \mathcal{A}_k(\Gamma)$ has a well defined divisor $\sum \nu_x(f)x$ on $X(\Gamma)$ even though f itself is not well defined on $X(\Gamma)$. However, formula (3.2) shows that the coefficients in this divisor need not be integers at the elliptic points. To allow for divisors with rational coefficients, make the definition

$$\mathrm{Div}_{\mathbb{Q}}(X(\varGamma)) = \left\{ \sum n_x x, \quad n_x \in \mathbb{Q} \text{ for all } x, \, n_x = 0 \text{ for almost all } x \right\}.$$

This is the *free* $\mathbb{Q}$-*module* (or *free vector space*) *on the set of points of* $X(\varGamma)$. Thus $\mathrm{div}(f) \in \mathrm{Div}_{\mathbb{Q}}(X(\varGamma))$. The discussion of $\mathrm{Div}(X)$ as an Abelian group with a relation "$\geq$" and a degree function extends to $\mathrm{Div}_{\mathbb{Q}}(X)$, but the Riemann–Roch Theorem requires integral divisors.

Now let the even integer k satisfy $k \geq 2$, and let f be any nonzero element of $\mathcal{A}_k(\varGamma)$ (Exercise 3.2.3 showed that such f exists). Then (Exercise 3.5.1)

$$\mathcal{A}_k(\varGamma) = \mathbb{C}(X(\varGamma))f = \{f_0 f : f_0 \in \mathbb{C}(X(\varGamma))\}.$$

The holomorphic subspace $\mathcal{M}_k(\varGamma)$ of modular forms can be described in terms of divisors,

$$\mathcal{M}_k(\varGamma) = \{f_0 f \in \mathcal{A}_k(\varGamma) : f_0 f = 0 \text{ or } \mathrm{div}(f_0 f) \geq 0\}$$
$$\cong \{f_0 \in \mathbb{C}(X(\varGamma)) : f_0 = 0 \text{ or } \mathrm{div}(f_0) + \mathrm{div}(f) \geq 0\},$$

where the isomorphism is a complex vector space isomorphism. If $\mathrm{div}(f)$ were integral then we could now compute $\dim(\mathcal{M}_k(\varGamma))$ with the Riemann–Roch Theorem, but it isn't. To approximate $\mathrm{div}(f)$ by an integral divisor, define the greatest integer function $\lfloor \; \rfloor$ on rational divisors by applying it to the coefficients,

$$\left\lfloor \sum n_x x \right\rfloor = \sum \lfloor n_x \rfloor x \quad \text{for } \sum n_x x \in \mathrm{Div}_{\mathbb{Q}}(X(\varGamma)).$$

On the other hand, $\mathrm{div}(f_0)$ is integral since f_0 is meromorphic on $X(\varGamma)$, giving the equivalence

$$\mathrm{div}(f_0) + \mathrm{div}(f) \geq 0 \quad \Longleftrightarrow \quad \mathrm{div}(f_0) + \lfloor \mathrm{div}(f) \rfloor \geq 0.$$

This shows that $\mathcal{M}_k(\varGamma) \cong L(\lfloor \mathrm{div}(f) \rfloor)$ and so $\dim(\mathcal{M}_k(\varGamma)) = \ell(\lfloor \mathrm{div}(f) \rfloor)$.

To study the integral divisor $\lfloor \mathrm{div}(f) \rfloor$ let $\omega \in \varOmega^{\otimes k/2}(X(\varGamma))$ be the meromorphic differential $\omega(f)$ from Theorem 3.3.1 that pulls back to $f(\tau)(d\tau)^{k/2}$ on $\mathcal{H}$. Let $\{x_{2,i}\}$ be the period 2 elliptic points of $X(\varGamma)$, $\{x_{3,i}\}$ the period 3 elliptic points, and $\{x_i\}$ the cusps. As in Section 3.1, let ε_2, ε_3, and ε_∞ be the sizes of these sets. Make the formal definition

$$\mathrm{div}(d\tau) = -\sum_i \frac{1}{2} x_{2,i} - \sum_i \frac{2}{3} x_{3,i} - \sum_i x_i,$$

an element of $\mathrm{Div}_{\mathbb{Q}}(X(\varGamma))$ even though $d\tau$ is not well defined on $X(\varGamma)$. Then formulas (3.8) and (3.9), relating the vanishing orders of ω and f, show that compatibly with the identification of ω and $f(\tau)(d\tau)^{k/2}$ and with the rule that the divisor of a product is the sum of the divisors (Exercise 3.5.2(a)),

$$\mathrm{div}(\omega) = \mathrm{div}(f) + (k/2)\mathrm{div}(d\tau). \tag{3.10}$$

Here the left side is integral, the right side a sum of rational expressions. It follows that (Exercise 3.5.2(b))

$$\lfloor \mathrm{div}(f) \rfloor = \mathrm{div}(\omega) + \sum_i \left\lfloor \frac{k}{4} \right\rfloor x_{2,i} + \sum_i \left\lfloor \frac{k}{3} \right\rfloor x_{3,i} + \sum_i \frac{k}{2} x_i. \qquad (3.11)$$

Therefore by the discussion immediately after Corollary 3.4.2 showing that $\deg(\mathrm{div}(\omega)) = k(g-1)$,

$$\deg(\lfloor \mathrm{div}(f) \rfloor) = k(g-1) + \left\lfloor \frac{k}{4} \right\rfloor \varepsilon_2 + \left\lfloor \frac{k}{3} \right\rfloor \varepsilon_3 + \frac{k}{2} \varepsilon_\infty$$

$$\geq \frac{k}{2}(2g-2) + \frac{k-2}{4} \varepsilon_2 + \frac{k-2}{3} \varepsilon_3 + \frac{k}{2} \varepsilon_\infty \quad \text{(Exercise 3.5.2(c))}$$

$$= 2g - 2 + \frac{k-2}{2} \left(2g - 2 + \frac{\varepsilon_2}{2} + \frac{2\varepsilon_3}{3} + \varepsilon_\infty \right) + \varepsilon_\infty$$

$$\geq 2g - 2 + \varepsilon_\infty \quad \text{by Corollary 3.1.2, since } k \geq 2$$

$$> 2g - 2.$$

So $\lfloor \mathrm{div}(f) \rfloor$ has large enough degree for Corollary 3.4.2(d) to apply and the simpler form of the Riemann–Roch Theorem says that for even $k \geq 2$,

$$\dim(\mathcal{M}_k(\Gamma)) = \ell(\lfloor \mathrm{div}(f) \rfloor) = (k-1)(g-1) + \left\lfloor \frac{k}{4} \right\rfloor \varepsilon_2 + \left\lfloor \frac{k}{3} \right\rfloor \varepsilon_3 + \frac{k}{2} \varepsilon_\infty.$$

For the cusp forms $\mathcal{S}_k(\Gamma)$, a similar argument shows that $\mathcal{S}_k(\Gamma) \cong L(\lfloor \mathrm{div}(f) - \sum_i x_i \rfloor)$ and that (Exercise 3.5.2(d))

$$\dim(\mathcal{S}_k(\Gamma)) = \ell(\lfloor \mathrm{div}(f) - \sum_i x_i \rfloor) = \ell(\lfloor \mathrm{div}(f) \rfloor) - \varepsilon_\infty \quad \text{for } k \geq 4. \qquad (3.12)$$

For $k = 2$ the divisor $\lfloor \mathrm{div}(f) - \sum_i x_i \rfloor$ works out to a canonical divisor $\mathrm{div}(\lambda)$, cf. Exercise 3.4.2, and so its linear space has dimension g.

Finally we dispense with the cases where the weight k is nonpositive. A modular form of weight 0 with respect to Γ is a meromorphic function on $X(\Gamma)$ with no poles. Since any holomorphic map between compact Riemann surfaces either surjects or maps to a single point (this was discussed in Section 1.3), this shows that $\mathcal{M}_0(\Gamma) = \mathbb{C}$ (the constant functions) and $\mathcal{S}_0(\Gamma) = \{0\}$. It follows that $\mathcal{M}_k(\Gamma) = \{0\}$ for $k < 0$, for if $f \in \mathcal{M}_k(\Gamma)$ then $f^{12} \Delta^{-k}$ (where Δ is the weight 12 cusp form from Chapter 1) lies in $\mathcal{S}_0(\Gamma) = \{0\}$. This argument works whether k is even or odd.

Assembling the results of this section gives

Theorem 3.5.1. *Let k be an even integer. Let Γ be a congruence subgroup of $\mathrm{SL}_2(\mathbb{Z})$, g the genus of $X(\Gamma)$, ε_2 the number of elliptic points with period 2, ε_3 the number of elliptic points with period 3, and ε_∞ the number of cusps. Then*

$$\dim(\mathcal{M}_k(\Gamma)) = \begin{cases} (k-1)(g-1) + \lfloor \frac{k}{4} \rfloor \varepsilon_2 + \lfloor \frac{k}{3} \rfloor \varepsilon_3 + \frac{k}{2} \varepsilon_\infty & \text{if } k \geq 2, \\ 1 & \text{if } k = 0, \\ 0 & \text{if } k < 0, \end{cases}$$

and

$$\dim(\mathcal{S}_k(\Gamma)) = \begin{cases} (k-1)(g-1) + \lfloor \frac{k}{4} \rfloor \varepsilon_2 + \lfloor \frac{k}{3} \rfloor \varepsilon_3 + (\frac{k}{2} - 1)\varepsilon_\infty & \text{if } k \geq 4, \\ g & \text{if } k = 2, \\ 0 & \text{if } k \leq 0. \end{cases}$$

In concert with the results from Exercises 3.1.4 through 3.1.6 this gives dimension formulas for $\Gamma_0(p)$, $\Gamma_1(p)$, and $\Gamma(p)$. When Γ is the full modular group $\mathrm{SL}_2(\mathbb{Z})$, the theorem is

Theorem 3.5.2. *The modular forms of weight 0 are $\mathcal{M}_0(\mathrm{SL}_2(\mathbb{Z})) = \mathbb{C}$. For any nonzero even integer $k < 4$, $\mathcal{M}_k(\mathrm{SL}_2(\mathbb{Z})) = \{0\}$. For any even integer $k < 4$, $\mathcal{S}_k(\mathrm{SL}_2(\mathbb{Z})) = \{0\}$. For any even integer $k \geq 4$,*

$$\mathcal{M}_k(\mathrm{SL}_2(\mathbb{Z})) = \mathcal{S}_k(\mathrm{SL}_2(\mathbb{Z})) \oplus \mathbb{C}E_k$$

where E_k is the normalized weight k Eisenstein series from Chapter 1, and

$$\dim(\mathcal{S}_k(\mathrm{SL}_2(\mathbb{Z}))) = \begin{cases} \lfloor \frac{k}{12} \rfloor - 1 & \text{if } k \equiv 2 \pmod{12}, \\ \lfloor \frac{k}{12} \rfloor & \text{otherwise.} \end{cases}$$

The ring of modular forms $\mathcal{M}(\mathrm{SL}_2(\mathbb{Z})) = \bigoplus_{k \in \mathbb{Z}} \mathcal{M}_k(\mathrm{SL}_2(\mathbb{Z}))$ and the ideal of cusp forms $\mathcal{S}(\mathrm{SL}_2(\mathbb{Z})) = \bigoplus_{k \in \mathbb{Z}} \mathcal{S}_k(\mathrm{SL}_2(\mathbb{Z}))$ are a polynomial ring in two variables and a principal ideal,

$$\mathcal{M}(\mathrm{SL}_2(\mathbb{Z})) = \mathbb{C}[E_4, E_6], \quad \mathcal{S}(\mathrm{SL}_2(\mathbb{Z})) = \Delta \cdot \mathcal{M}(\mathrm{SL}_2(\mathbb{Z}))$$

where Δ is the discriminant function from Chapter 1.

The proof is Exercise 3.5.3. Exercises 3.5.4 and 3.5.5 use the results of this section to describe some spaces of modular forms explicitly.

Exercises

3.5.1. Show that $\mathcal{A}_k(\Gamma) = \mathbb{C}(X(\Gamma))f$ for any nonzero $f \in \mathcal{A}_k(\Gamma)$.

3.5.2. (a) Prove formula (3.10).
  (b) Prove formula (3.11).
  (c) Prove that for $k \geq 2$ even, $\lfloor k/4 \rfloor \geq (k-2)/4$ and $\lfloor k/3 \rfloor \geq (k-2)/3$.
  (d) Prove formula (3.12). Where does the condition $k \geq 4$ arise?

3.5.3. Prove Theorem 3.5.2. Write bases of $\mathcal{M}_k(\mathrm{SL}_2(\mathbb{Z}))$ for all $k \leq 24$, expressing the basis elements as polynomials in E_4 and E_6. (A hint for this exercise is at the end of the book.).

3.5.4. (a) Let p be prime. Exercise 3.1.4(b) showed that $X_0(p)$ has exactly two cusps. Use this to show that

$$\mathcal{M}_2(\Gamma_0(p)) = \mathcal{S}_2(\Gamma_0(p)) \oplus \mathbb{C}G_{2,p}$$

where $G_{2,p}$ was defined in Section 1.2.

(b) Let k and N be positive with $k(N+1) = 24$ and k even. Thus (k, N) is one of $(2, 11)$, $(4, 5)$, $(6, 3)$, $(8, 2)$, or $(12, 1)$. Show that $\dim(\mathcal{S}_k(\Gamma_0(N))) = 1$. It follows from Proposition 3.2.2(b) that $\mathcal{S}_k(\Gamma_0(N)) = \mathcal{S}_k(\Gamma_1(N)) = \mathbb{C}\varphi_k$ where $\varphi_k(\tau) = \eta(\tau)^k \eta(N\tau)^k$.

(c) What do (a) and (b) combine to say about $\mathcal{M}_2(\Gamma_0(11))$?

3.5.5. Combine Exercise 3.1.4 and Theorem 3.5.2 to show that letting $k = p+1$ for p an odd prime, $\mathcal{S}_k(\mathrm{SL}_2(\mathbb{Z}))$ and $\mathcal{S}_2(\Gamma_0(p))$ are isomorphic as complex vector spaces. What is the isomorphism when $p = 11$? (A hint for this exercise is at the end of the book.)

3.6 Dimension formulas for odd k

Throughout this section k is an odd integer. Computing dimension formulas for $\mathcal{M}_k(\Gamma)$ and $\mathcal{S}_k(\Gamma)$ is essentially similar to the case where k is even, but it is a bit more technical. The formula $(f[-I]_k)(\tau) = -f(\tau)$ shows that $\mathcal{A}_k(\Gamma) = \{0\}$ whenever $-I \in \Gamma$, so throughout this section $-I \notin \Gamma$ as well. This implies that Γ has no elliptic points of period 2, as shown in Exercise 2.3.7.

Since $\mathcal{M}_k(\Gamma) = \{0\}$ for all negative k, assume that the odd integer k is positive. The argument in the preceding section that $\dim(\mathcal{M}_k(\Gamma)) = \ell(\lfloor \mathrm{div}(f) \rfloor)$ for any nonzero $f \in \mathcal{A}_k(\Gamma)$ didn't depend on k being even and thus remains valid here. On the other hand, the existence of such f has been established only for k even. Assume for now that f exists for odd k as well; this will be proved at the end of the section. Since k is odd and $-I \notin \Gamma$, the modular curve $X(\Gamma)$ has two possible types of cusps as discussed in Section 3.2, the *regular* cusps where $\nu_{\pi(s)}(f)$ is integral and the *irregular* cusps where $\nu_{\pi(s)}(f)$ is half-integral. For example, Exercise 3.2.1 showed that $1/2$ is an irregular cusp of $\Gamma_1(4)$.

This time let $\omega \in \Omega^{\otimes k}(X(\Gamma))$ be the differential $\omega(f^2)$ from Theorem 3.3.1, pulling back to $f(\tau)^2 (d\tau)^k$ on $\mathcal{H}$. Retain the notations $\{x_{3,i}\}$ and ε_3 for the period 3 elliptic points, and let $\{x_i\}$ and $\{x_i'\}$ be the regular and irregular cusps, with corresponding $\varepsilon_\infty^{\mathrm{reg}}$ and $\varepsilon_\infty^{\mathrm{irr}}$. The formal definition

$$\mathrm{div}(d\tau) = -\sum_i \frac{2}{3} x_{3,i} - \sum_i x_i - \sum_i x_i'$$

is consistent with the definition of $\operatorname{div}(d\tau)$ in Section 3.5 since there are no period 2 elliptic points now. By formulas like (3.8) and (3.9) but with f^2 and k replacing f and $k/2$ it gives the appropriate divisor identity $\operatorname{div}(\omega) = 2\operatorname{div}(f) + k\operatorname{div}(d\tau)$ (Exercise 3.6.1), implying

$$\operatorname{div}(f) = \frac{1}{2}\operatorname{div}(\omega) + \sum_i \frac{k}{3}x_{3,i} + \sum_i \frac{k}{2}x_i + \sum_i \frac{k}{2}x_i'. \qquad (3.13)$$

To study $\lfloor \operatorname{div}(f) \rfloor$, consider equation (3.13). At any point $x \notin \{x_{3,i}\} \cup \{x_i\} \cup \{x_i'\}$ of $X(\Gamma)$, the order $\nu_x(f)$ is integral, and so $\frac{1}{2}\nu_x(\omega)$ must be integral as well. At a point $x = x_{3,i}$, write $\nu_x(f) = m + j/3$, with m and j integers, cf. equation (3.2). Then $\frac{1}{2}\nu_x(\omega) = m + (j - k)/3$. Since $\nu_x(\omega)$ is integral this shows that $j \equiv k \pmod 3$ and $\frac{1}{2}\nu_x(\omega) \in \mathbb{Z}$. Thus the integral part of $\nu_x(f)$ is $\frac{1}{2}\nu_x(\omega) + \lfloor k/3 \rfloor$. Similarly, at a point $x = x_i$, the integral part of $\nu_x(f)$ is $\frac{1}{2}\nu_x(\omega) + \frac{k}{2}$, a sum of half-integers (Exercise 3.6.2(a)), and at a point $x = x_i'$, the integral part of $\nu_x(f)$ is $\frac{1}{2}\nu_x(\omega) + \frac{k-1}{2}$, a sum of integers (Exercise 3.6.2(b)). Thus

$$\lfloor \operatorname{div}(f) \rfloor = \frac{1}{2}\operatorname{div}(\omega) + \sum_i \left\lfloor \frac{k}{3} \right\rfloor x_{3,i} + \sum_i \frac{k}{2}x_i + \sum_i \frac{k-1}{2}x_i'.$$

The degree satisfies (Exercise 3.6.3(a))

$$
\begin{aligned}
\deg(\lfloor \operatorname{div}(f) \rfloor) &= k(g-1) + \left\lfloor \frac{k}{3} \right\rfloor \varepsilon_3 + \frac{k}{2}\varepsilon_\infty^{\mathrm{reg}} + \frac{k-1}{2}\varepsilon_\infty^{\mathrm{irr}} \\
&> (k-2)\left(g-1+\frac{\varepsilon_3}{3}+\frac{\varepsilon_\infty}{2}\right) + 2g - 2 \\
&> 2g - 2 \quad \text{if } k \geq 3.
\end{aligned}
$$

So when $k \geq 3$, Corollary 3.4.2(d) of the Riemann–Roch Theorem applies, giving

$$\ell(\lfloor \operatorname{div}(f) \rfloor) = (k-1)(g-1) + \left\lfloor \frac{k}{3} \right\rfloor \varepsilon_3 + \frac{k}{2}\varepsilon_\infty^{\mathrm{reg}} + \frac{k-1}{2}\varepsilon_\infty^{\mathrm{irr}}.$$

To study cusp forms, recall the general element $f_0 f \in \mathcal{A}_k(\Gamma)$ and note that at a regular cusp x,

$$\nu_x(f_0 f) > 0 \iff \nu_x(f_0 f) \geq 1 \iff \nu_x(f) + \nu_x(f_0) - 1 \geq 0,$$

while at an irregular cusp x',

$$\nu_{x'}(f_0 f) > 0 \iff \nu_{x'}(f_0 f) \geq 1/2 \iff \nu_{x'}(f) + \nu_{x'}(f_0) - 1/2 \geq 0.$$

Thus the relevant dimension is $\ell(\lfloor \operatorname{div}(f) - \sum x_i - \sum(1/2)x_i' \rfloor)$. Similar to the calculation above, this works out for $k \geq 3$ to (Exercise 3.6.3(b))

$$\ell(\lfloor \operatorname{div}(f) - \sum_i x_i - \sum_i \tfrac{1}{2} x_i' \rfloor) =$$

$$(k-1)(g-1) + \left\lfloor \frac{k}{3} \right\rfloor \varepsilon_3 + \frac{k-2}{2}\varepsilon_\infty^{\mathrm{reg}} + \frac{k-1}{2}\varepsilon_\infty^{\mathrm{irr}}.$$

This proves

Theorem 3.6.1. *Let k be an odd integer. Let Γ be a congruence subgroup of $\mathrm{SL}_2(\mathbb{Z})$. If Γ contains the negative identity matrix $-I$ then $\mathcal{M}_k(\Gamma) = \mathcal{S}_k(\Gamma) = \{0\}$. If $-I \notin \Gamma$, let g the genus of $X(\Gamma)$, ε_3 the number of elliptic points with period 3, $\varepsilon_\infty^{\mathrm{reg}}$ the number of regular cusps, and $\varepsilon_\infty^{\mathrm{irr}}$ the number of irregular cusps. Then*

$$\dim(\mathcal{M}_k(\Gamma)) = \begin{cases} (k-1)(g-1) + \lfloor \tfrac{k}{3} \rfloor \varepsilon_3 + \tfrac{k}{2}\varepsilon_\infty^{\mathrm{reg}} + \tfrac{k-1}{2}\varepsilon_\infty^{\mathrm{irr}} & \text{if } k \geq 3, \\ 0 & \text{if } k < 0, \end{cases}$$

and

$$\dim(\mathcal{S}_k(\Gamma)) = \begin{cases} (k-1)(g-1) + \lfloor \tfrac{k}{3} \rfloor \varepsilon_3 + \tfrac{k-2}{2}\varepsilon_\infty^{\mathrm{reg}} + \tfrac{k-1}{2}\varepsilon_\infty^{\mathrm{irr}} & \text{if } k \geq 3, \\ 0 & \text{if } k < 0. \end{cases}$$

If $\varepsilon_\infty^{\mathrm{reg}} > 2g - 2$ then $\dim(\mathcal{M}_1(\Gamma)) = \varepsilon_\infty^{\mathrm{reg}}/2$ and $\dim(\mathcal{S}_1(\Gamma)) = 0$. If $\varepsilon_\infty^{\mathrm{reg}} \leq 2g - 2$ then $\dim(\mathcal{M}_1(\Gamma)) \geq \varepsilon_\infty^{\mathrm{reg}}/2$ and $\dim(\mathcal{S}_1(\Gamma)) = \dim(\mathcal{M}_1(\Gamma)) - \varepsilon_\infty^{\mathrm{reg}}/2$.

The results for $k = 1$ are Exercise 3.6.4.

Regular and irregular cusps will be discussed further in Section 3.9. Here it remains to prove that some nonzero $f \in \mathcal{A}_k(\Gamma)$ exists when the positive integer k is odd and $-I \notin \Gamma$. The argument cites some results from Riemann surface theory that should be taken for granted if they are unfamiliar.

Let λ be any nonzero element of $\Omega^1(X(\Gamma))$, for example the divisor that pulls back to $j'(\tau)d\tau$ on $\mathcal{H}$ where j' is the derivative of the modular invariant j. Let x_0 be any point of $X(\Gamma)$. Then Corollary 3.4.2(b) of the Riemann–Roch Theorem says that $\operatorname{div}(\lambda) - 2(g-1)x_0 \in \mathrm{Div}^0$, where as before $\mathrm{Div}^0 = \mathrm{Div}^0(X(\Gamma))$ denotes the divisors of degree 0. Recall also that Div^ℓ denotes the divisors of nonzero meromorphic functions on $X(\Gamma)$, a subgroup of Div^0. Under the map $\mathrm{Div}^0/\mathrm{Div}^\ell \longrightarrow \mathbb{C}^g/\Lambda_g$ of Abel's Theorem (described in Section 3.4), let $z + \Lambda_g$ be the image of $\operatorname{div}(\lambda) - 2(g-1)x_0 + \mathrm{Div}^\ell$. Any divisor $D \in \mathrm{Div}^0$ such that $D + \mathrm{Div}^\ell$ maps to $z/2 + \Lambda_g$ satisfies $2D = \operatorname{div}(\lambda) - 2(g-1)x_0 + \operatorname{div}(f)$ for some nonzero $f \in \mathbb{C}(X(\Gamma))$. So $\operatorname{div}(f\lambda) = 2(D + (g-1)x_0)$.

Let $\tilde{f}$ be the function such that the differential $f\lambda$ on $X(\Gamma)$ pulls back to $\tilde{f}(\tau)d\tau$ on $\mathcal{H}$. Then

$$\operatorname{div}(\tilde{f}) = \operatorname{div}(f\lambda) - \operatorname{div}(d\tau)$$

$$= 2(D + (g-1)x_0) + \sum_i \frac{2}{3}x_{3,i} + \sum_i x_i + \sum_i x_i'.$$

This shows that $\nu_\tau(\tilde{f})$ is even for all $\tau \in \mathcal{H}$, since $\nu_\tau(\tilde{f}) = 3\nu_{\pi(\tau)}(\tilde{f})$ when $\pi(\tau)$ is an elliptic point $x_{3,i}$, and the cusps are not in $\mathcal{H}$. So there exists a function f_1 on $\mathcal{H}$ such that $f_1^2 = \tilde{f}$. Since $\tilde{f}$ is weight-2 invariant under Γ, f_1 transforms by the rule $(f_1[\gamma]_1) = \chi(\gamma)f_1$ for all $\gamma \in \Gamma$, where $\chi : \Gamma \longrightarrow \{\pm 1\}$ is a character. Let $\Gamma' = \{\gamma \in \Gamma : \chi(\gamma) = 1\}$, a subgroup of index 1 or 2. If the index is 1 then $f_1 \in \mathcal{A}_1(\Gamma)$.

If the index is 2 then $\Gamma = \Gamma' \cup \Gamma'\alpha$ where $f[\alpha]_1 = -f_1$. The surjection of Riemann surfaces $\pi : X(\Gamma') \longrightarrow X(\Gamma)$ pulls back to an inclusion of their fields of meromorphic functions, $\pi^* : \mathbb{C}(X(\Gamma)) \longrightarrow \mathbb{C}(X(\Gamma'))$, where the pullback of a function $f \in \mathbb{C}(X(\Gamma))$ is the function $\pi^*f = f \circ \pi \in \mathbb{C}(X(\Gamma'))$. After identifying $\mathbb{C}(X(\Gamma))$ with its image, the extension of function fields has the same degree degree as the original surjection, namely 2, and the Galois group Γ/Γ' acts as $\{\pm 1\}$ on generators of the field extension. So there exists some $f_2 \in \mathbb{C}(X(\Gamma'))$ such that $f_2 \circ \alpha = -f_2$ and thus $(f_1f_2)[\alpha]_1 = f_1f_2$, showing that $f_1f_2 \in \mathcal{A}_1(\Gamma)$.

Whichever of f_1 or f_1f_2 lies in $\mathcal{A}_1(\Gamma)$, its kth power is a nonzero element of $\mathcal{A}_k(\Gamma)$ as desired.

Exercises

3.6.1. Establish the formula $\text{div}(f) = 2\,\text{div}(f) + k\,\text{div}(d\tau)$ in the context of the section.

3.6.2. (a) Show that at a point $x = x_i$ the integral part of $\nu_x(f)$ is $\frac{1}{2}\nu_x(\omega) + k/2$, a sum of half-integers.

(b) Show that at a point $x = x_i'$ the integral part of $\nu_x(f)$ is $\frac{1}{2}\nu_x(\omega) + (k-1)/2$, a sum of integers.

3.6.3. (a) Justify the calculation that $\deg(\lfloor\text{div}(f)\rfloor) > 2g - 2$ if $k \geq 3$.

(b) Establish the formula for $\ell(\lfloor\text{div}(f) - \sum_i x_i - \sum_i (1/2)x_i'\rfloor)$ when $k \geq 3$.

3.6.4. Prove the remaining statements in Theorem 3.6.1.

3.6.5. Show that $\dim(\mathcal{S}_3(\Gamma_1(7)) = 1$, so Proposition 3.2.2(a) shows that this space is spanned by the function $\varphi_3(\tau) = \eta(\tau)^3\eta(7\tau)^3$.

3.7 More on elliptic points

Exercise 2.3.7 found the elliptic points of the groups $\Gamma(N)$ and $\Gamma_1(N)$, and it obtained some information about the elliptic points of $\Gamma_0(N)$. Exercise 3.1.4 of this chapter found the elliptic points of $\Gamma_0(p)$ when p is prime. This section finds the elliptic points of $\Gamma_0(N)$ for general N.

A point $\tau \in \mathcal{H}$ is an elliptic point of $\Gamma_0(N)$ when its isotropy subgroup takes the form $\Gamma_0(N)_\tau = \langle\gamma\rangle$, the matrix γ having order 4 or 6, twice the period of τ. For any $\alpha \in \Gamma_0(N)$, the translated point $\alpha(\tau)$ has the corresponding conjugate isotropy subgroup $\Gamma_0(N)_{\alpha(\tau)} = \alpha\langle\gamma\rangle\alpha^{-1}$. An elliptic point $\Gamma_0(N)\tau$ of the modular curve $X_0(N) = X(\Gamma_0(N))$ thus corresponds to a conjugacy class of isotropy subgroups of $\Gamma_0(N)$,

$$\Gamma_0(N)\tau \longleftrightarrow \{\alpha\langle\gamma\rangle\alpha^{-1} : \alpha \in \Gamma_0(N)\} \quad \text{where } \Gamma_0(N)_\tau = \langle\gamma\rangle.$$

Conjugacy classes of elements are more convenient than conjugacy classes of subgroups, so it is tempting to replace the subgroup $\langle\gamma\rangle$ by its generator γ in the correspondence. But this is ill defined since each nontrivial isotropy subgroup $\langle\gamma\rangle$ is also generated by γ^{-1}. Instead each elliptic point corresponds to a pair of conjugacy classes,

$$\Gamma_0(N)\tau \longleftrightarrow \left\{\{\alpha\gamma\alpha^{-1} : \alpha \in \Gamma_0(N)\}, \{\alpha\gamma^{-1}\alpha^{-1} : \alpha \in \Gamma_0(N)\}\right\},$$

where again $\Gamma_0(N)_\tau = \langle\gamma\rangle$. The inverse pair of generators γ and γ^{-1} are not conjugate in $\Gamma_0(N)$ (Exercise 3.7.1(a)), so the two conjugacy classes are distinct and in fact disjoint (Exercise 3.7.1(b)). Thus the number of period 2 elliptic points is half the number of conjugacy classes of order 4 elements, and similarly for period 3 and order 6.

To correct the factor of a half, introduce the group

$$\Gamma_0^{\pm}(N) = \left\{ \begin{bmatrix} a & b \\ c & d \end{bmatrix} \in \mathrm{GL}_2(\mathbb{Z}) : \begin{bmatrix} a & b \\ c & d \end{bmatrix} \equiv \begin{bmatrix} * & * \\ 0 & * \end{bmatrix} \pmod{N} \right\}$$

(where $\mathrm{GL}_2(\mathbb{Z})$ is the group of invertible 2-by-2 matrices with integer entries), similar to $\Gamma_0(N)$ but its elements can have determinant ± 1. Since the map $\det : \Gamma_0^{\pm}(N) \longrightarrow \{\pm 1\}$ is a surjective homomorphism, $\Gamma_0(N)$ is normal in $\Gamma_0^{\pm}(N)$ and $[\Gamma_0^{\pm}(N) : \Gamma_0(N)] = 2$. Thus for any $\gamma \in \Gamma_0(N)$ the extended conjugacy class

$$\{\alpha\gamma\alpha^{-1} : \alpha \in \Gamma_0^{\pm}(N)\}$$

lies in $\Gamma_0(N)$. This is the union of two conjugacy classes under $\Gamma_0(N)$ (Exercise 3.7.1(c)), so the number of period 2 (or period 3) elliptic points of $\Gamma_0(N)$ is the number of extended conjugacy classes of order 4 (or order 6) elements of $\Gamma_0(N)$. The extended conjugacy class of γ under $\Gamma_0^{\pm}(N)$ is not in general the union of the conjugacy classes of γ and γ^{-1} under $\Gamma_0(N)$ (see Exercise 3.7.1(d)).

Counting the elliptic points of $\Gamma_0(N)$ by counting these conjugacy classes is done in the same environment as the proof of Proposition 2.3.3. Consider elliptic points τ of period 3 with $\Gamma_0(N)_\tau = \langle\gamma\rangle$, $\gamma^6 = I$. Let $\mu_6 = e^{2\pi i/6}$ be the complex sixth root of unity and let $A = \mathbb{Z}[\mu_6]$, a principal ideal domain. Each extended conjugacy class will correspond to an ideal J of A such that $A/J \cong \mathbb{Z}/N\mathbb{Z}$ as an Abelian group,

$$\{\alpha\gamma\alpha^{-1} : \alpha \in \Gamma_0^{\pm}(N)\} \longleftrightarrow J_\gamma.$$

To construct the ideal J_γ from γ, note that the lattice $L = \mathbb{Z}^2$ is an A-module with the multiplication $\odot_\gamma : A \times L \longrightarrow L$ given by

$$(a + b\mu_6) \odot_\gamma l = (aI + b\gamma)l \quad \text{for } a, b \in \mathbb{Z} \text{ and } l \in L.$$

Define $L_0(N) = \{[\begin{smallmatrix} x \\ y \end{smallmatrix}] \in L : y \equiv 0 \ (\text{mod } N)\}$, a subgroup of L such that the quotient $L/L_0(N)$ is isomorphic to $\mathbb{Z}/N\mathbb{Z}$. Then $L_0(N)$ is also an A-submodule of L since $\gamma \in \Gamma_0(N)$ (Exercise 3.7.2), making $L/L_0(N)$ an A-module. Let J_γ be its annihilator,

$$J_\gamma = \text{Ann}(L/L_0(N)) = \{a + b\mu_6 \in A : (a + b\mu_6) \odot_\gamma L/L_0(N) = 0\}.$$

Thus $A/J_\gamma \cong A \odot_\gamma L/L_0(N) = L/L_0(N) \cong \mathbb{Z}/N\mathbb{Z}$ as desired.

To see that the ideal J_γ depends only on the extended conjugacy class of γ, let $\gamma' = \alpha\gamma\alpha^{-1}$ with $\alpha \in \Gamma_0^\pm(N)$. Then γ' is also an order-6 element of $\Gamma_0(N)$, and L is also an A-module associated to γ' with multiplication

$$(a + b\mu_6) \odot_{\gamma'} l = (aI + b\gamma')l.$$

Let L' denote L as the A-module associated to γ' in this fashion, and similarly for $L_0'(N)$. For any $a, b \in \mathbb{Z}$ and $l \in L$ compute that

$$\begin{aligned} \alpha((a + b\mu_6) \odot_\gamma l) &= \alpha(aI + b\gamma)l = (aI + b\gamma')\alpha l \quad \text{since } \alpha\gamma = \gamma'\alpha \\ &= (aI + b\mu_6) \odot_{\gamma'} \alpha l. \end{aligned}$$

This shows that multiplication by α gives an A-module isomorphism $L \xrightarrow{\sim} L'$, and since $\alpha L_0(N) \subset L_0'(N)$ because $\alpha \in \Gamma_0^\pm(N)$ this induces an isomorphism of quotients $L/L_0(N) \xrightarrow{\sim} L'/L_0'(N)$. Compute that for any $j = a + b\mu_6 \in A$,

$$\begin{aligned} j \in J_\gamma &\iff j \odot_\gamma L/L_0(N) = 0_{L/L_0(N)} \\ &\iff \alpha(j \odot_\gamma L/L_0(N)) = 0_{L'/L_0'(N)} \\ &\iff j \odot_{\gamma'} \alpha(L/L_0(N)) = 0_{L'/L_0'(N)} \\ &\iff j \odot_{\gamma'} L'/L_0'(N) = 0_{L'/L_0'(N)} \\ &\iff j \in J_{\gamma'}, \end{aligned}$$

showing that indeed J_γ depends only on the extended conjugacy class.

Conversely to the map from conjugacy classes to ideals, start now from an ideal J of A with $A/J \cong \mathbb{Z}/N\mathbb{Z}$. By the structure theorem for modules over a principal ideal domain again, there exists a basis (u, v) of A over $\mathbb{Z}$ such that (u, Nv) is a basis of J over $\mathbb{Z}$. Since (u, v) is a basis of A over $\mathbb{Z}$,

$$(\mu_6 u, \mu_6 v) = (ua + vc, ub + vd) \quad \text{for some integers } a, b, c, d,$$

or

$$\mu_6(u, v) = (u, v)\gamma_J, \quad \text{for some } \gamma_J = [\begin{smallmatrix} a & b \\ c & d \end{smallmatrix}] \in M_2(\mathbb{Z}).$$

So the matrix γ_J has order 6 and has the same minimal polynomial $X^2 - X + 1$ as μ_6, showing that $\gamma_J \in \text{SL}_2(\mathbb{Z})$ (Exercise 3.7.3(a)). And since $u \in J$ and J is an A-submodule of A, the formula $\mu_6 u = ua + vc$ shows that $c \equiv 0 \ (\text{mod } N)$, i.e., $\gamma_J \in \Gamma_0(N)$. Furthermore, for any $\alpha \in \Gamma_0^\pm(N)$ consider the $\mathbb{Z}$-basis of A

$$(u', v') = (u, v)\alpha.$$

A calculation (Exercise 3.7.3(b)) shows that (u', Nv') is again a $\mathbb{Z}$-basis of J. So now

$$\mu_6(u', v') = \mu_6(u, v)\alpha = (u, v)\gamma_J \alpha = (u', v')\alpha^{-1}\gamma_J \alpha,$$

and so the basis (u, v) of A and its $\Gamma_0^{\pm}(N)$-translates map back to a $\Gamma_0^{\pm}(N)$-conjugacy class of an order-6 element γ_J in $\Gamma_0(N)$. Finally, any two $\mathbb{Z}$-bases (u, v) and (u', v') of A such that (u, Nv) and (u', Nv') are $\mathbb{Z}$-bases of J satisfy the relation $(u', v') = (u, v)\alpha$ for some $\alpha \in \Gamma_0^{\pm}(N)$ (Exercise 3.7.3(c)), so the conjugacy class is the same for any such $\mathbb{Z}$-basis of A.

Thus we have a map from extended conjugacy classes to ideals and a map back from ideals to extended conjugacy classes. Suppose the class of $\gamma \in \Gamma_0(N)$ maps to the ideal J_γ of A. The proof of Proposition 2.3.3 shows that $\gamma = m^{-1} \begin{bmatrix} 0 & -1 \\ 1 & 1 \end{bmatrix} m$ for some $m = \begin{bmatrix} a & b \\ c & d \end{bmatrix} \in \mathrm{GL}_2(\mathbb{Z})$, and this computes out to

$$\gamma = \det m \cdot \begin{bmatrix} * & * \\ a^2 + ac + c^2 & ab + ad + cd \end{bmatrix}, \qquad a^2 + ac + c^2 \equiv 0 \pmod{N}.$$

Let $(u, v) = (1, \mu_6)m$, a $\mathbb{Z}$-basis of A. A calculation using the displayed description of γ shows that u annihilates $L/L_0(N)$ (Exercise 3.7.4), as Nv clearly does as well. Thus (u, Nv) is a $\mathbb{Z}$-basis of an A-submodule J of $\mathrm{Ann}(L/L_0(N)) = J_\gamma$ with $[A : J] = N = [A : J_\gamma]$, showing that $J = J_\gamma$. This ideal maps back to the extended conjugacy class of the matrix γ' such that $\mu_6(u, v) = (u, v)\gamma'$. Compute that

$$\mu_6(u, v) = \mu_6(1, \mu_6)m = (\mu_6, -1 + \mu_6)m$$
$$= (1, \mu_6) \begin{bmatrix} 0 & -1 \\ 1 & 1 \end{bmatrix} m = (u, v)m^{-1} \begin{bmatrix} 0 & -1 \\ 1 & 1 \end{bmatrix} m = (u, v)\gamma,$$

so J_γ maps back to the class of γ as desired. On the other hand, suppose the ideal J of A such that $A/J \cong \mathbb{Z}/N\mathbb{Z}$ maps to the extended conjugacy class of $\gamma \in \Gamma_0(N)$. This means that $\mu_6(u, v) = (u, v)\gamma$ for some $\mathbb{Z}$-basis (u, v) of A such that (u, Nv) is a $\mathbb{Z}$-basis of J. Note that $(u, v) = (1, \mu_6)m$ for some $m = \begin{bmatrix} a & b \\ c & d \end{bmatrix} \in \mathrm{GL}_2(\mathbb{Z})$. As above, $\mu_6(u, v) = (u, v)m^{-1} \begin{bmatrix} 0 & -1 \\ 1 & 1 \end{bmatrix} m$, showing that $\gamma = m^{-1} \begin{bmatrix} 0 & -1 \\ 1 & 1 \end{bmatrix} m$. Again the displayed description of such a γ shows that u annihilates $L/L_0(N)$ and consequently $J = J_\gamma$. That is, the class of γ maps back to J.

A virtually identical discussion with $A = \mathbb{Z}[i]$ applies to period 2 elliptic points (Exercise 3.7.5(a)), proving

Proposition 3.7.1. *The period 2 elliptic points of $\Gamma_0(N)$ are in bijective correspondence with the ideals J of $\mathbb{Z}[i]$ such that $\mathbb{Z}[i]/J \cong \mathbb{Z}/N\mathbb{Z}$. The period 3 elliptic points of $\Gamma_0(N)$ are in bijective correspondence with the ideals J of $\mathbb{Z}[\mu_6]$ (where $\mu_6 = e^{2\pi i/6}$) such that $\mathbb{Z}[\mu_6]/J \cong \mathbb{Z}/N\mathbb{Z}$.*

Counting the ideals gives

Corollary 3.7.2. *The number of elliptic points for $\Gamma_0(N)$ is*

$$\varepsilon_2(\Gamma_0(N)) = \begin{cases} \prod_{p|N}\left(1 + \left(\frac{-1}{p}\right)\right) & \text{if } 4 \nmid N, \\ 0 & \text{if } 4 \mid N, \end{cases}$$

where $(-1/p)$ is ± 1 if $p \equiv \pm 1 \pmod 4$ and is 0 if $p = 2$, and

$$\varepsilon_3(\Gamma_0(N)) = \begin{cases} \prod_{p|N}\left(1 + \left(\frac{-3}{p}\right)\right) & \text{if } 9 \nmid N, \\ 0 & \text{if } 9 \mid N, \end{cases}$$

where $(-3/p)$ is ± 1 if $p \equiv \pm 1 \pmod 3$ and is 0 if $p = 3$.

These formulas extend Exercise 2.3.7(c) and Exercise 3.1.4(b,c).

Proof. This is an application of beginning algebraic number theory; see for example Chapter 9 of [IR92] for the results to quote. For period 3, the ring $A = \mathbb{Z}[\mu_6]$ is a principal ideal domain and its maximal ideals are

- for each prime $p \equiv 1 \pmod 3$, two ideals $J_p = \langle a + b\mu_6 \rangle$ and $\bar{J}_p = \langle a + b\bar{\mu}_6 \rangle$ such that $\langle p \rangle = J_p\bar{J}_p$ and the quotients A/J_p^e and $A/\bar{J}_p^e$ are group-isomorphic to $\mathbb{Z}/p^e\mathbb{Z}$ for all $e \in \mathbb{N}$,
- for each prime $p \equiv -1 \pmod 3$, the ideal $J_p = \langle p \rangle$ such that the quotient A/J_p^e is group-isomorphic to $(\mathbb{Z}/p^e\mathbb{Z})^2$ for all $e \in \mathbb{N}$,
- for $p = 3$, the ideal $J_3 = \langle 1 + \mu_6 \rangle$ such that $\langle 3 \rangle = J_3^2$ and the quotient A/J_3^e is group-isomorphic to $(\mathbb{Z}/3^{e/2}\mathbb{Z})^2$ for even $e \in \mathbb{N}$ and is group-isomorphic to $\mathbb{Z}/3^{(e+1)/2}\mathbb{Z} \oplus \mathbb{Z}/3^{(e-1)/2}\mathbb{Z}$ for odd $e \in \mathbb{N}$.

The formula for $\varepsilon_3(\Gamma_0(N))$ now follows from Proposition 3.7.1 and the Chinese Remainder Theorem. Counting the period 2 elliptic points is left as Exercise 3.7.5(b), similarly citing the theory of the ring $A = \mathbb{Z}[i]$. $\qquad\square$

The elliptic points of $\Gamma_0(N)$ can be written down easily now that they are counted. Consider the set of translates in $\mathcal{H}$

$$\left\{ \begin{bmatrix} 1 & 0 \\ n & 1 \end{bmatrix}(\mu_3) : 0 \le n < N \right\}.$$

The corresponding isotropy subgroup generators $\begin{bmatrix} 1 & 0 \\ n & 1 \end{bmatrix}\begin{bmatrix} 0 & -1 \\ 1 & 1 \end{bmatrix}\begin{bmatrix} 1 & 0 \\ -n & 1 \end{bmatrix}$ are

$$\left\{ \begin{bmatrix} n & -1 \\ n^2 - n + 1 & 1 - n \end{bmatrix} : 0 \le n < N \right\}.$$

The number of these that are elements of $\Gamma_0(N)$ is the number of solutions to the congruence $x^2 - x + 1 \equiv 0 \pmod N$, and this number is given by the formula for $\varepsilon_3(\Gamma_0(N))$ in Corollary 3.7.2 (Exercise 3.7.6(a)). The cosets $\{\Gamma_0(N)\begin{bmatrix} 1 & 0 \\ n & 1 \end{bmatrix} : 0 \le n < N\}$ are distinct in the quotient space $\Gamma_0(N)\backslash\mathrm{SL}_2(\mathbb{Z})$, though they do not constitute the entire quotient space, and the corresponding orbits $\Gamma_0(N) \left[\begin{smallmatrix} 1 & 0 \\ n & 1 \end{smallmatrix}\right] (\mu_3)$ for such n such that $n^2 - n + 1 \equiv 0 \pmod{N}$ are distinct in $X_0(N)$ (Exercise 3.7.6(b)). Thus we have found all the period 3 elliptic points of $\Gamma_0(N)$ (Exercise 3.7.6(c)),

$$\Gamma_0(N) \frac{n + \mu_3}{n^2 - n + 1}, \quad n^2 - n + 1 \equiv 0 \pmod{N}. \tag{3.14}$$

Similarly, the period 2 elliptic points are (Exercise 3.7.6(d))

$$\Gamma_0(N) \frac{n + i}{n^2 + 1}, \quad n^2 + 1 \equiv 0 \pmod{N}. \tag{3.15}$$

Exercises

3.7.1. (a) Show that if γ generates a nontrivial isotropy subgroup in $\mathrm{SL}_2(\mathbb{Z})$ then γ and γ^{-1} are not conjugate in $\mathrm{GL}_2^+(\mathbb{Q})$. (Hints for this exercise are at the end of the book.)

(b) Show that any two conjugacy classes in a group are either equal or disjoint.

(c) Show that the $\Gamma_0^\pm(N)$-conjugacy class of $\gamma \in \Gamma_0(N)$ is the union of the $\Gamma_0(N)$-conjugacy classes of γ and $\left[\begin{smallmatrix} 1 & 0 \\ 0 & -1 \end{smallmatrix}\right] \gamma \left[\begin{smallmatrix} 1 & 0 \\ 0 & -1 \end{smallmatrix}\right]$. Show that if γ has order 4 or 6 then this union is disjoint.

(d) Let $\gamma = \left[\begin{smallmatrix} 1 & 1 \\ 1 & 2 \end{smallmatrix}\right] \left[\begin{smallmatrix} 0 & -1 \\ 1 & 0 \end{smallmatrix}\right] \left[\begin{smallmatrix} 1 & 1 \\ 1 & 2 \end{smallmatrix}\right]^{-1} = \left[\begin{smallmatrix} 3 & -2 \\ 5 & -3 \end{smallmatrix}\right]$, an order-4 element of $\Gamma_0(5)$. Show that γ is not conjugate to its inverse in $\Gamma_0^\pm(5)$.

3.7.2. In the context of mapping a matrix conjugacy class to an ideal, show that $L_0(N)$ is an A-submodule of L.

3.7.3. (a) In the context of mapping an ideal J of A such that $A/J \cong \mathbb{Z}/N\mathbb{Z}$ back to a matrix conjugacy class, retain the notation (u, v) for a $\mathbb{Z}$-basis of A such that (u, Nv) is a $\mathbb{Z}$-basis of J. Show that the matrix $\gamma_J \in \mathrm{M}_2(\mathbb{Z})$ such that $\mu_6(u, v) = (u, v)\gamma_J$ lies in $\mathrm{SL}_2(\mathbb{Z})$.

(b) For any $\alpha \in \Gamma_0^\pm(N)$ consider the $\mathbb{Z}$-basis $(u', v') = (u, v)\alpha$ of A. Show that (u', Nv') is again a $\mathbb{Z}$-basis of J.

(c) Show that any two such $\mathbb{Z}$-bases (u, v) and (u', v') of A satisfy the relation $(u', v') = (u, v)\alpha$ for some $\alpha \in \Gamma_0^\pm(N)$.

3.7.4. In the context of checking that the maps between conjugacy classes and ideals invert each other, let $\gamma \in \Gamma_0(N)$ of order 6 be given and define $(u, v) = (1, \mu_6)m$ as in the section. Show that $u \odot_\gamma L \subset L_0(N)$, so that u annihilates $L/L_0(N)$. (A hint for this exercise is at the end of the book.)

3.7.5. (a) Similarly to the methods of the section, check that the first half of Proposition 3.7.1 holds.

(b) Prove the first half of Corollary 3.7.2. (A hint for this exercise is at the end of the book.)

3.7.6. (a) Show that the number of solutions to the congruence $x^2 - x + 1 \equiv 0 \pmod{N}$ is given by the formula for $\varepsilon_3(\Gamma_0(N))$ in Corollary 3.7.2. (A hint for this exercise is at the end of the book.)

(b) Show that the orbits $\Gamma_0(N) \left[\begin{smallmatrix} 1 & 0 \\ n & 1 \end{smallmatrix}\right](\mu_3)$ for $n = 0, \ldots, N - 1$ such that $n^2 - n + 1 \equiv 0 \pmod{N}$ are distinct in $X_0(N)$.

(c) Confirm formula (3.14).

(d) Similarly show that the period 2 elliptic points of $\Gamma_0(N)$ are given by formula (3.15).

3.7.7. Let p^e and M be positive integers with p prime, $e \geq 1$, and $p \nmid M$. Let $m = p^{-1} \pmod{M}$, i.e., $mp \equiv 1 \pmod{M}$ and $0 \leq m < M$. Consider the matrices

$$\alpha_j = \begin{bmatrix} 1 & 0 \\ Mj & 1 \end{bmatrix}, \quad 0 \leq j < p^e$$

and

$$\beta_j = \begin{bmatrix} 1 & (m_j p - 1)/M \\ M & m_j p \end{bmatrix}, \quad m_j = m + jM, \ 0 \leq j < p^{e-1}.$$

Show that

$$\Gamma_0(p^e M) \backslash \Gamma_0(M) = \left(\bigcup_j \Gamma_0(p^e M)\alpha_j \right) \cup \left(\bigcup_j \Gamma_0(p^e M)\beta_j \right).$$

Iterating this construction gives a set of representatives for $\Gamma_0(N) \backslash \mathrm{SL}_2(\mathbb{Z})$ where $N = \prod p^e$, including the representatives used at the end of the section to find the elliptic points. For example, if representatives for $\Gamma_0(p^e q^f) \backslash \Gamma_0(q^f)$ are $\{\alpha_0, \ldots, \alpha_{p^e-1}, \beta_0, \ldots, \beta_{p^{e-1}-1}\}$ and representatives for $\Gamma_0(q^f) \backslash \mathrm{SL}_2(\mathbb{Z})$ are $\{\alpha'_0, \ldots, \alpha'_{q^f-1}, \beta'_0, \ldots, \beta'_{q^{f-1}-1}\}$ then the $p^e(1 + 1/p)q^f(1 + 1/q)$ products $\{\alpha_j \alpha'_{j'}, \alpha_j \beta'_{j'}, \beta_j \alpha'_{j'}, \beta_j \beta'_{j'}\}$ are a set of representatives for $\Gamma_0(p^e q^f) \backslash \mathrm{SL}_2(\mathbb{Z})$.

3.8 More on cusps

This section describes the cusps of the congruence subgroups $\Gamma(N)$, $\Gamma_1(N)$, and $\Gamma_0(N)$. When $N = 1$, these groups are $\mathrm{SL}_2(\mathbb{Z})$ and the only cusp is $\mathrm{SL}_2(\mathbb{Z})\infty$, so throughout the section let $N > 1$.

The first result states that the familiar action of $\mathrm{SL}_2(\mathbb{Z})/\{\pm I\}$ on $\mathbb{Q} \cup \{\infty\}$ comes from the action of $\mathrm{SL}_2(\mathbb{Z})$ on $\mathbb{Q}^2$.

Lemma 3.8.1. *Let* $s = a/c$ *and* $s' = a'/c'$ *be elements of* $\mathbb{Q} \cup \{\infty\}$, *with* $\gcd(a, c) = \gcd(a', c') = 1$. *Then for any* $\gamma \in \mathrm{SL}_2(\mathbb{Z})$,

$$s' = \gamma(s) \iff \begin{bmatrix} a' \\ c' \end{bmatrix} = \pm\gamma \begin{bmatrix} a \\ c \end{bmatrix}.$$

Proof. Assume that $c \neq 0$ and $c' \neq 0$. Then for $\gamma = \begin{bmatrix} p & q \\ r & t \end{bmatrix} \in \text{SL}_2(\mathbb{Z})$,

$$s' = \gamma(s) \iff \frac{a'}{c'} = \frac{pa + qc}{ra + tc}.$$

Both fractions on the right side are in lowest terms (Exercise 3.8.1(a)), and so the result follows in this case. The cases when at least one of c, c' is zero are left as Exercise 3.8.1(b). □

The notation from the lemma will remain in effect throughout the section. Thus $s = a/c$ and $s' = a'/c'$ are elements of $\mathbb{Q} \cup \{\infty\}$ expressed in lowest terms, so that $\begin{bmatrix} a \\ c \end{bmatrix}$ and $\begin{bmatrix} a' \\ c' \end{bmatrix}$ are elements of $\mathbb{Z}^2$ with $\gcd(a, c) = \gcd(a', c') = 1$. The gcd condition automatically excludes the zero vector. If $\begin{bmatrix} a \\ c \end{bmatrix} = \gamma \begin{bmatrix} a' \\ c' \end{bmatrix}$ for some $\gamma \in \Gamma(N)$ then taking the relation modulo N gives the congruence $\begin{bmatrix} a \\ c \end{bmatrix} \equiv \begin{bmatrix} a' \\ c' \end{bmatrix} \pmod{N}$. The next result, that the converse holds as well, is the small technical point needed to describe the cusps.

Lemma 3.8.2. *Let* $\begin{bmatrix} a \\ c \end{bmatrix}$ *and* $\begin{bmatrix} a' \\ c' \end{bmatrix}$ *be as above. Then*

$$\begin{bmatrix} a' \\ c' \end{bmatrix} = \gamma \begin{bmatrix} a \\ c \end{bmatrix} \text{ for some } \gamma \in \Gamma(N) \iff \begin{bmatrix} a' \\ c' \end{bmatrix} \equiv \begin{bmatrix} a \\ c \end{bmatrix} \pmod{N}.$$

Proof. " $\Longrightarrow$ " is immediate, as just noted. For " $\Longleftarrow$ " first assume $\begin{bmatrix} a \\ c \end{bmatrix} = \begin{bmatrix} 1 \\ 0 \end{bmatrix}$. Then $a' \equiv 1 \pmod{N}$. Take integers β and δ such that $a'\delta - c'\beta = (1 - a')/N$ and let $\gamma = \begin{bmatrix} a' & \beta N \\ c' & 1 + \delta N \end{bmatrix}$. Then $\gamma \in \Gamma(N)$ and $\begin{bmatrix} a' \\ c' \end{bmatrix} = \gamma \begin{bmatrix} a \\ c \end{bmatrix}$ in this case.

In the general case there exist integers b and d such that $ad - bc = 1$. Let $\alpha = \begin{bmatrix} a & b \\ c & d \end{bmatrix} \in \text{SL}_2(\mathbb{Z})$. Since $\alpha \begin{bmatrix} 1 \\ 0 \end{bmatrix} = \begin{bmatrix} a \\ c \end{bmatrix}$, it follows that $\alpha^{-1} \begin{bmatrix} a' \\ c' \end{bmatrix} \equiv \alpha^{-1} \begin{bmatrix} a \\ c \end{bmatrix} = \begin{bmatrix} 1 \\ 0 \end{bmatrix}$ working modulo N. Thus $\alpha^{-1} \begin{bmatrix} a' \\ c' \end{bmatrix} = \gamma'\alpha^{-1} \begin{bmatrix} a \\ c \end{bmatrix}$ for some $\gamma' \in \Gamma(N)$, as in the preceding paragraph. Since $\Gamma(N)$ is normal in $\text{SL}_2(\mathbb{Z})$, setting $\gamma = \alpha\gamma'\alpha^{-1}$ completes the proof. □

Proposition 3.8.3. *Let* $s = a/c$ *and* $s' = a'/c'$ *be elements of* $\mathbb{Q} \cup \{\infty\}$ *with* $\gcd(a, c) = \gcd(a', c') = 1$. *Then*

$$\Gamma(N)s' = \Gamma(N)s \iff \begin{bmatrix} a' \\ c' \end{bmatrix} \equiv \pm \begin{bmatrix} a \\ c \end{bmatrix} \pmod{N},$$

and

$$\Gamma_1(N)s' = \Gamma_1(N)s \iff \begin{bmatrix} a' \\ c' \end{bmatrix} \equiv \pm \begin{bmatrix} a + jc \\ c \end{bmatrix} \pmod{N} \text{ for some } j,$$

and

$$\Gamma_0(N)s' = \Gamma_0(N)s \iff \begin{bmatrix} ya' \\ c' \end{bmatrix} \equiv \begin{bmatrix} a + jc \\ yc \end{bmatrix} \pmod{N} \text{ for some } j, y.$$

In the second and third equivalences j can be any integer, and in the third equivalence y is any integer relatively prime to N.

Proof. For the first equivalence,

$$\Gamma(N)s' = \Gamma(N)s \iff s' = \gamma(s) \text{ for some } \gamma \in \Gamma(N)$$
$$\iff \left[\begin{smallmatrix} a' \\ c' \end{smallmatrix}\right] = \pm\gamma\left[\begin{smallmatrix} a \\ c \end{smallmatrix}\right] \text{ for some } \gamma \in \Gamma(N), \text{ by Lemma 3.8.1}$$
$$\iff \left[\begin{smallmatrix} a' \\ c' \end{smallmatrix}\right] \equiv \pm\left[\begin{smallmatrix} a \\ c \end{smallmatrix}\right] \pmod{N}, \text{ by Lemma 3.8.2.}$$

The decomposition $\Gamma_1(N) = \bigcup_j \Gamma(N)\left[\begin{smallmatrix} 1 & j \\ 0 & 1 \end{smallmatrix}\right]$ reduces the second equivalence to the first, cited at the last step of the computation

$$\Gamma_1(N)s' = \Gamma_1(N)s \iff s' \in \Gamma_1(N)s$$
$$\iff s' \in \Gamma(N)\left[\begin{smallmatrix} 1 & j \\ 0 & 1 \end{smallmatrix}\right]s \text{ for some } j$$
$$\iff \Gamma(N)s' = \Gamma(N)(s+j) \text{ for some } j$$
$$\iff \left[\begin{smallmatrix} a' \\ c' \end{smallmatrix}\right] \equiv \pm\left[\begin{smallmatrix} a+jc \\ c \end{smallmatrix}\right] \pmod{N} \text{ for some } j.$$

Similarly, the decomposition $\Gamma_0(N) = \bigcup_y \Gamma_1(N)\left[\begin{smallmatrix} x & k \\ N & y \end{smallmatrix}\right]$, taken over y relatively prime to N and $xy - kN = 1$, reduces the third equivalence to the second, cited at the third step of the computation

$$\Gamma_0(N)s' = \Gamma_0(N)s \iff s' \in \Gamma_1(N)\left[\begin{smallmatrix} x & k \\ N & y \end{smallmatrix}\right]s \text{ for some } y$$
$$\iff \Gamma_1(N)s' = \Gamma_1(N)\frac{xa+kc}{Na+yc} \text{ for some } y$$
$$\iff \left[\begin{smallmatrix} a' \\ c' \end{smallmatrix}\right] \equiv \pm\left[\begin{smallmatrix} xa+kc+jyc \\ yc \end{smallmatrix}\right] \pmod{N} \text{ for some } j, y$$
$$\iff \left[\begin{smallmatrix} ya' \\ c' \end{smallmatrix}\right] \equiv \left[\begin{smallmatrix} a+jc \\ yc \end{smallmatrix}\right] \pmod{N} \text{ for some } j, y,$$

where the "$\pm$" has been absorbed into y and some constants into j at the last step. $\qquad\square$

Thus the cusps of $\Gamma(N)$ are $\Gamma(N)s$, $s = a/c$, for all pairs $\pm\left[\begin{smallmatrix} a \\ c \end{smallmatrix}\right] \pmod{N}$ where $\gcd(a,c) = 1$. Condition (3) of the next lemma provides a characterization purely modulo N for these representatives.

Lemma 3.8.4. *Let the integers a, c have images $\bar{a}, \bar{c}$ in $\mathbb{Z}/N\mathbb{Z}$. Then the following are equivalent:*

(1) $\left[\begin{smallmatrix} \bar{a} \\ \bar{c} \end{smallmatrix}\right]$ *has a lift* $\left[\begin{smallmatrix} a' \\ c' \end{smallmatrix}\right] \in \mathbb{Z}^2$ *with* $\gcd(a', c') = 1$,
(2) $\gcd(a, c, N) = 1$,
(3) $\left[\begin{smallmatrix} \bar{a} \\ \bar{c} \end{smallmatrix}\right]$ *has order N in the additive group* $(\mathbb{Z}/N\mathbb{Z})^2$.

Proof. If condition (1) holds then $k(a+sN) + l(c+tN) = 1$ for some integers k, l, s, t, and so $ka + lc + (ks + lt)N = 1$, giving condition (2).

If condition (2) holds then $ad - bc + kN = 1$ for some b, d, and k, and the matrix $\gamma = \left[\begin{smallmatrix} a & b \\ c & d \end{smallmatrix}\right] \in M_2(\mathbb{Z})$ reduces modulo N into $\mathrm{SL}_2(\mathbb{Z}/N\mathbb{Z})$. Since $\mathrm{SL}_2(\mathbb{Z})$ surjects to $\mathrm{SL}_2(\mathbb{Z}/N\mathbb{Z})$, there is a lift $\left[\begin{smallmatrix} a' & b' \\ c' & d' \end{smallmatrix}\right] \in \mathrm{SL}_2(\mathbb{Z})$, giving condition (1). (The reader may remember this argument from the proof of Theorem 1.5.1.)

Condition (3) is the implication $(k\left[\begin{smallmatrix}a\\c\end{smallmatrix}\right] = 0$ in $(\mathbb{Z}/N\mathbb{Z})^2 \implies N \mid k)$, or equivalently $(N \mid ka$ and $N \mid kc \implies N \mid k)$, or equivalently $(N \mid k\gcd(a,c) \implies N \mid k)$, or equivalently $\gcd(\gcd(a,c),N) = 1$, or equivalently $\gcd(a,c,N) = 1$, i.e., condition (2). $\qquad\square$

By the equivalence of conditions (1) and (3), the cusps of $\Gamma(N)$ are now described by the pairs $\pm\left[\begin{smallmatrix}a\\c\end{smallmatrix}\right]$ of order N in $(\mathbb{Z}/N\mathbb{Z})^2$. The bijection is

$$\pm\left[\tfrac{a}{c}\right] \mapsto \Gamma(N)(a'/c'), \text{ where } \left[\begin{smallmatrix}a'\\c'\end{smallmatrix}\right] \text{ is a lift of } \left[\tfrac{a}{c}\right] \text{ with } \gcd(a',c') = 1.$$

Using condition (2) of Lemma 3.8.4 we can count the cusps of $\Gamma(N)$. For each $c \in \{0,1,\ldots,N-1\}$ let $d = \gcd(c,N)$. Then the number of values $a \in \{0,1,\ldots,N-1\}$ such that $\gcd(a,c,N) = 1$ is the number of values such that $\gcd(a,d) = 1$, or $(N/d)\phi(d)$ (Exercise 3.8.2(a)). So the number of elements of order N in $(\mathbb{Z}/N\mathbb{Z})^2$ is (considering only positive divisors here and elsewhere in this section)

$$\sum_{d\mid N}(N/d)\phi(d) \cdot |\{c : 0 \le c < N, \ \gcd(c,N) = d\}|$$

$$= \sum_{d\mid N}(N/d)\phi(d)\phi(N/d)$$

$$= N^2 \prod_{p\mid N}(1 - 1/p^2) \quad \text{(Exercise 3.8.2(b))}.$$

For $N > 2$ each pair $\pm\left[\tfrac{a}{c}\right]$ of order N elements in $(\mathbb{Z}/N\mathbb{Z})^2$ has two distinct members since an element equal to its negative has order 2. For $N = 2$ the "$\pm$" has no effect. Thus the number of cusps of $\Gamma(N)$ is

$$\varepsilon_\infty(\Gamma(N)) = \begin{cases}(1/2)N^2 \prod_{p\mid N}(1 - 1/p^2) & \text{if } N > 2, \\ 3 & \text{if } N = 2.\end{cases}$$

Deriving this formula doesn't really require first describing the cusps of $\Gamma(N)$ explicitly. Since $\Gamma(N)$ is normal in $\mathrm{SL}_2(\mathbb{Z})$, its cusps all have the same ramification degree over $\mathrm{SL}_2(\mathbb{Z})\infty$, most easily computed by applying the formula from Section 3.1 at ∞,

$$e_{\Gamma(N)\infty} = [\mathrm{SL}_2(\mathbb{Z})_\infty : \{\pm I\}\Gamma(N)_\infty] = [\pm \langle[\begin{smallmatrix}1&1\\0&1\end{smallmatrix}]\rangle : \pm\langle[\begin{smallmatrix}1&N\\0&1\end{smallmatrix}]\rangle] = N.$$

Since $-I \in \mathrm{SL}_2(\mathbb{Z})$ and $-I \in \Gamma(N)$ only for $N = 2$ (recall that $N > 1$), and since $[\mathrm{SL}_2(\mathbb{Z}) : \Gamma(N)] = N^3 \prod_{p\mid N}(1 - 1/p^2)$, the projection of modular curves $X(N) \longrightarrow X(1)$ has degree

$$d_N = [\mathrm{SL}_2(\mathbb{Z}) : \{\pm I\}\Gamma(N)] = \begin{cases}(1/2)N^3 \prod_{p\mid N}(1 - 1/p^2) & \text{if } N > 2, \\ 6 & \text{if } N = 2,\end{cases}$$

cf. Exercise 1.2.2(b). Now the formula $\varepsilon_\infty(\Gamma(N)) = d_N/N$ reproduces the result.

But since $\Gamma_1(N)$ is not normal in $\mathrm{SL}_2(\mathbb{Z})$, its cusps must be counted directly. The representatives are pairs $\pm\begin{bmatrix} a \\ c \end{bmatrix}$ of order N vectors modulo N, but now by Proposition 3.8.3 a cusp determines the upper entry a modulo the lower entry c, so in fact a is determined modulo $\gcd(c, N)$. This time, for each $c \in \{0, 1, \ldots, N-1\}$, letting $d = \gcd(c, N)$, the number of values $a \pmod d$ such that $\gcd(a, d) = 1$ is just $\phi(d)$. So the number of elements of in $(\mathbb{Z}/N\mathbb{Z})^2$ that pair to describe cusps of $\Gamma_1(N)$ is $\sum_{d|N} \phi(d)\phi(N/d)$.

$a =$	0	1	2	3	4	5	6	7	8	9	10	11
$c = 0$	·	⊗	·	·	·	⊗	·	⊗	·	·	·	⊗
1	⊗	×	×	×	×	×	×	×	×	×	×	×
2	·	⊗	·	×	·	×	·	×	·	×	·	×
3	·	⊗	⊗	·	×	×	·	×	×	·	×	×
4	·	⊗	·	⊗	·	×	·	×	·	×	·	×
5	⊗	×	×	×	×	×	×	×	×	×	×	×
$c = 6$	·	⊗	·	·	·	×	·	×	·	·	·	⊗
7	⊗	×	×	×	×	×	×	×	×	×	×	×
8	·	×	·	×	·	×	·	×	·	⊗	·	⊗
9	·	×	×	·	×	×	·	×	×	·	⊗	⊗
10	·	×	·	×	·	×	·	×	·	×	·	⊗
11	⊗	×	×	×	×	×	×	×	×	×	×	×

Figure 3.2. The cusps of $\Gamma(12)$ and of $\Gamma_1(12)$

The table in Figure 3.2 shows all of this for $N = 12$. The elements $\begin{bmatrix} a \\ c \end{bmatrix}$ of order 12, indicated by "×", pair under negation modulo 12 to represent the cusps of $\Gamma(12)$. In each row, one element is circled for each value of a modulo $\gcd(c, N)$, describing the cusps of $\Gamma_1(12)$. The circled elements from the top row, where $c = 0$, pair in an obvious way, and similarly for the left column, where $a = 0$. When $ac \neq 0$ the circled elements are taken from the left cell of length $\gcd(c, N)$ on row c when $c < 6$ and from the right cell when $c > 6$, and these visibly pair. On the middle row, where $c = 6$ (and more generally where $c = N/2$ when N is even), the pairing of circled elements is a bit more subtle: the left $\phi(N/2)/2$ elements in the left half of the row pair with the right $\phi(N/2)/2$ in the right half unless $\phi(N/2)$ is odd. But $\phi(N/2)$ is odd only for $N = 2$, when the whole pairing process collapses anyway, and for $N = 4$, when only four of the five "×"'s in the relevant table pair off, leaving three representatives (Exercise 3.8.3). Thus the number of cusps of $\Gamma_1(N)$ is

$$\varepsilon_\infty(\Gamma_1(N)) = \begin{cases} 2 & \text{if } N = 2, \\ 3 & \text{if } N = 4, \\ \frac{1}{2}\sum_{d|N} \phi(d)\phi(N/d) & \text{if } N = 3 \text{ or } N > 4. \end{cases}$$

The sum doesn't have as tidy a product form as the corresponding sum for $\Gamma(N)$ despite appearing simpler (this reflects that $\Gamma_1(N)$ is not normal in $\mathrm{SL}_2(\mathbb{Z})$), so we leave it as is.

To count the cusps of $\Gamma_0(N)$ recall from Proposition 3.8.3 that for this group, vectors $\begin{bmatrix} a \\ c \end{bmatrix}$ and $\begin{bmatrix} a' \\ c' \end{bmatrix}$ with $\gcd(a,c) = \gcd(a',c') = 1$ represent the same cusp when $\begin{bmatrix} ya' \\ c' \end{bmatrix} \equiv \begin{bmatrix} a+jc \\ yc \end{bmatrix}$ (mod N) for some j and y with $\gcd(y,N) = 1$. The bottom condition, $c' \equiv yc$ (mod N) for some such y, is equivalent to $\gcd(c',N) = \gcd(c,N)$, in which case letting $d = \gcd(c,N)$ and letting $y_0 \in \mathbb{Z}$ satisfy $y_0 \equiv c'c^{-1}$ (mod N/d) makes the condition equivalent to $y \equiv y_0 + iN/d$ (mod N) for some i (confirming the calculations in the paragraph is Exercise 3.8.4). For any divisor d of N, pick one value c modulo N such that $\gcd(c,N) = d$. Then any cusp of $\Gamma_0(N)$ represented by some vector $\begin{bmatrix} a' \\ c' \end{bmatrix}$ with $\gcd(c',N) = d$ is also represented by $\begin{bmatrix} a \\ c \end{bmatrix}$ whenever $(y_0 + iN/d)a' \equiv a + jc$ (mod N) for some i and j, or $a \equiv y_0 a'$ (mod $\gcd(c,N,a'N/d)$), or $a \equiv y_0 a'$ (mod $\gcd(d,N/d)$). Also, a is relatively prime to d since $\gcd(a,d) \mid \gcd(a,c) = 1$, so a is relatively prime to $\gcd(d,N/d)$. Thus for each divisor d of N there are $\phi(\gcd(d,N/d))$ cusps, and the number of cusps of $\Gamma_0(N)$ is

$$\varepsilon_\infty(\Gamma_0(N)) = \sum_{d|N} \phi(\gcd(d,N/d)).$$

The reader should use this discussion to find the cusps of $\Gamma_0(12)$ with the help of Figure 3.2 (Exercise 3.8.5).

Let N be any positive integer and let k be odd. All cusps of $\Gamma_0(N)$ are regular since $-I \in \Gamma_0(N)$. To study the cusps of $\Gamma_1(N)$, take any $s \in \mathbb{Q} \cup \{\infty\}$ and let $\gamma \in \Gamma_1(N)_s$, so if $\alpha \in \mathrm{SL}_2(\mathbb{Z})$ takes ∞ to s then $\alpha^{-1}\gamma\alpha \in (\alpha^{-1}\Gamma\alpha)_\infty$. If $\alpha^{-1}\gamma\alpha$ takes the form $-\begin{bmatrix} 1 & h \\ 0 & 1 \end{bmatrix}$ then γ has trace -2. But also, $\mathrm{trace}(\gamma) \equiv 2$ (mod N), so this can happen only if $N \mid 4$. Since $-I \in \Gamma_1(2)$, the only case where an irregular cusp might occur is $\Gamma_1(4)$. This argument shows that all cusps of $\Gamma(N)$, $N \neq 4$, are regular as well. In fact, the only irregular cusp in this context turns out to be the example we have already seen, $s = 1/2$ for $\Gamma_1(4)$ (Exercise 3.8.7).

Finally, the cusps of any congruence subgroup of $\mathrm{SL}_2(\mathbb{Z})$ have a purely group-theoretic description. Let G be an arbitrary group and let H_1 and H_2 be subgroups. A *double coset of G* is a subset of G of the form H_1gH_2. The space of double cosets is denoted $H_1\backslash G/H_2$, that is,

$$H_1\backslash G/H_2 = \{H_1gH_2 : g \in G\}.$$

As with one-sided cosets, any two double cosets are disjoint or equal, so there is a disjoint set decomposition

$$G = \bigcup_{g \in R} H_1gH_2 \quad \text{where R is a set of representatives.}$$

A specific case of all this describes the cusps of any congruence subgroup.

Proposition 3.8.5. *Let Γ be a congruence subgroup of $\mathrm{SL}_2(\mathbb{Z})$ and let P be the parabolic subgroup of $\mathrm{SL}_2(\mathbb{Z})$, $P = \{\pm \begin{bmatrix} 1 & j \\ 0 & 1 \end{bmatrix} : j \in \mathbb{Z}\} = \mathrm{SL}_2(\mathbb{Z})_\infty$, the fixing subgroup of ∞. Then the map*

$$\Gamma\backslash\mathrm{SL}_2(\mathbb{Z})/P \longrightarrow \{\text{cusps of } \Gamma\}$$

given by

$$\Gamma\alpha P \mapsto \Gamma\alpha(\infty)$$

is a bijection. Specifically, the map is $\Gamma \begin{bmatrix} a & b \\ c & d \end{bmatrix} P \mapsto \Gamma(a/c)$.

Proof. The map is well defined since if $\Gamma\alpha P = \Gamma\alpha' P$ then $\alpha' = \gamma\alpha\delta$ for some $\gamma \in \Gamma$ and $\delta \in P$, so that $\Gamma\alpha'(\infty) = \Gamma\gamma\alpha\delta(\infty) = \Gamma\alpha(\infty)$, the last equality because $\gamma \in \Gamma$ and δ fixes ∞.

The map is injective since the condition $\Gamma\alpha'(\infty) = \Gamma\alpha(\infty)$ is equivalent to $\alpha'(\infty) = \gamma\alpha(\infty)$ for some $\gamma \in \Gamma$, or $\alpha^{-1}\gamma^{-1}\alpha' \in P$ for some $\gamma \in \Gamma$, or $\alpha' \in \Gamma\alpha P$, meaning $\Gamma\alpha' P = \Gamma\alpha P$.

The map is clearly surjective. $\square$

Another proof, essentially identical, is to show first that $\mathrm{SL}_2(\mathbb{Z})/P$ identifies with $\mathbb{Q} \cup \{\infty\}$, so that the double coset space $\Gamma\backslash\mathrm{SL}_2(\mathbb{Z})/P$ identifies with the cusps $\Gamma\backslash(\mathbb{Q} \cup \{\infty\})$.

In particular, the double coset space $\Gamma(N)\backslash\mathrm{SL}_2(\mathbb{Z})/P$ is naturally viewed as $\mathrm{SL}_2(\mathbb{Z}/N\mathbb{Z})/\overline{P}$ where $\overline{P}$ denotes the projected image of P in $\mathrm{SL}_2(\mathbb{Z}/N\mathbb{Z})$, that is, $\overline{P} = \{\pm \begin{bmatrix} 1 & j \\ 0 & 1 \end{bmatrix} : j \in \mathbb{Z}/N\mathbb{Z}\}$. Thus double cosets $\Gamma(N)\alpha P$ with $\alpha \in \mathrm{SL}_2(\mathbb{Z})$ can be viewed as ordinary cosets $\alpha\overline{P}$ with $\alpha \in \mathrm{SL}_2(\mathbb{Z}/N\mathbb{Z})$. Since $\overline{P}$ acts from the right by

$$\begin{bmatrix} a & b \\ c & d \end{bmatrix} \mapsto \pm \begin{bmatrix} a & b + ja \\ c & d + jc \end{bmatrix},$$

the cosets are represented by pairs $\pm \begin{bmatrix} a \\ c \end{bmatrix}$ where the vector is the left column of some $\alpha \in \mathrm{SL}_2(\mathbb{Z}/N\mathbb{Z})$ as before. Similarly, when $\Gamma = \Gamma_1(N) = \bigcup_j \begin{bmatrix} 1 & j \\ 0 & 1 \end{bmatrix} \Gamma(N)$, the double coset space is naturally viewed as $\overline{P}_+\backslash\mathrm{SL}_2(\mathbb{Z}/N\mathbb{Z})/\overline{P}$ where $P_+ = \{\begin{bmatrix} 1 & j \\ 0 & 1 \end{bmatrix} : j \in \mathbb{Z}\}$ is the "positive" half of P and again the overbar signifies reduction modulo N. Since $\begin{bmatrix} 1 & j \\ 0 & 1 \end{bmatrix}\begin{bmatrix} a & * \\ c & * \end{bmatrix} = \begin{bmatrix} a+jc & * \\ c & * \end{bmatrix}$ this recovers the description of the cusps of $\Gamma_1(N)$. The double coset decomposition $\Gamma_0(N)\backslash\mathrm{SL}_2(\mathbb{Z})/P$ recovers the cusps of $\Gamma_0(N)$ in the same way. See Section 1.6 of [Shi73] for an elegant enumeration of the cusps of $\Gamma_0(N)$ using double cosets.

In Chapter 5 of this book double cosets will define the Hecke operators mentioned in the preface. For future reference in Chapter 5 we now state that for any prime p,

$$\Gamma_1(N) \begin{bmatrix} 1 & 0 \\ 0 & p \end{bmatrix} \Gamma_1(N) = \left\{\gamma \in \mathrm{M}_2(\mathbb{Z}) : \gamma \equiv \begin{bmatrix} 1 & * \\ 0 & p \end{bmatrix} \pmod{N},\ \det\gamma = p\right\}.$$
$$(3.16)$$

Supply details as necessary to the following argument (Exercise 3.8.8(a)). One containment is clear. For the other containment, let $L = \mathbb{Z}^2$ and let $L_0 = L_0(N) = \{[\begin{smallmatrix} x \\ y \end{smallmatrix}] \in L : y \equiv 0 \pmod{N}\}$. Then $M_2(\mathbb{Z})$ acts on L by left multiplication. Take $\gamma \in M_2(\mathbb{Z})$ such that $\gamma \equiv [\begin{smallmatrix} 1 & * \\ 0 & p \end{smallmatrix}] \pmod{N}$ and $\det \gamma = p$. Because $\gamma \in M_2(\mathbb{Z})$ and $\gamma \equiv [\begin{smallmatrix} * & * \\ 0 & * \end{smallmatrix}] \pmod{N}$, $\gamma L_0 \subset L_0$. Because $\det \gamma = p$ is positive, $[L : \gamma L_0] = [L : L_0][L_0 : \gamma L_0] = Np$. By the theory of Abelian groups there exists a basis $\{u, v\}$ of L such that $\det[u\ v] = 1$ and $\gamma L_0 = mu\mathbb{Z} \oplus nv\mathbb{Z}$ where $0 < m \mid n$ and $mn = Np$. The first column of γ is $[\begin{smallmatrix} a_\gamma \\ 0 \end{smallmatrix}] \pmod{N}$, but it is also γe_1 (where e_1 is the first standard basis vector), an element of γL_0 and therefore $[\begin{smallmatrix} 0 \\ 0 \end{smallmatrix}] \pmod{m}$. Because $\gcd(a_\gamma, N) = 1$, also $\gcd(m, N) = 1$. Because p is prime it follows that $m = 1$ and $n = Np$,

$$\gamma L_0 = u\mathbb{Z} \oplus Npv\mathbb{Z}.$$

The right side has unique supergroups of index p and N inside $L = u\mathbb{Z} \oplus v\mathbb{Z}$, so now

$$L_0 = u\mathbb{Z} \oplus Nv\mathbb{Z}, \qquad \gamma L = u\mathbb{Z} \oplus pv\mathbb{Z}, \qquad \gamma L_0 = u\mathbb{Z} \oplus Npv\mathbb{Z}.$$

Let $\gamma_1 = [u\ v]$. The condition $u \in L_0$ shows that $\gamma_1 \in \Gamma_0(N)$. Let $\gamma_2 = (\gamma_1 [\begin{smallmatrix} 1 & 0 \\ 0 & p \end{smallmatrix}])^{-1} \gamma$, an element $[\begin{smallmatrix} a & b \\ c & d \end{smallmatrix}]$ of $GL_2^+(\mathbb{Q})$ with determinant 1 such that as desired,

$$\gamma = \gamma_1 \begin{bmatrix} 1 & 0 \\ 0 & p \end{bmatrix} \gamma_2. \tag{3.17}$$

The condition $\gamma e_1 \in \gamma L_0$ is $au + cpv \in u\mathbb{Z} \oplus Npv\mathbb{Z}$, or $a \in \mathbb{Z}$ and $c \in N\mathbb{Z}$. The condition $\gamma e_2 \in \gamma L$ is $ub + pvd \in u\mathbb{Z} \oplus pv\mathbb{Z}$, or $b, d \in \mathbb{Z}$. Thus $\gamma_2 \in \Gamma_0(N)$ as well. Because $a_\gamma \equiv 1 \pmod{N}$, $\gamma e_1 \equiv e_1 \pmod{N}$ and thus (3.17) shows that $au_1 \equiv 1 \pmod{N}$ using only that $\gamma_1, \gamma_2 \in \Gamma_0(N)$ where $\gamma_2 = [\begin{smallmatrix} a & b \\ c & d \end{smallmatrix}]$. So if γ satisfies (3.17) for $\gamma_1, \gamma_2 \in \Gamma_0(N)$ such that either γ_1 or γ_2 lies in $\Gamma_1(N)$ then both do. To complete the argument it suffices to show that

$$\Gamma_0(N) \begin{bmatrix} 1 & 0 \\ 0 & p \end{bmatrix} \Gamma_0(N) = \Gamma_1(N) \begin{bmatrix} 1 & 0 \\ 0 & p \end{bmatrix} \Gamma_0(N).$$

For the nontrivial containment, since $\Gamma_1(N) \backslash \Gamma_0(N)$ is represented by matrices of the form $[\begin{smallmatrix} a & k \\ N & d \end{smallmatrix}] \in SL_2(\mathbb{Z})$, it suffices to show that for each such matrix there exists a matrix $\delta \in \Gamma_1(N)$ such that $[\begin{smallmatrix} a & k \\ N & d \end{smallmatrix}][\begin{smallmatrix} 1 & 0 \\ 0 & p \end{smallmatrix}] \Gamma_0(N) = \delta [\begin{smallmatrix} 1 & 0 \\ 0 & p \end{smallmatrix}] \Gamma_0(N)$, or a matrix $\delta' \in \Gamma_1(N)$ such that

$$\begin{bmatrix} 1 & 0 \\ 0 & p \end{bmatrix}^{-1} \delta' \begin{bmatrix} a & k \\ N & d \end{bmatrix} \begin{bmatrix} 1 & 0 \\ 0 & p \end{bmatrix} \in \Gamma_0(N).$$

If $p \mid N$ then $\delta' = [\begin{smallmatrix} dN+1 & -1 \\ -dN & 1 \end{smallmatrix}]$ works. If $p \nmid N$ then any $\delta' = [\begin{smallmatrix} * & * \\ N & d' \end{smallmatrix}]$ where $d' \equiv 1 \pmod{N}$ and $d' \equiv -a \pmod{p}$ will do. This completes the proof. Exercise 3.8.8(b) requests a similar description of the double coset $\Gamma_0(N) [\begin{smallmatrix} 1 & 0 \\ 0 & p \end{smallmatrix}] \Gamma_0(N)$.

Exercises

3.8.1. (a) In the proof of Lemma 3.8.1 show that the fraction $(pa+qc)/(ra+tc)$ is in lowest terms. (A hint for this exercise is at the end of the book.)
(b) Complete the proof of Lemma 3.8.1.

3.8.2. (a) Let d be a divisor of N. Show that the number of values a modulo N such that $\gcd(a, d) = 1$ is $(N/d)\phi(d)$.
(b) Show that $\sum_{d|N}(N/d)\phi(d)\phi(N/d) = N^2 \prod_{p|N}(1 - 1/p^2)$. (A hint for this exercise is at the end of the book.)

3.8.3. Make a table counting the cusps of $\Gamma(4)$, of $\Gamma_1(4)$, and of $\Gamma_0(4)$.

3.8.4. Confirm the calculations that count the cusps of $\Gamma_0(N)$.

3.8.5. Find the cusps of $\Gamma_0(12)$.

3.8.6. Find the cusps of $\Gamma(10)$, of $\Gamma_1(10)$, and of $\Gamma_0(10)$.

3.8.7. Show that then when k is odd, all cusps of $\Gamma(4)$ are regular and only the cusp $s = 1/2$ is irregular for $\Gamma_1(4)$. (A hint for this exercise is at the end of the book.)

3.8.8. (a) Supply details as necessary to the proof of (3.16) in the section.
(b) Analogously to (3.16), describe the double coset $\Gamma_0(N) \left[\begin{smallmatrix} 1 & 0 \\ 0 & p \end{smallmatrix}\right] \Gamma_0(N)$ as a subset of $M_2(\mathbb{Z})$.

3.9 More dimension formulas

This section computes dimension formulas for $\mathcal{M}_k(\Gamma)$ and $\mathcal{S}_k(\Gamma)$ when Γ is $\Gamma_0(N)$, $\Gamma_1(N)$, or $\Gamma(N)$. Since we already have formulas for $\mathrm{SL}_2(\mathbb{Z})$, let $N > 1$ throughout this section.

Recall that $X(N) = X(\Gamma(N))$, $X_1(N) = X(\Gamma_1(N))$, and $X_0(N) = X(\Gamma_0(N))$. Projection $X(N) \longrightarrow X(1)$ has the degree computed in the previous section,

$$d(\Gamma(N)) = d_N = \begin{cases} (1/2)N^3 \prod_{p|N}(1 - 1/p^2) & \text{if } N > 2, \\ 6 & \text{if } N = 2. \end{cases}$$

Since $-I \in \Gamma_1(N)$ if and only if $-I \in \Gamma(N)$, and since the index $[\Gamma_1(N) : \Gamma(N)]$ is N, projection $X_1(N) \longrightarrow X(1)$ has degree

$$d(\Gamma_1(N)) = d_N/N \quad \text{for } N \geq 2.$$

In discussing $\Gamma_0(N)$ we may take $N > 2$ since $\Gamma_0(2) = \Gamma_1(2)$. So $-I \notin \Gamma_1(N)$ but $-I \in \Gamma_0(N)$. Along with the index $[\Gamma_0(N) : \Gamma_1(N)] = \phi(N)$ this shows that projection $X_0(N) \longrightarrow X(1)$ has degree

$$d(\Gamma_0(N)) = 2d_N/(N\phi(N)) \quad \text{for } N > 2.$$

By Exercise 2.3.7(a), $\varepsilon_2(\Gamma(N)) = \varepsilon_3(\Gamma(N)) = 0$ for $N > 1$. By Exercise 2.3.7(b),

$$\varepsilon_2(\Gamma_1(N)) = \begin{cases} 1 & \text{if } N = 2, \\ 0 & \text{if } N > 2, \end{cases} \qquad \varepsilon_3(\Gamma_1(N)) = \begin{cases} 1 & \text{if } N = 3, \\ 0 & \text{if } N = 2 \text{ or } N > 3. \end{cases}$$

Finally, $\varepsilon_2(\Gamma_0(N))$ and $\varepsilon_3(\Gamma_0(N))$ were computed in Section 3.7. The data are summarized in Figure 3.3.

Γ	d	ε_2	ε_3	ε_∞
$\Gamma_0(N)$ $(N > 2)$	$\dfrac{2d_N}{N\phi(N)}$	$\displaystyle\prod_{p\mid N}(1 + \left(\frac{-1}{p}\right))$ if $4 \nmid N$ 0 if $4 \mid N$	$\displaystyle\prod_{p\mid N}(1 + \left(\frac{-3}{p}\right))$ if $9 \nmid N$ 0 if $9 \mid N$	$\displaystyle\sum_{d\mid N}\phi(\gcd(d, \tfrac{N}{d}))$
$\Gamma_1(2)$ $(\Gamma_0(2))$	3	1	0	2
$\Gamma_1(3)$	4	0	1	2
$\Gamma_1(4)$	6	0	0	3
$\Gamma_1(N)$ $(N > 4)$	$\dfrac{d_N}{N}$	0	0	$\dfrac{1}{2}\displaystyle\sum_{d\mid N}\phi(d)\phi(\tfrac{N}{d})$
$\Gamma(N)$ $(N > 1)$	d_N	0	0	$\dfrac{d_N}{N}$

Figure 3.3. Data for the dimension formulas

For $\Gamma_0(N)$, the formula for the genus g from Theorem 3.1.1 and the dimension formulas for $\mathcal{M}_k(\Gamma_0(N))$ and $\mathcal{S}_k(\Gamma_0(N))$ when k is even from Theorem 3.5.1 don't simplify and are best left in their given forms. (Since $-I \in \Gamma_0(N)$ the dimensions are 0 for k odd.) But for $\Gamma_1(N)$ and $\Gamma(N)$, see Figure 3.4. In the table, each "$\pm$" refers to the dimensions for the modular forms and the cusp forms respectively. The last column is split when the formulas differ for k even and k odd. Since $\dim(\mathcal{S}_2(\Gamma)) = g$ and $\dim(\mathcal{M}_0(\Gamma)) = 1$ and $\dim(\mathcal{S}_0(\Gamma)) = 0$ and $\dim(\mathcal{M}_k(\Gamma)) = 0$ when $k < 0$, the only case not resolved by the table is $k = 1$.

Γ	g	$\dim(\mathcal{M}_k(\Gamma))$ for $k \geq 2$ and $\dim(\mathcal{S}_k(\Gamma))$ for $k \geq 3$	
		(k even)	(k odd)
$\Gamma_1(2)$	0	$\left\lfloor \dfrac{k}{4} \right\rfloor \pm 1$	0
$\Gamma_1(3)$	0		$\left\lfloor \dfrac{k}{3} \right\rfloor \pm 1$
$\Gamma_1(4)$	0	$\dfrac{k-1 \pm 3}{2}$	$\dfrac{k-1 \pm 2}{2}$
$\Gamma_1(N)$ $(N > 4)$	$1 + \dfrac{d_N}{12N} - \dfrac{1}{4}\sum_{d\mid N}\phi(d)\phi(\tfrac{N}{d})$	$\dfrac{(k-1)d_N}{12N} \pm \dfrac{1}{4}\sum_{d\mid N}\phi(d)\phi(\tfrac{N}{d})$	
$\Gamma(2)$	0	$\dfrac{k-1 \pm 3}{2}$	0
$\Gamma(N)$ $(N > 2)$	$1 + \dfrac{d_N(N-6)}{12N}$	$\dfrac{(k-1)d_N}{12} \pm \dfrac{d_N}{2N}$	

Figure 3.4. Genera and dimensions for most values of k

Exercises

3.9.1. Verify all the entries in Figure 3.4. (A hint for this exercise is at the end of the book.)

3.9.2. Theorem 3.6.1 provides the dimension formulas $\dim(\mathcal{M}_1(\Gamma)) = \varepsilon_\infty^{\text{reg}}/2$ and $\dim(\mathcal{S}_1(\Gamma)) = 0$ when $\varepsilon_\infty^{\text{reg}} > 2g - 2$. Show that this condition holds for $\Gamma(N)$ if and only if $N < 12$.

3.9.3. Show that $\Gamma_0(N) = \{\pm I\}\Gamma_1(N)$ for $N = 2, 3, 4$. What does this show about $\dim \mathcal{M}_k(\Gamma_0(N))$ and $\dim \mathcal{S}_k(\Gamma_0(N))$ for these values of N? As a special case of this, note the formula $\dim(\mathcal{M}_2(\Gamma_0(4))) = 2$ cited in the context of the four squares problem in Section 1.2.

3.9.4. Show that $\Gamma(2) = \begin{bmatrix} 0 & 1/2 \\ -1 & 0 \end{bmatrix}^{-1} \{\pm I\}\Gamma_1(4) \begin{bmatrix} 0 & 1/2 \\ -1 & 0 \end{bmatrix}$. What does this explain about Figure 3.4?

3.9.5. For what values of N does the genus $g(X(N))$ equal 0? 1?

4

Eisenstein Series

For any congruence subgroup Γ of $\mathrm{SL}_2(\mathbb{Z})$, the space $\mathcal{M}_k(\Gamma)$ of modular forms naturally decomposes into its subspace of cusp forms $\mathcal{S}_k(\Gamma)$ and the corresponding quotient space $\mathcal{M}_k(\Gamma)/\mathcal{S}_k(\Gamma)$, the *Eisenstein space* $\mathcal{E}_k(\Gamma)$. This chapter gives bases of $\mathcal{E}_k(\Gamma(N))$, $\mathcal{E}_k(\Gamma_1(N))$, and subspaces of $\mathcal{E}_k(\Gamma_1(N))$ called *eigenspaces*, including $\mathcal{E}_k(\Gamma_0(N))$. The basis elements are variants of the Eisenstein series from Chapter 1. For $k \geq 3$ they are straightforward to write down, but for $k = 2$ and $k = 1$ the process is different.

Aside from demonstrating explicit examples of modular forms, computing the Fourier expansions of these Eisenstein series leads naturally to related subjects that are appealing in their own right: Dirichlet characters, zeta and L-functions, their analytic continuations and functional equations, Bernoulli numbers and Bernoulli polynomials, Fourier analysis, theta functions, and Mellin transforms. These ideas are presented in context as the need arises. They help to show that a more general kind of Eisenstein series, augmented by a complex parameter, satisfies a functional equation as well. At the end of the chapter they are used to construct a modular form that is related to the equation $x^3 = d$ and to the Cubic Reciprocity Theorem as the motivating example in the book's preface is related to $x^2 = d$ and Quadratic Reciprocity.

Related reading: Various parts of this material are covered in the texts [Gun62], [Hid93], [Kob93], [Miy89], [Lan76], [Sch74], and in sections of the papers [Hec27] and [Hid86].

4.1 Eisenstein series for $\mathrm{SL}_2(\mathbb{Z})$

The Eisenstein series $G_k(\tau)$ for even $k \geq 4$ were defined in Chapter 1, as were the normalized Eisenstein series $E_k(\tau) = G_k(\tau)/(2\zeta(k))$ with rational Fourier coefficients and leading coefficient 1. Recalling that a primed summation sign means to sum over nonzero elements, compute that

$$G_k(\tau) = \sideset{}{'}\sum_{(c,d)\in\mathbb{Z}^2} \frac{1}{(c\tau+d)^k} = \sum_{n=1}^{\infty} \sum_{\substack{(c,d) \\ \gcd(c,d)=n}} \frac{1}{(c\tau+d)^k}$$

$$= \sum_{n=1}^{\infty} \frac{1}{n^k} \sum_{\substack{(c,d) \\ \gcd(c,d)=1}} \frac{1}{(c\tau+d)^k} = \zeta(k) \sum_{\substack{(c,d) \\ \gcd(c,d)=1}} \frac{1}{(c\tau+d)^k}.$$

These calculations are valid since the series $G_k(\tau)$ converges absolutely for all integers $k \geq 3$, as shown in Exercise 1.1.4. It follows that

$$E_k(\tau) = \tfrac{1}{2} \sum_{\substack{(c,d)\in\mathbb{Z}^2 \\ \gcd(c,d)=1}} \frac{1}{(c\tau+d)^k}. \tag{4.1}$$

The series defining G_k and E_k cancel to zero for odd k.

Define $P_+ = \{ [\begin{smallmatrix} 1 & n \\ 0 & 1 \end{smallmatrix}] : n \in \mathbb{Z} \}$, the positive part of the parabolic subgroup of $\mathrm{SL}_2(\mathbb{Z})$. This allows an intrinsic description of the normalized Eisenstein series (Exercise 4.1.1),

$$E_k(\tau) = \tfrac{1}{2} \sum_{\gamma \in P_+ \backslash \mathrm{SL}_2(\mathbb{Z})} j(\gamma,\tau)^{-k}. \tag{4.2}$$

To show intrinsically that E_k is weakly modular of weight k, compute that for any $\gamma \in \mathrm{SL}_2(\mathbb{Z})$

$$(E_k[\gamma]_k)(\tau) = \tfrac{1}{2} j(\gamma,\tau)^{-k} \sum_{\gamma' \in P_+ \backslash \mathrm{SL}_2(\mathbb{Z})} j(\gamma',\gamma(\tau))^{-k},$$

so the relations $j(\gamma',\gamma(\tau)) = j(\gamma'\gamma,\tau)/j(\gamma,\tau)$ and $P_+\backslash\mathrm{SL}_2(\mathbb{Z})\gamma = P_+\backslash\mathrm{SL}_2(\mathbb{Z})$ show that the right side is again $E_k(\tau)$. This argument isn't really different from the nonintrinsic proof from Section 1.1 (and they both require the absolute convergence shown there in Exercise 1.1.2 to rearrange the sum), but the intrinsic method is being shown for elegance here and in the next section and because it remains tractable in more general situations.

This chapter will construct Eisenstein series for congruence subgroups of $\mathrm{SL}_2(\mathbb{Z})$. For any congruence subgroup Γ and any integer k, define the *weight k Eisenstein space* of Γ to be the quotient space of the modular forms by the cusp forms,

$$\mathcal{E}_k(\Gamma) = \mathcal{M}_k(\Gamma)/\mathcal{S}_k(\Gamma).$$

Recall from Chapter 3 that ε_∞ denotes the number of cusps of the compact modular curve $X(\Gamma)$ and $\varepsilon_\infty^{\mathrm{reg}}$ denotes the number of regular cusps. By the dimension formulas from Chapter 3 (Exercise 4.1.2),

$$\dim(\mathcal{E}_k(\Gamma)) = \begin{cases} \varepsilon_\infty & \text{if } k \geq 4 \text{ is even,} \\ \varepsilon_\infty^{\text{reg}} & \text{if } k \geq 3 \text{ is odd and } -I \notin \Gamma, \\ \varepsilon_\infty - 1 & \text{if } k = 2, \\ \varepsilon_\infty^{\text{reg}}/2 & \text{if } k = 1 \text{ and } -I \notin \Gamma, \\ 1 & \text{if } k = 0, \\ 0 & \text{if } k < 0 \text{ or if } k > 0 \text{ is odd and } -I \in \Gamma. \end{cases} \quad (4.3)$$

At the end of the next chapter the Eisenstein space $\mathcal{E}_k(\Gamma)$ will be redefined as a subspace of $\mathcal{M}_k(\Gamma)$ complementary to $\mathcal{S}_k(\Gamma)$, meaning the subspaces are linearly disjoint and their sum is the full space. In the meantime it suffices to think of $\mathcal{E}_k(\Gamma)$ as a quotient space.

Exercises

4.1.1. Show that the intrinsic sum for E_k in (4.2) agrees with the sum in (4.1).

4.1.2. Confirm formula (4.3).

4.2 Eisenstein series for $\Gamma(N)$ when $k \geq 3$

Let N be a positive integer and let $\overline{v} \in (\mathbb{Z}/N\mathbb{Z})^2$ be a row vector of order N. (In this chapter an overline generally denotes reduction modulo N, so for example v is any lift of $\overline{v}$ to $\mathbb{Z}^2$. Occasionally the overline will denote complex conjugation instead, but context will make the distinction clear.) Let $\delta = \begin{bmatrix} a & b \\ c_v & d_v \end{bmatrix} \in \mathrm{SL}_2(\mathbb{Z})$ with (c_v, d_v) a lift of $\overline{v}$ to $\mathbb{Z}^2$, and let $k \geq 3$ be an integer. Let ϵ_N be $1/2$ if $N \in \{1, 2\}$ and 1 if $N > 2$. Define (Exercise 4.2.1)

$$E_k^{\overline{v}}(\tau) = \epsilon_N \sum_{\substack{(c,d) \equiv v \,(N) \\ \gcd(c,d)=1}} (c\tau + d)^{-k} = \epsilon_N \sum_{\gamma \in (P_+ \cap \Gamma(N)) \backslash \Gamma(N)\delta} j(\gamma, \tau)^{-k}. \quad (4.4)$$

When $N = 1$ there is only one choice of $\overline{v}$ and $E_k^{\overline{v}}$ is E_k from before. Note that $E_k^{-\overline{v}} = (-1)^k E_k^{\overline{v}}$. This is a special case of

Proposition 4.2.1. *For any* $\gamma \in \mathrm{SL}_2(\mathbb{Z})$,

$$(E_k^{\overline{v}}[\gamma]_k)(\tau) = E_k^{\overline{v\gamma}}(\tau).$$

Proof. Compute for any $\gamma \in \mathrm{SL}_2(\mathbb{Z})$, using the fact that $\Gamma(N)$ is normal in $\mathrm{SL}_2(\mathbb{Z})$ at the second and fourth steps, that

$$(E_k^{\overline{v}}[\gamma]_k)(\tau) = \epsilon_N j(\gamma,\tau)^{-k} \sum_{\gamma' \in (P_+ \cap \Gamma(N)) \backslash \Gamma(N)\delta} j(\gamma', \gamma(\tau))^{-k}$$

$$= \epsilon_N \sum_{\gamma' \in (P_+ \cap \Gamma(N)) \backslash \delta \Gamma(N)} j(\gamma'\gamma, \tau)^{-k}$$

$$= \epsilon_N \sum_{\gamma'' \in (P_+ \cap \Gamma(N)) \backslash \delta \Gamma(N)\gamma} j(\gamma'', \tau)^{-k}$$

$$= \epsilon_N \sum_{\gamma'' \in (P_+ \cap \Gamma(N)) \backslash \Gamma(N)\delta\gamma} j(\gamma'', \tau)^{-k} = E_k^{\overline{v\gamma}}(\tau).$$

$\square$

Corollary 4.2.2. $E_k^{\overline{v}} \in \mathcal{M}_k(\Gamma(N))$.

Proof. As a subseries of the series E_k analyzed in Exercise 1.1.4, $E_k^{\overline{v}}$ is holomorphic on $\mathcal{H}$, meeting condition (1) of Definition 1.2.3. Since every $\gamma \in \Gamma(N)$ reduces to the identity matrix modulo N, Proposition 4.2.1 here shows that $\overline{v\gamma} = \overline{v}$ and $E_k^{\overline{v}}$ is weight-k invariant with respect to $\Gamma(N)$, i.e., it satisfies condition (2) in the definition. Finally, the Fourier expansion calculation to follow will produce coefficients meeting condition (3') of Proposition 1.2.4, showing that $E_k^{\overline{v}}$ satisfies condition (3) in the definition as well. $\square$

With the corollary established, one can symmetrize the Eisenstein series $E_k^{\overline{v}}$ to create modular forms for any congruence subgroup of level N (Exercise 4.2.2).

It is straightforward (Exercise 4.2.3) to compute that

$$\lim_{\mathrm{Im}(\tau) \to \infty} E_k^{\overline{v}}(\tau) = \begin{cases} (\pm 1)^k & \text{if } \overline{v} = \pm \overline{(0,1)}, \text{ unless } k \text{ is odd and } N \in \{1,2\}, \\ 0 & \text{otherwise.} \end{cases}$$

When k is odd and $N \in \{1,2\}$, the Eisenstein space $\mathcal{E}_k(\Gamma(N))$ has dimension 0 since $-I \in \Gamma(N)$. Excluding these cases for the rest of this paragraph, $E_k^{\overline{v}}$ is nonvanishing at ∞ if $\overline{v} = \pm \overline{(0,1)}$ and vanishes at ∞ otherwise. Next take any $\overline{v} = \overline{(c,d)} \in (\mathbb{Z}/N\mathbb{Z})^2$ of order N with corresponding $\delta = \begin{bmatrix} a & b \\ c & d \end{bmatrix} \in \mathrm{SL}_2(\mathbb{Z})$, and take any cusp $s = a'/c' \in \mathbb{Q} \cup \{\infty\}$, so that a matrix $\alpha = \begin{bmatrix} a' & b' \\ c' & d' \end{bmatrix} \in \mathrm{SL}_2(\mathbb{Z})$ takes ∞ to s. The behavior of $E_k^{\overline{v}}$ at s is described by the Fourier series of $E_k^{\overline{v}}[\alpha]_k$. By Proposition (4.2.1)

$$E_k^{\overline{v}}[\alpha]_k = E_k^{\overline{v\alpha}} = E_k^{\overline{(0,1)\delta\alpha}}.$$

Thus $E_k^{\overline{v}}[\alpha]_k$ is nonvanishing at ∞ only when $\overline{(0,1)\delta\alpha} = \pm\overline{(0,1)}$, or $\overline{(0,1)\delta} = \pm\overline{(0,1)}\alpha^{-1}$, or $\begin{bmatrix} a' \\ c' \end{bmatrix} \equiv \pm \begin{bmatrix} -d \\ c \end{bmatrix} \pmod{N}$. By Proposition 3.8.3, this is equivalent to $\Gamma(N)s = \Gamma(N)(-d/c)$, showing that $E_k^{\overline{v}}$ is nonvanishing at $\Gamma(N)(-d/c)$ and vanishes at the other cusps of $\Gamma(N)$.

If k is even or $N > 2$, pick a set of vectors $\{\overline{v}\} = \{\overline{(c,d)}\}$ such that the quotients $-d/c$ represent the cusps of $\Gamma(N)$, cf. Section 3.8. The preceding paragraph shows that the corresponding set $\{E_k^{\overline{v}}\}$ of Eisenstein series is linearly independent. The set contains $\varepsilon_\infty(\Gamma(N))$ elements, so it represents a basis of $\mathcal{E}_k(\Gamma(N))$ according to formula (4.3) since all cusps of $\Gamma(N)$ are regular. Since $\mathcal{E}_k(\Gamma(N)) = \{0\}$ when k is odd and $N \in \{1,2\}$, we have a basis of $\mathcal{E}_k(\Gamma(N))$ in all cases when $k \geq 3$. The basis elements are cosets $E_k^{\overline{v}} + \mathcal{S}_k(\Gamma(N))$ since $\mathcal{E}_k(\Gamma(N))$ is the quotient space $\mathcal{M}_k(\Gamma(N))/\mathcal{S}_k(\Gamma(N))$. Once $\mathcal{E}_k(\Gamma(N))$ is redefined as a subspace of $\mathcal{M}_k(\Gamma(N))$ in the next chapter, the basis elements will be the Eisenstein series themselves.

The series $E_k^{\overline{v}}$ are normalized, but in Chapter 1 it was the nonnormalized series G_k whose Fourier series was readily calculable. Analogously, define for any point $\overline{v} \in (\mathbb{Z}/N\mathbb{Z})^2$ of order N

$$G_k^{\overline{v}}(\tau) = \sideset{}{'}\sum_{(c,d)\equiv v\,(N)} \frac{1}{(c\tau + d)^k}.$$

Any (c,d) in this sum satisfies $\gcd(\gcd(c,d),N) = 1$. Thus

$$
\begin{aligned}
G_k^{\overline{v}}(\tau) &= \sum_{\substack{n=1 \\ \gcd(n,N)=1}}^{\infty} \sum_{\substack{(c,d)\equiv v\,(N) \\ \gcd(c,d)=n}} \frac{1}{(c\tau + d)^k} \\
&= \sum_{\substack{n=1 \\ \gcd(n,N)=1}}^{\infty} \frac{1}{n^k} \sum_{\substack{(c',d')\equiv n^{-1}v\,(N) \\ \gcd(c',d')=1}} \frac{1}{(c'\tau + d')^k} \qquad (4.5) \\
&= \frac{1}{\epsilon_N} \sum_{n\in(\mathbb{Z}/N\mathbb{Z})^*} \zeta_+^n(k) E_k^{n^{-1}\overline{v}}(\tau)
\end{aligned}
$$

where the modified Riemann zeta function (essentially the *Hurwitz zeta function* to be defined in Section 4.7) is

$$\zeta_+^n(k) = \sum_{\substack{m=1 \\ m\equiv n\,(N)}}^{\infty} \frac{1}{m^k}, \quad n \in (\mathbb{Z}/N\mathbb{Z})^*.$$

(Note that $(\mathbb{Z}/1\mathbb{Z})^* = \{\overline{0}\}$ since $0 \equiv 1 \pmod 1$, i.e., $\overline{0}$ is multiplicatively invertible for once. This is consistent with the definition $\phi(1) = 1$ for the Euler totient function. So ζ_+^n is the usual zeta function when $N = 1$.) Also, letting μ denote the Möbius function from elementary number theory, Möbius inversion shows that (Exercise 4.2.4)

$$E_k^{\overline{v}}(\tau) = \epsilon_N \sum_{n\in(\mathbb{Z}/N\mathbb{Z})^*} \zeta_+^n(k,\mu) G_k^{n^{-1}\overline{v}}(\tau) \qquad (4.6)$$

where as before ϵ_N is $1/2$ if $N \in \{1,2\}$ and 1 if $N > 2$, and where the coefficient in the sum is another modified zeta function,

$$\zeta_+^n(k,\mu) = \sum_{\substack{m=1 \\ m \equiv n \, (N)}}^{\infty} \frac{\mu(m)}{m^k}, \quad n \in (\mathbb{Z}/N\mathbb{Z})^*.$$

The set $\{G_k^{\bar{v}}\}$ for the same vectors $\bar{v}$ as before also represents a basis of $\mathcal{E}_k(\Gamma(N))$. We will reformulate the modified zeta functions in Section 4.4.

To compute the Fourier expansion of $G_k^{\bar{v}}$ when $N > 1$, start from

$$G_k^{\bar{v}}(\tau) = \sum_{c \equiv c_v} \sum_{d \in \mathbb{Z}} \frac{1}{(c\tau + d_v + Nd)^k} = \frac{1}{N^k} \sum_{c \equiv c_v} \sum_{d \in \mathbb{Z}} \frac{1}{\left(\frac{c\tau+d_v}{N} + d\right)^k}, \quad (4.7)$$

where all congruence conditions in this calculation are modulo N and all sums are taken over positive and negative and possibly zero values unless otherwise specified. Examining the first sum in (4.7) shows that the constant term of $G_k^{\bar{v}}(\tau)$ is (Exercise 4.2.5(a))

$$\begin{Bmatrix} 1 & \text{if } \bar{c}_v = \bar{0} \\ 0 & \text{otherwise} \end{Bmatrix} \cdot \sum_{d \equiv d_v}' \frac{1}{d^k}. \quad (4.8)$$

Next recall equation (1.2),

$$\sum_{d \in \mathbb{Z}} \frac{1}{(\tau + d)^k} = C_k \sum_{m=1}^{\infty} m^{k-1} q^m \quad \text{for } \tau \in \mathcal{H} \text{ and } k \geq 2,$$

where $C_k = (-2\pi i)^k/(k-1)!$ (this symbol will be used repeatedly for the rest of the chapter) and $q = e^{2\pi i \tau}$. If $c > 0$ in the second sum in (4.7) then $(c\tau + d_v)/N \in \mathcal{H}$ and so letting this be the τ in (1.2), part of (4.7) is (Exercise 4.2.5(b))

$$\frac{C_k}{N^k} \sum_{\substack{c \equiv c_v \\ c>0}} \sum_{m=1}^{\infty} m^{k-1} \mu_N^{d_v m} q_N^{cm} = \frac{C_k}{N^k} \sum_{n=1}^{\infty} \sum_{\substack{m|n \\ n/m \equiv c_v \\ m>0}} m^{k-1} \mu_N^{d_v m} q_N^n, \quad (4.9)$$

where $\mu_N = e^{2\pi i/N}$ and $q_N = e^{2\pi i \tau/N}$. Note that the inner sum is zero unless $n \equiv c_v$. Similarly, if $c < 0$ in (4.7) then $-(c\tau + d_v)/N \in \mathcal{H}$ and so the rest of (4.7) is (Exercise 4.2.5(c))

$$(-1)^k \frac{C_k}{N^k} \sum_{c \equiv c_v} \sum_{m=1}^{\infty} m^{k-1} \mu_N^{-d_v m} q_N^{-cm} = \frac{C_k}{N^k} \sum_{n=1}^{\infty} \sum_{\substack{m|n \\ n/m \equiv c_v \\ m<0}} -m^{k-1} \mu_N^{d_v m} q_N^n.$$

$$(4.10)$$

Assembling the results of this calculation gives

Theorem 4.2.3. *The Fourier expansion of $G_k^{\bar{v}}(\tau)$ for $k \geq 3$ and $\bar{v} \in (\mathbb{Z}/N\mathbb{Z})^2$ a point of order N is*

$$G_k^{\bar{v}}(\tau) = \delta(\bar{c}_v)\zeta^{\bar{d}_v}(k) + \frac{C_k}{N^k} \sum_{n=1}^{\infty} \sigma_{k-1}^{\bar{v}}(n)q_N^n, \quad q_N = e^{2\pi i \tau/N},$$

where

$$\delta(\bar{c}_v) = \begin{cases} 1 & \text{if } \bar{c}_v = \bar{0}, \\ 0 & \text{otherwise} \end{cases}, \quad \zeta^{\bar{d}_v}(k) = \sum_{d \equiv d_v \, (N)}' \frac{1}{d^k},$$

and

$$C_k = \frac{(-2\pi i)^k}{(k-1)!}, \quad \sigma_{k-1}^{\bar{v}}(n) = \sum_{\substack{m|n \\ n/m \equiv c_v \, (N)}} \text{sgn}(m)m^{k-1}\mu_N^{d_v m}.$$

The sums for $\zeta^{\bar{d}_v}$ and $\sigma_{k-1}^{\bar{v}}$ are taken over positive and negative values of d and m.

Any set $\{G_k^{\bar{v}}(\tau)\}$ with one $\bar{v}$ corresponding to each cusp of $\Gamma(N)$ represents a basis of $\mathcal{E}_k(\Gamma(N))$.

When $N = 1$, the Fourier expansion agrees with $G_k(\tau)$ from Chapter 1 for k even and it cancels to zero when k is odd (Exercise 4.2.6). Clearly the nth Fourier coefficient is bounded by Cn^k, completing the proof that $E_k^{\bar{v}}$ is a modular form.

Exercises

4.2.1. Show that the two sums in (4.4) agree.

4.2.2. Let N, $\bar{v}$, and k be as in the section. Let Γ be a congruence subgroup of $\text{SL}_2(\mathbb{Z})$ of level N, so that $\Gamma(N) \subset \Gamma \subset \text{SL}_2(\mathbb{Z})$. Define

$$E_{k,\Gamma}^{\bar{v}} = \sum_{\gamma_j \in \Gamma(N) \backslash \Gamma} E_k^{\bar{v}}[\gamma_j]_k,$$

the sum taken over a set of coset representatives. Show that $E_{k,\Gamma}^{\bar{v}}$ is well defined and lies in $\mathcal{M}_k(\Gamma)$.

4.2.3. Establish the formula in the section for $\lim_{\text{Im}(\tau) \to \infty} E_k^{\bar{v}}(\tau)$.

4.2.4. (a) The Möbius function $\mu : \mathbb{Z}^+ \longrightarrow \mathbb{Z}$ is defined by

$$\mu(n) = \begin{cases} (-1)^g & \text{if } n = p_1 \cdots p_g \text{ for distinct primes } p_1, \ldots, p_g, \\ 0 & \text{if } p^2 \mid n \text{ for some prime } p. \end{cases}$$

This includes the case $\mu(1) = 1$. Show that for $n \in \mathbb{Z}^+$,

$$\sum_{\substack{d|n \\ d>0}} \mu(d) = \begin{cases} 1 & \text{if } n = 1, \\ 0 & \text{if } n > 1. \end{cases}$$

(b) Let $\bar{v} \in (\mathbb{Z}/N\mathbb{Z})^2$ be a point of order N. It follows from part (a) that

$$E_k^{\bar{v}}(\tau) = \epsilon_N \sum_{\substack{(c,d)\equiv v \, (N)}}' \sum_{\substack{n|\gcd(c,d) \\ n>0}} \frac{\mu(n)}{(c\tau + d)^k}.$$

Rearrange the sum to obtain formula (4.6). (A hint for this exercise is at the end of the book.)

4.2.5. (a) Confirm that the constant term of (4.7) is given by (4.8).
(b) Show that the terms of (4.7) with $c > 0$ sum to (4.9).
(c) Show that the terms of (4.7) with $c < 0$ sum to (4.10).

4.2.6. Confirm that when $N = 1$ in Theorem 4.2.3, the series becomes the Fourier expansion of $G_k(\tau)$ from Chapter 1 when k is even and cancels to zero when k is odd.

4.3 Dirichlet characters, Gauss sums, and eigenspaces

Computing the Eisenstein series for the groups $\Gamma_1(N)$ and $\Gamma_0(N)$ requires some machinery, to be given over the next two sections. For any positive integer N, let G_N denote the multiplicative group $(\mathbb{Z}/N\mathbb{Z})^*$, of order $\phi(N)$ where ϕ is the Euler totient, when this doesn't conflict with the notation for Eisenstein series. A *Dirichlet character modulo N* is a homomorphism of multiplicative groups,

$$\chi : G_N \longrightarrow \mathbb{C}^*.$$

For any two Dirichlet characters χ and ψ modulo N, the product character defined by the rule $(\chi\psi)(n) = \chi(n)\psi(n)$ for $n \in G_N$ is again a Dirichlet character modulo N. In fact, the set of Dirichlet characters modulo N is again a multiplicative group, called the *dual group* of G_N, denoted $\widehat{G}_N$, whose identity element is the *trivial character modulo N* mapping every element of G_N to 1, denoted $\mathbf{1}_N$ or just $\mathbf{1}$ when N is clear. Since G_N is a finite group the values taken by any Dirichlet character are complex roots of unity, and so the inverse of a Dirichlet character is its complex conjugate, defined by the rule $\bar{\chi}(n) = \overline{\chi(n)}$ for all $n \in G_N$. (So here the overline denotes complex conjugation, not reduction modulo N.) As explained in the previous section, $G_1 = \{\bar{0}\}$, and so the only Dirichlet character modulo 1 is the trivial character $\mathbf{1}_1$.

For any prime p the group G_p is cyclic of order $p - 1$. Let g be a generator and let μ_{p-1} be a primitive $(p - 1)$st complex root of unity. Then the group of Dirichlet characters modulo p is again cyclic of order $p - 1$, generated by the character taking g to μ_{p-1}. In general (see, for example, [Ser73]),

Proposition 4.3.1. *Let $\widehat{G}_N$ be the dual group of G_N. Then $\widehat{G}_N$ is isomorphic to G_N. In particular, the number of Dirichlet characters modulo N is $\phi(N)$.*

The two groups are noncanonically isomorphic, meaning that constructing an actual isomorphism from G_N to $\widehat{G}_N$ involves arbitrary choices of which elements map to which characters. The groups G_N and $\widehat{G}_N$ satisfy the *orthogonality relations* (Exercise 4.3.1),

$$\sum_{n \in G_N} \chi(n) = \begin{cases} \phi(N) & \text{if } \chi = 1, \\ 0 & \text{if } \chi \neq 1, \end{cases} \qquad \sum_{\chi \in \widehat{G}_N} \chi(n) = \begin{cases} \phi(N) & \text{if } n = 1, \\ 0 & \text{if } n \neq 1. \end{cases}$$

Let N be a positive integer and let d be a positive divisor of N. Every Dirichlet character χ modulo d lifts to a Dirichlet character χ_N modulo N, defined by the rule $\chi_N(n \ (\text{mod } N)) = \chi(n \ (\text{mod } d))$ for all $n \in \mathbb{Z}$ relatively prime to N. That is, $\chi_N = \chi \circ \pi_{N,d}$ where $\pi_{N,d} : G_N \longrightarrow G_d$ is natural projection. For example, the Dirichlet character modulo 4 taking 1 to 1 and 3 to -1 lifts to the Dirichlet character modulo 12 taking 1 and 5 to 1 and 7 and 11 to -1. Going in the other direction, from modulus N to modulus d, isn't always possible. Every Dirichlet character χ modulo N has a *conductor*, the smallest positive divisor d of N such that $\chi = \chi_d \circ \pi_{N,d}$ for some character χ_d modulo d, or, equivalently, such that χ is trivial on the normal subgroup

$$K_{N,d} = \ker(\pi_{N,d}) = \{n \in G_N : n \equiv 1 \ (\text{mod } d)\}.$$

For example, the Dirichlet character modulo 12 taking 1 and 7 to 1 and 5 and 11 to -1 has conductor 3. A Dirichlet character modulo N is *primitive* if its conductor is N. The only character modulo N with conductor 1 is the trivial character $\mathbf{1}_N$, and the trivial character $\mathbf{1}_N$ modulo N is primitive only for $N = 1$.

Every Dirichlet character χ modulo N extends to a function $\chi : \mathbb{Z}/N\mathbb{Z} \longrightarrow \mathbb{C}$ where $\chi(n) = 0$ for noninvertible elements n of the ring $\mathbb{Z}/N\mathbb{Z}$, and then extends further to a function $\chi : \mathbb{Z} \longrightarrow \mathbb{C}$ where (abusing notation) $\chi(n) = \chi(n \ (\text{mod } N))$ for all $n \in \mathbb{Z}$. Thus $\chi(n) = 0$ for all n such that $\gcd(n, N) > 1$. The extended function is no longer a homomorphism, but it still satisfies $\chi(nm) = \chi(n)\chi(m)$ for all n, m. In particular, the trivial character modulo N extends to the function

$$\mathbf{1}_N(n) = \begin{cases} 1 & \text{if } \gcd(n, N) = 1, \\ 0 & \text{if } \gcd(n, N) > 1. \end{cases}$$

Thus the extended trivial character is no longer identically 1 unless $N = 1$. The extension of any Dirichlet character χ modulo N satisfies

$$\chi(0) = \begin{cases} 1 & \text{if } N = 1, \\ 0 & \text{if } N > 1. \end{cases}$$

Summing over $n = 0$ to $N - 1$ in the first orthogonality relation and taking $n \in \mathbb{Z}$ in the second gives modified versions,

$$\sum_{n=0}^{N-1} \chi(n) = \begin{cases} \phi(N) & \text{if } \chi = 1, \\ 0 & \text{if } \chi \neq 1, \end{cases} \qquad \sum_{\chi \in \widehat{G}_N} \chi(n) = \begin{cases} \phi(N) & \text{if } n \equiv 1 \ (N), \\ 0 & \text{if } n \not\equiv 1 \ (N). \end{cases}$$

The *Gauss sum* of a Dirichlet character χ modulo N is the complex number

$$g(\chi) = \sum_{n=0}^{N-1} \chi(n) \mu_N^n, \quad \mu_N = e^{2\pi i / N}.$$

If χ is primitive modulo N then (Exercise 4.3.2) for any integer m,

$$\sum_{n=0}^{N-1} \chi(n) \mu_N^{nm} = \bar{\chi}(m) g(\chi). \tag{4.11}$$

It follows that the Gauss sum of a primitive character is nonzero. Indeed, the square of its absolute value is

$$g(\chi)\overline{g(\chi)} = \sum_{m=0}^{N-1} \bar{\chi}(m) g(\chi) \mu_N^{-m} = \sum_{m=0}^{N-1} \sum_{n=0}^{N-1} \chi(n) \mu_N^{nm} \mu_N^{-m} \quad \text{by (4.11)} \quad (4.12)$$

$$= \sum_{n=0}^{N-1} \chi(n) \sum_{m=0}^{N-1} \mu_N^{(n-1)m} = N, \tag{4.13}$$

the last equality holding because the inner sum is N when $n = 1$ and 0 otherwise. Formula (4.11) and the following lemma will be used in the next section.

Lemma 4.3.2. *Let N be a positive integer. If $N = 1$ or $N = 2$ then every Dirichlet character χ modulo N satisfies $\chi(-1) = 1$. If $N > 2$ then the number of Dirichlet characters modulo N is even, half of them satisfying $\chi(-1) = 1$ and the other half satisfying $\chi(-1) = -1$.*

Proof. The result for $N = 1$ and $N = 2$ is clear. If $N > 2$ then $4 \mid N$ or $p \mid N$ for some odd prime p. The nontrivial character modulo 4 takes $-1 \pmod 4$ to -1, and for every odd prime p the character modulo p taking a generator g of G_p to a primitive $(p-1)$st complex root of unity takes $-1 \pmod p$ to -1 since $-1 \equiv g^{(p-1)/2} \pmod p$. In either case the character lifts to a character modulo N taking $-1 \pmod N$ to -1.

Let $\widehat{G}_N$ denote the group of Dirichlet characters modulo N. The map $\widehat{G}_N \longrightarrow \{\pm 1\}$ taking each character χ to $\chi(-1)$ is a homomorphism. We have just seen that the homomorphism surjects if $N > 2$, and so the result follows from the First Isomorphism Theorem of group theory. $\qquad \square$

We are interested in Dirichlet characters because they decompose the vector space $\mathcal{M}_k(\Gamma_1(N))$ into a direct sum of subspaces that we can analyze independently. For each Dirichlet character χ modulo N define the χ-*eigenspace* of $\mathcal{M}_k(\Gamma_1(N))$,

$$\mathcal{M}_k(N,\chi) = \{f \in \mathcal{M}_k(\Gamma_1(N)) : f[\gamma]_k = \chi(d_\gamma)f \text{ for all } \gamma \in \Gamma_0(N)\}.$$

(Here d_γ denotes the lower right entry of γ.) In particular the eigenspace $\mathcal{M}_k(N,\mathbf{1})$ is $\mathcal{M}_k(\Gamma_0(N))$ (Exercise 4.3.3(a)). Also note that $\mathcal{M}_k(N,\chi)$ is just $\{0\}$ unless $\chi(-1) = (-1)^k$ (Exercise 4.3.3(b)). The vector space $\mathcal{M}_k(\Gamma_1(N))$ decomposes as the direct sum of the eigenspaces (Exercise 4.3.4(a)),

$$\mathcal{M}_k(\Gamma_1(N)) = \bigoplus_\chi \mathcal{M}_k(N,\chi),$$

and the same result holds for cusp forms (Exercise 4.3.4(b)), so it holds for the quotients as well (Exercise 4.3.4(c)),

$$\mathcal{E}_k(\Gamma_1(N)) = \bigoplus_\chi \mathcal{E}_k(N,\chi).$$

Exercises

4.3.1. Prove the orthogonality relations. (A hint for this exercise is at the end of the book.)

4.3.2. Let χ be a primitive Dirichlet character modulo N and let m be an integer. Prove formula (4.11) as follows.

(a) First assume that $\gcd(m, N) = 1$. Use the fact that $\bar{\chi}(m)\chi(m) = 1$ to prove the formula.

(b) Now assume that $\gcd(m, N) > 1$. Let $g = \gcd(m, N)$, so that $m = m'g$ for some integer m' and $N = N'g$ for some positive integer N'. Show that

$$\sum_{n=0}^{N-1} \chi(n)\mu_N^{nm} = \sum_{n'=0}^{N'-1} \left(\sum_{\substack{n=0 \\ n \equiv n'\,(N')}}^{N-1} \chi(n) \right) \mu_{N'}^{n'm'}. \tag{4.14}$$

Let

$$K = K_{N,N'} = \{n \in (\mathbb{Z}/N\mathbb{Z})^* : n \equiv 1 \pmod{N'}\},$$

the subgroup of $(\mathbb{Z}/N\mathbb{Z})^*$ defined in the section. Use the fact that χ is primitive to show that

$$\sum_{n \in K} \chi(n) = 0.$$

Since $(\mathbb{Z}/N\mathbb{Z})^* = \bigcup_{n'} n'K$ where the coset representatives n' taken modulo N' run through $(\mathbb{Z}/N'\mathbb{Z})^*$, show that the inner sum in (4.14) is $\sum_{n \in n'K} \chi(n)$, and that this is 0. Explain why formula (4.11) follows.

4.3.3. (a) Show that $\mathcal{M}_k(N, \mathbf{1}) = \mathcal{M}_k(\Gamma_0(N))$.
  (b) Show that $\mathcal{M}_k(N, \chi) = \{0\}$ unless $\chi(-1) = (-1)^k$.

4.3.4. (a) Show that $\mathcal{M}_k(\Gamma_1(N)) = \bigoplus_\chi \mathcal{M}_k(N, \chi)$. (A hint for this exercise is at the end of the book.)
  (b) Show the same result for the cusp form spaces $\mathcal{S}_k(N, \chi)$.
  (c) Show the same result for the Eisenstein spaces $\mathcal{E}_k(N, \chi)$.

4.4 Gamma, zeta, and L-functions

This section gives a brief discussion of three functions from complex analysis. Some results are stated without proof since the arguments are written in many places, see for example [Kob93]. For any complex number s with positive real part, the *gamma function* of s is defined as an integral,

$$\Gamma(s) = \int_{t=0}^\infty e^{-t} t^s \frac{dt}{t}, \quad s \in \mathbb{C}, \ \mathrm{Re}(s) > 0.$$

Since $|t^s| = t^{\mathrm{Re}(s)}$ for $t > 0$, the condition on s ensures that the integral converges at $t = 0$, and since the integrand decays exponentially as $t \to \infty$, the integral converges at the other end as well. It is easy to show that $\Gamma(1) = 1$, that $\Gamma(1/2) = \sqrt{\pi}$, and that Γ satisfies the *functional equation*

$$\Gamma(s + 1) = s\Gamma(s)$$

(Exercise 4.4.1(a–c)). Consequently $\Gamma(n) = (n - 1)!$ for $n \in \mathbb{Z}^+$. (This tidies the notation for the ubiquitous constant C_k of this chapter, which is now $(-2\pi i)^k / \Gamma(k)$.) Rewriting the functional equation as $\Gamma(s) = \Gamma(s+1)/s$ defines the left side as a meromorphic function of s for s-values with $\mathrm{Re}(s) > -1$ and thus extends the domain of Γ one unit leftward. Repeatedly applying the functional equation this way extends the domain of Γ to all of $\mathbb{C}$, and now the functional equation holds for all s. For any $n \in \mathbb{N}$, it gives

$$\lim_{s \to -n} (s + n)\Gamma(s) = \lim_{s \to -n} \frac{\Gamma(s + n + 1)}{s(s + 1)\cdots(s + n - 1)} = \frac{(-1)^n}{n!},$$

showing that the extended Γ function has a simple pole at each nonpositive integer with residue $\mathrm{Res}_{s=-n}\Gamma(s) = (-1)^n/n!$. We will need the formula (Exercise 4.4.1(d))

$$\Gamma(s)\Gamma(1 - s) = \frac{\pi}{\sin(\pi s)} \tag{4.15}$$

in Section 4.7. This shows that the extended Gamma function has no zeros, making its reciprocal $1/\Gamma(s)$ entire. We will need the formula (Exercise 4.4.2)

$$\frac{\pi^{-\frac{1-k}{2}} \Gamma\left(\frac{1-k}{2}\right)}{\pi^{-\frac{k}{2}} \Gamma\left(\frac{k}{2}\right)} = \frac{C_k}{2} \quad \text{for even integers } k \geq 2 \tag{4.16}$$

in the next section.

The *Riemann zeta function* of the complex variable s is

$$\zeta(s) = \sum_{n=1}^{\infty} \frac{1}{n^s}, \quad \text{Re}(s) > 1.$$

The sum converges absolutely. An elegant restatement due to Euler of the Fundamental Theorem of Arithmetic (positive integers factor uniquely into products of primes) is

$$\zeta(s) = \prod_{p \in \mathcal{P}} (1 - p^{-s})^{-1}, \quad \text{Re}(s) > 1,$$

where $\mathcal{P}$ is the set of primes. Define

$$\xi(s) = \pi^{-s/2} \Gamma(s/2) \zeta(s), \quad \text{Re}(s) > 1.$$

Section 4.9 will prove that the function ξ has a meromorphic continuation to the entire s-plane satisfying the functional equation

$$\xi(s) = \xi(1 - s) \quad \text{for all } s \in \mathbb{C}$$

and having simple poles at $s = 0$ and $s = 1$ with respective residues -1 and 1. By the properties of the gamma function, this shows that the Riemann zeta function has a meromorphic continuation to the entire s-plane, with one simple pole at $s = 1$ having residue 1 and with simple zeros at $s = -2, -4, -6, \ldots$ It also shows that $\zeta(0) = -1/2$ and gives the values of $\zeta(-1)$, $\zeta(-3)$, $\zeta(-5)$, $\ldots$ in terms of $\zeta(2)$, $\zeta(4)$, $\zeta(6)$, $\ldots$, which were computed in Exercise 1.1.7(b). See Exercise 4.4.3 for an appealing heuristic argument due to Euler in support of these values $\zeta(1 - k)$ for even $k \geq 2$.

Every Dirichlet character χ modulo N has an associated *Dirichlet L-function* similar to the Riemann zeta function,

$$L(s, \chi) = \sum_{n=1}^{\infty} \frac{\chi(n)}{n^s} = \prod_{p \in \mathcal{P}} (1 - \chi(p) p^{-s})^{-1}, \quad \text{Re}(s) > 1. \tag{4.17}$$

Again this extends meromorphically to the s-plane and the extension is entire unless $\chi = \mathbf{1}_N$, in which case the L-function is essentially the Riemann zeta function (Exercise 4.4.4) and again has a simple pole at $s = 1$. When $\chi(-1) = 1$, the functional equation satisfied by $L(s, \chi)$ is

$$\pi^{-s/2} \Gamma(\tfrac{s}{2}) N^s L(s, \chi) = \pi^{-(1-s)/2} \Gamma(\tfrac{1-s}{2}) g(\chi) L(1 - s, \bar{\chi}), \tag{4.18}$$

and when $\chi(-1) = -1$ it is

$$\pi^{-(s+1)/2} \Gamma(\tfrac{s+1}{2}) N^s L(s, \chi) = -i \pi^{-(2-s)/2} \Gamma(\tfrac{2-s}{2}) g(\chi) L(1 - s, \bar{\chi}). \tag{4.19}$$

We will not prove these, but see the comment at the end of Section 4.7.

The modified zeta functions from Section 4.2,

$$\zeta_+^n(k) = \sum_{\substack{m=1 \\ m\equiv n\,(N)}}^{\infty} \frac{1}{m^k}, \quad \zeta_+^n(k,\mu) = \sum_{\substack{m=1 \\ m\equiv n\,(N)}}^{\infty} \frac{\mu(m)}{m^k}, \quad n \in (\mathbb{Z}/N\mathbb{Z})^*,$$

are expressible in terms of L-functions. For any $n \in (\mathbb{Z}/N\mathbb{Z})^*$ the second orthogonality relation shows that summing over Dirichlet characters picks out the desired terms of the full zeta function,

$$\frac{1}{\phi(N)} \sum_{\chi \in \widehat{G}_N} \chi(n^{-1}) L(s,\chi) = \sum_{m=1}^{\infty} \frac{1}{\phi(N)} \sum_{\chi \in \widehat{G}_N} \chi(mn^{-1} \,(\mathrm{mod}\,N)) \frac{1}{m^s}$$

$$= \sum_{\substack{m=1 \\ m\equiv n\,(N)}}^{\infty} \frac{1}{m^s} = \zeta_+^n(s), \quad \mathrm{Re}(s) > 1.$$

Since the sum of L-functions extends meromorphically to the full s-plane with a simple pole at $s = 1$, so does $\zeta_+^n(s)$. Substituting the sum for $\zeta_+^n(s)$ in relation (4.5) gives

$$G_k^{\overline{v}}(\tau) = \frac{1}{\epsilon_N \phi(N)} \sum_{\chi,n} \chi(n) L(k,\chi) E_k^{n\overline{v}}(\tau), \qquad (4.20)$$

with the sum taken over $\chi \in (\widehat{\mathbb{Z}/N\mathbb{Z}})^*$ and $n \in (\mathbb{Z}/N\mathbb{Z})^*$. To invert this relation, take an arbitrary linear combination of series $G_k^{m^{-1}\overline{v}}$ over $m \in (\mathbb{Z}/N\mathbb{Z})^*$,

$$\sum_m a_m G_k^{m^{-1}\overline{v}} = \frac{1}{\epsilon_N \phi(N)} \sum_{m,\chi,n} a_m \chi(n) L(k,\chi) E_k^{nm^{-1}\overline{v}}$$

$$= \frac{1}{\epsilon_N \phi(N)} \sum_{m,\chi,n} a_m \chi(mn) L(k,\chi) E_k^{n\overline{v}}$$

$$= \frac{1}{\epsilon_N \phi(N)} \sum_{m,\chi,n} a_m \chi(m) \chi(n) L(k,\chi) E_k^{n\overline{v}}.$$

The left side is the inner product (without complex conjugation) $\langle a, G \rangle$ where a is the vector with entries a_m and G is the vector with entries $G_k^{m^{-1}\overline{v}}$. The right side is the inner product $1/\epsilon_N \langle aAB_k, E \rangle$ where A and B_s ($s \in \mathbb{C}$) are the matrices

$$A = \left[\frac{\chi(m)}{(\phi(N))^{1/2}}\right]_{(m,\chi)\in G_N \times \widehat{G}_N} \qquad B_s = \left[\frac{\chi(n)L(s,\chi)}{(\phi(N))^{1/2}}\right]_{(\chi,n)\in \widehat{G}_N \times G_N}$$

and E is the vector with entries $E_k^{n\overline{v}}$. By the second orthogonality relation, $A^*A = I$ where A^* is the adjoint (transpose-conjugate), and so A is invertible.

By the first orthogonality relation, $B_s B_s^*$ (where the adjoint operator conjugates complex scalars but not the variable s) is the diagonal matrix with diagonal entries $L(s, \chi) L(s, \bar{\chi})$, and so B_s is invertible as a matrix of meromorphic functions on $\mathbb{C}$. The product formula (4.17) shows that $L(s, \chi) \neq 0$ when $\mathrm{Re}(s) > 1$, and so the meromorphic functions in the inverse matrix have no poles when $\mathrm{Re}(s) > 1$, in particular when s is an integer $k \geq 3$. Letting e_1 denote the first standard basis vector and choosing $a(k) = e_1 (AB_k)^{-1}$ gives

$$E_k^{\bar{v}} = \epsilon_N \sum_m a_m(k) G_k^{m^{-1}\bar{v}}. \tag{4.21}$$

Comparing this to (4.6) shows that $\zeta_+^n(k, \mu) = a_n(k)$ for integers $k \geq 3$, and in fact the argument establishes the same relation replacing k by any complex s with $\mathrm{Re}(s) > 2$. Thus $\zeta_+^n(s, \mu)$ is also a meromorphic function that continues to the full s-plane.

The meromorphic continuation of $\zeta_+^n(s)$ quickly gives an analytic continuation of the function $\zeta^n(s) = \sum'_{m \equiv n \,(N)} 1/m^s$ occurring in the constant term of $G_k^{\bar{v}}(\tau)$, cf. Theorem 4.2.3 (Exercise 4.4.5(c)). We will need the formula (Exercise 4.4.5(d))

$$\zeta^n(1) = \frac{\pi i}{N} + \frac{\pi}{N} \cot\left(\frac{\pi n}{N}\right), \quad \gcd(n, N) = 1. \tag{4.22}$$

in Section 4.8.

Exercises

4.4.1. (a) Show that $\Gamma(1) = 1$.

(b) Show that $\Gamma(1/2) = \sqrt{\pi}$. (Hints for this exercise are at the end of the book.)

(c) Show that $\Gamma(s + 1) = s\Gamma(s)$ when $\mathrm{Re}(s) > 0$.

(d) For any positive integer n consider the integral

$$I_n(s) = \int_{t=0}^n \left(1 - \frac{t}{n}\right)^n t^s \frac{dt}{t}, \quad s \in \mathbb{C}, \ \mathrm{Re}(s) > 0.$$

Thus $\lim_{n \to \infty} I_n(s) = \Gamma(s)$. Change variables to get

$$I_n(s) = n^s \int_{t=0}^1 (1 - t)^n t^{s-1} dt,$$

integrate by parts to show that $I_n(s) = (1/s)(n/(n - 1))^{s+1} I_{n-1}(s + 1)$, and evaluate $I_1(s + n - 1)$ to conclude that

$$I_n(s) = \frac{n^s}{s \prod_{m=1}^n (1 + s/m)}.$$

Therefore

$$I_n(s)I_n(-s) = \frac{1}{-s^2 \prod_{m=1}^n (1 - s^2/m^2)}.$$

Letting $n \to \infty$, use the formula $\sin(\pi s) = \pi s \prod_{m=1}^\infty (1 - s^2/m^2)$ and the identity $-s\Gamma(-s) = \Gamma(1-s)$ to prove (4.15). If you have the relevant background, make this heuristic argument rigorous by citing appropriate convergence theorems for integrals and infinite products.

4.4.2. Prove formula (4.16). (A hint for this exercise is at the end of the book.)

4.4.3. This exercise presents an argument due to Euler for the functional equation of $\zeta(s)$.

(a) Let t be a formal variable. Starting from the identity

$$t + t^2 + t^3 + t^4 + \cdots = (t - t^2 + t^3 - t^4 + \cdots) + 2(t^2 + t^4 + t^6 + t^8 + \cdots),$$

show that applying the operator $t\frac{d}{dt}$ (i.e., differentiation and then multiplication by t) n times gives

$$1^n t + 2^n t^2 + 3^n t^3 + 4^n t^4 + \cdots$$
$$= \left(t\frac{d}{dt}\right)^n \left(\frac{t}{1+t}\right) + 2^{n+1}(1^n t^2 + 2^n t^4 + 3^n t^6 + 4^n t^8 + \cdots).$$

Formally, when $t = 1$ this is $\zeta(-n) = \left(t\frac{d}{dt}\right)^n \left(\frac{t}{1+t}\right)_{t=1} + 2^{n+1}\zeta(-n)$, giving a heuristic value for $\zeta(-n)$. Thus for example, according to Euler, $1 + 1 + 1 + 1 + \cdots = -1/2$ and $1 + 2 + 3 + 4 + \cdots = -1/12$.

(b) Let $t = e^X$ and note that $t\frac{d}{dt} = \frac{d}{dX}$. Now we have

$$(1 - 2^{n+1})\zeta(-n) = \left[\frac{d^n}{dX^n}\left(\frac{e^X}{e^X + 1}\right)\right]_{X=0} \quad \text{for } n \in \mathbb{N},$$

giving the Taylor series

$$\frac{e^X}{e^X + 1} = \sum_{n=0}^\infty \frac{(1 - 2^{n+1})\zeta(-n)}{n!} X^n.$$

Thus the function of a complex variable

$$F(z) = \frac{e^{2\pi i z}}{e^{2\pi i z} + 1} = \sum_{n=0}^\infty \frac{(1 - 2^{n+1})\zeta(-n)(2\pi i)^n}{n!} z^n$$

generates (in the sense of generating function, cf. Section 1.2) the values $\zeta(-n)$ for all natural numbers n.

Let $G(z) = \pi \cot \pi z$. Recall from equations (1.1) that

$$G(z) = \pi i \frac{e^{2\pi i z}+1}{e^{2\pi i z}-1} = \frac{1}{z} - 2\sum_{k=1}^{\infty} \zeta(2k) z^{2k-1}.$$

The rational forms of F and G look similar, suggesting a relation between $\zeta(1-k)$ and $\zeta(k)$ for even $k \geq 2$. Use the rational forms of F and G to show that in fact

$$\frac{1}{\pi i}(G(z) - 2G(2z)) = -F(z) + F(-z),$$

and then equate coefficients to obtain

$$\zeta(1-k) = \frac{2\Gamma(k)}{(2\pi i)^k}\zeta(k) \quad \text{for even } k \geq 2.$$

The value of $\zeta(k)$ computed in Exercise 1.1.7 shows that $\zeta(1-k) = -B_k/k$ for even $k \geq 2$ where B_k is the Bernoulli number.

(c) Use formula (4.16) to show that

$$\pi^{-k/2}\Gamma(\tfrac{k}{2})\zeta(k) = \pi^{-(1-k)/2}\Gamma(\tfrac{1-k}{2})\zeta(1-k) \quad \text{for even } k \geq 2,$$

giving a partial version of the functional equation.

The ideas here can be turned into a rigorous proof of the meromorphic continuation and the functional equation of ζ, cf. [Hid93].

4.4.4. Show that $L(s, 1_N) = \zeta(s)\prod_{p|N}(1 - p^{-s})$.

4.4.5. Take the nonpositive imaginary axis as a branch cut in the z-plane, so that $\arg(z) \in (-\pi/2, 3\pi/2)$ on the remaining set and the function $\log(z) = \ln|z| + i\arg(z)$ is single-valued and analytic there. For any z off the branch cut define

$$z^s = e^{s\log(z)}, \quad s \in \mathbb{C}.$$

(a) Show that $(-r)^s = (-1)^s r^s$ for any positive real number r and any $s \in \mathbb{C}$, even though the rule $(zw)^s = z^s w^s$ does not hold in general.

(b) Show that for any $n \in G_N$,

$$\zeta^n(s) = \zeta_+^n(s) + (-1)^{-s}\zeta_+^{-n}(s), \quad \mathrm{Re}(s) > 1.$$

(c) It follows that

$$\zeta^n(s) = \frac{1}{\phi(N)}\sum_{\chi \in \widehat{G}_N} (\chi(n^{-1}) + (-1)^{-s}\chi((-n)^{-1}))L(s, \chi), \quad \mathrm{Re}(s) > 1.$$

This continues meromorphically to the full s-plane with the only possible pole coming from the term where $\chi = 1_N$. Show that

$$\lim_{s \to 1} \frac{1}{\phi(N)}(1_N(n^{-1}) + (-1)^{-s}1_N((-n)^{-1}))L(s, 1_N) = \frac{\pi i}{N}.$$

Thus the continuation of $\zeta^n(s)$ is entire. (Hints for this exercise are at the end of the book.)

(d) For $\chi \neq 1_N$, the analytic continuation of $L(s, \chi)$ is bounded at $s = 1$. Show that for $N > 1$,

$$\zeta^n(1) = \frac{\pi i}{N} + \frac{1}{\phi(N)} \sum_{\chi \neq 1_N} (\chi(n^{-1}) - \chi((-n)^{-1}))L(1, \chi)$$

$$= \frac{\pi i}{N} + \frac{\pi}{N} \cot\left(\frac{\pi n}{N}\right).$$

4.4.6. Substitute $u^2 = t$ in the definition of Γ to get

$$\Gamma(s) = 2 \int_{u=0}^{\infty} e^{-u^2} u^{2s} \frac{du}{u}, \quad s \in \mathbb{C}, \ \mathrm{Re}(s) > 0.$$

Show that therefore for $a, b > 0$,

$$\Gamma(a)\Gamma(b) = \Gamma(a + b) \int_{x=0}^{1} x^{a-1}(1 - x)^{b-1} dx.$$

This last integral is a *Beta integral*. Show also that for any positive integer m,

$$\int_{t=0}^{1} \frac{dt}{\sqrt{1 - t^m}} = \frac{1}{m} \int_{x=0}^{1} x^{\frac{1}{m}-1}(1 - x)^{-\frac{1}{2}} dx.$$

Use these results to explain the last equalities in the formulas for ϖ_4 and ϖ_3 at the end of Section 1.1. (A hint for this exercise is at the end of the book.)

4.5 Eisenstein series for the eigenspaces when $k \geq 3$

The Eisenstein series $G_k^{\bar{v}}$ from Section 4.2 are linear combinations of the series $E_k^{\bar{v}}$, so the transformation rule of Proposition 4.2.1 applies to them as well, $G_k^{\bar{v}}[\gamma]_k = G_k^{\bar{v}\gamma}$ for all $\gamma \in \mathrm{SL}_2(\mathbb{Z})$. Note that vectors modulo N of the form $v = \overline{(0, d)}$ satisfy

$$\overline{(0, d)}\gamma = \overline{(0, dd_\gamma)} \quad \text{for all } \gamma \in \Gamma_0(N),$$

where d_γ is the lower right entry of γ. Thus, symmetrizing over d gives a sum of Eisenstein series

$$\sum_{d \in (\mathbb{Z}/N\mathbb{Z})^*} G_k^{(0,d)}$$

lying in $\mathcal{M}_k(\Gamma_0(N))$ (Exercise 4.5.1(a)). Similarly, if χ is a Dirichlet character modulo N then the sum

$$\sum_{d \in (\mathbb{Z}/N\mathbb{Z})^*} \bar{\chi}(d) G_k^{(0,d)}$$

(here $\bar{\chi}$ is the complex conjugate of χ) lies in $\mathcal{M}_k(N, \chi)$ (Exercise 4.5.1(b)). This section generalizes these ideas to construct bases of the spaces $\mathcal{E}_k(N, \chi)$.

For any two Dirichlet characters ψ modulo u and φ modulo v such that $uv = N$ and $(\psi\varphi)(-1) = (-1)^k$ (here the characters are raised to level uv so their product makes sense) and φ is primitive, consider a linear combination of the Eisenstein series for $\Gamma(N)$,

$$G_k^{\psi,\varphi}(\tau) = \sum_{c=0}^{u-1}\sum_{d=0}^{v-1}\sum_{e=0}^{u-1} \psi(c)\bar{\varphi}(d)G_k^{\overline{(cv, d+ev)}}(\tau).$$

Let γ be an element of $\Gamma_0(N)$ and consider the transformed series $G_k^{\psi,\varphi}[\gamma]_k$,

$$(G_k^{\psi,\varphi}[\gamma]_k)(\tau) = \sum_{c=0}^{u-1}\sum_{d=0}^{v-1}\sum_{e=0}^{u-1} \psi(c)\bar{\varphi}(d)G_k^{\overline{(cv, d+ev)\gamma}}(\tau).$$

Writing $\gamma = \begin{bmatrix} a_\gamma & b_\gamma \\ c_\gamma & d_\gamma \end{bmatrix} \in \Gamma_0(N)$, it is straightforward to compute (Exercise 4.5.1(c)) that $\overline{(cv, d+ev)\gamma} = \overline{(c'v, d'+e'v)}$, where $c', e' \in \{0, \ldots, u-1\}$ and $d' \in \{0, \ldots, v-1\}$ are defined by the conditions $c' \equiv ca_\gamma \pmod{u}$ and $d' \equiv dd_\gamma \pmod{v}$ and $e' \equiv (e+c'b_\gamma)d_\gamma - q \pmod{u}$, where $q = (d' - dd_\gamma)/v$. Also, $\psi(c)\bar{\varphi}(d) = (\psi\varphi)(d_\gamma)\psi(c')\bar{\varphi}(d')$. It follows that

$$(G_k^{\psi,\varphi}[\gamma]_k)(\tau) = (\psi\varphi)(d_\gamma)G_k^{\psi,\varphi}(\tau),$$

and $G_k^{\psi,\varphi} \in \mathcal{M}_k(N, \psi\varphi)$.

Now we compute the Fourier expansion of $G_k^{\psi,\varphi}$. Since the nonconstant part of $G_k^{\bar{v}}(\tau)$ rewrites as

$$\frac{C_k}{N^k} \sum_{\substack{mn>0 \\ n \equiv c_v\,(N)}} \operatorname{sgn}(m)m^{k-1}\mu_N^{d_v m}q_N^{mn},$$

the nonconstant part of $G_k^{\psi,\varphi}(\tau)$ is (remembering that v denotes a vector in the previous expression and denotes the conductor of φ in the next one)

$$\frac{C_k}{N^k} \sum_{c=0}^{u-1}\sum_{d=0}^{v-1}\sum_{e=0}^{u-1} \psi(c)\bar{\varphi}(d) \sum_{\substack{mn>0 \\ n \equiv cv\,(N)}} \operatorname{sgn}(m)m^{k-1}\mu_N^{(d+ev)m}q_N^{mn}$$

$$= \frac{C_k}{N^k} \sum_{c=0}^{u-1}\sum_{d=0}^{v-1} \psi(c)\bar{\varphi}(d) \sum_{\substack{mn>0 \\ n \equiv cv\,(N)}} \operatorname{sgn}(m)m^{k-1}\mu_N^{dm}\left(\sum_{e=0}^{u-1}\mu_u^{em}\right)q_N^{mn}.$$

The inner sum is 0 unless $u \mid m$, in which case it is u. The third sum is empty unless $v \mid n$. So, replacing m by um and n by vn and letting $\mu_v = e^{2\pi i/v}$, this is

$$\frac{C_k}{v^k} \sum_{c=0}^{u-1} \sum_{d=0}^{v-1} \psi(c)\bar{\varphi}(d) \sum_{\substack{mn>0 \\ n\equiv c\,(u)}} \mathrm{sgn}(m)m^{k-1}\mu_v^{dm}q^{mn}$$

$$= \frac{C_k}{v^k} \sum_{c=0}^{u-1} \psi(c) \sum_{\substack{mn>0 \\ n\equiv c\,(u)}} \mathrm{sgn}(m)m^{k-1}\left(\sum_{d=0}^{v-1}\bar{\varphi}(d)\mu_v^{dm}\right)q^{mn}.$$

The inner sum is $\varphi(m)g(\bar{\varphi})$ by formula (4.11). (This is where the calculation requires φ to be primitive. Note also that since φ is primitive the Gauss sum is nonzero.) So now the sum is

$$\frac{C_k g(\bar{\varphi})}{v^k} \sum_{c=0}^{u-1} \psi(c) \sum_{\substack{mn>0 \\ n\equiv c\,(u)}} \mathrm{sgn}(m)\varphi(m)m^{k-1}q^{mn}$$

$$= \frac{C_k g(\bar{\varphi})}{v^k} \sum_{mn>0} \psi(n)\mathrm{sgn}(m)\varphi(m)m^{k-1}q^{mn}.$$

Using the fact that $(\psi\varphi)(-1) = (-1)^k$ and absorbing one more constant shows that the nonconstant part of $G_k^{\psi,\varphi}(\tau)$ is

$$\frac{2C_k g(\bar{\varphi})}{v^k} \sum_{m,n>0} \psi(n)\varphi(m)m^{k-1}q^{mn}$$

$$= \frac{C_k g(\bar{\varphi})}{v^k}2\sum_{n=1}^{\infty}\left(\sum_{\substack{m|n \\ m>0}} \psi(n/m)\varphi(m)m^{k-1}\right)q^n.$$

Meanwhile, the constant part of $G_k^{\psi,\varphi}(\tau)$ is

$$\sum_{c=0}^{u-1}\sum_{d=0}^{v-1}\sum_{e=0}^{u-1} \psi(c)\bar{\varphi}(d)\delta(\overline{cv})\zeta^{\overline{d+ev}}(k) = \psi(0)\sum_{d=0}^{v-1}\sum_{e=0}^{u-1}\bar{\varphi}(d)\zeta^{\overline{d+ev}}(k).$$

Since $\psi(0)$ is 0 unless $u = 1$, we may take $u = 1$ in continuing, and the constant term is

$$\psi(0)\sum_{d=0}^{v-1}\bar{\varphi}(d)\zeta^{\bar{d}}(k) = \psi(0)\sum_{d=0}^{v-1}\bar{\varphi}(d)\sum_{\substack{m\equiv d\,(v) \\ m\neq 0}}\frac{1}{m^k} = \psi(0)\sum_{m\neq 0}\frac{\bar{\varphi}(m)}{m^k}$$

$$= 2\psi(0)L(k,\bar{\varphi}),$$

where for the last step, $\varphi(-1) = (-1)^k$ when ψ is trivial and both sides are 0 otherwise. When k is even, the functional equation (4.18) with v and

$\bar{\varphi}$ in place of N and χ and formula (4.16) show that the constant term is (Exercise 4.5.2(a))

$$\frac{C_k g(\bar{\varphi})}{v^k} \psi(0) L(1-k, \varphi).$$

When k is odd, the functional equation (4.19) and formula (4.16) show the same relation (Exercise 4.5.2(b)). So finally,

Theorem 4.5.1. *The Eisenstein series* $G_k^{\psi,\varphi}$ *takes the form*

$$G_k^{\psi,\varphi}(\tau) = \frac{C_k g(\bar{\varphi})}{v^k} E_k^{\psi,\varphi}(\tau),$$

where $E_k^{\psi,\varphi}$ *has Fourier expansion*

$$E_k^{\psi,\varphi}(\tau) = \delta(\psi) L(1-k, \varphi) + 2 \sum_{n=1}^{\infty} \sigma_{k-1}^{\psi,\varphi}(n) q^n, \quad q = e^{2\pi i \tau}.$$

Here $\delta(\psi)$ *is 1 if* $\psi = 1_1$ *and is 0 otherwise, and the generalized power sum in the Fourier coefficient is*

$$\sigma_{k-1}^{\psi,\varphi}(n) = \sum_{\substack{m|n \\ m>0}} \psi(n/m) \varphi(m) m^{k-1}.$$

For any positive integer N and any integer $k \geq 3$, let $A_{N,k}$ be the set of triples (ψ, φ, t) such that ψ and φ are primitive Dirichlet characters modulo u and v with $(\psi\varphi)(-1) = (-1)^k$, and t is a positive integer such that $tuv \mid N$. Then (Exercise 4.5.3) $|A_{N,k}| = \dim(\mathcal{E}_k(\Gamma_1(N)))$. For any triple $(\psi, \varphi, t) \in A_{N,k}$ define

$$E_k^{\psi,\varphi,t}(\tau) = E_k^{\psi,\varphi}(t\tau).$$

Now that N no longer necessarily equals uv, we need to recognize that our calculations actually showed that $E_k^{\psi,\varphi} \in \mathcal{M}_k(\Gamma_1(uv))$ and all the congruences were taken modulo uv. By the observation at the end of Section 1.2, it follows that $E_k^{\psi,\varphi,t} \in \mathcal{M}_k(\Gamma_1(tuv))$, and since $tuv \mid N$ it follows that $E_k^{\psi,\varphi,t} \in \mathcal{M}_k(\Gamma_1(N))$.

Theorem 4.5.2. *Let* N *be a positive integer and let* $k \geq 3$. *The set*

$$\{E_k^{\psi,\varphi,t} : (\psi, \varphi, t) \in A_{N,k}\}$$

represents a basis of $\mathcal{E}_k(\Gamma_1(N))$. *For any character* χ *modulo* N, *the set*

$$\{E_k^{\psi,\varphi,t} : (\psi, \varphi, t) \in A_{N,k}, \ \psi\varphi = \chi\}$$

represents a basis of $\mathcal{E}_k(N, \chi)$.

As a special case, when $\chi = \mathbf{1}_N$ we have a basis of $\mathcal{E}_k(\Gamma_0(N))$. We skip the details of proving this theorem and its pending analogs for $k = 2$ and $k = 1$. Again the actual basis elements are cosets for now, but once $\mathcal{E}_k(\Gamma(N))$ is redefined in the next chapter as a subspace of $\mathcal{M}_k(\Gamma(N))$ the Eisenstein series themselves will make up the basis.

The Eisenstein series of Theorem 4.5.1 will be revisited at the end of the book in a context encompassing the Modularity Theorem.

Exercises

4.5.1. (a) Check that $\sum_{d \in (\mathbb{Z}/N\mathbb{Z})^*} E_k^{(0,d)}$ is an element of $\mathcal{M}_k(\Gamma_0(N))$.

(b) Check that $\sum_{d \in (\mathbb{Z}/N\mathbb{Z})^*} \bar{\chi}(d) E_k^{(0,d)} \in \mathcal{M}_k(N, \chi)$.

(c) Verify the calculations showing that $G_k^{\psi,\varphi} \in \mathcal{M}_k(N, \psi\varphi)$.

4.5.2. (a) Show that when k is even, the constant term of $G_k^{\psi,\varphi}(\tau)$ is $(C_k g(\bar{\varphi})/v^k)\psi(0)L(1-k,\varphi)$.

(b) Show that when k is odd, the constant term of $G_k^{\psi,\varphi}(\tau)$ is given by the same formula.

4.5.3. Let $A_{N,k}$ be the set of triples (ψ,φ,t) defined in the section and let $B_{N,k}$ be the set of pairs (ψ',φ') such that ψ' and φ' are Dirichlet characters modulo u' and v', not necessarily primitive, with $(\psi'\varphi')(-1) = (-1)^k$ and $u'v' = N$. For such a pair, let u be the conductor of ψ', let ψ be the corresponding primitive character modulo u, and similarly for v and φ. Conversely, for a primitive character ψ modulo u and u' a multiple of u, let $\psi'_{u'}$ be the corresponding character modulo u', and similarly for $\varphi'_{v'}$.

(a) Show that the map from $A_{N,k}$ to $B_{N,k}$ taking (ψ,φ,t) to $(\psi'_{tu}, \varphi'_{N/(tu)})$ and the map from $B_{N,k}$ to $A_{N,k}$ taking (ψ',φ') to $(\psi,\varphi,u'/u)$ invert each other. Thus $|A_{N,k}| = |B_{N,k}|$.

(b) Show that the number of pairs (ψ',φ') satisfying the membership conditions for $B_{N,k}$ except the parity condition on $\psi\varphi$ is $\sum_{d|N} \phi(d)\phi(N/d)$.

(c) Use Lemma 4.3.2 to show that $|B_{N,k}| = \dim(\mathcal{E}_k(\Gamma_1(N)))$. (A hint for this exercise is at the end of the book.)

4.6 Eisenstein series of weight 2

This section describes $\mathcal{E}_2(\Gamma(N))$, $\mathcal{E}_2(\Gamma_1(N))$, and the eigenspaces $\mathcal{E}_2(N,\chi)$, including $\mathcal{E}_2(\Gamma_0(N))$. Recall from Section 1.5 that for any $\bar{v} \in (\mathbb{Z}/N\mathbb{Z})^2$ of order N the function

$$f_2^{\bar{v}}(\tau) = \frac{1}{N^2}\wp_\tau\left(\frac{c_v\tau + d_v}{N}\right)$$

$$= \frac{1}{(c_v\tau + d_v)^2} + \frac{1}{N^2}\sideset{}{'}\sum_{(c,d)\in\mathbb{Z}^2}\left(\frac{1}{\left(\frac{c_v\tau+d_v}{N} - c\tau - d\right)^2} - \frac{1}{(c\tau + d)^2}\right)$$

is weakly modular of weight 2 with respect to $\Gamma(N)$. The methods of Section 4.2 give the Fourier series, so briefly, letting $0 \le c_v < N$ for convenience, the leading term and the terms for $c = 0$ sum to

$$\frac{1}{(c_v\tau + d_v)^2} + \frac{1}{N^2}{\sum_{d\in\mathbb{Z}}}' \left(\frac{1}{\left(\frac{c_v\tau+d_v}{N} - d\right)^2} - \frac{1}{d^2} \right)$$

$$= \delta(\bar{c}_v)\zeta^{\bar{d}_v}(2) + (1 - \delta(\bar{c}_v))\frac{C_2}{N^2}\sum_{m=1}^{\infty} m\mu_N^{d_v m}q_N^{c_v m} - \frac{1}{N^2}2\zeta(2),$$

and the terms with $c > 0$ sum to

$$\frac{1}{N^2}\sum_{c>0}\sum_{d\in\mathbb{Z}} \left(\frac{1}{\left(\frac{c_v\tau+d_v}{N} - c\tau - d\right)^2} - \frac{1}{(c\tau + d)^2} \right)$$

$$= \frac{C_2}{N^2}\sum_{n=1}^{\infty}\left(\sum_{\substack{m|n \\ \frac{n}{m}\equiv c_v \\ m<0}} \mathrm{sgn}(m)m\mu_N^{d_v m} \right)q_N^n - \frac{C_2}{N^2}\sum_{n=1}^{\infty}\sigma(n)q^n,$$

and the terms with $c < 0$ sum to

$$\frac{1}{N^2}\sum_{c<0}\sum_{d\in\mathbb{Z}} \left(\frac{1}{\left(\frac{c_v\tau+d_v}{N} - c\tau - d\right)^2} - \frac{1}{(c\tau + d)^2} \right)$$

$$= \frac{C_2}{N^2}\sum_{n=1}^{\infty}\left(\sum_{\substack{m|n \\ \frac{n}{m}\equiv c_v \\ m>0}} \mathrm{sgn}(m)m\mu_N^{d_v m} \right)q_N^n - \frac{C_2}{N^2}\sum_{n=1}^{\infty}\sigma(n)q^n$$

$$- (1 - \delta(\bar{c}_v))\frac{C_2}{N^2}\sum_{m=1}^{\infty} m\mu_N^{d_v m}q_N^{c_v m}.$$

In these calculations, $C_2 = (-2\pi i)^2/\Gamma(2)$ and $\sigma(n) = \sum_{0<d|n} d$. In total, $f_2^{\bar{v}}(\tau) = G_2^{\bar{v}}(\tau) - G_2(\tau)/N^2$ where

$$G_2^{\bar{v}}(\tau) = \delta(\bar{c}_v)\zeta^{\bar{d}_v}(2) + \frac{C_2}{N^2}\sum_{n=1}^{\infty}\sigma_1^{\bar{v}}(n)q_N^n, \quad q_N = e^{2\pi i\tau/N}, \qquad (4.23)$$

the terms having the same meaning as in Theorem 4.2.3, and $G_2(\tau)$ is the weight 2 Eisenstein series from Section 1.2,

$$G_2(\tau) = 2\zeta(2) + 2C_2\sum_{n=1}^{\infty}\sigma(n)q^n, \quad q = e^{2\pi i\tau}.$$

Recall from Section 1.2 that the function $G_2(\tau) - \pi/\mathrm{Im}(\tau)$ is weight-2 invariant with respect to $\mathrm{SL}_2(\mathbb{Z})$. So instead of $f_2^{\bar{v}}$, the function

$$g_2^{\overline{v}}(\tau) = G_2^{\overline{v}}(\tau) - \frac{\pi}{N^2 \mathrm{Im}(\tau)}, \qquad \overline{v} \in (\mathbb{Z}/N\mathbb{Z})^2 \text{ of order } N \qquad (4.24)$$

is weight-2 invariant with respect to $\Gamma(N)$. Thus $g_2^{\overline{v}}$ is the series of Theorem 4.2.3 extended to $k = 2$ except that now it also includes a nonholomorphic correction term. The Fourier coefficients of $g_2^{\overline{v}}(\tau)$ are bounded by Cn^2 (Exercise 4.6.2).

The formula $\dim(\mathcal{E}_2(\Gamma)) = \varepsilon_\infty - 1$ in (4.3) at the beginning of this chapter shows that as $\overline{v}$ runs through a set of cusp representatives for $\Gamma(N)$, the set $\{g_2^{\overline{v}}\}$ has one too many elements to represent a basis. The set of differences $\{g_2^{\overline{v}_1} - g_2^{\overline{v}_2}, g_2^{\overline{v}_2} - g_2^{\overline{v}_3}, \ldots, g_2^{\overline{v}_{\varepsilon-1}} - g_2^{\overline{v}_\varepsilon}\}$, where $\varepsilon = \varepsilon_\infty$ and $v_1, \ldots, v_\varepsilon$ represent the cusps of $\Gamma(N)$, does represent a basis though we omit the proof. These differences are modular forms since they are holomorphic and weakly modular and their Fourier coefficients are small enough. Thus $\mathcal{E}_2(\Gamma(N))$ is the set of linear combinations of $g_2^{\overline{v}}$ whose coefficients sum to 0, canceling the occurrences of the correction so that we can use the $G_2^{\overline{v}}$ instead, subject to the most symmetrical constraint,

Theorem 4.6.1.

$$\mathcal{E}_2(\Gamma(N)) = \left\{ \sum_{\overline{v}} a_{\overline{v}} G_2^{\overline{v}} : \sum_{\overline{v}} a_{\overline{v}} = 0 \right\},$$

where the sums are taken over vectors $\overline{v}$ of order N in $(\mathbb{Z}/N\mathbb{Z})^2$.

Moving on to $\mathcal{E}_2(\Gamma_1(N))$ and its eigenspaces, let ψ and φ be Dirichlet characters modulo u and v (so v is no longer a vector) with $uv = N$ and $(\psi\varphi)(-1) = 1$ and φ primitive, and consider the sums

$$G_2^{\psi,\varphi}(\tau) = \sum_{c=0}^{u-1} \sum_{d=0}^{v-1} \sum_{e=0}^{u-1} \psi(c)\overline{\varphi}(d) G_2^{\overline{(cv,d+ev)}}(\tau),$$

$$E_2^{\psi,\varphi}(\tau) = \delta(\psi)L(-1,\varphi) + 2\sum_{n=1}^{\infty} \sigma_1^{\psi,\varphi}(n)q^n, \qquad q = e^{2\pi i \tau}.$$

Here $G_2^{\overline{(cv,d+ev)}}(\tau)$ is from (4.23), and $\delta(\psi)$ and $\sigma_1^{\psi,\varphi}(n)$ have the same meanings as in Theorem 4.5.1. If either ψ or φ is nontrivial then the coefficients sum to 0 in $G_2^{\psi,\varphi}(\tau)$, and as in Section 4.5

$$G_2^{\psi,\varphi} \in \mathcal{M}_2(N, \psi\varphi), \qquad G_2^{\psi,\varphi}(\tau) = \frac{C_2 g(\overline{\varphi})}{v^2} E_2^{\psi,\varphi}(\tau).$$

When ψ and φ are both trivial no sum $G_{2,1_u,1_v}(\tau)$ is a modular form. But for any positive integer t (Exercise 4.6.3),

$$
\begin{aligned}
G_{2,1_1,1_1}(\tau) - tG_{2,1_1,1_1}(t\tau) &= \frac{C_2}{N^2}(E_{2,1_1,1_1}(\tau) - tE_{2,1_1,1_1}(t\tau)) \\
&= \frac{1}{N^2} G_{2,t}(\tau)
\end{aligned}
\qquad (4.25)
$$

where $G_{2,t} \in \mathcal{M}_2(\Gamma_0(t))$ is the modular form from Section 1.2.

Let $A_{N,2}$ be the set of triples (ψ, φ, t) such that ψ and φ are primitive Dirichlet characters modulo u and v with $(\psi\varphi)(-1) = 1$, and t is an integer such that $1 < tuv \mid N$. Note that the triple $(\mathbf{1}_1, \mathbf{1}_1, 1)$ is excluded from $A_{N,2}$. Checking formula (4.3) shows that this makes $|A_{N,2}| = \dim(\mathcal{E}_2(\Gamma_1(N)))$. For any triple $(\psi, \varphi, t) \in A_{N,2}$ define

$$E_2^{\psi, \varphi, t}(\tau) = \begin{cases} E_2^{\psi, \varphi}(t\tau) & \text{unless } \psi = \varphi = \mathbf{1}_1, \\ E_2^{\mathbf{1}_1, \mathbf{1}_1}(\tau) - tE_2^{\mathbf{1}_1, \mathbf{1}_1}(t\tau) & \text{if } \psi = \varphi = \mathbf{1}_1. \end{cases}$$

Theorem 4.6.2. *Let N be a positive integer. The set*

$$\{E_2^{\psi, \varphi, t} : (\psi, \varphi, t) \in A_{N,2}\}$$

represents a basis of $\mathcal{E}_2(\Gamma_1(N))$. For any character χ modulo N, the set

$$\{E_2^{\psi, \varphi, t} : (\psi, \varphi, t) \in A_{N,2}, \ \psi\varphi = \chi\}$$

represents a basis of $\mathcal{E}_2(N, \chi)$.

Exercises

4.6.1. Confirm the Fourier expansion of $f_2^{\overline{v}}$.

4.6.2. Show that the nth Fourier coefficient of $g_2^{\overline{v}}(\tau)$ is bounded by Cn^2 for some constant C.

4.6.3. Prove equations (4.25).

4.6.4. Use results from Chapter 3 to show that $S_2(\Gamma_0(4)) = 0$ and that $\dim(\mathcal{M}_2(\Gamma_0(4))) = 2$. This section shows that $E_2^{\mathbf{1}_1, \mathbf{1}_1, 2}$ and $E_2^{\mathbf{1}_1, \mathbf{1}_1, 4}$ form a basis of $\mathcal{M}_2(\Gamma_0(4))$; the function $\theta(\tau, 4)$ from the beginning of Section 1.2 lies in $\mathcal{M}_2(\Gamma_0(4))$ as well. Show that $\theta(\tau, 4)$ is a scalar multiple of $E_2^{\mathbf{1}_1, \mathbf{1}_1, 4}$. Show that $E_2^{\mathbf{1}_1, \mathbf{1}_1, 4} - 3E_2^{\mathbf{1}_1, \mathbf{1}_1, 2}$ is a scalar multiple of the function

$$f(\tau) = \sum_{\substack{n \geq 1 \\ \text{odd}}} \sigma_1(n)q^n$$

(which is not a cusp form despite vanishing at infinity). Thus $\theta(\tau, 4)$ and $f(\tau)$ form a basis of $\mathcal{M}_2(\Gamma_0(4))$. (A hint for this exercise is at the end of the book.)

4.7 Bernoulli numbers and the Hurwitz zeta function

Calculating the weight 1 Eisenstein series requires a generalization of the Bernoulli numbers introduced in Exercise 1.1.7. The Bernoulli numbers arise naturally in the context of computing the power sums

$$1^0 + 2^0 + \cdots + n^0 = n,$$

$$1^1 + 2^1 + \cdots + n^1 = \frac{1}{2}(n^2 + n),$$

$$1^2 + 2^2 + \cdots + n^2 = \frac{1}{6}(2n^3 + 3n^2 + n),$$

etc.

To study these, let n be a positive integer and let the kth power sum up to $n-1$ be

$$S_k(n) = \sum_{m=0}^{n-1} m^k, \quad k \in \mathbb{N}.$$

Thus $S_0(n) = n$ while for $k > 0$ the term 0^k is 0. (Having the sum start at 0 and stop at $n-1$ neatens the ensuing calculation.) The power series having these sums as its coefficients is their generating function, an idea we encountered back in Section 1.2,

$$\mathbb{S}(n, t) = \sum_{k=0}^{\infty} S_k(n) \frac{t^k}{k!}.$$

Rearranging the double sum gives (Exercise 4.7.1(a))

$$\mathbb{S}(n, t) = \frac{e^{nt} - 1}{t} \frac{t}{e^t - 1}. \tag{4.26}$$

The second term is independent of n. Its coefficients are the Bernoulli numbers by definition, constants that can be computed once and for all,

$$\frac{t}{e^t - 1} = \sum_{k=0}^{\infty} B_k \frac{t^k}{k!}.$$

Now the generating function rearranges to (Exercise 4.7.1(b))

$$\mathbb{S}(n, t) = \sum_{l=1}^{\infty} n^l \frac{t^{l-1}}{l!} \sum_{j=0}^{\infty} B_j \frac{t^j}{j!}$$

$$= \sum_{k=0}^{\infty} \left(\frac{1}{k+1} \sum_{j=0}^{k} \binom{k+1}{j} B_j n^{k+1-j} \right) \frac{t^k}{k!}. \tag{4.27}$$

Thus, defining the kth *Bernoulli polynomial* as

$$B_k(X) = \sum_{j=0}^{k} \binom{k}{j} B_j X^{k-j},$$

which again can be computed once and for all, comparing the first and last expressions for $\mathbb{S}(n, t)$ shows that the kth power sum is

$$S_k(n) = \frac{1}{k+1}(B_{k+1}(n) - B_{k+1}).$$

As in Exercise 1.1.7, $B_0 = 1$, $B_1 = -1/2$, and $B_2 = 1/6$, so the first few Bernoulli polynomials are $B_0(X) = 1$ and

$$B_1(X) = X - \frac{1}{2}, \quad B_2(X) = X^2 - X + \frac{1}{6}, \quad B_3(X) = X^3 - \frac{3}{2}X^2 + \frac{1}{2}X.$$

For example, $1^2 + 2^2 + \cdots + n^2 = S_2(n+1) = (B_3(n+1) - B_3)/3$ works out to $(2n^3 + 3n^2 + n)/6$ as it should.

To define more general Bernoulli numbers, note that the Bernoulli polynomials also have a generating function (Exercise 4.7.2(a)),

$$\frac{te^{tX}}{e^t - 1} = \sum_{k=0}^{\infty} B_k(X)\frac{t^k}{k!}. \tag{4.28}$$

Let u be a positive integer and let $\psi : \mathbb{Z}/u\mathbb{Z} \longrightarrow \mathbb{C}$ be any function, not necessarily a Dirichlet character. The *Bernoulli numbers of* ψ are determined by a generating function,

$$\sum_{c=0}^{u-1} \psi(c)\frac{te^{ct}}{e^{ut} - 1} = \sum_{k=0}^{\infty} B_{k,\psi}\frac{t^k}{k!}. \tag{4.29}$$

In particular, $B_{k,1_1} = B_k$. Substituting the generating function of the Bernoulli polynomials in the left side of this relation shows that (Exercise 4.7.2(b))

$$B_{k,\psi} = u^{k-1}\sum_{c=0}^{u-1} \psi(c)B_k(c/u). \tag{4.30}$$

In the next section we need (4.30) specialized to $k = 1$,

$$\sum_{c=0}^{u-1} \psi(c)\left(\frac{c}{u} - \frac{1}{2}\right) = B_{1,\psi}. \tag{4.31}$$

The *Hurwitz zeta function* is

$$\zeta(s,r) = \sum_{n=0}^{\infty} \frac{1}{(r+n)^s}, \quad 0 < r \leq 1, \; \mathrm{Re}(s) > 1.$$

In particular, $\zeta(s,1)$ is the Riemann zeta function, and if r takes the form $r = d/N$ with $0 < d < N$ (both integers) then $\zeta(s,r) = N^s\zeta_+^d(s)$ where $\zeta_+^d(s)$ is the modified zeta function of Sections 4.2 and 4.4.

Let $f_r(t) = \frac{e^{-rt}}{1 - e^{-t}} = \sum_{n=0}^{\infty} e^{-(r+n)t}$ for $t > 0$ and consider the integral

$$g_r(s) = \int_{t=0}^{\infty} f_r(t) t^s \frac{dt}{t}.$$

Since $f_r(t)$ grows like $1/t$ as $t \to 0^+$ the integral converges at its left end when $\mathrm{Re}(s) > 1$, and since $f_r(t)$ decays exponentially as $t \to \infty$ the integral converges at its right end for all values of s. The rapidly converging sum passes through the integral (Exercise 4.7.3), and since dt/t and the limits of integration are invariant when t is replaced by $t/(r+n)$, the integral works out to (Exercise 4.7.4)

$$g_r(s) = \Gamma(s)\zeta(s,r), \quad \mathrm{Re}(s) > 1.$$

This integral representation of a well understood factor times the zeta function is a *Mellin transform*, an idea that will be developed further in Section 4.9. To connect the Hurwitz zeta function with Bernoulli numbers, consider the function

$$\tilde{f}_r(t) = -t f_r(-t) = \frac{t e^{rt}}{e^t - 1}, \quad t < 0.$$

If ψ is a Dirichlet character modulo u then $\tilde{f}_{c/u}(ut)/u$ is the summand in the generating function (4.29) for $\{B_{k,\psi}\}$. Returning to a general value of r, since dt/t is preserved when t is replaced by $-t$ and since $f_r(-t) = \tilde{f}_r(t)/(-t)$ for $t < 0$, the Mellin transform is also

$$g_r(s) = \int_{t=0}^{-\infty} \tilde{f}_r(t)(-t)^{s-1} \frac{dt}{t}, \quad \mathrm{Re}(s) > 1. \tag{4.32}$$

Similarly to Exercise 4.4.5, take the nonpositive real axis as a branch cut in the z-plane so that $\arg(z) \in (-\pi, \pi)$ on the remaining set and the function $\log(z) = \ln|z| + i\arg(z)$ is single-valued and analytic there. For any z off the branch cut and any $s \in \mathbb{C}$ define $z^s = e^{s\log(z)}$. For any positive real number ε, let γ_ε be the complex contour traversing the "underside" of the branch cut (where $\arg(z) = -\pi$) from $-\infty$ to $-\varepsilon$, and then a counterclockwise circle of radius ε about the origin, and finally the "upperside" of the branch cut (where $\arg(z) = \pi$) from $-\varepsilon$ to $-\infty$. Consider the complex contour integral

$$\int_{\gamma_\varepsilon} \tilde{f}_r(z) z^{s-1} \frac{dz}{z}.$$

This integral converges, and the rapid decay of the integrand as $z \to -\infty$ shows that the integral defines an entire function of s. Comparing the z^{s-1} in the integrand against the $(-t)^{s-1}$ in (4.32) as z traverses the negative axis back and forth shows that the integral is $-2i\sin(\pi s)\Gamma(s)\zeta(s,r)$ when $\mathrm{Re}(s) > 1$ (Exercise 4.7.5). Thus the function

$$-\frac{1}{2i\sin(\pi s)\Gamma(s)} \int_{\gamma_\varepsilon} \tilde{f}_r(z) z^{s-1} \frac{dz}{z}, \quad s \in \mathbb{C}$$

is a meromorphic continuation of $\zeta(s,r)$ to the full s-plane. By formula (4.15) the continuation is

$$\zeta(s,r) = -\frac{\Gamma(1-s)}{2\pi i}\int_{\gamma_\varepsilon} \tilde{f}_r(z)z^{s-1}\frac{dz}{z}, \quad s \in \mathbb{C}.$$

This lets us evaluate L-functions at nonnegative integers. Let $s = 1-k$ for a positive integer k and let $\psi \neq 1_1$ be a Dirichlet character modulo u. Then on the one hand,

$$\sum_{c=1}^{u} \psi(c)\zeta(1-k,c/u) = u^{1-k}\sum_{c=1}^{u}\psi(c)\zeta_+^c(1-k) = u^{1-k}L(1-k,\psi).$$

But on the other hand,

$$\sum_{c=1}^{u} \psi(c)\zeta(1-k,c/u) = -\sum_{c=1}^{u}\psi(c)\frac{\Gamma(k)}{2\pi i}\int_{\gamma_\varepsilon}\tilde{f}_{c/u}(z)z^{-k}\frac{dz}{z}$$

$$= -\frac{\Gamma(k)}{2\pi i}\int_{\gamma_\varepsilon}\sum_{c=1}^{u}\psi(c)\tilde{f}_{c/u}(z)\frac{dz}{z^{k+1}}.$$

The integrals along the two straight segments cancel. Replace z by uz and apply Cauchy's integral formula to the remaining integral around the circle to get

$$\sum_{c=1}^{u}\psi(c)\zeta(1-k,c/u) = -u^{1-k}\frac{\Gamma(k)}{2\pi i}\int_{\gamma_\varepsilon}\left(\sum_{c=1}^{u}\psi(c)\frac{\tilde{f}_{c/u}(uz)}{u}\right)\frac{dz}{z^{k+1}}$$

$$= -\frac{u^{1-k}}{k}\left(\sum_{c=1}^{u}\psi(c)\frac{\tilde{f}_{c/u}(uz)}{u}\right)^{(k)}\Bigg|_{z=0} = -u^{1-k}\frac{B_{k,\psi}}{k}.$$

Therefore $L(1-k,\psi) = -B_{k,\psi}/k$ for $k \geq 1$ just as we saw that $\zeta(1-k) = -B_k/k$ for $k \geq 2$ earlier. We need the result for $k = 1$ in the next section,

$$B_{1,\psi} = -L(0,\psi) \quad \text{if } \psi \neq 1_1. \tag{4.33}$$

For more on the topics of this section see [Lan76], Chapters 13 and 14. In particular, the functional equations (4.18) and (4.19) for L-functions follow quickly. These are also presented as a series of exercises in [Kob93].

Exercises

4.7.1. (a) Establish formula (4.26).
　(b) Establish formula (4.27).

4.7.2. (a) Establish formula (4.28).
　(b) Establish formula (4.30).

4.7.3. Let $0 < r \le 1$ and let $\mathrm{Re}(s) > 1$. Show that the sum $\sum_{n=0}^{\infty} e^{-(r+n)t}t^{s-1}$ converges absolutely. Show that the sum $\sum_n \int_{t=0}^{\infty} e^{-(r+n)t}t^{s-1}dt$ converges. Show that the integral of the sum $\int_t \sum_n e^{-(r+n)t}t^{s-1}dt$ is equal to the sum of the integrals $\sum_n \int_t e^{-(r+n)t}t^{s-1}dt$. (A hint for this exercise is at the end of the book.)

4.7.4. Compute that $g_r(s) = \Gamma(s)\zeta(s,r)$ when $\mathrm{Re}(s) > 1$.

4.7.5. (a) Show that the contour integral in the section is independent of ε when $0 < \varepsilon < 1$.

(b) For $\mathrm{Re}(s) > 1$, show that as $\varepsilon \to 0^+$ the sum of the two linear pieces of the contour integral in the section converges to $-2i\sin(\pi s)\Gamma(s)\zeta(s,r)$.

(c) For $\mathrm{Re}(s) > 1$, show that as $\varepsilon \to 0^+$ the circular piece of the contour integral converges to 0. (A hint for this exercise is at the end of the book.)

4.8 Eisenstein series of weight 1

Just as the Weierstrass $\wp$-function defines weight 2 Eisenstein series, another function associated with a lattice Λ leads naturally to series of weight 1. The natural starting point is the *Weierstrass σ-function*

$$\sigma_\Lambda(z) = z\prod_{\omega \in \Lambda}{}' \left(1 - \frac{z}{\omega}\right)e^{z/\omega + \frac{1}{2}(z/\omega)^2}, \quad z \in \mathbb{C}.$$

Since this function has simple zeros at the 2-dimensional lattice $\Lambda \subset \mathbb{C}$ just as $\sin \pi x$ has simple zeros at the 1-dimensional lattice $\mathbb{Z} \subset \mathbb{R}$, it is named "$\sigma$" by analogy. The exponential factors are needed to make the infinite product converge (Exercise 4.8.1(a)).

The logarithmic derivative σ'/σ (Exercise 4.8.1(b)) is the *Weierstrass zeta function*, denoted Z here to avoid confusion with the Riemann zeta,

$$Z_\Lambda(z) = \frac{1}{z} + \sum_{\omega \in \Lambda}{}' \left(\frac{1}{z - \omega} + \frac{1}{\omega} + \frac{z}{\omega^2}\right), \quad z \in \mathbb{C}.$$

This function has simple poles with residue 1 at the lattice points, analogously to the logarithmic derivative $\pi \cot \pi x$ of $\sin \pi x$, but it isn't quite periodic with respect to Λ. Instead, since $Z'_\Lambda = -\wp_\Lambda$ is periodic, if $\Lambda = \omega_1\mathbb{Z} \oplus \omega_2\mathbb{Z}$ then the quantities

$$\eta_1(\Lambda) = Z_\Lambda(z + \omega_1) - Z_\Lambda(z) \quad \text{and} \quad \eta_2(\Lambda) = Z_\Lambda(z + \omega_2) - Z_\Lambda(z)$$

are lattice constants such that

$$Z_\Lambda(z + n_1\omega_1 + n_2\omega_2) = Z_\Lambda(z) + n_1\eta_1(\Lambda) + n_2\eta_2(\Lambda), \quad n_1, n_2 \in \mathbb{Z}.$$

Under the normalizing convention $\omega_1/\omega_2 \in \mathcal{H}$ the lattice constants satisfy the *Legendre relation* $\eta_2(\Lambda)\omega_1 - \eta_1(\Lambda)\omega_2 = 2\pi i$ (Exercise 4.8.2). The second lattice constant appears in the q-product expansion of σ specialized to $\Lambda = \Lambda_\tau$ (proved in [Lan73], Chapter 18),

$$\sigma_{\Lambda_\tau}(z) = \frac{1}{2\pi i} e^{\frac{1}{2}\eta_2(\Lambda_\tau)z^2}(e^{\pi i z} - e^{-\pi i z}) \prod_{n=1}^{\infty} \frac{(1 - e^{2\pi i z}q^n)(1 - e^{-2\pi i z}q^n)}{(1 - q^n)^2},$$

where as always, $q = e^{2\pi i \tau}$. The logarithmic derivative is therefore

$$Z_{\Lambda_\tau}(z) = \eta_2(\Lambda_\tau)z - \pi i \frac{1 + e^{2\pi i z}}{1 - e^{2\pi i z}} - 2\pi i \sum_{n=1}^{\infty} \left(\frac{e^{2\pi i z}q^n}{1 - e^{2\pi i z}q^n} - \frac{e^{-2\pi i z}q^n}{1 - e^{-2\pi i z}q^n} \right).$$

[Lan73] also shows that $\eta_2(\Lambda_\tau) = G_2(\tau)$ (see Exercise 4.8.3), and so the Legendre relation shows that $\eta_1(\Lambda_\tau) = \tau G_2(\tau) - 2\pi i$.

For any vector $\bar{v} \in (\mathbb{Z}/N\mathbb{Z})^2$ of order N, the function of modular points

$$F_1^{\bar{v}}(\mathbb{C}/\Lambda, (\omega_1/N + \Lambda, \omega_2/N + \Lambda))$$
$$= Z_\Lambda \left(\frac{c_v \omega_1 + d_v \omega_2}{N} \right) - \frac{c_v \eta_1(\Lambda) + d_v \eta_2(\Lambda)}{N}$$

is well defined (Exercise 4.8.4(a)) and degree-1 homogeneous with respect to $\Gamma(N)$ (Exercise 4.8.4(b)). The corresponding function (divided by N for convenience)

$$g_1^{\bar{v}}(\tau) = \frac{1}{N} Z_{\Lambda_\tau} \left(\frac{c_v \tau + d_v}{N} \right) - \frac{c_v \eta_1(\Lambda_\tau) + d_v \eta_2(\Lambda_\tau)}{N^2}$$

is weakly modular of weight 1 with respect to $\Gamma(N)$. Letting $z = (c_v \tau + d_v)/N$ in the last expression for Z_{Λ_τ} and taking $0 \le c_v < N$ for convenience, a calculation similar to the weight 2 case now shows that the terms of $g_1^{\bar{v}}(\tau)$ are (Exercise 4.8.5)

$$\frac{\eta_2(\Lambda_\tau)z}{N} - \frac{c_v \eta_1(\Lambda_\tau) + d_v \eta_2(\Lambda_\tau)}{N^2} = \frac{2\pi i c_v}{N^2}$$

and

$$-\frac{\pi i}{N} \frac{1 + e^{2\pi i z}}{1 - e^{2\pi i z}}$$
$$= \delta(\bar{c}_v) \frac{\pi}{N} \cot \left(\frac{\pi d_v}{N} \right) + (1 - \delta(\bar{c}_v)) \left(-\frac{\pi i}{N} + \frac{C_1}{N} \sum_{m=1}^{\infty} \mu_N^{d_v m} q_N^{c_v m} \right)$$

and

$$-\frac{2\pi i}{N} \sum_{n=1}^{\infty} \frac{e^{2\pi i z}q^n}{1 - e^{2\pi i z}q^n}$$
$$= \frac{C_1}{N} \sum_{n=1}^{\infty} \left(\sum_{\substack{m|n \\ \frac{n}{m} \equiv c_v \\ m>0}} \text{sgn}(m)\mu_N^{d_v m} \right) q_N^n - (1 - \delta(\bar{c}_v)) \frac{C_1}{N} \sum_{m=1}^{\infty} \mu_N^{d_v m} q_N^{c_v m}$$

and

$$\frac{2\pi i}{N} \sum_{n=1}^{\infty} \frac{e^{-2\pi i z} q^n}{1 - e^{-2\pi i z} q^n} = \frac{C_1}{N} \sum_{n=1}^{\infty} \Big(\sum_{\substack{m|n \\ \frac{n}{m} \equiv c_v \\ m<0}} \operatorname{sgn}(m) \mu_N^{d_v m} \Big) q_N^n.$$

Assembling the results gives

$$g_1^{\overline{v}}(\tau) = G_1^{\overline{v}}(\tau) - \frac{C_1}{N}\left(\frac{c_v}{N} - \frac{1}{2}\right), \qquad 0 \le c_v < N, \tag{4.34}$$

where by using formula (4.22) and defining $\sigma_0^{\overline{v}}(n)$ as in Theorem 4.2.3,

$$G_1^{\overline{v}}(\tau) = \delta(\overline{c}_v)\zeta^{\overline{d}_v}(1) + \frac{C_1}{N}\sum_{n=1}^{\infty}\sigma_0^{\overline{v}}(n)q_N^n. \tag{4.35}$$

Thus the series $G_1^{\overline{v}}$ is again analogous to $G_k^{\overline{v}}$ for $k \ge 3$. Since $g_1^{\overline{v}}(\tau)$ is holomorphic and weakly modular with respect to $\Gamma(N)$ and its nth Fourier coefficient grows as Cn, it is a weight 1 modular form with respect to $\Gamma(N)$.

Note that $G_1^{\overline{(N-c_v,d_v)}}(\tau) = -G_1^{\overline{(c_v,d_v)}}(\tau)$ when $\overline{c}_v \ne 0$, and $G_1^{\overline{(0,-d_v)}}(\tau) = -G_1^{\overline{(0,d_v)}}(\tau)$ (Exercise 4.8.6), consistently with the dimension of $\mathcal{E}_1(\Gamma(N))$ being $\varepsilon_\infty/2$ in formula (4.3) rather than ε_∞.

Let ψ and φ be Dirichlet characters modulo u and v (again v is no longer a vector) with $uv = N$ and φ primitive and $(\psi\varphi)(-1) = -1$. As in Sections 4.5 and 4.6, consider the sums

$$G_1^{\psi,\varphi}(\tau) = \sum_{c=0}^{u-1}\sum_{d=0}^{v-1}\sum_{e=0}^{u-1}\psi(c)\overline{\varphi}(d)g_1^{\overline{(cv,d+ev)}}(\tau),$$

$$E_1^{\psi,\varphi}(\tau) = \delta(\varphi)L(0,\psi) + \delta(\psi)L(0,\varphi) + 2\sum_{n=1}^{\infty}\sigma_0^{\psi,\varphi}(n)q^n.$$

Here $\delta(\psi)$ and $\sigma_0^{\psi,\varphi}(n)$ have the same meanings as in Theorem 4.5.1. Part of the constant term of $G_1^{\psi,\varphi}(\tau)$ comes from summing the correction term from (4.34),

$$-\frac{C_1}{N}\sum_{c=0}^{u-1}\psi(c)\left(\frac{cv}{N} - \frac{1}{2}\right)\sum_{d=0}^{v-1}\overline{\varphi}(d)\sum_{e=0}^{u-1}1.$$

This is 0 unless $\varphi = \mathbf{1}_1$, in which case $u = N$ and ψ is nontrivial and $g(\overline{\varphi})/v = 1$, and so it works out to

$$-\frac{C_1 g(\overline{\varphi})}{v}\delta(\varphi)\sum_{c=0}^{u-1}\psi(c)\left(\frac{c}{u} - \frac{1}{2}\right) = -\frac{C_1 g(\overline{\varphi})}{v}\delta(\varphi)B_{1,\psi} \qquad \text{by (4.31)}$$

$$= \frac{C_1 g(\overline{\varphi})}{v}\delta(\varphi)L(0,\psi) \qquad \text{by (4.33).}$$

As in the calculation for Theorem 4.5.1, the rest of the constant term of $G_1^{\psi,\varphi}(\tau)$ is

$$\sum_{c=0}^{u-1}\sum_{d=0}^{v-1}\sum_{e=0}^{u-1}\psi(c)\bar{\varphi}(d)\delta(cv)\overline{\zeta^{d+ev}}(1) = \psi(0)\sum_{d=0}^{v-1}\bar{\varphi}(d)\bar{\zeta^d}(1)$$

where in the second step we may assume $\psi = \mathbf{1}_1$. This makes $\varphi(-1) = -1$ and the sum is the analytic continuation as $s \to 1^+$ of

$$\sum_{d=0}^{v-1}\bar{\varphi}(d)\bar{\zeta^d}(s) = \sum_{d}\sideset{}{'}\sum_{n\equiv d\,(v)}\bar{\varphi}(n)n^{-s} = (1-(-1)^{-s})L(s,\bar{\varphi}).$$

So the rest of the constant term is $2\delta(\psi)L(1,\bar{\varphi})$, and as in Section 4.5 this is $(C_1 g(\bar{\varphi})/v)\delta(\psi)L(0,\varphi)$. The constant term thus totals

$$\frac{C_1 g(\bar{\varphi})}{v}(\delta(\varphi)L(0,\psi)+\delta(\psi)L(0,\varphi)).$$

The rest of the calculations in Section 4.5 go as before, giving

$$G_1^{\psi,\varphi} \in \mathcal{M}_1(N,\psi\varphi), \qquad G_1^{\psi,\varphi}(\tau) = \frac{C_1 g(\bar{\varphi})}{v}E_1^{\psi,\varphi}(\tau).$$

Let $A_{N,1}$ be the set of triples $(\{\psi,\varphi\},t)$ such that ψ and φ, taken this time as an unordered pair, are primitive Dirichlet characters modulo u and v satisfying the parity condition $(\psi\varphi)(-1) = -1$, and t is a positive integer such that $tuv \mid N$. The parity condition shows that $A_{N,1}$ contains no triples $(\{\psi,\psi\},t)$ with the same character twice, so taking the characters in unordered pairs means that $A_{N,1}$ contains half as many elements as it would otherwise, and formula (4.3) shows that again $|A_{N,1}| = \dim(\mathcal{E}_1(\Gamma_1(N)))$. Since the constant term and the Fourier coefficients of $E_1^{\psi,\varphi}$ are symmetric in ψ and φ, the series depends on the two characters only as an unordered pair, and it makes sense to define for each triple $(\{\psi,\varphi\},t) \in A_{N,1}$

$$E_1^{\psi,\varphi,t}(\tau) = E_1^{\psi,\varphi}(t\tau).$$

Theorem 4.8.1. *Let N be a positive integer. The set*

$$\{E_1^{\psi,\varphi,t} : (\{\psi,\varphi\},t) \in A_{N,1}\}$$

represents a basis of $\mathcal{E}_1(\Gamma_1(N))$. For any character χ modulo N, the set

$$\{E_1^{\psi,\varphi,t} : (\{\psi,\varphi\},t) \in A_{N,1},\ \psi\varphi = \chi\}$$

represents a basis of $\mathcal{E}_1(N,\chi)$.

This theorem combines with Theorem 4.5.2 to give a solution to the s squares problem (cf. Section 1.2) for $s = 2, 6, 8$ along with the solution we already have for $s = 4$. For any even $s \geq 10$ the same methods give an asymptotic estimate $\tilde{r}(n, s)$ of the solution $r(n, s)$, meaning that $\lim_{n \to \infty} \tilde{r}(n, s)/r(n, s) = 1$. See Exercise 4.8.7 for the details.

A weight 1 Eisenstein series $E_1^{\psi, 1}$ figures in the proof of the Modularity Theorem. This will be explained in Section 9.6.

Exercises

4.8.1. This exercise touches briefly on properties of the infinite product $\sigma_\Lambda(z)$. Since the general theory of infinite products is not assumed and will not be explained here (see, for example, [JS87]), the results are necessarily informal.

(a) Using the principal branch of the complex logarithm, show that for $0 < |z| < \min\{|\omega| : \omega \in \Lambda,\ \omega \neq 0\}$, the sum

$$\sum_{\omega \in \Lambda}{}' \log\left((1 - z/\omega)e^{z/\omega + \frac{1}{2}(z/\omega)^2}\right)$$

converges absolutely. It follows that the infinite product $\sigma_\Lambda(z)$ converges to a holomorphic function.

(b) Show that if $f = \prod_{n=1}^N f_n$ then $f'/f = \sum_{n=1}^N f_n'/f_n$. The analogous result for infinite products holds for the functions σ and Z.

4.8.2. Prove the Legendre relation $\eta_2(\Lambda)\omega_1 - \eta_1(\Lambda)\omega_2 = 2\pi i$. (A hint for this exercise is at the end of the book.)

4.8.3. The actual formula for $\eta_2(\Lambda_\tau)$ in [Lan73] is

$$\eta_2(\Lambda_\tau) = \frac{(2\pi i)^2}{12}\left(-1 + 24\sum_{m=1}^\infty \frac{mq^m}{1 - q^m}\right).$$

Prove that the right side is $G_2(\tau)$.

4.8.4. (a) Verify that the function $F_1^{\bar{v}}$ is is well defined.

(b) Verify that the function is degree-1 homogeneous with respect to $\Gamma(N)$.

4.8.5. Verify the calculations leading to formula (4.34).

4.8.6. Show that $G_1^{\overline{(N-c_v, d_v)}}(\tau) = -G_1^{\overline{(c_v, d_v)}}(\tau)$ when $\bar{c}_v \neq 0$, and show that $G_1^{\overline{(0, -d_v)}}(\tau) = -G_1^{\overline{(0, d_v)}}(\tau)$.

4.8.7. (a) Recall the function $\theta(\tau, k)$ from Section 1.2. Similarly to that section's argument that $\theta(\tau, 4) \in \mathcal{M}_2(\Gamma_0(4))$, show that $\theta(\tau, 2) \in \mathcal{M}_1(\Gamma_1(4))$ and consequently $\theta(\tau, s) \in \mathcal{M}_{s/2}(\Gamma_1(4))$ for all even $s \geq 2$.

(b) Use results from Chapter 3 to show that $\mathcal{S}_{s/2}(\Gamma_1(4)) = \{0\}$ for $s = 2, 6, 8$. It follows that $\theta(\tau, s)$ is a linear combination of Eisenstein series for these values of s. Use results from Chapter 3 to compute $\dim(\mathcal{M}_{s/2}(\Gamma_1(4)))$ in each case. (Hints for this exercise are at the end of the book.)

(c) Recall from Section 1.2 the notation $r(n, k)$ for the number of ways of representing the positive integer n as a sum of k squares. Use Theorem 4.8.1 to show that

$$r(n, 2) = 4 \sum_{\substack{0 < m | n \\ m \text{ odd}}} (-1)^{(m-1)/2}.$$

Similarly, use Theorem 4.5.2 to find formulas for $r(n, 6)$ and $r(n, 8)$.

(d) The proof of Proposition 5.9.1 to follow will establish that the Fourier coefficients of any cusp form $f \in \mathcal{S}_k(\Gamma)$ satisfy $|a_n(f)| \leq Cn^{k/2}$ for some C. Let $s \geq 10$ be even. Argue that the Fourier coefficients of the Eisenstein series in $\mathcal{M}_{s/2}(\Gamma_1(4))$ dominate those of the cusp forms in $\mathcal{S}_{s/2}(\Gamma_1(4))$ as $n \to \infty$, and so the methods of this exercise give an asymptotic solution to the s squares problem by ignoring the cusp forms.

4.9 The Fourier transform and the Mellin transform

The next section will explain an idea originally due to Hecke to get around the conditional convergence of $G_2(\tau)$ and the divergence of the analogous series $G_1(\tau)$ in a much farther-reaching way than the methods so far. This will require a bit more analysis. As a warmup, the ideas are first used in this section to prove the meromorphic continuation and functional equation of the Riemann zeta function as discussed in Section 4.4.

For any positive integer l, the space of measurable and absolutely integrable functions on $\mathbb{R}^l$ is

$$\mathcal{L}^1(\mathbb{R}^l) = \{\text{measurable } f : \mathbb{R}^l \longrightarrow \mathbb{C} : \int_{x \in \mathbb{R}^l} |f(x)| dx < \infty\}.$$

Any $f \in \mathcal{L}^1(\mathbb{R}^l)$ has a *Fourier transform* $\hat{f} : \mathbb{R}^l \longrightarrow \mathbb{C}$ given by

$$\hat{f}(x) = \int_{y \in \mathbb{R}^l} f(y) e^{-2\pi i \langle y, x \rangle} dy,$$

where $\langle \, , \, \rangle$ is the usual Euclidean inner product. Although the Fourier transform is continuous (Exercise 4.9.1), it need not belong to $\mathcal{L}^1(\mathbb{R}^l)$. But if also $\int_{x \in \mathbb{R}^l} |f(x)|^2 dx < \infty$ then $\int_{x \in \mathbb{R}^l} |\hat{f}(x)|^2 dx < \infty$, see for example [Rud74].

The l-*dimensional theta function* $\vartheta(\, , l) : \mathcal{H} \longrightarrow \mathbb{C}$ is

$$\vartheta(\tau, l) = \sum_{n \in \mathbb{Z}^l} e^{\pi i |n|^2 \tau}, \qquad \tau \in \mathcal{H}$$

where $|\ |$ is the usual Euclidean absolute value. The sum converges very rapidly away from the real axis, making absolute and uniform convergence on compact subsets of $\mathcal{H}$ easy to show (Exercise 4.9.2) and thus defining a holomorphic function. Specializing to $\tau = it$ with $t > 0$ and letting $f \in \mathcal{L}^1(\mathbb{R}^l)$ be the Gaussian function $f(x) = e^{-\pi|x|^2}$ gives $\vartheta(it, l) = \sum_{n \in \mathbb{Z}^l} f(nt^{1/2})$, a sum of quickly decreasing functions whose graphs narrow as $|n|$ grows. (See Figure 4.1.)

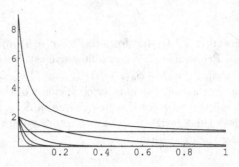

Figure 4.1. The first few terms of $\vartheta(it, 1)$ and their sum

We need the Fourier transform of the Gaussian $f(x)$. First set $l = 1$, call the function φ in this case, and compute

$$\hat{\varphi}(x) = \int_{y=-\infty}^{\infty} e^{-\pi(y^2 + 2iyx - x^2)} e^{-\pi x^2} \, dy = e^{-\pi x^2} \int_{y=-\infty}^{\infty} e^{-\pi(y+ix)^2} \, dy.$$

Complex contour integration shows that the integral is just the Gaussian integral $\int_{-\infty}^{\infty} e^{-\pi y^2} \, dy$ (Exercise 4.9.3(a)), and this is 1 (Exercise 4.9.3(b)). Thus $\hat{f} = f$ when $l = 1$. To prove the same result for $l > 1$, note that f takes the product form $f(x) = \varphi(x_1) \cdots \varphi(x_l)$ for any vector $x = (x_1, \ldots, x_l)$, and consequently the Fourier transform is $\hat{f}(x) = \hat{\varphi}(x_1) \cdots \hat{\varphi}(x_l)$, i.e., $\hat{f} = f$ as desired (Exercise 4.9.3(c)). Also, for any function $h \in \mathcal{L}^1(\mathbb{R}^l)$ and any positive number r, the Fourier transform of the function $h(xr)$ is $r^{-l}\hat{h}(x/r)$ (Exercise 4.9.3(d)), so in particular the Fourier transform of $f(xt^{1/2})$ is $t^{-l/2} f(xt^{-1/2})$.

For any function $h \in \mathcal{L}^1(\mathbb{R}^l)$ such that the sum $\sum_{d \in \mathbb{Z}^l} h(x + d)$ converges absolutely and uniformly on compact sets and is infinitely differentiable as a function of x, the *Poisson summation formula* is

$$\sum_{d \in \mathbb{Z}^l} h(x + d) = \sum_{m \in \mathbb{Z}^l} \hat{h}(m) e^{2\pi i \langle m, x \rangle}.$$

For example, formula (1.2), which we have used to compute the Fourier expansions of Eisenstein series, is a special case of one-dimensional Poisson summation known as the Lipschitz formula (Exercise 4.9.4). Letting

$h(x)$ be the Gaussian $f(xt^{1/2})$, Poisson summation with $x = 0$ shows that $\sum f(nt^{1/2}) = t^{-l/2} \sum f(nt^{-1/2})$, i.e.,

$$\vartheta(i/t, l) = t^{l/2} \vartheta(it, l), \quad t > 0. \tag{4.36}$$

By the Uniqueness Theorem from complex analysis this relation extends to

$$\vartheta(-1/\tau, l) = (-i\tau)^{l/2} \vartheta(\tau, l), \quad \tau \in \mathcal{H}.$$

Here $-i\tau$ lies in the right half plane, and the square root is defined for all complex numbers except the negative reals, extending the real positive square root function of positive real numbers. This transformation law should look familiar from Section 1.2 since the function that arose there from the four squares problem is $\theta(\tau, l) = \vartheta(2\tau, l)$.

The properties of the Riemann zeta function are established by examining the *Mellin transform* of (essentially) the theta function. In general, the Mellin transform of a function $f : \mathbb{R}^+ \longrightarrow \mathbb{C}$ is the integral

$$g(s) = \int_{t=0}^{\infty} f(t) t^s \frac{dt}{t}$$

for s-values such that the integral converges absolutely. (See Exercise 4.9.5 for the inverse Mellin transform.) For example, the Mellin transform of e^{-t} is $\Gamma(s)$. Setting $l = 1$ and writing $\vartheta(\tau)$ rather than $\vartheta(\tau, 1)$ for the duration of this section, consider the Mellin transform of the function $\sum_{n=1}^{\infty} e^{-\pi n^2 t} = 1/2(\vartheta(it) - 1)$ for $t > 0$,

$$g(s) = \int_{t=0}^{\infty} \sum_{n=1}^{\infty} e^{-\pi n^2 t} t^s \frac{dt}{t} = \frac{1}{2} \int_{t=0}^{\infty} (\vartheta(it) - 1) t^s \frac{dt}{t}. \tag{4.37}$$

Since $\vartheta(it)$ converges to 1 as $t \to \infty$, the transformation law (4.36) shows that as $t \to 0$, $\vartheta(it)$ grows at the same rate as $t^{-1/2}$, and therefore the integral $g(s)$ converges at its left endpoint if $\operatorname{Re}(s) > 1/2$. And since the convergence of $\vartheta(it)$ to 1 as $t \to \infty$ is rapid, the integral converges at its right end for all values of s. Rapid convergence lets the sum pass through the integral in (4.37) to yield, after a change of variable,

$$g(s) = \sum_{n=1}^{\infty} (\pi n^2)^{-s} \int_{t=0}^{\infty} e^{-t} t^s \frac{dt}{t} = \pi^{-s} \Gamma(s) \zeta(2s), \quad \operatorname{Re}(s) > 1/2.$$

Thus the integral $g(s/2)$ is the function $\xi(s) = \pi^{-s/2} \Gamma(s/2) \zeta(s)$ from Section 4.4 when $\operatorname{Re}(s) > 1$. That is, the Mellin transform of the 1-dimensional theta function is the Riemann zeta function multiplied by a well understood factor. Thinking in these terms, the factor $\pi^{-s/2} \Gamma(s/2)$ is intrinsically associated to $\zeta(s)$, making $\xi(s)$ the natural function to consider.

The second integral in (4.37) provides the meromorphic continuation and functional equation of ξ. Computing part of $g(s/2)$ by splitting off a term, replacing t by $1/t$, using the transformation law (4.36) for $\vartheta(it)$ with $l = 1$, and splitting off another term gives (Exercise 4.9.6)

$$
\begin{aligned}
\frac{1}{2}\int_{t=0}^{1}(\vartheta(it)-1)t^{s/2}\frac{dt}{t} &= \frac{1}{2}\int_{t=0}^{1}\vartheta(it)t^{s/2}\frac{dt}{t}-\frac{1}{s} \\
&= \frac{1}{2}\int_{t=1}^{\infty}\vartheta(i/t)t^{-s/2}\frac{dt}{t}-\frac{1}{s} \\
&= \frac{1}{2}\int_{t=1}^{\infty}\vartheta(it)t^{(1-s)/2}\frac{dt}{t}-\frac{1}{s} \\
&= \frac{1}{2}\int_{t=1}^{\infty}(\vartheta(it)-1)t^{(1-s)/2}\frac{dt}{t}-\frac{1}{s}-\frac{1}{1-s}.
\end{aligned}
$$

And combining this with the remainder of the integral $g(s/2)$,

$$
\xi(s) = \frac{1}{2}\int_{t=1}^{\infty}(\vartheta(it)-1)(t^{s/2}+t^{(1-s)/2})\frac{dt}{t}-\frac{1}{s}-\frac{1}{1-s}, \quad \mathrm{Re}(s) > 1. \quad (4.38)
$$

But the integral in (4.38) is entire in s, making the right side holomorphic everywhere in the s-plane except for simple poles at $s = 0$ and $s = 1$. And the right side is clearly invariant under $s \mapsto 1 - s$. Thus the function $\xi(s)$ has a meromorphic continuation to the full s-plane and satisfies the functional equation $\xi(1 - s) = \xi(s)$. The properties of the Riemann zeta function as stated in Section 4.4 now follow.

Exercises

4.9.1. Show that the Fourier transform is continuous. (A hint for this exercise is at the end of the book.)

4.9.2. This exercise shows that the series $\vartheta(\tau, l) = \sum_{n\in\mathbb{Z}^l} e^{\pi i |n|^2 \tau}$ is holomorphic.

(a) Let $S_m = \{n \in \mathbb{Z}^l : |n| = m\}$ for each natural number m. Show that $|S_m| \le (2m+1)^l$.

(b) Let $K \subset \mathcal{H}$ be compact. Thus there exists some $y_0 > 0$ such that $\mathrm{Im}(\tau) \ge y_0$ for all $\tau \in K$. Show that for any $\tau \in K$ and $n \in S_m$,

$$
|e^{\pi i |n|^2 \tau}| \le e^{-\pi m^2 y_0}.
$$

(c) There exists a natural number M_0 such that if $m \ge M_0$ then $(2m+1)^l < e^m$ and $\pi m^2 y_0 > 2m$. Show that for any $\tau \in K$ and any $M \ge M_0$, the corresponding tail of the series $\vartheta(\tau, l)$ satisfies

$$
\sum_{\substack{n\in\mathbb{Z}^l \\ |n|\ge M}} |e^{\pi i |n|^2 \tau}| \le \sum_{m=M}^{\infty} e^{-m} = e^{-M}(1-e^{-1})^{-1}.
$$

This is less than arbitrary $\varepsilon > 0$ for large enough M independently of τ, showing that the series $\vartheta(\tau, l)$ converges absolutely and uniformly on compact subsets of $\mathcal{H}$. Since each summand is holomorphic, so is the sum.

4.9.3. (a) Use complex contour integration to show that $\int_{t=-\infty}^{\infty} e^{-\pi(t+ix)^2} dt$ equals $\int_{-\infty}^{\infty} e^{-\pi t^2} dt$.

  (b) Show that this Gaussian integral is 1. (A hint for this exercise is at the end of the book.)

  (c) Show that if $f \in L^1(R^l)$ takes the form of a product $f(x) = f_1(x_1) \cdots f_l(x_l)$ then its Fourier transform is the corresponding product $\hat{f}(x) = \hat{f}_1(x_1) \cdots \hat{f}_l(x_l)$.

  (d) Show that for any function $h \in L^1(\mathbb{R}^l)$ and any positive number r, the Fourier transform of the function $h(xr)$ is $r^{-l}\hat{h}(x/r)$.

4.9.4. Formula (1.2) is a special case of the *Lipschitz formula* for $\tau \in \mathcal{H}$ and $q = e^{2\pi i \tau}$,

$$\sum_{d \in \mathbb{Z}} \frac{1}{(\tau + d)^s} = \frac{(-2\pi i)^s}{\Gamma(s)} \sum_{m=1}^{\infty} m^{s-1} q^m, \quad \mathrm{Re}(s) > 1.$$

Prove the Lipschitz formula by applying Poisson summation to the function $f \in L^1(\mathbb{R})$ given by

$$f(x) = \begin{cases} x^{s-1} e^{2\pi i x \tau} & \text{if } x > 0, \\ 0 & \text{if } x \leq 0. \end{cases}$$

(A hint for this exercise is at the end of the book.)

4.9.5. If g is a holomorphic function of the complex variable s in some right half plane then its *inverse Mellin transform* is

$$f(t) = \frac{1}{2\pi i} \int_{s=\sigma-i\infty}^{\sigma+i\infty} g(s)t^{-s} ds$$

for positive t-values such that the integral converges absolutely. Complex contour integration shows that this is independent of σ.

  (a) Let $t = e^x$, $s = \sigma + 2\pi i y$, $f(t) = e^{-\sigma x} f_\sigma(x)$, and $g(s) = g_\sigma(y)$. Show that

$$g(s) = \int_{t=0}^{\infty} f(t)t^s \frac{dt}{t} \iff g_\sigma(y) = \int_{x=-\infty}^{\infty} f_\sigma(x)e^{2\pi i x y} dx$$

and that

$$f(t) = \frac{1}{2\pi i} \int_{s=\sigma-i\infty}^{\sigma+i\infty} g(s)t^{-s} ds \iff f_\sigma(x) = \int_{y=-\infty}^{\infty} g_\sigma(y)e^{-2\pi i y x} dy.$$

The right side conditions are equivalent by Fourier inversion, and so the left side conditions are equivalent as well. This is the *Mellin inversion formula*.

  (b) Evaluate the integral $\int_{s=\sigma-i\infty}^{\sigma+i\infty} \Gamma(s)x^{-s} ds$ for any $\sigma > 0$. (A hint for this exercise is at the end of the book.)

4.9.6. Verify the calculations leading up to formula (4.38).

4.10 Nonholomorphic Eisenstein series

Hecke put the Eisenstein series of low weights 1 and 2 into a larger context by augmenting Eisenstein series with a complex parameter s. We will study a variant of Hecke's original series. Let N be a positive integer, let $\overline{v}$ be a vector in $(\mathbb{Z}/N\mathbb{Z})^2$ of order N, let k be any integer, and let ϵ_N be $1/2$ if $N \in \{1, 2\}$ and 1 if $N > 2$. Define

$$E_k^{\overline{v}}(\tau, s) = \epsilon_N \sum_{\substack{(c,d) \equiv v\,(N) \\ \gcd(c,d)=1}}' \frac{y^s}{(c\tau + d)^k |c\tau + d|^{2s}}, \quad \tau = x + iy \in \mathcal{H}.$$

This converges absolutely on the right half plane $\{s : \operatorname{Re}(k + 2s) > 2\}$. The convergence is uniform on compact subsets, making $E_k^{\overline{v}}(\tau, s)$ an analytic function of s on this half plane. To rewrite the series in intrinsic form, let $\delta = \left[\begin{smallmatrix} a & b \\ c_v & d_v \end{smallmatrix}\right] \in \mathrm{SL}_2(\mathbb{Z})$ with (c_v, d_v) a lift of $\overline{v}$ to $\mathbb{Z}^2$, and recall the positive part of the parabolic subgroup of $\mathrm{SL}_2(\mathbb{Z})$, $P_+ = \{\left[\begin{smallmatrix} 1 & n \\ 0 & 1 \end{smallmatrix}\right] : n \in \mathbb{Z}\}$. Extend the weight-$k$ operator to functions of τ and s by the definition

$$(f[\gamma]_k)(\tau, s) = j(\gamma, \tau)^{-k} f(\gamma(\tau), s), \quad \gamma \in \mathrm{SL}_2(\mathbb{Z}).$$

Then

$$E_k^{\overline{v}}(\tau, s) = \epsilon_N \sum_{\gamma \in (P_+ \cap \Gamma(N)) \backslash \Gamma(N)\delta} \operatorname{Im}(\tau)^s [\gamma]_k.$$

This formula makes it clear that $(E_k^{\overline{v}}[\gamma]_k)(\tau, s) = E_k^{\overline{v}\gamma}(\tau, s)$ for all $\gamma \in \mathrm{SL}_2(\mathbb{Z})$ as in Proposition 4.2.1, so that in particular

$$E_k^{\overline{v}}[\gamma]_k = E_k^{\overline{v}} \quad \text{for all } \gamma \in \Gamma(N).$$

As earlier, the corresponding nonnormalized series

$$G_k^{\overline{v}}(\tau, s) = \sum_{(c,d) \equiv v\,(N)}' \frac{y^s}{(c\tau + d)^k |c\tau + d|^{2s}} \tag{4.39}$$

will be easier to analyze. Relations like (4.5) and (4.6) expressing each $G_k^{\overline{v}}$ as a linear combination of the $E_k^{\overline{v}}$ and vice versa still hold (Exercise 4.10.1(a)), so $G_k^{\overline{v}}$ transforms under $\mathrm{SL}_2(\mathbb{Z})$ in the same way as $E_k^{\overline{v}}$ (Exercise 4.10.1(b)). The discussion leading to relations (4.20) and (4.21) generalizes to show that meromorphically continuing either sort of Eisenstein series meromorphically continues the other (Exercise 4.10.1(c)). This section will show that $G_k^{\overline{v}}$ as a function of s has a meromorphic continuation to the entire s-plane. For s-values to the left of the original half plane of convergence $\operatorname{Re}(k + 2s) > 2$, the continuation is no longer defined by the sum, but the transformation law continues to hold by the Uniqueness Theorem from complex analysis. The natural Eisenstein series to consider is

$$G_k^{\overline{v}}(\tau, 0) = G_k^{\overline{v}}(\tau, s)|_{s=0}.$$

For $k \geq 3$ the parameter s simply appears and disappears, but for $k \leq 2$ it gives something new. Hecke's method for weights 1 and 2, obtaining the functions $g_1^{\overline{v}}(\tau)$ and $g_2^{\overline{v}}(\tau)$ from before, is presented in [Sch74], and a generalization of Hecke's method due to Shimura is presented in [Miy89] and [Hid93]. The method here uses a Mellin transform and Poisson summation, extending the ideas of the previous section. See for example Rankin [Ran39], Selberg [Sel40], and Godement [God66].

Working in dimension $l = 2$, view elements of $\mathbb{R}^2$ as row vectors and define a modified theta function whose argument is now a matrix,

$$\vartheta(\gamma) = \sum_{n \in \mathbb{Z}^2} e^{-\pi |n\gamma|^2}, \quad \gamma \in \mathrm{GL}_2(\mathbb{R}).$$

Again the second variable from the general notation for ϑ is omitted since l is fixed at 2. When $\gamma = It^{1/2}$ where I is the identity matrix and $t > 0$, this is $\vartheta(it)$ from the previous section. To see how a more general matrix γ affects the transformation law for ϑ, compute that for any function $f \in \mathcal{L}^1(\mathbb{R}^2)$, any matrix $\gamma \in \mathrm{SL}_2(\mathbb{R})$, and any positive number r, the Fourier transform of the function $\varphi(x) = f(x\gamma r)$ is $\hat{\varphi}(x) = r^{-2}\hat{f}(x\gamma^{-T}/r)$ where γ^{-T} denotes the inverse-transpose of γ (Exercise 4.10.2). In particular, again letting $f(x) = e^{-\pi |x|^2}$ be the Gaussian, Poisson summation shows that for $\gamma \in \mathrm{SL}_2(\mathbb{R})$, $r \sum f(n\gamma r) = r^{-1} \sum f(n\gamma^{-T}/r)$. Let $S = \left[\begin{smallmatrix} 0 & -1 \\ 1 & 0 \end{smallmatrix}\right]$. Then $S\gamma^{-T} = \gamma S$ for all $\gamma \in \mathrm{SL}_2(\mathbb{R})$ and $|xS| = |x|$ for all $x \in \mathbb{R}^2$, so $f(nS\gamma^{-T}/r) = f(n\gamma/r)$. As n varies through $\mathbb{Z}^2$ so does nS, and thus the Poisson summation is $r \sum f(n\gamma r) = r^{-1} \sum f(n\gamma/r)$. That is,

$$r\vartheta(\gamma r) = (1/r)\vartheta(\gamma/r), \quad \gamma \in \mathrm{SL}_2(\mathbb{R}), \ r > 0. \tag{4.40}$$

For any $\gamma \in \mathrm{SL}_2(\mathbb{R})$, the Mellin transform of $\sum'_{n \in \mathbb{Z}^2} e^{-\pi |n\gamma|^2 t} = \vartheta(\gamma t^{1/2}) - 1$ for $t > 0$ is

$$g(s, \gamma) = \int_{t=0}^{\infty} \sum_{n \in \mathbb{Z}^2}' e^{-\pi |n\gamma|^2 t} t^s \frac{dt}{t} = \int_{t=0}^{\infty} (\vartheta(\gamma t^{1/2}) - 1) t^s \frac{dt}{t}. \tag{4.41}$$

Again $\vartheta(\gamma t^{1/2})$ converges to 1 as $t \to \infty$, so the transformation law (4.40) shows that as $t \to 0$, $\vartheta(\gamma t^{1/2})$ grows as $1/t$, and therefore the integral converges at its left endpoint if $\mathrm{Re}(s) > 1$. Also as before, the integral converges at its right end for all values of s, and rapid convergence of the sum lets it pass through the integral to yield after a change of variable,

$$g(s, \gamma) = \sum_{n \in \mathbb{Z}^2}' (\pi |n\gamma|^2)^{-s} \int_{t=0}^{\infty} e^{-t} t^s \frac{dt}{t} = \pi^{-s} \Gamma(s) \sum_{n \in \mathbb{Z}^2}' |n\gamma|^{-2s}$$

for $\mathrm{Re}(s) > 1$. To connect this with Eisenstein series, for any point $\tau = x + iy \in \mathcal{H}$ let γ_τ be the corresponding matrix

$$\gamma_\tau = \frac{1}{\sqrt{y}} \begin{bmatrix} y & x \\ 0 & 1 \end{bmatrix} \in \mathrm{SL}_2(\mathbb{R}).$$

(This matrix, which transforms i to τ, appeared in Chapter 2, in Exercise 2.1.3(c) and again in the proof of Corollary 2.3.4.) For $n = (c, d) \in \mathbb{Z}^2$ compute that $|n\gamma_\tau|^2 = |(cy, cx + d)/\sqrt{y}|^2 = |c\tau + d|^2/y$. Therefore the Mellin transform is essentially the Eisenstein series,

$$g(s, \gamma_\tau) = \pi^{-s}\Gamma(s) \sideset{}{'}\sum_{(c,d)} \frac{y^s}{|c\tau + d|^{2s}} = \pi^{-s}\Gamma(s)G_0(\tau, s), \quad \mathrm{Re}(s) > 1,$$

where $G_0(\tau, s)$ is (4.39) with weight $k = 0$ and level $N = 1$. (In this case there is only one $\overline{v} \in (\mathbb{Z}/N\mathbb{Z})^2$, so it is suppressed from the notation $G_0^{\overline{v}}$.)

The second integral in (4.41) provides the meromorphic continuation and functional equation of G_0. As in the previous section, the transformation law for ϑ shows that part of the integral is

$$\begin{aligned}
\int_{t=0}^1 (\vartheta(\gamma t^{1/2}) - 1)t^s \frac{dt}{t} &= \int_{t=0}^1 \vartheta(\gamma t^{1/2})t^s \frac{dt}{t} - \frac{1}{s} \\
&= \int_{t=1}^\infty \vartheta(\gamma t^{-1/2})t^{-s} \frac{dt}{t} - \frac{1}{s} \\
&= \int_{t=1}^\infty \vartheta(\gamma t^{1/2})t^{1-s} \frac{dt}{t} - \frac{1}{s} \\
&= \int_{t=1}^\infty (\vartheta(\gamma t^{1/2}) - 1)t^{1-s} \frac{dt}{t} - \frac{1}{s} - \frac{1}{1-s}.
\end{aligned}$$

Combining this with the remainder of the integral gives

$$g(s, \gamma) = \int_{t=1}^\infty (\vartheta(\gamma t^{1/2}) - 1)(t^s + t^{1-s}) \frac{dt}{t} - \frac{1}{s} - \frac{1}{1-s}, \quad \mathrm{Re}(s) > 1.$$

This integral is entire in s, making the right side holomorphic everywhere in the s-plane except for simple poles at $s = 0$ and $s = 1$. And the right side is invariant under $s \mapsto 1-s$. Specializing to $\gamma = \gamma_\tau$, the function $\pi^{-s}\Gamma(s)G_0(\tau, s)$ for $\mathrm{Re}(s) > 1$ has a meromorphic continuation to the full s-plane that is invariant under $s \mapsto 1 - s$.

Suitably modified, this argument extends to higher weights and levels. Let $N \geq 1$ and let G be the group $(\mathbb{Z}/N\mathbb{Z})^2$. Rather than taking one vector $\overline{v} \in G$, consider any function

$$a : G \longrightarrow \mathbb{C}.$$

Of course, the function a could simply pick off a vector, but thinking about the entire group symmetrizes the result nicely. Let $\mu_N = e^{2\pi i/N}$, let $S = \begin{bmatrix} 0 & -1 \\ 1 & 0 \end{bmatrix}$, and let $\langle \, , \rangle$ be the usual inner product on $\mathbb{R}^2$. The *Fourier transform* of a is a function $\hat{a} : G \longrightarrow \mathbb{C}$,

$$\hat{a}(\overline{v}) = \frac{1}{N} \sum_{\overline{w} \in G} a(\overline{w}) \mu_N^{-\langle w, vS \rangle}, \quad \overline{v} \in G.$$

The resulting Fourier series reproduces the original function (Exercise 4.10.3)

$$a(\overline{u}) = \frac{1}{N} \sum_{\overline{v} \in G} \hat{a}(\overline{v}) \mu_N^{\langle u, vS \rangle}, \quad \overline{u} \in G,$$

and $\hat{\hat{a}} = a$ (Exercise 4.10.3 again).

Let k be a positive integer and let h_k be the harmonic polynomial

$$h_k(c, d) = (-i)^k (c + id)^k, \quad (c, d) \in \mathbb{R}^2.$$

Associate a theta function to each vector $\overline{v} \in (\mathbb{Z}/N\mathbb{Z})^2$,

$$\vartheta_k^{\overline{v}}(\gamma) = \sum_{n \in \mathbb{Z}^2} h_k((v/N + n)\gamma) e^{-\pi |(v/N+n)\gamma|^2}, \quad \gamma \in \mathrm{GL}_2(\mathbb{R}). \tag{4.42}$$

Again letting $f(x) = e^{-\pi|x|^2}$, define the *Schwartz function* $f_k(x)$ to be a modified Gaussian,

$$f_k(x) = h_k(x) f(x).$$

Thus $\vartheta_k^{\overline{v}}(\gamma) = \sum_{n \in \mathbb{Z}^2} f_k((v/N + n)\gamma)$. For $\gamma \in \mathrm{SL}_2(\mathbb{R})$ and $r > 0$, compute that $r\vartheta_k^{\overline{v}}(\gamma r) = r \sum f_k((v/N + n)\gamma r) = r \sum \varphi_k(v/N + n)$ where $\varphi_k(x) = f_k(x\gamma r)$. The Schwartz function has Fourier transform $\hat{f}_k = (-i)^k f_k$ (Exercise 4.10.4(a)), and so $\hat{\varphi}_k(x) = (-i)^k r^{-2} f_k(x\gamma^{-T} r^{-1})$. Poisson summation over nS and the relations $S\gamma^{-T} = \gamma S$ and $S^T = -S$ give

$$r \vartheta_k^{\overline{v}}(\gamma r) = (-i)^k r^{-1} \sum_{n \in \mathbb{Z}^2} f_k(nS\gamma^{-T} r^{-1}) e^{2\pi i \langle nS, v/N \rangle}$$

$$= (-i)^k r^{-1} \sum_{n \in \mathbb{Z}^2} f_k(n\gamma S r^{-1}) e^{2\pi i \langle n, -vS/N \rangle}.$$

Letting $z(x) = c + id \in \mathbb{C}$ for any $x = (c, d) \in \mathbb{R}^2$ makes $h_k(x) = (-i)^k (z(x))^k = (z(xS))^k$ and thus $h_k(xS) = (-i)^k h_k(x)$ (Exercise 4.10.4(b)). Consequently $f_k(xS) = (-i)^k f_k(x)$ as well, and now

$$r \vartheta_k^{\overline{v}}(\gamma r) = (-1)^k r^{-1} \sum_{n \in \mathbb{Z}^2} f_k(n\gamma r^{-1}) \mu_N^{-\langle n, vS \rangle}$$

$$= (-1)^k r^{-1} \sum_{\overline{w} \in G} \sum_{\substack{n \in \mathbb{Z}^2 \\ n \equiv w \,(N)}} f_k(n\gamma r^{-1}) \mu_N^{-\langle w, vS \rangle}$$

$$= (-1)^k r^{-1} \sum_{\overline{w} \in G} \sum_{n \in \mathbb{Z}^2} f_k((w/N + n)\gamma N r^{-1}) \mu_N^{-\langle w, vS \rangle}$$

$$= (-1)^k r^{-1} \sum_{\overline{w} \in G} \vartheta_k^{\overline{w}}(\gamma N r^{-1}) \mu_N^{-\langle w, vS \rangle}.$$

Thinking here of $\vartheta^{\overline{v}}$ as a function of $\overline{v}$, the sum is N times this function's Fourier transform, giving $r\,\vartheta_k^{\overline{v}}(\gamma r) = (-1)^k N r^{-1} \widehat{\vartheta_k^{\overline{v}}}(\gamma N r^{-1})$. Since $r > 0$ is arbitrary, replace it by $N^{1/2}r$ to get

$$r\,\vartheta_k^{\overline{v}}(\gamma N^{1/2}r) = (-1)^k r^{-1} \widehat{\vartheta_k^{\overline{v}}}(\gamma N^{1/2}r^{-1}), \quad \gamma \in \mathrm{SL}_2(\mathbb{R}),\; r > 0. \tag{4.43}$$

Thus (Exercise 4.10.5)

Proposition 4.10.1. *For any function* $a : G \longrightarrow \mathbb{C}$, *the associated sum of theta functions*

$$\Theta_k^a(\gamma) = \sum_{\overline{v} \in G} (a(\overline{v}) + (-1)^k \hat{a}(-\overline{v}))\vartheta_k^{\overline{v}}(\gamma N^{1/2}), \quad \gamma \in \mathrm{GL}_2(\mathbb{R})$$

satisfies the transformation law

$$r\,\Theta_k^a(\gamma r) = r^{-1}\Theta_k^a(\gamma r^{-1}), \quad \gamma \in \mathrm{SL}_2(\mathbb{R}),\; r > 0.$$

The functional equation for the associated sum of Eisenstein series follows as before. The constant terms of the series $\vartheta_k^{\overline{v}}(\gamma)$ are 0, so that $\Theta_k^a(\gamma r)$ converges to 0 rapidly as $r \to \infty$. Consider the Mellin transform

$$g_k^a(s, \gamma) = \int_{t=0}^{\infty} \Theta_k^a(\gamma t^{1/2})t^s\,\frac{dt}{t}. \tag{4.44}$$

The transformation law shows that the integral converges at $t = 0$ for all values of s, and as before it converges at its right end for all s as well. Passing the sums through the integral, noting that $h_k(xr) = h_k(x)r^k$ for $r \in \mathbb{R}$, and changing variable gives

$$g_k^a(s, \gamma) = \pi^{-k/2-s}\Gamma(k/2 + s)N^s$$
$$\cdot \sum_{\overline{v} \in G} (a(\overline{v}) + (-1)^k \hat{a}(-\overline{v})) \sum_{n \equiv v\,(N)}^{\prime} h_k(n\gamma)|n\gamma|^{-k-2s}.$$

Setting $\gamma = \gamma_\tau$ gives $h_k(n\gamma) = (z(n\gamma S))^k = \overline{(c\tau + d)}^k/y^{k/2}$ and gives $|n\gamma|^{-k-2s} = y^{k/2+s}/|c\tau + d|^{k+2s}$, with product $y^s/((c\tau + d)^k|c\tau + d|^{2(s-k/2)})$. Thus for $\mathrm{Re}(s) > 1$,

$$g_k^a(s, \gamma_\tau) = \pi^{-k/2-s}\Gamma(k/2 + s)N^s y^{k/2} G_k^a(\tau, s - k/2) \tag{4.45}$$

where G_k^a is the associated sum of Eisenstein series (4.39),

$$G_k^a(\tau, s) = \sum_{\overline{v}} (a(\overline{v}) + (-1)^k \hat{a}(-\overline{v}))G_k^{\overline{v}}(\tau, s), \quad \mathrm{Re}(k/2 + s) > 1. \tag{4.46}$$

Proposition 4.10.1 shows that part of the integral (4.44) is

$$\int_{t=0}^{1} \Theta_k^a(\gamma t^{1/2})t^s\,\frac{dt}{t} = \int_{t=1}^{\infty} \Theta_k^a(\gamma t^{-1/2})t^{-s}\,\frac{dt}{t} = \int_{t=1}^{\infty} \Theta_k^a(\gamma t^{1/2})t^{1-s}\,\frac{dt}{t}.$$

Combining this with the remainder of the integral gives

$$g_k^{\overline{v}}(s,\gamma) = \int_{t=1}^{\infty} \Theta_k^a(\gamma t^{1/2})(t^s + t^{1-s})\frac{dt}{t}.$$

The integral is entire in s and invariant under $s \mapsto 1 - s$. Similar arguments for $k = 0$ and $k < 0$ (Exercise 4.10.6) show that in all cases,

Theorem 4.10.2. *Let N be a positive integer and let $G = (\mathbb{Z}/N\mathbb{Z})^2$. For any function $a : G \longrightarrow \mathbb{C}$, let $G_k^a(\tau, s)$ be the associated sum (4.46) of Eisenstein series. Then for any integer k and any point $\tau = x + iy \in \mathcal{H}$, the function*

$$(\pi/N)^{-s}\Gamma(|k|/2 + s)G_k^a(\tau, s - k/2), \quad \mathrm{Re}(s) > 1$$

has a continuation to the full s-plane that is invariant under $s \mapsto 1 - s$. The continuation is analytic for $k \neq 0$ and has simple poles at $s = 0, 1$ for $k = 0$.

The function in the theorem comes from canceling the s-independent terms in the Mellin transform (4.45). Exercise 4.10.8(b–c) shows that sums of only two Eisenstein series satisfy functional equations for the larger group $\Gamma_1(N)$.

The meromorphically continued Eisenstein series of this section have several applications. As already mentioned, continuing the series for weights $k = 1$ and $k = 2$ to $s = 0$ recovers the Eisenstein series from earlier in the chapter. Also, the *Rankin–Selberg method* integrates an Eisenstein series against the square of the absolute value of a cusp form, obtaining an L-function satisfying a functional equation. This idea was originally used to estimate the Fourier coefficients of the cusp form, but L-functions of the sort obtained by this method are now understood to be important in their own right. For more on this topic, see [Bum97].

A more subtle use of the meromorphically continued Eisenstein series is in the study of square-integrable functions called *automorphic forms* on the quotient space $\mathrm{SL}_2(\mathbb{Z})\backslash\mathrm{SL}_2(\mathbb{R})$, the *arithmetic quotient*. The space of such functions decomposes as a discrete part and a continuous part, the former like Fourier series and the latter like Fourier transforms. The discrete part contains cusp forms and a little more, the *residual spectrum*, so called because it consists of residues of Eisenstein series in the half plane to the right of the line of symmetry for the functional equation. The continuous part consists of integrals of Eisenstein series on the critical line against square-integrable functions on that line. Thus the explicit spectral decomposition cannot be stated at all without knowing the meromorphic continuation of Eisenstein series. These ideas extend in interesting ways to matrix groups beyond $\mathrm{SL}_2(\mathbb{R})$, where their original proofs often proceeded by establishing the meromorphic continuation en route.

Exercises

4.10.1. (a) Find relations like (4.5) and (4.6) expressing each $G_k^{\overline{v}}$ as a linear combination of the $E_k^{\overline{v}}$ and vice versa.

(b) Verify that $G_k^{\overline{v}}$ transforms under $\mathrm{SL}_2(\mathbb{Z})$ the same way as $E_k^{\overline{v}}$.

(c) Show that meromorphic continuation of either sort of Eisenstein series gives meromorphic continuation of the other.

4.10.2. Let $f(x)$ be a function in $\mathcal{L}^1(\mathbb{R}^2)$. For any $\gamma \in \mathrm{SL}_2(\mathbb{R})$, and $r > 0$, show that the function $\varphi(x) = f(x\gamma r)$ has Fourier transform $\hat{\varphi}(x) = r^{-2}\hat{f}(x\gamma^{-T}/r)$ where γ^{-T} denotes the inverse-transpose of γ. (A hint for this exercise is at the end of the book.)

4.10.3. Show that $a(\overline{u}) = \dfrac{1}{N}\sum_{\overline{v}\in G}\hat{a}(\overline{v})\mu_N^{\langle u,vS\rangle}$ for $\overline{u}\in G$ and that $\hat{\hat{a}} = a$. (A hint for this exercise is at the end of the book.)

4.10.4. (a) Let $f \in \mathcal{L}^1(\mathbb{R}^l)$, let $j \in \{1,\ldots,l\}$, and define

$$g : \mathbb{R}^l \longrightarrow \mathbb{C}, \quad g(x) = x_j f(x)$$

where x_j is the jth component of x. Suppose that $g \in \mathcal{L}^1(\mathbb{R}^l)$. Show that

$$\hat{g}(x) = -\frac{1}{2\pi i}\frac{\partial}{\partial x_j}\hat{f}(x).$$

Applying this identity repeatedly shows that the Schwartz function $f_k(x,y) = (-i)^k(x+iy)^k e^{-\pi(x^2+y^2)}$ has Fourier transform

$$\hat{f}_k(x,y) = (-i)^k\left(-\frac{1}{2\pi i}\right)^k\left(\frac{\partial}{\partial x}+i\frac{\partial}{\partial y}\right)^k e^{-\pi(x^2+y^2)}.$$

Switching to complex notation $f_k(z) = (-i)^k z^k e^{-\pi z\overline{z}}$, this is

$$\hat{f}_k(z) = (-i)^k\left(-\frac{1}{\pi i}\right)^k\left(\frac{\partial}{\partial\overline{z}}\right)^k e^{-\pi z\overline{z}}.$$

Compute that this relation is $\hat{f}_k = (-i)^k f_k$ as desired.

(b) Show that $h_k(x) = (z(xS))^k$ and $h_k(xS) = (-i)^k h_k(x)$.

4.10.5. (a) Show that for all $\gamma \in \mathrm{GL}_2(\mathbb{R})$,

$$\sum_{\overline{v}\in G}a(\overline{v})\vartheta_k^{\overline{v}}(\gamma) = \sum_{\overline{v}\in G}\hat{a}(-\overline{v})\widehat{\vartheta_k^{\overline{v}}}(\gamma) \text{ and } \sum_{\overline{v}\in G}\hat{a}(-\overline{v})\vartheta_k^{\overline{v}}(\gamma) = \sum_{\overline{v}\in G}a(\overline{v})\widehat{\vartheta_k^{\overline{v}}}(\gamma).$$

(A hint for this exercise is at the end of the book.)

(b) Prove Proposition 4.10.1.

4.10.6. (a) Define $h_0(x) = 1$ for all $x \in \mathbb{R}^2$. For $N \geq 1$ and $k = 0$ define theta series $\{\vartheta_0^{\overline{v}} : \overline{v}\in G\}$ by the formula in the section. Prove Theorem 4.10.2 when $k = 0$.

(b) When $k < 0$ define $h_k(x) = \overline{h}_{-k}(x)$ for $x \in \mathbb{R}^2$, where the overbar denotes complex conjugation. Prove Theorem 4.10.2 when $k < 0$. (A hint for this exercise is at the end of the book.)

4.10.7. What does Theorem 4.10.2 say when the function a picks off one vector, i.e., $a = a_{\overline{v}}$ where $a_{\overline{v}}(\overline{w})$ is 1 if $\overline{w} = \overline{v}$ and 0 otherwise?

4.10.8. (a) Let ψ and φ be primitive Dirichlet characters modulo u and v with $(\psi\varphi)(-1) = (-1)^k$ and $uv = N$. If $a : (\mathbb{Z}/N\mathbb{Z})^2 \longrightarrow \mathbb{C}$ is the function

$$a(\overline{cv}, \overline{d + ev}) = \psi(c)\overline{\varphi}(d), \quad a(\overline{x}, \overline{y}) = 0 \text{ otherwise,}$$

show that its Fourier transform is

$$\hat{a}(-\overline{cu}, -\overline{(d + eu)}) = (\varphi(-1)g(\overline{\varphi})/v)g(\psi)\varphi(c)\overline{\psi}(d), \quad \hat{a}(\overline{x}, \overline{y}) = 0 \text{ otherwise.}$$

(Hints for this exercise are at the end of the book.)
  (b) Define an Eisenstein series with parameter,

$$G_k^{\psi,\varphi}(\tau, s) = \sum_{c=0}^{u-1} \sum_{d=0}^{v-1} \sum_{e=0}^{u-1} \psi(c)\overline{\varphi}(d) G_k^{\overline{(cv, d+ev)}}(\tau, s).$$

Show that the sum (4.46) of Eisenstein series for the function a in this problem is

$$G_k^a(\tau, s) = G_k^{\psi,\varphi}(\tau, s) + \psi(-1)(g(\overline{\varphi})/v)g(\psi)G_k^{\varphi,\psi}(\tau, s).$$

Thus the functional equations for the eigenspaces $\mathcal{E}_k(N, \chi)$ involve only two series at a time.
  (c) Define a function $b : (\mathbb{Z}/N\mathbb{Z})^2 \longrightarrow \mathbb{C}$ and a series $E_k^{\psi,\varphi}(\tau, s)$ by the conditions

$$a = \frac{g(\overline{\varphi})}{v}b, \qquad G_k^{\psi,\varphi} = \frac{g(\overline{\varphi})}{v}E_k^{\psi,\varphi}.$$

Show that the sum (4.46) of Eisenstein series for b is more nicely symmetrized than the one for a,

$$G_k^b(\tau, s) = E_k^{\psi,\varphi}(\tau, s) + E_k^{\varphi,\psi}(\tau, s).$$

4.11 Modular forms via theta functions

This chapter ends by using theta functions to construct a modular form that both connects back to the preface and adumbrates the ideas at the end of the book. The construction is one case of a general method due to Hecke [Hec26].
  Recall that the preface used Quadratic Reciprocity to motivate the Modularity Theorem via a simple analog, counting the solutions modulo p to the quadratic equation $x^2 = d$. Now consider a cubic equation instead,

$$C : x^3 = d, \qquad d \in \mathbb{Z}^+, \ d \text{ cubefree,}$$

and for each prime p let

$$a_p(C) = (\text{the number of solutions modulo } p \text{ of equation } C) - 1.$$

Results from elementary number theory show that (Exercise 4.11.1)

$$a_p(C) = \begin{cases} 2 & \text{if } p \equiv 1 \ (\text{mod } 3) \text{ and } d \text{ is a nonzero cube modulo } p, \\ -1 & \text{if } p \equiv 1 \ (\text{mod } 3) \text{ and } d \text{ is not a cube modulo } p, \\ 0 & \text{if } p \equiv 2 \ (\text{mod } 3) \text{ or } p \mid 3d. \end{cases} \qquad (4.47)$$

This section will use Poisson summation and the Cubic Reciprocity Theorem from number theory to construct a modular form θ_χ with Fourier coefficients $a_p(\theta_\chi) = a_p(C)$. Section 5.9 will show that these Fourier coefficients are eigenvalues. That is, the solution-counts of the cubic equation C are a system of eigenvalues arising from a modular form. Chapter 9 will further place this example in the context of Modularity.

Introduce the notation

$$\mathbf{e}(z) = e^{2\pi i z}, \quad z \in \mathbb{C}.$$

Let $A = \mathbb{Z}[\mu_3]$, let $\alpha = i\sqrt{3}$, and let $B = \frac{1}{\alpha}A$. Thus $A \subset B \subset \frac{1}{3}A$. Note that

$$|x|^2 = x_1^2 - x_1 x_2 + x_2^2 \quad \text{for any } x = x_1 + x_2\mu_3 \in \mathbb{R}[\mu_3].$$

We will frequently use the formula $|x+y|^2 = |x|^2 + \mathrm{tr}\,(xy^*) + |y|^2$ for $x, y \in \mathbb{C}$, where y^* is the complex conjugate of y and $\mathrm{tr}\,(z) = z + z^*$. For any positive integer N and any $\overline{u}$ in the quotient group $\frac{1}{3}A/NA$ define a theta function,

$$\theta^{\overline{u}}(\tau, N) = \sum_{n \in A} \mathbf{e}\left(N|u/N + n|^2\tau\right), \quad \tau \in \mathcal{H}. \qquad (4.48)$$

An argument similar to Exercise 4.9.2 shows that $\theta^{\overline{u}}$ is holomorphic. The following lemma establishes its basic transformation properties. From now until near the end of the section the symbol d is unrelated to the d of the cubic equation C.

Lemma 4.11.1. *Let N be a positive integer. Then*

$$\theta^{\overline{u}}(\tau + 1, N) = \mathbf{e}\left(\frac{|u|^2}{N}\right)\theta^{\overline{u}}(\tau, N), \qquad\qquad \overline{u} \in B/NA,$$

$$\theta^{\overline{u}}(\tau, N) = \sum_{\substack{\overline{v} \in B/dNA \\ \overline{v} \equiv \overline{u}\ (NA)}} \theta^{\overline{v}}(d\tau, dN), \qquad\qquad \overline{u} \in B/NA,\ d \in \mathbb{Z}^+,$$

$$\theta^{\overline{v}}(-1/\tau, N) = \frac{-i\tau}{N\sqrt{3}} \sum_{\overline{w} \in B/NA} \mathbf{e}\left(-\frac{\mathrm{tr}\,(vw^*)}{N}\right)\theta^{\overline{w}}(\tau, N), \quad \overline{v} \in B/NA.$$

Proof. For the first statement compute that for $u \in B$ and $n \in A$,

$$N\left|\frac{u}{N} + n\right|^2 \equiv \frac{|u|^2}{N} \ (\text{mod } \mathbb{Z}).$$

For the second statement note that $\theta^{\overline{u}}(\tau, N) = \sum_{n \in A} \mathbf{e}\left(dN | \frac{u/N+n}{d}|^2 d\tau\right)$. Let $n = r + dm$, making the fraction $\frac{u+Nr}{dN} + m$. Thus

$$\theta^{\overline{u}}(\tau, N) = \sum_{\substack{\overline{r} \in A/dA \\ m \in A}} \mathbf{e}\left(dN\left|\frac{u+Nr}{dN} + m\right|^2 d\tau\right) = \sum_{\overline{r} \in A/dA} \theta^{\overline{u+Nr}}(d\tau, dN),$$

where the reduction $\overline{u + Nr}$ is taken modulo dN. This gives the result.

The third statement is shown by Poisson summation. Recall that the defining equation (4.42) from the previous section was extended to $k = 0$ in Exercise 4.10.6(a),

$$\vartheta_0^{\overline{v}}(\gamma) = \sum_{n \in \mathbb{Z}^2} e^{-\pi|(v/N+n)\gamma|^2}, \quad \overline{v} \in (\mathbb{Z}/N\mathbb{Z})^2, \; \gamma \in \mathrm{GL}_2(\mathbb{R}).$$

To apply this let $\gamma \in \mathrm{SL}_2(\mathbb{R})$ be the positive square root of $\frac{1}{\sqrt{3}}\begin{bmatrix} 2 & -1 \\ -1 & 2 \end{bmatrix}$, satisfying

$$|(x_1, x_2)\gamma|^2 = \tfrac{2}{\sqrt{3}}|x_1 + x_2\mu_3|^2, \quad x_1, x_2 \in \mathbb{R}.$$

Note that $B/NA \subset \frac{1}{3}A/NA \cong A/3NA$. Identify A with $\mathbb{Z}^2$ so that if $v \in B$ then its multiple $3v \in \alpha A \subset A$ can also be viewed as an element of $\mathbb{Z}^2$. Compute with $3N$ in place of N and with γ as above that for any $t \in \mathbb{R}^+$,

$$\vartheta_0^{\overline{3v}}(\gamma(3N)^{1/2}(t/\sqrt{3})^{1/2}) = \sum_{n \in \mathbb{Z}^2} e^{-\pi\sqrt{3}N\left|\left(\frac{3v}{3N}+n\right)\gamma\right|^2 t}$$

$$= \sum_{n \in A} e^{-2\pi N|\frac{v}{N}+n|^2 t} = \theta^{\overline{v}}(it, N).$$

This shows that the identity (4.43) with $k = 0$, with $3N$ in place of N, with γ as above, and with $r = (\sqrt{3}\,t)^{-1/2}$ is (Exercise 4.11.2)

$$\theta^{\overline{v}}(-1/(it), N) = \frac{t}{N\sqrt{3}} \sum_{\overline{u} \in \frac{1}{3}A/NA} \mathbf{e}\left(-\frac{3(vu^*)_2}{N}\right) \theta^{\overline{u}}(3it, N), \quad \overline{v} \in B/NA,$$

where $vu^* = (vu^*)_1 + (vu^*)_2\mu_3$. Generalize from $t \in \mathbb{R}^+$ to $-i\tau$ for $\tau \in \mathcal{H}$ by the Uniqueness Theorem of complex analysis to get

$$\theta^{\overline{v}}(-1/\tau, N) = \frac{-i\tau}{N\sqrt{3}} \sum_{\overline{u} \in \frac{1}{3}A/NA} \mathbf{e}\left(-\frac{(\alpha v(\alpha u)^*)_2}{N}\right) \theta^{\overline{u}}(3\tau, N), \quad \overline{v} \in B/NA.$$

The sum on the right side is

$$\sum_{\overline{w} \in B/NA} \mathbf{e}\left(-\frac{(\alpha v w^*)_2}{N}\right) \sum_{\substack{\overline{u} \in \frac{1}{3}A/NA \\ \overline{\alpha u} = \overline{w} \,(NA)}} \theta^{\overline{u}}(3\tau, N).$$

To simplify the inner sum note that $\theta^{\overline{w}}(\tau, N) = \sum_{n \in A} e\left(N|\frac{w/N+n}{\alpha}|^2 3\tau\right)$ for any $\overline{w} \in B/NA$, and similarly to the proof of the second statement this works out to

$$\theta^{\overline{w}}(\tau, N) = \sum_{\substack{\overline{r} \in A/\alpha A \\ m \in A}} e\left(N \left|\frac{(w+Nr)/\alpha}{N} + m\right|^2 3\tau\right) = \sum_{\overline{r} \in A/\alpha A} \theta^{\overline{(w+Nr)/\alpha}}(3\tau, N).$$

That is, the inner sum is

$$\sum_{\substack{\overline{u} \in \frac{1}{3} A/NA \\ \overline{\alpha u} \equiv \overline{w} \ (NA)}} \theta^{\overline{u}}(3\tau, N) = \theta^{\overline{w}}(\tau, N), \quad \overline{w} \in B/NA.$$

The proof is completed by noting that $(\alpha v w^*)_2 = \text{tr}(v w^*)$.   □

The next result shows how the theta function transforms under the group

$$\Gamma_0(3N, N) = \left\{ \begin{bmatrix} a & b \\ c & d \end{bmatrix} \in SL_2(\mathbb{Z}) : b \equiv 0 \ (\text{mod } N), \ c \equiv 0 \ (\text{mod } 3N) \right\}.$$

Proposition 4.11.2. *Let N be a positive integer. Then*

$$(\theta^{\overline{u}}[\gamma]_1)(\tau, N) = \left(\frac{d}{3}\right) \theta^{\overline{au}}(\tau, N), \quad \overline{u} \in A/NA, \ \gamma = \begin{bmatrix} a & b \\ c & d \end{bmatrix} \in \Gamma_0(3N, N).$$

Here $(d/3)$ is the Legendre symbol.

Proof. Since $\theta^{-\overline{au}} = \theta^{\overline{au}}$ we may assume $d > 0$ by replacing γ with $-\gamma$ if necessary. Write

$$\frac{a\tau + b}{c\tau + d} = \frac{1}{d}\left(\frac{1}{d/\tau + c} + b\right).$$

Apply the second statement of the lemma and then the first statement to get

$$\theta^{\overline{u}}(\gamma(\tau), N) = \sum_{\substack{\overline{v} \in B/dNA \\ \overline{v} \equiv \overline{u} \ (NA)}} e\left(\frac{b|v|^2}{dN}\right) \theta^{\overline{v}}\left(-\frac{1}{-d/\tau - c}, dN\right).$$

The third statement and again the first now give

$$\theta^{\overline{u}}(\gamma(\tau), N) = \frac{i(d/\tau + c)}{dN\sqrt{3}} \sum_{\substack{\overline{v}, \overline{w} \in B/dNA \\ \overline{v} \equiv \overline{u} \ (NA)}} e\left(\frac{b|v|^2 - \text{tr}(vw^*)}{dN}\right) \theta^{\overline{w}}(-d/\tau - c, dN)$$

$$= \frac{i(c\tau + d)}{dN\sqrt{3}\,\tau} \sum_{\substack{\overline{v}, \overline{w} \in B/dNA \\ \overline{v} \equiv \overline{u} \ (NA)}} e\left(\frac{b|v|^2 - \text{tr}(vw^*) - c|w|^2}{dN}\right) \theta^{\overline{w}}(-d/\tau, dN).$$

Note that $cw \in NA$ for $w \in B$ since $c \equiv 0 \ (\text{mod } 3N)$. It follows that

$$\sum_{\substack{\overline{v}\in B/dNA \\ \overline{v}\equiv\overline{u}\,(NA)}} \mathbf{e}\left(\frac{b|v|^2-\operatorname{tr}(vw^*)-c|w|^2}{dN}\right) = \sum_{\substack{\overline{v}\in B/dNA \\ \overline{v}\equiv\overline{u}\,(NA)}} \mathbf{e}\left(\frac{b|v-cw|^2-\operatorname{tr}((v-cw)w^*)-c|w|^2}{dN}\right)$$

$$= \mathbf{e}\left(-\frac{\operatorname{tr}(auw^*)}{N}\right) \sum_{\substack{\overline{v}\in B/dNA \\ \overline{v}\equiv\overline{u}\,(NA)}} \mathbf{e}\left(\frac{b|v|^2}{dN}\right),$$

where the last equality uses the relation $ad - bc = 1$. Since $b \equiv 0 \pmod{N}$ the summand $\mathbf{e}\left(b|v|^2/(dN)\right)$ depends only on $v \pmod{dA}$, and since $(d, N) = 1$ and $\overline{u} \in A/NA$ the sum is $\sum_{\overline{v}\in A/dA} \mathbf{e}\left(b|v|^2/(dN)\right)$. This takes the value $(d/3)d$ (Exercise 4.11.3). Substitute this into the transformation formula and apply the second and third statements of the lemma to continue,

$$\theta^{\overline{u}}(\gamma(\tau), N) = \frac{i(c\tau + d)}{N\sqrt{3}\,\tau}\left(\frac{d}{3}\right) \sum_{\overline{w}\in B/dNA} \mathbf{e}\left(-\frac{\operatorname{tr}(auw^*)}{N}\right) \theta^{\overline{w}}(d(-1/\tau), dN)$$

$$= \frac{i(c\tau + d)}{N\sqrt{3}\,\tau}\left(\frac{d}{3}\right) \sum_{\overline{v}\in B/NA} \mathbf{e}\left(-\frac{\operatorname{tr}(auv^*)}{N}\right) \theta^{\overline{v}}(-1/\tau, N)$$

$$= \frac{c\tau + d}{3N^2}\left(\frac{d}{3}\right) \sum_{\overline{v},\overline{w}\in B/NA} \mathbf{e}\left(-\frac{\operatorname{tr}(vw^*+auv^*)}{N}\right) \theta^{\overline{w}}(\tau, N).$$

Exercise 4.11.3(b) shows that the inner sum is

$$\sum_{\overline{v}\in B/NA} \mathbf{e}\left(-\frac{\operatorname{tr}(v(w^*+au^*))}{N}\right) = \begin{cases} 3N^2 & \text{if } \overline{w} = -\overline{au}, \\ 0 & \text{if } \overline{w} \neq -\overline{au}, \end{cases}$$

completing the proof since $\theta^{-\overline{au}} = \theta^{\overline{au}}$. $\qquad\qquad\square$

To construct a modular form from the theta functions we need to conjugate and then symmetrize. To conjugate, let $\delta = \left[\begin{smallmatrix} N & 0 \\ 0 & 1 \end{smallmatrix}\right]$ so that

$$\delta\Gamma_0(3N^2)\delta^{-1} = \Gamma_0(3N, N)$$

and the conjugation preserves matrix entries on the diagonal. Recall that the weight-k operator was extended to $\mathrm{GL}_2^+(\mathbb{Q})$ in Exercise 1.2.11. Thus for any $\gamma = \left[\begin{smallmatrix} a & b \\ c & d \end{smallmatrix}\right] \in \Gamma_0(3N^2)$,

$$(\theta^{\overline{u}}[\delta\gamma]_1)(\tau, N) = (\theta^{\overline{u}}[\gamma'\delta]_1)(\tau, N) \qquad \text{where } \gamma' = \delta\gamma\delta^{-1} \in \Gamma_0(3N, N)$$

$$= (d/3)(\theta^{\overline{au}}[\delta]_1)(\tau, N) \qquad \text{since } d = d_{\gamma'}.$$

$$\text{(4.49)}$$

The construction is completed by symmetrizing:

Theorem 4.11.3. *Let N be a positive integer and let $\chi : (A/NA)^* \longrightarrow \mathbb{C}^*$ be a character, extended multiplicatively to A. Define*

$$\theta_\chi(\tau) = \tfrac{1}{6} \sum_{\overline{u} \in A/NA} \chi(u)\theta^{\overline{u}}(N\tau, N).$$

Then

$$\theta_\chi[\gamma]_1 = \chi(d)(d/3)\theta_\chi, \quad \gamma = \left[\begin{smallmatrix} a & b \\ c & d \end{smallmatrix}\right] \in \Gamma_0(3N^2).$$

Therefore

$$\theta_\chi \in \mathcal{M}_1(3N^2, \psi), \qquad \psi(d) = \chi(d)(d/3).$$

The desired transformation of θ_χ under $\Gamma_0(3N^2)$ follows from (4.49) since $\theta_\chi = \sum_{\overline{u}} \chi(u)\theta^{\overline{u}}[\delta]_1$ (Exercise 4.11.4). To finish proving the theorem note that

$$\theta_\chi(\tau) = \tfrac{1}{6} \sum_{n \in A} \chi(n)\mathbf{e}\left(|n|^2\tau\right) = \sum_{m=0}^{\infty} a_m(\theta_\chi)e^{2\pi i m\tau}$$

where

$$a_m(\theta_\chi) = \tfrac{1}{6} \sum_{\substack{n \in A \\ |n|^2 = m}} \chi(n). \tag{4.50}$$

This shows that $a_n(\theta) = \mathcal{O}(n)$, i.e., the Fourier coefficients are small enough to satisfy the condition in Proposition 1.2.4.

For example, when $N = 1$ the theta function

$$\theta_1(\tau) = \tfrac{1}{6} \sum_{n \in A} e^{2\pi i |n|^2 \tau}, \quad \tau \in \mathcal{H}$$

is a constant multiple of $E_1^{\psi,1}$, the Eisenstein series mentioned at the end of Section 4.8 (Exercise 4.11.5). This fact is equivalent to a representation number formula like those in Exercise 4.8.7.

Along with Poisson summation, the other ingredient for constructing a modular form to match the cubic equation C from the beginning of the section is the Cubic Reciprocity Theorem. The unit group of A is $A^* = \{\pm 1, \pm \mu_3, \pm \mu_3^2\}$. Note that formula (4.50) shows that $\theta_\chi = 0$ unless χ is trivial on A^*. Let p be a rational prime, $p \equiv 1 \pmod 3$. Then there exists an element $\pi = a + b\mu_3 \in A$ such that

$$\{n \in A : |n|^2 = p\} = A^*\pi \cup A^*\overline{\pi}.$$

The choice of π can be normalized, e.g., to $\pi = a + b\mu_3$ where $a \equiv 2 \pmod 3$ and $b \equiv 0 \pmod 3$. On the other hand a rational prime $p \equiv 2 \pmod 3$ does not take the form $p = |n|^2$ for any $n \in A$, as is seen by checking $|n|^2$ modulo 3. (See 9.1–9.6 of [IR92] for more on the arithmetic of A.) A weak form of Cubic Reciprocity is: *Let $d \in \mathbb{Z}^+$ be cubefree and let $N = 3\prod_{p|d} p$. Then there exists a character*

$$\chi : (A/NA)^* \longrightarrow \{1, \mu_3, \mu_3^2\}$$

such that the multiplicative extension of χ to all of A is trivial on A^* and on primes $p \nmid N$, while on elements π of A such that $\pi\bar{\pi}$ is a prime $p \nmid N$ it is trivial if and only if d is a cube modulo p. See Exercise 4.11.6 for simple examples.

For this character, $\theta_\chi(\tau, N) \in \mathcal{M}_1(3N^2, \psi)$ where ψ is the quadratic character with conductor 3. Formula (4.50) shows that the Fourier coefficients of prime index are

$$a_p(\theta_\chi) = \begin{cases} 2 & \text{if } p \equiv 1 \pmod 3 \text{ and } d \text{ is a nonzero cube modulo } p, \\ -1 & \text{if } p \equiv 1 \pmod 3 \text{ and } d \text{ is not a cube modulo } p, \\ 0 & \text{if } p \equiv 2 \pmod 3 \text{ or } p \mid 3d. \end{cases} \qquad (4.51)$$

That is, the Fourier coefficients are the solution-counts (4.47) of equation C as anticipated at the beginning of the section.

Exercises

4.11.1. Show that for any prime p the map $x \mapsto x^3$ is an endomorphism of the multiplicative group $(\mathbb{Z}/p\mathbb{Z})^*$. Show that the map is 3-to-1 if $p \equiv 1 \pmod 3$ and is 1-to-1 if $p \equiv 2 \pmod 3$. Use this to establish (4.47).

4.11.2. Confirm that under the identification $(x_1, x_2) \leftrightarrow x_1 + x_2\mu_3$, the exponent $-\langle u, vS \rangle$ from Section 4.10 becomes $-(vu^*)_2$. Use this to verify the application of Poisson summation with $3u$, $3v$, and $3N$ in the proof of Lemma 4.11.1.

4.11.3. For $b, d \in \mathbb{Z}$ with $(3b, d) = 1$ let $\check{\varphi}_{b,d} = \sum_{\bar{v} \in A/dA} \mathbf{e}\left(b|v|^2/d\right)$. This exercise proves the formula $\varphi_{b,d} = (d/3)d$.

(a) Prove the formula when $d = p$ where $p \neq 3$ is prime. For $p = 2$ compute directly. For $p > 3$ use the isomorphism $\mathbb{Z}[\sqrt{-3}]/p\mathbb{Z}[\sqrt{-3}] \longrightarrow A/pA$ to show that

$$\varphi_{b,d} = \sum_{r_1, r_2 \in \mathbb{Z}/p\mathbb{Z}} \mathbf{e}\left(b(r_1^2 + 3r_2^2)/p\right).$$

Show that if m is not divisible by p then

$$\sum_{r \in \mathbb{Z}/p\mathbb{Z}} \mathbf{e}\left(mr^2/p\right) = \sum_{s \in \mathbb{Z}/p\mathbb{Z}} \left(1 + \left(\tfrac{ms}{p}\right)\right)\mathbf{e}\left(s/p\right) = \left(\tfrac{m}{p}\right)g(\chi)$$

where $g(\chi)$ is the Gauss sum associated to the character $\chi(s) = (s/p)$. Show that $g(\chi)^2 = \chi(-1)|g(\chi)|^2 = (-1/p)p$ similarly to (4.12). Use Quadratic Reciprocity to complete the proof.

(b) Before continuing show that for any $x \in N^{-1}B$,

$$\sum_{\bar{w} \in A/NA} \mathbf{e}\left(\text{tr}\left(xw^*\right)\right) = \begin{cases} N^2 & \text{if } x \in B, \\ 0 & \text{if } x \notin B, \end{cases}$$

and

$$\sum_{\overline{v} \in B/NA} \mathbf{e}\left(\mathrm{tr}\left(vx^*\right)\right) = \begin{cases} 3N^2 & \text{if } x \in A, \\ 0 & \text{if } x \notin A. \end{cases}$$

(c) Prove the formula for $d = p^t$ inductively by showing that for $t \geq 2$,

$$\varphi_{b,p^t} = \sum_{\overline{v} \in A/p^{t-1}A} \sum_{\overline{w} \in A/pA} \mathbf{e}\left(b|v + p^{t-1}w|^2/p^t\right)$$

$$= \sum_{\overline{v} \in A/p^{t-1}A} \mathbf{e}\left(b|v|^2/p^t\right) \sum_{\overline{w} \in A/pA} \mathbf{e}\left(\mathrm{tr}\left(bvw^*\right)/p\right) = \varphi_{b,p^{t-2}} p^2.$$

(d) For arbitrary $d > 0$ suppose that $d = d_1 d_2$ with $(d_1, d_2) = 1$ and $d_1, d_2 > 0$. Use the bijection

$$A/d_1 A \times A/d_2 A \longrightarrow A/dA, \qquad (\overline{u_1}, \overline{u_2}) \mapsto \overline{d_2 u_1 + d_1 u_2}$$

to show that $\varphi_{b,d} = \varphi_{bd_2,d_1} \varphi_{bd_1,d_2}$. Deduce that this holds for $d < 0$ as well. Complete the proof of the formula.

4.11.4. Verify the transformation law in Theorem 4.11.3.

4.11.5. (a) Use results from Chapter 3 to show that $\dim(\mathcal{S}_1(\Gamma_0(3))) = 0$, so that if ψ is the quadratic character modulo 3 then $\mathcal{M}_1(3, \psi) = \mathbb{C} E_1^{\psi,1}$.

(b) Let $\theta_1(\tau)$ denote the theta function in Theorem 4.11.3 specialized to $N = 1$, making the character trivial. Thus $\theta_1 \in \mathcal{M}_1(3, \psi)$ and so θ_1 is a constant multiple of $E_1^{\psi,1}$. What is the constant?

(c) The Fourier coefficients $a_m(\theta_1)$ are (up to a constant multiple) representation numbers for the quadratic form $n_1^2 - n_1 n_2 + n_2^2$. Thus the representation number is a constant multiple of the arithmetic function $\sigma_0^{\psi,1}(m) = \sum_{d|m} \psi(d)$ for $m \geq 1$. Check the relation between $r(p)$ and $\sigma_0^{\psi,1}(p)$ for prime p by using the information about this ring given in the proof of Corollary 3.7.2. Indeed, the reader with background in number theory can work this exercise backwards by deriving the representation numbers and thus the identity $\theta_1 = c E_1^{\psi,1}$ arithmetically.

4.11.6. (a) Describe the character χ provided by Cubic Reciprocity for $d = 1$.

(b) To describe χ for $d = 2$, first determine the conditions modulo 2 on $a, b \in \mathbb{Z}$ that make $a + b\mu_3 \in A$ invertible modulo $2A$, and similarly for 3. Use these to show that the multiplicative group $G = (A/6A)^*$ has order 18. Show that A^* reduces to a cyclic subgroup H of order 6 in G and that the quotient G/H is generated by $g = \overline{1 + 3\mu_3}$. Explain why χ is defined on the quotient and why up to complex conjugation it is

$$\chi(H) = 1, \qquad \chi(gH) = \zeta_3, \qquad \chi(g^2 H) = \zeta_3^2.$$

(Here the symbol ζ_3 is being used to distinguish the cube root of unity in the codomain $\mathbb{C}^*$ of χ from the cube root of unity in A.)

(c) Describe the character χ provided by Cubic Reciprocity for $d = 3$.

5

Hecke Operators

This chapter addresses the question of finding a canonical basis for the space of cusp forms $\mathcal{S}_k(\Gamma_1(N))$. Since cusp forms are not easy to write explicitly like Eisenstein series, specifying a basis requires more sophisticated methods than the direct calculations of the preceding chapter.

For any congruence subgroups Γ_1 and Γ_2 of $\mathrm{SL}_2(\mathbb{Z})$, a family of *double coset operators* takes $\mathcal{M}_k(\Gamma_1)$ to $\mathcal{M}_k(\Gamma_2)$, taking cusp forms to cusp forms. These operators are linear. Specializing to $\Gamma_1 = \Gamma_2 = \Gamma_1(N)$, particular double coset operators $\langle n \rangle$ and T_n for all $n \in \mathbb{Z}^+$ are the *Hecke operators*, commuting endomorphisms of the vector space $\mathcal{M}_k(\Gamma_1(N))$ and the subspace $\mathcal{S}_k(\Gamma_1(N))$. The *Petersson inner product* makes $\mathcal{S}_k(\Gamma_1(N))$ an inner product space, and the Hecke operators $\langle n \rangle$ and T_n for n relatively prime to the level N are normal with respect to this inner product. Thus by linear algebra the space $\mathcal{S}_k(\Gamma_1(N))$ has an orthogonal basis whose elements are simultaneous eigenfunctions for the Hecke operators away from N. Further decomposing $\mathcal{S}_k(\Gamma_1(N))$ into *old* and *new* subspaces partially eliminates the restriction: the new subspace has an orthogonal basis of eigenfunctions for all the Hecke operators, and after normalizing this basis is canonical. The old subspace is the image of new subspaces of lower levels.

The canonical basis elements for the new space, and more generally all normalized eigenfunctions for the Hecke operators, correspond naturally to Dirichlet series having *Euler product expansions* and satisfying *functional equations*. These Dirichlet series, the *L-functions of eigenfunctions*, will later express the connection between modular forms and elliptic curves.

Related reading: [Lan76], [Miy89], and Chapter 3 of [Shi73].

5.1 The double coset operator

Let Γ_1 and Γ_2 be congruence subgroups of $\mathrm{SL}_2(\mathbb{Z})$. Then Γ_1 and Γ_2 are subgroups of $\mathrm{GL}_2^+(\mathbb{Q})$, the group of 2-by-2 matrices with rational entries and positive determinant. For each $\alpha \in \mathrm{GL}_2^+(\mathbb{Q})$ the set

© Springer Science+Business Media New York 2005
F. Diamond, J. Shurman, *A First Course in Modular Forms*,
Graduate Texts in Mathematics 228, DOI 10.1007/978-0-387-27226-9_5

$$\Gamma_1 \alpha \Gamma_2 = \{\gamma_1 \alpha \gamma_2 : \gamma_1 \in \Gamma_1, \gamma_2 \in \Gamma_2\}$$

is a *double coset* in $\mathrm{GL}_2^+(\mathbb{Q})$. Under a definition to be developed in this section, such double cosets transform modular forms with respect to Γ_1 into modular forms with respect to Γ_2.

The group Γ_1 acts on the double coset $\Gamma_1 \alpha \Gamma_2$ by left multiplication, partitioning it into orbits. A typical orbit is $\Gamma_1 \beta$ with representative $\beta = \gamma_1 \alpha \gamma_2$, and the orbit space $\Gamma_1 \backslash \Gamma_1 \alpha \Gamma_2$ is thus a disjoint union $\bigcup \Gamma_1 \beta_j$ for some choice of representatives β_j. The next two lemmas combine to show that this union is finite.

Lemma 5.1.1. *Let Γ be a congruence subgroup of $\mathrm{SL}_2(\mathbb{Z})$ and let α be an element of $\mathrm{GL}_2^+(\mathbb{Q})$. Then $\alpha^{-1} \Gamma \alpha \cap \mathrm{SL}_2(\mathbb{Z})$ is again a congruence subgroup of $\mathrm{SL}_2(\mathbb{Z})$.*

Proof. There exists $\tilde{N} \in \mathbb{Z}^+$ satisfying the conditions $\Gamma(\tilde{N}) \subset \Gamma$, $\tilde{N}\alpha \in \mathrm{M}_2(\mathbb{Z})$, $\tilde{N}\alpha^{-1} \in \mathrm{M}_2(\mathbb{Z})$. Set $N = \tilde{N}^3$. The calculation

$$\alpha \Gamma(N) \alpha^{-1} \subset \alpha(I + \tilde{N}^3 \mathrm{M}_2(\mathbb{Z}))\alpha^{-1}$$
$$= I + \tilde{N} \cdot \tilde{N}\alpha \cdot \mathrm{M}_2(\mathbb{Z}) \cdot \tilde{N}\alpha^{-1} \subset I + \tilde{N}\mathrm{M}_2(\mathbb{Z})$$

and the observation that $\alpha\Gamma(N)\alpha^{-1}$ consists of determinant-1 matrices combine to show that $\alpha\Gamma(N)\alpha^{-1} \subset \Gamma(\tilde{N})$. Thus $\Gamma(N) \subset \alpha^{-1}\Gamma(\tilde{N})\alpha \subset \alpha^{-1}\Gamma\alpha$, and intersecting with $\mathrm{SL}_2(\mathbb{Z})$ completes the proof. $\square$

Lemma 5.1.2. *Let Γ_1 and Γ_2 be congruence subgroups of $\mathrm{SL}_2(\mathbb{Z})$, and let α be an element of $\mathrm{GL}_2^+(\mathbb{Q})$. Set $\Gamma_3 = \alpha^{-1}\Gamma_1\alpha \cap \Gamma_2$, a subgroup of Γ_2. Then left multiplication by α,*

$$\Gamma_2 \longrightarrow \Gamma_1 \alpha \Gamma_2 \qquad \text{given by} \qquad \gamma_2 \mapsto \alpha \gamma_2,$$

induces a natural bijection from the coset space $\Gamma_3 \backslash \Gamma_2$ to the orbit space $\Gamma_1 \backslash \Gamma_1 \alpha \Gamma_2$. In concrete terms, $\{\gamma_{2,j}\}$ is a set of coset representatives for $\Gamma_3 \backslash \Gamma_2$ if and only if $\{\beta_j\} = \{\alpha\gamma_{2,j}\}$ is a set of orbit representatives for $\Gamma_1 \backslash \Gamma_1 \alpha \Gamma_2$.

Proof. The map $\Gamma_2 \longrightarrow \Gamma_1 \backslash \Gamma_1 \alpha \Gamma_2$ taking γ_2 to $\Gamma_1 \alpha \gamma_2$ clearly surjects. It takes elements γ_2, γ_2' to the same orbit when $\Gamma_1 \alpha \gamma_2 = \Gamma_1 \alpha \gamma_2'$, i.e., $\gamma_2' \gamma_2^{-1} \in \alpha^{-1}\Gamma_1\alpha$, and of course $\gamma_2' \gamma_2^{-1} \in \Gamma_2$ as well. So the definition $\Gamma_3 = \alpha^{-1}\Gamma_1\alpha \cap \Gamma_2$ gives a bijection $\Gamma_3 \backslash \Gamma_2 \longrightarrow \Gamma_1 \backslash \Gamma_1 \alpha \Gamma_2$ from cosets $\Gamma_3 \gamma_2$ to orbits $\Gamma_1 \alpha \gamma_2$. The last statement of the lemma follows immediately. $\square$

Any two congruence subgroups G_1 and G_2 of $\mathrm{SL}_2(\mathbb{Z})$ are *commensurable*, meaning that the indices $[G_1 : G_1 \cap G_2]$ and $[G_2 : G_1 \cap G_2]$ are finite (Exercise 5.1.2). In particular, since $\alpha^{-1}\Gamma_1\alpha \cap \mathrm{SL}_2(\mathbb{Z})$ is a congruence subgroup of $\mathrm{SL}_2(\mathbb{Z})$ by Lemma 5.1.1, the coset space $\Gamma_3 \backslash \Gamma_2$ in Lemma 5.1.2 is finite and hence so is the orbit space $\Gamma_1 \backslash \Gamma_1 \alpha \Gamma_2$. With finiteness of the orbit space established, the double coset $\Gamma_1 \alpha \Gamma_2$ can act on modular forms. Recall that for

$\beta \in \mathrm{GL}_2^+(\mathbb{Q})$ and $k \in \mathbb{Z}$, the *weight-k β operator* on functions $f : \mathcal{H} \longrightarrow \mathbb{C}$ is given by

$$(f[\beta]_k)(\tau) = (\det \beta)^{k-1} j(\beta, \tau)^{-k} f(\beta(\tau)), \quad \tau \in \mathcal{H}.$$

Definition 5.1.3. *For congruence subgroups Γ_1 and Γ_2 of $\mathrm{SL}_2(\mathbb{Z})$ and $\alpha \in \mathrm{GL}_2^+(\mathbb{Q})$, the* **weight-k $\Gamma_1 \alpha \Gamma_2$ operator** *takes functions $f \in \mathcal{M}_k(\Gamma_1)$ to*

$$f[\Gamma_1 \alpha \Gamma_2]_k = \sum_j f[\beta_j]_k$$

where $\{\beta_j\}$ are orbit representatives, i.e., $\Gamma_1 \alpha \Gamma_2 = \bigcup_j \Gamma_1 \beta_j$ is a disjoint union.

The double coset operator is well defined, i.e., it is independent of how the β_j are chosen (Exercise 5.1.3). Seeing that it takes modular forms with respect to Γ_1 to modular forms with respect to Γ_2,

$$[\Gamma_1 \alpha \Gamma_2]_k : \mathcal{M}_k(\Gamma_1) \longrightarrow \mathcal{M}_k(\Gamma_2),$$

means showing that for each $f \in \mathcal{M}_k(\Gamma_1)$, the transformed $f[\Gamma_1 \alpha \Gamma_2]_k$ is Γ_2-invariant and is holomorphic at the cusps. Seeing that the double coset operator takes cusp forms to cusp forms,

$$[\Gamma_1 \alpha \Gamma_2]_k : \mathcal{S}_k(\Gamma_1) \longrightarrow \mathcal{S}_k(\Gamma_2),$$

means showing that for each $f \in \mathcal{S}_k(\Gamma_1)$, the transformed $f[\Gamma_1 \alpha \Gamma_2]_k$ vanishes at the cusps.

To show invariance, note that any $\gamma_2 \in \Gamma_2$ permutes the orbit space $\Gamma_1 \backslash \Gamma_1 \alpha \Gamma_2$ by right multiplication. That is, the map $\gamma_2 : \Gamma_1 \backslash \Gamma_1 \alpha \Gamma_2 \longrightarrow \Gamma_1 \backslash \Gamma_1 \alpha \Gamma_2$ given by $\Gamma_1 \beta \mapsto \Gamma_1 \beta \gamma_2$ is well defined and bijective. So if $\{\beta_j\}$ is a set of orbit representatives for $\Gamma_1 \backslash \Gamma_1 \alpha \Gamma_2$ then $\{\beta_j \gamma_2\}$ is a set of orbit representatives as well. Thus

$$(f[\Gamma_1 \alpha \Gamma_2]_k)[\gamma_2]_k = \sum_j f[\beta_j \gamma_2]_k = f[\Gamma_1 \alpha \Gamma_2]_k,$$

and $f[\Gamma_1 \alpha \Gamma_2]_k$ is weight-k invariant under Γ_2 as claimed.

To show holomorphy at the cusps, first note that for any $f \in \mathcal{M}_k(\Gamma_1)$ and for any $\gamma \in \mathrm{GL}_2^+(\mathbb{Q})$, the function $g = f[\gamma]_k$ is holomorphic at infinity, meaning it has a Fourier expansion

$$g(\tau) = \sum_{n \geq 0} a_n(g) e^{2\pi i n \tau / h}$$

for some period $h \in \mathbb{Z}^+$ (this was Exercise 1.2.11(b)). Second, note that if functions $g_1, ..., g_d : \mathcal{H} \longrightarrow \mathbb{C}$ are holomorphic at infinity, meaning that each g_j has a Fourier expansion

$$g_j(\tau) = \sum_{n \geq 0} a_n(g_j) e^{2\pi i n \tau / h_j},$$

then so is their sum $g_1 + \cdots + g_d$ (Exercise 5.1.4). For any $\delta \in \mathrm{SL}_2(\mathbb{Z})$, the function $(f[\Gamma_1 \alpha \Gamma_2]_k)[\delta]_k$ is a sum of functions $g_j = f[\gamma_j]_k$ with $\gamma_j = \beta_j \delta \in \mathrm{GL}_2^+(\mathbb{Q})$, so it is holomorphic at infinity by the two facts just noted. Since δ is arbitrary this is the condition for holomorphy at the cusps.

For any $f \in \mathcal{S}_k(\Gamma_1)$ and for any $\gamma \in \mathrm{GL}_2^+(\mathbb{Q})$, the function $g = f[\gamma]_k$ vanishes at infinity (this was also Exercise 1.2.11(b)), and the previous paragraph now shows that $f[\Gamma_1 \alpha \Gamma_2]_k \in \mathcal{S}_k(\Gamma_2)$, i.e., the double coset operator takes cusp forms to cusp forms as claimed.

Special cases of the double coset operator $[\Gamma_1 \alpha \Gamma_2]_k$ arise when

(1) $\Gamma_1 \supset \Gamma_2$. Taking $\alpha = I$ makes the double coset operator be $f[\Gamma_1 \alpha \Gamma_2]_k = f$, the natural inclusion of the subspace $\mathcal{M}_k(\Gamma_1)$ in $\mathcal{M}_k(\Gamma_2)$, an injection.

(2) $\alpha^{-1} \Gamma_1 \alpha = \Gamma_2$. Here the double coset operator is $f[\Gamma_1 \alpha \Gamma_2]_k = f[\alpha]_k$, the natural translation from $\mathcal{M}_k(\Gamma_1)$ to $\mathcal{M}_k(\Gamma_2)$, an isomorphism.

(3) $\Gamma_1 \subset \Gamma_2$. Taking $\alpha = I$ and letting $\{\gamma_{2,j}\}$ be a set of coset representatives for $\Gamma_1 \backslash \Gamma_2$ makes the double coset operator be $f[\Gamma_1 \alpha \Gamma_2]_k = \sum_j f[\gamma_{2,j}]_k$, the natural *trace map* that projects $\mathcal{M}_k(\Gamma_1)$ onto its subspace $\mathcal{M}_k(\Gamma_2)$ by symmetrizing over the quotient, a surjection.

In fact, any double coset operator is a composition of these. Given Γ_1, Γ_2, and α, set $\Gamma_3 = \alpha^{-1} \Gamma_1 \alpha \cap \Gamma_2$ as usual and set $\Gamma_3' = \alpha \Gamma_3 \alpha^{-1} = \Gamma_1 \cap \alpha \Gamma_2 \alpha^{-1}$. Then $\Gamma_1 \supset \Gamma_3'$ and $\alpha^{-1} \Gamma_3' \alpha = \Gamma_3$ and $\Gamma_3 \subset \Gamma_2$, giving the three cases. The corresponding composition of double coset operators is

$$f \mapsto f \mapsto f[\alpha]_k \mapsto \sum_j f[\alpha \gamma_{2,j}]_k,$$

which by Lemma 5.1.2 is the general $[\Gamma_1 \alpha \Gamma_2]_k$.

The process of transferring functions forward from $\mathcal{M}_k(\Gamma_1)$ to $\mathcal{M}_k(\Gamma_2)$ by the double coset operator also has a geometric interpretation in terms of transferring points back between the corresponding modular curves. This leads to an algebraic interpretation of the double coset as a homomorphism of divisor groups. Recall that every congruence subgroup Γ has a modular curve $X(\Gamma) = \Gamma \backslash \mathcal{H}^*$ consisting of orbits $\Gamma \tau$. The configuration of groups is

$$
\begin{array}{ccc}
\Gamma_3 & \xrightarrow{\ \sim\ } & \Gamma_3' \\
\downarrow & & \downarrow \\
\Gamma_2 & & \Gamma_1
\end{array}
$$

where the group isomorphism is $\gamma \mapsto \alpha \gamma \alpha^{-1}$ and the vertical arrows are inclusions. The corresponding configuration of modular curves is

$$X_3 \xrightarrow{\sim} X_3'$$

$$\pi_2 \downarrow \qquad \downarrow \pi_1 \qquad\qquad (5.1)$$

$$X_2 \qquad X_1$$

where the modular curve isomorphism is $\Gamma_3\tau \mapsto \Gamma_3'\alpha(\tau)$, denoted α (Exercise 5.1.5). Again letting $\Gamma_3\backslash\Gamma_2 = \bigcup_j \Gamma_3\gamma_{2,j}$ and $\beta_j = \alpha\gamma_{2,j}$ for each j so that $\Gamma_1\alpha\Gamma_2 = \bigcup_j \Gamma_1\beta_j$, each point of X_2 is taken back by $\pi_1 \circ \alpha \circ \pi_2^{-1}$ to a set of points of X_1,

$$\{\Gamma_3\gamma_{2,j}(\tau)\} \xrightarrow{\alpha} \{\Gamma_3'\beta_j(\tau)\}$$

$$\pi_2^{-1} \uparrow \qquad\qquad \downarrow \pi_1$$

$$\Gamma_2\tau \qquad\qquad \{\Gamma_1\beta_j(\tau)\}.$$

Here π_2^{-1} takes each point $x \in X_2$ to the multiset (meaning elements can repeat) of overlying points $y \in X_3$ each with multiplicity according to its ramification degree, $\pi_2^{-1}(x) = \{e_y \cdot y : y \in X_3, \pi_2(y) = x\}$. To place the composition in the right environment for counting with multiplicity let $\mathrm{Div}(X)$ denote the divisor group of any modular curve X, where as in Chapter 3 the divisor group of a set is the free Abelian group on its points. In terms of divisors the composition is

$$[\Gamma_1\alpha\Gamma_2]_k : X_2 \longrightarrow \mathrm{Div}(X_1), \qquad \Gamma_2\tau \mapsto \sum_j \Gamma_1\beta_j(\tau),$$

and this has a unique $\mathbb{Z}$-linear extension to a divisor group homomorphism,

$$[\Gamma_1\alpha\Gamma_2]_k : \mathrm{Div}(X_2) \longrightarrow \mathrm{Div}(X_1).$$

In this context the special cases $\Gamma_1 \supset \Gamma_2$, $\alpha^{-1}\Gamma_1\alpha = \Gamma_2$, and $\Gamma_1 \subset \Gamma_2$ from before lead respectively to a surjection, an isomorphism, and an injection (Exercise 5.1.6).

This chapter will focus on double coset operators acting on modular forms, but the emphasis later in the book will move toward the divisor group interpretation.

Exercises

5.1.1. In Lemma 5.1.1, need $\alpha^{-1}\Gamma\alpha$ lie in $\mathrm{SL}_2(\mathbb{Z})$?

5.1.2. Show that for congruence subgroups G_1 and G_2 of $\mathrm{SL}_2(\mathbb{Z})$, the indices $[G_1 : G_1 \cap G_2]$ and $[G_2 : G_1 \cap G_2]$ are finite. (A hint for this exercise is at the end of the book.)

5.1.3. Show that the weight-k double coset operator $[\Gamma_1\alpha\Gamma_2]_k$ is independent of how the orbit representatives β_j are chosen. (A hint for this exercise is at the end of the book.)

5.1.4. Show that if $g_1, ..., g_d : \mathcal{H} \longrightarrow \mathbb{C}$ are holomorphic at infinity, then so is their sum $g_1 + \cdots + g_d$. (A hint for this exercise is at the end of the book.)

5.1.5. Check that the map $\alpha : X_3 \longrightarrow X_3'$ given by $\Gamma_3 \tau \mapsto \Gamma_3' \alpha(\tau)$ is well defined.

5.1.6. Show that in the special cases $\Gamma_1 \supset \Gamma_2$, $\alpha^{-1} \Gamma_1 \alpha = \Gamma_2$, and $\Gamma_1 \subset \Gamma_2$ the divisor group interpretation of $[\Gamma_1 \alpha \Gamma_2]_k$ gives a surjection, an isomorphism, and an injection.

5.2 The $\langle d \rangle$ and T_p operators

Recall the congruence subgroups

$$\Gamma_0(N) = \left\{ \begin{bmatrix} a & b \\ c & d \end{bmatrix} \in \mathrm{SL}_2(\mathbb{Z}) : \begin{bmatrix} a & b \\ c & d \end{bmatrix} \equiv \begin{bmatrix} * & * \\ 0 & * \end{bmatrix} \pmod{N} \right\}$$

and

$$\Gamma_1(N) = \left\{ \begin{bmatrix} a & b \\ c & d \end{bmatrix} \in \mathrm{SL}_2(\mathbb{Z}) : \begin{bmatrix} a & b \\ c & d \end{bmatrix} \equiv \begin{bmatrix} 1 & * \\ 0 & 1 \end{bmatrix} \pmod{N} \right\},$$

where as always N is a positive integer. The smaller group has more modular forms, i.e., $\mathcal{M}_k(\Gamma_1(N)) \supset \mathcal{M}_k(\Gamma_0(N))$. This section introduces two operators on the larger vector space $\mathcal{M}_k(\Gamma_1(N))$.

The map $\Gamma_0(N) \longrightarrow (\mathbb{Z}/N\mathbb{Z})^*$ taking $\begin{bmatrix} a & b \\ c & d \end{bmatrix}$ to $d \pmod{N}$ is a surjective homomorphism with kernel $\Gamma_1(N)$. This shows that $\Gamma_1(N)$ is normal in $\Gamma_0(N)$ and induces an isomorphism

$$\Gamma_0(N)/\Gamma_1(N) \xrightarrow{\sim} (\mathbb{Z}/N\mathbb{Z})^* \qquad \text{where} \qquad \begin{bmatrix} a & b \\ c & d \end{bmatrix} \mapsto d \pmod{N}.$$

To define the first type of Hecke operator, take any $\alpha \in \Gamma_0(N)$, set $\Gamma_1 = \Gamma_2 = \Gamma_1(N)$, and consider the weight-k double coset operator $[\Gamma_1 \alpha \Gamma_2]_k$. Since $\Gamma_1(N) \triangleleft \Gamma_0(N)$ this operator is case (2) from the list in Section 5.1, translating each function $f \in \mathcal{M}_k(\Gamma_1(N))$ to

$$f[\Gamma_1(N) \alpha \Gamma_1(N)]_k = f[\alpha]_k, \qquad \alpha \in \Gamma_0(N),$$

again in $\mathcal{M}_k(\Gamma_1(N))$. Thus the group $\Gamma_0(N)$ acts on $\mathcal{M}_k(\Gamma_1(N))$, and since its subgroup $\Gamma_1(N)$ acts trivially, this is really an action of the quotient $(\mathbb{Z}/N\mathbb{Z})^*$. The action of $\alpha = \begin{bmatrix} a & b \\ c & d \end{bmatrix}$, determined by $d \pmod{N}$ and denoted $\langle d \rangle$, is

$$\langle d \rangle : \mathcal{M}_k(\Gamma_1(N)) \longrightarrow \mathcal{M}_k(\Gamma_1(N))$$

given by

$$\langle d \rangle f = f[\alpha]_k \quad \text{for any } \alpha = \begin{bmatrix} a & b \\ c & \delta \end{bmatrix} \in \Gamma_0(N) \text{ with } \delta \equiv d \pmod{N}.$$

This is the first type of Hecke operator, also called a *diamond operator*. For any character $\chi : (\mathbb{Z}/N\mathbb{Z})^* \longrightarrow \mathbb{C}$, the space $\mathcal{M}_k(N, \chi)$ from Section 4.3 is precisely the χ-eigenspace of the diamond operators,

$$\mathcal{M}_k(N, \chi) = \{f \in \mathcal{M}_k(\Gamma_1(N)) : \langle d \rangle f = \chi(d)f \text{ for all } d \in (\mathbb{Z}/N\mathbb{Z})^* \}.$$

That is, the diamond operator $\langle d \rangle$ respects the decomposition $\mathcal{M}_k(\Gamma_1(N)) = \bigoplus_\chi \mathcal{M}_k(N, \chi)$, operating on the eigenspace associated to each character χ as multiplication by $\chi(d)$.

The second type of Hecke operator is also a weight-k double coset operator $[\Gamma_1 \alpha \Gamma_2]_k$ where again $\Gamma_1 = \Gamma_2 = \Gamma_1(N)$, but now

$$\alpha = \begin{bmatrix} 1 & 0 \\ 0 & p \end{bmatrix}, \quad p \text{ prime}.$$

This operator is denoted T_p. Thus

$$T_p : \mathcal{M}_k(\Gamma_1(N)) \longrightarrow \mathcal{M}_k(\Gamma_1(N)), \quad p \text{ prime}$$

is given by

$$T_p f = f[\Gamma_1(N) \begin{bmatrix} 1 & 0 \\ 0 & p \end{bmatrix} \Gamma_1(N)]_k.$$

From (3.16), the double coset here is

$$\Gamma_1(N) \begin{bmatrix} 1 & 0 \\ 0 & p \end{bmatrix} \Gamma_1(N) = \left\{ \gamma \in \mathrm{M}_2(\mathbb{Z}) : \gamma \equiv \begin{bmatrix} 1 & * \\ 0 & p \end{bmatrix} \pmod{N}, \det \gamma = p \right\},$$

so in fact $\begin{bmatrix} 1 & 0 \\ 0 & p \end{bmatrix}$ can be replaced by any matrix in this double coset in the definition of T_p.

The two kinds of Hecke operator commute. To see this, continue to let $\alpha = \begin{bmatrix} 1 & 0 \\ 0 & p \end{bmatrix}$ and check that $\gamma \alpha \gamma^{-1} \equiv \begin{bmatrix} 1 & * \\ 0 & p \end{bmatrix} \pmod{N}$ for any $\gamma \in \Gamma_0(N)$. If

$$\Gamma_1(N) \alpha \Gamma_1(N) = \bigcup_j \Gamma_1(N) \beta_j$$

then the last sentence of the preceding paragraph and the fact that $\Gamma_1(N)$ is normal in $\Gamma_0(N)$ show that the double coset is also

$$\Gamma_1(N) \alpha \Gamma_1(N) = \Gamma_1(N) \gamma \alpha \gamma^{-1} \Gamma_1(N) = \gamma \Gamma_1(N) \alpha \Gamma_1(N) \gamma^{-1}$$

$$= \gamma \bigcup_j \Gamma_1(N) \beta_j \gamma^{-1} = \bigcup_j \Gamma_1(N) \gamma \beta_j \gamma^{-1}.$$

Comparing the two decompositions of the double coset gives $\bigcup_j \Gamma_1(N) \gamma \beta_j = \bigcup_j \Gamma_1(N) \beta_j \gamma$, even though it need not be true that $\Gamma_1(N) \gamma \beta_j = \Gamma_1(N) \beta_j \gamma$

for each j. Thus for any $f \in \mathcal{M}_k(\Gamma_1(N))$ and any $\gamma \in \Gamma_0(N)$ with lower right entry $\delta \equiv d \pmod{N}$,

$$\langle d \rangle T_p f = \sum_j f[\beta_j \gamma]_k = \sum_j f[\gamma \beta_j]_k = T_p \langle d \rangle f, \qquad f \in \mathcal{M}_k(\Gamma_1(N)).$$

To find an explicit representation of T_p, recall that it is specified by orbit representatives for $\Gamma_1 \backslash \Gamma_1 \alpha \Gamma_2$, and these are coset representatives for $\Gamma_3 \backslash \Gamma_2$ left multiplied by α, where $\Gamma_3 = \alpha^{-1} \Gamma_1 \alpha \cap \Gamma_2$. For the particular Γ_1, Γ_2, and α in play here, recall from Exercise 1.5.6 the groups

$$\Gamma^0(p) = \left\{ \begin{bmatrix} a & b \\ c & d \end{bmatrix} \in \mathrm{SL}_2(\mathbb{Z}) : \begin{bmatrix} a & b \\ c & d \end{bmatrix} \equiv \begin{bmatrix} * & 0 \\ * & * \end{bmatrix} \pmod{p} \right\}$$

and

$$\Gamma_1^0(N, p) = \Gamma_1(N) \cap \Gamma^0(p).$$

Then $\Gamma_3 = \Gamma_1^0(N, p)$ (Exercise 5.2.1). Since Γ_3 is Γ_2 subject to the additional condition $b \equiv 0 \pmod{p}$, the obvious candidates for coset representatives are

$$\gamma_{2,j} = \begin{bmatrix} 1 & j \\ 0 & 1 \end{bmatrix}, \qquad 0 \le j < p.$$

Given $\gamma_2 = \begin{bmatrix} a & b \\ c & d \end{bmatrix} \in \Gamma_2$, we have $\gamma_2 \in \Gamma_3 \gamma_{2,j}$ if $\gamma_2 \gamma_{2,j}^{-1} \in \Gamma_3 = \Gamma_2 \cap \Gamma^0(p)$. Certainly $\gamma_2 \gamma_{2,j}^{-1} \in \Gamma_2$ for any j since γ_2 and $\gamma_{2,j}$ are, but also we need the upper right entry $b - ja$ of $\gamma_2 \gamma_{2,j}^{-1} = \begin{bmatrix} a & b \\ c & d \end{bmatrix} \begin{bmatrix} 1 & -j \\ 0 & 1 \end{bmatrix}$ to be 0 (mod p).

If $p \nmid a$ then setting $j = ba^{-1} \pmod{p}$ does the job. But if $p \mid a$ then $b - ja$ can't be 0 (mod p) for any j, for then $p \mid b$ and so $p \mid ad - bc = 1$. Instances of $\gamma_2 \in \Gamma_2$ with $p \mid a$ occur if and only if $p \nmid N$, and when this happens $\gamma_{2,0}, \dots, \gamma_{2,p-1}$ fail to represent $\Gamma_3 \backslash \Gamma_2$. To complete the set of coset representatives in this case, set

$$\gamma_{2,\infty} = \begin{bmatrix} mp & n \\ N & 1 \end{bmatrix} \qquad \text{where } mp - nN = 1.$$

Now given $\gamma_2 = \begin{bmatrix} a & b \\ c & d \end{bmatrix} \in \Gamma_2$ with $p \mid a$, it is easy to show that $\gamma_2 \gamma_{2,\infty}^{-1} \in \Gamma_3$ as needed (Exercise 5.2.2). Thus $\gamma_{2,0}, \dots, \gamma_{2,p-1}$ are a complete set of coset representatives when $p \mid N$, but $\gamma_{2,\infty}$ is required as well when $p \nmid N$. In any case, it is easy to show that the $\gamma_{2,j}$ represent distinct cosets (Exercise 5.2.3). The reader may recognize these matrices from Exercise 1.5.6(b).

The corresponding orbit representatives $\beta_j = \alpha \gamma_{2,j}$ for $\Gamma_1 \backslash \Gamma_1 \alpha \Gamma_2$, needed to compute the double coset operator, work out to

$$\beta_j = \begin{bmatrix} 1 & j \\ 0 & p \end{bmatrix} \text{ for } 0 \le j < p, \quad \beta_\infty = \begin{bmatrix} m & n \\ N & p \end{bmatrix} \begin{bmatrix} p & 0 \\ 0 & 1 \end{bmatrix} \text{ if } p \nmid N. \qquad (5.2)$$

This proves

Proposition 5.2.1. *Let $N \in \mathbb{Z}^+$, let $\Gamma_1 = \Gamma_2 = \Gamma_1(N)$, and let $\alpha = \left[\begin{smallmatrix} 1 & 0 \\ 0 & p \end{smallmatrix}\right]$ where p is prime. The operator $T_p = [\Gamma_1 \alpha \Gamma_2]_k$ on $\mathcal{M}_k(\Gamma_1(N))$ is given by*

$$
T_p f = \begin{cases}
\displaystyle\sum_{j=0}^{p-1} f\left[\left[\begin{smallmatrix} 1 & j \\ 0 & p \end{smallmatrix}\right]\right]_k & \text{if } p \mid N, \\[2em]
\displaystyle\sum_{j=0}^{p-1} f\left[\left[\begin{smallmatrix} 1 & j \\ 0 & p \end{smallmatrix}\right]\right]_k + f\left[\left[\begin{smallmatrix} m & n \\ N & p \end{smallmatrix}\right]\left[\begin{smallmatrix} p & 0 \\ 0 & 1 \end{smallmatrix}\right]\right]_k & \text{if } p \nmid N, \text{ where } mp - nN = 1.
\end{cases}
$$

Letting $\Gamma_1 = \Gamma_2 = \Gamma_0(N)$ instead and keeping $\alpha = \left[\begin{smallmatrix} 1 & 0 \\ 0 & p \end{smallmatrix}\right]$ gives the same orbit representatives for $\Gamma_1 \backslash \Gamma_1 \alpha \Gamma_2$ (Exercise 5.2.4), but in this case the last representative can be replaced by $\beta_\infty = \left[\begin{smallmatrix} p & 0 \\ 0 & 1 \end{smallmatrix}\right]$ since $\left[\begin{smallmatrix} m & n \\ N & p \end{smallmatrix}\right] \in \Gamma_1$.

The next result describes the effect of T_p on Fourier coefficients.

Proposition 5.2.2. *Let $f \in \mathcal{M}_k(\Gamma_1(N))$. Since $\left[\begin{smallmatrix} 1 & 1 \\ 0 & 1 \end{smallmatrix}\right] \in \Gamma_1(N)$, f has period 1 and hence has a Fourier expansion*

$$
f(\tau) = \sum_{n=0}^{\infty} a_n(f) q^n, \quad q = e^{2\pi i \tau}.
$$

Then:

(a) *Let $\mathbf{1}_N : (\mathbb{Z}/N\mathbb{Z})^* \longrightarrow \mathbb{C}^*$ be the trivial character modulo N. Then $T_p f$ has Fourier expansion*

$$
(T_p f)(\tau) = \sum_{n=0}^{\infty} a_{np}(f) q^n + \mathbf{1}_N(p) p^{k-1} \sum_{n=0}^{\infty} a_n(\langle p \rangle f) q^{np}
$$

$$
= \sum_{n=0}^{\infty} \left(a_{np}(f) + \mathbf{1}_N(p) p^{k-1} a_{n/p}(\langle p \rangle f) \right) q^n.
$$

That is,

$$
a_n(T_p f) = a_{np}(f) + \mathbf{1}_N(p) p^{k-1} a_{n/p}(\langle p \rangle f) \quad \text{for } f \in \mathcal{M}_k(\Gamma_1(N)). \quad (5.3)
$$

(Here $a_{n/p} = 0$ when $n/p \notin \mathbb{N}$. As in Chapter 4, $\mathbf{1}_N(p) = 1$ when $p \nmid N$ and $\mathbf{1}_N(p) = 0$ when $p \mid N$.)

(b) *Let $\chi : (\mathbb{Z}/N\mathbb{Z})^* \longrightarrow \mathbb{C}^*$ be a character. If $f \in \mathcal{M}_k(N, \chi)$ then also $T_p f \in \mathcal{M}_k(N, \chi)$, and now its Fourier expansion is*

$$
(T_p f)(\tau) = \sum_{n=0}^{\infty} a_{np}(f) q^n + \chi(p) p^{k-1} \sum_{n=0}^{\infty} a_n(f) q^{np}
$$

$$
= \sum_{n=0}^{\infty} \left(a_{np}(f) + \chi(p) p^{k-1} a_{n/p}(f) \right) q^n.
$$

That is,

$$
a_n(T_p f) = a_{np}(f) + \chi(p) p^{k-1} a_{n/p}(f) \quad \text{for } f \in \mathcal{M}_k(N, \chi). \quad (5.4)
$$

Proof. For part (a), take $0 \leq j < p$ and compute

$$f[\begin{bmatrix} 1 & j \\ 0 & p \end{bmatrix}]_k(\tau) = p^{k-1}(0\tau + p)^{-k} f\left(\frac{\tau + j}{p}\right) = \frac{1}{p} \sum_{n=0}^{\infty} a_n(f)e^{2\pi i n(\tau + j)/p}$$

$$= \frac{1}{p} \sum_{n=0}^{\infty} a_n(f)q_p^n \mu_p^{nj}$$

where $q_p = e^{2\pi i \tau/p}$ and $\mu_p = e^{2\pi i/p}$. Since the geometric sum $\sum_{j=0}^{p-1} \mu_p^{nj}$ is p when $p \mid n$ and 0 when $p \nmid n$, summing over j gives

$$\sum_{j=0}^{p-1} f[\begin{bmatrix} 1 & j \\ 0 & p \end{bmatrix}]_k(\tau) = \sum_{n \equiv 0 \,(p)} a_n(f)q_p^n = \sum_{n=0}^{\infty} a_{np}(f)q^n.$$

This is $(T_p f)(\tau)$ when $p \mid N$. When $p \nmid N$, $(T_p f)(\tau)$ also includes the term

$$f[\begin{bmatrix} m & n \\ N & p \end{bmatrix}\begin{bmatrix} p & 0 \\ 0 & 1 \end{bmatrix}]_k(\tau) = (\langle p \rangle f)[\begin{bmatrix} p & 0 \\ 0 & 1 \end{bmatrix}]_k(\tau)$$

$$= p^{k-1}(0\tau + 1)^{-k}(\langle p \rangle f)(p\tau) = p^{k-1} \sum_{n=0}^{\infty} a_n(\langle p \rangle f)q^{np}.$$

In either case, the Fourier series of $T_p f$ is as claimed.

The first statement of (b) follows from the relation $\langle d \rangle (T_p f) = T_p(\langle d \rangle f)$. Formula (5.4) is immediate from (5.3). □

Since double coset operators take cusp forms to cusp forms, the T_p operator restricts to the subspace $\mathcal{S}_k(\Gamma_1(N))$ of $\mathcal{M}_k(\Gamma_1(N))$. In particular the weight 12 cusp form Δ, the discriminant function from Chapter 1, is an eigenvector of the T_p operators for $SL_2(\mathbb{Z})$ since $\mathcal{S}_{12}(SL_2(\mathbb{Z}))$ is 1-dimensional, and similarly for the generator of any other 1-dimensional space such as the functions $\varphi_k(\tau) = \eta(\tau)^k \eta(N\tau)^k$ from Proposition 3.2.2. Part (b) of the proposition here shows that T_p further restricts to the subspace $\mathcal{S}_k(N, \chi)$ for every character χ modulo N.

Eisenstein series are also eigenvectors of the Hecke operators. Recall the series $E_k^{\psi,\varphi}$ from Chapter 4 for any pair ψ, φ of Dirichlet characters modulo u and v such that $uv \mid N$ and $(\psi\varphi)(-1) = (-1)^k$,

$$E_k^{\psi,\varphi}(\tau) = \delta(\psi)L(1-k, \varphi) + 2\sum_{n=1}^{\infty} \sigma_{k-1}^{\psi,\varphi}(n)q^n, \quad q = e^{2\pi i \tau},$$

$$\delta(\psi) = \begin{cases} 1 & \text{if } \psi = \mathbf{1}_1, \\ 0 & \text{otherwise,} \end{cases} \qquad \sigma_{k-1}^{\psi,\varphi}(n) = \sum_{\substack{m \mid n \\ m > 0}} \psi(n/m)\varphi(m)m^{k-1}.$$

Recall also the series $E_k^{\psi,\varphi,t}(\tau) = E_k^{\psi,\varphi}(t\tau)$ where ψ and φ are primitive and t is a positive integer such that $tuv \mid N$. From Chapter 4, as (ψ, φ, t) runs through a set of such triples such that $\psi\varphi = \chi$ at level N these Eisenstein series represent a basis of $\mathcal{E}_k(N, \chi)$. (At weight $k = 2$ when $\psi = \varphi = \mathbf{1}_1$ the definition is different, $E_2^{\mathbf{1}_1, \mathbf{1}_1, t}(\tau) = E_2^{\mathbf{1}_1, \mathbf{1}_1}(\tau) - tE_2^{\mathbf{1}_1, \mathbf{1}_1}(t\tau)$, so the triple $(\mathbf{1}_1, \mathbf{1}_1, 1)$ contributes nothing and is excluded.) By definition of $\mathcal{M}_k(N, \chi)$ as an eigenspace, $\langle d \rangle E_k^{\psi, \varphi, t} = \chi(d) E_k^{\psi, \varphi, t}$ for all d relatively prime to N, but also

Proposition 5.2.3. *Let χ modulo N, ψ, φ, and t be as above. Let p be prime. Excluding the case $k = 2$, $\psi = \varphi = \mathbf{1}_1$,*

$$T_p E_k^{\psi, \varphi, t} = (\psi(p) + \varphi(p)p^{k-1}) E_k^{\psi, \varphi, t} \quad \text{if } uv = N \text{ or if } p \nmid N.$$

Also,

$$T_p E_2^{\mathbf{1}_1, \mathbf{1}_1, t} = (1 + \mathbf{1}_N(p)p) E_2^{\mathbf{1}_1, \mathbf{1}_1, t} \quad \begin{array}{l} \text{if } t \text{ is prime and } N \text{ is a power of } t \\ \text{or if } p \nmid N. \end{array}$$

This is a direct calculation (Exercise 5.2.5). When $uv < N$ (excluding the special case $k = 2$, $\psi = \varphi = \mathbf{1}_1$) the series $E_k^{\psi, \varphi, t}$ is "old" at level N in the sense that it comes from a lower level, being the series $E_k^{\psi, \varphi}$ at level uv raised to level tuv by multiplying the variable by t and then viewed at level N since $tuv \mid N$. On the other hand, when $uv = N$ (or in the special case if also t is prime and $N = t$) the series is "new" at level N, not arising from a lower level. The proposition shows that Eisenstein series at level N are eigenvectors for the Hecke operators away from the level, and new Eisenstein series at level N are eigenvectors for all the Hecke operators. We will see this same phenomenon for cusp forms later in the chapter.

The Hecke operators commute.

Proposition 5.2.4. *Let d and e be elements of $(\mathbb{Z}/N\mathbb{Z})^*$, and let p and q be prime. Then*

(a) $\langle d \rangle T_p = T_p \langle d \rangle$,
(b) $\langle d \rangle \langle e \rangle = \langle e \rangle \langle d \rangle = \langle de \rangle$,
(c) $T_p T_q = T_q T_p$.

Proof. Part (a) has already been shown. Since the $\langle d \rangle$ and T_p operators preserve the decomposition $\mathcal{M}_k(\Gamma_1(N)) = \bigoplus \mathcal{M}_k(N, \chi)$, it suffices to check (b) and (c) on an arbitrary $f \in \mathcal{M}_k(N, \chi)$. Now (b) is immediate (Exercise 5.2.7). As for (c), applying formula (5.4) twice gives

$$\begin{aligned} a_n(T_p(T_q f)) &= a_{np}(T_q f) + \chi(p)p^{k-1} a_{n/p}(T_q f) \\ &= a_{npq}(f) + \chi(q)q^{k-1} a_{np/q}(f) \\ &\quad + \chi(p)p^{k-1}(a_{nq/p}(f) + \chi(q)q^{k-1} a_{n/pq}(f)) \\ &= a_{npq}(f) + \chi(q)q^{k-1} a_{np/q}(f) + \chi(p)p^{k-1} a_{nq/p}(f) \\ &\quad + \chi(pq)(pq)^{k-1} a_{n/pq}(f), \end{aligned}$$

and this is symmetric in p and q. $\qquad\square$

The modular curve interpretation of T_p as at the end of Section 5.1 is

$$T_p : \mathrm{Div}(X_1(N)) \longrightarrow \mathrm{Div}(X_1(N)), \qquad \Gamma_1(N)\tau \mapsto \sum_j \Gamma_1(N)\beta_j(\tau), \quad (5.5)$$

with the matrices β_j from (5.2), excluding β_∞ when $p \mid N$. (Strictly speaking the map on the left is the $\mathbb{Z}$-linear extension of the description on the right, whose domain is only $X_1(N)$, but we use this notation from now on without comment.) There is a corresponding interpretation of T_p in terms of the moduli space $S_1(N)$ from Section 1.5. Recall that $S_1(N)$ consists of equivalence classes of enhanced elliptic curves (E, Q) where E is a complex elliptic curve and Q is a point of order N. Let $\mathrm{Div}(S_1(N))$ denote its divisor group. Then the moduli space interpretation is

$$T_p : \mathrm{Div}(S_1(N)) \longrightarrow \mathrm{Div}(S_1(N)), \qquad [E, Q] \mapsto \sum_C [E/C, Q+C], \quad (5.6)$$

where the sum is taken over all order p subgroups $C \subset E$ such that $C \cap \langle Q \rangle = \{0_E\}$ and where the square brackets denote equivalence class. To see where this comes from, recall from Chapter 1 the lattice $\Lambda_\tau = \tau\mathbb{Z} \oplus \mathbb{Z}$ and the complex elliptic curve $E_\tau = \mathbb{C}/\Lambda_\tau$ for $\tau \in \mathcal{H}$. Associate to each β_j (including β_∞ when $p \nmid N$) a subgroup $C = C_j = c_j \Lambda_{\beta_j(\tau)}$ of E_τ where $c_j \in \mathbb{C}$, working out to $C = \langle (\tau + j)/p \rangle + \Lambda_\tau$ for $0 \leq j < p$ and to $C = \langle 1/p \rangle + \Lambda_\tau$ for $j = \infty$ (Exercise 5.2.8(a)) and therefore satisfying the conditions

$$C \cong \mathbb{Z}/p\mathbb{Z} \text{ in } E_\tau, \qquad C \cap (\langle 1/N \rangle + \Lambda_\tau) = \{0\} \text{ in } E_\tau. \quad (5.7)$$

The groups $C_0, \ldots, C_{p-1}, C_\infty$ are subgroups of $E_\tau[p]$ and pairwise disjoint except for 0, making their union a subset of $E_\tau[p]$ totaling $1 + (p+1)(p-1) = p^2$ elements, i.e., their union is all of $E_\tau[p]$. (See Figure 5.1 and Exercise 5.2.8(b,c).) Any subgroup C of E_τ satisfying the first condition of (5.7) must lie in $E_\tau[p]$, making it one of the C_j. The group C_∞ fails the second condition of (5.7) when $p \mid N$. Thus the matrices β appearing in T_p describe the subgroups C in the moduli space interpretation (5.6). This discussion elaborates on Exercise 1.5.6(a).

The moduli space $S_1(N)$ is in bijective correspondence with the noncompact modular curve $Y_1(N) = \Gamma_1(N)\backslash\mathcal{H}$ by Theorem 1.5.1. The relation between descriptions (5.5) and (5.6) of T_p is summarized in the following commutative diagram (Exercise 5.2.8(d)):

$$
\begin{array}{ccc}
\mathrm{Div}(S_1(N)) & \xrightarrow{\;T_p\;} & \mathrm{Div}(S_1(N)) \\
\Big\downarrow{\psi_1} & & \Big\downarrow{\psi_1} \\
\mathrm{Div}(Y_1(N)) & \xrightarrow{\;T_p\;} & \mathrm{Div}(Y_1(N)).
\end{array}
\qquad (5.8)
$$

Here the vertical map ψ_1 is the bijection from Theorem 1.5.1. Elementwise the mappings are

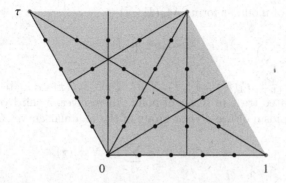

Figure 5.1. $E_\tau[5]$ as a union of six 5-cyclic subgroups

$$[E_\tau, \tfrac{1}{N} + \Lambda_\tau] \longmapsto \sum_C [E_\tau/C, \tfrac{1}{N} + C]$$

$$\Gamma_1(N)\tau \longmapsto \sum_j \Gamma_1(N)\beta_j(\tau).$$

When $p \nmid N$ the term $[E_\tau/C_\infty, \tfrac{1}{N} + C_\infty]$ works out to $[E_{p\tau}, \tfrac{p}{N} + \Lambda_{p\tau}]$ (Exercise 5.2.8(e)).

Similarly and more easily for the diamond operator, there is a commutative diagram

$$
\begin{array}{ccc}
S_1(N) & \xrightarrow{\langle d \rangle} & S_1(N) \\
\psi_1 \downarrow & & \downarrow \psi_1 \\
Y_1(N) & \xrightarrow{\langle d \rangle} & Y_1(N),
\end{array}
\tag{5.9}
$$

where if $\alpha = \left[\begin{smallmatrix} a & b \\ c & \delta \end{smallmatrix} \right] \in \Gamma_0(N)$ with $\delta \equiv d \pmod{N}$ then the mappings are given by

$$[E_\tau, \tfrac{1}{N} + \Lambda_\tau] \longmapsto [E_\tau, \tfrac{d}{N} + \Lambda_\tau]$$

$$\Gamma_1(N)\tau \longmapsto \Gamma_1(N)\alpha(\tau).$$

This was Exercise 1.5.3. Naturally, $\langle d \rangle$ extends $\mathbb{Z}$-linearly to divisors.

To summarize, we have four compatible notions of the Hecke operator T_p, starting from the double coset $\Gamma_1(N) \left[\begin{smallmatrix} 1 & 0 \\ 0 & p \end{smallmatrix} \right] \Gamma_1(N)$. This works out to

$$\left\{ \gamma \in M_2(\mathbb{Z}) : \gamma \equiv \begin{bmatrix} 1 & * \\ 0 & p \end{bmatrix} \pmod{N}, \ \det \gamma = p \right\},$$

similar to the definitions of congruence subgroups by conditions on integer matrices. The double coset gives the second version of T_p, a linear operator on the space of modular forms $\mathcal{M}_k(\Gamma_1(N))$,

$$T_p : f \mapsto \sum_j f[\beta_j]_k,$$

where $\Gamma_1(N) \begin{bmatrix} 1 & 0 \\ 0 & p \end{bmatrix} \Gamma_1(N) = \bigcup_j \Gamma_1(N)\beta_j$. Thus $T_p f(\tau)$ evaluates f at a set of points associated to τ. In terms of points themselves, a third version of T_p is the endomorphism of the divisor group of the modular curve $X_1(N)$ induced by

$$T_p : \Gamma_1(N)\tau \mapsto \sum_j \Gamma_1(N)\beta_j(\tau).$$

By the general configuration of the double coset operator from Section 5.1, this lifts each point of $X_1(N)$ to its overlying points on the modular curve $X_3 = X_1^0(N,p)$ of the group $\Gamma_3 = \Gamma_1^0(N,p) = \Gamma_1(N) \cap \Gamma^0(p)$, translates them to another modular curve X_3' over $X_1(N)$ by dividing them by p (since the definition uses the matrix $\begin{bmatrix} 1 & 0 \\ 0 & p \end{bmatrix}$), and then projects them back down. That is, T_p factors as

$$T_p : \Gamma_1(N)\tau \mapsto \sum_j \Gamma_3 \gamma_{2,j}(\tau) \mapsto \sum_j \Gamma_1(N)\gamma_{2,j}(\tau)/p,$$

where $\begin{bmatrix} 1 & 0 \\ 0 & p \end{bmatrix} \gamma_{2,j} = \beta_j$ for each j (see Exercise 5.2.10). The net effect is to take a point at level N to a formal sum of level N points associated to it in a manner depending on p. Section 6.3 will revisit this description of T_p, and Exercise 7.9.3 will show that X_3' is the modular curve $X_{1,0}(N,p)$ of the group $\Gamma_{1,0}(N,p) = \Gamma_1(N) \cap \Gamma_0(Np)$. Fourth, T_p is an endomorphism of the divisor group of the moduli space $S_1(N)$, induced by

$$T_p : [E,Q] \mapsto \sum_C [E/C, Q+C].$$

Similarly to the third version (recall (1.11) and see Exercise 5.2.10), this factors as

$$T_p : [E,Q] \mapsto \sum_C [E,C,Q] \mapsto \sum_C [E/C, Q+C].$$

Exercises

5.2.1. Show that when $\Gamma_1 = \Gamma_2 = \Gamma_1(N)$ and $\alpha = \begin{bmatrix} 1 & 0 \\ 0 & p \end{bmatrix}$, the group $\Gamma_3 = \alpha^{-1}\Gamma_1\alpha \cap \Gamma_2$ works out to $\Gamma_1^0(N,p)$.

5.2.2. When $\gamma_2 = \begin{bmatrix} a & b \\ c & d \end{bmatrix}$ with $p \mid a$, show that $\gamma_2 \gamma_{2,\infty}^{-1} \in \Gamma_3$ as needed.

5.2.3. Show that the $\gamma_{2,j}$ represent distinct cosets.

5.2.4. Show that letting $\Gamma_1 = \Gamma_2 = \Gamma_0(N)$ and keeping $\alpha = \begin{bmatrix} 1 & 0 \\ 0 & p \end{bmatrix}$ gives the same orbit representatives $\{\beta_j\}$ for $\Gamma_1 \backslash \Gamma_1 \alpha \Gamma_2$.

5.2.5. This exercise proves Proposition 5.2.3.

(a) Show that the generalized divisor sum $\sigma_{k-1}^{\psi,\varphi}$ is multiplicative, i.e.,
$\sigma_{k-1}^{\psi,\varphi}(nm) = \sigma_{k-1}^{\psi,\varphi}(n)\sigma_{k-1}^{\psi,\varphi}(m)$ when $\gcd(n,m) = 1$.

(b) Let p be prime and let $n \geq 1$. Write $n = n'p^e$ with $p \nmid n'$ and $e \geq 0$.
Use part (a) to show that

$$\sigma_{k-1}^{\psi,\varphi}(np) = \psi(p)\sigma_{k-1}^{\psi,\varphi}(n) + \varphi(p^{e+1})(p^{e+1})^{k-1}\sigma_{k-1}^{\psi,\varphi}(n').$$

Also use part (a) to show that when $e \geq 1$ and $p \nmid N$,

$$\chi(p)p^{k-1}\sigma_{k-1}^{\psi,\varphi}(n/p) = \varphi(p)p^{k-1}\sigma_{k-1}^{\psi,\varphi}(n) - \varphi(p^{e+1})(p^{e+1})^{k-1}\sigma_{k-1}^{\psi,\varphi}(n').$$

Use the two formulas to show that for all $e \geq 0$ and $p \nmid N$,

$$\sigma_{k-1}^{\psi,\varphi}(np) + \chi(p)p^{k-1}\sigma_{k-1}^{\psi,\varphi}(n/p) = (\psi(p) + \varphi(p)p^{k-1})\sigma_{k-1}^{\psi,\varphi}(n).$$

Show that this formula also holds for all $e \geq 0$ and $p \mid uv$.

(c) Excluding the case $k = 2$, $\psi = \varphi = 1_1$, use part (b) and formula (5.4)
to show that if $uv = N$ or $p \nmid N$ then

$$a_n(T_p E_k^{\psi,\varphi,t}) = (\psi(p) + \varphi(p)p^{k-1})a_n(E_k^{\psi,\varphi,t}), \quad n \geq 1.$$

Show the same result for $n = 0$. (Hints for this exercise are at the end of the
book.)

(d) Complete the proof by computing $a_n(E_2^{1_1,1_1,t})$ and $a_n(T_p E_2^{1_1,1_1,t})$ for
all n.

5.2.6. Let N be a positive integer. Renormalize the weight 2 Eisenstein series
to $E_2(\tau) = -1/24 + \sum_{n \geq 1} \sigma(n)q^n$ where $\tau \in \mathcal{H}$ and $q = e^{2\pi i \tau}$. For any
divisor $t > 1$ of N define $f_t \in \mathcal{M}_2(\Gamma_0(N))$ to be

$$f_t(\tau) = E_2(\tau) - tE_2(t\tau), \quad \tau \in \mathcal{H}.$$

(a) Show that for any prime p,

$$T_p(f_t) = \begin{cases} (1+p)f_t & \text{if } p \nmid N, \\ (1+p)f_t + f_p - f_{pt} & \text{if } p \mid N, \ p \nmid t, \\ f_p + pf_{t/p} & \text{if } p \mid t. \end{cases}$$

(A hint for this exercise is at the end of the book.)

(b) Let q be prime (here q is not $e^{2\pi i \tau}$). Use Theorem 4.6.2 and part (a)
to show that the two functions f_q and $(1+q)f_q - f_{q^2}$ represent a Hecke
eigenfunction basis of $\mathcal{E}_2(\Gamma_0(q^2))$.

(c) Let q and r be distinct primes. Use Theorem 4.6.2 and part (a) to show
that the three functions $f_q + f_r - f_{qr}$ and $f_q + qf_r - f_{qr}$ and $rf_q + f_r - f_{qr}$
represent a Hecke eigenfunction basis of $\mathcal{E}_2(\Gamma_0(qr))$.

(d) Note that the definition of f_t extends to $f_1 = 0$. Let $p \mid N$, $t \mid N$, $p \nmid t$. Show that $f_t - f_{pt}$ is a Hecke eigenfunction with T_p-eigenvalue 1. Use this to show that if q, r, u are distinct primes then the function $f = f_1 - f_q - f_r - f_u + f_{qr} + f_{qu} + f_{ru} - f_{qru}$ is a Hecke eigenfunction with eigenvalues $\lambda_p = 1$ for $p = q, r, u$ (this generalizes the first function in part (c)), and similarly for the function $f = \sum_{d \mid N} (-1)^{\varepsilon(d)} f_d$ where N is squarefree and $\varepsilon(d)$ is the number of prime factors of d.

5.2.7. Verify part (b) of Proposition 5.2.4.

5.2.8. (a) Show that for each $\tau \in \mathcal{H}$ and each matrix $\beta_j = \begin{bmatrix} 1 & j \\ 0 & p \end{bmatrix}$ where $0 \leq j < p$, the group $C = \Lambda_{\beta_j(\tau)}$ is $\langle (\tau + j)/p \rangle + \Lambda_\tau$. Show that this group satisfies conditions (5.7). If $p \nmid N$ show that for each $\tau \in \mathcal{H}$ and the matrix $\beta_\infty = \begin{bmatrix} m & n \\ N & p \end{bmatrix} \begin{bmatrix} p & 0 \\ 0 & 1 \end{bmatrix}$, the group $C = C_\infty = (N\tau + 1)\Lambda_{\beta_\infty(\tau)}$ is $\langle 1/p \rangle + \Lambda_\tau$ and this group satisfies conditions (5.7). (A hint for this exercise is at the end of the book.)

(b) How do the subgroups of $E_\tau[5]$ in Figure 5.1 correspond to the matrices $\beta_0, \ldots, \beta_5, \beta_\infty$?

(c) As an alternative to the counting argument in the section, consider any subgroup $C \subset E_\tau$ satisfying (5.7). By the first condition $[E_\tau[p] : C] = p$ and so a basis of C is

$$\beta \begin{bmatrix} \tau/p + \Lambda_\tau \\ 1/p + \Lambda_\tau \end{bmatrix}, \quad \beta \in M_2(\mathbb{Z}), \quad \det(\beta) = p.$$

Two such matrices β and β' specify the same group exactly when $\beta' = \gamma\beta$ for some $\gamma \in SL_2(\mathbb{Z})$. If $\beta = \begin{bmatrix} a & b \\ c & d \end{bmatrix}$ then $\gcd(a, c)$ is 1 or p. Show that repeated left multiplication by elements of $SL_2(\mathbb{Z})$ to carry out the Euclidean algorithm on a and c and then further left multiplication if necessary reduce β to exactly one of $\beta_0, \ldots, \beta_{p-1}, \beta_\infty$.

(d) Show diagram (5.8) commutes by showing that $\psi_1([E/C_j, 1/N + C_j]) = \Gamma_1(N)\beta_j(\tau)$ for all j, including $j = \infty$ when $p \nmid N$.

(e) Show that $[E_\tau/C_\infty, \frac{1}{N} + C_\infty] = [E_{p\tau}, \frac{p}{N} + \Lambda_{p\tau}]$ when $p \nmid N$.

5.2.9. Evaluate $T_3(SL_2(\mathbb{Z})i) \in \mathrm{Div}(SL_2(\mathbb{Z}) \backslash \mathcal{H}^*)$, viewing T_3 in the sense of (5.5) and keeping an eye out for ramification. (A hint for this exercise is at the end of the book.)

5.2.10. Specializing the general double coset operator to the data for T_p, i.e., $\Gamma_1 = \Gamma_2 = \Gamma_1(N)$ and $\alpha = \begin{bmatrix} 1 & 0 \\ 0 & p \end{bmatrix}$, so that $\Gamma_3 = \Gamma_1^0(N, p)$, the corresponding modular curve X_3 is denoted $X_1^0(N, p)$. Show that in the configuration (5.1) the maps π_2 and $\pi_1 \circ \alpha$ from $X_1^0(N, p)$ to $X_1(N)$ are

$$\pi_2(\Gamma_1^0(N, p)\tau) = \Gamma_1(N)\tau, \qquad (\pi_1 \circ \alpha)(\Gamma_1^0(N, p)\tau) = \Gamma_1(N)\tau/p.$$

The moduli space versions of these maps were given in (1.11) and derived in Exercise 1.5.6. Show that their descriptions and the commutative diagram

$$S_1(N) \longleftarrow S_1^0(N,p) \longrightarrow S_1(N)$$

$$\psi_1 \downarrow \qquad\qquad \downarrow \psi_1^0 \qquad\qquad \downarrow \psi_1$$

$$X_1(N) \xleftarrow{\;\pi_2\;} X_1^0(N,p) \xrightarrow{\;\pi_1 \circ \alpha\;} X_1(N)$$

give another derivation of (5.6).

5.3 The $\langle n \rangle$ and T_n operators

So far the Hecke operators $\langle d \rangle$ and T_p are defined for $d \in (\mathbb{Z}/N\mathbb{Z})^*$ and p prime. This section extends the definitions to $\langle n \rangle$ and T_n for all $n \in \mathbb{Z}^+$.

For $n \in \mathbb{Z}^+$ with $(n, N) = 1$, $\langle n \rangle$ is determined by $n \pmod{N}$. For $n \in \mathbb{Z}^+$ with $(n, N) > 1$, define $\langle n \rangle = 0$, the zero operator on $\mathcal{M}_k(\Gamma_1(N))$. The mapping $n \mapsto \langle n \rangle$ is totally multiplicative, i.e., $\langle nm \rangle = \langle n \rangle \langle m \rangle$ for all $n, m \in \mathbb{Z}^+$.

To define T_n, set $T_1 = 1$ (the identity operator); T_p is already defined for primes p. For prime powers, define inductively

$$T_{p^r} = T_p T_{p^{r-1}} - p^{k-1} \langle p \rangle T_{p^{r-2}}, \quad \text{for } r \geq 2, \tag{5.10}$$

and note that inductively on r and s starting from Proposition 5.2.4(c), $T_{p^r} T_{q^s} = T_{q^s} T_{p^r}$ for distinct primes p and q. Extend the definition multiplicatively to T_n for all n,

$$T_n = \prod T_{p_i^{r_i}} \quad \text{where } n = \prod p_i^{r_i}, \tag{5.11}$$

so that the T_n all commute by Proposition 5.2.4 and

$$T_{nm} = T_n T_m \quad \text{if } (n, m) = 1.$$

Exercise 5.3.1 gives a more direct description of T_n on $\mathcal{M}_k(\Gamma_0(N))$. To motivate the mysterious-looking prime power definition, let $g(s)$ be a generating function of the T_n,

$$g(s) = \sum_{n=1}^{\infty} T_n n^{-s}.$$

Then definitions (5.10) and (5.11), determining all the T_n in terms of T_p and $\langle p \rangle$, are encapsulated as a product expression for g (Exercise 5.3.2),

$$g(s) = \prod_p (1 - T_p p^{-s} + \langle p \rangle p^{k-1-2s})^{-1}, \tag{5.12}$$

the product taken over all primes. This idea will be pursued further in Section 5.9. Also we remark that for any congruence subgroup Γ of $SL_2(\mathbb{Z})$, the double cosets $\Gamma \alpha \Gamma$ for $\alpha \in GL_2^+(\mathbb{Q})$ generate an algebra $\mathbb{Z}[\Gamma \backslash GL_2^+(\mathbb{Q})/\Gamma]$.

Especially when Γ is $\Gamma_0(N)$ or $\Gamma_1(N)$, the Hecke operators $\langle n \rangle$ and T_n for any $n \in \mathbb{Z}^+$ can be defined as elements of the algebra,

$$\langle n \rangle = \begin{cases} \Gamma \left[\begin{smallmatrix} * & * \\ * & n \end{smallmatrix} \right] \Gamma \text{ where } \left[\begin{smallmatrix} * & * \\ * & n \end{smallmatrix} \right] \in \Gamma_0(N) & \text{if } (n, N) = 1, \\ 0 & \text{if } (n, N) > 1, \end{cases}$$

and

$$T_n = \sum_{\substack{ad=n \\ a|d}} \langle a \rangle \Gamma \begin{bmatrix} a & 0 \\ 0 & d \end{bmatrix} \Gamma,$$

where it is tacit that a and d are positive in the sum. Doing so makes the recursive expression (5.10) for T_{p^r} and the multiplicative property (5.11) of T_n calculations in $\mathbb{Z}[\Gamma \backslash \mathrm{GL}_2^+(\mathbb{Q})/\Gamma]$ rather than definitions. For more on this subject see Chapter 3 of [Shi73].

The Fourier coefficient formulas in Proposition 5.2.2 generalize to

Proposition 5.3.1. *Let $f \in \mathcal{M}_k(\Gamma_1(N))$ have Fourier expansion*

$$f(\tau) = \sum_{m=0}^{\infty} a_m(f) q^m \quad \text{where } q = e^{2\pi i \tau}.$$

Then for all $n \in \mathbb{Z}^+$, $T_n f$ has Fourier expansion

$$(T_n f)(\tau) = \sum_{m=0}^{\infty} a_m(T_n f) q^m,$$

where

$$a_m(T_n f) = \sum_{d | (m,n)} d^{k-1} a_{mn/d^2}(\langle d \rangle f). \tag{5.13}$$

In particular, if $f \in \mathcal{M}_k(N, \chi)$ then

$$a_m(T_n f) = \sum_{d | (m,n)} \chi(d) d^{k-1} a_{mn/d^2}(f). \tag{5.14}$$

Proof. As usual, we may take $f \in \mathcal{M}_k(N, \chi)$ and thus formula (5.13) reduces to (5.14). Formula (5.14) checks trivially when $n = 1$ and it reduces to (5.4) when $n = p$ is prime (Exercise 5.3.3). Let $r \geq 2$ and assume (5.14) holds for $n = 1, p, p^2, \ldots, p^{r-1}$. By definition of T_{p^r} and formula (5.4),

$$\begin{aligned} a_m(T_{p^r} f) &= a_m(T_p(T_{p^{r-1}} f)) - p^{k-1} a_m(\langle p \rangle (T_{p^{r-2}} f)) \\ &= a_{mp}(T_{p^{r-1}} f) + \chi(p) p^{k-1} a_{m/p}(T_{p^{r-1}} f) \\ &\quad - \chi(p) p^{k-1} a_m(T_{p^{r-2}} f). \end{aligned}$$

By induction on r this is

$$a_m(T_{p^r}f) = \sum_{d|(mp,p^{r-1})} \chi(d)d^{k-1}a_{mp^r/d^2}(f)$$

$$+ \chi(p)p^{k-1} \sum_{d|(m/p,p^{r-1})} \chi(d)d^{k-1}a_{mp^{r-2}/d^2}(f)$$

$$- \chi(p)p^{k-1} \sum_{d|(m,p^{r-2})} \chi(d)d^{k-1}a_{mp^{r-2}/d^2}(f).$$

The first term is

$$a_{mp^r}(f) + \sum_{\substack{d|(mp,p^{r-1}) \\ d>1}} \chi(d)d^{k-1}a_{mp^r/d^2}(f),$$

and the sum here cancels the third term. Thus

$$a_m(T_{p^r}f) = a_{mp^r}(f) + \chi(p)p^{k-1} \sum_{d|(m/p,p^{r-1})} \chi(d)d^{k-1}a_{mp^{r-2}/d^2}(f).$$

A little inspection shows that this regroups to the desired formula (5.14) with $n = p^r$ (Exercise 5.3.4).

Finally, take $n_1, n_2 \in \mathbb{Z}^+$ with $(n_1, n_2) = 1$. Then

$$a_m(T_{n_1}(T_{n_2}f)) = \sum_{d|(m,n_1)} \chi(d)d^{k-1}a_{mn_1/d^2}(T_{n_2}f)$$

$$= \sum_{d|(m,n_1)} \chi(d)d^{k-1} \sum_{e|(mn_1/d^2,n_2)} \chi(e)e^{k-1}a_{mn_1n_2/d^2e^2}(f),$$

and this regroups to formula (5.14) (Exercise 5.3.4 again). $\square$

Exercises

5.3.1. Let $f \in \mathcal{M}_k(\Gamma_0(N))$. For each $n \in \mathbb{Z}^+$ define

$$S_n = \{d \in \mathbb{Z}^+ : d \mid n, \gcd(n/d, N) = 1\}, \quad M_n = \bigcup_{d \in S_n} \bigcup_{j=0}^{d-1} \begin{bmatrix} n/d & j \\ 0 & d \end{bmatrix}.$$

Show that $T_n f = \sum_{\gamma \in M_n} f[\gamma]_k$. (A hint for this exercise is at the end of the book.)

5.3.2. Show that definitions (5.10) and (5.11) imply relation (5.12).

5.3.3. Check that Proposition 5.3.1 reduces to formula (5.4) when $n = p$ is prime.

5.3.4. Convince yourself that the two expressions computed in the proof of Proposition 5.3.1 regroup to formula (5.14) as claimed.

5.4 The Petersson inner product

To study the space of cusp forms $\mathcal{S}_k(\Gamma_1(N))$ further, we make it into an inner product space. The inner product will be defined as an integral. The first few results will establish that the integral in question converges and is well defined. It does not converge on the larger space $\mathcal{M}_k(\Gamma_1(N))$, so the inner product structure is restricted to the cusp forms.

Define the *hyperbolic measure* on the upper half plane,

$$d\mu(\tau) = \frac{dx\,dy}{y^2}, \quad \tau = x + iy \in \mathcal{H}.$$

This is invariant under the automorphism group $\mathrm{GL}_2^+(\mathbb{R})$ of $\mathcal{H}$, meaning $d\mu(\alpha(\tau)) = d\mu(\tau)$ for all $\alpha \in \mathrm{GL}_2^+(\mathbb{R})$ and $\tau \in \mathcal{H}$ (Exercise 5.4.1(a)), and thus in particular $d\mu$ is $\mathrm{SL}_2(\mathbb{Z})$-invariant. Since the set $\mathbb{Q} \cup \{\infty\}$ is countable it has measure zero, and so $d\mu$ suffices for integrating over the extended upper half plane $\mathcal{H}^*$. Recall from Chapter 2 that a fundamental domain of $\mathcal{H}^*$ under the action of $\mathrm{SL}_2(\mathbb{Z})$ is

$$\mathcal{D}^* = \{\tau \in \mathcal{H} : |\mathrm{Re}(\tau)| \leq 1/2, |\tau| \geq 1\} \cup \{\infty\}.$$

That is, every point $\tau' \in \mathcal{H}$ transforms under $\mathrm{SL}_2(\mathbb{Z})$ into the connected set $\mathcal{D}$, and barring certain cases on the boundary of $\mathcal{D}$ the transformation is unique; and every point $s \in \mathbb{Q} \cup \{\infty\}$ transforms under $\mathrm{SL}_2(\mathbb{Z})$ to ∞. For any continuous, bounded function $\varphi : \mathcal{H} \longrightarrow \mathbb{C}$ and any $\alpha \in \mathrm{SL}_2(\mathbb{Z})$, the integral $\int_{\mathcal{D}^*} \varphi(\alpha(\tau))d\mu(\tau)$ converges (Exercise 5.4.1(b)).

Let $\Gamma \subset \mathrm{SL}_2(\mathbb{Z})$ be a congruence subgroup and let $\{\alpha_j\} \subset \mathrm{SL}_2(\mathbb{Z})$ represent the coset space $\{\pm I\}\Gamma\backslash\mathrm{SL}_2(\mathbb{Z})$, meaning that the union

$$\mathrm{SL}_2(\mathbb{Z}) = \bigcup_j \{\pm I\}\Gamma\alpha_j$$

is disjoint. If the function φ is Γ-invariant then the sum $\sum_j \int_{\mathcal{D}^*} \varphi(\alpha_j(\tau))d\mu(\tau)$ is independent of the choice of coset representatives α_j. Since $d\mu$ is $\mathrm{SL}_2(\mathbb{Z})$-invariant the sum is $\int_{\bigcup\alpha_j(\mathcal{D}^*)} \varphi(\tau)d\mu(\tau)$. And since $\bigcup\alpha_j(\mathcal{D}^*)$ represents the modular curve $X(\Gamma)$ up to some boundary identification (cf. the end of Section 2.3), this quantity is naturally denoted $\int_{X(\Gamma)}$. Thus we have made the definition

$$\int_{X(\Gamma)} \varphi(\tau)d\mu(\tau) = \int_{\bigcup\alpha_j(\mathcal{D}^*)} \varphi(\tau)d\mu(\tau) = \sum_j \int_{\mathcal{D}^*} \varphi(\alpha_j(\tau))d\mu(\tau).$$

In particular, setting $\varphi = 1$, the *volume* of $X(\Gamma)$ is

$$V_\Gamma = \int_{X(\Gamma)} d\mu(\tau).$$

The volume and index of a congruence subgroup are related by

$$V_\Gamma = [\mathrm{SL}_2(\mathbb{Z}) : \{\pm I\}\Gamma] V_{\mathrm{SL}_2(\mathbb{Z})}. \tag{5.15}$$

To construct the Petersson inner product, take any $f, g \in \mathcal{S}_k(\Gamma)$ and let

$$\varphi(\tau) = f(\tau)\overline{g(\tau)}(\mathrm{Im}(\tau))^k \qquad \text{for } \tau \in \mathcal{H}.$$

This is clearly continuous. It is Γ-invariant because for any $\gamma \in \Gamma$,

$$\varphi(\gamma(\tau)) = f(\gamma(\tau))\overline{g(\gamma(\tau))}(\mathrm{Im}(\gamma(\tau)))^k$$

$$= (f[\gamma]_k)(\tau)j(\gamma,\tau)^k \overline{(g[\gamma]_k)(\tau)} \, \overline{j(\gamma,\tau)}^k (\mathrm{Im}(\tau))^k |j(\gamma,\tau)|^{-2k}$$

$$= (f[\gamma]_k)(\tau)\overline{(g[\gamma]_k)(\tau)}(\mathrm{Im}(\tau))^k$$

$$= \varphi(\tau) \quad \text{since } f \text{ and } g \text{ are weight-}k \text{ invariant under } \Gamma.$$

With Γ-invariance established, showing that φ is bounded on $\mathcal{H}$ reduces to showing that φ is bounded on $\bigcup \alpha_i(\mathcal{D})$, and since the union is finite it suffices to show that for any $\alpha \in \mathrm{SL}_2(\mathbb{Z})$, $\varphi \circ \alpha$ is bounded on $\mathcal{D}$. Being continuous, $\varphi \circ \alpha$ is certainly bounded on any compact subset of $\mathcal{D}$. As for neighborhoods $\{\mathrm{Im}(\tau) > y\}$ of $i\infty$, first note the Fourier expansions

$$(f[\alpha]_k)(\tau) = \sum_{n=1}^\infty a_n(f[\alpha]_k)q_h^n, \qquad (g[\alpha]_k)(\tau) = \sum_{n=1}^\infty a_n(g[\alpha]_k)q_h^n,$$

where $q_h = e^{2\pi i \tau/h}$ for some $h \in \mathbb{Z}^+$. Each of these is of the order q_h (written $\mathcal{O}(q_h)$) as $\mathrm{Im}(\tau) \to \infty$. Thus as above,

$$\varphi(\alpha(\tau)) = (f[\alpha]_k)(\tau)\overline{(g[\alpha]_k)(\tau)}(\mathrm{Im}(\tau))^k = \mathcal{O}(q_h)^2(\mathrm{Im}(\tau))^k$$

by the Fourier expansions. Since $|q_h| = e^{-2\pi \mathrm{Im}(\tau)/h}$ and exponential decay dominates polynomial growth, $\varphi(\alpha(\tau)) \to 0$ as $\mathrm{Im}(\tau) \to \infty$ and $\varphi \circ \alpha$ is bounded on $\mathcal{D}$ as desired. This shows that in the next definition the integral is well defined and convergent.

Definition 5.4.1. *Let $\Gamma \subset \mathrm{SL}_2(\mathbb{Z})$ be a congruence subgroup. The* **Petersson inner product,**

$$\langle,\rangle_\Gamma : \mathcal{S}_k(\Gamma) \times \mathcal{S}_k(\Gamma) \longrightarrow \mathbb{C},$$

is given by

$$\langle f, g \rangle_\Gamma = \frac{1}{V_\Gamma} \int_{X(\Gamma)} f(\tau)\overline{g(\tau)}(\mathrm{Im}(\tau))^k d\mu(\tau).$$

Clearly this product is linear in f, conjugate linear in g, Hermitian-symmetric, and positive definite. The normalizing factor $1/V_\Gamma$ ensures that if $\Gamma' \subset \Gamma$ then $\langle,\rangle_{\Gamma'} = \langle,\rangle_\Gamma$ on $\mathcal{S}_k(\Gamma)$ (Exercise 5.4.3). When the subgroup Γ is clear from context, the inner product is written $\langle,\rangle$ without the subscript.

Inspecting the argument that the integral defining the Petersson inner product converges shows that in fact this holds so long as the product fg vanishes at each cusp. In particular it holds when only one of f and g is a cusp form. Phrasing Definition 5.4.1 and the various results in the next section to encompass this additional generality is a bit cumbersome, so we don't bother since our main objects of study via the Petersson inner product are cusp forms. At the end of this chapter, however, we will revisit Eisenstein series briefly, showing among other things that the Petersson inner product of an Eisenstein series and a cusp form is always 0. That is, Eisenstein series and cusp forms are in some sense orthogonal, but this statement is an abuse of language since the Petersson inner product does not converge on all of $\mathcal{M}_k(\Gamma)$, possibly diverging for two noncusp forms.

Exercises

5.4.1. (a) Show that the hyperbolic measure, $d\mu(\tau) = dx\,dy/y^2$ for $\tau = x + iy \in \mathcal{H}$, is $\mathrm{GL}_2^+(\mathbb{R})$-invariant. (A hint for this exercise is at the end of the book.)

(b) Let $\varphi : \mathcal{H} \longrightarrow \mathbb{C}$ be continuous and bounded, and let $\alpha \in \mathrm{SL}_2(\mathbb{Z})$. Show that the integral $\int_{\mathcal{D}^*} \varphi(\alpha(\tau))d\mu(\tau)$ converges.

5.4.2. What is $V_{\mathrm{SL}_2(\mathbb{Z})}$?

5.4.3. Show that if $\Gamma' \subset \Gamma \subset \mathrm{SL}_2(\mathbb{Z})$ are congruence subgroups then $\langle\,,\,\rangle_{\Gamma'} = \langle\,,\,\rangle_\Gamma$ on $\mathcal{S}_k(\Gamma)$. (A hint for this exercise is at the end of the book.)

5.4.4. Let $\Gamma' \subset \Gamma \subset \mathrm{SL}_2(\mathbb{Z})$ be congruence subgroups with $-I \in \Gamma'$. Suppose that $f \in \mathcal{S}_k(\Gamma) \subset \mathcal{S}_k(\Gamma')$ and that $g \in \mathcal{S}_k(\Gamma')$. Letting $\Gamma = \bigcup_i \Gamma'\alpha_i$, recall the trace of g from Section 5.1, $\mathrm{tr}\,g = \sum_i g[\alpha_i]_k \in \mathcal{S}_k(\Gamma)$. Show that

$$V_{\Gamma'}\langle f, g\rangle_{\Gamma'} = V_\Gamma\langle f, \mathrm{tr}\,g\rangle_\Gamma.$$

(A hint for this exercise is at the end of the book.)

5.5 Adjoints of the Hecke Operators

Recall that if V is an inner product space and T is a linear operator on V, then the *adjoint* T^* is the linear operator on V defined by the condition

$$\langle Tv, w\rangle = \langle v, T^*w\rangle \qquad \text{for all } v, w \in V.$$

Recall also that the operator T is called *normal* when it commutes with its adjoint.

Continuing to work in the space $\mathcal{S}_k(\Gamma_1(N))$ endowed with the Petersson inner product, the next task is to show that the Hecke operators $\langle n\rangle$ and T_n

for n relatively prime to N are normal. To do so, we need to compute their adjoints. This first requires a few technical points. If $\Gamma \subset \mathrm{SL}_2(\mathbb{Z})$ is a congruence subgroup and $\mathrm{SL}_2(\mathbb{Z}) = \bigcup_j \{\pm I\} \Gamma \alpha_j$ and $\alpha \in \mathrm{GL}_2^+(\mathbb{Q})$ then the map $\mathcal{H} \longrightarrow \mathcal{H}$ given by $\tau \mapsto \alpha(\tau)$ induces a bijection $\alpha^{-1}\Gamma\alpha \backslash \mathcal{H}^* \longrightarrow X(\Gamma)$. Thus the union $\bigcup_j \alpha^{-1}\alpha_j(\mathcal{D}^*)$ represents the quotient space $\alpha^{-1}\Gamma\alpha \backslash \mathcal{H}^*$ up to some boundary identification. Analogously to the last section, for continuous, bounded, $\alpha^{-1}\Gamma\alpha$-invariant functions $\varphi : \mathcal{H} \longrightarrow \mathbb{C}$ define

$$\int_{\alpha^{-1}\Gamma\alpha\backslash\mathcal{H}^*} \varphi(\tau)d\mu(\tau) = \sum_j \int_{\mathcal{D}^*} \varphi(\alpha^{-1}\alpha_j(\tau))d\mu(\tau).$$

Lemma 5.5.1. *Let $\Gamma \subset \mathrm{SL}_2(\mathbb{Z})$ be a congruence subgroup, and let $\alpha \in \mathrm{GL}_2^+(\mathbb{Q})$.*

(a) *If $\varphi : \mathcal{H} \longrightarrow \mathbb{C}$ is continuous, bounded, and Γ-invariant, then*

$$\int_{\alpha^{-1}\Gamma\alpha\backslash\mathcal{H}^*} \varphi(\alpha(\tau))d\mu(\tau) = \int_{X(\Gamma)} \varphi(\tau)d\mu(\tau).$$

(b) *If $\alpha^{-1}\Gamma\alpha \subset \mathrm{SL}_2(\mathbb{Z})$ then $V_{\alpha^{-1}\Gamma\alpha} = V_\Gamma$ and $[\mathrm{SL}_2(\mathbb{Z}) : \alpha^{-1}\Gamma\alpha] = [\mathrm{SL}_2(\mathbb{Z}) : \Gamma]$.*

(c) *There exist $\beta_1, \ldots, \beta_n \in \mathrm{GL}_2^+(\mathbb{Q})$, where $n = [\Gamma : \alpha^{-1}\Gamma\alpha \cap \Gamma] = [\Gamma : \alpha\Gamma\alpha^{-1} \cap \Gamma]$, such that*

$$\Gamma\alpha\Gamma = \bigcup \Gamma\beta_j = \bigcup \beta_j\Gamma,$$

with both unions disjoint.

Proof. Part (a) is immediate.

The first formula in (b) follows from (a) and the definition of volume. The rest follows from formula (5.15) and the fact that $-I \in \alpha^{-1}\Gamma\alpha \iff -I \in \Gamma$.

For (c), apply (b) with $\alpha\Gamma\alpha^{-1} \cap \Gamma$ in place of Γ to get

$$[\mathrm{SL}_2(\mathbb{Z}) : \alpha^{-1}\Gamma\alpha \cap \Gamma] = [\mathrm{SL}_2(\mathbb{Z}) : \alpha\Gamma\alpha^{-1} \cap \Gamma],$$

and the same formula holds with Γ replacing $\mathrm{SL}_2(\mathbb{Z})$. Thus there exist $\gamma_1, \ldots, \gamma_n$ and $\tilde{\gamma}_1, \ldots, \tilde{\gamma}_n$ in Γ such that

$$\Gamma = \bigcup (\alpha^{-1}\Gamma\alpha \cap \Gamma)\gamma_j = \bigcup (\alpha\Gamma\alpha^{-1} \cap \Gamma)\tilde{\gamma}_j^{-1},$$

both unions disjoint. Recall that Lemma 5.1.2 relates coset representatives in the quotient space to orbit representatives in the double coset. Setting Γ_1 and Γ_2 in the lemma to Γ here, and setting α in the lemma first to α here and then to α^{-1} here gives

$$\Gamma\alpha\Gamma = \bigcup \Gamma\alpha\gamma_j \quad \text{and} \quad \Gamma\alpha^{-1}\Gamma = \bigcup \Gamma\alpha^{-1}\tilde{\gamma}_j^{-1},$$

both unions disjoint. The second formula is also $\Gamma\alpha\Gamma = \bigcup \tilde{\gamma}_j \alpha\Gamma$. For each j, the left and right cosets $\Gamma\alpha\gamma_j$ and $\tilde{\gamma}_j\alpha\Gamma$ have nonempty intersection, for otherwise $\Gamma\alpha\gamma_j \subset \bigcup_{i\neq j} \tilde{\gamma}_i\alpha\Gamma$, and multiplying from the right by Γ gives $\Gamma\alpha\Gamma \subset \bigcup_{i\neq j} \tilde{\gamma}_i\alpha\Gamma$, a contradiction. So for each j, pick some $\beta_j \in \Gamma\alpha\gamma_j \cap \tilde{\gamma}_j\alpha\Gamma$. Then $\Gamma\alpha\Gamma = \bigcup \Gamma\beta_j = \bigcup \beta_j\Gamma$ as desired. $\square$

With the technicalities established, the next proposition shows how to compute adjoints.

Proposition 5.5.2. *Let $\Gamma \subset \mathrm{SL}_2(\mathbb{Z})$ be a congruence subgroup, and let $\alpha \in \mathrm{GL}_2^+(\mathbb{Q})$. Set $\alpha' = \det(\alpha)\alpha^{-1}$. Then*

(a) *If $\alpha^{-1}\Gamma\alpha \subset \mathrm{SL}_2(\mathbb{Z})$ then for all $f \in \mathcal{S}_k(\Gamma)$ and $g \in \mathcal{S}_k(\alpha^{-1}\Gamma\alpha)$,*

$$\langle f[\alpha]_k, g\rangle_{\alpha^{-1}\Gamma\alpha} = \langle f, g[\alpha']_k\rangle_\Gamma.$$

(b) *For all $f, g \in \mathcal{S}_k(\Gamma)$,*

$$\langle f[\Gamma\alpha\Gamma]_k, g\rangle = \langle f, g[\Gamma\alpha'\Gamma]_k\rangle.$$

In particular, if $\alpha^{-1}\Gamma\alpha = \Gamma$ then $[\alpha]_k^* = [\alpha']_k$, and in any case $[\Gamma\alpha\Gamma]_k^* = [\Gamma\alpha'\Gamma]_k$.

Proof. For part (a), expand the $[\alpha]_k$ operator, note that α' acts as α^{-1} on $\mathcal{H}^*$, and apply Lemma 5.5.1(a) to get

$$\int_{\alpha^{-1}\Gamma\alpha\backslash\mathcal{H}^*} (f[\alpha]_k)(\tau)\overline{g(\tau)}(\mathrm{Im}(\tau))^k d\mu(\tau)$$

$$= \int_{\alpha^{-1}\Gamma\alpha\backslash\mathcal{H}^*} (\det\alpha)^{k-1} f(\alpha(\tau))j(\alpha,\tau)^{-k}\overline{g(\tau)}(\mathrm{Im}(\tau))^k d\mu(\tau)$$

$$= \int_{X(\Gamma)} (\det\alpha)^{k-1} f(\tau)j(\alpha,\alpha'(\tau))^{-k}\overline{g(\alpha'(\tau))}(\mathrm{Im}(\alpha'(\tau)))^k d\mu(\tau).$$

The cocycle condition $j(\alpha\alpha',\tau) = j(\alpha,\alpha'(\tau))j(\alpha',\tau)$, the upper half plane identity $\mathrm{Im}(\alpha'(\tau)) = (\det\alpha')\mathrm{Im}(\tau)|j(\alpha',\tau)|^{-2}$, and the observation $\det\alpha' = \det\alpha$ reduce the integral to

$$\int_{X(\Gamma)} f(\tau)\overline{(g[\alpha']_k)(\tau)}(\mathrm{Im}(\tau))^k d\mu(\tau).$$

Since also $V_{\alpha^{-1}\Gamma\alpha} = V_\Gamma$ by Lemma 5.5.1(b), the result follows.

For part (b), the relation $\Gamma\alpha\Gamma = \bigcup \Gamma\beta_j$ from Lemma 5.5.1(c) shows that the $\{\beta_j\}$ can serve in Definition 5.1.3 of the operator $[\Gamma\alpha\Gamma]_k$. And the relation $\Gamma\alpha\Gamma = \bigcup \beta_j\Gamma$ gives $\Gamma\alpha'\Gamma = \bigcup \Gamma\beta'_j$ where $\beta'_j = \det(\beta_j)\beta_j^{-1}$, showing that the $\{\beta'_j\}$ can serve in the definition of $[\Gamma\alpha'\Gamma]_k$. Now the result is immediate from (a) with each $\Gamma \cap \beta_j\Gamma\beta_j^{-1}$ in place of Γ,

$$\langle f[\Gamma \alpha \Gamma]_k, g \rangle_\Gamma = \sum_j \langle f[\beta_j]_k, g \rangle_{\beta_j^{-1}\Gamma\beta_j \cap \Gamma}$$

$$= \sum_j \langle f, g[\beta_j']_k \rangle_{\Gamma \cap \beta_j \Gamma \beta_j^{-1}} = \langle f, g[\Gamma \alpha' \Gamma]_k \rangle_\Gamma.$$

$\square$

The adjoints of the Hecke operators follow from the proposition.

Theorem 5.5.3. *In the inner product space* $\mathcal{S}_k(\Gamma_1(N))$, *the Hecke operators* $\langle p \rangle$ *and* T_p *for* $p \nmid N$ *have adjoints*

$$\langle p \rangle^* = \langle p \rangle^{-1} \quad \text{and} \quad T_p^* = \langle p \rangle^{-1} T_p.$$

Thus the Hecke operators $\langle n \rangle$ *and* T_n *for* n *relatively prime to* N *are normal.*

Proof. Take any $f, g \in \mathcal{S}_k(\Gamma_1(N))$. Recall that $\Gamma_1(N)$ is normal in $\Gamma_0(N)$, so Proposition 5.5.2(a) gives

$$\langle p \rangle^* = [\alpha]_k^* \quad \text{for any } \alpha \in \Gamma_0(N) \text{ such that } \alpha \equiv [\begin{smallmatrix} * & * \\ 0 & p \end{smallmatrix}] \pmod{N}$$

$$= [\alpha^{-1}]_k = \langle p \rangle^{-1}.$$

For the second formula, Proposition 5.5.2(b) gives

$$T_p^* = [\Gamma_1(N) [\begin{smallmatrix} 1 & 0 \\ 0 & p \end{smallmatrix}] \Gamma_1(N)]_k^* = [\Gamma_1(N) [\begin{smallmatrix} p & 0 \\ 0 & 1 \end{smallmatrix}] \Gamma_1(N)]_k.$$

To study this double coset, note that for m and n with $mp - nN = 1$, the matrix in its middle satisfies $[\begin{smallmatrix} p & 0 \\ 0 & 1 \end{smallmatrix}] = [\begin{smallmatrix} 1 & n \\ N & mp \end{smallmatrix}]^{-1} [\begin{smallmatrix} 1 & 0 \\ 0 & p \end{smallmatrix}] [\begin{smallmatrix} p & n \\ N & m \end{smallmatrix}]$. The first matrix of this triple product lies in $\Gamma_1(N)$, the third in $\Gamma_0(N)$. Thus, substituting the triple product for $[\begin{smallmatrix} p & 0 \\ 0 & 1 \end{smallmatrix}]$ and recalling again that $\Gamma_1(N)$ is normal in $\Gamma_0(N)$ shows that

$$\Gamma_1(N) [\begin{smallmatrix} p & 0 \\ 0 & 1 \end{smallmatrix}] \Gamma_1(N) = \Gamma_1(N) [\begin{smallmatrix} 1 & 0 \\ 0 & p \end{smallmatrix}] \Gamma_1(N) [\begin{smallmatrix} p & n \\ N & m \end{smallmatrix}].$$

Now if $\Gamma_1(N) [\begin{smallmatrix} 1 & 0 \\ 0 & p \end{smallmatrix}] \Gamma_1(N) = \bigcup \Gamma_1(N)\beta_j$ is the decomposition of the double coset describing T_p, then the display shows that $\Gamma_1(N) [\begin{smallmatrix} p & 0 \\ 0 & 1 \end{smallmatrix}] \Gamma_1(N) = \bigcup \Gamma_1(N)\beta_j [\begin{smallmatrix} p & n \\ N & m \end{smallmatrix}]$ is the decomposition for T_p^*. Since $m \equiv p^{-1} \pmod{N}$, the result $T_p^* = \langle p \rangle^{-1} T_p$ follows. $\square$

From the Spectral Theorem of linear algebra, given a commuting family of normal operators on a finite-dimensional inner product space, the space has an orthogonal basis of simultaneous eigenvectors for the operators. Since each such vector is a modular form we say *eigenform* instead, and the result is

Theorem 5.5.4. *The space* $\mathcal{S}_k(\Gamma_1(N))$ *has an orthogonal basis of simultaneous eigenforms for the Hecke operators* $\{\langle n \rangle, T_n : (n, N) = 1\}$.

The next few sections will partly eliminate the restriction that $(n, N) = 1$. Note that when $(n, N) > 1$, we have $\langle n \rangle^* = 0^* = 0$ (the zero operator). And Exercise 5.5.1 shows that for any Hecke operator $T = T_n$ or $T = \langle n \rangle$, whether n shares factors with N or not,

$$T^* = w_N T w_N^{-1} \quad \text{where } w_N \text{ is the operator } [[\begin{smallmatrix} 0 & -1 \\ N & 0 \end{smallmatrix}]]_k. \tag{5.16}$$

Exercise

5.5.1. (a) Let $\gamma = \begin{bmatrix} 0 & -1 \\ N & 0 \end{bmatrix}$. Establish the normalization formula

$$\gamma^{-1} \begin{bmatrix} a & b \\ Nc & d \end{bmatrix} \gamma = \begin{bmatrix} d & -c \\ -Nb & a \end{bmatrix}.$$

Use this to show that $\gamma^{-1}\Gamma_1(N)\gamma = \Gamma_1(N)$, and so the operator $w_N = [\gamma]_k$ is the double coset operator $[\Gamma_1(N)\gamma\Gamma_1(N)]_k$ on $\mathcal{S}_k(\Gamma_1(N))$. Show that $w_N \langle n \rangle w_N^{-1} = \langle n \rangle^*$ for all n such that $(n, N) = 1$ and thus for all n.

(b) Let $\Gamma_1(N) \begin{bmatrix} 1 & 0 \\ 0 & p \end{bmatrix} \Gamma_1(N) = \bigcup \Gamma_1(N)\beta_j$ be a distinct union, i.e., $T_p = \sum [\beta_j]_k$. From part (a), $\gamma\Gamma_1(N) = \Gamma_1(N)\gamma$ and similarly for γ^{-1}. Use this and the formula $\begin{bmatrix} p & 0 \\ 0 & 1 \end{bmatrix} = \gamma^{-1} \begin{bmatrix} 1 & 0 \\ 0 & p \end{bmatrix} \gamma$ to find coset representatives for $\Gamma_1(N) \begin{bmatrix} p & 0 \\ 0 & 1 \end{bmatrix} \Gamma_1(N)$. Use Proposition 5.5.2 and the coset representatives to show that $T_p^* = w_N T_p w_N^{-1}$, and so $T_n^* = w_N T_n w_N^{-1}$ for all n.

(c) Show that $w_N^* = (-1)^k w_N$ and that $i^k w_N T_n$ is self-adjoint. (A hint for this exercise is at the end of the book.)

5.6 Oldforms and Newforms

So far the theory has all taken place at one generic level N. This section begins results that move between levels, taking forms from lower levels $M \mid N$ up to level N, mostly with $M = Np^{-1}$ where p is some prime factor of N.

The most trivial way to move between levels is to observe that if $M \mid N$ then $\mathcal{S}_k(\Gamma_1(M)) \subset \mathcal{S}_k(\Gamma_1(N))$ (Exercise 5.6.1).

Another way to embed $\mathcal{S}_k(\Gamma_1(M))$ into $\mathcal{S}_k(\Gamma_1(N))$ is by composing with the multiply-by-d map where d is any factor of N/M. For any such d, let

$$\alpha_d = \begin{bmatrix} d & 0 \\ 0 & 1 \end{bmatrix}$$

so that $(f[\alpha_d]_k)(\tau) = d^{k-1} f(d\tau)$ for $f : \mathcal{H} \longrightarrow \mathbb{C}$. By Exercise 1.2.11, the injective linear map $[\alpha_d]_k$ takes $\mathcal{S}_k(\Gamma_1(M))$ to $\mathcal{S}_k(\Gamma_1(N))$, lifting the level from M to N.

Combining the observations so far, it is natural to distinguish the part of $\mathcal{S}_k(\Gamma_1(N))$ coming from lower levels.

Definition 5.6.1. *For each divisor d of N, let i_d be the map*

$$i_d : (\mathcal{S}_k(\Gamma_1(Nd^{-1})))^2 \longrightarrow \mathcal{S}_k(\Gamma_1(N))$$

given by

$$(f, g) \mapsto f + g[\alpha_d]_k.$$

*The subspace of **oldforms** at level N is*

$$\mathcal{S}_k(\Gamma_1(N))^{\text{old}} = \sum_{\substack{p \mid N \\ \text{prime}}} i_p((\mathcal{S}_k(\Gamma_1(Np^{-1})))^2)$$

and the subspace of **newforms at level** N is the orthogonal complement with respect to the Petersson inner product,

$$\mathcal{S}_k(\Gamma_1(N))^{\text{new}} = (\mathcal{S}_k(\Gamma_1(N))^{\text{old}})^{\perp}.$$

Exercise 5.6.2 shows that changing the definition of oldforms to sum over all divisors of N, rather than just the prime divisors, makes no difference. For newspace dimension formulas see [Mar]. The term "newform" will be given a more specific meaning in Definition 5.8.1 to follow.

The Hecke operators respect the decomposition of $\mathcal{S}_k(\Gamma_1(N))$ into old and new.

Proposition 5.6.2. *The subspaces $\mathcal{S}_k(\Gamma_1(N))^{\text{old}}$ and $\mathcal{S}_k(\Gamma_1(N))^{\text{new}}$ are stable under the Hecke operators T_n and $\langle n \rangle$ for all $n \in \mathbb{Z}^+$.*

Proof. Let $p \mid N$. The argument breaks into cases. First, let $T = \langle d \rangle$ with $(d, N) = 1$ or let $T = T_{p'}$ for p' a prime other than p. Then the diagram

$$
\begin{array}{ccc}
\mathcal{S}_k(\Gamma_1(Np^{-1}))^2 & \xrightarrow{\left[\begin{smallmatrix} T & 0 \\ 0 & T \end{smallmatrix}\right]} & \mathcal{S}_k(\Gamma_1(Np^{-1}))^2 \\
{\scriptstyle i_p}\downarrow & & \downarrow{\scriptstyle i_p} \\
\mathcal{S}_k(\Gamma_1(N)) & \xrightarrow{\quad T \quad} & \mathcal{S}_k(\Gamma_1(N))
\end{array}
$$

commutes (Exercise 5.6.3(a)); note that the operator T means one thing at level Np^{-1} in the top line of the diagram and something else at level N in the bottom line.

Second, continuing to take $p \mid N$, the diagram

$$
\begin{array}{ccc}
\mathcal{S}_k(\Gamma_1(Np^{-1}))^2 & \xrightarrow{\left[\begin{smallmatrix} T_p & p^{k-1} \\ -\langle p \rangle & 0 \end{smallmatrix}\right]} & \mathcal{S}_k(\Gamma_1(Np^{-1}))^2 \\
{\scriptstyle i_p}\downarrow & & \downarrow{\scriptstyle i_p} \\
\mathcal{S}_k(\Gamma_1(N)) & \xrightarrow{\quad T_p \quad} & \mathcal{S}_k(\Gamma_1(N))
\end{array}
$$

commutes as well (Exercise 5.6.3(b)). The two diagrams combine to show that $\mathcal{S}_k(\Gamma_1(N))^{\text{old}}$ is stable under all T_n and $\langle n \rangle$ (Exercise 5.6.3(c)).

To establish the result for $\mathcal{S}_k(\Gamma_1(N))^{\text{new}}$ it is enough to show that $\mathcal{S}_k(\Gamma_1(N))^{\text{old}}$ is stable under the adjoints of T_n and $\langle n \rangle$ for all n (Exercise 5.6.3(d)). Since $T_n^* = \langle n \rangle^{-1} T_n$ and $\langle n \rangle^* = \langle n \rangle^{-1}$ when $(n, N) = 1$, and since $\langle n \rangle^* = 0$ when $(n, N) > 1$, the result is clear in these cases. But discussing T_n^* when $(n, N) > 1$ requires Exercise 5.5.1: $T_n^* = w T_n w^{-1}$ where

$w = [[\begin{smallmatrix} 0 & 1 \\ -N & 0 \end{smallmatrix}]]_k$. Thus it suffices to show that the oldforms are preserved under the injective linear map w. This follows from the commutative diagram

$$
\begin{array}{ccc}
\mathcal{S}_k(\Gamma_1(Np^{-1}))^2 & \xrightarrow{\left[\begin{smallmatrix} 0 & p^{k-2}w \\ w & 0 \end{smallmatrix}\right]} & \mathcal{S}_k(\Gamma_1(Np^{-1}))^2 \\
\Big\downarrow{\scriptstyle i_p} & & \Big\downarrow{\scriptstyle i_p} \\
\mathcal{S}_k(\Gamma_1(N)) & \xrightarrow{\quad w \quad} & \mathcal{S}_k(\Gamma_1(N))
\end{array}
$$

(Exercise 5.6.3(e); note that like the Hecke operators, the w operator means different things at different levels: in the top row of the diagram it is $w = w_{Np^{-1}} = [[\begin{smallmatrix} 0 & 1 \\ -Np^{-1} & 0 \end{smallmatrix}]]_k)$.   □

Corollary 5.6.3. *The spaces $\mathcal{S}_k(\Gamma_1(N))^{\text{old}}$ and $\mathcal{S}_k(\Gamma_1(N))^{\text{new}}$ have orthogonal bases of eigenforms for the Hecke operators away from the level, $\{T_n, \langle n \rangle : (n, N) = 1\}$.*

As we will see, the condition $(n, N) = 1$ can be removed for the newforms.

Exercises

5.6.1. Show that if $M \mid N$ then $\mathcal{S}_k(\Gamma_1(M)) \subset \mathcal{S}_k(\Gamma_1(N))$.

5.6.2. If $pd \mid N$ then what relation holds between $i_{pd}(\mathcal{S}_k(\Gamma_1(Np^{-1}d^{-1}))^2)$ and $i_p(\mathcal{S}_k(\Gamma_1(Np^{-1})) \times i_d(\mathcal{S}_k(\Gamma_1(Np^{-1}d^{-1}))^2))$? (Here i_{pd} is from level $Np^{-1}d^{-1}$ to level N, i_p is from level Np^{-1} to level N, and i_d is from level $Np^{-1}d^{-1}$ to level Np^{-1}.) Show that changing the definition of $\mathcal{S}_k(\Gamma_1(N))^{\text{old}}$ to sum over all divisors of N doesn't add anything to the space of oldforms.

5.6.3. (a) Show that the first diagram commutes. (Hints for this exercise are at the end of the book.)
  (b) Show that the second diagram commutes.
  (c) Explain how the two diagrams show that $\mathcal{S}_k(\Gamma_1(N))^{\text{old}}$ is stable under all T_n and $\langle n \rangle$.
  (d) How does showing that $\mathcal{S}_k(\Gamma_1(N))^{\text{old}}$ is also stable under the adjoints of T_n and $\langle n \rangle$ for all n establish the rest of the proposition?
  (e) Show that the third diagram commutes.

5.7 The Main Lemma

Let $M \mid N$ and let $d \mid (N/M)$, $d > 1$. Thus $\Gamma_1(M) \supset \Gamma_1(N)$.
  From the previous section, recall two maps from $\mathcal{S}_k(\Gamma_1(M))$ to $\mathcal{S}_k(\Gamma_1(N))$: inclusion and the weight-k operator $[\alpha_d]_k$, where $\alpha_d = [\begin{smallmatrix} d & 0 \\ 0 & 1 \end{smallmatrix}]$ so that the operator is—up to a scalar multiple—composition with the multiply-by-d map. To

normalize the scalar to 1, define a variant ι_d of the map i_d from the previous section,

$$\iota_d = d^{1-k}[\alpha_d]_k : \mathcal{S}_k(\Gamma_1(M)) \longrightarrow \mathcal{S}_k(\Gamma_1(N)), \quad (\iota_d f)(\tau) = f(d\tau),$$

acting on Fourier expansions as

$$\iota_d : \sum_{n=1}^{\infty} a_n q^n \mapsto \sum_{n=1}^{\infty} a_n q^{dn}, \quad \text{where } q = e^{2\pi i \tau}.$$

This shows that if $f \in \mathcal{S}_k(\Gamma_1(N))$ takes the form $f = \sum_{p|N} \iota_p f_p$ with each $f_p \in \mathcal{S}_k(\Gamma_1(N/p))$, and if the Fourier expansion of f is $f(\tau) = \sum a_n(f)q^n$, then $a_n(f) = 0$ for all n such that $(n, N) = 1$. The main lemma in the theory of newforms is that the converse holds as well.

Theorem 5.7.1 (Main Lemma). *If $f \in \mathcal{S}_k(\Gamma_1(N))$ has Fourier expansion $f(\tau) = \sum a_n(f)q^n$ with $a_n(f) = 0$ whenever $(n, N) = 1$, then f takes the form $f = \sum_{p|N} \iota_p f_p$ with each $f_p \in \mathcal{S}_k(\Gamma_1(N/p))$.*

The Main Lemma is due to Atkin and Lehner [AL70]. The elegant proof presented here is due to David Carlton [Car99, Car01], using some basic results about group representations and tensor products. Explaining these from scratch would take the exposition too far afield, so the reader should consult a text such as [FH91] (Chapters 1 and 2) as necessary. A simpler instance of the tensor product will be presented in Chapter 6. An elementary proof of the Main Lemma is in [Lan76], and see also [Miy89].

The first step is to change congruence subgroups, moving the congruence conditions from the lower left matrix entry to the upper right one. This simplifies the Main Lemma by converting the map ι_p to inclusion. Analogously to $\Gamma_1(N)$, define for any positive integer N

$$\Gamma^1(N) = \left\{ \begin{bmatrix} a & b \\ c & d \end{bmatrix} \in \mathrm{SL}_2(\mathbb{Z}) : \begin{bmatrix} a & b \\ c & d \end{bmatrix} \equiv \begin{bmatrix} 1 & 0 \\ * & 1 \end{bmatrix} \pmod{N} \right\}.$$

Lemma 5.7.2. $\alpha_M \Gamma_1(M) \alpha_M^{-1} = \Gamma^1(M)$, *and the same formula holds with M replaced by N.*

The proof is Exercise 5.7.1.

Thus the two maps $M^{k-1}[\alpha_M^{-1}]_k : \mathcal{S}_k(\Gamma_1(M)) \longrightarrow \mathcal{S}_k(\Gamma^1(M))$ and $N^{k-1}[\alpha_N^{-1}]_k : \mathcal{S}_k(\Gamma_1(N)) \longrightarrow \mathcal{S}_k(\Gamma^1(N))$ are isomorphisms. The first map takes $f(\tau)$ to $f(\tau/M)$ and thus in terms of Fourier expansions it takes $\sum a_n q^n$ to $\sum a_n q_M^n$ where $q_M = e^{2\pi i \tau/M} = q^{1/M}$. Similarly for the second map with N replacing M.

Since $\Gamma^1(M) \supset \Gamma^1(N)$, we have $\mathcal{S}_k(\Gamma^1(M)) \subset \mathcal{S}_k(\Gamma^1(N))$, so inclusion makes sense as the bottom row in the (not necessarily commutative) diagram

$$\mathcal{S}_k(\Gamma_1(M)) \xrightarrow{\ \iota_d\ } \mathcal{S}_k(\Gamma_1(N))$$

$$\downarrow \qquad\qquad\qquad \downarrow$$

$$\mathcal{S}_k(\Gamma^1(M)) \xrightarrow{\ \text{incl}\ } \mathcal{S}_k(\Gamma^1(N)).$$

In terms of Fourier expansions, the top row and the columns are

$$\sum a_n q^n \longmapsto \sum a_n q^{dn}$$

$$\downarrow \qquad\qquad\qquad \downarrow$$

$$\sum a_n q_M^n \qquad \sum a_n q_N^{dn}$$

and so the diagram commutes when $N = dM$. This condition holds when $d = p$ and $M = N/p$, reducing the Main Lemma to

Theorem 5.7.3 (Main Lemma, second version). *If $f \in \mathcal{S}_k(\Gamma^1(N))$ has Fourier expansion $f(\tau) = \sum a_n(f) q_N^n$ with $a_n(f) = 0$ whenever $(n, N) = 1$, then $f = \sum_{p|N} f_p$ with each $f_p \in \mathcal{S}_k(\Gamma^1(N/p))$.*

The second step is to translate the Main Lemma's hypothesis into linear algebra. This is done by constructing a suitable projection operator. For any positive integer m recall the group

$$\Gamma^0(m) = \left\{ \begin{bmatrix} a & b \\ c & d \end{bmatrix} \in \mathrm{SL}_2(\mathbb{Z}) : \begin{bmatrix} a & b \\ c & d \end{bmatrix} \equiv \begin{bmatrix} * & 0 \\ * & * \end{bmatrix} \ (\mathrm{mod}\ m) \right\}.$$

For any $d \mid N$, let $\Gamma_d = \Gamma_1(N) \cap \Gamma^0(N/d)$, a congruence subgroup of level N.

Lemma 5.7.4. *A set of representatives for the quotient space $\Gamma(N) \backslash \Gamma_d$ is*

$$\left\{ \begin{bmatrix} 1 & bN/d \\ 0 & 1 \end{bmatrix} : 0 \leq b < d \right\}.$$

Proof. Exercise 5.7.2.   □

Averaging over the coset representatives gives the trace operator,

$$\pi_d : \mathcal{S}_k(\Gamma(N)) \longrightarrow \mathcal{S}_k(\Gamma(N)), \quad \pi_d(f) = \frac{1}{d} \cdot \sum_{b=0}^{d-1} f[[\begin{smallmatrix} 1 & bN/d \\ 0 & 1 \end{smallmatrix}]]_k,$$

a projection to $\mathcal{S}_k(\Gamma_d)$. (Recall that a projection is an idempotent linear map, i.e., $\pi_d^2 = \pi_d$.) In terms of Fourier series, a geometric sum calculation (Exercise 5.7.3) shows that π_d preserves the terms whose indices are multiples of d and kills everything else; that is,

$$\pi_d : \sum_{n=1}^{\infty} a_n q_N^n \mapsto \sum_{n\ :\ d|n} a_n q_N^n.$$

It follows that $\pi_{d_1}\pi_{d_2} = \pi_{d_2}\pi_{d_1}$ when $d_1d_2 \mid N$.

Now define $\pi : \mathcal{S}_k(\Gamma(N)) \longrightarrow \mathcal{S}_k(\Gamma(N))$ to be

$$\pi = 1 - \sum_{p|N}\pi_p + \sum_{\substack{p_1|N,\ p_2|N\\ p_1<p_2}}\pi_{p_1p_2} - \cdots = \prod_{p|N}(1-\pi_p),$$

where the product denotes composition. By nature of the π_d and the additive expression for π, the Inclusion-Exclusion Principle shows that π preserves the part of f away from N and kills the rest,

$$\pi : \sum_{n=1}^{\infty}a_nq_N^n \mapsto \sum_{n\ :\ (n,N)=1}a_nq_N^n.$$

Thus the hypothesis of the Main Lemma is that $f \in \mathcal{S}_k(\Gamma^1(N)) \cap \ker(\pi)$. Since the π_p's are commuting projections, facts from linear algebra about commuting projections (Exercise 5.7.4) show that

$$\ker(\pi) = \ker(\prod_{p|N}(1-\pi_p)) = \sum_{p|N}\ker(1-\pi_p) = \sum_{p|N}\operatorname{im}(\pi_p).$$

And recall that π_p is a projection of $\mathcal{S}_k(\Gamma(N))$ to $\mathcal{S}_k(\Gamma_p)$,

$$\operatorname{im}(\pi_p) = \mathcal{S}_k(\Gamma_p) = \mathcal{S}_k(\Gamma_1(N) \cap \Gamma^0(N/p)).$$

So $\ker(\pi) = \sum_{p|N}\mathcal{S}_k(\Gamma_1(N)\cap\Gamma^0(N/p))$. Now it suffices to prove "$\subset$" (though we'll prove equality, no harder since the other inclusion is obvious) in

Theorem 5.7.5 (Main Lemma, third version).

$$\mathcal{S}_k(\Gamma^1(N)) \cap \sum_{p|N}\mathcal{S}_k(\Gamma_1(N)\cap\Gamma^0(N/p)) = \sum_{p|N}\mathcal{S}_k(\Gamma^1(N/p)).$$

This version of the Main Lemma reduces to group theory. The group $G = \operatorname{SL}_2(\mathbb{Z}/N\mathbb{Z})$ acts on the complex vector space $\mathcal{S}_k(\Gamma(N))$ from the right via the weight-k operator. (The full modular group $\operatorname{SL}_2(\mathbb{Z})$ acts and its normal subgroup $\Gamma(N)$ acts trivially, so the quotient acts and the quotient is naturally isomorphic to G.) If the prime factorization of N is $N = \prod_{i=1}^n p_i^{e_i}$ then G is naturally identified with a direct product,

$$G = \prod_{i=1}^n G_i, \quad \text{each } G_i = \operatorname{SL}_2(\mathbb{Z}/p_i^{e_i}\mathbb{Z}).$$

For $i = 1,\ldots,n$ define subgroups of G_i,

$$H_i = \Gamma^1(p_i^{e_i})/\Gamma(p_i^{e_i}),$$
$$K_i = (\Gamma_1(p_i^{e_i}) \cap \Gamma^0(p_i^{e_i-1}))/\Gamma(p_i^{e_i}).$$

Lemma 5.7.6. *For p prime and $e \geq 1$,*

$$\langle \Gamma^1(p^e), \Gamma_1(p^e) \cap \Gamma^0(p^{e-1}) \rangle = \Gamma^1(p^{e-1}).$$

Proof. To stay with the main argument, this will be done afterward.   □

Let $H = \prod_{i=1}^n H_i$. Using superscripts to denote subspaces fixed by subgroups of G, Lemma 5.7.6 shows that Theorem 5.7.5 rewrites as

$$\mathcal{S}_k(\Gamma(N))^H \cap \sum_{i=1}^n \mathcal{S}_k(\Gamma(N))^{K_i} = \sum_{i=1}^n \mathcal{S}_k(\Gamma(N))^{\langle H, K_i \rangle} \qquad (5.17)$$

(Exercise 5.7.5). We now quote a result from representation theory: the vector space $\mathcal{S}_k(\Gamma(N))$ is a direct sum of subspaces irreducible under the G-action. So (Exercise 5.7.6) condition (5.17) follows from a purely group-theoretical result,

Proposition 5.7.7. *Let V be an irreducible representation of the group $G = \prod_{i=1}^n G_i$. Let $H = \prod_{i=1}^n H_i$ and $K = \prod_{i=1}^n K_i$ be subgroups. Then*

$$V^H \cap \sum_{i=1}^n V^{K_i} = \sum_{i=1}^n V^{\langle H, K_i \rangle}. \qquad (5.18)$$

Quoting another result from representation theory, under the action of $G = \prod_{i=1}^n G_i$ the irreducible vector space V decomposes as a corresponding tensor product $V = \bigotimes_{i=1}^n V_i$ with each G_i acting on V_i. Establishing Proposition 5.7.7 is just a matter of decomposing each component V_i according to the various groups involved. Specifically (Exercise 5.7.7), each V_i has pairwise linearly disjoint subspaces V_{i1}, V_{i2}, V_{i3} such that

$$V_i^{\langle H_i, K_i \rangle} = V_{i1}, \quad V_i^{H_i} = V_{i1} \oplus V_{i2}, \quad V_i^{K_i} = V_{i1} \oplus V_{i3}. \qquad (5.19)$$

Now proving the proposition reduces to inspecting the various terms in (5.18). Conditions (5.19) show that

$$V^H = (V_{11} \oplus V_{12}) \otimes \cdots \otimes (V_{i1} \oplus V_{i2}) \otimes \cdots \otimes (V_{n1} \oplus V_{n2}),$$

$$\sum_{i=1}^n V^{K_i} = \sum_{i=1}^n V_1 \otimes \cdots \otimes (V_{i1} \oplus V_{i3}) \otimes \cdots \otimes V_n, \qquad (5.20)$$

$$\sum_{i=1}^n V^{\langle H, K_i \rangle} = \sum_{i=1}^n (V_{11} \oplus V_{12}) \otimes \cdots \otimes V_{i1} \otimes \cdots \otimes (V_{n1} \oplus V_{n2}).$$

The last result we need to quote is that tensor products and direct sums of vector spaces satisfy the distributive law. Now (5.20) shows that the tensor products belonging to V^H and contributing to $\sum V^{K_i}$ are precisely the products contributing to $\sum V^{\langle H, K_i \rangle}$. That is, multiplying the three right sides out and converting the second and third ones to direct sums by eliminating redundant factors shows that the third is the intersection of the first two (Exercise 5.7.8). The proposition follows and the Main Lemma is proved.

Finally, we need to go back and establish Lemma 5.7.6, that

$$\langle \Gamma^1(p^e), \Gamma_1(p^e) \cap \Gamma^0(p^{e-1}) \rangle = \Gamma^1(p^{e-1}).$$

The containment "$\subset$" is clear, so only "$\supset$" is needed. Let Γ denote the group on the left side. Let $m = \begin{bmatrix} a & b \\ c & d \end{bmatrix} \in \Gamma^1(p^{e-1})$. It suffices to show that some element of $\Gamma m \Gamma$ lies in Γ, so the proof will repeatedly replace m by $\gamma m \gamma'$ with $\gamma, \gamma' \in \Gamma$.

First eliminate the cases $p \mid a$ and $p \mid d$, which can occur only if $e = 1$. If $p \mid a$ then $p \nmid b$ and so $m \begin{bmatrix} 1 & 0 \\ 1 & 1 \end{bmatrix} = \begin{bmatrix} a+b & b \\ c+d & d \end{bmatrix}$ satisfies $p \nmid a + b$. Since $\begin{bmatrix} 1 & 0 \\ 1 & 1 \end{bmatrix} \in \Gamma$ we may take $p \nmid a$. The argument for $p \nmid d$ is similar.

Next reduce to the case $b \equiv 0 \pmod{p^e}$ and $c \equiv 0 \pmod{p^e}$. Let $\beta = -bd^{-1} \pmod{p^e}$ (which makes sense now that $p \nmid d$), satisfying $b + d\beta \equiv 0 \pmod{p^e}$ and $\beta \equiv 0 \pmod{p^{e-1}}$. Then $\begin{bmatrix} 1 & \beta \\ 0 & 1 \end{bmatrix} \in \Gamma$ and $\begin{bmatrix} 1 & \beta \\ 0 & 1 \end{bmatrix} m = \begin{bmatrix} a+c\beta & b+d\beta \\ c & d \end{bmatrix}$, so we may take $b \equiv 0 \pmod{p^e}$. The argument for $c \equiv 0 \pmod{p^e}$ is similar and slightly easier, left multiplying m by suitable $\begin{bmatrix} 1 & 0 \\ \gamma & 1 \end{bmatrix}$.

So now $m = \begin{bmatrix} a & b \\ c & d \end{bmatrix}$ with $a \equiv d \equiv 1 \pmod{p^{e-1}}$, $b \equiv c \equiv 0 \pmod{p^e}$. Since $\det m = 1$, we have $ad \equiv 1 \pmod{p^e}$. Consider the matrix

$$\gamma = \begin{bmatrix} 1 & 1-a \\ 0 & 1 \end{bmatrix} \begin{bmatrix} 1 & 0 \\ -1 & 1 \end{bmatrix} \begin{bmatrix} 1 & 1-d \\ 0 & 1 \end{bmatrix} \begin{bmatrix} 1 & 0 \\ a & 1 \end{bmatrix} = \begin{bmatrix} a + a(1-ad) & 1-ad \\ ad-1 & d \end{bmatrix}.$$

Inspecting the multiplicands shows that $\gamma \in \Gamma$. Inspecting the product shows that $\gamma \equiv m \pmod{p^e}$, so $m\gamma^{-1} \equiv \begin{bmatrix} 1 & 0 \\ 0 & 1 \end{bmatrix} \pmod{p^e}$ and $m\gamma^{-1} \in \Gamma$ as well. Thus $m \in \Gamma$ and the proof is complete.

Exercises

5.7.1. Prove Lemma 5.7.2.

5.7.2. Prove Lemma 5.7.4. (A hint for this exercise is at the end of the book.)

5.7.3. Show that $\pi_d \left(\sum_{n=0}^{\infty} a_n q_N^n \right) = \sum_{n:d|n} a_n q_N^n$.

5.7.4. State and prove the facts about commuting projections needed in the calculation $\ker(\pi) = \sum_{p|N} \operatorname{im}(\pi_p)$. (A hint for this exercise is at the end of the book.)

5.7.5. Use Lemma 5.7.6 to explain why Theorem 5.7.5 rephrases as condition (5.17). (A hint for this exercise is at the end of the book.)

5.7.6. Show that Proposition 5.7.7 implies condition (5.17). (A hint for this exercise is at the end of the book.)

5.7.7. In the proof of Proposition 5.7.7, explain why each V_i has pairwise linearly disjoint subspaces satisfying conditions (5.19).

5.7.8. Letting $n = 2$, multiply out the right sides of (5.20), express the results as direct sums, and show that $V^H \cap \sum V^{K_i} = \sum V^{\langle H, K_i \rangle}$.

5.8 Eigenforms

Recall from Corollary 5.6.3 that the spaces $\mathcal{S}_k(\Gamma_1(N))^{\mathrm{old}}$ and $\mathcal{S}_k(\Gamma_1(N))^{\mathrm{new}}$ have orthogonal bases of eigenforms for the Hecke operators $\{T_n, \langle n \rangle : (n, N) = 1\}$. Let f be such an eigenform. Using the Main Lemma, this section will show that if $f \in \mathcal{S}_k(\Gamma_1(N))^{\mathrm{new}}$ then in fact f is an eigenform for all T_n and $\langle n \rangle$. If $(n, N) > 1$ then $\langle n \rangle = 0$ (the zero operator) and so $\langle n \rangle f = 0$ (the zero function), meaning f is an eigenform for all $\langle n \rangle$, with eigenvalues $d_n = 0$ when $(n, N) > 1$. Thus the only operators in question are the T_n.

Proposition 5.2.3 showed a similar result for Eisenstein series. The series $E_k^{\psi,\varphi,t}(\tau)$ where $\psi\varphi = \chi$ modulo N is an eigenform for the Hecke operators $\langle d \rangle$ and T_p where $p \nmid N$, and when $t = 1$ the series is an eigenform for all T_p. To allow the discussion here to cover Eisenstein series as well as cusp forms, definitions and results are given for all modular forms when appropriate.

Definition 5.8.1. *A nonzero modular form $f \in \mathcal{M}_k(\Gamma_1(N))$ that is an eigenform for the Hecke operators T_n and $\langle n \rangle$ for all $n \in \mathbb{Z}^+$ is a **Hecke eigenform** or simply an **eigenform**. The eigenform $f(\tau) = \sum_{n=0}^{\infty} a_n(f)q^n$ is **normalized** when $a_1(f) = 1$. A **newform** is a normalized eigenform in $\mathcal{S}_k(\Gamma_1(N))^{\mathrm{new}}$.*

The definition precludes the zero form consistently with the general definition of eigenvector from linear algebra. In the parlance of this definition, we are about to show that $\mathcal{S}_k(\Gamma_1(N))^{\mathrm{new}}$ has an orthogonal basis of newforms.

Let $f \in \mathcal{S}_k(\Gamma_1(N))$ be an eigenform for the Hecke operators T_n and $\langle n \rangle$ with $(n, N) = 1$. Thus for all such n there exist eigenvalues $c_n, d_n \in \mathbb{C}$ such that $T_n f = c_n f$ and $\langle n \rangle f = d_n f$. The map $n \mapsto d_n$ defines a Dirichlet character $\chi : (\mathbb{Z}/N\mathbb{Z})^* \longrightarrow \mathbb{C}^*$ (Exercise 5.8.1), and $f \in \mathcal{S}_k(N, \chi)$. Consequently formula (5.14) applies and says

$$a_1(T_n f) = a_n(f) \quad \text{for all } n \in \mathbb{Z}^+. \tag{5.21}$$

Since f is an eigenform away from the level, also

$$a_1(T_n f) = c_n a_1(f) \quad \text{when } (n, N) = 1,$$

so

$$a_n(f) = c_n a_1(f) \quad \text{when } (n, N) = 1.$$

Thus if $a_1(f) = 0$ then $a_n(f) = 0$ when $(n, N) = 1$ and so $f \in \mathcal{S}_k(\Gamma_1(N))^{\mathrm{old}}$ by the Main Lemma.

Now assume that $f \in \mathcal{S}_k(\Gamma_1(N))^{\text{new}}$, $f \neq 0$. Then $f \notin \mathcal{S}_k(\Gamma_1(N))^{\text{old}}$ and the preceding paragraph shows that $a_1(f) \neq 0$, so we may assume f is normalized to $a_1(f) = 1$. For any $m \in \mathbb{Z}^+$ let $g_m = T_m f - a_m(f)f$, an element of $\mathcal{S}_k(\Gamma_1(N))^{\text{new}}$ and an eigenform for the Hecke operators T_n and $\langle n \rangle$ for $(n, N) = 1$ (Exercise 5.8.2). Compute that its first coefficient is

$$
\begin{aligned}
a_1(g_m) &= a_1(T_m f) - a_1(a_m(f)f) \\
&= a_m(f) - a_m(f) \qquad \text{by (5.21) and because } a_1(f) = 1 \\
&= 0,
\end{aligned}
$$

showing that $g_m \in \mathcal{S}_k(\Gamma_1(N))^{\text{old}}$ by the argument of the preceding paragraph. So $g_m \in \mathcal{S}_k(\Gamma_1(N))^{\text{new}} \cap \mathcal{S}_k(\Gamma_1(N))^{\text{old}} = \{0\}$, i.e., $T_m f = a_m(f)f$. Since m is arbitrary this discussion proves most of

Theorem 5.8.2. *Let $f \in \mathcal{S}_k(\Gamma_1(N))^{\text{new}}$ be a nonzero eigenform for the Hecke operators T_n and $\langle n \rangle$ for all n with $(n, N) = 1$. Then*

(a) *f is a Hecke eigenform, i.e., an eigenform for T_n and $\langle n \rangle$ for all $n \in \mathbb{Z}^+$. A suitable scalar multiple of f is a newform.*
(b) *If $\tilde{f}$ satisfies the same conditions as f and has the same T_n-eigenvalues, then $\tilde{f} = cf$ for some constant c.*

The set of newforms in the space $\mathcal{S}_k(\Gamma_1(N))^{\text{new}}$ is an orthogonal basis of the space. Each such newform lies in an eigenspace $\mathcal{S}_k(N, \chi)$ and satisfies $T_n f = a_n(f)f$ for all $n \in \mathbb{Z}^+$. That is, its Fourier coefficients are its T_n-eigenvalues.

Proof. All that remains to prove is that the set of newforms in the space $\mathcal{S}_k(\Gamma_1(N))^{\text{new}}$ is linearly independent. To see this, suppose there is a nontrivial linear relation

$$
\sum_{i=1}^{n} c_i f_i = 0, \qquad c_i \in \mathbb{C}
$$

with all c_i nonzero and with as few terms as possible, necessarily at least two. For any prime p, applying $T_p - a_p(f_1)$ to the relation gives

$$
\sum_{i=2}^{n} c_i(a_p(f_1) - a_p(f_i))f_i = 0.
$$

This relation must be trivial since it has fewer terms, so $a_p(f_i) = a_p(f_1)$ for all i. Since p is arbitrary this means that $f_i = f_1$ for all i, giving a contradiction since the original relation has at least two terms. $\square$

The newforms in each diamond operator eigenspace $\mathcal{S}_k(N, \chi)^{\text{new}}$ therefore are an orthogonal basis of the eigenspace. Part (b) in the theorem is the *Multiplicity One* property of newforms, showing that the basis of $\mathcal{S}_k(\Gamma_1(N))^{\text{new}}$

contains one element per eigenvalue where "eigenvalue" means a set of T_n-eigenvalues $\{c_n : n \in \mathbb{Z}^+\}$. That is, each eigenspace for the T_n operators is 1-dimensional. Exercise 5.8.3 demonstrates a subspace of $\mathcal{S}_k(\Gamma_1(N))$ without a basis of eigenforms. If $f \in \mathcal{S}_k(\Gamma_1(N))$ is an eigenform then f is old or new, never a hybrid $f = g + h$ with g old and h new and both nonzero (Exercise 5.8.4).

The functions $\varphi_k(\tau) = \eta(\tau)^k \eta(N\tau)^k$ from Proposition 3.2.2 are newforms, each spanning $\mathcal{S}_k(\Gamma_1(N))^{\text{new}}$ and also spanning $\mathcal{S}_k(\Gamma_0(N))^{\text{new}}$ when k is even. The reader is encouraged to browse the online modular forms database LMFDB to see numerical examples of newforms. The database also displays examples of many other constructs associated to modular forms, some of which will be explained in the following chapters here.

Theorem 5.8.3. *The set*

$$\mathcal{B}_k(N) = \{f(n\tau) : f \text{ is a newform of level } M \text{ and } nM \mid N\}$$

is a basis of $\mathcal{S}_k(\Gamma_1(N))$.

Proof. (Partial.) Consider the decomposition

$$\mathcal{S}_k(\Gamma_1(N)) = \mathcal{S}_k(\Gamma_1(N))^{\text{new}} \oplus \sum_{p \mid N} i_p((\mathcal{S}_k(\Gamma_1(N/p)))^2).$$

The first term on the right side is spanned by the elements of $\mathcal{B}_k(N)$ with $M = N$ and $n = 1$, and by induction on the level each summand in the second term is spanned by $f(\tau)$ and $f(p\tau)$ for newforms f of level dividing N/p. Thus $\mathcal{B}_k(N)$ spans $\mathcal{S}_k(\Gamma_1(N))$.

To show that $\mathcal{B}_k(N)$ is linearly independent, suppose there is a nontrivial relation

$$\sum_{i,j} c_{i,j} f_i(n_{i,j}\tau) = 0, \quad c_{i,j} \in \mathbb{C}. \tag{5.22}$$

Here each f_i lies in a space $\mathcal{S}_k(M_i, \chi_i)$ with $M_i \mid N$ and χ_i a Dirichlet character modulo M_i, and each $n_{i,j}$ divides N/M_i. Assume that each $c_{i,j}$ in (5.22) is nonzero and that the relation has as few terms as possible. It must involve more than one f_i since for a given i each jth function $f_i(n_{i,j}\tau) = q^{n_{i,j}} + \cdots$ starts at a different index. Each χ_i lifts to a Dirichlet character $\widetilde{\chi}_i$ modulo N, and then $f_i \in \mathcal{S}_k(N, \widetilde{\chi}_i)$. In fact all χ_i must lift to the same character modulo N, for if (say) $\widetilde{\chi}_1(d) \neq \widetilde{\chi}_2(d)$ for some $d \in (\mathbb{Z}/N\mathbb{Z})^*$ then applying $\langle d \rangle - \widetilde{\chi}_1(d)$ to (5.22) yields a nontrivial relation with fewer terms (Exercise 5.8.5(a)). Similarly all f_i have the same Fourier coefficients away from N, for if (say) $a_p(f_1) \neq a_p(f_2)$ for some $p \nmid N$ then applying $T_p - a_p(f_1)$ to (5.22) also yields a nontrivial relation with fewer terms (Exercise 5.8.5(b)). Thus (5.22) involves distinct forms $f_i \in \mathcal{S}_k(M_i, \chi_i)$ with all χ_i lifting to the same character modulo N and all $a_p(f_i)$ equal for $p \nmid N$. By a result known as *Strong Multiplicity One*, these conditions make all of the f_i equal, contradicting the possibility of linear dependence. $\qquad\square$

Strong Multiplicity One also plays a role in the proof (Exercise 5.8.6(a)) of

Proposition 5.8.4. *Let* $g \in \mathcal{S}_k(\Gamma_1(N))$ *be a normalized eigenform. Then there is a newform* $f \in \mathcal{S}_k(\Gamma_1(M))^{\mathrm{new}}$ *for some* $M \mid N$ *such that* $a_p(f) = a_p(g)$ *for all* $p \nmid N/M$. *Let* χ *be the character of* g, *i.e.,* $g \in \mathcal{S}_k(N, \chi)$. *Then* $f \in \mathcal{S}_k(M, \chi_M)$ *where* χ_M *lifts to* χ *modulo* N.

The level M of the newform f in the proposition is called the *conductor of* f. The result is analogous to the fact that any Dirichlet character χ modulo N has a corresponding primitive character χ_{prim} modulo the conductor M of χ such that $\chi_{\mathrm{prim}}(p) = \chi(p)$ for all $p \nmid N$.

We do not prove Strong Multiplicity One. See for example [Miy89] for the proof. The argument here that $\mathcal{B}_k(N)$ spans $\mathcal{S}_k(\Gamma_1(N))$ is complete, and this suffices for most purposes.

The last result of this section holds for all modular forms.

Proposition 5.8.5. *Let* $f \in \mathcal{M}_k(N, \chi)$. *Then* f *is a normalized eigenform if and only if its Fourier coefficients satisfy the conditions*

(1) $a_1(f) = 1$,
(2) $a_{p^r}(f) = a_p(f)a_{p^{r-1}}(f) - \chi(p)p^{k-1}a_{p^{r-2}}(f)$ *for all* p *prime and* $r \geq 2$,
(3) $a_{mn}(f) = a_m(f)a_n(f)$ *when* $(m, n) = 1$.

The proposition does not say that any function $f(\tau) = \sum_n a_n(f)q^n$ with coefficients satisfying conditions (1), (2), and (3) is a normalized eigenform. The function need not be a modular form at all.

Proof. The forward implication ($\Longrightarrow$) follows from the definition of T_n in Section 5.3 (Exercise 5.8.7). For the reverse implication ($\Longleftarrow$), suppose f satisfies the three conditions. Then f is normalized, and to be an eigenform for all the Hecke operators it need only satisfy $a_m(T_p f) = a_p(f)a_m(f)$ for all p prime and $m \in \mathbb{Z}^+$. If $p \nmid m$ then formula (5.14) gives $a_m(T_p f) = a_{pm}(f)$ and by the third condition this is $a_p(f)a_m(f)$ as desired. On the other hand, if $p \mid m$ write $m = p^r m'$ with $r \geq 1$ and $p \nmid m'$. This time

$$\begin{aligned}
a_m(T_p f) &= a_{p^{r+1}m'}(f) + \chi(p)p^{k-1}a_{p^{r-1}m'}(f) && \text{by formula (5.14)} \\
&= (a_{p^{r+1}}(f) + \chi(p)p^{k-1}a_{p^{r-1}}(f))a_{m'}(f) && \text{by the third condition} \\
&= a_p(f)a_{p^r}(f)a_{m'}(f) && \text{by the second condition} \\
&= a_p(f)a_m(f) && \text{by the third condition.}
\end{aligned}$$

$\square$

In particular, Proposition 5.2.3 shows that this result applies to certain Eisenstein series $E_k^{\psi,\varphi}/2$. That is, the Fourier coefficients $a_n(f) = \sigma_{k-1}^{\psi,\varphi}(n)$ satisfy the three relations of Proposition 5.8.5. Indeed, the proof of Proposition 5.2.3 essentially was to verify this.

Exercises

5.8.1. Show that if f is an eigenform for the operators $\langle n \rangle$ with eigenvalues d_n, then the map $n \mapsto d_n$ defines a Dirichlet character $\chi : (\mathbb{Z}/N\mathbb{Z})^* \longrightarrow \mathbb{C}^*$.

5.8.2. Show that the function g_m in the section is an eigenfunction for the Hecke operators T_n and $\langle n \rangle$ when $(n, N) = 1$.

5.8.3. Let $f_0 \in \mathcal{S}_2(\Gamma_0(11))$ be the unique normalized eigenform, cf. Exercise 3.5.4.

(a) Compute that $a_2(f_0) = -2$.

(b) Define $f_1(\tau) = f_0(2\tau)$, $f_2(\tau) = f_0(4\tau)$, and $f_3(\tau) = f_0(8\tau)$. Show that the set $\beta = \{f_0, f_1, f_2, f_3\}$ is linearly independent in $\mathcal{S}_2(\Gamma_0(88))$.

(c) Let $V = \text{span}(\beta)$. Show that V is stable under the Hecke operators, computing the matrix $[T_p]_\beta$ describing the action of T_p on V with respect to β for each prime p. (A hint for this exercise is at the end of the book.)

(d) Show that there are exactly three normalized eigenforms in V. Thus this problem exhibits a space without a basis of eigenforms.

5.8.4. Show that if $f \in \mathcal{S}_k(\Gamma_1(N))$ is an eigenform then f is old or new. (A hint for this exercise is at the end of the book.)

5.8.5. (a) Show that if $\tilde{\chi}_1(d) \neq \tilde{\chi}_2(d)$ for some $d \in (\mathbb{Z}/N\mathbb{Z})^*$ then applying $\langle d \rangle - \tilde{\chi}_1(d)$ to (5.22) yields a nontrivial relation with fewer terms.

(b) Show that if $a_p(f_1) \neq a_p(f_2)$ for some $p \nmid N$ then applying $T_p - a_p(f_1)$ to (5.22) yields a nontrivial relation with fewer terms. Why does this require $p \nmid N$?

5.8.6. (a) Prove Proposition 5.8.4 by supplying details to this argument: Suppose that for each newform f_i of level dividing N there exists a prime $p_i \nmid N/M_i$ such that $a_{p_i}(f_i) \neq a_{p_i}(g)$. Write g in terms of the basis provided by Theorem 5.8.3,

$$g = \sum_i \sum_{n:nM_i \mid N} c_{i,n} f_{i,n} \quad \text{where } f_{i,n}(\tau) = f_i(n\tau),$$

and then apply the level N operator $\prod_i (T_{p_i} - a_{p_i}(f_i))$ to obtain a contradiction. (A hint for this exercise is at the end of the book.)

(b) According to Strong Multiplicity One, the newform f for which the equality in the proposition holds is unique. Deduce that g is in the span of $\{f(n\tau) : nM \mid N\}$ where M is the level of f.

5.8.7. Prove ($\Longrightarrow$) in Proposition 5.8.5.

5.9 The connection with L-functions

Each modular form $f \in \mathcal{M}_k(\Gamma_1(N))$ has an associated Dirichlet series, its L-function. Let $f(\tau) = \sum_{n=0}^{\infty} a_n q^n$, let $s \in \mathbb{C}$ be a complex variable, and write formally

$$L(s, f) = \sum_{n=1}^{\infty} a_n n^{-s}.$$

Convergence of $L(s, f)$ in a half plane of s-values follows from estimating the Fourier coefficients of f.

Proposition 5.9.1. *If $f \in \mathcal{M}_k(\Gamma_1(N))$ is a cusp form then $L(s, f)$ converges absolutely for all s with $\mathrm{Re}(s) > k/2+1$. If f is not a cusp form then $L(s, f)$ converges absolutely for all s with $\mathrm{Re}(s) > k$.*

Proof. First assume f is a cusp form. Let $g(q) = \sum_{n=1}^{\infty} a_n q^n$, a holomorphic function on the unit disk $\{q : |q| < 1\}$. Then by Cauchy's formula,

$$a_n = \frac{1}{2\pi i} \int_{|q|=r} g(q) q^{-n} dq/q \qquad \text{for any } r \in (0, 1)$$

$$= \int_{x=0}^{1} f(x + iy) e^{-2\pi i n(x+iy)} dx \quad \text{for any } y > 0, \text{ where } q = e^{2\pi i(x+iy)}$$

$$= e^{2\pi} \int_{x=0}^{1} f(x + i/n) e^{-2\pi i n x} dx \quad \text{letting } y = 1/n.$$

Since f is a cusp form, $\mathrm{Im}(\tau)^{k/2} |f(\tau)|$ is bounded on the upper half plane $\mathcal{H}$ (Exercise 5.9.1(a)), and so estimating this last integral shows that $|a_n| \leq C n^{k/2}$. The result for a cusp form f now follows since $|a_n n^{-s}| = \mathcal{O}(n^{k/2 - \mathrm{Re}(s)})$.

If E is an Eisenstein series in $\mathcal{M}_k(\Gamma_1(N))$ then by direct inspection its Fourier coefficients satisfy $|a_n| \leq C n^{k-1}$ (Exercise 5.9.1(b)) and now $|a_n n^{-s}| = \mathcal{O}(n^{k-1-\mathrm{Re}(s)})$. Since any modular form is the sum of a cusp form and an Eisenstein series the rest of the proposition follows. $\square$

The estimate $|a_n(f)| \leq C n^{k/2}$ for $f \in \mathcal{S}_k(\Gamma_1(N))$ readily extends to $\mathcal{S}_k(\Gamma(N))$ and therefore to $\mathcal{S}_k(\Gamma)$ for any congruence subgroup Γ of $SL_2(\mathbb{Z})$. Similarly for the estimate $|a_n(E)| \leq C n^{k-1}$ for Eisenstein series $E \in \mathcal{M}_k(\Gamma(N))$. The upshot is that every modular form with respect to a congruence subgroup satisfies condition (3′) in Proposition 1.2.4,

(3′) f is holomorphic at ∞, and in the Fourier expansion $f(\tau) = \sum_{n=0}^{\infty} a_n q_N^n$
the coefficients for $n > 0$ satisfy the condition

$$|a_n| \leq C n^r \quad \text{for some positive constants } C \text{ and } r.$$

So finally the converse to that proposition holds as well: if f is holomorphic and weight-k invariant under Γ then f is a modular form if and only if it satisfies condition (3′).

The condition of f being a normalized eigenform is equivalent to its L-function series having an *Euler product*.

Theorem 5.9.2. *Let $f \in \mathcal{M}_k(N, \chi)$, $f(\tau) = \sum_{n=0}^{\infty} a_n q^n$. The following are equivalent:*

- *f is a normalized eigenform.*
- *$L(s, f)$ has an Euler product expansion*

$$L(s, f) = \prod_p (1 - a_p p^{-s} + \chi(p) p^{k-1-2s})^{-1},$$

where the product is taken over all primes.

Note that the Euler product here is the Hecke operator generating function product (5.12).

Proof. By Proposition 5.8.5, the first item here is equivalent to three conditions on the coefficients a_n, so it suffices to show that those conditions are equivalent to the second item here.

Fix a prime p. Multiplying condition (2) in Proposition 5.8.5 by p^{-rs} and summing over $r \geq 2$ shows, after a little algebra, that it is equivalent to

$$\sum_{r=0}^{\infty} a_{p^r} p^{-rs} \cdot (1 - a_p p^{-s} + \chi(p) p^{k-1-2s}) = a_1 + (1 - a_1) a_p p^{-s}. \qquad (5.23)$$

If also condition (1) in Proposition 5.8.5 holds then this becomes

$$\sum_{r=0}^{\infty} a_{p^r} p^{-rs} \cdot (1 - a_p p^{-s} + \chi(p) p^{k-1-2s}) = 1. \qquad (5.24)$$

Conversely, suppose (5.24) holds. Letting $s \to +\infty$ shows $a_1 = 1$ so condition (1) in Proposition 5.8.5 holds, and so does (5.23), implying condition (2) in Proposition 5.8.5. So conditions (1) and (2) in Proposition 5.8.5 are equivalent to

$$\sum_{r=0}^{\infty} a_{p^r} p^{-rs} = (1 - a_p p^{-s} + \chi(p) p^{k-1-2s})^{-1} \quad \text{for } p \text{ prime.} \qquad (5.25)$$

Before continuing, note that the Fundamental Theorem of Arithmetic (positive integers factor uniquely into prime powers) implies that for a function g of prime powers (Exercise 5.9.2),

$$\prod_p \sum_{r=0}^{\infty} g(p^r) = \sum_{n=1}^{\infty} \prod_{p^r \| n} g(p^r). \qquad (5.26)$$

The notation $p^r \| n$ means that p^r is the highest power of p that divides n, and we are assuming that g is small enough to justify formal rearrangements.

Now, if (5.25) holds along with condition (3) of Proposition 5.8.5 then compute

$$L(s,f) = \sum_{n=1}^{\infty} a_n n^{-s} = \sum_{n=1}^{\infty} \left(\prod_{p^r \| n} a_{p^r} \right) n^{-s} \quad \text{by the third condition}$$

$$= \sum_{n=1}^{\infty} \prod_{p^r \| n} a_{p^r} p^{-rs} = \prod_{p} \sum_{r=0}^{\infty} a_{p^r} p^{-rs} \quad \text{by (5.26)}$$

$$= \prod_{p} (1 - a_p p^{-s} + \chi(p) p^{k-1-2s})^{-1} \quad \text{by (5.25),}$$

giving the Euler product expansion.

Conversely, given the Euler product expansion, compute (using the geometric series formula and (5.26))

$$L(s,f) = \prod_{p} (1 - a_p p^{-s} + \chi(p) p^{1-k-2s})^{-1}$$

$$= \prod_{p} \sum_{r=0}^{\infty} b_{p,r} p^{-rs} \quad \text{for some } \{b_{p,r}\}$$

$$= \sum_{n=1}^{\infty} \prod_{p^r \| n} b_{p,r} p^{-rs} = \sum_{n=1}^{\infty} \left(\prod_{p^r \| n} b_{p,r} \right) n^{-s}.$$

So $a_n = \prod_{p^r \| n} b_{p,r}$, giving condition (3) of Proposition 5.8.5 and showing in particular that $b_{p,r} = a_{p^r}$. This in turn implies (5.25), implying conditions (1) and (2) of Proposition 5.8.5. $\square$

As an example, the L-function of the Eisenstein series $E_k^{\psi,\varphi}/2$ works out to (Exercise 5.9.3)

$$L(s, E_k^{\psi,\varphi}/2) = L(s,\psi) L(s-k+1,\varphi) \tag{5.27}$$

where the L-functions on the right side are as defined in Chapter 4. For another example see Exercise 5.9.5.

Let N be a positive integer and let A be the ring $\mathbb{Z}[\mu_3]$. For any character $\chi : (A/NA)^* \longrightarrow \mathbb{C}^*$, Section 4.11 constructed a modular form $\theta_\chi \in \mathcal{M}_1(3N^2, \psi)$ where $\psi(d) = \chi(d)(d/3)$. Recall that χ needs to be trivial on A^* for θ_χ to be nonzero, so assume this. The arithmetic of A and Theorem 5.9.2 show that θ_χ is a normalized eigenform, as follows. The relevant facts about A were invoked in the proof of Corollary 3.7.2 and in Section 4.11. To reiterate, A is a principal ideal domain. For each prime $p \equiv 1 \pmod 3$ there exists an element $\pi_p \in A$ such that $\pi_p \bar{\pi}_p = p$, but there is no such element if $p \equiv 2 \pmod 3$. The maximal ideals of A are

- for each prime $p \equiv 1 \pmod 3$, the two ideals $\langle \pi_p \rangle$ and $\langle \bar{\pi}_p \rangle$,
- for each prime $p \equiv 2 \pmod 3$, the ideal $\langle p \rangle$,
- for $p = 3$, the ideal $\langle \sqrt{-3} \rangle$.

Let $\pi_p = p$ for each prime $p \equiv 2 \pmod 3$, let $\pi_3 = \sqrt{-3}$, and take the set of generators of the maximal ideals,

$$S = \{\pi_p, \overline{\pi}_p : p \equiv 1 \pmod 3\} \cup \{\pi_p : p \equiv 2 \pmod 3\} \cup \{\pi_3\}.$$

Then each nonzero $n \in A$ can be written uniquely as

$$n = u \prod_{\pi \in S} \pi^{a_\pi}, \quad u \in A^*, \text{ each } a_\pi \in \mathbb{N}, \, a_\pi = 0 \text{ for all but finitely many } \pi.$$

Correspondingly $\chi(n) = \prod_{\pi \in S} \chi(\pi)^{a_\pi}$. The Fourier coefficients of θ_χ were given in (4.50),

$$a_m(\theta_\chi) = \tfrac{1}{6} \sum_{\substack{n \in A \\ |n|^2 = m}} \chi(n).$$

Compute that therefore

$$L(s, \theta_\chi) = \tfrac{1}{6} \sum_{\substack{n \in A \\ n \neq 0}} \chi(n)|n|^{-2s} = \prod_{\pi \in S}(1 - \chi(\pi)|\pi|^{-2s})^{-1} = \prod_p L_p(s, \theta_\chi),$$

where (Exercise 5.9.4)

$$L_p(s, \theta_\chi) = \begin{cases} (1 - (\chi(\pi_p) + \chi(\overline{\pi}_p))p^{-s} + \chi(p)p^{-2s})^{-1} & \text{if } p \equiv 1 \pmod 3, \\ (1 - \chi(p)p^{-2s})^{-1} & \text{if } p \equiv 2 \pmod 3, \\ (1 - \chi(\sqrt{-3})3^{-s})^{-1} & \text{if } p = 3. \end{cases}$$

$$(5.28)$$

Since $L_p(s, \theta_\chi) = (1 - a_p(\theta_\chi)p^{-s} + \psi(p)p^{-2s})^{-1}$ in all cases, Theorem 5.9.2 shows that θ_χ is a normalized eigenform.

Let N be a positive integer and let χ be a Dirichlet character modulo N. Let $g \in S_k(N, \chi)$ be a normalized Hecke eigenform. By Proposition 5.8.4 there is a newform $f \in S_k(M, \chi_M)$ for some $M \mid N$, with χ_M lifting to χ modulo N, such that $a_p(f) = a_p(g)$ for all $p \nmid N/M$. Consequently the Euler factors $L_p(s, f) = (1 - a_p(f)p^{-s} + \chi_M(p)p^{k-1-2s})^{-1}$ and $L_p(s, g) = (1 - a_p(g)p^{-s} + \chi(p)p^{k-1-2s})^{-1}$ are equal for all $p \nmid N/M$. We discuss the relation between the Euler factors of f and g for $p \mid N/M$. For such p the Euler factor of f, with an indeterminate X in place of p^{-s}, is

$$L_p(X, f) = (1 - a_p(f)X + \chi_M(p)p^{k-1}X^2)^{-1}$$
$$= \begin{cases} (1 - \alpha X)^{-1}(1 - \beta X)^{-1} & \text{if } p \nmid M, \\ (1 - a_p(f)X)^{-1} & \text{if } p \mid M, \end{cases}$$

where in the first case α and β are the reciprocal roots of the quadratic polynomial. Because $p \mid N/M$ the corresponding Euler factor of g is

$$L_p(X, g) = (1 - a_p(g)X)^{-1}.$$

We will show that $L_p(X, g)^{-1}$ divides $L_p(X, f)^{-1}$.

By Exercise 5.8.6(b), g lies in the subspace of $\mathcal{S}_k(\Gamma_1(N))$ having basis $\{f_d : dM \mid N\}$ where $f_d(\tau) = f(d\tau)$. Each f_d has Fourier coefficients $a_n(f_d) = a_{n/d}(f)$ and lies in $\mathcal{S}_k(\Gamma_1(N))$. Fix a prime $p \mid N/M$. Define the function $f_{p^{-1}}(\tau) = \sum_{n \geq 1} a_{np}(f)q^n$. (This function is not $f(p^{-1}\tau)$ despite the Fourier coefficient formula $a_n(f_{p^{-1}}) = a_{np}(f)$.) Working at level M, we have $a_n(T_p f) = a_{np}(f) + \chi_M(p)p^{k-1}a_{n/p}(f)$ for all positive integers n, showing that $T_p f = f_{p^{-1}} + \chi_M(p)p^{k-1}f_p$ even though neither summand on the right side is claimed to lie in $\mathcal{S}_k(\Gamma_1(M))$. But also $T_p f = a_p(f)f$, and the previous two equalities combine to give

$$f_{p^{-1}} = a_p(f)f - \chi_M(p)p^{k-1}f_p.$$

The right side of the previous display lies in $\mathcal{S}_k(\Gamma_1(N))$ because $p \mid N/M$. More generally the argument in this paragraph shows that if $pm \mid N/M$ and $p \nmid m$, and if $f_{mp^{-1}}$ is defined in terms of f_m similarly to $f_{p^{-1}}$ in terms of f, then $f_{mp^{-1}} = a_p(f)f_m - \chi_M(p)p^{k-1}f_{mp}$ (Exercise 5.9.5(a)).

Retaining the fixed prime $p \mid N/M$, now work at level N. For any d such that $dM \mid N$, for all positive integers n,

$$a_n(T_p f_d) = a_{np}(f_d) = a_{np/d}(f).$$

If $p \mid d$ then this says that $a_n(T_p f_d) = a_n(f_{d/p})$ for all n. If $p \nmid d$ then the previous display says that $a_n(T_p f_d) = a_{n/d}(T_p f)$ regardless of whether n is a multiple of d (Exercise 5.9.5(b)). But $T_p f$ has Fourier coefficients $a_n(T_p f) = a_{np}(f)$, identifying it as the function $f_{p^{-1}} = a_p(f)f - \chi_M(p)p^{k-1}f_p$ from the previous paragraph; this does not contradict the level M relation $T_p f = a_p(f)f$ because here we are working at level N. So if $p \nmid d$ then $a_n(T_p f_d) = a_p(f)a_{n/d}(f) - \chi_M(p)p^{k-1}a_{n/d}(f_p) = a_p(f)a_n(f_d) - \chi_M(p)p^{k-1}a_n(f_{pd})$. Thus altogether:

$$\text{For } p \mid N/M, \quad T_p f_d = \begin{cases} f_{d/p} & \text{if } p \mid d, \\ a_p(f)f_d - \chi_M(p)p^{k-1}f_{pd} & \text{if } p \nmid d. \end{cases}$$

Suppose that $N = p^e M$ with p prime and $e \geq 1$. Thus g lies in the subspace of $\mathcal{S}_k(\Gamma_1(N))$ having basis $\{f, f_p, \cdots, f_{p^e}\}$. The previous two paragraphs have shown that this space is T_p-stable and the matrix of T_p on this space with respect to this basis is

$$A = \begin{bmatrix} a_p(f) & 1 & 0 & \cdots & 0 \\ -\chi_M(p)p^{k-1} & 0 & 1 & \cdots & 0 \\ 0 & 0 & 0 & \cdots & 0 \\ \vdots & & \vdots & & \vdots \\ 0 & 0 & 0 & \cdots & 1 \\ 0 & 0 & 0 & \cdots & 0 \end{bmatrix},$$

with characteristic polynomial $X^{e-1}(X^2 - a_p(f)X + \chi_M(p)p^{k-1})$. (The reader may recognize the matrix A from Exercise 5.8.3(c).) Therefore the Euler factor $L_p(X, g) = (1 - a_p(g)X)^{-1}$ is $(1 - \lambda X)^{-1}$ for some root λ of the characteristic polynomial. If $p \nmid M$ then either $\lambda = 0$, giving $L_p(X, g) = 1$ (this is possible only if $e > 1$), or $1 - a_p(f)\lambda^{-1} + \chi_M(p)p^{k-1}\lambda^{-2} = 0$, which is to say that λ is a reciprocal root of $L_p(X, f)^{-1} = (1 - \alpha X)(1 - \beta X)$ and so $L_p(X, g) = (1 - \alpha X)^{-1}$ or $L_p(X, g) = (1 - \beta X)^{-1}$. If $p \mid M$ then either $\lambda = 0$, giving $L_p(X, g) = 1$, or $\lambda = a_p(f)$, giving $L_p(X, g) = (1 + a_p(f)X)^{-1} = L_p(X, f)$. In all cases $L_p(X, g)^{-1}$ divides $L_p(X, f)^{-1}$ as claimed above. See Exercise 5.9.5(c) for an example of these ideas.

The general case is $N = np^e M$ with p prime and $e \geq 1$ and $p \nmid n$. Now g lies in the subspace of $\mathcal{S}_k(\Gamma_1(N))$ having basis $\bigcup_{m|n}\{f_m, f_{mp}, \cdots, f_{mp^e}\}$. This space is T_p-stable and the matrix of T_p on this space with respect to this basis is a direct sum of copies of A over the divisors of n (Exercise 5.9.5(d)). Consequently the eigenvalues of g admit the same analysis as in the previous paragraph, and hence so does the Euler factor $L_p(X, g)$.

The ideas that have just been discussed for cusp forms also work for Eisenstein series in consequence of Theorems 4.5.2 and 4.6.2 and 4.8.1. For example, take $k \geq 3$ and recall the set $A_{N,k}$ of triples (ψ, φ, t) introduced before Theorem 4.5.2, i.e., ψ and φ are primitive Dirichlet characters modulo u and v with $(\psi\varphi)(-1) = (-1)^k$, and t is a positive integer such that $tuv \mid N$. Working in $\mathcal{M}_k(\Gamma_1(N))$, for each pair (ψ, φ) let

$$\mathcal{E}_k(N, \psi, \varphi) = \bigoplus_{t : tuv | N} \mathbb{C}E_k^{\psi,\varphi,t},$$

and for each Dirichlet character χ modulo N with $\chi(-1) = (-1)^k$ let

$$\mathcal{E}_k(N, \chi) = \bigoplus_{(\psi,\varphi) : \psi\varphi = \chi \bmod N} \mathcal{E}_k(N, \psi, \varphi),$$

and let

$$\mathcal{E}_k(\Gamma_1(N)) = \bigoplus_{\chi} \mathcal{E}_k(N, \chi).$$

Thus $\mathcal{E}_k(N, \psi, \varphi)$ is the subspace of $\mathcal{M}_k(\Gamma_1(N))$ having as basis the Eisenstein series $E_k^{\psi,\varphi,t}$ indexed by the triples in $A_{N,k}$ whose first two entries are the specified ψ and φ, and similarly for $\mathcal{E}_k(N, \chi)$ but with all ψ and φ such that $\psi\varphi$ lifts to χ modulo N, and similarly for $\mathcal{E}_k(\Gamma_1(N))$ with all of $A_{N,k}$. Chapter 4 defined $\mathcal{E}_k(\Gamma_1(N))$ as the quotient space $\mathcal{M}_k(\Gamma_1(N))/\mathcal{S}_k(\Gamma_1(N))$; the redefinition here is naturally isomorphic to the quotient as a vector space, but we need to consider the Hecke action as well. Because the Hecke operators act on $\mathcal{M}_k(\Gamma_1(N))$ and the action restricts to $\mathcal{S}_k(\Gamma_1(N))$ they act on the quotient, so what needs to be shown is that they act on the complement $\mathcal{E}_k(\Gamma_1(N))$ in the previous display. By the same argument as for cusp forms, for each pair

(ψ, φ) the action of T_p for $p \nmid N/(uv)$ on $\mathcal{E}_k(N, \psi, \varphi)$ is diagonal, and the action of T_p for $p \mid N/(uv)$ is given by copies of the matrix A from above with its upper left corner specialized to $\begin{bmatrix} \psi(p) + \varphi(p)p^{k-1} & 1 \\ -(\psi\varphi)(p)p^{k-1} & 0 \end{bmatrix}$, the copies indexed by the divisors of $N/(p^e uv)$ where $p^e \| N/(uv)$. In this context we take the Eisenstein series $E_k^{\psi,\varphi}$ to be rescaled so that $a_1(E_k^{\psi,\varphi}) = 1$.

Now let $g \in \mathcal{E}_k(N, \chi)$ be a normalized eigenform. Its Euler factors are $L_p(X, g) = (1 - a_p(g)X + \chi(p)p^{k-1}X^2)^{-1}$. Similarly to Proposition 5.8.4, there exists some pair (ψ, φ) with $\psi\varphi = \chi$ modulo N such that $a_p(E_k^{\psi,\varphi}) = a_p(g)$ for all primes $p \nmid N/(uv)$ (Exercise 5.9.6(a)), and so $L_p(X, E_k^{\psi,\varphi}) = L_p(X, g)$ for such p. Moreover, Dirichlet's Theorem on Arithmetic Progressions shows that the pair of characters is unique (Exercise 5.9.6(b)). The Fourier coefficients $a_p(E_k^{\psi,\varphi}) = \sigma_{k-1}^{\psi,\varphi}(p) = \psi(p) + \varphi(p)p^{k-1}$ show that the Euler factors of $E_k^{\psi,\varphi}$ are $(1 - \psi(p)X)^{-1}(1 - \varphi(p)p^{k-1}X)^{-1}$ and so $L(s, E_k^{\psi,\varphi}) = L(s, \psi)L(s - k + 1, \varphi)$, as already noted earlier in this section. The normalized eigenform g lies in the space $\mathcal{E}_k(N, \psi, \varphi)$ (Exercise 5.9.6(c)), and so as in the discussion for cusp forms it follows that $L_p(X, g)^{-1}$ divides $L_p(X, E_k^{\psi,\varphi})^{-1}$.

The discussions for $k = 2$ (excluding $\psi = \varphi = 1$) and for $k = 1$ are essentially identical to the discussion just given for $k \geq 3$, except that for $k = 1$ the uniqueness of ψ and φ follows from the linear independence of characters rather than from the argument for $k \geq 2$. Even in the exceptional case $k = 2$ and $\psi = \varphi = 1$ the discussion goes through notwithstanding that the Eisenstein series E_2 is not a modular form, with Hecke stability a consequence of Exercise 5.2.6(a) rather than the method of this section. Exercise 5.9.7(a) asks the reader to check all of this to his or her satisfaction. In the exceptional case of $\mathcal{E}_2(N, 1, 1)$, the natural basic L-function is $\zeta(s)\zeta(s - 1)$ and the L-functions of eigenforms are related to it in the same way as this section describes in other settings. Furthermore in this case, the L-functions of the basis functions $f_t(\tau) = E_2(\tau) - tE_2(t\tau)$ for $t \mid N$ are holomorphic at $s = 1$ (exercise 5.9.7(b)) and hence so are the L-functions of the eigenforms. The reader can work out the L-functions of the eigenforms from Exercise 5.2.6(b–d) and confirm that they are holomorphic at $s = 1$ (Exercise 5.9.7(c)).

Exercises

5.9.1. (a) For any cusp form $f \in \mathcal{S}_k(\Gamma_1(N))$ show that the function $\varphi(\tau) = \text{Im}(\tau)^{k/2}|f(\tau)|$ is bounded on the upper half plane $\mathcal{H}$. (A hint for this exercise is at the end of the book.)

   (b) Establish the relation $1 \leq \sigma_{k-1}(n)/n^{k-1} < \zeta(k - 1)$ where ζ is the Riemann zeta function. Show that the Fourier coefficients a_n of any Eisenstein series satisfy $|a_n| \leq Cn^{k-1}$.

5.9.2. Prove formula (5.26). (A hint for this exercise is at the end of the book.)

5.9.3. Prove formula (5.27). What is a half plane of convergence?

5.9.4. Establish formula (5.28).

5.9.5. (a) Generalize the argument in the section that $f_{p^{-1}} = a_p(f)f - \chi_M(p)p^{k-1}f_p$ for $p \nmid N/M$ to show that $f_{mp^{-1}} = a_p(f)f_m - \chi_M(p)p^{k-1}f_{mp}$ where $f_{mp^{-1}}$ is defined similarly to $f_{p^{-1}}$. (Note that the work will take place at level dM and so it will involve the corresponding character χ_{dM}. Note also that the coefficients of f_m and f_{mp} on the right side of the equality to be shown are independent of m.)

(b) Prove casewise that the displayed relation $a_n(T_p f_d) = a_{np}(f_d) = a_{np/d}(f)$ in the section says when $p \nmid d$ that $a_n(T_p f_d) = a_{n/d}(T_p f)$ regardless of whether n is a multiple of d.

(c) Recall the functions f, f_1, f_2, and f_3 from Exercise 5.8.3. The exercise showed that the 4-dimensional space spanned by these functions contains only three normalized eigenforms. How do the L-functions of the three eigenforms relate to $L(s, f)$?

(d) Let $N = np^e M$ with p prime and $e \geq 1$ and $p \nmid n$. The normalized Hecke eigenform g lies in the subspace of $S_k(\Gamma_1(N))$ having basis $\bigcup_{m|n}\{f_m, f_{mp}, \cdots, f_{mp^e}\}$. Show that the matrix of T_p on this space with respect to this basis is a direct sum of copies of the matrix A in the section, taken over the divisors of n.

5.9.6. (a) Let $g \in \mathcal{E}_k(N, \chi)$ be a normalized eigenform, with $k \geq 3$. Use the method of Exercise 5.8.6(a) to show that there exists some pair (ψ, φ) with $\psi\varphi = \chi$ modulo N such that $a_p(E_k^{\psi,\varphi}) = a_p(g)$ for all primes $p \nmid N/(uv)$; here $E_k^{\psi,\varphi}$ is scaled so that $a_1(E_k^{\psi,\varphi}) = 1$.

(b) Prove that the pair ψ, φ in part (a) is unique. (A hint for this exercise is at the end of the book.)

(c) Use the method of Exercise 5.8.6(b) to show that the normalized eigenform g lies in the space $\mathcal{E}_k(N, \psi, \varphi)$.

5.9.7. (a) Convince yourself that parts (a–c) of the previous exercise hold for $k = 2$ and for $k = 1$, including the exceptional case $k = 2$ and $\psi = \varphi = 1$.

(b) Recall that the Riemann zeta function $\zeta(s)$ has a meromorphic continuation to all of $\mathbb{C}$, with $\zeta(s) \sim 1/(s-1)$ near $s = 1$ and $\zeta(0) = -1/2$. Let $f_t(\tau) = E_2(\tau) - tE_2(t\tau)$. Show that $L(s, f_t) = (1 - t^{1-s})\zeta(s)\zeta(s-1)$, and that this function is holomorphic at $s = 1$ with $L(1, f_t) = -\frac{1}{2}\log t$.

(c) Determine the L-functions of the eigenforms from Exercise 5.2.6(b–d). Confirm that these L-functions are holomorphic at $s = 1$.

5.10 Functional equations

Let $f(\tau) = \sum_{n=1}^{\infty} a_n q^n \in S_k(\Gamma_1(N))$ be a cusp form of weight k. Its associated L-function is $L(s, f) = \sum_{n=1}^{\infty} a_n n^{-s}$, convergent for $\text{Re}(s) > k/2 + 1$. As in Chapter 4, the Mellin transform of f is

$$g(s) = \int_{t=0}^{\infty} f(it)t^s \frac{dt}{t}$$

for values of s such that the integral converges absolutely.

Proposition 5.10.1. *The Mellin transform of f is*

$$g(s) = (2\pi)^{-s}\Gamma(s)L(s,f), \quad \mathrm{Re}(s) > k/2 + 1.$$

This is shown exactly as in Chapter 4 (Exercise 5.10.1). This section shows that the function

$$\Lambda_N(s) = N^{s/2}g(s)$$

satisfies a functional equation. Define an operator

$$W_N : \mathcal{S}_k(\Gamma_1(N)) \longrightarrow \mathcal{S}_k(\Gamma_1(N)), \quad f \mapsto i^k N^{1-k/2} f[[\begin{smallmatrix} 0 & -1 \\ N & 0 \end{smallmatrix}]]_k.$$

That is, $(W_N f)(\tau) = i^k N^{-k/2}\tau^{-k} f(-1/(N\tau))$. This is essentially the w_N operator from the end of Section 5.5. The operator W_N is an involution, meaning W_N^2 is the identity operator (Exercise 5.10.2), and similarly to Exercise 5.5.1(c) W_N is self-adjoint,

$$\langle W_N f_1, f_2 \rangle = \langle f_1, W_N f_2 \rangle, \quad f_1, f_2 \in \mathcal{S}_k(\Gamma_1(N)).$$

Letting $\mathcal{S}_k(\Gamma_1(N))^+$ and $\mathcal{S}_k(\Gamma_1(N))^-$ denote the eigenspaces

$$\mathcal{S}_k(\Gamma_1(N))^{\pm} = \{f \in \mathcal{S}_k(\Gamma_1(N)) : W_N f = \pm f\}$$

gives an orthogonal decomposition of $\mathcal{S}_k(\Gamma_1(N))$,

$$\mathcal{S}_k(\Gamma_1(N)) = \mathcal{S}_k(\Gamma_1(N))^+ \oplus \mathcal{S}_k(\Gamma_1(N))^-.$$

Theorem 5.10.2. *Suppose $f \in \mathcal{S}_k(\Gamma_1(N))^{\pm}$. Then the Mellin transform $\Lambda_N(s)$ extends to an entire function satisfying the functional equation*

$$\Lambda_N(s) = \pm\Lambda_N(k - s).$$

Consequently, $L(s, f)$ has an analytic continuation to the full s-plane.

Proof. Take $f \in \mathcal{S}_k(\Gamma_1(N))^{\pm}$ and compute

$$\Lambda_N(s) = N^{s/2}\int_{t=0}^{\infty} f(it)t^s \frac{dt}{t} = \int_{t=0}^{\infty} f(it/\sqrt{N})t^s \frac{dt}{t}.$$

Since $f(it/\sqrt{N})$ is of order $e^{-2\pi t/\sqrt{N}}$ as $t \to \infty$, the piece of this integral from $t = 1$ to ∞ converges to an entire function of s. For the other piece compute that $(W_N f)(i/(\sqrt{N}\,t)) = t^k f(it/\sqrt{N})$, so that

$$\int_{t=0}^{1} f(it/\sqrt{N})t^s \frac{dt}{t} = \int_{t=0}^{1} (W_N f)(i/(\sqrt{N}\,t))t^{s-k} \frac{dt}{t}$$

$$= \int_{t=1}^{\infty} (W_N f)(it/\sqrt{N})t^{k-s} \frac{dt}{t}$$

Since $W_N f = \pm f$ this also converges to an entire function of s. Thus $\Lambda_N(s)$ has an analytic continuation to the full s-plane. To obtain the functional equation note that in total the integral is

$$\Lambda_N(s) = \int_{t=1}^{\infty} \left(f(it/\sqrt{N})t^s + (W_N f)(it/\sqrt{N})t^{k-s} \right) \frac{dt}{t}.$$

Since $W_N f = \pm f$ this is $\pm\Lambda_N(k-s)$, completing the proof of the functional equation. $\qquad\square$

It is natural to wonder about a converse to Theorem 5.10.2, a result that Dirichlet series with analytic continuations and functional equations come from modular forms. Such a theorem exists due to Hecke and Weil, though it is not quite as straightforward as one might guess. For a summary discussion see [Kob93], and for the full story see [Ogg69], [Bum97], or [Miy89].

Exercises

5.10.1. Prove Proposition 5.10.1.

5.10.2. Show that W_N is an involution.

5.11 Eisenstein series again

Recall from Section 5.4 that the integral defining the Petersson inner product converges when at least one of f and g is a cusp form, not necessarily both. In particular it is meaningful to take the inner product of an Eisenstein series and a cusp form. This section will show that such an inner product always vanishes, meaning that the Eisenstein series are orthogonal to the cusp forms. (As mentioned before, this usage of "orthogonal" is a slight abuse of language.) Thus the Eisenstein spaces can be redefined as complements rather than as quotient spaces.

Specifically, recall the Eisenstein series with parameter associated to the group $\Gamma(N)$ in Section 4.10,

$$E_k^{\overline{v}}(\tau, s) = \epsilon_N \sum_{\gamma \in (P_+ \cap \Gamma(N)) \backslash \Gamma(N) \delta} \mathrm{Im}(\tau)^s [\gamma]_k, \quad \mathrm{Re}(k+2s) > 2.$$

Here ϵ_N is $1/2$ if $N \in \{1,2\}$ and 1 if $N > 2$, $\overline{v} = \overline{(c_v, d_v)}$ is a vector in $(\mathbb{Z}/N\mathbb{Z})^2$, $\delta = \begin{bmatrix} a & b \\ c_v & d_v \end{bmatrix}$ is a matrix in $\mathrm{SL}_2(\mathbb{Z})$ whose bottom row (c_v, d_v) is a lift of $\overline{v}$

to $\mathbf{Z}^2$, and $P_+ = \{[\begin{smallmatrix} 1 & n \\ 0 & 1 \end{smallmatrix}] : n \in \mathbf{Z}\}$ is the positive part of the parabolic subgroup of $\mathrm{SL}_2(\mathbf{Z})$. The weight-$k$ operator in this context is

$$(f[\gamma]_k)(\tau, s) = j(\gamma, \tau)^{-k} f(\gamma(\tau), s) \quad \text{for } \gamma \in \mathrm{SL}_2(\mathbf{Z}).$$

Analytically continuing $E_k^{\overline{v}}(\tau, s)$ in the s-plane to $s = 0$ recovered the Eisenstein series $E_k^{\overline{v}}(\tau)$ from earlier in Chapter 4. At weight $k = 2$ the continued series are nonholomorphic at $s = 0$ but linear combinations cancel away the nonholomorphic terms. Bearing all of this in mind, redefine

$$\mathcal{E}_k(\Gamma(N)) = \text{the holomorphic functions in span}(\{E_k^{\overline{v}}(\tau, 0) : v \in (\mathbf{Z}/N\mathbf{Z})^2\}).$$

So now $\mathcal{E}_k(\Gamma(N))$ is a subspace of $\mathcal{M}_k(\Gamma(N))$ linearly disjoint from the cusp forms $\mathcal{S}_k(\Gamma(N))$, replacing its earlier definition as the quotient space $\mathcal{M}_k(\Gamma(N))/\mathcal{S}_k(\Gamma(N))$. That is,

$$\mathcal{M}_k(\Gamma(N)) = \mathcal{S}_k(\Gamma(N)) \oplus \mathcal{E}_k(\Gamma(N)).$$

To show that this decomposition is orthogonal introduce the notation

$$P_+(N) = P_+ \cap \Gamma(N) = \left\{ \begin{bmatrix} 1 & nN \\ 0 & 1 \end{bmatrix} : n \in \mathbf{Z} \right\},$$

so a fundamental domain for the orbit space $P_+(N)\backslash\mathcal{H}^*$ is the set

$$\mathcal{D}_N^* = \{\tau \in \mathcal{H}^* \cap \mathbf{C} : 0 \le \mathrm{Re}(\tau) \le N\} \cup \{\infty\}.$$

Let

$$P_+(N)\backslash\Gamma(N) = \bigcup_i P_+(N)\alpha_i \quad \text{and} \quad \Gamma(N)\backslash\mathrm{SL}_2(\mathbf{Z}) = \bigcup_{i'} \Gamma(N)\beta_{i'}$$

so that

$$P_+(N)\backslash\mathrm{SL}_2(\mathbf{Z}) = \bigcup_{i,i'} P_+(N)\alpha_i\beta_{i'} \quad \text{and} \quad \mathcal{D}_N^* = \bigcup_{i,i'} \alpha_i\beta_{i'}(\mathcal{D}^*).$$

(For instance, the decomposition $\mathcal{D}_1^* = \bigcup_i \alpha_i(\mathcal{D}^*)$ is partially illustrated in Figure 2.4.) For any cusp form $f \in \mathcal{S}_k(\Gamma(N))$ the relation $a_0(f) = 0$ is

$$\int_{x=0}^N f(x + iy)dx = 0 \quad \text{for any } y > 0, \tag{5.29}$$

similar to Exercise 1.1.6 (Exercise 5.11.1(a)). Consequently for any $s \in \mathbf{C}$ with $\mathrm{Re}(k + 2s) \ge 0$ we can evaluate the absolutely convergent double integral

$$\int_{y=0}^\infty \int_{x=0}^N f(x + iy)y^{k+s-2}dx\,dy = 0, \tag{5.30}$$

or in terms of the complex variable τ, recalling the measure $d\mu(\tau) = dx\,dy/y^2$,

$$\int_{\mathcal{D}_N^*} f(\tau)(\mathrm{Im}(\tau))^{k+s} d\mu(\tau) = 0 \qquad (5.31)$$

(Exercise 5.11.1(b–c)). This relation rewrites as

$$0 = \sum_{i,i'} \int_{\mathcal{D}^*} f(\alpha_i \beta_{i'}(\tau))(\mathrm{Im}(\alpha_i \beta_{i'}(\tau)))^{k+s} d\mu(\tau)$$

$$= \sum_{i,i'} \int_{\mathcal{D}^*} f(\beta_{i'}(\tau)) j(\alpha_i, \beta_{i'}(\tau))^k \frac{(\mathrm{Im}(\beta_{i'}(\tau)))^{k+s}}{|j(\alpha_i, \beta_{i'}(\tau))|^{2(k+s)}}\, d\mu(\tau).$$

If $\mathrm{Re}(k+2s) > 2$ then the sum over i converges absolutely and passes through the integral, giving

$$0 = \sum_{i'} \int_{\mathcal{D}^*} f(\beta_{i'}(\tau)) \overline{E_k^{(0,1)}}(\beta_{i'}(\tau), s)/\epsilon_N \cdot (\mathrm{Im}(\beta_{i'}(\tau)))^k d\mu(\tau).$$

This sum of translated integrals is by definition the integral over the modular curve $X(\Gamma(N))$, cf. Section 5.4, and in fact it is a constant multiple of the Petersson inner product of the cusp form $f(\tau)$ and $\overline{E_k^{(0,1)}}(\tau, s)$. Thus, the inner product is 0 for all s in a right half plane, and the relation analytically continues to $s = 0$,

$$\langle f, \overline{E_k^{(0,1)}} \rangle_{\Gamma(N)} = 0.$$

Consequently for any $\gamma \in \mathrm{SL}_2(\mathbb{Z})$,

$$\langle f, \overline{E_k^{(0,1)\gamma}} \rangle_{\Gamma(N)} = \langle f, \overline{E_k^{(0,1)}}[\gamma]_k \rangle_{\Gamma(N)} = \langle f[\gamma^{-1}]_k, \overline{E_k^{(0,1)}} \rangle_{\Gamma(N)}$$

by Proposition 5.5.2(a) generalized to the Petersson inner product of a cusp form and a general modular form. Since also $f[\gamma^{-1}]_k \in \mathcal{S}_k(\Gamma(N))$ the product is 0. Thus the cusp forms and the Eisenstein series are orthogonal at level N as claimed.

For any congruence subgroup Γ at level N redefine

$$\mathcal{E}_k(\Gamma) = \mathcal{E}_k(\Gamma(N)) \cap \mathcal{M}_k(\Gamma).$$

Already section 5.9 has introduced this redefinition for $\Gamma = \Gamma_1(N)$, as for any Dirichlet character χ modulo N it has redefined

$$\mathcal{E}_k(N, \chi) = \mathcal{E}_k(\Gamma_1(N)) \cap \mathcal{M}_k(N, \chi)$$

and shown that $\mathcal{E}_k(N, \chi)$ and $\mathcal{E}_k(\Gamma_1(N))$ are stable under the action of the Hecke operators. For any congruence subgroup Γ the Eisenstein space is linearly disjoint from the cusp forms and similarly for the eigenspaces,

$$\mathcal{M}_k(\Gamma) = \mathcal{S}_k(\Gamma) \oplus \mathcal{E}_k(\Gamma) \quad \text{and} \quad \mathcal{M}_k(N, \chi) = \mathcal{S}_k(N, \chi) \oplus \mathcal{E}_k(N, \chi),$$

and the decompositions are orthogonal. The sets of Eisenstein series specified in Chapter 4 as coset representatives for bases of the Eisenstein spaces as quotients are now actual bases of the Eisenstein spaces as complements.

Using Proposition 5.8.5, it is easy (except when $k = 2$ and $\chi = 1$) to prove that $\mathcal{S}_k(\Gamma_1(N)) \perp \mathcal{E}_k(N, \chi)$ with Hecke theory rather than the previous integral calculation for the larger space $\mathcal{M}_k(\Gamma(N))$. Exercise 5.11.2 gives the argument.

Exercises

5.11.1. (a) Let the modular form $f \in \mathcal{M}_k(\Gamma(N))$ have Fourier expansion $f(\tau) = \sum_{n=0}^{\infty} a_n q_N^n$ where $q_N = e^{2\pi i \tau / N}$. Show that for all $n \geq 0$,

$$a_n = \frac{1}{N} \int_{x=0}^{N} f(x + iy) e^{-2\pi i n (x+iy)/N} dx \quad \text{for any } y > 0,$$

so in particular (5.29) holds.

(b) Assuming absolute convergence, show that the integral in (5.30) is 0.

(c) For any cusp $s = \alpha(\infty)$ with $\alpha \in \text{SL}_2(\mathbb{Z})$ show by changing variable that the integral in (5.31) over a neighborhood of s takes the form

$$\int_{\text{Im}(\tau') > y_0} (f[\alpha]_k)(\tau') j(\alpha, \tau')^k \text{Im}(\tau')^{k+s} |j(\alpha, \tau')|^{-2(k+s)} d\mu(\tau').$$

Explain why the integrand decays exponentially as $\text{Im}(\tau') \to \infty$.

5.11.2. (a) Let $f \in \mathcal{S}_k(\Gamma_1(N))$ be a normalized eigenform with eigenvalue a_p under the Hecke operator T_p for each prime p. Show that f has eigenvalue $\bar{a}_p$ under the adjoint T_p^*. (Hints for this exercise are at the end of the book.)

(b) Let χ be a Dirichlet character modulo N. Let ψ and φ be primitive Dirichlet characters modulo u and v such that $uv \mid N$ and $(\psi\varphi)(-1) = (-1)^k$ and $\psi\varphi = \chi$ at level N, excluding the case $k = 2$, $\psi = \varphi = 1_1$. Recall that $E_k^{\psi, \varphi}/2$ is a normalized eigenform. Let its T_p-eigenvalues be denoted b_p. Show that $b_p \neq a_p$ for some prime p.

(c) Show that $E_k^{\psi, \varphi} \perp f$.

(d) Let t be a positive integer such that $tuv \mid N$. Let $\alpha = \begin{bmatrix} t^{-1} & 0 \\ 0 & 1 \end{bmatrix}$ so that $\alpha' = (1/t) \begin{bmatrix} t & 0 \\ 0 & 1 \end{bmatrix}$ in Proposition 5.5.2. We know that $\alpha^{-1} \Gamma_1(N) \alpha = \Gamma_1(N/t) \cap \Gamma^0(t)$. Note that $E_k^{\psi, \varphi, t}$ is a nonzero constant multiple of $E_k^{\psi, \varphi}[\alpha']_k$ since the exceptional case is excluded. For any $f \in \mathcal{S}_k(\Gamma_1(N))$ show that

$$\langle f, E_k^{\psi, \varphi}[\alpha']_k \rangle_{\Gamma_1(N)} = \langle f[\alpha]_k, E_k^{\psi, \varphi} \rangle_{\Gamma_1(N/t)}.$$

Explain why this is 0 and why consequently all of $\mathcal{E}_k(N, \chi)$ is orthogonal to $\mathcal{S}_k(\Gamma_1(N))$.

5.11.3. Let A be the ring $\mathbb{Z}[\mu_3]$. Recall the statement of Cubic Reciprocity in Section 4.11: Let $d \in \mathbb{Z}^+$ be cubefree and let $N = 3 \prod_{p|d} p$. Then there exists a character

$$\chi : (A/NA)^* \longrightarrow \{1, \mu_3, \mu_3^2\}$$

such that the multiplicative extension of χ to all of A is trivial on A^* and on primes $p \nmid N$, while on elements π of A such that $\pi\bar{\pi}$ is a prime $p \nmid N$ it is trivial if and only if d is a cube modulo p. The character χ is nontrivial if $d > 1$. Define $\psi(n) = \chi(n)(n/3)$ (Legendre symbol) for $\bar{n} \in (\mathbb{Z}/3N^2\mathbb{Z})^*$ and let $\theta_\chi \in \mathcal{M}_1(3N^2, \psi)$ be the corresponding modular form constructed in Section 4.11, shown to be a normalized eigenform at the end of Section 5.9. This exercise shows that θ_χ is a cusp form if $d > 1$. The proof is by contradiction. Thus take $d > 1$ and suppose that θ_χ is not a cusp form.

(a) For any positive integers t, u, v such that $tuv \mid 3N^2$ and any Dirichlet characters ϕ and φ modulo u and v with $\phi\varphi = \psi$ at level $3N^2$ recall the Eisenstein series $E_1^{\phi, \varphi, t} \in \mathcal{E}_1(3N^2, \psi)$. According to Theorem 4.8.1 the set of such Eisenstein series is a basis. Use the idea of Exercise 5.8.6(a) to show that there exist such u, v, ϕ, and φ such that

$$a_p(E_1^{\phi, \varphi}/2) = a_p(\theta_\chi) \quad \text{for all } p \nmid 3N^2.$$

(b) Show that $\phi(n) = -\varphi(n)$ for all $\bar{n} \in (\mathbb{Z}/3N^2\mathbb{Z})^*$ such that $n \equiv 2 \pmod 3$. Deduce that

$$\phi(n) = \varphi(n) \left(\frac{n}{3}\right), \quad \bar{n} \in (\mathbb{Z}/3N^2\mathbb{Z})^*.$$

(c) Show that ϕ^2 is trivial and so $a_p(\theta_\chi) \in \{\pm 2\}$ for all $p \equiv 1 \pmod 3$ such that $p \nmid N$. Deduce that $\chi(\pi) \in \{\pm 1\}$ for all primes π of A not dividing N (cf. Section 5.9), so χ is trivial or quadratic. This contradiction shows that θ_χ is a cusp form.

6

Jacobians and Abelian Varieties

Let X be a compact Riemann surface of genus $g \geq 1$ and fix a point x_0 in X. Letting x vary over points of X and viewing path integration as a function of holomorphic differentials ω on X, the map

$$x \mapsto \left(\omega \mapsto \int_{x_0}^{x} \omega \right)$$

is an injection

$$X \longrightarrow \left\{ \begin{array}{c} \text{linear functions of holomorphic differentials on } X \\ \text{modulo integration over loops in } X \end{array} \right\}.$$

When $g = 1$ this is an isomorphism of Abelian groups. When $g > 1$ the domain X is no longer a group, but the codomain still is. The codomain is the *Jacobian* of X, complex analytically a g-dimensional torus $\mathbb{C}^g / \Lambda_g$ where $\Lambda_g \cong \mathbb{Z}^{2g}$. This chapter presents the Jacobian and states a version of the Modularity Theorem mapping the Jacobian of a modular curve holomorphically to a given elliptic curve. The map is also a homomorphism, incorporating group structure into the Modularity Theorem whereas the first version, back in Chapter 2, was solely complex analytic. This chapter then uses the Jacobian to prove number-theoretic results about weight 2 eigenforms of the Hecke operators. It ends with another version of Modularity replacing the Jacobian with an *Abelian variety*, a quotient of the Jacobian. The Abelian variety comes from a weight 2 eigenform, so this version of the Modularity Theorem associates an eigenform to an elliptic curve. From now on, virtually all modular forms in this book will be of weight 2.

Related reading: The main results of this chapter are from Chapters 3, 7, and 8 of [Shi73]. For the basics on Jacobians of Riemann surfaces, see [FK80]. For material on complex tori and complex Abelian varieties, see [Swi74]. There are many introductory texts to algebraic number theory, such as [Mar89] and [Sam72].

© Springer Science+Business Media New York 2005   215
F. Diamond, J. Shurman, *A First Course in Modular Forms*,
Graduate Texts in Mathematics 228, DOI 10.1007/978-0-387-27226-9_6

6.1 Integration, homology, the Jacobian, and Modularity

We begin by describing the situation sketched in the chapter introduction when $g = 1$. Let X be a complex elliptic curve $\mathbb{C}/\Lambda$. For any point $z \in \mathbb{C}$ consider the family of integrals in the complex plane $\{\int_0^{z+\lambda} d\zeta : \lambda \in \Lambda\} = z + \Lambda$. Each such integral can be viewed as the integral over any path from 0 to $z + \lambda$, and two such integrals differ by the integral over any path between lattice points λ_1 and λ_2 in $\mathbb{C}$. Since $d\zeta$ is translation-invariant it makes sense as a holomorphic differential on X and so path integrals in $\mathbb{C}$ project to path integrals in X. Two projected path integrals in X from $0 + \Lambda$ to the same endpoint $z + \Lambda$ can take different values, coming from plane integrals whose difference projects to an integral $\int_\alpha d\zeta$ where α is a loop in X. Thus, viewing the cosets $z + \Lambda$ in two ways gives a bijection from the elliptic curve to a quotient of its path integrals,

$$X \xrightarrow{\sim} \Big\{\text{path integrals } \int_{0+\Lambda}^{z+\Lambda} d\zeta\Big\} \Big/ \Big\{\text{integrals } \int_\alpha d\zeta \text{ over loops } \alpha\Big\}.$$

The bijection is a group isomorphism since computing modulo integrals over loops and using the translation-invariance of $d\zeta$,

$$\int_{0+\Lambda}^{z_1+\Lambda} d\zeta + \int_{0+\Lambda}^{z_2+\Lambda} d\zeta = \int_{0+\Lambda}^{z_1+\Lambda} d\zeta + \int_{z_1+\Lambda}^{z_1+z_2+\Lambda} d\zeta = \int_{0+\Lambda}^{z_1+z_2+\Lambda} d\zeta.$$

To generalize these ideas to higher genus $g > 1$, let X now be any compact Riemann surface. The first thing to show is that path integrals on X, similar to the path integrals ubiquitous in complex analysis, are well defined. That is, if $\gamma : [0,1] \longrightarrow X$ is a continuous function and ω is a holomorphic differential on X then there is a meaningful notion of the integral $\int_\gamma \omega$.

To see this, begin by supposing that the image of γ lies in one neighborhood U with local coordinate $\varphi : U \longrightarrow V \subset \mathbb{C}$ and with $\omega|_V = f(q)dq$. In this case the integral is evaluated in local coordinates,

$$\int_\gamma \omega = \int_{\varphi \circ \gamma} \omega|_V = \int_{\varphi \circ \gamma} f(q)dq.$$

(See Exercise 6.1.1 for the definition of the last integral here.) This definition seems ambiguous if the image of γ lies in the intersection $U_1 \cap U_2$ of two coordinate neighborhoods, but both coordinates give the integral the same value. Indeed, let $V_1 = \varphi_1(U_1 \cap U_2)$, let $V_2 = \varphi_2(U_1 \cap U_2)$, and let $\varphi_{2,1} : V_1 \longrightarrow V_2$ be the transition function between the two coordinate systems. By definition of the transition function, by the change of variable formula, and by the compatibility condition $\varphi_{2,1}^*(\omega|_{V_2}) = \omega|_{V_1}$,

$$\int_{\varphi_2 \circ \gamma} \omega|_{V_2} = \int_{\varphi_{2,1} \circ \varphi_1 \circ \gamma} \omega|_{V_2} = \int_{\varphi_1 \circ \gamma} \varphi_{2,1}^*(\omega|_{V_2}) = \int_{\varphi_1 \circ \gamma} \omega|_{V_1}.$$

So the local integral is the same on overlapping coordinate systems.

In general, the domain $[0, 1]$ of γ partitions into finitely many intervals each having image in one coordinate neighborhood of X (Exercise 6.1.2(a)). The integral over γ is defined as the corresponding sum of local integrals, and it is easy to check that the result is well defined (Exercise 6.1.2(b)).

With path integration understood we now want to define as path-independent an integral as possible. For arbitrary points $x, x' \in X$, the notion of integrating holomorphic differentials from x to x',

$$\int_x^{x'} \omega, \quad \omega \in \Omega_{\text{hol}}^1(X),$$

is not well defined since different paths from x to x' might give different values for the integral. But if γ and $\tilde{\gamma}$ are paths from x to x' then letting α be the loop obtained by traversing forward along γ and then back along $\tilde{\gamma}$, the integrals along the two paths differ by the integral around the loop,

$$\int_\gamma \omega = \int_{\tilde{\gamma}} \omega + \int_\alpha \omega, \quad \omega \in \Omega_{\text{hol}}^1(X).$$

That is, for path-independent integration of holomorphic differentials from x to x' we need to quotient away integration over loops.

Viewing X as a sphere with g handles where g is the genus of X, let $A_1, \ldots, A_g$ be longitudinal loops around each handle like arm-bands, and let $B_1, \ldots, B_g$ be latitudinal loops around each handle like equators. We state without proof that for any nonnegative integer N, any integers $l_1, \ldots, l_N$, and any loops $\alpha_1, \ldots, \alpha_N$ there are unique integers $m_1, \ldots, m_g, n_1, \ldots, n_g$ such that

$$\sum_{i=1}^N l_i \int_{\alpha_i} \omega = \sum_{i=1}^g m_i \int_{A_i} \omega + \sum_{i=1}^g n_i \int_{B_i} \omega, \quad \omega \in \Omega_{\text{hol}}^1(X).$$

(See a Riemann surface theory book such as [FK80] for the unproved assertions in this section.) Thus the group of integer sums of integrations over loops is the free Abelian group generated by integration over the A_i and the B_i, called the *(first) homology group* of X,

$$\text{H}_1(X, \mathbb{Z}) = \mathbb{Z} \int_{A_1} \oplus \cdots \oplus \mathbb{Z} \int_{A_g} \oplus \mathbb{Z} \int_{B_1} \oplus \cdots \oplus \mathbb{Z} \int_{B_g} \cong \mathbb{Z}^{2g}.$$

(Since we deal only with this homology group we won't bother saying "first" any more.) Elements of $\text{H}_1(X, \mathbb{Z})$ are maps taking the holomorphic differentials on X to complex numbers, and in fact the homology group is a subgroup of the dual space $\Omega_{\text{hol}}^1(X)^\wedge = \text{Hom}_{\mathbb{C}}(\Omega_{\text{hol}}^1(X), \mathbb{C})$, the vector space of $\mathbb{C}$-linear maps from $\Omega_{\text{hol}}^1(X)$ to $\mathbb{C}$. We also state without proof that

$$\mathbb{R} \int_{A_1} \oplus \cdots \oplus \mathbb{R} \int_{A_g} \oplus \mathbb{R} \int_{B_1} \oplus \cdots \oplus \mathbb{R} \int_{B_g} = \Omega_{\text{hol}}^1(X)^\wedge.$$

Recall from Section 3.4 that $\dim_{\mathbb{C}}(\Omega^1_{\text{hol}}(X)) = g$. Thus $H_1(X, \mathbb{Z})$ is a lattice in the dual space. The quotient, complex analytically a g-dimensional complex torus $\mathbb{C}^g / \Lambda_g$, is the Jacobian.

Definition 6.1.1. *The **Jacobian** of X is the quotient group*

$$\text{Jac}(X) = \Omega^1_{\text{hol}}(X)^\wedge / H_1(X, \mathbb{Z}).$$

Abel's Theorem relates the Jacobian of X to the divisors on X, as discussed in Chapter 3. Let $\mathbb{C}(X)$ be the function field of the compact Riemann surface X. The degree-0 divisor group of X is

$$\text{Div}^0(X) = \left\{ \sum_{x \in X} n_x x : n_x \in \mathbb{Z}, \ n_x = 0 \text{ for almost all } x, \ \sum_x n_x = 0 \right\},$$

and the subgroup of principal divisors is all the divisors linearly equivalent to 0,

$$\text{Div}^\ell(X) = \{ \delta \in \text{Div}^0(X) : \delta = \text{div}(f) \text{ for some } f \in \mathbb{C}(X)^* \}.$$

The quotient group of degree-0 divisors modulo principal divisors on X is the *(degree-0) divisor class group of X* or the *(degree-0) Picard group of X*,

$$\text{Pic}^0(X) = \text{Div}^0(X) / \text{Div}^\ell(X).$$

The full Picard group is divisors of all degrees modulo principal divisors, but we use only $\text{Pic}^0(X)$ and so we simply call it the Picard group from now on. The Picard group measures the extent to which degree-0 divisors fail to be the divisors of meromorphic functions on X. If X has genus $g > 0$ and x_0 is a *base point* in X then X embeds in its Picard group under the map

$$X \longrightarrow \text{Pic}^0(X), \qquad x \mapsto [x - x_0],$$

where $[x - x_0]$ denotes the equivalence class $x - x_0 + \text{Div}^\ell(X)$, the subtraction occurring in $\text{Div}^0(X)$. (In general there is no subtraction on X.) The map is an embedding because there are no functions $f \in \mathbb{C}(X)$ of degree 1 (Exercise 6.1.3). The map from degree-0 divisors to the Jacobian

$$\text{Div}^0(X) \longrightarrow \text{Jac}(X), \qquad \sum_x n_x x \mapsto \sum_x n_x \int_{x_0}^x \qquad (6.1)$$

is well defined.

Theorem 6.1.2 (Abel's Theorem). *The map (6.1) descends to divisor classes, inducing an isomorphism*

$$\text{Pic}^0(X) \xrightarrow{\sim} \text{Jac}(X), \qquad \left[\sum_x n_x x \right] \mapsto \sum_x n_x \int_{x_0}^x.$$

Thus the degree-0 divisors describing meromorphic functions on X are those that map to trivial integration on $\Omega_{\text{hol}}^1(X)$ modulo integration over loops. If X has genus $g > 0$ then its embedding into its Picard group followed by the isomorphism of Abel's Theorem shows that the map

$$X \longrightarrow \text{Jac}(X), \qquad x \mapsto \int_{x_0}^x$$

embeds the Riemann surface in its Jacobian. When X is an elliptic curve this map is the isomorphism from the beginning of the chapter. Abel's Theorem also shows that the finite integer sums of path integrals in X—in fact even just the sums whose coefficients sum to 0 (add a suitable multiple of $\int_{x_0}^{x_0}$)—make up the entire dual space of $\Omega_{\text{hol}}^1(X)$,

$$\Omega_{\text{hol}}^1(X)^\wedge = \left\{ \sum_\gamma n_\gamma \int_\gamma : \sum_\gamma n_\gamma = 0 \right\}. \tag{6.2}$$

We will use this freely from now on.

For example, when $X = \widehat{\mathbb{C}}$ the Jacobian is trivial because $g = 0$, and Abel's Theorem states that every degree-0 divisor on the Riemann sphere is principal. This is familiar from complex analysis: if the divisor is $\delta = \sum z_i - \sum p_i + n_\infty \infty$ where the sums are over the desired zeros and poles in $\mathbb{C}$, allowing repetition, then the suitable rational function $\prod(z - z_i)/\prod(z - p_i)$ has this divisor since its order at ∞ is minus the sum of its orders in $\mathbb{C}$.

For another example, when $X = \mathbb{C}/\Lambda$ the Jacobian $\text{Jac}(X)$ identifies with X as explained at the beginning of the section, and the Abel map takes divisors $\sum n_x x$ to the corresponding sums in X. Abel's Theorem states in this case that degree-0 divisors on X such that $\sum n_x x \in \Lambda$ are the divisors of elliptic functions with respect to Λ. This is the constraint obtained in Exercise 1.4.1(c) and used in Section 1.4 to describe the group law on elliptic curves in $\mathbb{C}^2$.

The Modularity Theorem can be stated using the Jacobian. Recall that $X_0(N)$ is the compact modular curve associated to the congruence subgroup $\Gamma_0(N)$. Let $J_0(N)$ denote its Jacobian,

$$J_0(N) = \text{Jac}(X_0(N)).$$

Theorem 6.1.3 (Modularity Theorem, Version $J_{\mathbb{C}}$). *Let E be a complex elliptic curve with $j(E) \in \mathbb{Q}$. Then for some positive integer N there exists a surjective holomorphic homomorphism of complex tori*

$$J_0(N) \longrightarrow E.$$

The word "homomorphism" here introduces algebraic structure into the Modularity Theorem but it comes essentially for free. View $J_0(N)$ as a complex torus $\mathbb{C}^g / \Lambda_g$. Proposition 1.3.2 extends to higher dimension:

Proposition 6.1.4. *Let* $\varphi : \mathbb{C}^g/\Lambda_g \longrightarrow \mathbb{C}^h/\Lambda_h$ *be a holomorphic map of complex tori. Then*

$$\varphi(z + \Lambda_g) = Mz + b + \Lambda_h$$

where $M \in \mathrm{M}_{h,g}(\mathbb{C})$ *is an h-by-g matrix with complex entries such that* $M\Lambda_g \subset \Lambda_h$, *and* $b \in \mathbb{C}^h$.

This is proved the same way as in Chapter 1. By topology the map φ lifts to a map $\tilde{\varphi} : \mathbb{C}^g \longrightarrow \mathbb{C}^h$ of universal covers, and it follows that $\tilde{\varphi}'$ is bounded. Applying Liouville's Theorem in each direction makes $\tilde{\varphi}'$ constant, and now everything proceeds as before. In particular, if the Modularity Theorem supplied only a holomorphic map then this would take the form $z + \Lambda_g \mapsto m \cdot z + b + \Lambda$ where now $m \in \mathbb{C}^g$ is a vector such that $m \cdot \Lambda_g \subset \Lambda$. Translating by $-b$ gives a homomorphism as claimed.

Version $J_{\mathbb{C}}$ of the Modularity Theorem implies Version $X_{\mathbb{C}}$ (from Section 2.5), that every complex elliptic curve E with $j(E) \in \mathbb{Q}$ is the holomorphic image of a modular curve $X_0(N)$. Indeed, if E is the image of $J_0(N)$ then $J_0(N)$ is nontrivial, so $X_0(N)$ itself has genus $g > 0$ and embeds in its Jacobian as discussed. The embedding $X_0(N) \longrightarrow J_0(N)$ composes with the map $J_0(N) \longrightarrow E$ of Version $J_{\mathbb{C}}$ to show Version $X_{\mathbb{C}}$ (Exercise 6.1.4 is to show that the composition surjects). The converse, that Version $X_{\mathbb{C}}$ of the Modularity Theorem implies Version $J_{\mathbb{C}}$, will be shown at the end of the next section.

Exercises

6.1.1. This exercise defines the integral of a holomorphic differential over a continuous path in $\mathbb{C}$, using integration over polygonal paths (which is easy) and Cauchy's Theorem for such integration. Let V be an open subset of $\mathbb{C}$, let $\gamma : [0,1] \longrightarrow V$ be a continuous function, and let $\omega = f(q)dq$ be a holomorphic differential on V. For each $t \in [0,1]$, the image point $\gamma(t)$ lies in some disk $D_t \subset V$. Show that $\{\gamma^{-1}(D_t) : t \in [0,1]\}$ is a collection of open subsets of $[0,1]$ such that $[0,1] = \bigcup_t \gamma^{-1}(D_t)$. (Intervals $[0,b)$, $(a,1]$, and $[0,1]$ are open as subsets of $[0,1]$.) Every open subset of $[0,1]$ is a union of open intervals by definition. Show that since $[0,1]$ is compact, it is covered by finitely many open intervals each taken by γ into a disk D of V, and therefore it partitions into finitely many closed intervals with the same property. Define $\int_\gamma \omega$ to be the integral of ω along the polygonal path connecting the images of the partition points. Show that this is well defined by using a common refinement and Cauchy's Theorem.

6.1.2. Let X be a compact Riemann surface and let $\gamma : [0,1] \longrightarrow X$ be continuous.

(a) If $\{U_j : j \in J\}$ is a collection of coordinate neighborhoods in X such that $X = \bigcup_j U_j$, show that $\{\gamma^{-1}(U_j) : j \in J\}$ is a collection of open subsets of $[0,1]$ such that $[0,1] = \bigcup_j \gamma^{-1}(U_j)$. Show that $[0,1]$ is covered by finitely many open intervals each taken by γ into one coordinate neighborhood U of X, and therefore it partitions into finitely many closed intervals with the same property. Thus, integrating along the path γ can be defined as a finite sum of local integrals.

(b) Consider two partitions of the domain $[0,1]$ as in part (a). Show that the two corresponding sums of local integrals are equal.

6.1.3. Use the Riemann–Hurwitz formula from Section 3.1 to show that if X has genus $g > 0$ then there are no functions $f \in \mathbb{C}(X)$ of degree 1. (A hint for this exercise is at the end of the book.)

6.1.4. In general, if a map φ is surjective this need not imply that a composition $\varphi \circ f$ is surjective as well. Yet the argument at the end of the section asserts that if a holomorphic homomorphism $\varphi : J_0(N) \longrightarrow E$ surjects then so does $\varphi \circ f$ where $f : X_0(N) \longrightarrow J_0(N)$ is an embedding $x \mapsto \int_{x_0}^x$. Explain. (A hint for this exercise is at the end of the book.)

6.2 Maps between Jacobians

Let $h : X \longrightarrow Y$ be a nonconstant holomorphic map of compact Riemann surfaces. This section defines corresponding forward and reverse holomorphic homomorphisms of Jacobians, $h_J : \mathrm{Jac}(X) \longrightarrow \mathrm{Jac}(Y)$ and $h^J : \mathrm{Jac}(Y) \longrightarrow \mathrm{Jac}(X)$. Since the Jacobian and the Picard group are isomorphic under Abel's Theorem, there also are homomorphisms h_P and h^P of Picard groups. The Jacobian maps h_J and h^J are defined in terms of transferring holomorphic differentials between X and Y and have interpretations in terms of transferring integration between X and Y. The Picard group maps h_P and h^P come from transferring meromorphic functions between X and Y, independently of Abel's Theorem. In this chapter, where our methods are still complex analytic in keeping with the book so far, the Picard group and its maps don't play a large role, but from the next chapter onwards, when we move from complex analysis to algebraic geometry and begin working over fields other than $\mathbb{C}$, the Picard group formulation is the one that will continue to make sense.

To define the forward map of Jacobians in terms of differentials, begin with the *pullback map* induced by h, embedding the function field of Y in the function field of X,

$$h^* : \mathbb{C}(Y) \longrightarrow \mathbb{C}(X), \qquad h^*g = g \circ h.$$

Recall from Section 3.1 that each point $x \in X$ has a ramification degree $e_x \in \mathbb{Z}^+$ such that h is locally e_x-to-1 at x. The orders of vanishing of a nonzero function and its pullback are related by

$$\nu_x(h^*g) = e_x \nu_{h(x)}(g), \quad g \in \mathbb{C}(Y)^*.$$

(Exercise 6.2.1(a)). Since ramification degree is the same thing as order in local coordinates, this says that the order of the composition is the product of the orders. In particular, h^*g is holomorphic at x if g is holomorphic at $h(x)$. The pullback extends to a linear map of holomorphic differentials,

$$h^* : \Omega^1_{\mathrm{hol}}(Y) \longrightarrow \Omega^1_{\mathrm{hol}}(X).$$

To define this, let $\varphi_j : U_j \longrightarrow V_j$ and $\tilde{\varphi}_j : \tilde{U}_j \longrightarrow \tilde{V}_j$ be local coordinates on X and Y, where $h(U_j) = \tilde{U}_j$ so that $\tilde{\varphi}_j h \varphi_j^{-1} = h_j : V_j \longrightarrow \tilde{V}_j$, and let λ be a holomorphic differential on Y. Then the pullback is defined locally in terms of the local pullback as

$$(h^*\lambda)_j = h_j^*(\lambda_j) \in \Omega^1_{\mathrm{hol}}(V_j), \quad \lambda_j \in \Omega^1_{\mathrm{hol}}(\tilde{V}_j).$$

The required compatibility condition $\varphi^*_{k,j}((h^*\lambda)_k|_{V_{k,j}}) = (h^*\lambda)_j|_{V_{j,k}}$ from Section 3.3 follows from the known compatibility condition $\tilde{\varphi}^*_{k,j}(\lambda_k|_{\tilde{V}_{k,j}}) = \lambda_j|_{\tilde{V}_{j,k}}$ (Exercise 6.2.1(b)). In coordinates, if $\lambda_j = g(\tilde{q})d\tilde{q}$ then $(h^*\lambda)_j = (h_j^*g)(q)h_j'(q)dq$. The pullback on differentials dualizes to a linear map of dual spaces, denoted h_* rather than $(h^*)^\wedge$,

$$h_* : \Omega^1_{\mathrm{hol}}(X)^\wedge \longrightarrow \Omega^1_{\mathrm{hol}}(Y)^\wedge, \qquad h_*\varphi = \varphi \circ h^*.$$

The *forward change of variable formula* says that for any path γ in X,

$$\int_\gamma h^*\lambda = \int_{h\circ\gamma} \lambda, \quad \lambda \in \Omega^1_{\mathrm{hol}}(Y). \tag{6.3}$$

Showing this reduces to finitely many applications of the usual change of variable formula on coordinate patches. Especially if α is a loop in X and $\varphi \in \Omega^1_{\mathrm{hol}}(X)^\wedge$ is $\int_\alpha$ then $h_*\varphi \in \Omega^1_{\mathrm{hol}}(Y)^\wedge$ is $\int_\alpha h^* = \int_{h(\alpha)}$. Since $h(\alpha)$ is a loop in Y this shows that h_* takes homology to homology and therefore descends to a map of Jacobians, a holomorphic homomorphism because it is induced by a linear map of the ambient complex spaces. This defines the forward map of Jacobians in terms of differentials. For any $\varphi \in \Omega^1_{\mathrm{hol}}(X)^\wedge$ let $[\varphi] = \varphi + H_1(X, \mathbb{Z})$ denote its equivalence class in $\mathrm{Jac}(X)$, and similarly for Y.

Definition 6.2.1. *The forward map of Jacobians is the holomorphic homomorphism induced by composition with the pullback,*

$$h_J : \mathrm{Jac}(X) \longrightarrow \mathrm{Jac}(Y), \qquad h_J[\varphi] = [h_*\varphi] = [\varphi \circ h^*].$$

Letting x_0 be a base point in X as before and writing elements of $\mathrm{Jac}(X)$ as sums of integrations as in Abel's Theorem, forward change of variable shows that the forward map of Jacobians transfers integration modulo homology from X to Y by pushing forward the limits of integration,

$$h_J\left(\sum_x n_x \int_{x_0}^x\right) = \sum_x n_x \int_{h(x_0)}^{h(x)}. \tag{6.4}$$

For example, let $X = \mathbb{C}/\Lambda$ and $Y = \mathbb{C}/\Lambda'$ be complex elliptic curves, so that their embeddings in their Jacobians, $z + \Lambda \mapsto \int_{0+\Lambda}^{z+\Lambda}$ and $z + \Lambda' \mapsto \int_{0+\Lambda'}^{z+\Lambda'}$, are isomorphisms. By Proposition 1.3.2 or its generalization Proposition 6.1.4 any nonconstant holomorphic map between them takes the form

$$h : X \longrightarrow Y, \qquad h(z + \Lambda) = mz + b + \Lambda',$$

where $m, b \in \mathbb{C}$, $m \neq 0$, and $m\Lambda \subset \Lambda'$. Thus by (6.4),

$$h_J\left(\int_{0+\Lambda}^{z+\Lambda}\right) = \int_{b+\Lambda'}^{mz+b+\Lambda'} = \int_{0+\Lambda'}^{mz+\Lambda'}.$$

This shows that under the isomorphisms between the complex elliptic curves and their Jacobians the map retains its linear part but drops its translation,

$$h_J : X \longrightarrow Y, \qquad h_J(z + \Lambda) = mz + \Lambda'.$$

In particular, if h is an isogeny then $h_J = h$. Let $\tilde{h} : X \longrightarrow Y$ be another isogeny. Then $h + \tilde{h} : X \longrightarrow Y$ is an isogeny as well so long as $\tilde{h} \neq -h$, and so

$$(h + \tilde{h})_J = h + \tilde{h} = h_J + \tilde{h}_J \quad \text{if } h + \tilde{h} \neq 0. \tag{6.5}$$

That is, taking forward maps of isogenies of complex elliptic curves is additive.

Returning to the general situation $h : X \longrightarrow Y$, there is also a natural forward map of Picard groups. The *norm* map transfers functions forward from $\mathbb{C}(X)$ to $\mathbb{C}(Y)$ (Exercise 6.2.2(a)),

$$\mathrm{norm}_h : \mathbb{C}(X) \longrightarrow \mathbb{C}(Y), \qquad (\mathrm{norm}_h f)(y) = \prod_{x \in h^{-1}(y)} f(x)^{e_x}.$$

The orders of vanishing of a nonzero function and its norm are related by

$$\nu_y(\mathrm{norm}_h f) = \sum_{x \in h^{-1}(y)} \nu_x(f), \quad f \in \mathbb{C}(X)^*$$

(Exercise 6.2.2(b)). Consequently

$$\mathrm{div}(\mathrm{norm}_h f) = \sum_y \left(\sum_{x \in h^{-1}(y)} \nu_x(f) \right) y = \sum_x \nu_x(f) h(x).$$

That is, the norm takes $\sum_x \nu_x(f) x$ to $\sum_x \nu_x(f) h(x)$ at the level of principal divisors. The map on general divisors that extends this,

$$h_D : \mathrm{Div}(X) \longrightarrow \mathrm{Div}(Y), \qquad h_D\left(\sum_x n_x x\right) = \sum_x n_x h(x),$$

is a homomorphism taking degree-0 divisors to degree-0 divisors (Exercise 6.2.3), taking principal divisors to principal divisors since by its definition $h_D(\mathrm{div}(f)) = \mathrm{div}(\mathrm{norm}_h f)$, and therefore descending to divisor classes.

Definition 6.2.2. *The forward map of Picard groups is the homomorphism*

$$h_{\hat{P}} : \mathrm{Pic}^0(X) \longrightarrow \mathrm{Pic}^0(Y), \qquad h_P\Big[\sum_x n_x x\Big] = \Big[\sum_x n_x h(x)\Big].$$

This map of Picard groups corresponds to the map (6.4) of Jacobians under the isomorphism of Abel's Theorem,

$$\mathrm{Pic}^0(X) \longrightarrow \mathrm{Jac}(X), \qquad \Big[\sum_x n_x x\Big] \mapsto \sum_x n_x \int_{x_0}^x,$$

and similarly for Y using the base point $y_0 = h(x_0)$. That is, the following diagram commutes:

$$
\begin{array}{ccc}
\mathrm{Pic}^0(X) & \xrightarrow{\ h_P\ } & \mathrm{Pic}^0(Y) \\
\downarrow & & \downarrow \\
\mathrm{Jac}(X) & \xrightarrow{\ h_J\ } & \mathrm{Jac}(Y).
\end{array}
$$

We could have defined h_P from h_J via this diagram but then it would be dependent on complex analysis. Assuming the norm will continue to make sense in other contexts, we could have defined h_J from h_P via this diagram but doing so would lose the information that h_J is holomorphic.

Defining the reverse map h^J in terms of differentials is more delicate than doing so for the forward map. Recall from Section 3.1 that h is a surjection of finite degree d, that h is locally e_x-to-1 at each $x \in X$ as already mentioned here, and that the set of exceptional points in X is defined as $\mathcal{E} = \{x \in X : e_x > 1\}$, the finite set of points where h is ramified. Let $Y' = Y - h(\mathcal{E})$ and $X' = h^{-1}(Y')$ be the Riemann surfaces obtained by removing the images of the exceptional points from Y and their preimages from X. The restriction of h away from ramification and its image,

$$h : X' \longrightarrow Y',$$

is a *d-fold covering map* where $d = \deg(h)$. This means that every point $y \in Y'$ has a neighborhood $\widetilde{U}$ whose inverse image is a disjoint union of neighborhoods $U_1, \ldots, U_d$ in X' such that each restriction $h_i : U_i \longrightarrow \widetilde{U}$ of h is invertible.

A theorem from topology says that given a path δ in Y' and a point $x \in h^{-1}(\delta(0))$ in X', there exists a unique *lift* γ of δ to X' starting at x, i.e., a unique path γ in X' such that $\gamma(0) = x$ and $h \circ \gamma = \delta$. If δ is a path in Y and only its endpoints might lie in $h(\mathcal{E})$ then the Local Mapping Theorem of complex analysis shows that for each $x \in h^{-1}(\delta(0))$ there exist e_x lifts of δ starting at x, for a total of d lifts γ in X. If β is a loop in Y' then the map taking the initial point of each of its lifts to the terminal point is a permutation of the d-element set $h^{-1}(\beta(0))$, and so the lifts concatenate to loops α in X' corresponding to the permutation's cycles. Any path in X can be perturbed locally to avoid $h(\mathcal{E})$ away from its endpoints, and by Cauchy's Theorem this doesn't affect integration of holomorphic differentials. Thus any path integral of holomorphic differentials on Y can be taken over a path δ such that only its endpoints might lie in $h(\mathcal{E})$. In particular, loop integrals in Y can always be taken in Y'.

The *trace* map induced by h is a linear map transferring differentials from X to Y,

$$\text{tr}_h : \Omega^1_{\text{hol}}(X) \longrightarrow \Omega^1_{\text{hol}}(Y).$$

If δ is a path in Y' lifting to a path in X' and h_i^{-1} is a local inverse of h about $\delta(0)$ taking $\delta(0)$ to the initial point of the lift then h_i^{-1} has an *analytic continuation along* δ, the chaining together of overlapping local inverses culminating in the local inverse about $\delta(1)$ taking $\delta(1)$ to the terminal point of the lift. To define the trace, let $\omega \in \Omega^1_{\text{hol}}(X)$. Suppose y is a point in Y', so that h has local inverses $h_i^{-1} : \widetilde{U} \longrightarrow U_i$, $i = 1, \ldots, d$. The trace is defined on $\widetilde{U}$ as the sum of local pullbacks,

$$(\text{tr}_h \omega)|_{\widetilde{U}} = \sum_{i=1}^{d} (h_i^{-1})^* (\omega|_{U_i}).$$

This local definition pieces together to a well defined global trace on Y' because analytically continuing the local inverses h_i^{-1} along any loop in Y' back to y permutes them, leaving the trace unaltered. The trace extends holomorphically from Y' to all of Y (Exercise 6.2.4). The trace dualizes to a linear map of dual spaces,

$$\text{tr}_h^\wedge : \Omega^1_{\text{hol}}(Y)^\wedge \longrightarrow \Omega^1_{\text{hol}}(X)^\wedge, \qquad \text{tr}_h^\wedge \psi = \psi \circ \text{tr}_h.$$

The *reverse change of variable formula* for any path δ in Y',

$$\int_\delta (h^{-1})^* \omega = \int_{h^{-1} \circ \delta} \omega, \quad \omega \in \Omega^1_{\text{hol}}(X),$$

is meaningful so long as h^{-1} is understood to be some local inverse of h at $\delta(0)$ analytically continued along δ, making $h^{-1} \circ \delta$ the lift of δ starting at $h^{-1}(\delta(0))$. This reduces to a finite sum of the same result on coordinate patches, where it is just forward change of variable (6.3) with h^{-1} in place of h. Summing the reverse change of variable formula over local inverses gives for paths δ in Y'

$$\int_\delta \text{tr}_h \omega = \sum_{\text{lifts } \gamma} \int_\gamma \omega, \quad \omega \in \Omega^1_{\text{hol}}(X). \tag{6.6}$$

This formula extends continuously to paths δ in Y such that only their endpoints might lie in $h(\mathcal{E})$. Especially if β is a loop in Y' and $\psi \in \Omega^1_{\text{hol}}(Y)^\wedge$ is $\int_\beta$ then $\text{tr}_h^\wedge \psi \in \Omega^1_{\text{hol}}(X)^\wedge$ is $\int_\beta \text{tr}_h = \sum_{\text{lifts } \gamma} \int_\gamma$. Since β lifts to a concatenation

of loops in X this shows that $\mathrm{tr}_h^\wedge$ takes homology to homology and therefore
descends to a map of Jacobians, again a holomorphic homomorphism because
it is induced by a linear map of the ambient complex spaces. This defines the
reverse map of Jacobians in terms of differentials.

Definition 6.2.3. *The reverse map of Jacobians is the holomorphic homo-*
morphism induced by composition with the trace,

$$h^J : \mathrm{Jac}(Y) \longrightarrow \mathrm{Jac}(X), \qquad h^J[\psi] = [\psi \circ \mathrm{tr}_h].$$

Writing elements of $\mathrm{Jac}(Y)$ as sums of integrations per Abel's Theorem,
the summed reverse change of variable formula (6.6) shows that the reverse
maps of Jacobians transfers integration modulo homology from Y to X by
pulling back the limits of integration with suitable multiplicity,

$$h^J\left(\sum_y n_y \int_{y_0}^y\right) = \sum_y n_y \sum_{x \in h^{-1}(y)} e_x \int_{x_0}^x. \qquad (6.7)$$

We return to the example where $X = \mathbb{C}/\Lambda$ and $Y = \mathbb{C}/\Lambda'$ are complex
elliptic curves and h is a nonconstant holomorphic map, $h(z+\Lambda) = mz+b+\Lambda'$
with $m \neq 0$ and $m\Lambda \subset \Lambda'$. The map is unramified and has degree $\deg(h) =$
$[m^{-1}\Lambda' : \Lambda]$. For any $w \in \mathbb{C}$ the point $w + \Lambda' \in Y$ has inverse image

$$h^{-1}(w + \Lambda') = \{m^{-1}(w - b) + t + \Lambda : t \in m^{-1}\Lambda'/\Lambda\}.$$

By (6.7),

$$h^J\left(\int_{0+\Lambda'}^{w+\Lambda'}\right) = \sum_t \int_{-m^{-1}b+\Lambda}^{m^{-1}(w-b)+t+\Lambda} = \sum_t \int_{0+\Lambda}^{m^{-1}w+t+\Lambda} = \int_{0+\Lambda}^{\sum_t m^{-1}w+t+\Lambda}.$$

and again under the isomorphisms $X \overset{\sim}{\longrightarrow} \mathrm{Jac}(X)$ and $Y \overset{\sim}{\longrightarrow} \mathrm{Jac}(Y)$ this is

$$h^J(w + \Lambda') = \sum_t m^{-1}w + t + \Lambda.$$

Following this by the forward map $h_J(z + \Lambda) = mz + \Lambda'$ from before shows
that $(h_J \circ h^J)(w + \Lambda') = [m^{-1}\Lambda' : \Lambda](w + \Lambda')$. That is, the composition is
multiplication by the degree of h on Y,

$$h_J \circ h^J = [\deg(h)].$$

Thus h_J is the dual isogeny of h^J, making h^J the dual isogeny of h_J in turn
as shown in Section 1.3. Consequently if h is an isogeny and $\tilde{h} : X \longrightarrow Y$ is
an isogeny as well, then using a result from earlier in the section and a result
from Section 1.3 that taking duals of isogenies is additive,

$$(h + \tilde{h})^J = \widehat{(h + \tilde{h})}_J = \widehat{h_J + \tilde{h}_J} \quad \text{by (6.5)}$$
$$= \widehat{h_J} + \widehat{\tilde{h}_J} \quad \text{by (1.7)} \quad \Big\} \quad \text{if } h + \tilde{h} \neq 0.$$
$$= h^J + \tilde{h}^J$$

This shows that taking reverse maps of isogenies of complex elliptic curves is additive as well.

The result $h_J \circ h^J = [\deg(h)]$ is general: for any nonconstant holomorphic map $h : X \longrightarrow Y$ of compact Riemann surfaces, pullback followed by trace multiplies differentials by the degree of the map (Exercise 6.2.5),

$$(\mathrm{tr}_h \circ h^*)(\lambda) = \deg(h)\lambda, \qquad \lambda \in \Omega^1_{\mathrm{hol}}(Y). \tag{6.8}$$

It follows that $h_J \circ h^J$ is multiplication by $\deg(h)$ in $\mathrm{Jac}(Y)$. We will need another consequence in Section 6.6:

Proposition 6.2.4. *Let $h : X \longrightarrow Y$ be a nonconstant holomorphic map of compact Riemann surfaces. Let $h_* = (h^*)^\wedge|_{\mathrm{H}_1(X,\mathbb{Z})}$ be the induced forward homomorphism of homology groups,*

$$h_* : \mathrm{H}_1(X, \mathbb{Z}) \longrightarrow \mathrm{H}_1(Y, \mathbb{Z}), \qquad h_*\Big(\sum_\alpha n_\alpha \int_\alpha \Big) = \sum_\alpha n_\alpha \int_{h \circ \alpha}.$$

Then $h_*(\mathrm{H}_1(X, \mathbb{Z}))$ is a subgroup of finite index in $\mathrm{H}_1(Y, \mathbb{Z})$. If V is a subspace of $\Omega^1_{\mathrm{hol}}(Y)$ then the restriction $h_*(\mathrm{H}_1(X, \mathbb{Z}))|_V$ is a subgroup of finite index in $\mathrm{H}_1(Y, \mathbb{Z})|_V$.

Proof. Using (6.8) at the last step, compute that

$$h_*(\mathrm{H}_1(X, \mathbb{Z})) \supset ((h^*)^\wedge \circ \mathrm{tr}_h^\wedge)(\mathrm{H}_1(Y, \mathbb{Z}))$$
$$= (\mathrm{tr}_h \circ h^*)^\wedge(\mathrm{H}_1(Y, \mathbb{Z})) = \deg(h)\mathrm{H}_1(Y, \mathbb{Z}).$$

Therefore $\mathrm{H}_1(Y, \mathbb{Z})/h_*(\mathrm{H}_1(X, \mathbb{Z}))$ is a quotient of $(\mathbb{Z}/\deg(h)\mathbb{Z})^{2g}$. If V is a subspace of $\Omega^1_{\mathrm{hol}}(Y)$ then the restriction $\mathrm{H}_1(Y, \mathbb{Z}) \longrightarrow \mathrm{H}_1(Y, \mathbb{Z})|_V$ is a surjection, and (Exercise 6.2.6) $[\mathrm{H}_1(Y, \mathbb{Z})|_V : h_*(\mathrm{H}_1(X, \mathbb{Z}))|_V]$ is at most $[\mathrm{H}_1(Y, \mathbb{Z}) : h_*(\mathrm{H}_1(X, \mathbb{Z}))]$. $\qquad \square$

To complete the program of this section we need to describe the reverse map of Picard groups. This is easy. The pullback $h^* : \mathbb{C}(Y) \longrightarrow \mathbb{C}(X)$ transfers functions backwards. The earlier formula $\nu_x(h^*g) = e_x \nu_{h(x)}(g)$ for $g \in \mathbb{C}(Y)$ gives

$$\mathrm{div}(h^*g) = \sum_x e_x \nu_{h(x)}(g)x = \sum_y \nu_y(g) \sum_{x \in h^{-1}(y)} e_x x.$$

That is, the pullback takes $\sum_y \nu_y(g)y$ to $\sum_y \nu_y(g) \sum_{x \in h^{-1}(y)} e_x x$ at the level of principal divisors. The map on general divisors that extends this,

$$h^D : \mathrm{Div}(Y) \longrightarrow \mathrm{Div}(X), \qquad h^D\Big(\sum_y n_y y\Big) = \sum_y n_y \sum_{x \in h^{-1}(y)} e_x x,$$

is again a homomorphism taking degree-0 divisors to degree-0 divisors (Exercise 6.2.7(a)), taking principal divisors to principal divisors since by its definition $h^D(\mathrm{div}(g)) = \mathrm{div}(h^*g)$, and therefore descending to divisor classes.

Definition 6.2.5. *The reverse map of Picard groups is the homomorphism*

$$h^P : \mathrm{Pic}^0(Y) \longrightarrow \mathrm{Pic}^0(X), \qquad h^P\Big[\sum_y n_y y\Big] = \Big[\sum_y n_y \sum_{x \in h^{-1}(y)} e_x x\Big].$$

As with forward maps this correlates via Abel's Theorem to the description (6.7) of h^J so that a diagram commutes:

$$
\begin{array}{ccc}
\mathrm{Pic}^0(Y) & \xrightarrow{\ h^P\ } & \mathrm{Pic}^0(X) \\
\downarrow & & \downarrow \\
\mathrm{Jac}(Y) & \xrightarrow{\ h^J\ } & \mathrm{Jac}(X).
\end{array}
$$

Again this means that we could have defined either of h^J and h^P in terms of the other, but doing so would either leave h^P dependent on complex analysis or lose the information that h^J is holomorphic. It is easy to show that the composite $h_D \circ h^D$ multiplies by $\deg(h)$ in $\mathrm{Div}(Y)$ (Exercise 6.2.7(b)), and so the result that $h_P \circ h^P$ multiplies by $\deg(h)$ in $\mathrm{Pic}^0(Y)$ holds without reference to Jacobians.

The compact Riemann surfaces in this book other than complex elliptic curves are the modular curves. The next section will discuss maps between their Jacobians as well.

Returning to Versions $X_{\mathbb{C}}$ and $J_{\mathbb{C}}$ of the Modularity Theorem, we already know that Version $J_{\mathbb{C}}$ implies Version $X_{\mathbb{C}}$. Conversely, let E be a complex elliptic curve with $j(E) \in \mathbb{Q}$ and let $h : X_0(N) \longrightarrow E$ be the map guaranteed by Version $X_{\mathbb{C}}$. Then $h_J : J_0(N) \longrightarrow \mathrm{Jac}(E)$ is the necessary map for Version $J_{\mathbb{C}}$, remembering that E is naturally isomorphic to its Jacobian. The map surjects since $h_J \circ h^J$ is multiplication by $\deg(h)$, a surjection.

The myriad functorial adornments of the letter h in this section (see Exercise 6.2.8) serve only as irritants once the ideas are clear. When we move from complex analysis to algebraic geometry in later chapters, forward and reverse maps will just be denoted h_* and h^*.

Exercises

6.2.1. (a) Show that if $h : X \longrightarrow Y$ is a nonconstant holomorphic map of compact Riemann surfaces and $g \in \mathbb{C}(Y)$ then $\nu_x(h^*g) = e_x \nu_{h(x)}(g)$ at any point $x \in X$.

(b) Show that the pullback $h^*\lambda$ of a holomorphic differential λ on Y satisfies the compatibility condition on X.

6.2.2. (a) Show that the norm function defined in the section is meromorphic. (A hint for this exercise is at the end of the book.)

(b) Show that if $f \in \mathbb{C}(X)$ then $\nu_y(\mathrm{norm}_h f) = \sum_{x \in h^{-1}(y)} \nu_x(f)$ at any point $y \in Y$.

6.2.3. Show that the map h_D is a homomorphism and takes degree-0 divisors to degree-0 divisors.

6.2.4. This exercise extends the trace holomorphically from Y' to all of Y. Part (a) gives an intrinsic argument and part (b) gives a lower-powered argument using local coordinates. Fill in details.

(a) Every path δ in Y' has d lifts γ in X'. By definition of the trace, $\int_\delta \mathrm{tr}_h \omega = \sum_\gamma \int_\gamma \omega$ for $\omega \in \Omega^1_{\mathrm{hol}}(X)$. Let $\tilde{U} \subset Y$ be homeomorphic to a disk and contain only one point $y \in h(\mathcal{E})$, and let $\tilde{U}' = \tilde{U} - \{y\} \subset Y'$. Any loop β in $\tilde{U}'$ lifts to fewer than d loops α in $h^{-1}(\tilde{U})$. Each connected component of $h^{-1}(\tilde{U})$ is homeomorphic to a disk and so $\int_\beta \mathrm{tr}_h \omega = \sum_\alpha \int_\alpha \omega = 0$. This shows that integrating $\mathrm{tr}_h \omega$ is path-independent in $\tilde{U}'$. Let $w(y) = \int_{\tilde{y}}^y \mathrm{tr}_h(\omega)$ where $\tilde{y}$ is any point in $\tilde{U}'$. Then w is holomorphic on $\tilde{U}'$ and $dw = \mathrm{tr}_h \omega$. The expression $w(y) = \sum_\gamma \int_\gamma \omega$ shows that w is bounded on $\tilde{U}'$ and hence extends holomorphically to $\tilde{U}$. Therefore so does tr_h.

(b) Let $y \in h(\mathcal{E})$ be the image of an exceptional point. Suppose first that the ramification of h over y is total, i.e., $h^{-1}(y)$ is a single point x with $e_x = \deg(h)$. In suitable local coordinates the map about x is $r = h(q) = q^e$ (where $e = e_x$) on some disk V about $q = 0$. For any nonzero $r_0 \in h(V)$ let

$$(\ )^{1/e} : \{r : |r - r_0| < |r_0|\} \longrightarrow V$$

denote a holomorphic branch of the eth root function so that the local inverses of h about r_0 are all of the branches, $h_i^{-1} = \mu_e^i (\ )^{1/e}$ for $i = 0, \ldots, e-1$. Let $\omega = f(q)dq$ on V. Then the trace on $h(V) \backslash \{0\}$ is

$$\mathrm{tr}_h \omega = \sum_{i=0}^{e-1} f(\mu_e^i r^{1/e}) d(\mu_e^i r^{1/e}) = \frac{1}{e} \sum_{i=0}^{e-1} f(\mu_e^i r^{1/e}) \mu_e^i r^{1/e-1} dr.$$

If f has power series expansion $f(q) = \sum_{n=0}^\infty a_n q^n$ then rearranging a double sum shows that locally

$$\mathrm{tr}_h \omega = \sum_{m=0}^\infty a_{(m+1)e-1} r^m dr,$$

and this is holomorphic at $r = 0$. In the general case where $h^{-1}(y)$ contains more than one point x, each with its ramification degree e_x, carrying out this process at each x and summing the results shows that $\mathrm{tr}_h \omega$ extends holomorphically to y.

6.2.5. Prove formula (6.8).

6.2.6. Let $\pi : G \longrightarrow \pi(G)$ be a group homomorphism and let H be a subgroup of G. Show that $[\pi(G) : \pi(H)] \leq [G : H]$.

6.2.7. (a) Show that the map h^D is a homomorphism and takes degree-0 divisors to degree-0 divisors.
    (b) Show that the composition $h_D \circ h^D$ multiplies by $\deg(h)$ in $\mathrm{Div}(Y)$.

6.2.8. Briefly summarize the relations among h, h_*, h^*, norm_h, tr_h, h_J, h^J, h_D, h^D, h_P, and h^P.

6.3 Modular Jacobians and Hecke operators

The double coset operators from the beginning of Chapter 5 lead naturally to maps between Jacobians of modular curves. In particular, letting $J_1(N)$ denote the Jacobian associated to the group $\Gamma_1(N)$,

$$J_1(N) = \mathrm{Jac}(X_1(N)),$$

the Hecke operators act naturally on $J_1(N)$. This section explains these ideas.

Recall the double coset operator from Section 5.1: Let Γ_1 and Γ_2 be congruence subgroups of $\mathrm{SL}_2(\mathbb{Z})$ and let $\alpha \in \mathrm{GL}_2^+(\mathbb{Q})$. Set $\Gamma_3 = \alpha^{-1}\Gamma_1\alpha \cap \Gamma_2$, and take representatives $\{\gamma_{2,j}\}$ such that $\Gamma_3 \backslash \Gamma_2 = \bigcup_j \Gamma_3\gamma_{2,j}$. Then $\{\beta_j\} = \{\alpha\gamma_{2,j}\}$ are in turn representatives such that $\Gamma_1\alpha\Gamma_2 = \bigcup_j \Gamma_1\beta_j$. Also letting $\Gamma_3' = \alpha\Gamma_3\alpha^{-1} = \Gamma_1 \cap \alpha\Gamma_2\alpha^{-1}$ gives the configuration

$$\Gamma_2 \longleftarrow \Gamma_3 \xrightarrow{\;\sim\;} \Gamma_3' \longrightarrow \Gamma_1$$

where the group isomorphism is $\gamma \mapsto \alpha\gamma\alpha^{-1}$ and the other arrows are inclusions. The groups have modular curves X_1, X_2, X_3, and X_3', and their corresponding configuration is

$$X_2 \xleftarrow{\;\pi_2\;} X_3 \xrightarrow{\;\sim\;} X_3' \xrightarrow{\;\pi_1\;} X_1$$

where the modular curve isomorphism is $\Gamma_3\tau \mapsto \Gamma_3'\alpha(\tau)$, denoted α. Each point of X_2 is taken back by $\pi_1 \circ \alpha \circ \pi_2^{-1}$ to a multiset of points of X_1,

$$\Gamma_2\tau \xmapsto{\;\pi_2^{-1}\;} \{\Gamma_3\gamma_{2,j}(\tau)\} \xrightarrow{\;\alpha\;} \{\Gamma_3'\beta_j(\tau)\} \xrightarrow{\;\pi_1\;} \{\Gamma_1\beta_j(\tau)\},$$

where π_2^{-1} takes a point to its overlying points, each with multiplicity according to its ramification degree, $\pi_2^{-1}(x) = \{e_y \cdot y : y \in X_3, \pi_2(y) = x\}$. This leads to an interpretation of the double coset operator as a reverse map of divisor groups,

$$[\Gamma_1 \alpha \Gamma_2]_2 : \mathrm{Div}(X_2) \longrightarrow \mathrm{Div}(X_1),$$

the $\mathbb{Z}$-linear extension of the map $\Gamma_2 \tau \mapsto \sum_j \Gamma_1 \beta_j(\tau)$ from X_2 to $\mathrm{Div}(X_1)$. Using the language of the previous section, we now recognize all of this as saying that the double coset operator on divisor groups is a composition of forward and reversed induced maps, $[\Gamma_1 \alpha \Gamma_2]_2 = (\pi_1)_D \circ \alpha_D \circ \pi_2^D$. It therefore descends to the corresponding map of Picard groups, denoted by the same symbol,

$$[\Gamma_1 \alpha \Gamma_2]_2 = (\pi_1)_P \circ \alpha_P \circ \pi_2^P : \mathrm{Pic}^0(X_2) \longrightarrow \mathrm{Pic}^0(X_1),$$

given by

$$[\Gamma_1 \alpha \Gamma_2]_2 \Big[\sum_\tau n_\tau \Gamma_2 \tau \Big] = \Big[\sum_\tau n_\tau \sum_j \Gamma_1 \beta_j(\tau) \Big].$$

To rephrase this in terms of Jacobians and modular forms, let Γ be a congruence subgroup of $\mathrm{SL}_2(\mathbb{Z})$. By Section 3.3 the holomorphic differentials $\Omega^1_{\mathrm{hol}}(X(\Gamma))$ and the weight 2 cusp forms $\mathcal{S}_2(\Gamma)$ are naturally identified—every cusp form f describes a holomorphic differential $\omega(f)$ (essentially $f(\tau)d\tau$ but suitably defined on $X(\Gamma)$) and every holomorphic differential takes the form $\omega(f)$ for some cusp form f. That is, $\omega : \mathcal{S}_2(\Gamma) \longrightarrow \Omega^1_{\mathrm{hol}}(X(\Gamma))$ is a linear isomorphism. The dual spaces are consequently identified as well under $\omega^\wedge$,

$$\mathcal{S}_2(\Gamma)^\wedge = \omega^\wedge(\Omega^1_{\mathrm{hol}}(X(\Gamma))^\wedge).$$

In the modular context, let $\mathrm{H}_1(X(\Gamma), \mathbb{Z})$ denote the corresponding subgroup of $\mathcal{S}_2(\Gamma)^\wedge$, what would be denoted $\omega^\wedge(\mathrm{H}_1(X(\Gamma), \mathbb{Z}))$ in the language of differentials. It is natural to redefine the Jacobian of $X(\Gamma)$ as a quotient of the dual space of weight 2 cusp forms,

Definition 6.3.1. *Let Γ be a congruence subgroup of $\mathrm{SL}_2(\mathbb{Z})$. The Jacobian of the corresponding modular curve $X(\Gamma)$ is*

$$\mathrm{Jac}(X(\Gamma)) = \mathcal{S}_2(\Gamma)^\wedge / \mathrm{H}_1(X(\Gamma), \mathbb{Z}).$$

Under this change of terminology the induced maps of the previous section need to be rephrased in terms of functions rather than differentials. Let X and Y be the modular curves associated to congruence subgroups Γ_X and Γ_Y of $\mathrm{SL}_2(\mathbb{Z})$. Let $\alpha \in \mathrm{GL}_2^+(\mathbb{Q})$ be such that $\alpha \Gamma_X \alpha^{-1} \subset \Gamma_Y$, and consider the corresponding holomorphic map,

$$h : X \longrightarrow Y, \qquad h(\Gamma_X \tau) = \Gamma_Y \alpha(\tau). \tag{6.9}$$

The weight-2 operator on functions is compatible with the pullback on differentials in the sense that the following diagram commutes (Exercise 6.3.1(a)):

$$\begin{array}{ccc}
\mathcal{S}_2(\Gamma_Y) & \xrightarrow{[\alpha]_2} & \mathcal{S}_2(\Gamma_X) \\
\omega_Y \downarrow & & \downarrow \omega_X \\
\Omega^1_{\text{hol}}(Y) & \xrightarrow{h^*} & \Omega^1_{\text{hol}}(X).
\end{array} \qquad (6.10)$$

The induced forward map on dual spaces correspondingly becomes

$$h_* : \mathcal{S}_2(\Gamma_X)^\wedge \longrightarrow \mathcal{S}_2(\Gamma_Y)^\wedge, \qquad h_*\varphi = \varphi \circ [\alpha]_2.$$

Similarly, if $\alpha \Gamma_X \alpha^{-1} \backslash \Gamma_Y = \bigcup_j \alpha \Gamma_X \alpha^{-1} \gamma_{Y,j}$ then the following diagram commutes (Exercise 6.3.1(b)):

$$\begin{array}{ccc}
\mathcal{S}_2(\Gamma_X) & \xrightarrow{\sum_j [\alpha^{-1} \gamma_{Y,j}]_2} & \mathcal{S}_2(\Gamma_Y) \\
\omega_X \downarrow & & \downarrow \omega_Y \\
\Omega^1_{\text{hol}}(X) & \xrightarrow{\text{tr}_h} & \Omega^1_{\text{hol}}(Y).
\end{array} \qquad (6.11)$$

Letting tr_h also denote the top map, the induced reverse map on dual spaces is thus

$$\text{tr}_h^\wedge : \mathcal{S}_2(\Gamma_Y)^\wedge \longrightarrow \mathcal{S}_2(\Gamma_X)^\wedge, \qquad \text{tr}_h^\wedge \psi = \psi \circ \sum_j [\alpha^{-1} \gamma_{Y,j}]_2.$$

By the compatibilities shown here, h_* and $\text{tr}_h^\wedge$ descend to maps of Jacobians, denoted h_J and h^J as before.

Recall from Section 5.1 that the double coset operator on divisor groups corresponds to the double coset operator on modular forms, in our case of weight 2 cusps forms

$$[\Gamma_1 \alpha \Gamma_2]_2 : \mathcal{S}_2(\Gamma_1) \longrightarrow \mathcal{S}_2(\Gamma_2), \qquad f[\Gamma_1 \alpha \Gamma_2]_2 = \sum_j f[\beta_j]_2.$$

Its pullback is denoted by the same symbol,

$$[\Gamma_1 \alpha \Gamma_2]_2 : \mathcal{S}_2(\Gamma_2)^\wedge \longrightarrow \mathcal{S}_2(\Gamma_1)^\wedge, \qquad [\Gamma_1 \alpha \Gamma_2]_2 \psi = \psi \circ [\Gamma_1 \alpha \Gamma_2]_2.$$

As with the divisor map, we now recognize this as a composition of induced maps, $[\Gamma_1 \alpha \Gamma_2]_2 = (\pi_1)_* \circ \alpha_* \circ \text{tr}_{\pi_2}^\wedge$, and this operator correspondingly descends to a composition of maps of Jacobians. That is, the double coset operator on Jacobians is

$$[\Gamma_1 \alpha \Gamma_2]_2 = (\pi_1)_J \circ \alpha_J \circ \pi_2^J : \text{Jac}(X_2) \longrightarrow \text{Jac}(X_1),$$

where again letting brackets denote equivalence classes modulo homology,

$$[\Gamma_1 \alpha \Gamma_2]_2[\psi] = [\psi \circ [\Gamma_1 \alpha \Gamma_2]_2] \text{ for } \psi \in \mathcal{S}_2(\Gamma_2)^\wedge.$$

To summarize, the double coset operator acts on Jacobians as composition with its action on modular forms in the other direction.

The Hecke operators are special cases of the double coset operator. For p prime the Hecke operator T_p on $\mathcal{S}_2(\Gamma_1(N))$ is $[\Gamma_1(N) \left[\begin{smallmatrix} 1 & 0 \\ 0 & p \end{smallmatrix}\right] \Gamma_1(N)]_2$, and similarly for $\Gamma_0(N)$. For d relatively prime to N the Hecke operator $\langle d \rangle$ on $\mathcal{S}_2(\Gamma_1(N))$ is $[\Gamma_1(N)\alpha\Gamma_1(N)]_2$ for any $\alpha = \left[\begin{smallmatrix} a & b \\ c & \delta \end{smallmatrix}\right] \in \Gamma_0(N)$ with $\delta \equiv d \pmod N$. This proves

Proposition 6.3.2. *The Hecke operators $T = T_p$ and $T = \langle d \rangle$ act by composition on the Jacobian associated to $\Gamma_1(N)$,*

$$T : J_1(N) \longrightarrow J_1(N), \qquad [\varphi] \mapsto [\varphi \circ T] \text{ for } \varphi \in \mathcal{S}_2(\Gamma_1(N))^{\wedge},$$

and similarly for T_p on $J_0(N)$.

Recall from Section 1.5 and Section 5.2 that when the double coset operator is specialized to T_p the group Γ_3 becomes $\Gamma_1^0(N,p) = \Gamma_1(N) \cap \Gamma^0(p)$, its modular curve X_3 is denoted $X_1^0(N,p)$, and the maps π_2 and $\pi_1 \circ \alpha$ from $X_1^0(N,p)$ to $X_1(N)$ are

$$\pi_2(\Gamma_1^0(N,p)\tau) = \Gamma_1(N)\tau, \qquad (\pi_1 \circ \alpha)(\Gamma_1^0(N,p)\tau) = \Gamma_1(N)(\tau/p).$$

The discussion here thus describes T_p explicitly as a pullback followed by a pushforward,

$$T_p = (\pi_1 \circ \alpha)_J \circ \pi_2^J : J_1(N) \longrightarrow J_1(N).$$

Similarly for the diamond operator, letting any $\alpha \in \Gamma_0(N)$ also denote the self-map $\Gamma_1(N)\tau \mapsto \Gamma_1(N)\alpha(\tau)$ of $X_1(N)$,

$$\langle d \rangle = \alpha_J : J_1(N) \longrightarrow J_1(N), \qquad \alpha = \left[\begin{smallmatrix} a & b \\ c & \delta \end{smallmatrix}\right] \in \Gamma_0(N), \ \delta \equiv d \pmod N.$$

Again as in Section 5.1, when $\Gamma_1 \supset \Gamma_2$ and $\alpha = I$, the maps of curves in the double coset operator configuration collapse to

$$\pi : X_2 \longrightarrow X_1.$$

The discussion earlier shows that the forward map on dual spaces is restriction,

$$\pi_* : \mathcal{S}_2(\Gamma_2)^{\wedge} \longrightarrow \mathcal{S}_2(\Gamma_1)^{\wedge}, \qquad \pi_*\varphi = \varphi|_{\mathcal{S}_2(\Gamma_1)}.$$

This descends to Jacobians,

$$\pi_J : \mathrm{Jac}(X_2) \longrightarrow \mathrm{Jac}(X_1), \qquad \pi_J[\varphi] = [\varphi|_{\mathcal{S}_2(\Gamma_1)}].$$

Especially when $\Gamma_1 = \Gamma_0(N)$ and $\Gamma_2 = \Gamma_1(N)$ this shows that restriction induces a natural surjection from $J_1(N)$ to $J_0(N)$. We will cite this fact at the end of Section 6.6.

Exercise

6.3.1. (a) Show that diagram (6.10) commutes. (A hint for this exercise is at the end of the book.)
(b) Show that diagram (6.11) commutes.

6.4 Algebraic numbers and algebraic integers

To prepare for applying the results of the previous section, this section sketches the requisite basics of algebraic number theory for the reader who is not familiar with them.

A complex number α is an *algebraic number* if α satisfies some monic polynomial with rational coefficients,

$$p(\alpha) = 0, \quad p(x) = x^n + c_1 x^{n-1} + \cdots + c_n, \quad c_1, \ldots, c_n \in \mathbb{Q}.$$

Every rational number r is algebraic since it satisfies the polynomial $x - r$, but not every algebraic number is rational, for example either complex square root of a rational number r satisfies the polynomial $x^2 - r$. Complex numbers expressible over $\mathbb{Q}$ in radicals are algebraic, for example $\mu_n = e^{2\pi i/n}$, but the converse is not true.

The algebraic numbers form a field, denoted $\overline{\mathbb{Q}}$. This is shown as follows.

Theorem 6.4.1. *Let α be a complex number. The following conditions on α are equivalent:*

(1) α *is an algebraic number, i.e.,* $\alpha \in \overline{\mathbb{Q}}$,
(2) *The ring* $\mathbb{Q}[\alpha]$ *is a finite-dimensional vector space over* $\mathbb{Q}$,
(3) α *belongs to a ring R in $\mathbb{C}$ that is a finite-dimensional vector space over* $\mathbb{Q}$.

Proof. (1) $\implies$ (2): Let α satisfy the polynomial $x^n + c_1 x^{n-1} + \cdots + c_n$ with $c_1, \ldots, c_n \in \mathbb{Q}$. Then $\alpha^n = -\sum_{i=0}^{n-1} c_{n-i}\alpha^i$, so the complex vector space generated by the powers $\{1, \alpha, \ldots, \alpha^{n-1}\}$ also contains α^n. Similarly $\alpha^{n+1} = -\sum_{i=0}^{n-1} c_{n-i}\alpha^{i+1}$, showing that α^{n+1} is in the space, and so on by induction for all higher powers of α.
(2) $\implies$ (3) is immediate.
(3) $\implies$ (1): Let the ring R have basis $g_1, \ldots, g_n$ as a vector space over $\mathbb{Q}$. Then multiplying each g_i by α gives a rational linear combination of the generators, $\alpha g_i = \sum_{j=1}^n c_{ij} g_j$. Letting g denote the column vector with entries g_i, this means that $\alpha g = Mg$ where M is the n-by-n rational matrix with entries c_{ij}. Thus α is an eigenvalue of M, meaning it satisfies the characteristic polynomial of M, a monic polynomial with rational coefficients. $\square$

Condition (3) in the theorem easily proves

Corollary 6.4.2. *The algebraic numbers $\overline{\mathbb{Q}}$ form a field.*

Proof. Let α and β be algebraic numbers. Then the rings $\mathbb{Q}[\alpha]$ and $\mathbb{Q}[\beta]$ have respective bases $\{\alpha^i : 0 \le i < m\}$ and $\{\beta^j : 0 \le j < n\}$ as vector spaces over $\mathbb{Q}$. Let $R = \mathbb{Q}[\alpha, \beta]$, spanned as a vector space over $\mathbb{Q}$ by the set $\{\alpha^i \beta^j : 0 \le i < m, 0 \le j < m\}$. Then $\alpha + \beta$ and $\alpha\beta$ belong to R, making them algebraic numbers by condition (3) of the theorem. If $\alpha \ne 0$ then its polynomial $p(x)$ can be taken to have a nonzero constant term c_n after dividing through by its lowest power of x. The relation $p(\alpha) = 0$ is $\alpha^{-1} = (p(\alpha) - c_n)/(-c_n\alpha) \in \mathbb{Q}[\alpha]$, making α^{-1} an algebraic number by condition (3) as well.   □

If α and β are algebraic numbers satisfying the monic rational polynomials $p(x)$ and $q(x)$ then the proofs of Corollary 6.4.2 and of (3) $\implies$ (1) in Theorem 6.4.1 combine to produce the polynomials satisfied by $\alpha + \beta$ and $\alpha\beta$ and $1/\alpha$ if $\alpha \ne 0$. The theory of resultants provides an efficient algorithm to find these polynomials (see Exercise 6.4.1).

One can now consider complex numbers α satisfying monic polynomials with coefficients in $\overline{\mathbb{Q}}$. But in fact $\overline{\mathbb{Q}}$ is *algebraically closed*, meaning any such α is already in $\overline{\mathbb{Q}}$. The proof again uses condition (3) in the theorem.

Corollary 6.4.3. *The field $\overline{\mathbb{Q}}$ of algebraic numbers is algebraically closed.*

Proof. (Sketch.) Consider a monic polynomial $x^n + c_1 x^{n-1} + \cdots + c_n$ with coefficients $c_i \in \overline{\mathbb{Q}}$ and let α be one of its roots. Since each ring $\mathbb{Q}[c_i]$ is a finite-dimensional vector space over $\mathbb{Q}$, so is the ring $R = \mathbb{Q}[c_1, \ldots, c_n]$. Let $R' = R[\alpha]$. If $\{v_i : 1 \le i \le m\}$ is a basis for R over $\mathbb{Q}$ then

$$\{v_i \alpha^j : 1 \le i \le m, 0 \le j < n\}$$

is a spanning set for R' as a vector space over $\mathbb{Q}$. (This set is not necessarily a basis since α might satisfy a polynomial of lower degree.) Now condition (3) of the theorem shows that $\alpha \in \overline{\mathbb{Q}}$.   □

Definition 6.4.4. *A **number field** is a field $\mathbb{K} \subset \overline{\mathbb{Q}}$ such that the degree $[\mathbb{K} : \mathbb{Q}] = \dim_{\mathbb{Q}}(\mathbb{K})$ is finite.*

The ring of integers $\mathbb{Z}$ in the rational number field $\mathbb{Q}$ has a natural analog in the field of algebraic numbers $\overline{\mathbb{Q}}$. A complex number α is an *algebraic integer* if α satisfies some monic polynomial with integer coefficients. The algebraic integers are denoted $\overline{\mathbb{Z}}$. The algebraic integers in the rational number field $\mathbb{Q}$ are the usual integers $\mathbb{Z}$ (Exercise 6.4.3), now called the *rational integers*. Every algebraic number takes the form of an algebraic integer divided by a rational integer (Exercise 6.4.4). Similarly to Theorem 6.4.1 and its corollaries (Exercise 6.4.5),

Theorem 6.4.5. *Let α be a complex number. The following conditions on α are equivalent:*

(1) α is an algebraic integer, i.e., $\alpha \in \overline{\mathbb{Z}}$,
(2) The ring $\mathbb{Z}[\alpha]$ is finitely generated as an Abelian group,
(3) α belongs to a ring R in $\mathbb{C}$ that is finitely generated as an Abelian group.

Corollary 6.4.6. The algebraic integers $\overline{\mathbb{Z}}$ form a ring.

Corollary 6.4.7. The algebraic integers form an **integrally closed** ring, meaning that every monic polynomial with coefficients in $\overline{\mathbb{Z}}$ factors down to linear terms over $\overline{\mathbb{Z}}$, i.e., its roots lie in $\overline{\mathbb{Z}}$.

Definition 6.4.8. Let $\mathbb{K}$ be a number field. The **number ring** of $\mathbb{K}$ is the ring of algebraic integers in $\mathbb{K}$,

$$\mathcal{O}_{\mathbb{K}} = \overline{\mathbb{Z}} \cap \mathbb{K}.$$

We need a few more results from algebraic number theory, stated here without proof. Every number ring $\mathcal{O}_{\mathbb{K}}$ is a free Abelian group of rank $[\mathbb{K} : \mathbb{Q}]$. That is, letting $d = [\mathbb{K} : \mathbb{Q}]$, there is a basis $\{\alpha_1, \ldots, \alpha_d\} \subset \mathcal{O}_{\mathbb{K}}$ such that $\mathcal{O}_{\mathbb{K}}$ is the free Abelian group

$$\mathcal{O}_{\mathbb{K}} = \mathbb{Z}\alpha_1 \oplus \cdots \oplus \mathbb{Z}\alpha_d.$$

There exist d nonzero field homomorphisms from $\mathbb{K}$ to $\mathbb{C}$. These have trivial kernel, making them embeddings $\sigma_1, \ldots, \sigma_d : \mathbb{K} \hookrightarrow \mathbb{C}$, and they fix the rational field $\mathbb{Q}$ elementwise. The d-by-d matrix

$$A = \left[\alpha_i^{\sigma_j}\right] = \begin{bmatrix} \alpha_1^{\sigma_1} & \cdots & \alpha_1^{\sigma_d} \\ \vdots & \ddots & \vdots \\ \alpha_d^{\sigma_1} & \cdots & \alpha_d^{\sigma_d} \end{bmatrix}$$

has nonzero determinant.

Every embedding $\sigma : \mathbb{K} \hookrightarrow \mathbb{C}$ extends (in many ways) to an automorphism $\sigma : \mathbb{C} \longrightarrow \mathbb{C}$. Conversely every such automorphism restricts to an automorphism $\sigma : \overline{\mathbb{Q}} \longrightarrow \overline{\mathbb{Q}}$ of the algebraic numbers that takes the algebraic integers $\overline{\mathbb{Z}}$ to $\overline{\mathbb{Z}}$, and this automorphism further restricts to an embedding of $\mathbb{K}$ in $\overline{\mathbb{Q}}$ that injects $\mathcal{O}_{\mathbb{K}}$ into $\overline{\mathbb{Z}}$. The only algebraic numbers that are fixed under all automorphisms of $\mathbb{C}$ are the rational numbers $\mathbb{Q}$. After this we will be casual about distinguishing between maps σ as embeddings of $\mathbb{K}$, as automorphisms of $\mathbb{C}$, or as automorphisms of $\overline{\mathbb{Q}}$.

Exercises

6.4.1. Recall from algebra that if $f(x) = a_0 x^m + a_1 x^{n-1} + \cdots + a_m$ and $g(x) = b_0 x^n + b_1 x^{n-1} + \cdots + b_n$ then their *resultant* $R(f(x), g(x); x)$ is the determinant of their *Sylvester matrix*

$$M = \begin{bmatrix} a_0 & a_1 & \cdots & \cdots & \cdots & a_m \\ & \ddots & \ddots & & & & \ddots \\ & & a_0 & a_1 & \cdots & \cdots & a_m \\ b_0 & b_1 & \cdots & b_n \\ & b_0 & b_1 & \cdots & b_n \\ & & \ddots & \ddots & & \ddots \\ & & & b_0 & b_1 & \cdots & b_n \end{bmatrix}$$

(n staggered rows of a_i's, m staggered rows of b_j's, all other entries 0). The resultant eliminates x, leaving a polynomial in the coefficients that vanishes if and only if f and g share a root. Let $p(x)$ and $q(x)$ be monic polynomials in $\mathbb{Q}[x]$. Consider the resultants

$$\tilde{q}(t, u) = R(p(s), u - s - t; s), \qquad r(u) = R(\tilde{q}(t, u), q(t); t).$$

Show that if α and β satisfy the polynomials p and q then $\alpha + \beta$ satisfies the polynomial r. Similarly, find polynomials satisfied by $\alpha\beta$ and $1/\alpha$ if $\alpha \neq 0$. (A hint for this exercise is at the end of the book.)

6.4.2. Use the methods of the section or of Exercise 6.4.1 to find a monic integer polynomial satisfied by $\sqrt{2} + \mu_3$.

6.4.3. Show that $\overline{\mathbb{Z}} \cap \mathbb{Q} = \mathbb{Z}$. (A hint for this exercise is at the end of the book.)

6.4.4. Show that every algebraic number takes the form of an algebraic integer divided by a rational integer. (A hint for this exercise is at the end of the book.)

6.4.5. Prove Theorem 6.4.5 and its corollaries.

6.5 Algebraic eigenvalues

Returning to the material of Section 6.3, recall the action of the weight-2 Hecke operators $T = T_p$ and $T = \langle d \rangle$ on the dual space as composition from the right,

$$T : \mathcal{S}_2(\Gamma_1(N))^\wedge \longrightarrow \mathcal{S}_2(\Gamma_1(N))^\wedge, \qquad \varphi \mapsto \varphi \circ T,$$

and recall that the action descends to the quotient $J_1(N)$. Thus the operators act as endomorphisms on the kernel $H_1(X_1(N), \mathbb{Z})$, a finitely generated Abelian group. In particular the characteristic polynomial $f(x)$ of T_p acting on $H_1(X_1(N), \mathbb{Z})$ has integer coefficients, and being a characteristic polynomial it is monic. Since an operator satisfies its characteristic polynomial, $f(T_p) = 0$ on $H_1(X_1(N), \mathbb{Z})$. Since T_p is $\mathbb{C}$-linear, also $f(T_p) = 0$ on $\mathcal{S}_2(\Gamma_1(N))^\wedge$ and so $f(T_p) = 0$ on $\mathcal{S}_2(\Gamma_1(N))$. Therefore the minimal polynomial of T_p on $\mathcal{S}_2(\Gamma_1(N))$ divides $f(x)$ and the eigenvalues of T_p satisfy $f(x)$, making them algebraic integers. Since p is arbitrary this proves

Theorem 6.5.1. *Let $f \in \mathcal{S}_2(\Gamma_1(N))$ be a normalized eigenform for the Hecke operators T_p. Then the eigenvalues $a_n(f)$ are algebraic integers.*

To refine this result we need to view the Hecke operators as lying within an algebraic structure, not merely as a set.

Definition 6.5.2. *The **Hecke algebra over $\mathbb{Z}$** is the algebra of endomorphisms of $\mathcal{S}_2(\Gamma_1(N))$ generated over $\mathbb{Z}$ by the Hecke operators,*

$$\mathbb{T}_{\mathbb{Z}} = \mathbb{Z}[\{T_n, \langle n \rangle : n \in \mathbb{Z}^+\}].$$

*The **Hecke algebra** $\mathbb{T}_{\mathbb{C}}$ **over $\mathbb{C}$** is defined similarly.*

Each level has its own Hecke algebra, but N is omitted from the notation since it is usually written somewhere nearby. Clearly any $f \in \mathcal{S}_2(\Gamma_1(N))$ is an eigenform for all of $\mathbb{T}_{\mathbb{C}}$ if and only if f is an eigenform for all Hecke operators T_p and $\langle d \rangle$.

For the remainder of this chapter the methods will shift to working with algebraic structure rather than thinking about objects such as Hecke operators one at a time. In particular modules will figure prominently, and so in this context Abelian groups will often be called $\mathbb{Z}$-modules. For example, viewing the $\mathbb{Z}$-module $\mathbb{T}_{\mathbb{Z}}$ as a ring of endomorphisms of the finitely generated free $\mathbb{Z}$-module $\mathrm{H}_1(X_1(N), \mathbb{Z})$ shows that it is finitely generated as well (Exercise 6.5.1). Again letting $f(\tau) = \sum_{n=1}^{\infty} a_n(f)q^n$ be a normalized eigenform, the eigenvalue homomorphism (defined by its characterizing property in the next display)

$$\lambda_f : \mathbb{T}_{\mathbb{Z}} \longrightarrow \mathbb{C}, \quad Tf = \lambda_f(T)f$$

therefore has as its image a finitely generated $\mathbb{Z}$-module. Since the image is $\mathbb{Z}[\{a_n(f) : n \in \mathbb{Z}^+\}]$ this shows that even though there are infinitely many eigenvalues $a_n(f)$, the ring they generate has finite rank as a $\mathbb{Z}$-module. More specifically, letting

$$I_f = \ker(\lambda_f) = \{T \in \mathbb{T}_{\mathbb{Z}} : Tf = 0\}$$

gives a ring and $\mathbb{Z}$-module isomorphism (Exercise 6.5.2)

$$\mathbb{T}_{\mathbb{Z}}/I_f \xrightarrow{\sim} \mathbb{Z}[\{a_n(f)\}]. \tag{6.12}$$

The image ring sits inside some finite-degree extension field of $\mathbb{Q}$, i.e., a number field. The rank of $\mathbb{T}_{\mathbb{Z}}/I_f$ is the degree of this number field as an extension of $\mathbb{Q}$.

Definition 6.5.3. *Let $f \in \mathcal{S}_2(\Gamma_1(N))$ be a normalized eigenform, $f(\tau) = \sum_{n=1}^{\infty} a_n q^n$. The field $\mathbb{K}_f = \mathbb{Q}(\{a_n\})$ generated by the Fourier coefficients of f is called **the number field of f.***

The reader is referred to the online LMFDB database for examples.

Any embedding $\sigma : \mathbb{K}_f \hookrightarrow \mathbb{C}$ conjugates f by acting on its coefficients. That is, if $f(\tau) = \sum_{n=1}^{\infty} a_n q^n$ then noting the action with a superscript,

$$f^\sigma(\tau) = \sum_{n=1}^\infty a_n^\sigma q^n.$$

In fact this action produces another eigenform.

Theorem 6.5.4. *Let f be a weight 2 normalized eigenform of the Hecke operators, so that $f \in S_2(N, \chi)$ for some N and χ. Let $\mathbb{K}_f$ be its number field. For any embedding $\sigma : \mathbb{K}_f \hookrightarrow \mathbb{C}$ the conjugated f^σ is also a normalized eigenform in $S_2(N, \chi^\sigma)$ where $\chi^\sigma(n) = \chi(n)^\sigma$. If f is a newform then so is f^σ.*

The proof will require two beginning results from commutative algebra, so these are stated first.

Proposition 6.5.5 (Nakayama's Lemma). *Suppose that A is a commutative ring with unit and $J \subset A$ is an ideal contained in every maximal ideal of A, and suppose that M is a finitely generated A-module such that $JM = M$. Then $M = \{0\}$.*

Proof. Suppose that $M \neq \{0\}$ and let $m_1, \ldots, m_n$ be a minimal set of generators for M over A. Since $JM = M$, in particular $m_n \in JM$, giving it the form $m_n = a_1 m_1 + \cdots + a_n m_n$ with all $a_i \in J$. Thus

$$(1 - a_n)m_n = a_1 m_1 + \cdots + a_{n-1} m_{n-1}.$$

But $1 - a_n$ is invertible in A, else it sits in a maximal ideal, which necessarily contains a_n as well and therefore is all of A, impossible. Thus $m_1, \ldots, m_{n-1}$ is a smaller generating set, and this contradiction proves the lemma.   □

Again suppose that A is a commutative ring with unit and $J \subset A$ is an ideal, and suppose that M is an A-module and a finite-dimensional vector space over some field $\mathbf{k}$. The dual space $M^\wedge = \mathrm{Hom}_{\mathbf{k}}(M, \mathbf{k})$ is an A-module in the natural way, $a\varphi = \varphi \circ a$ for $a \in A$ and $\varphi \in M^\wedge$ (i.e., $(a\varphi)(m) = \varphi(am)$ for $m \in M$), and similarly for $(M/JM)^\wedge$. Let $M[J]$ be the elements of M annihilated by J, and similarly for $M^\wedge[J]$. Then there exist natural isomorphisms of A-modules (Exercise 6.5.3)

$$(M/JM)^\wedge \cong M^\wedge[J], \qquad M^\wedge/JM^\wedge \cong M[J]^\wedge. \tag{6.13}$$

Now we can prove Theorem 6.5.4.

Proof. The Fourier coefficients $\{a_n\}$ are a system of eigenvalues for the operators T_n acting on $S_2(\Gamma_1(N))$. We need to show that the conjugated coefficients $\{a_n^\sigma\}$ are again a system of eigenvalues.

As explained at the beginning of the section, the action of $\mathbb{T}_{\mathbb{Z}}$ on $S_2(\Gamma_1(N))$ transfers to the dual space $S_2(\Gamma_1(N))^\wedge$ as composition on the right and then descends to the Jacobian $\mathrm{J}_1(N)$, inducing an action on the homology group $\mathrm{H}_1(X_1(N), \mathbb{Z})$. This $\mathbb{Z}$-module is free of rank $2g$ where g is the genus of $X_1(N)$ and the dimension of $\mathcal{S}_2(\Gamma_1(N))$. Take a homology basis $\{\varphi_1, \ldots, \varphi_{2g}\} \subset \mathcal{S}_2(\Gamma_1(N))^\wedge$, so that

$$H_1(X_1(N), \mathbb{Z}) = \mathbb{Z}\varphi_1 \oplus \cdots \oplus \mathbb{Z}\varphi_{2g}.$$

With respect to this basis, each group element $\sum_{j=1}^{2g} n_j \varphi_j$ is represented as an integral row vector $\vec{v} = [n_j] \in \mathbb{Z}^{2g}$ and each $T \in \mathbb{T}_\mathbb{Z}$ is represented by an integral $2g$-by-$2g$ matrix $[T] \in \mathrm{M}_{2g}(\mathbb{Z})$, so that the action of T as composition from the right is multiplication by $[T]$,

$$T : \vec{v} \mapsto \vec{v}[T]. \tag{6.14}$$

This action of $\mathbb{T}_\mathbb{Z}$ extends linearly to the free $\mathbb{C}$-module generated by the set $\{\varphi_1, \ldots, \varphi_{2g}\}$, the $2g$-dimensional complex vector space

$$V = \mathbb{C}\varphi_1 \oplus \cdots \oplus \mathbb{C}\varphi_{2g}.$$

Each element $(z_1\varphi_1, \ldots, z_{2g}\varphi_{2g})$ is represented by a complex row vector $\vec{v} = [z_j] \in \mathbb{C}^{2g}$, and the action of each T is still described by (6.14). Suppose $\{\lambda(T) : T \in \mathbb{T}_\mathbb{Z}\}$ is a system of eigenvalues of $\mathbb{T}_\mathbb{Z}$ on V, i.e., some nonzero $\vec{v} \in \mathbb{C}^{2g}$ satisfies

$$\vec{v}[T] = \lambda(T)\vec{v}, \quad T \in \mathbb{T}_\mathbb{Z}.$$

Let $\sigma : \mathbb{C} \longrightarrow \mathbb{C}$ be any automorphism extending the given embedding $\sigma : \mathbb{K}_f \hookrightarrow \mathbb{C}$. Then σ acts on $\vec{v}$ elementwise and fixes the elements of each matrix $[T]$ since it fixes $\mathbb{Q}$. Thus

$$\vec{v}^\sigma[T] = (\vec{v}[T])^\sigma = (\lambda(T)\vec{v})^\sigma = \lambda(T)^\sigma \vec{v}^\sigma, \quad T \in \mathbb{T}_\mathbb{Z},$$

showing that if $\{\lambda(T) : T \in \mathbb{T}_\mathbb{Z}\}$ is a system of eigenvalues on V then so is $\{\lambda(T)^\sigma : T \in \mathbb{T}_\mathbb{Z}\}$. To prove the theorem, this result needs to be transferred from V to $\mathcal{S}_2(\Gamma_1(N))$.

For convenience, abbreviate $\mathcal{S}_2(\Gamma_1(N))$ to $\mathcal{S}_2$ for the duration of this proof. The space $\mathcal{S}_2$ is isomorphic as a complex vector space to its dual space

$$\mathcal{S}_2^\wedge = \mathbb{C}\varphi_1 + \cdots + \mathbb{C}\varphi_{2g}.$$

The dual space is not V but a g-dimensional quotient of V under the map $(z_1\varphi_1, \ldots, z_{2g}\varphi_{2g}) \mapsto \sum z_j \varphi_j$. The map has a g-dimensional kernel because the $\{\varphi_j\}$ are linearly independent over $\mathbb{R}$ but dependent over $\mathbb{C}$. The proof will construct a complementary space $\overline{\mathcal{S}_2^\wedge}$ isomorphic to $\mathcal{S}_2$ such that V is isomorphic to the direct sum $\mathcal{S}_2^\wedge \oplus \overline{\mathcal{S}_2^\wedge}$ as a $\mathbb{T}_\mathbb{Z}$-module and therefore the desired result transfers to the sum, and then the proof will show that the systems of eigenvalues on the sum are the systems of eigenvalues on $\mathcal{S}_2$, transferring the result to $\mathcal{S}_2$ as necessary.

To construct the complementary space, recall the operator $w_N = [[\begin{smallmatrix} 0 & 1 \\ -N & 0 \end{smallmatrix}]]_2$ from Section 5.5, satisfying the relation $w_N T = T^* w_N$ for all $T \in \mathbb{T}_\mathbb{Z}$ where as usual T^* is the adjoint of T. For any cusp form $g \in \mathcal{S}_2$ consider an associated map ψ_g from cusp forms to scalars,

$$\psi_g : \mathcal{S}_2 \longrightarrow \mathbb{C}, \qquad \psi_g(h) = \langle w_N g, h \rangle.$$

Then $\psi_g(h + \tilde{h}) = \psi_g(h) + \psi_g(\tilde{h})$ but since the Petersson inner product is conjugate-linear in its second factor,

$$\psi_g(zh) = \bar{z}\psi_g(h), \quad z \in \mathbb{C}.$$

Thus ψ_g belongs to the set $\overline{\mathcal{S}_2^\wedge}$ of conjugate-linear functions on $\mathcal{S}_2$, the complex conjugates of the dual space $\mathcal{S}_2^\wedge$ (Exercise 6.5.4). This set forms a complex vector space with the obvious operations. It is immediate that $\psi_{g+\tilde{g}} = \psi_g + \psi_{\tilde{g}}$ and $\psi_{zg} = z\psi_g$ for $g, \tilde{g} \in \mathcal{S}_2$ and $z \in \mathbb{C}$, and therefore the map

$$\Psi : \mathcal{S}_2 \longrightarrow \overline{\mathcal{S}_2^\wedge}, \qquad g \mapsto \psi_g$$

is $\mathbb{C}$-linear. It has trivial kernel, making it an isomorphism. (The fact that each vector in $\overline{\mathcal{S}_2^\wedge}$ is itself a conjugate-linear function is a bit of a red herring here—what matters here is that altogether $\overline{\mathcal{S}_2^\wedge}$ forms a complex vector space isomorphic to $\mathcal{S}_2$.) The vector space $\overline{\mathcal{S}_2^\wedge}$ is a $\mathbb{T}_{\mathbb{Z}}$-module with the Hecke operators acting from the right as composition, and the linear isomorphism Ψ is also $\mathbb{T}_{\mathbb{Z}}$-linear since

$$\psi_{Tg}(h) = \langle w_N Tg, h \rangle = \langle T^* w_N g, h \rangle = \langle w_N g, Th \rangle = (\psi_g \circ T)(h).$$

That is, $\mathcal{S}_2$ and $\overline{\mathcal{S}_2^\wedge}$ are isomorphic as complex vector spaces and as $\mathbb{T}_{\mathbb{Z}}$-modules. In particular, every system of eigenvalues $\{\lambda(T) : T \in \mathbb{T}_{\mathbb{Z}}\}$ on $\mathcal{S}_2$ is a system of eigenvalues on $\overline{\mathcal{S}_2^\wedge}$ and conversely.

Also, every system of eigenvalues on $\mathcal{S}_2$ is a system of eigenvalues on the dual space $\mathcal{S}_2^\wedge$ and conversely. To see this, let $f \in \mathcal{S}_2$ be a normalized eigenform. Similarly to before there is an eigenvalue map

$$\lambda_f : \mathbb{T}_{\mathbb{C}} \longrightarrow \mathbb{C}, \qquad Tf = \lambda_f(T)f.$$

(We need the complex Hecke algebra $\mathbb{T}_{\mathbb{C}}$ in this paragraph.) Let $J_f = \ker(\lambda_f) = \{T \in \mathbb{T}_{\mathbb{C}} : Tf = 0\}$, a prime ideal of $\mathbb{T}_{\mathbb{C}}$. An application of Nakayama's Lemma shows that $J_f \mathcal{S}_2 \neq \mathcal{S}_2$ (Exercise 6.5.5), making the quotient $\mathcal{S}_2 / J_f \mathcal{S}_2$ nontrivial. It follows that the subspace of the dual space annihilated by J_f,

$$\mathcal{S}_2^\wedge[J_f] = \{\varphi \in \mathcal{S}_2^\wedge : \varphi \circ T = 0 \text{ for all } T \in J_f\},$$

is nonzero since it is isomorphic to $(\mathcal{S}_2 / J_f \mathcal{S}_2)^\wedge$ by the first isomorphism in (6.13). Since T_1 is the identity operator, $T - \lambda_f(T)T_1 \in J_f$ for any $T \in \mathbb{T}_{\mathbb{C}}$, and so any nonzero $\varphi \in \mathcal{S}_2^\wedge[J_f]$ satisfies

$$\varphi \circ T = \varphi \circ (T - \lambda_f(T)T_1) + \lambda_f(T)\varphi = \lambda_f(T)\varphi, \quad T \in \mathbb{T}_{\mathbb{C}}.$$

Restricting our attention to $\mathbb{T}_\mathbb{Z}$ again, this shows that $\{\lambda_f(T) : T \in \mathbb{T}_\mathbb{Z}\}$ is a system of eigenvalues on $\mathcal{S}_2^\wedge$ as claimed. The converse follows by replacing $\mathcal{S}_2$ and $\mathcal{S}_2^\wedge$ with their duals, since the finite-dimensional vector space $\mathcal{S}_2$ is naturally isomorphic to its double dual as a $\mathbb{T}_\mathbb{Z}$-module. Thus the cusp forms $\mathcal{S}_2$ and the sum $\mathcal{S}_2^\wedge \oplus \overline{\mathcal{S}_2^\wedge}$ have the same systems of eigenvalues.

Consider the $\mathbb{C}$-linear map

$$V \longrightarrow \mathcal{S}_2^\wedge \oplus \overline{\mathcal{S}_2^\wedge}, \qquad (z_1\varphi_1, \ldots, z_{2g}\varphi_{2g}) \mapsto \left(\sum z_j\varphi_j, \sum z_j\bar{\varphi}_j \right).$$

This is also a $\mathbb{T}_\mathbb{Z}$-module map since $\overline{\varphi_j \circ T} = \bar{\varphi}_j \circ T$. The map has trivial kernel since if $\sum z_j\varphi_j = 0$ in $\mathcal{S}_2^\wedge$ and $\sum z_j\bar{\varphi}_j = 0$ in $\overline{\mathcal{S}_2^\wedge}$ then both $\sum z_j\varphi_j = 0$ and $\sum \bar{z}_j\varphi_j = 0$ in $\mathcal{S}_2^\wedge$, i.e., $\sum \mathrm{Re}(z_j)\varphi_j = 0$ and $\sum \mathrm{Im}(z_j)\varphi_j = 0$ in $\mathcal{S}_2^\wedge$; but the $\{\varphi_j\}$ are linearly independent over $\mathbb{R}$, so this implies $z_j = 0$ for all j. Since the domain and codomain have the same dimension the map is a linear isomorphism of $\mathbb{T}_\mathbb{Z}$-modules. The result that if $\{\lambda(T) : T \in \mathbb{T}_\mathbb{Z}\}$ is a system of eigenvalues then so is $\{\lambda(T)^\sigma : T \in \mathbb{T}_\mathbb{Z}\}$ now transfers from V to $\mathcal{S}_2^\wedge \oplus \overline{\mathcal{S}_2^\wedge}$ and then to $\mathcal{S}_2$. Thus if $f(\tau) = \sum a_n q^n$ is a normalized eigenform in $\mathcal{S}_2(N, \chi)$ then its conjugate $f^\sigma(\tau) = \sum a_n^\sigma q^n$ is a normalized eigenform in $\mathcal{S}_2(N, \chi^\sigma)$ as desired.

It remains to prove the last statement of the theorem, that if f is a newform then so is f^σ. By Theorem 5.8.3, f^σ takes the form $f^\sigma(\tau) = \sum_i a_i f_i(n_i\tau)$ where each f_i is a newform at level M_i with $n_i M_i \mid N$. (Note that this uses only the part of that theorem that we have proved, that the set of such f_i spans $\mathcal{S}_2(\Gamma_1(N))$.) Let $\tau = \sigma^{-1} : \mathbb{C} \longrightarrow \mathbb{C}$, an extension of another embedding $\tau : \mathbb{K}_f \hookrightarrow \mathbb{C}$. Then $f = (f^\sigma)^\tau = \sum_i a_i^\tau f_i^\tau(n_i\tau)$. If f^σ is not new then by Exercise 5.8.4 it is old and all M_i are strictly less than N. Since each f_i^τ is also a modular form at level M_i this shows that f is old as well. The result follows by contraposition. $\square$

Linearly combining the normalized eigenforms gives modular forms with coefficients in $\mathbb{Z}$.

Corollary 6.5.6. *The space $\mathcal{S}_2(\Gamma_1(N))$ has a basis of forms with rational integer coefficients.*

Proof. Let f be any newform at level M where $M \mid N$. Let $\mathbb{K} = \mathbb{K}_f$ be the number field of f. Let $\{\alpha_1, \ldots, \alpha_d\}$ be a basis of $\mathcal{O}_\mathbb{K}$ as a $\mathbb{Z}$-module and let $\{\sigma_1, \ldots, \sigma_d\}$ be the embeddings of $\mathbb{K}$ into $\mathbb{C}$. Consider the matrix from the end of the previous section and the vector

$$A = \begin{bmatrix} \alpha_1^{\sigma_1} & \cdots & \alpha_1^{\sigma_d} \\ \vdots & \ddots & \vdots \\ \alpha_d^{\sigma_1} & \cdots & \alpha_d^{\sigma_d} \end{bmatrix}, \qquad \vec{f} = \begin{bmatrix} f^{\sigma_1} \\ \vdots \\ f^{\sigma_d} \end{bmatrix},$$

and let $\vec{g} = A\vec{f}$, i.e.,

$$g_i = \sum_{j=1}^{d} \alpha_i^{\sigma_j} f^{\sigma_j}, \quad i = 1, \ldots, d.$$

Then $\mathrm{span}(\{g_1, \ldots, g_d\}) = \mathrm{span}(\{f^{\sigma_1}, \ldots, f^{\sigma_d}\})$ since A is invertible. Each g_i takes the form $g_i(\tau) = \sum_n a_n(g_i) q^n$ with all $a_n(g_i) \in \overline{\mathbb{Z}}$. For any automorphism $\sigma : \mathbb{C} \longrightarrow \mathbb{C}$, as σ_j runs through the embeddings of $\mathbb{K}_f$ into $\mathbb{C}$ so does $\sigma_j\sigma$ (composing left to right), and so

$$g_i^{\sigma} = \sum_{j=1}^{d} \alpha_i^{\sigma_j\sigma} f^{\sigma_j\sigma} = g_i.$$

That is, each $a_n(g_i)$ is fixed by all automorphisms of $\mathbb{C}$, showing that each $a_n(g_i)$ lies in $\overline{\mathbb{Z}} \cap \mathbb{Q} = \mathbb{Z}$. Repeating this argument for each newform f whose level divides N gives the result. $\square$

Exercises

6.5.1. Let M be a free $\mathbb{Z}$-module of rank r. Show that the ring of endomorphisms of M is a free $\mathbb{Z}$-module of rank r^2, and so any subring is a free $\mathbb{Z}$-module of finite rank.

6.5.2. Let $f \in \mathcal{S}_2(\Gamma_1(N))$ be a normalized eigenform. Thus $f \in \mathcal{S}_2(N, \chi)$ for some Dirichlet character $\chi : (\mathbb{Z}/N\mathbb{Z})^* \longrightarrow \mathbb{C}^*$ and $\lambda_f(\langle d \rangle) = \chi(d)$ for all $d \in (\mathbb{Z}/N\mathbb{Z})^*$. Show that there is a ring and $\mathbb{Z}$-module isomorphism $\mathbb{T}_{\mathbb{Z}}/I_f \xrightarrow{\sim} \mathbb{Z}[\{a_n(f), \chi(d)\}]$. Show that adjoining the $\chi(d)$ values is redundant, making (6.12) in the text correct. (A hint for this exercise is at the end of the book.)

6.5.3. Prove the isomorphisms (6.13). (A hint for this exercise is at the end of the book.)

6.5.4. Let V be any complex vector space with dual space $V^{\wedge}$. Show that the set $\overline{V^{\wedge}} = \{\bar{\varphi} : \varphi \in V^{\wedge}\}$ is the set of functions $\psi : V \longrightarrow \mathbb{C}$ such that $\psi(v + v') = \psi(v) + \psi(v')$ and $\psi(zv) = \bar{z}\psi(v)$ for all $v, v' \in V$ and $z \in \mathbb{C}$.

6.5.5. Let J_f, $\mathbb{T}_{\mathbb{C}}$, and $\mathcal{S}_2$ be as in the proof of Theorem 6.5.4. Show that J_f is a prime ideal. Define the *local ring of* $\mathbb{T}_{\mathbb{C}}$ *at* J_f as a set of equivalence classes of formal elements

$$A = \{T/U : T \in \mathbb{T}_{\mathbb{C}}, U \in \mathbb{T}_{\mathbb{C}} - J_f\}/ \sim$$

where the equivalence relation is

$$T/U \sim T'/U' \quad \text{if} \quad V(U'T - UT') = 0 \text{ for some nonzero } V \in \mathbb{T}_{\mathbb{C}} - J_f.$$

Show that "$\sim$" is indeed an equivalence relation and that consequently the local ring is a ring under the usual rules for manipulating fractions,

$$T/U + T'/U' = (U'T + UT')/(UU'), \qquad (T/U) \cdot (T'/U') = (TT')/(UU'),$$

taken at the level of equivalence classes. Show that the ideal $J = J_f A$ is the unique maximal ideal in the localization. Similarly define the *local module of* $\mathcal{S}_2$ *at* J_f as a set of equivalence classes for formal elements

$$M = \{g/U : g \in \mathcal{S}_2,\, U \in \mathbb{T}_{\mathbb{C}} - J_f\}/ \sim$$

where the equivalence relation is

$$g/U \sim g'/U' \quad \text{if} \quad V(U'g - Ug') = 0 \text{ for some nonzero } V \in \mathbb{T}_{\mathbb{C}} - J_f.$$

Show that this is an equivalence relation and that consequently the local module is an A-module under the rules

$$g/U + g'/U' = (U'g + Ug')/(UU'), \qquad (T/U) \cdot (g/U') = (Tg)/(UU'),$$

again at the level of equivalence classes. Show that $M \neq \{0\}$, so by Nakayama's Lemma $JM \neq M$. Show that therefore $J_f \mathcal{S}_2 \neq \mathcal{S}_2$.

6.6 Eigenforms, Abelian varieties, and Modularity

This section decomposes the Jacobian $\mathrm{J}_1(N)$ into a direct sum of complex tori associated to newforms, and similarly for $\mathrm{J}_0(N)$. The resulting restatement of the Modularity Theorem associates a newform to an elliptic curve.

Studying complex tori requires some preliminary results on tensor products of Abelian groups and fields. For the reader not familiar with this material, the following ad hoc definition of tensor product in this special case is all we need.

Definition 6.6.1. *Given a finitely generated Abelian group G and a field* **k**, *the* **tensor product** $G \otimes \mathbf{k}$ *is the vector space over* **k** *consisting of* **k**-*linear sums of elements $g \otimes k$ where $g \in G$ and $k \in \mathbf{k}$, subject to the relations*

$$g \otimes k + g' \otimes k = (g + g') \otimes k,$$
$$(g \otimes k) + (g \otimes k') = g \otimes (k + k'),$$
$$k(g \otimes k') = g \otimes (kk').$$

The fields of interest in this section are $\mathbb{C}$ and $\mathbb{R}$. For these fields, and for any field of characteristic 0, if g is a group element of finite order n then $g \otimes k = (ng) \otimes (k/n) = 0 \otimes (k/n) = 0$ for any $k \in \mathbf{k}$, and so $\mathbb{Z}g \otimes \mathbf{k} = 0$. If g has infinite order then $\mathbb{Z}g \otimes \mathbf{k}$ is the 1-dimensional vector space spanned by $g \otimes 1$ with the usual field operations carried out in the second component. For a direct sum of cyclic groups the tensor product is the corresponding direct sum of vector spaces,

$$\left(\bigoplus_i G_i \right) \otimes \mathbf{k} = \bigoplus_i (G_i \otimes \mathbf{k}).$$

If G is a subgroup of some vector space over $\mathbf{k}$ then tensoring it with $\mathbf{k}$ is different from taking the $\mathbf{k}$-linear combinations of its elements. The tensor product $G \otimes \mathbf{k}$, where linear independence over $\mathbb{Z}$ is preserved over $\mathbf{k}$, surjects to the vector space $\mathbf{k} \cdot G$, where linearly independent sets over $\mathbb{Z}$ can be dependent over $\mathbf{k}$, the surjection taking elements $\sum g_i \otimes k_i$ to $\sum k_i g_i$. For example, in the proof of Theorem 6.5.4 the vector space V was $H_1(X_1(N), \mathbb{Z}) \otimes \mathbb{C}$ and the proof made use of the surjection from V to $\mathcal{S}_2(\Gamma_1(N))^\wedge = \mathbb{C} \cdot H_1(X_1(N), \mathbb{Z})$. Similarly, $\mathbb{T}_\mathbb{Z} \otimes \mathbb{C}$ surjects to $\mathbb{T}_\mathbb{C}$.

Lemma 6.6.2. *Let G be a finitely generated Abelian group and let $\mathbf{k}$ be a field of characteristic 0. Then*

(a) $G \otimes \mathbf{k} \cong \mathbf{k}^{\mathrm{rank}(G)}$.

(b) $(G/K) \otimes \mathbf{k} \cong (G \otimes \mathbf{k})/(K \otimes \mathbf{k})$ *for any subgroup $K \subset G$.*

(c) *If A is a commutative ring with unit and G is an A-module then $G \otimes \mathbf{k}$ is an A-module under the rule $a(\sum_i g_i \otimes k_i) = \sum_i (ag_i) \otimes k_i$. If J is an ideal of A then $(JG) \otimes \mathbf{k} = J(G \otimes \mathbf{k})$.*

Proof. Exercise 6.6.1. □

Now we can proceed with the material. Let $f \in \mathcal{S}_2(\Gamma_1(M_f))$ be a newform at some level M_f and therefore an eigenform of the Hecke algebra $\mathbb{T}_\mathbb{Z}$. Recall the eigenvalue map

$$\lambda_f : \mathbb{T}_\mathbb{Z} \longrightarrow \mathbb{C}, \qquad Tf = \lambda_f(T)f$$

and its kernel

$$I_f = \ker(\lambda_f) = \{T \in \mathbb{T}_\mathbb{Z} : Tf = 0\}.$$

Since $\mathbb{T}_\mathbb{Z}$ acts on the Jacobian $J_1(M_f)$, the subgroup $I_f J_1(M_f)$ of $J_1(M_f)$ makes sense.

Definition 6.6.3. *The **Abelian variety associated to** f is defined as the quotient*

$$A_f = J_1(M_f)/I_f J_1(M_f).$$

As explained in the previous section, the map $T \mapsto \lambda_f(T)$ induces a $\mathbb{Z}$-module isomorphism from the quotient $\mathbb{T}_\mathbb{Z}/I_f$ to $\mathbb{Z}[\{a_n(f)\}]$ and the latter has rank $[\mathbb{K}_f : \mathbb{Q}]$. The quotient $\mathbb{T}_\mathbb{Z}/I_f$ acts on A_f and hence so does its isomorphic image $\mathbb{Z}[\{a_n(f)\}]$. Since $\lambda_f(T_p) = a_p(f)$ the following diagram commutes:

$$
\begin{array}{ccc}
J_1(M_f) & \xrightarrow{\ T_p\ } & J_1(M_f) \\
\downarrow & & \downarrow \\
A_f & \xrightarrow[\ a_p(f)\ ]{} & A_f.
\end{array}
\qquad (6.15)
$$

The bottom row here is not in general multiplication by $a_p(f)$ but rather $a_p(f)$ acting as T_p at the level of A_f, so for example if $\varphi \in A_f$ and $\sigma : \mathbb{K}_f \hookrightarrow \mathbb{C}$ is an embedding then $(a_p(f)\varphi)(f^\sigma) = a_p(f)^\sigma \varphi(f^\sigma)$ by Theorem 6.5.4. The submodule $\mathbb{Z}$ of $\mathbb{Z}[\{a_n\}]$, being the isomorphic image of the submodule $\mathbb{Z} + I_f$ of $\mathbb{T}_\mathbb{Z}/I_f$, does act as multiplication, so in particular if $a_p(f)$ lies in $\mathbb{Z}$ then indeed it acts as multiplication by itself.

To describe how the Jacobian decomposes into Abelian varieties A_f put an equivalence relation on newforms,

$$\tilde{f} \sim f \iff \tilde{f} = f^\sigma \quad \text{for some automorphism } \sigma : \mathbb{C} \longrightarrow \mathbb{C}.$$

Let $[f]$ denote the equivalence class of f,

$$[f] = \{f^\sigma : \sigma \text{ is an automorphism of } \mathbb{C}\}.$$

Its cardinality is the number of embeddings $\sigma : \mathbb{K}_f \hookrightarrow \mathbb{C}$, and each $f^\sigma \in [f]$ is a newform at level M_f by Theorem 6.5.4. The subspace of $\mathcal{S}_2(\Gamma_1(M_f))$ corresponding to f and its equivalence class is

$$V_f = \text{span}([f]) \subset \mathcal{S}_2(\Gamma_1(M_f)),$$

a subspace of dimension $[\mathbb{K}_f : \mathbb{Q}]$, the number of embeddings. Restricting the subgroup $\text{H}_1(X_1(M_f), \mathbb{Z})$ of $\mathcal{S}_2(\Gamma_1(M_f))^\wedge$ to functions on V_f gives a subgroup of the dual space $V_f^\wedge$,

$$\Lambda_f = \text{H}_1(X_1(M_f), \mathbb{Z})|_{V_f}.$$

Restricting to V_f gives a well-defined homomorphism

$$\text{J}_1(M_f) \longrightarrow V_f^\wedge / \Lambda_f, \qquad [\varphi] \mapsto \varphi|_{V_f} + \Lambda_f \text{ for } \varphi \in \mathcal{S}_2(\Gamma_1(M_f))^\wedge.$$

Proposition 6.6.4. *Let $f \in \mathcal{S}_2(\Gamma_1(M_f))$ be a newform with number field $\mathbb{K}_f$. Then restricting to V_f induces an isomorphism*

$$A_f \xrightarrow{\sim} V_f^\wedge / \Lambda_f, \qquad [\varphi] + I_f \text{J}_1(M_f) \mapsto \varphi|_{V_f} + \Lambda_f \text{ for } \varphi \in \mathcal{S}_2(\Gamma_1(M_f))^\wedge,$$

and the right side is a complex torus of dimension $[\mathbb{K}_f : \mathbb{Q}]$.

Proof. Let $\mathcal{S}_2 = \mathcal{S}_2(\Gamma_1(M_f))$ and let $H_1 = \text{H}_1(X_1(M_f), \mathbb{Z}) \subset \mathcal{S}_2^\wedge$. The Abelian variety is

$$A_f = \text{J}_1(M_f)/I_f \text{J}_1(M_f) = (\mathcal{S}_2^\wedge/H_1)/I_f(\mathcal{S}_2^\wedge/H_1)$$
$$\cong \mathcal{S}_2^\wedge/(I_f \mathcal{S}_2^\wedge + H_1) \cong (\mathcal{S}_2^\wedge/I_f \mathcal{S}_2^\wedge)/(\text{image of } H_1 \text{ in } \mathcal{S}_2^\wedge/I_f \mathcal{S}_2^\wedge).$$

But $\mathcal{S}_2^\wedge/I_f \mathcal{S}_2^\wedge \cong \mathcal{S}_2[I_f]^\wedge$ by the second isomorphism in (6.13), and in fact the isomorphism is the restriction map $\varphi + I_f \mathcal{S}_2^\wedge \mapsto \varphi|_{\mathcal{S}_2[I_f]}$ for $\varphi \in \mathcal{S}_2^\wedge$, so now

$$A_f \xrightarrow{\sim} \mathcal{S}_2[I_f]^\wedge/H_1|_{\mathcal{S}_2[I_f]}, \qquad [\varphi] + I_f \text{J}_1(M_f) \mapsto \varphi|_{\mathcal{S}_2[I_f]} + H_1|_{\mathcal{S}_2[I_f]}.$$

We need to show that $\mathcal{S}_2[I_f] = V_f$, and that $\Lambda_f = H_1|_{V_f}$ is a lattice.

Clearly $V_f \subset \mathcal{S}_2[I_f]$. To show equality, consider the pairing

$$\mathbb{T}_\mathbb{C} \times \mathcal{S}_2 \longrightarrow \mathbb{C}, \qquad (T, g) \mapsto a_1(Tg).$$

This pairing is linear and nondegenerate in each component, a so-called *perfect pairing*. Linearity in each component is clear. To show nondegeneracy in the first component, suppose $T \in \mathbb{T}_\mathbb{C}$ and $a_1(Tg) = 0$ for all $g \in \mathcal{S}_2$. Then also $0 = a_1(TT_ng) = a_1(T_nTg) = a_n(Tg)$ for all $n \in \mathbb{Z}^+$ and $g \in \mathcal{S}_2$, so $Tg = 0$ for all g, and $T = 0$. Nondegeneracy in the second component is similar (Exercise 6.6.2(a)). Thus the map $g \mapsto (T \mapsto a_1(Tg))$ is a vector space isomorphism $\mathcal{S}_2 \cong \mathbb{T}_\mathbb{C}^\wedge$. The map is also a $\mathbb{T}_\mathbb{Z}$-module isomorphism since $(T'T, g)$ and $(T, T'g)$ both pair to $a_1(T'Tg) = a_1(T(T'g))$. Now (Exercise 6.6.2(b))

$$\dim(\mathcal{S}_2[I_f]) = \dim(\mathcal{S}_2[I_f]^\wedge) = \dim(\mathcal{S}_2^\wedge / I_f \mathcal{S}_2^\wedge) = \dim(\mathbb{T}_\mathbb{C}/I_f \mathbb{T}_\mathbb{C}). \quad (6.16)$$

The surjection from $\mathbb{T}_\mathbb{Z} \otimes \mathbb{C}$ to $\mathbb{T}_\mathbb{C}$ takes elements $\sum_i U_i \otimes z_i$ to $\sum_i z_i U_i$. If each $U_i \in I_f$ then since $z_i = z_i T_1 \in \mathbb{T}_\mathbb{C}$ the image lies in $I_f \mathbb{T}_\mathbb{C}$. Thus the surjection descends to a surjection from $(\mathbb{T}_\mathbb{Z} \otimes \mathbb{C})/(I_f \otimes \mathbb{C})$ to $\mathbb{T}_\mathbb{C}/I_f \mathbb{T}_\mathbb{C}$, and so (Exercise 6.6.2(b) again)

$$\begin{aligned}
\dim(\mathcal{S}_2[I_f]) &\leq \dim((\mathbb{T}_\mathbb{Z} \otimes \mathbb{C})/(I_f \otimes \mathbb{C})) = \dim((\mathbb{T}_\mathbb{Z}/I_f) \otimes \mathbb{C}) \\
&= \operatorname{rank}(\mathbb{T}_\mathbb{Z}/I_f) = [\mathbb{K}_f : \mathbb{Q}] = \dim(V_f).
\end{aligned} \quad (6.17)$$

The containment $V_f \subset \mathcal{S}_2[I_f]$ and the relation $\dim(\mathcal{S}_2[I_f]) \leq \dim(V_f)$ combine to show that $\mathcal{S}_2[I_f] = V_f$ as desired.

To show that Λ_f is a lattice in $V_f^\wedge$, we need to establish that the $\mathbb{R}$-span of Λ_f is $V_f^\wedge$ and that $\operatorname{rank}(\Lambda_f) \leq \dim_\mathbb{R}(V_f^\wedge)$. The containment $\mathcal{S}_2 \supset V_f$ of vector spaces over $\mathbb{R}$ makes the restriction map of dual spaces $\pi : \mathcal{S}_2^\wedge \longrightarrow V_f^\wedge$ a surjection. Since the $\mathbb{R}$-span of H_1 is $\mathcal{S}_2^\wedge$, the $\mathbb{R}$-span of $\Lambda_f = \pi(H_1)$ is $V_f^\wedge = \pi(\mathcal{S}_2^\wedge)$ as desired. Furthermore, taking dimensions over $\mathbb{R}$ (Exercise 6.6.2(b) again),

$$\begin{aligned}
\dim(V_f^\wedge) &= \dim(\mathcal{S}_2[I_f]^\wedge) = \dim(\mathcal{S}_2^\wedge / I_f \mathcal{S}_2^\wedge) \\
&= \dim((H_1 \otimes \mathbb{R})/I_f(H_1 \otimes \mathbb{R})) \\
&= \dim((H_1 \otimes \mathbb{R})/(I_f H_1 \otimes \mathbb{R})) \\
&= \dim((H_1/I_f H_1) \otimes \mathbb{R}) = \operatorname{rank}(H_1/I_f H_1).
\end{aligned} \quad (6.18)$$

On the other hand, $\Lambda_f = \pi(H_1) \cong H_1/(H_1 \cap \ker(\pi))$, and I_f acts on H_1, taking it into $\ker(\pi)$ since $V_f = \mathcal{S}_2[I_f]$, so $I_f H_1 \subset H_1 \cap \ker(\pi)$. This shows that there is a surjection $H_1/I_f H_1 \longrightarrow \Lambda_f$, making $\operatorname{rank}(\Lambda_f) \leq \operatorname{rank}(H_1/I_f H_1) = \dim_\mathbb{R}(V_f^\wedge)$, and the proof is complete. $\qquad \square$

The useful equivalence relation between tori in our context is isogeny, as in Section 1.3. Extending the definition to more than one dimension,

Definition 6.6.5. *An **isogeny** is a holomorphic homomorphism between complex tori that surjects and has finite kernel.*

By Proposition 6.1.4, a holomorphic homomorphism between complex tori $\mathbb{C}^g/\Lambda_g$ and $\mathbb{C}^h/\Lambda_h$ takes the form

$$\varphi(z + \Lambda_g) = Mz + \Lambda_h, \qquad M \in M_{h,g}(\mathbb{C}), \; M\Lambda_g \subset \Lambda_h.$$

This is an isogeny when $h = g$ and M is invertible (Exercise 6.6.3(a)), and now the one-dimensional argument in Chapter 1 that isogeny is an equivalence relation extends easily to dimension g (Exercise 6.6.3(b)).

Theorem 6.6.6. *The Jacobian associated to $\Gamma_1(N)$ is isogenous to a direct sum of Abelian varieties associated to equivalence classes of newforms,*

$$J_1(N) \longrightarrow \bigoplus_f A_f^{m_f}.$$

Here the sum is taken over a set of representatives $f \in \mathcal{S}_2(\Gamma_1(M_f))$ at levels M_f dividing N, and each m_f is the number of divisors of N/M_f.

Proof. By Theorem 5.8.3 and Theorem 6.5.4, $\mathcal{S}_2(\Gamma_1(N))$ has basis

$$\mathcal{B}_2(N) = \bigcup_f \bigcup_n \bigcup_\sigma f^\sigma(n\tau)$$

where the first union is taken over equivalence class representatives, the second over divisors of N/M_f, and the third over embeddings of $\mathbb{K}_f$ in $\mathbb{C}$. (This relies on the full strength of Theorem 5.8.3. The part of that theorem that we have proved—that $\mathcal{B}_2(N)$ spans—shows that a subset of $\mathcal{B}_2(N)$ is a basis, and this is enough for the proof here except that the end of Theorem 6.6.6 is weakened to "each m_f is at most the number of divisors of N/M_f.") For each pair (f, n) let $d = [\mathbb{K}_f : \mathbb{Q}]$, let $\sigma_1, \ldots, \sigma_d$ be the embeddings of $\mathbb{K}_f$ in $\mathbb{C}$, and consider the map

$$\Psi_{f,n} : \mathcal{S}_2(\Gamma_1(N))^\wedge \longrightarrow V_f^\wedge$$

taking each $\varphi \in \mathcal{S}_2(\Gamma_1(N))^\wedge$ to $\psi \in V_f^\wedge$ where

$$\psi\Big(\sum_{j=1}^d z_j f^{\sigma_j}(\tau)\Big) = \sum_{j=1}^d z_j n\varphi(f^{\sigma_j}(n\tau)).$$

The map takes $H_1(X_1(N), \mathbb{Z})$ into $\Lambda_f = H_1(X_1(M_f), \mathbb{Z})|_{V_f}$. To see this, let $\varphi = \int_\alpha$ for some loop α in $X_1(N)$. Then correspondingly

$$\psi(f^\sigma(\tau)) = n \int_\alpha f^\sigma(n\tau)d\tau = \int_{\tilde\alpha} f^\sigma(\tau)d\tau \qquad \text{where } \tilde\alpha(t) = n\alpha(t).$$

Here the holomorphic differential $\omega(f^\sigma(n\tau))$ on $X_1(N)$ is identified with its pullback to $\mathcal{H}$ as in Chapter 3 and α is identified with some lift in $\mathcal{H}$. Consequently $\tilde\alpha$ is the lift of a loop in $X_1(M_f)$ (Exercise 6.6.4), and $\Psi_{f,n}$ takes $H_1(X_1(N), \mathbb{Z})$ to Λ_f as claimed.

Taking the product map over all pairs (f, n) gives

$$\Psi = \prod_{f,n} \Psi_{f,n} : S_2(\Gamma_1(N))^\wedge \longrightarrow \bigoplus_{f,n} V_f^\wedge = \bigoplus_f (V_f^\wedge)^{m_f}.$$

This surjects since if $\varphi \in S_2(\Gamma_1(N))^\wedge$ picks off a basis element $f^\sigma(n\tau)$, i.e., $\varphi(f^\sigma(n\tau)) = 1$ and $\varphi = 0$ on the rest of the basis, then its image ψ picks off the basis element $f^\sigma(\tau)$ in the copy of $V_f^\wedge$ corresponding to n. Counting dimensions shows that Ψ is a vector space isomorphism. It descends to an isomorphism of quotients

$$\overline{\Psi} : J_1(N) \xrightarrow{\sim} \bigoplus_f (V_f^\wedge)^{m_f} / \Psi(H_1(X_1(N), \mathbb{Z})).$$

Since $\Psi(H_1(X_1(N), \mathbb{Z})) \subset \bigoplus_f \Lambda_f^{m_f}$ is a containment of Abelian groups of the same rank, the natural surjection

$$\pi : \bigoplus_f (V_f^\wedge)^{m_f} / \Psi(H_1(X_1(N), \mathbb{Z})) \longrightarrow \bigoplus_f (V_f^\wedge / \Lambda_f)^{m_f}$$

has finite kernel. By Proposition 6.6.4 there is an isomorphism

$$i : \bigoplus_f (V_f^\wedge / \Lambda_f)^{m_f} \xrightarrow{\sim} \bigoplus_f A_f^{m_f},$$

and so $i \circ \pi \circ \overline{\Psi} : J_1(N) \longrightarrow \bigoplus_f A_f^{m_f}$ is the desired isogeny. $\qquad\square$

For future reference in Chapter 8 we make some observations in connection with Theorem 6.6.6 and its proof. The following diagram commutes, where p is a prime not dividing N, the vertical arrows are the isogeny of the theorem, and the sum on the bottom row is the sum in the theorem notated slightly differently (Exercise 6.6.5(a)):

$$
\begin{array}{ccc}
J_1(N) & \xrightarrow{\quad T_p \quad} & J_1(N) \\
\downarrow & & \downarrow \\
\bigoplus_{f,n} A_f & \xrightarrow{\quad \prod_{f,n} a_p(f) \quad} & \bigoplus_{f,n} A_f.
\end{array}
\qquad (6.19)
$$

Since isogeny is an equivalence relation there exists an isogeny from $\bigoplus_{f,n} A_f$ back to $J_1(N)$. The following diagram commutes as well, where the vertical arrows denote this isogeny (Exercise 6.6.5(b)):

$$\bigoplus_{f,n} A_f \xrightarrow{\ \Pi_{f,n}\ a_p(f)\ } \bigoplus_{f,n} A_f$$

$$\downarrow \qquad\qquad\qquad\qquad \downarrow$$

$$J_1(N) \xrightarrow{\quad T_p \quad} J_1(N). \tag{6.20}$$

We end this chapter with a version of the Modularity Theorem involving Abelian varieties. Since the group $\Gamma_0(N)$ contains $\Gamma_1(N)$, its associated objects $S_2(\Gamma_0(N))$, $X_0(N)$, and $J_0(N)$ are respectively a subspace, an image, and a quotient of their counterparts for $\Gamma_1(N)$, smaller objects in all cases. Our practice is to develop results in the larger context of $\Gamma_1(N)$ for generality but to state versions of the Modularity Theorem in the restricted context of $\Gamma_0(N)$, making them slightly sharper.

In particular, for each newform $f \in S_2(\Gamma_0(M_f))$ let

$$A'_f = J_0(M_f)/I_f J_0(M_f).$$

This is another Abelian variety associated to f. The proofs of Proposition 6.6.4 and Theorem 6.6.6 transfer to $\Gamma_0(N)$, showing that there is an isomorphism

$$A'_f \xrightarrow{\ \sim\ } V_f^\wedge/\Lambda'_f \quad \text{where } \Lambda'_f = H_1(X_0(M_f), \mathbb{Z})|_{V_f}$$

and an isogeny

$$J_0(N) \longrightarrow \bigoplus_f (A'_f)^{m_f}$$

where now the sum is taken over the equivalence classes of newforms $f \in S_2(\Gamma_0(M_f))$.

In fact A_f and A'_f are isogenous. By Proposition 6.2.4, Λ_f is a finite-index subgroup of Λ'_f, and thus there is a surjection from A_f to A'_f,

$$V_f^\wedge/\Lambda_f \longrightarrow V_f^\wedge/\Lambda'_f, \qquad \varphi + \Lambda_f \mapsto \varphi + \Lambda'_f,$$

whose kernel Λ'_f/Λ_f is finite, showing that there is an isogeny $A_f \longrightarrow A'_f$. The purpose of introducing A'_f is to phrase the following version of the Modularity Theorem entirely in terms of $\Gamma_0(N)$. By incorporating Abelian varieties this version associates a modular form, in fact a newform, to an elliptic curve. The distinction between using A'_f or A_f in the theorem for a newform f is only cosmetic; what matters is that f itself lies in the smaller space associated to $\Gamma_0(N)$ rather than the larger one associated to $\Gamma_1(N)$, making the theorem more specific.

Theorem 6.6.7 (Modularity Theorem, Version $A_\mathbb{C}$). *Let E be a complex elliptic curve with $j(E) \in \mathbb{Q}$. Then for some positive integer N and some newform $f \in S_2(\Gamma_0(N))$ there exists a surjective holomorphic homomorphism of complex tori*

$$A'_f \longrightarrow E.$$

Versions $J_{\mathbb{C}}$ and $A_{\mathbb{C}}$ are equivalent by the isogeny $J_0(N) \longrightarrow \bigoplus_f (A_f')^{m_f}$ since each restriction $A_f' \longrightarrow E$ of a map $\bigoplus_f (A_f')^{m_f} \longrightarrow E$ is trivial or surjects. Note that the integers N appearing in Versions $J_{\mathbb{C}}$ and $A_{\mathbb{C}}$ are not necessarily the same: if Version $J_{\mathbb{C}}$ holds for N then Version $A_{\mathbb{C}}$ holds for some $N' \mid N$, while if Version $A_{\mathbb{C}}$ holds for N then so does Version $J_{\mathbb{C}}$. Later versions of the Modularity Theorem will specify N precisely.

The versions of the Modularity Theorem so far are designated as variants of Type $\mathbb{C}$ because they all involve holomorphic maps of compact Riemann surfaces and complex tori. The versions in the next chapter will move from complex analysis to algebraic geometry and change the field of definition from $\mathbb{C}$ to $\mathbb{Q}$.

Exercises

6.6.1. Prove Lemma 6.6.2.

6.6.2. (a) In the proof of Proposition 6.6.4, prove that the pairing $\mathbb{T}_{\mathbb{C}} \times \mathcal{S}_2 \longrightarrow \mathbb{C}$ is nondegenerate in the second component.
(b) Justify the equalities in (6.16), (6.17), and (6.18).

6.6.3. (a) Show that a map $\varphi(z + \Lambda_g) = Mz + \Lambda_h$ where $M \in \mathrm{M}_{h,g}(\mathbb{C})$ and $M\Lambda_g \subset \Lambda_h$ is an isogeny when $h = g$ and M is invertible.
(b) Show that isogeny is an equivalence relation.

6.6.4. (a) For any $\gamma \in \Gamma_1(N)$ and any positive integer $n \mid N/M_f$, show that $\left[\begin{smallmatrix} n & 0 \\ 0 & 1 \end{smallmatrix}\right] \gamma \left[\begin{smallmatrix} n & 0 \\ 0 & 1 \end{smallmatrix}\right]^{-1} \in \Gamma_1(M_f)$.
(b) Suppose that the path $\alpha : [0,1] \longrightarrow \mathcal{H}$ is the lift of a loop in $X_1(N)$. Show that the path $\tilde{\alpha}(t) = n\alpha(t)$ is the lift of a loop in $X_1(M_f)$.

6.6.5. (a) Show that diagram (6.19) commutes. (Hints for this exercise are at the end of the book.)
(b) Show that diagram (6.20) commutes.

7

Modular Curves as Algebraic Curves

Recall that every complex elliptic curve is described as an algebraic curve, the solution set of a polynomial equation in two variables, via the Weierstrass $\wp$-function,

$$(\wp, \wp') : \mathbb{C}/\Lambda \longrightarrow \{(x,y) : y^2 = 4x^3 - g_2(\Lambda)x - g_3(\Lambda)\} \cup \{\infty\}.$$

Let N be a positive integer. The modular curves

$$X_0(N) = \Gamma_0(N)\backslash\mathcal{H}^*, \qquad X_1(N) = \Gamma_1(N)\backslash\mathcal{H}^*, \qquad X(N) = \Gamma(N)\backslash\mathcal{H}^*$$

can also be described as algebraic curves, solution sets of systems of polynomial equations in many variables. Since modular curves are compact Riemann surfaces, such polynomials with complex coefficients exist by a general theorem of Riemann surface theory, but $X_0(N)$ and $X_1(N)$ are in fact curves over the rational numbers, meaning the polynomials can be taken to have rational coefficients. The Modularity Theorem in its various guises rephrases with the relevant complex analytic objects replaced by their algebraic counterparts and with the relevant complex analytic mappings replaced by rational maps, i.e., maps defined by polynomials with rational coefficients. The methods of this chapter also show that $X(N)$ is defined by polynomials with coefficients in the field $\mathbb{Q}(\mu_N)$ obtained by adjoining the complex Nth roots of unity to the rational numbers.

The process of defining modular curves algebraically is less direct than writing a polynomial explicitly via the Weierstrass $\wp$-function as we can do for elliptic curves; it works with fields of functions on the curves rather than with the curves themselves. According to the moduli space point of view from Section 1.5, a modular curve is a set of equivalence classes of elliptic curves and associated torsion data, and that description will emerge again naturally during this chapter. Thus the moduli space description of modular curves carries over to the algebraic context. The descriptions of the Hecke operators in terms of modular curves and moduli spaces rephrase algebraically as well.

© Springer Science+Business Media New York 2005
F. Diamond, J. Shurman, *A First Course in Modular Forms*,
Graduate Texts in Mathematics 228, DOI 10.1007/978-0-387-27226-9_7

Related reading: The main topics of this chapter are in Chapters 4 and 6 of [Shi73]. Chapter 6 of [Lan73] also overlaps with this chapter's material on modular function fields. The presentation here is drawn from David Rohrlich's article in the proceedings volume from the 1995 Boston University conference on modular forms and Fermat's Last Theorem [CSS97] and from early parts of Joseph Silverman's book on elliptic curves [Sil86]. Other books on elliptic curves are [ST92], [Was03], [Kob93], [Kna93], [Hus04], and [Sil94], the first two especially inviting for beginners, the third emphasizing the connections between elliptic curves and modular forms, and the fourth discussing much of the material here. The notes at the end of [Kna93] give an overview of sources up through 1992 for much of the material in Chapters 7 through 9 of this book. For algebraic curves in general, [Ful69] is an excellent introduction.

7.1 Elliptic curves as algebraic curves

Seeing how the description of a modular curve as an algebraic curve reflects its description as a moduli space—essentially a family of elliptic curves and torsion data—first requires understanding elliptic curves as algebraic curves. Up to now we have worked mostly over the field $\mathbb{C}$, but this chapter will use function fields both over $\mathbb{C}$ and over $\mathbb{Q}$ to discuss algebraic curves, so let $\mathbf{k}$ denote any field of characteristic 0. Fields of positive characteristic will be considered in the next chapter.

Definition 7.1.1. *An element α of an extension field $\mathbb{K}$ of $\mathbf{k}$ is **algebraic over $\mathbf{k}$** if it satisfies some monic polynomial with coefficients in $\mathbf{k}$. Otherwise α is **transcendental over $\mathbf{k}$**. A field extension $\mathbb{K}/\mathbf{k}$ is **algebraic** if every $\alpha \in \mathbb{K}$ is algebraic over $\mathbf{k}$. A field $\mathbf{k}$ is **algebraically closed** if there is no proper algebraic extension $\mathbb{K}/\mathbf{k}$; that is, in any extension field $\mathbb{K}$ of $\mathbf{k}$, any element α that is algebraic over $\mathbf{k}$ in fact lies in $\mathbf{k}$. An **algebraic closure $\overline{\mathbf{k}}$** of $\mathbf{k}$ is a minimal algebraically closed extension field of $\mathbf{k}$.*

Every field has an algebraic closure. Any two algebraic closures $\overline{\mathbf{k}}$ and $\overline{\mathbf{k}}'$ of $\mathbf{k}$ are $\mathbf{k}$-isomorphic, meaning there exists an isomorphism $\overline{\mathbf{k}} \xrightarrow{\sim} \overline{\mathbf{k}}'$ fixing $\mathbf{k}$ pointwise, so we often refer imprecisely to "the" algebraic closure of $\mathbf{k}$. In Chapter 6 the algebraic closure $\overline{\mathbb{Q}}$ was unique because it was set inside the fixed ambient field $\mathbb{C}$.

A *Weierstrass equation over* $\mathbf{k}$ is any cubic equation of the form

$$E : y^2 = 4x^3 - g_2 x - g_3, \quad g_2, g_3 \in \mathbf{k}. \tag{7.1}$$

As in Section 1.1, define the *discriminant* of the equation to be

$$\Delta = g_2^3 - 27g_3^2 \in \mathbf{k},$$

and if $\Delta \neq 0$ define the *invariant* of the equation to be

$$j = 1728g_2^3/\Delta \in \mathbf{k}.$$

Definition 7.1.2. *Let* $\overline{\mathbf{k}}$ *be an algebraic closure of the field* **k**. *When a Weierstrass equation E has nonzero discriminant* Δ *it is called* **nonsingular** *and the set*

$$\mathcal{E} = \{(x, y) \in \overline{\mathbf{k}}^2 \text{ satisfying } E(x, y)\} \cup \{\infty\}$$

is called an **elliptic curve over k**.

Thus an elliptic curve always contains the point ∞. As the solution set of a polynomial equation in two variables, an elliptic curve as defined here is a special case of a plane algebraic curve. The term "algebraic curve" is being used casually so far, but it will be defined in the next section after some of the ideas are first demonstrated here in the specific context of elliptic curves.

Strictly speaking we are interested in equivalence classes of elliptic curves under *admissible changes of variable*,

$$x = u^2 x', \quad y = u^3 y', \quad u \in \mathbf{k}^*.$$

These transform Weierstrass equations to Weierstrass equations, taking g_2 to g_2/u^4, g_3 to g_3/u^6, the discriminant Δ to Δ/u^{12} (again nonzero) and preserving the invariant j (Exercise 7.1.1). Admissible changes of variable are special cases of isomorphisms between algebraic curves, a topic to be discussed more generally in the next section.

For example, consider an elliptic curve $\mathcal{E}$ over $\mathbb{C}$ whose invariant is rational. Supposing g_2 and g_3 are nonzero, the condition $j \in \mathbb{Q}$ is $g_2^3 = r g_3^2$ for some nonzero $r \in \mathbb{Q}$ (Exercise 7.1.2(a)). Let $u \in \mathbb{C}$ satisfy $g_2/u^4 = r$; then also $r^3 u^{12} = g_2^3 = r g_3^2$ and thus $g_3/u^6 = \pm r$, and we may take $g_3/u^6 = r$ after replacing u by iu if necessary. Since the admissible change of variable $x = u^2 x'$, $y = u^3 y'$ produces new Weierstrass coefficients $g_2' = g_2/u^4$ and $g_3' = g_3/u^6$, this shows that up to isomorphism $\mathcal{E}$ is defined over $\mathbb{Q}$ by a Weierstrass equation

$$y^2 = 4x^3 - gx - g, \quad g \in \mathbb{Q}.$$

A separate argument when one of g_2, g_3 is zero (Exercise 7.1.2(b)) shows that in all cases an elliptic curve over $\mathbb{C}$ has rational invariant $j \in \mathbb{Q}$ if and only if it is isomorphic over $\mathbb{C}$ to an elliptic curve over $\mathbb{Q}$. This argument works with $\mathbb{C}$ replaced by any algebraically closed field **k** of characteristic 0 and with $\mathbb{Q}$ replaced by any subfield **f** of **k**. It shows that any two elliptic curves over **k** with the same invariant j are isomorphic when **k** is algebraically closed.

Associated to each Weierstrass equation and also denoted E is a corresponding *Weierstrass polynomial*,

$$E(x, y) = y^2 - 4x^3 + g_2 x + g_3 \in \mathbf{k}[x, y].$$

The Weierstrass equation (7.1) is nonsingular, meaning as in Definition 7.1.2 that its discriminant Δ is nonzero, if and only if the corresponding curve $\mathcal{E}$ is *geometrically nonsingular*, meaning that at each point $(x, y) \in \mathcal{E}$ at least one of the partial derivatives $D_1 E(x, y)$, $D_2 E(x, y)$ of the Weierstrass polynomial is nonzero. To see this, note that the Weierstrass polynomial takes the form

$$E(x,y) = y^2 - 4(x - x_1)(x - x_2)(x - x_3), \quad x_1, x_2, x_3 \in \overline{\mathbf{k}}.$$

The third condition of

$$E(x,y) = 0, \qquad D_1 E(x,y) = 0, \qquad D_2 E(x,y) = 0$$

is $y = 0$ since char$(\mathbf{k}) = 0$. Now the first condition is $x \in \{x_1, x_2, x_3\}$, making the second condition impossible exactly when x_1, x_2, x_3 are distinct, i.e., when $\Delta \neq 0$. One can think of geometric nonsingularity as meaning that $\mathcal{E}$ has a tangent line at each point.

Many good texts on elliptic curves exist, cf. the references at the beginning of the chapter, so we state without proof the facts that we need. Most importantly,

> *every elliptic curve forms an Abelian group with the point* ∞ *as its additive identity.*

The point ∞ is therefore denoted $0_{\mathcal{E}}$ from now on. One can think of addition on $\mathcal{E}$ geometrically, algebraically, and analytically.

Geometrically the elliptic curve really sits in the *projective plane over* $\overline{\mathbf{k}}$, denoted $\mathbb{P}^2(\overline{\mathbf{k}})$, the usual plane $\overline{\mathbf{k}}^2$ (called the *affine* plane) along with some additional points conceptually out at infinity added to complete it. (The reader who is unfamiliar with this construction should work Exercise 7.1.3 for details of the following assertions.) The projective plane is the union of three overlapping affine planes and the elliptic curve is correspondingly the union of three affine pieces. The projective point ∞ of $\mathcal{E}$ is infinitely far in the y-direction, i.e., $0_{\mathcal{E}} = [0, 1, 0]$ in projective notation. The curve can be studied about $0_{\mathcal{E}}$ by working in a different affine piece of $\mathbb{P}^2(\overline{\mathbf{k}})$, and it is geometrically nonsingular at $0_{\mathcal{E}}$ as it is at its finite points. Projective space is a natural construct, e.g., the Riemann sphere $\mathbb{C} \cup \{\infty\}$ is $\mathbb{P}^1(\mathbb{C})$, and similarly the set $\{0, \ldots, p-1, \infty\}$ that often serves as an index set in the context of the Hecke operator T_p can be viewed as $\mathbb{P}^1(\mathbb{Z}/p\mathbb{Z})$.

Bézout's Theorem says that if C_1 and C_2 are plane curves over $\mathbf{k}$ whose defining polynomials have degrees d_1 and d_2 and are relatively prime in $\overline{\mathbf{k}}[x, y]$ (a unique factorization domain) then their intersection in $\mathbb{P}^2(\overline{\mathbf{k}})$ consists of $d_1 d_2$ points, suitably counting multiplicity. Thus cubic curves, and only cubic curves, naturally produce triples P, Q, R of collinear points, meaning projective points satisfying an equation $ax + by + cz = 0$ where $a, b, c \in \mathbf{k}$ are not all zero. If the line is tangent to the curve then two or three of P, Q, and R will coincide. The addition law on elliptic curves $\mathcal{E}$ is that *collinear triples sum to* $0_{\mathcal{E}}$. That is,

$$P + Q + R = 0_{\mathcal{E}} \iff P, Q, R \text{ are collinear.}$$

We saw this idea back in Section 1.4. In particular, since $P - P + 0_{\mathcal{E}} = 0_{\mathcal{E}}$ it follows that P, $-P$, and $0_{\mathcal{E}}$ are collinear. So far this only requires $0_{\mathcal{E}}$ to be any point of $\mathcal{E}$, but the condition $0_{\mathcal{E}} = [0, 1, 0]$ gives the addition law a pleasing geometry. Since $0_{\mathcal{E}}$ is infinitely far in the vertical direction, P and $-P$ have the same x-coordinate. At most two points with the same x-coordinate satisfy the Weierstrass equation (7.1), so any two points with the same x-coordinate are equal or opposite, possibly both. Since the y-values satisfying (7.1) for a given x sum to 0 the additive inverse of $P = (x_P, y_P)$ is the natural companion point

$$-P = (x_P, -y_P).$$

As remarked, this could well be P again. More generally, given points P and Q of $\mathcal{E}$, let R be their third collinear point. The addition law $P + Q = -R$ says that the sum is the companion point of R, its reflection through the x-axis,

$$\textit{if } P, Q, R \textit{ are collinear then } P + Q = (x_R, -y_R). \tag{7.2}$$

Figure 7.1 illustrates this for the elliptic curve $y^2 = 4x^3 - 4x$.

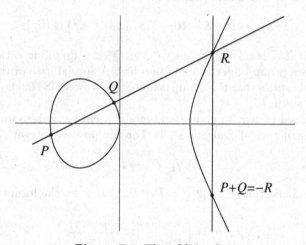

Figure 7.1. The addition law

Moving from geometry to algebra, the group law is defined by rational functions over the field $\mathbf{k}$. Let $P = (x_P, y_P)$ and $Q = (x_Q, y_Q)$ be nonzero points of the curve, and suppose their sum $P + Q = (x_{P+Q}, y_{P+Q})$ is nonzero as well. Then

$$x_{P+Q} = r(x_P, y_P, x_Q, y_Q) \quad \text{and} \quad y_{P+Q} = s(x_P, y_P, x_Q, y_Q)$$

where r and s are rational functions with coefficients in $\mathbf{k}$. Even more specifically since char$(\mathbf{k}) = 0$, r and s are rational functions over the field $\mathbb{Q}(g_2, g_3)$. If $x_Q = x_P$ and $y_Q = -y_P$ then $Q = -P$ and so $P + Q = 0_{\mathcal{E}}$. Otherwise $P + Q$ lies in the affine part of $\mathcal{E}$. Let

$$\lambda = \begin{cases} \dfrac{y_Q - y_P}{x_Q - x_P} & x_P \neq x_Q, \\[2mm] \dfrac{12x_P^2 - g_2}{2y_P} & x_P = x_Q, \end{cases} \qquad \mu = \begin{cases} \dfrac{x_Q y_P - x_P y_Q}{x_Q - x_P} & x_P \neq x_Q, \\[2mm] \dfrac{-4x_P^3 - g_2 x_P - 2g_3}{2y_P} & x_P = x_Q. \end{cases}$$

The line $y = \lambda x + \mu$ passes through P and Q when $P \neq Q$ and is the tangent line to $\mathcal{E}$ at P when $P = Q$ (Exercise 7.1.4(a)). The casewise nature of λ and μ will be discussed further in the next section. In either case, the rational functions giving x_{P+Q} and y_{P+Q} are

$$r(x_P, y_P, x_Q, y_Q) = \lambda^2/4 - x_P - x_Q,$$
$$s(x_P, y_P, x_Q, y_Q) = -\lambda\, r(x_P, y_P, x_Q, y_Q) - \mu. \tag{7.3}$$

This algebraic definition of addition corresponds to the geometric description (7.2) (Exercise 7.1.4(b)). For example, on the curve $y^2 = 4x^3 - 4x$ of Figure 7.1, $(0,0) + (2, 2\sqrt{6}) = (-1/2, \sqrt{6}/2)$.

Because the coordinates of $0_{\mathcal{E}} = [0, 1, 0]$ lie in $\mathbf{k}$ and because the group law is rational over $\mathbf{k}$, for any algebraic extension $\mathbb{K}/\mathbf{k}$ (i.e., $\mathbf{k} \subset \mathbb{K} \subset \overline{\mathbf{k}}$) the set of $\mathbb{K}$-*points of* $\mathcal{E}$ is a subgroup of $\mathcal{E}$,

$$\mathcal{E}(\mathbb{K}) = \{P \in \mathcal{E} - \{0_{\mathcal{E}}\} : (x_P, y_P) \in \mathbb{K}^2\} \cup \{0_{\mathcal{E}}\}.$$

In particular if $\mathcal{E}$ is an elliptic curve over $\mathbb{Q}$ then its affine rational points and $0_{\mathcal{E}}$ form a group under the addition law. A special case of the *Mordell–Weil Theorem* states that this group is finitely generated. [ST92] is an excellent text on this subject.

As explained earlier, to understand modular curves we also want to study the torsion structure of $\mathcal{E}$ algebraically. For any positive integer N let

$$[N] : \mathcal{E} \longrightarrow \mathcal{E}$$

denote N-fold addition, e.g., $[2]P = P + P$. This takes the form of a rational function,

$$[N](x, y) = \left(\frac{\phi_N(x, y)}{\psi_N(x, y)^2}, \frac{\omega_N(x, y)}{\psi_N(x, y)^3} \right).$$

Here the Nth *division polynomials* ϕ_N, ω_N, ψ_N lie in $\mathbb{Z}[g_2, g_3, x, y]$, and (∞, ∞) is understood to mean $0_{\mathcal{E}}$. Thus the condition $[N](x, y) = 0_{\mathcal{E}}$ is $\psi_N(x, y) = 0$. This works out to a polynomial condition on x alone,

$$[N](x, y) = 0_{\mathcal{E}} \iff \tilde{\psi}_N(x) = 0, \quad \tilde{\psi}_N \in \mathbb{Z}[g_2, g_3, x].$$

For instance, the condition $[2]P = 0$ for a nonzero point P is $y_P = 0$, or $4x_P^3 - g_2 x_P - g_3 = 0$. Similarly, the condition that $[3]P = 0$ for nonzero P is $[2]P = -P$, or $x([2]P) = x_P$, or $r(x_P, y_P, x_P, y_P) = x_P$, and this works out to $48x_P^4 - 24g_2 x_P^2 - 48g_3 x_P - g_2^2 = 0$ (Exercise 7.1.5(a)). The polynomial $\tilde{\psi}_N$ will be used in Section 7.5.

Let $\mathcal{E}[N]$ denote the group of N-*torsion points of* $\mathcal{E}$, the kernel of $[N]$,

$$\mathcal{E}[N] = \{P \in \mathcal{E} : [N]P = 0_{\mathcal{E}}\}.$$

We quote the structure theorem for $\mathcal{E}[N]$.

Theorem 7.1.3. *Let $\mathcal{E}$ be an elliptic curve over a field $\mathbf{k}$ of characteristic 0 and let N be a positive integer. Then $\mathcal{E}[N] \cong (\mathbb{Z}/N\mathbb{Z})^2$.*

Let $\mathbb{K}$ be a Galois extension field of $\mathbf{k}$ containing the x- and y-coordinates of $\mathcal{E}[N] - \{0_{\mathcal{E}}\}$. The relations $E(x^\sigma, y^\sigma) = E(x, y)^\sigma$ and $\tilde{\psi}_N(x^\sigma) = \tilde{\psi}_N(x)^\sigma$ for any $x, y \in \mathbb{K}$ and any automorphism $\sigma \in \mathrm{Gal}(\mathbb{K}/\mathbf{k})$ show that the Galois group acts on $\mathcal{E}[N]$. The action is an automorphism since the coefficients of the rational functions r and s lie in $\mathbb{Q}(g_2, g_3)$ and hence in $\mathbf{k}$. That is, $r(x_P^\sigma, y_P^\sigma, x_Q^\sigma, y_Q^\sigma) = r(x_P, y_P, x_Q, y_Q)^\sigma$ for all $P, Q \in \mathcal{E}[N]$, and similarly for s, making $P^\sigma + Q^\sigma = (P + Q)^\sigma$. Since $\mathcal{E}[N] \cong (\mathbb{Z}/N\mathbb{Z})^2$, once an ordered basis (P, Q) of $\mathcal{E}[N]$ over $\mathbb{Z}/N\mathbb{Z}$ is chosen this gives a representation

$$\rho : \mathrm{Gal}(\mathbb{K}/\mathbf{k}) \longrightarrow \mathrm{GL}_2(\mathbb{Z}/N\mathbb{Z}), \qquad \begin{bmatrix} P^\sigma \\ Q^\sigma \end{bmatrix} = \rho(\sigma) \begin{bmatrix} P \\ Q \end{bmatrix},$$

where as always $\mathrm{GL}_2(\mathbb{Z}/N\mathbb{Z})$ is the group of invertible 2-by-2 matrices with entries in $\mathbb{Z}/N\mathbb{Z}$. We will return to this idea later in the chapter and again more extensively in Chapter 9.

Analytically, when $\mathbf{k} \subset \mathbb{C}$ we may view the coefficients of the Weierstrass equation as complex numbers. As shown in Proposition 1.4.3, there exists a lattice $\Lambda \subset \mathbb{C}$ such that the Weierstrass equation (7.1) takes the form $y^2 = 4x^3 - g_2(\Lambda)x - g_3(\Lambda)$, and addition on the corresponding curve $\mathcal{E}$ is compatible with the natural addition on $\mathbb{C}/\Lambda$ via the Weierstrass function $\wp$. In particular, Theorem 7.1.3 here is clear and familiar when $\mathbf{k} = \mathbb{C}$. Also as shown in Section 1.4, holomorphic isomorphisms of complex tori correspond to admissible changes of variable in complex Weierstrass equations.

Exercises

7.1.1. Show that every admissible change of variable $x = u^2 x'$, $y = u^3 y'$ where $u \in \mathbf{k}^*$ transforms a Weierstrass equation E into another Weierstrass equation E' with

$$u^4 g_2' = g_2, \qquad u^6 g_3' = g_3, \qquad u^{12} \Delta' = \Delta, \qquad j' = j.$$

7.1.2. (a) Confirm that for a Weierstrass equation (7.1) with g_2 and g_3 nonzero the condition $j \in \mathbb{Q}$ is equivalent to the condition $g_2^3 = rg_3^2$ for some nonzero $r \in \mathbb{Q}$.

(b) Show that an elliptic curve over $\mathbb{C}$ with either of g_2, g_3 zero is isomorphic to an elliptic curve over $\mathbb{Q}$.

7.1.3. For any positive integer n and any field $\mathbb{K}$, n-*dimensional projective space over* $\mathbb{K}$ is the set of equivalence classes of nonzero $(n+1)$-tuples modulo scalar multiplication,

$$\mathbb{P}^n(\mathbb{K}) = (\mathbb{K}^{n+1} - \{\mathbf{0}\})/\sim$$

where $v \sim v'$ if $v' = cv$ for some nonzero $c \in \mathbb{K}$. Let $[v] \in \mathbb{P}^n(\mathbb{K})$ denote the equivalence class of the vector v.

(a) When $n = 1$ this construction gives the *projective line*. Show that

$$\mathbb{P}^1(\mathbb{K}) = \{[x, 1] : x \in \mathbb{K}\} \cup \{[1, y] : y \in \mathbb{K}\},$$

so the projective line is an overlapping union of two affine lines. Show also that

$$\mathbb{P}^1(\mathbb{K}) = \{[x, 1] : x \in \mathbb{K}\} \cup \{[1, 0]\},$$

so the projective line is a disjoint union of the line and a point at infinity. In particular, $\mathbb{P}^1(\mathbb{C})$ is the Riemann sphere $\widehat{\mathbb{C}}$.

(b) When $n = 2$ the construction gives the projective plane mentioned in the section. Show that

$$\mathbb{P}^2(\mathbb{K}) = \{[x, y, 1] : x, y \in \mathbb{K}\} \cup \{[x, 1, z] : x, z \in \mathbb{K}\} \cup \{[1, y, z] : y, z \in \mathbb{K}\},$$

so the projective plane is a union of three affine planes. Show also that

$$\mathbb{P}^2(\mathbb{K}) = \{[x, y, 1] : x, y \in \mathbb{K}\} \cup \{[x, 1, 0] : x \in \mathbb{K}\} \cup \{[1, 0, 0]\}$$
$$= \{[x, y, 1] : x, y \in \mathbb{K}\} \cup (\mathbb{P}^1(\mathbb{K}) \times \{0\}),$$

so the projective plane is a disjoint union of the plane and a projective line at infinity.

(c) *Homogenize* the Weierstrass polynomial by adding in powers of z to make each term cubic,

$$E_{\mathrm{hom}}(x, y, z) = y^2 z - 4x^3 + g_2 x z^2 + g_3 z^3.$$

Show that either all points or no points in each equivalence class $[x, y, z] \in \mathbb{P}^2(\overline{\mathbb{k}})$ satisfy E_{hom}. Show that $[0, 1, 0]$ satisfies E_{hom} and no other $[x, y, 0]$ does.

(d) Dehomogenize E_{hom} by setting $y = 1$ to obtain a second affine version of E,

$$E'(x, z) = z - 4x^3 + g_2 x z^2 + g_3 z^3.$$

In the (x, z) coordinate system, the infinite point $0_{\mathcal{E}}$ is $(0, 0)$. Working in this affine coordinate system, show that $\mathcal{E}$ is geometrically nonsingular at $0_{\mathcal{E}}$. Is the set of points $(x, z) \in \overline{\mathbb{k}}^2$ satisfying $E'(x, z)$ all of $\mathcal{E}$?

(e) Let $P = (x_P, y_P)$ be any (x, y)-point of $\mathcal{E}$, and homogenize it to $P = [x_P, y_P, 1]$. Show that the homogeneous equation $x - x_P z = 0$ is satisfied by P and by $0_{\mathcal{E}}$, thus defining the projective line containing them. Dehomogenize back to an (x, y)-equation to obtain a vertical line. Thus $0_{\mathcal{E}}$ is infinitely far in the vertical direction.

7.1.4. (a) Show that the casewise definitions of λ and μ make the line $y = \lambda x + \mu$ the secant line through P and Q when $P \neq Q$ and the tangent line to $\mathcal{E}$ through P when $P = Q$.

(b) Show that the algebraic definition (7.3) of elliptic curve addition corresponds to the geometric description (7.2). (A hint for this exercise is at the end of the book.)

(c) The points $(2, \pm 5)$ satisfy the equation $y^2 = 4x^3 - 7$. Find another point $(x, y) \in \mathbb{Q}^2$ that does so as well.

7.1.5. (a) Confirm that the condition $x([2]P) = x_P$ works out to the polynomial condition given in the section.

(b) Compute $\mathcal{E}[2]$ and $\mathcal{E}[3]$ for the elliptic curve with Weierstrass equation $y^2 = 4x^3 - 4x$.

7.2 Algebraic curves and their function fields

Describing modular curves as algebraic curves without writing specific polynomial equations requires some general theory. The idea is that curves are characterized by fields, and so to construct curves it suffices to construct appropriate fields instead. Similarly, mappings between curves are characterized by field extensions, and so to construct mappings it suffices to construct appropriate extensions. This section sketches some background and then states the theorems.

Let $\mathbf{k}$ be any field of characteristic 0. Let m, n be positive integers. Consider a set of m polynomials over $\mathbf{k}$ in n variables, $\varphi_1, \ldots, \varphi_m \in \mathbf{k}[x_1, \ldots, x_n]$. Let I be the ideal generated by the φ_i in the ring of polynomials over the algebraic closure of $\mathbf{k}$,

$$I = \langle \varphi_1, \ldots, \varphi_m \rangle \subset \overline{\mathbf{k}}[x_1, \ldots, x_n],$$

and let C be the set of simultaneous solutions of the polynomials, or equivalently the simultaneous solutions of all polynomials in the ideal,

$$C = \{P \in \overline{\mathbf{k}}^n : \varphi(P) = 0 \text{ for all } \varphi \in I\}.$$

Suppose that I is a prime ideal of $\overline{\mathbf{k}}[x_1, \ldots, x_n]$. The *coordinate ring of C over $\overline{\mathbf{k}}$* is the integral domain

$$\overline{\mathbf{k}}[C] = \overline{\mathbf{k}}[x_1, \ldots, x_n]/I.$$

An element of the coordinate ring is called a *polynomial function on C*. The *function field of C over $\overline{\mathbf{k}}$* is the quotient field of the coordinate ring,

$$\overline{\mathbf{k}}(C) = \{F = f/g : f, g \in \overline{\mathbf{k}}[C], \ g \neq 0\}.$$

The function field really consists of equivalence classes F of ordered pairs $(f, g) \in \overline{\mathbf{k}}[C]^2$ where the equivalence is determined by the usual rule for fractions, $(f, g) \sim (f', g')$ if $fg' = f'g$ (primes do not denote differentiation here or elsewhere in the chapter), but the notation $F = f/g$ is less cumbersome. An element of the function field is called a *rational function on C*.

Definition 7.2.1. *If* $\overline{\mathbf{k}}(C)$ *is a finite extension of a field* $\overline{\mathbf{k}}(t)$ *where* t *is transcendental over* $\overline{\mathbf{k}}$ *then* C *is an* ***affine algebraic curve over* k**. *If also for each point* $P \in C$ *the* m-*by*-n *derivative matrix* $[D_j\varphi_i(P)]$ *has rank* $n-1$ *then the curve* C *is* ***nonsingular***.

The definition shows that if C is an affine algebraic curve over $\mathbf{k}$ then $m \geq n-1$, i.e., the number of constraints on the variables is at least the geometrically obvious minimal number for a curve. After we introduce more ideas it will be an exercise to show that the nonsingularity condition is a maximality condition on the rank of the derivative matrix, i.e., the rank cannot be n.

Homogenize the polynomials φ_i with another variable x_0, consider the ideal

$$I_{\mathrm{hom}} = \langle \varphi_{i,\mathrm{hom}} \rangle \subset \overline{\mathbf{k}}[x_0, \dots, x_n]$$

generated by the homogeneous polynomials but also containing inhomogeneous ones (unless $I_{\mathrm{hom}} = \{0\}$), and consider the set

$$C_{\mathrm{hom}} = \left\{ P \in \mathbb{P}^n(\overline{\mathbf{k}}) : \varphi(P) = 0 \text{ for all homogeneous } \varphi \in I_{\mathrm{hom}} \right\}.$$

This is the projective version of the curve, a superset of C, the union of the affine curves obtained by dehomogenizing the $\varphi_{i,\mathrm{hom}}$ at each coordinate, cf. Exercise 7.1.3(d). The definition of nonsingularity extends to C_{hom} by dehomogenizing to appropriate coordinate systems at the points outside of $\overline{\mathbf{k}}^n$. In this chapter we are interested in nonsingular projective algebraic curves, but for convenience we usually work in the $(x_1, \dots, x_n)$ affine coordinate system, illustrating some projective ideas in examples and exercises.

Setting $m = 0$ and $n = 1$ shows that the projective line $\mathbb{P}^1(\overline{\mathbf{k}})$ is a nonsingular algebraic curve over $\mathbf{k}$ with function field $\overline{\mathbf{k}}(x)$. When $m = 1$ and $n = 2$ and the single polynomial defining C is a Weierstrass polynomial $E(x,y)$ then C is an elliptic curve $\mathcal{E}$ as in Section 7.1. The function field is $\overline{\mathbf{k}}(x,y)$ where x and y are related by a Weierstrass equation, or $\overline{\mathbf{k}}(x)[y]/\langle E(x,y) \rangle$, showing that it is quadratic over $\overline{\mathbf{k}}(x)$. Thus elliptic curves over $\mathbf{k}$ are indeed nonsingular projective algebraic curves over $\mathbf{k}$ as in Definition 7.2.1 and the preceding paragraph. The projective line can also be placed in the $m = 1$, $n = 2$ context by taking $\varphi(x,y) = y$. At the end of the section we will see why the definition of algebraic curve has not been limited to $m = 1$ and $n = 2$, i.e., to plane curves.

Let C be a nonsingular algebraic curve over $\mathbf{k}$. Since C is defined by the condition $\varphi(P) = 0$ for all $\varphi \in I$, an element $f + I$ of the coordinate ring takes a well defined value in $\overline{\mathbf{k}}$ at each point $P \in C$ even though its representative f is not unique. That is, a polynomial function on C, a formal algebraic object, also defines a mapping from the set C to $\overline{\mathbf{k}}$. *A priori* it is not clear that an element $F = f/g$ of the function field takes a well defined value in $\overline{\mathbf{k}} \cup \{\infty\}$ at each $P \in C$ since $f(P)/g(P)$ might evaluate to $0/0$, but we will see that it does and therefore a rational function $F \in \overline{\mathbf{k}}(C)$ defines a mapping from C to $\mathbb{P}^1(\overline{\mathbf{k}})$. In fact it defines a mapping on all of C_{hom}.

Let $P = (p_1, \ldots, p_n)$ be a point of C. To study rational functions on C at P start with the elements of the coordinate ring that vanish at P,

$$m_P = \{f \in \overline{k}[C] : f(P) = 0\} = \langle\{x_i - p_i + I : 1 \le i \le n\}\rangle.$$

This is a maximal ideal of $\overline{k}[C]$ (Exercise 7.2.2), and its second power is

$$m_P^2 = \langle\{(x_i - p_i)(x_j - p_j) + I : 1 \le i, j \le n\}\rangle.$$

Lemma 7.2.2. *Let C be a nonsingular algebraic curve. For any point $P \in C$, the quotient m_P/m_P^2 is a 1-dimensional vector space over $\overline{k}$.*

Proof. The *tangent line* to C at P is

$$T_P(C) = \{v \in \overline{k}^n : [D_j \varphi_i(P)]v = 0\},$$

a 1-dimensional vector space over $\overline{k}$ by the Rank-Nullity Theorem of linear algebra since C is nonsingular.

Although a function $f + I \in \overline{k}[C]$ has a well defined value at P, its form as a coset makes its gradient (vector of partial derivatives) at P ill defined. But changing coset representatives changes the gradient to

$$\nabla(f + \varphi)(P) = \nabla f(P) + \nabla\varphi(P), \quad \varphi \in I,$$

and since $\nabla\varphi(P)$ is in the row space of $[D_j\varphi_i(P)]$ (Exercise 7.2.3(a)) the definition of the tangent line shows that the inner product of the gradient $\nabla f(P)$ with any $v \in T_P(C)$ is well defined. Translating f by a constant has no effect on the inner product, and any function $f \in m_P^2$ satisfies $\nabla f(P) \cdot v = 0$. Thus there is a well defined pairing

$$m_P/m_P^2 \times T_P(C) \longrightarrow \overline{k}, \quad (f, v) \mapsto \nabla f(P) \cdot v.$$

This pairing is perfect, meaning that it is linear and nondegenerate in each component, cf. the proof of Proposition 6.6.4 (Exercise 7.2.3(b)). It follows that m_P/m_P^2 and $T_p(C)$ have the same dimension as vector spaces over $\overline{k}$ since each can be viewed as the dual space of the other. Since the tangent line $T_p(C)$ has dimension 1, so does m_P/m_P^2. $\qquad\square$

For example, let $\mathcal{E}$ be an elliptic curve defined by a Weierstrass polynomial E. The proof of Lemma 7.2.2 shows that $x - x_P + m_P^2$ spans m_P/m_P^2 if $D_2 E(P) \ne 0$, meaning the tangent line at P is not vertical, and $y - y_P + m_P^2$ spans m_P/m_P^2 if $D_1 E(P) \ne 0$, meaning the tangent line is not horizontal (Exercise 7.2.3(c)).

The *local ring of C over $\overline{k}$ at P* is

$$\overline{k}[C]_P = \{f/g \in \overline{k}(C) : g(P) \ne 0\},$$

a subring of the function field. The *local maximal ideal at P* is the ideal of functions that vanish at the point,

$$M_P = m_P \overline{\mathbf{k}}[C]_P = \{f/g \in \overline{\mathbf{k}}[C]_P : f(P) = 0\}.$$

This is a maximal ideal of $\overline{\mathbf{k}}[C]_P$, and since it consists of all the noninvertible elements it is the *unique* maximal ideal (this is not true for m_P and $\overline{\mathbf{k}}[C]$). Consider the natural map from the maximal ideal to the quotient of its localization,

$$i : m_P \longrightarrow M_P/M_P^2, \qquad f \mapsto f + M_P^2.$$

The kernel is $m_P \cap M_P^2 = m_P^2$ (Exercise 7.2.3(d)). The map surjects since for any $f/g \in M_P$ the subtraction $f/g(P) - f/g = f(g - g(P))/(g(P)g) \in M_P^2$ shows that $i(f/g(P)) = f/g + M_P^2$. Thus i induces an isomorphism

$$i : m_P/m_P^2 \xrightarrow{\sim} M_P/M_P^2.$$

Proposition 7.2.3. *For any nonsingular point $P \in C$ the ideal M_P is principal.*

Proof. By the lemma and the isomorphism i, some function $t \in \overline{\mathbf{k}}[C]$ generates M_P/M_P^2 as a vector space over $\overline{\mathbf{k}}$. This function will also generate M_P as an ideal of $\overline{\mathbf{k}}[C]_P$. Letting $N = t\overline{\mathbf{k}}[C]_P$, an ideal of M_P, it suffices to show that the quotient M_P/N is trivial. Since M_P is a $\overline{\mathbf{k}}[C]_P$-module, so is the quotient. Noting that $M_P \cdot (M_P/N) = (N + M_P^2)/N$ and that $N + M_P^2 = (\overline{\mathbf{k}}t + M_P^2)\overline{\mathbf{k}}[C]_P = M_P$ shows that $M_P \cdot (M_P/N) = M_P/N$. By Nakayama's Lemma (Proposition 6.5.5), $M_P/N = 0$ as desired. $\square$

The proposition requires the local ring to apply Nakayama's Lemma. The results so far in this section allow an argument (Exercise 7.2.4) that the condition for a curve to be nonsingular is a maximality condition on the rank of the derivative matrix, as stated after Definition 7.2.1.

A generator t of M_P is called a *uniformizer* at P. Any $t \in m_P - m_P^2$ is a uniformizer but t need not generate m_P. For example, let $\mathcal{E}$ be the elliptic curve over $\mathbb{Q}$ with Weierstrass equation $E : y^2 = 4x^3 - 4x$. The remark after the proof of Lemma 7.2.2 shows that y is a uniformizer at the point $P = (0,0)$. Indeed the Weierstrass relation $x = y^2/(4x^2 - 4)$ shows that x is redundant as a generator of M_P in $\overline{\mathbf{k}}[\mathcal{E}]_P$. On the other hand, x is needed along with y to generate m_P in $\overline{\mathbf{k}}[\mathcal{E}]$. To see this, note that $\overline{\mathbb{Q}}[\mathcal{E}]/\langle y \rangle = \overline{\mathbb{Q}}[x]/\langle x^3 - x \rangle$ and this is not a field, showing that $\langle y \rangle$ is not maximal in $\overline{\mathbb{Q}}[\mathcal{E}]$ and thus is not m_P.

Let F be any nonzero element of $\overline{\mathbf{k}}[C]_P$ and let t be a uniformizer at P. Either F is a unit or it takes the form $F = tF_1$ where F_1 is a nonzero element of $\overline{\mathbf{k}}[C]_P$. Either F_1 is a unit or it takes the form $F_1 = tF_2$, and so on. If this process doesn't terminate then it leads to an ascending chain of ideals in $\overline{\mathbf{k}}[C]_P$,

$$\langle F \rangle \subset \langle F_1 \rangle \subset \langle F_2 \rangle \subset \cdots,$$

where each containment is proper. But $\overline{\mathbf{k}}[C]_P$ is Noetherian (see Exercise 7.2.5), so this cannot happen. Therefore F takes the form $F = t^e u$ where $e \in \mathbb{N}$ and $u \in \overline{\mathbf{k}}[C]_P^*$. This representation of F is unique, for if $F = t^e u = t^{e'} u'$ with $e \geq e'$ then $t^{e-e'} u = u'$, showing that $e = e'$ and $u = u'$.

The *valuation at P* on the coordinate ring is the function

$$\nu_P : \overline{\mathbf{k}}[C] \longrightarrow \mathbb{N} \cup \{+\infty\}, \qquad \nu_P(f) = \begin{cases} +\infty & \text{if } f = 0, \\ e & \text{if } f = t^e u. \end{cases}$$

This extends to the function field,

$$\nu_P : \overline{\mathbf{k}}(C) \longrightarrow \mathbb{Z} \cup \{+\infty\}, \qquad \nu_P(F) = \nu_P(f) - \nu_P(g), \quad F = f/g,$$

where it is well defined and it satisfies (Exercise 7.2.6)

Proposition 7.2.4. *Let C be a nonsingular algebraic curve. For any point $P \in C$ the valuation $\nu_P : \overline{\mathbf{k}}(C) \longrightarrow \mathbb{Z} \cup \{+\infty\}$ surjects, and for all nonzero $F, G \in \overline{\mathbf{k}}(C)$ it satisfies the conditions*

(1) $\nu_P(FG) = \nu_P(F) + \nu_P(G)$,
(2) $\nu_P(F + G) \geq \min\{\nu_P(F), \nu_P(G)\}$ *with equality if* $\nu_P(F) \neq \nu_P(G)$.

Now we can describe the mapping from C to $\mathbb{P}^1(\overline{\mathbf{k}})$ defined by a rational function $F \in \overline{\mathbf{k}}(C)$. The zero function defines the zero mapping. Any nonzero function takes the form

$$F = t^{\nu_P(F)} \frac{f}{g}, \quad \nu_P(F) \in \mathbb{Z}, \; f, g \in \overline{\mathbf{k}}[C], \; f(P) \text{ and } g(P) \text{ nonzero}$$

at each point $P \in C$. The mapping it defines is

$$F(P) = \begin{cases} 0 & \text{if } \nu_P(F) > 0, \\ \infty & \text{if } \nu_P(F) < 0, \\ f(P)/g(P) & \text{if } \nu_P(F) = 0. \end{cases}$$

The valuation is the algebraic analog of the order of vanishing for meromorphic functions at a point P of a Riemann surface, and so one speaks of the zeros and poles of a rational function in the usual way.

Since these methods are local they work on all of C_{hom}, using an appropriate coordinate system at each point. It turns out that dehomogenizing the polynomials defining a nonsingular projective curve C_{hom} to define one of its affine pieces leads to the same function field up to isomorphism whenever the affine piece is nonempty, regardless of which variable is set to 1, and the isomorphism between two affine function fields associated to C_{hom} preserves valuations on the overlap of the two corresponding affine pieces of the curve. We will touch on this point, but not prove it completely, in Section 8.5. Also, the valuations on $\overline{\mathbf{k}}(C)$, meaning the surjections $\nu : \overline{\mathbf{k}}(C) \longrightarrow \mathbb{Z} \cup \{+\infty\}$ satisfying the conditions of Proposition 7.2.4, are precisely the valuations ν_P arising from the points of the projective curve C_{hom} (for the proof see Proposition 11.43 of [Kna93]). For these two reasons it is natural to conceive of the function field as being associated with the projective curve rather than with an affine piece. That is, even when C is projective we can write $\overline{\mathbf{k}}(C)$, interpreting it as the function field of an affine piece of C. The notation C_{hom} to distinguish a projective curve will no longer be used.

For example, let $\mathcal{E}$ be an elliptic curve over $\mathbf{k}$ and consider the rational functions x and y, elements of $\overline{\mathbf{k}}(\mathcal{E})$. These make sense as mappings on the affine part of $\mathcal{E}$. To study them at the infinite point $0_{\mathcal{E}} = [0, 1, 0]$, homogenize them to x/z and y/z and then dehomogenize to $\tilde{x}/\tilde{z}$ and $1/\tilde{z}$, introducing the tildes to emphasize the different coordinate system. In these coordinates the Weierstrass polynomial is $E'(\tilde{x}, \tilde{z}) = \tilde{z} - 4\tilde{x}^3 + g_2\tilde{x}\tilde{z}^2 + g_3\tilde{z}^3$. This has nonzero second partial derivative at $(0, 0)$, showing that $\tilde{x}$ is a uniformizer at $0_{\mathcal{E}}$, i.e., $\nu_{0_{\mathcal{E}}}(\tilde{x}) = 1$. The Weierstrass relation becomes

$$\tilde{z}(1 + g_2\tilde{x}\tilde{z} + g_3\tilde{z}^2) = 4\tilde{x}^3,$$

and so Proposition 7.2.4 shows that $\nu_{0_{\mathcal{E}}}(\tilde{z}) = 3$. Thus the rational functions x and y have valuations $\nu_{0_{\mathcal{E}}}(x) = \nu_{0_{\mathcal{E}}}(\tilde{x}/\tilde{z}) = 1 - 3 = -2$ and $\nu_{0_{\mathcal{E}}}(y) = \nu_{0_{\mathcal{E}}}(1/\tilde{z}) = -3$, making x/y a uniformizer at $0_{\mathcal{E}}$. Working over the complex numbers and thinking of x and y as $\wp$ and $\wp'$, this analysis reflects their double and triple poles at the origin.

To summarize the flow of ideas so far: Lemma 7.2.2 translates the geometric data that a nonsingular algebraic curve has a one-dimensional tangent line at each point into algebraic information that a vector space associated to each point is one-dimensional. Proposition 7.2.3 says that after localizing, this makes a maximal ideal at each point principal. The Noetherian property of the local ring then leads to a valuation at each point. The uniformizer and the valuation show that the curve's rational functions give well defined mappings on the entire curve.

Different representations of a rational function F may be required to evaluate it at different points. The idea is that an expression giving $0/0$ at a point yields the correct value after its numerator and denominator are divided by their common power of a uniformizer. For example, consider the function x/y on the elliptic curve over $\mathbb{Q}$ with Weierstrass equation $E : y^2 = 4x^3 - 4x$. To determine its behavior at the point $P = (0, 0)$, recall the discussion earlier that y is a uniformizer at P. This shows that $x = y^2/(4x^2 - 4)$ has order 2, giving a simple zero of x/y at P. Equivalently, x/y rewrites as $y/(4x^2 - 4)$ (here is where the numerator and denominator get divided by the uniformizer), again giving a simple zero at P. On the other hand, the value of the new expression $y/(4x^2 - 4)$ for x/y is no longer immediately clear at the points $(\pm 1, 0)$ of the curve.

For a more elaborate example, let $\mathcal{E}$ be any elliptic curve over any field $\mathbf{k}$ as in Section 7.1, and let P be any nonzero point of $\mathcal{E}$. The translation function

$$\tau_P : \mathcal{E} \longrightarrow \mathcal{E}, \qquad \tau_P(X) = X + P$$

has as its components the rational functions $F(x, y) = r(x_P, y_P, x, y)$ and $G(x, y) = s(x_P, y_P, x, y)$ from Section 7.1, using the general case expressions for the slope and intercept λ and μ,

$$\lambda = \frac{y - y_P}{x - x_P}, \quad \mu = \frac{xy_P - x_P y}{x - x_P}.$$

These expressions for λ and μ are not well defined when $x = x_P$, but multiplying the numerators and denominators by $y + y_P$ and then using the Weierstrass equation (7.1) to substitute for y^2 and y_P^2 and then canceling $x - x_P$ shows that the general expressions for λ and μ are also (Exercise 7.2.7)

$$\lambda = \frac{4(x_P^2 + x_P x + x^2) - g_2}{y_P + y}, \quad \mu = \frac{-4(x_P^2 x + x_P x^2) - g_3 + y_P y}{y_P + y}.$$

These are defined when $x = x_P$ provided $y \neq -y_P$, when if also $y = y_P$ then they are the special case expressions for λ and μ from Section 7.1. The new expressions are no panacea, however, since they are not well defined when $y = -y_P$ even when $x \neq x_P$. Nonetheless, the point is that despite their casewise expressions, the components F and G of τ_P are single rational functions on $\mathcal{E}$.

As explained at the beginning of this section, we need a general result from algebraic geometry that curves are determined by their function fields. The precise statement requires appropriate equivalence relations both on curves and on fields.

To define the equivalence relation on curves, let C be a nonsingular projective algebraic curve over $\mathbf{k}$. Let r be a positive integer. Consider any element of $\mathbb{P}^r(\overline{\mathbf{k}}(C))$,

$$h = [F_0, \ldots, F_r].$$

For each point $P \in C$ let t_P be a uniformizer at P and let $\nu_P(h) = \min\{\nu_P(F_0), \ldots, \nu_P(F_r)\}$. Then h defines a map from C to $\mathbb{P}^r(\overline{\mathbf{k}})$,

$$h(P) = [(t_P^{-\nu_P(h)} F_0)(P), \ldots, (t_P^{-\nu_P(h)} F_r)(P)], \quad P \in C.$$

Let $C' \subset \mathbb{P}^r(\overline{\mathbf{k}})$ be another nonsingular algebraic curve over $\mathbf{k}$. Then h is a *morphism from C to C'* if $h(P) \in C'$ for all $P \in C$. The affine notation is $h = (F_1/F_0, \ldots, F_r/F_0) \in \overline{\mathbf{k}}(C)^r$ when $F_0 \neq 0$, but this needn't map points P of the affine piece of C back to affine space when $F_0(P) = 0$. For instance, every element of $\overline{\mathbf{k}}(C)$ defines a morphism from C to $\mathbb{P}^1(\overline{\mathbf{k}})$, and the example τ_P from a moment ago is a morphism from $\mathcal{E}$ to $\mathcal{E}$. A morphism h from C to C' is an *isomorphism* if there exists a morphism h' from C' to C such that the composites $h' \circ h$ and $h \circ h'$ are the respective identity maps on C and C', in which case C and C' are *isomorphic*.

Isomorphism is an equivalence relation but it is not fine enough for our purposes. Again letting C be a nonsingular projective algebraic curve over $\mathbf{k}$, an element F of $\overline{\mathbf{k}}(C)$ is *defined over* $\mathbf{k}$ if it has a representation $F = f/g$ with f and g from $\mathbf{k}[C] = \mathbf{k}[x_1, \ldots, x_n]/I_{\mathbf{k}}$ (where $I_{\mathbf{k}} = I \cap \mathbf{k}[x_1, \ldots, x_n]$) rather than $\overline{\mathbf{k}}[C]$. Such functions form a subfield of $\overline{\mathbf{k}}(C)$,

$$\mathbf{k}(C) = \{F \in \overline{\mathbf{k}}(C) : F \text{ is defined over } \mathbf{k}\}.$$

A morphism $h = (F_1, \ldots, F_r)$ of curves over $\mathbf{k}$ is *defined over* $\mathbf{k}$ if each F_i is. For instance, every element of $\mathbf{k}(C)$ defines a morphism over $\mathbf{k}$ from C to $\mathbb{P}^1(\overline{\mathbf{k}})$, and τ_P is a morphism over $\mathbf{k}$. The equivalence relation we need is that two nonsingular projective curves C and C' over $\mathbf{k}$ are *isomorphic over* $\mathbf{k}$ if they are isomorphic via morphisms h and h' over $\mathbf{k}$. See Exercise 7.2.8 for an example. In the case of two elliptic curves the only possible isomorphisms over $\mathbf{k}$ taking $0_{\mathcal{E}}$ to $0_{\mathcal{E}'}$ turn out to be the admissible changes of variable from Section 7.1. As affine maps these take lines to lines, making them group isomorphisms as well. Two curves defined over a field $\mathbf{k}$ can be isomorphic over $\overline{\mathbf{k}}$ without being isomorphic over $\mathbf{k}$ itself. Exercise 7.2.9 gives an example.

To define the needed equivalence relation on function fields, call a field $\mathbb{K}$ a *function field over* $\mathbf{k}$ if

(1) $\mathbb{K} \cap \overline{\mathbf{k}} = \mathbf{k}$,
(2) $\mathbb{K}$ is a finite extension of a field $\mathbf{k}(t)$ where t is transcendental over $\mathbf{k}$.

Two function fields $\mathbb{K}$ and $\mathbb{K}'$ over $\mathbf{k}$ are *conjugate over* $\mathbf{k}$ if there exists a $\mathbf{k}$-isomorphism $\mathbb{K} \xrightarrow{\sim} \mathbb{K}'$, meaning an isomorphism that fixes $\mathbf{k}$ pointwise.

The relation between curves and their function fields is

Theorem 7.2.5 (Curves–Fields Correspondence, Part 1). *The map*

$$C \mapsto \mathbf{k}(C)$$

induces a bijection from the set of isomorphism classes over $\mathbf{k}$ *of nonsingular projective algebraic curves over* $\mathbf{k}$ *to the set of conjugacy classes over* $\mathbf{k}$ *of function fields over* $\mathbf{k}$.

The importance of Theorem 7.2.5 is that rather than constructing algebraic curves directly it now suffices to construct suitable fields instead.

The theorem gives the map from curve-classes to field-classes explicitly. To describe the map from field-classes to curve-classes, let $\mathbb{K}$ be a function field over $\mathbf{k}$. Since the extension $\mathbb{K}/\mathbf{k}(t)$ is finite and we are in characteristic 0 the Primitive Element Theorem of field theory says that $\mathbb{K} = \mathbf{k}(t, u)$ for some u. The generator u satisfies an irreducible polynomial over $\mathbf{k}(t)$, and clearing denominators gives a polynomial relation between t and u over $\mathbf{k}$,

$$\varphi(t, u) = 0, \qquad \varphi \in \mathbf{k}[x, y].$$

Because $\mathbb{K} \cap \overline{\mathbf{k}} = \mathbf{k}$, the polynomial φ is irreducible over $\overline{\mathbf{k}}$ (Exercise 7.2.10). The set of points $(x, y) \in \overline{\mathbf{k}}^2$ such that $\varphi(x, y) = 0$ gives a plane curve C', but this curve can have finitely many singular points, and at these points rational functions need not define mappings. A process called *desingularizing* (see Chapter 7 of [Ful69]) produces a nonsingular curve C with function field $\mathbb{K}$. The map (t, u) takes C to the plane curve C', but this map is guaranteed only to be a *birational equivalence*, not necessarily an isomorphism. In general given two curves C and C', possibly containing singular points, a *rational map* from C to C' is an element h of $\mathbb{P}^r(\overline{\mathbf{k}}(C))$ like a morphism but required only to define an actual map from all but finitely many points of C to C'. The rational map is a birational equivalence if there exists a rational map h' from C' to C such that the two composites are the identity maps on C and on C' wherever they are defined. See Exercise 7.2.11 for an example of birational equivalence between a singular curve and a nonsingular curve. A nonsingular curve corresponding to the field $\mathbb{K}$ need not lie in the plane, and this is why Definition 7.2.1 is not restricted to plane curves. In any case, with Theorem 7.2.5 in hand we can think of curves in terms of their function fields.

Theorem 7.2.5 extends from curves and fields to maps between curves and maps between fields. If $h : C \longrightarrow C'$ is a nonconstant morphism over $\mathbf{k}$ of nonsingular projective curves over $\mathbf{k}$ then as in the compact Riemann surface environment, h is surjective. Its pullback is a map of function fields,

$$h^* : \mathbf{k}(C') \longrightarrow \mathbf{k}(C), \qquad h^*G = G \circ h.$$

This isn't quite as simple as it looks, the fact that h^* is well defined relying on a result about polynomial ideals and their solution sets (Exercise 7.2.12(a)). Another such result shows that if h is assumed only to be a nonconstant rational map then its pullback still makes sense (Exercise 7.2.12(b)).

Theorem 7.2.6 (Curves–Fields Correspondence, Part 2). *Let C and C' be nonsingular projective algebraic curves over $\mathbf{k}$. Then the map*

$$h \mapsto h^*$$

is a bijection from the set of nonconstant morphisms over $\mathbf{k}$ from C to C' to the set of $\mathbf{k}$-injections of $\mathbf{k}(C')$ in $\mathbf{k}(C)$.

Theorem 7.2.6 says that rather than constructing morphisms between curves it now suffices to construct suitable field injections instead. For example, since any nonconstant rational map h between nonsingular projective curves C and C' defines a corresponding $\mathbf{k}$-injection h^* of $\mathbf{k}(C')$ in $\mathbf{k}(C)$, the theorem shows that in fact h is a morphism. In particular a birational equivalence between nonsingular projective curves is an isomorphism. For another example, consider the injection $i : \mathbf{k}(x) \longrightarrow \mathbf{k}(x, y)$ where x and y satisfy a Weierstrass equation. Letting $\mathcal{E}$ denote the relevant elliptic curve, the morphism corresponding to i is $h : \mathcal{E} \longrightarrow \mathbb{P}^1(\mathbf{k})$ where $h(x, y) = x$.

The theorem gives the map from morphisms to field injections explicitly. To describe the map from field injections to morphisms, let $i : \mathbb{K}' \longrightarrow \mathbb{K}$ be a $\mathbf{k}$-injection of function fields. By Theorem 7.2.5 there exist nonsingular projective curves C' and C such that $\mathbf{k}(C') \cong \mathbb{K}'$ and $\mathbf{k}(C) \cong \mathbb{K}$ ($\mathbf{k}$-isomorphisms), and i can be viewed instead as a $\mathbf{k}$-injection

$$i : \mathbf{k}(C') \longrightarrow \mathbf{k}(C).$$

Define two points P and Q of C to be $\mathbf{k}(C')$-equivalent if the embedded image of $\mathbf{k}(C')$ in $\mathbf{k}(C)$ doesn't separate them, i.e., $(iG)(P) = (iG)(Q)$

for all $G \in \mathbf{k}(C')$. Then the quotient set of $\mathbf{k}(C')$-equivalence classes in C bijects to the curve C', the equivalence class of $P \in C$ mapping to the unique point $P' \in C'$ such that $G(P') = (iG)(P)$ for all G. The morphism $h : C \longrightarrow C'$ corresponding to i is the quotient map followed by the bijection. For instance, in the example ending the previous paragraph, pairs $(x, \pm y)$ in $\mathcal{E}$ are $\mathbf{k}(x)$-equivalent and the morphism $h(x, y) = x$ identifies them in mapping $\mathcal{E}$ to $\mathbb{P}^1(\overline{\mathbf{k}})$. We will use Theorem 7.2.6 and the quotient morphism in Section 7.8 to revisit isogenies of elliptic curves in algebraic terms.

Exercises

7.2.1. The polynomial $\varphi(x, y) = x^2 - 2y^2$ is irreducible over $\mathbb{Q}$. Does it define an algebraic curve over $\mathbb{Q}$? Describe its solution set in $\mathbb{P}^2(\overline{\mathbb{Q}})$.

7.2.2. Show that m_P is a maximal ideal of $\overline{\mathbf{k}}[C]$.

7.2.3. (a) In the proof of Lemma 7.2.2, show that $\nabla\varphi(P)$ is in the row space of $[D_j \varphi_i(P)]$. (Hints for this exercise are at the end of the book.)
  (b) In the proof of Lemma 7.2.2, show that the pairing is perfect.
  (c) Let $\mathcal{E}$ be an elliptic curve defined by a Weierstrass polynomial E, and let $P = (x, y)$ be a point of $\mathcal{E}$. Explain how the proof of Lemma 7.2.2 shows that $x - x_P + m_P^2$ spans m_P/m_P^2 if $D_2 E(P) \neq 0$ and $y - y_P + m_P^2$ spans m_P/m_P^2 if $D_1 E(P) \neq 0$.
  (d) In the discussion before Proposition 7.2.3, show that $m_P \cap M_P^2 = m_P^2$.

7.2.4. Let C be an affine algebraic curve as in Definition 7.2.1. Supply details as necessary in the following argument that the condition $\mathrm{rank}([D_j\varphi_i(P)]) = n$ for some $P \in C$ is impossible: If the rank is n then the method of proving Lemma 7.2.2 shows that $m_P = m_P^2$ and so $M_P = M_P^2$. By Nakayama's Lemma $M_P = 0$ and so $m_P = 0$. This forces $\overline{\mathbf{k}}[C] = \overline{\mathbf{k}}$ and then $\overline{\mathbf{k}}(C) = \overline{\mathbf{k}}$, contradiction.

7.2.5. A special case of the *Hilbert Basis Theorem* (see, for example, [Ful69]) says that the ring $\overline{\mathbf{k}}[x_1, \ldots, x_n]$ is Noetherian, meaning that all ideals are finitely generated, or equivalently that there is no infinite chain of proper increasing ideal containments $I_1 \subset I_2 \subset I_3 \subset \cdots$ in $\overline{\mathbf{k}}[x_1, \ldots, x_n]$.
  (a) Show that the two conditions are equivalent.
  (b) Let I be an ideal of $\overline{\mathbf{k}}[x_1, \ldots, x_n]$. Show that the ideals of the quotient $\overline{\mathbf{k}}[x_1, \ldots, x_n]/I$ are in bijective correspondence with the ideals of $\overline{\mathbf{k}}[x_1, \ldots, x_n]$ that contain I. Conclude that $\overline{\mathbf{k}}[x_1, \ldots, x_n]/I$ is Noetherian. In particular $\overline{\mathbf{k}}[C]$ is Noetherian for an algebraic curve C.
  (c) For any ideal J of $\overline{\mathbf{k}}[C]_P$ the ideal $J \cap \overline{\mathbf{k}}[C]$ of $\overline{\mathbf{k}}[C]$ is finitely generated by part (a). Show that its generators also generate J and therefore the local ring $\overline{\mathbf{k}}[C]_P$ is Noetherian.

7.2.6. Show that the valuation at P on $\overline{\mathbf{k}}(C)$ is well defined and prove Proposition 7.2.4.

7.2.7. Derive the new expressions for λ and μ in the section.

7.2.8. Working over $\mathbb{Q}$, let $\varphi(x, y) = xy - 1$ and $\varphi'(z, w) = z^2 + w^2 - 1$, and let C and C' be the corresponding curves. Let

$$(F, G)(x, y) = \left(\frac{2x}{x^2 + 1}, \frac{x^2 - 1}{x^2 + 1} \right), \quad (F', G')(z, w) = \left(\frac{z}{1 - w}, \frac{1 - w}{z} \right).$$

The polynomials homogenize to $\varphi(x, y, s) = xy - s^2$ and to $\varphi'(z, w, t) = z^2 + w^2 - t^2$. Explain why the maps homogenize to $(F, G)[x, y, s] = [2xs, x^2 - s^2, x^2 + s^2]$ and $(F', G')[z, w, t] = [z^2, (t - w)^2, z(t - w)]$, and why these maps are

$$(F, G)[x, y, s] = [2s, x - y, x + y], \quad (F', G')[z, w, t] = [t + w, t - w, z].$$

Show that (F, G) takes C to C' and that (F', G') takes C' to C, including infinite points. Show that the maps invert each other. Thus the curves are isomorphic over $\mathbb{Q}$. In particular, the sets of rational points on the hyperbola and the circle are isomorphic as algebraic curves.

7.2.9. Consider the curves $\mathcal{E}$ and $\mathcal{E}'$ defined by the Weierstrass relations $y^2 = 4x^3 - 4x$ and $y^2 = 4x^3 + 4x$. Show that $\mathcal{E}$ and $\mathcal{E}'$ are isomorphic as curves over $\overline{\mathbb{Q}}$ by finding an admissible change of variable over $\overline{\mathbb{Q}}$ between them. Show that $\mathcal{E}$ and $\mathcal{E}'$ are not isomorphic as curves over $\mathbb{Q}$ either by showing that no appropriate admissible change of variable exists or by comparing their subgroups of 2-torsion points with rational coordinates, $\mathcal{E}[2](\mathbb{Q})$ and $\mathcal{E}'[2](\mathbb{Q})$.

7.2.10. Working inside a fixed algebraic closure of $\mathbf{k}(t)$, let $\mathbb{K} = \mathbf{k}(t, u)$ be a field as in Theorem 7.2.5 and let $f \in \mathbf{k}(t)[y]$ be the minimal polynomial of u over $\mathbf{k}(t)$. After clearing denominators the relation $f(u) = 0$ rewrites as a polynomial relation between t and u over $\mathbf{k}$, $\varphi(t, u) = 0$ where $\varphi \in \mathbf{k}[x, y]$. The text asserts that φ is irreducible over $\overline{\mathbf{k}}$ because $\mathbb{K} \cap \overline{\mathbf{k}} = \mathbf{k}$. Supply details as necessary for the following argument that this is so: If $\varphi = \varphi_1 \varphi_2$ over $\overline{\mathbf{k}}$ then this factorization occurs over a finite extension field of $\mathbf{k}$, necessarily of the form $\mathbf{k}(v)$ where $g(v) = 0$, $g \in \mathbf{k}[z]$ is irreducible. Also g is irreducible in $\mathbf{k}(t)[z]$ since its roots lie in $\overline{\mathbf{k}}$, so that the coefficients of any factorization would lie in $\mathbf{k}(t) \cap \overline{\mathbf{k}} = \mathbf{k}$. Let $\mathbf{l} = \mathbf{k}(t, v)$. Thus $[\mathbb{K} : \mathbf{k}(t)] = \deg(f)$ and $[\mathbf{l} : \mathbf{k}(t)] = \deg(g)$. Consider the composite field $\mathbb{K}\mathbf{l} = \mathbf{k}(t, u, v)$. The factorization of φ over $\mathbf{k}(v)$ gives a factorization of f over $\mathbf{l}$, and this makes the extension degree $[\mathbb{K}\mathbf{l} : \mathbf{l}]$ less than $\deg(f)$. Thus $[\mathbb{K}\mathbf{l} : \mathbf{k}(t)] < \deg(f) \deg(g)$ and $[\mathbb{K}\mathbf{l} : \mathbb{K}] < \deg(g)$, and consequently $g = g_1 g_2$ in $\mathbb{K}[z]$. Since the roots of g lie in $\overline{\mathbf{k}}$, the factorization of g occurs in $(\mathbb{K} \cap \overline{\mathbf{k}})[z] = \mathbf{k}[z]$, giving a contradiction.

7.2.11. Let $\varphi(x, y) = y^2 - x^3 - x^2$, and let C be the corresponding curve. Show that C is an algebraic curve in the sense of Definition 7.2.1 but that it has a singular point at $(0, 0)$. Let

$$F(x, y) = y/x, \quad (G, H)(t) = (t^2 - 1, t^3 - t).$$

Show that F and (G, H) are birational equivalences between C and $\overline{k}$ but not isomorphisms.

7.2.12. (a) Let $I \subset \overline{k}[x_1, \ldots, x_n]$ and $I' \subset \overline{k}[y_1, \ldots, y_r]$ be prime ideals defining nonsingular projective algebraic curves C and C'. Using projective notation, let

$$h = [h_0 + I, \ldots, h_r + I], \quad \text{each } h_i + I \in \overline{k}[C]$$

be a nonconstant (and hence surjective) morphism from C to C', and let

$$G = [g_0 + I', g_1 + I'], \quad \text{each } g_i + I' \in \overline{k}[C'], \, g_0 + I' \neq I'$$

be an element of the function field $\overline{k}(C')$. The pullback $h^*G = G \circ h \in \overline{k}(C)$ is

$$h^*G = [g_0 \circ (h_0, \ldots, h_r) + I, g_1 \circ (h_0, \ldots, h_r) + I].$$

But to make sense this must be independent of how the various coset representatives are chosen. Show that replacing any h_i by $h_i + \varphi$ where $\varphi \in I$ doesn't change the definition of h^*G. Show that replacing either g_i by $g_i + \psi'$ where $\psi' \in I'$ doesn't change the definition of h^*G provided that for any polynomial $\varphi \in \overline{k}[x_1, \ldots, x_n]$,

$$\varphi(P) = 0 \text{ for all } P \in C \implies \varphi \in I.$$

The definition of C as the solution set of I immediately gives the opposite implication, but this also follows using the fact that I is prime. For the details see the first few pages of an algebraic geometry text.

(b) Now make the weaker assumption that h is a nonconstant rational map from C to C'. Show that verifying that h^*G is well defined now requires the additional implication

$$\varphi(P) = 0 \text{ for all but finitely many } P \in C \implies \varphi(P) = 0 \text{ for all } P \in C.$$

Again, see an algebraic geometry text.

7.3 Divisors on curves

This section begins by citing some results from Riemann surface theory in the new context of algebraic curves. Every compact Riemann surface has a description as a nonsingular projective algebraic curve over $\mathbb{C}$. The meromorphic functions on a compact Riemann surface are the rational functions on the surface viewed as a curve, and the holomorphic functions between compact Riemann surfaces are the morphisms between the surfaces viewed as curves. Many results about Riemann surfaces rephrase algebraically and then apply to algebraic curves over an arbitrary field k, although sometimes the characteristic of k needs to be taken into account when it is not 0 as we assume it is here.

For instance, as mentioned in the previous section, a morphism of nonsingular projective algebraic curves is either constant or surjective. If $h : C \longrightarrow C'$ is a nonconstant morphism then its pullback is a $\overline{\mathbf{k}}$-injection of function fields, $h^* : \overline{\mathbf{k}}(C') \longrightarrow \overline{\mathbf{k}}(C)$, and the *degree* of h is defined as the field extension degree

$$\deg(h) = [\overline{\mathbf{k}}(C) : h^*(\overline{\mathbf{k}}(C'))].$$

Thus a morphism is an isomorphism if and only if its degree is 1 (Exercise 7.3.1(a)). The formula shows that the degree of a composition of nonconstant morphisms is the product of the degrees, and this extends to all morphisms if constant morphisms are given degree 0.

The degree of h is the number of inverse images under h of each point of C' suitably counting an algebraically defined ramification. Specifically the *ramification degree* of h at a point $P \in C$ is the positive integer

$$e_P(h) = \nu_P(t' \circ h) \qquad \text{where } t' \text{ is a uniformizer at } h(P).$$

Thus $t' \circ h = t^{e_P(h)} F$ where t is a uniformizer at P and $F \in \overline{\mathbf{k}}(C)$ with $\nu_P(F) = 0$. The resulting degree formula is analogous to the result in Riemann surface theory,

$$\sum_{P \in h^{-1}(Q)} e_P(h) = \deg(h) \quad \text{for any } Q \in C'. \tag{7.4}$$

(As a special case, a rational function has only finitely many zeros and poles.) For morphisms $h : C \longrightarrow C'$ over $\mathbf{k}$ of curves over $\mathbf{k}$, the methods of Exercise 7.2.10 show that (Exercise 7.3.1(b))

$$[\mathbf{k}(C) : h^*(\mathbf{k}(C'))] = [\overline{\mathbf{k}}(C) : h^*(\overline{\mathbf{k}}(C'))],$$

so that $\deg(h)$ is defined by either of these.

In particular, let $\mathcal{E}$ be an elliptic curve over $\mathbf{k}$ and let N be a positive integer. We assert without proof that since $\mathbf{k}$ has characteristic 0 the map $[N] : \mathcal{E} \longrightarrow \mathcal{E}$ is unramified and has degree N^2,

$$[\overline{\mathbf{k}}(\mathcal{E}) : [N]^*(\overline{\mathbf{k}}(\mathcal{E}))] = N^2. \tag{7.5}$$

The terminology of divisors carries over from Riemann surfaces to curves. Let C be a nonsingular algebraic curve over $\mathbf{k}$. The *divisor group of C* is the free Abelian group on its points,

$$\mathrm{Div}(C) = \left\{ \sum_{P \in C} n_P(P) : n_P \in \mathbb{Z}, \, n_P = 0 \text{ for almost all } P \right\}.$$

Since both an elliptic curve and its divisor group have group laws, we have adopted the notational convention of parenthesizing points in divisors, thus the divisor (P) versus the point P and more generally in the case of an elliptic curve

$$\sum n_P(P) \in \mathrm{Div}(\mathcal{E}) \quad \text{versus} \quad \sum [n_P]P \in \mathcal{E}.$$

Coefficients n_P in the divisor on the left can be negative, so the addition in $\mathcal{E}$ on the right is using the definition $[n_P]P = -[-n_P]P$ when $n_P < 0$. Returning to a general curve, the subgroup of *degree-0 divisors* is

$$\mathrm{Div}^0(C) = \left\{ \sum n_P(P) \in \mathrm{Div}(C) : \sum n_P = 0 \right\}.$$

The divisor of a nonzero element of the function field of C is

$$\mathrm{div}(F) = \sum \nu_P(F)(P), \quad F \in \overline{\mathbf{k}}(C)^*.$$

Divisors of the form $\mathrm{div}(F)$ are called *principal* and the set of principal divisors is denoted $\mathrm{Div}^{\ell}(C)$. Principal divisors have degree 0. This follows from another formula analogous to Riemann surface theory,

$$\mathrm{div}(F) = \sum_{P \in F^{-1}(0)} e_P(F)(P) - \sum_{P \in F^{-1}(\infty)} e_P(F)(P),$$

and from (7.4), combining to show $\deg(\mathrm{div}F) = \deg(F) - \deg(F) = 0$. Part (1) of Proposition 7.2.4 shows that the map

$$\mathrm{div} : \overline{\mathbf{k}}(C)^* \longrightarrow \mathrm{Div}^0(C)$$

is a homomorphism. Thus $\mathrm{Div}^{\ell}(C) = \mathrm{div}(\overline{\mathbf{k}}(C)^*)$ is a subgroup of $\mathrm{Div}^0(C)$. As in Chapter 6 the (degree zero) *Picard group* of C is the (degree zero) divisor class group, the quotient

$$\mathrm{Pic}^0(C) = \mathrm{Div}^0(C)/\mathrm{Div}^{\ell}(C).$$

We will need the following result in Chapter 8.

Proposition 7.3.1. *Let C be a nonsingular projective algebraic curve over a field $\mathbf{k}$. Let $S \subset C$ be a finite subset. Then every divisor class in $\mathrm{Pic}^0(C)$ has a representative $\sum n_P(P)$ such that $n_P = 0$ for all $P \in S$.*

Proof. We may assume that all of S lies in affine space $\overline{\mathbf{k}}^n$. To see this, note that any matrix $A \in \mathrm{GL}_{n+1}(\overline{\mathbf{k}})$ defines an automorphism of $\mathbb{P}^n(\overline{\mathbf{k}})$ just as 2-by-2 matrices act on the Riemann sphere $\widehat{\mathbb{C}} = \mathbb{P}^1(\mathbb{C})$. Each such automorphism restricts to an isomorphism from C to another curve C' taking S to a set S'. Choose vectors $v_P \in \overline{\mathbf{k}}^{n+1}$ that represent the points $P \in \mathbb{P}^n(\overline{\mathbf{k}})$ of S. There is a vector $a_{n+1} \in \overline{\mathbf{k}}^{n+1}$ that is not orthogonal to any v_P (Exercise 7.3.2(a)). Extending to a basis $\{a_1, \ldots, a_{n+1}\}$ of $\overline{\mathbf{k}}^{n+1}$ and then making these the rows of A gives an isomorphism of C taking S into $\overline{\mathbf{k}}^n$ as desired. With this done, all points of S fall under the purview of an affine description of the function field $\overline{\mathbf{k}}(C)$ even though C is projective.

Recall from Section 7.2 the coordinate ring of C and its localization at P, $\overline{\mathbf{k}}[C] \subset \overline{\mathbf{k}}[C]_P$, and recall their maximal ideals of functions that vanish at P, $m_P \subset M_P$. For each point P of S define an intermediate ring and maximal ideal to avoid poles in S,

$$\overline{\mathbf{k}}[C]_S = \{f/g \in \overline{\mathbf{k}}(C) : g(Q) \neq 0 \text{ for all } Q \in S\},$$
$$M_{S,P} = \{f/g \in \overline{\mathbf{k}}[C]_S : f(P) = 0\}.$$

Thus $m_P \subset M_{S,P} \subset M_P$, and so there is a natural map

$$m_P/m_P^2 \longrightarrow M_{S,P}/M_{S,P}^2 \longrightarrow M_P/M_P^2.$$

Section 7.2 showed that the composite is an isomorphism of nontrivial vector spaces, so the middle quotient is nonzero. Choose any $G_P \in M_{S,P} - M_{S,P}^2$. Then $G_P \in M_P$. The argument given in the hint to Exercise 7.2.3(d) modifies easily to show that $M_{S,P} \cap M_P^2 = M_{S,P}^2$ (Exercise 7.3.2(b)), so $G_P \notin M_P^2$. This shows that $\nu_P(G_P) = 1$. By the Chinese Remainder Theorem there is a surjective map

$$\overline{\mathbf{k}}[C]_S \longrightarrow \overline{\mathbf{k}}[C]_S/M_{S,P}^2 \times \prod_{\substack{Q \in S \\ Q \neq P}} \overline{\mathbf{k}}[C]_S/M_{S,Q}.$$

In particular, some $F_P \in \overline{\mathbf{k}}[C]_S$ maps to $(G_P, 1, \ldots, 1)$. This function F_P therefore has valuations

$$\nu_P(F_P) = 1, \qquad \nu_Q(F_P) = 0 \quad \text{for all } Q \in S - \{P\}.$$

Now the result is clear. Take any divisor in $\mathrm{Div}^0(C)$. Adding the divisor of a suitable product $\prod_{P \in S} F_P^{e_P}$ gives an equivalent divisor at the level of $\mathrm{Pic}^0(C)$ but with $n_P = 0$ for all $P \in S$. $\qquad\square$

If $h : C \longrightarrow C'$ is a nonconstant morphism then it induces forward and reverse maps of Picard groups,

$$h_* : \mathrm{Pic}^0(C) \longrightarrow \mathrm{Pic}^0(C') \quad \text{and} \quad h^* : \mathrm{Pic}^0(C') \longrightarrow \mathrm{Pic}^0(C),$$

given by

$$h_*\left(\left[\sum_P n_P(P)\right]\right) = \left[\sum_P n_P(h(P))\right]$$

and

$$h^*\left(\left[\sum_Q n_Q(Q)\right]\right) = \left[\sum_Q n_Q \sum_{P \in h^{-1}(Q)} e_P(h)(P)\right].$$

Here the square brackets denote equivalence class modulo $\mathrm{Div}^\ell(C)$ or $\mathrm{Div}^\ell(C')$ as appropriate. These maps were denoted h_P and h^P in Section 6.2. They continue to make sense in the algebraic geometry environment because at the level of divisors they take degree-0 divisors to degree-0 divisors and they satisfy $h_*(\operatorname{div} f) = \operatorname{div}(\operatorname{norm}_h f)$ where norm is defined as in Section 6.2 (cf. Exercise 6.2.4) and $h^*(\operatorname{div} g) = \operatorname{div}(h^* g)$ where this last h^* is the pullback. Also as in Section 6.2, mapping backwards and then forwards multiplies by the degree,

$$h_* \circ h^* = \deg(h) \qquad \text{on } \operatorname{Pic}^0(C').$$

For example, let $\mathcal{E}$ be an elliptic curve over $\mathbf{k}$ and consider a first degree function in $\overline{\mathbf{k}}(\mathcal{E})$,

$$F(x, y) = ax + by + c, \quad a, b, c \in \mathbf{k}, \ a \text{ or } b \text{ nonzero}.$$

This function has zeros at the affine points where the line with equation $F(x, y) = 0$ meets $\mathcal{E}$, and it has a pole at $0_{\mathcal{E}}$ of the right order for its divisor to have degree 0. Letting P, Q, R denote the points where the line meets $\mathcal{E}$, allowing repeats and allowing $0_{\mathcal{E}}$, in all cases

$$\operatorname{div}(F) = (P) + (Q) + (R) - 3(0_{\mathcal{E}}).$$

Specializing to $F(x, y) = x$, the vertical line meets the affine part of $\mathcal{E}$ twice, showing again that $\nu_{0_{\mathcal{E}}}(x) = -2$ as in the previous section; specializing to $F(x, y) = y$, the horizontal line meets the affine part of $\mathcal{E}$ three times, showing again that $\nu_{0_{\mathcal{E}}}(y) = -3$. Thus a uniformizer at $0_{\mathcal{E}}$ is x/y as before.

In particular, let $\mathcal{E}$ be the elliptic curve over $\mathbb{Q}$ with Weierstrass equation $E : y^2 = 4x^3 - 4x$. Consider the rational function

$$F(x, y) = \frac{y}{x} \in \overline{\mathbb{Q}}(\mathcal{E}).$$

The numerator y of F vanishes at the points $(\pm 1, 0)$ of $\mathcal{E}$, where the denominator x is nonzero and so F has zeros. To compute their orders, note that y is a uniformizer at these points since $D_1 E(\pm 1, 0) \neq 0$, making $\nu_{(\pm 1, 0)}(F) = 1$. The numerator y is also a uniformizer at the point $(0, 0)$ of $\mathcal{E}$ where the denominator x vanishes, but as in the previous section $F(x, y) = (4x^2 - 4)/y$ as well and F has a pole at $(0, 0)$ with $\nu_{(0,0)}(F) = -1$. There are no other zeros or poles at affine points of $\mathcal{E}$. At the infinite point $0_{\mathcal{E}}$ the previous paragraph shows that F has a pole and $\nu_{0_{\mathcal{E}}}(F) = -1$. Thus

$$\operatorname{div}(F) = -(0, 0) + (1, 0) + (-1, 0) - (0_{\mathcal{E}}).$$

Section 1.4, especially Exercise 1.4.3, shows that this is essentially the same as the divisor of the meromorphic function $F(z) = \wp_i'(z)/\wp_i(z)$ on the complex elliptic curve $\mathbb{C}/\Lambda_i$ as shown in Section 3.4. See Exercise 7.3.3 for a similar example.

The next section will redefine the Weil pairing for an elliptic curve $\mathcal{E}$ over a field $\mathbf{k}$. Doing so requires a characterization of principal divisors on the curve, to be derived from a small calculation. Consider two distinct points P and Q of $\mathcal{E}$. The claim is that there is no rational function $F \in \overline{\mathbf{k}}(\mathcal{E})$ with divisor $(P) - (Q)$. This can be shown using general machinery since such a map would have to be an isomorphism from the genus 1 curve $\mathcal{E}$ to the genus 0 curve $\mathbb{P}^1(\overline{\mathbf{k}})$ and genus is an algebraic invariant (or cf. Exercise 6.1.3), but we give an elementary proof here. Given such a function F, considering $F \circ \tau_Q$ (where again $\tau_Q(X) = X + Q$) shows that we may assume $Q = 0_{\mathcal{E}}$, i.e., $\mathrm{div}(F) = (P) - (0_{\mathcal{E}})$. The function F takes the form (Exercise 7.3.4(a))

$$F(x, y) = F_1(x)y + F_2(x), \quad F_1, F_2 \in \overline{\mathbf{k}}(x). \tag{7.6}$$

At an affine point R of $\mathcal{E}$ whose y-coordinate is nonzero, (7.6) gives the second equality in the calculation

$$\nu_R(F_1(x)) = \nu_R(F_1(x)y) = \nu_R((F(x, y) - F(x, -y))/2)$$
$$\geq \min\{\nu_R(F(x, y)), \nu_R(F(x, -y))\} \geq 0.$$

At an affine point R whose y-coordinate is 0, $\nu_R(y) = 1$ and $\nu_R(x)$ is 0 or 2, making $\nu_R(F_i(x))$ even for $i = 1, 2$ (Exercise 7.3.4(b)) and so $\nu_R(F_1(x)y)$ and $\nu_R(F_2(x))$ are distinct. This gives

$$0 \leq \nu_R(F) = \min\{\nu_R(F_1(x)) + 1, \nu_R(F_2(x))\} \leq \nu_R(F_1(x)) + 1,$$

and again $\nu_R(F_1(x)) \geq 0$ since it is even. At the projective point $0_{\mathcal{E}}$, $\nu_{0_{\mathcal{E}}}(x) = -2$ and $\nu_{0_{\mathcal{E}}}(y) = -3$ as shown above, so $\nu_{0_{\mathcal{E}}}(F_i(x))$ is even for $i = 1, 2$ (Exercise 7.3.4(b) again) and the relation $\nu_{0_{\mathcal{E}}}(F) = -1$ is

$$\min\{\nu_{0_{\mathcal{E}}}(F_1(x)) - 3, \nu_{0_{\mathcal{E}}}(F_2(x))\} = -1,$$

forcing $\nu_{0_{\mathcal{E}}}(F_1(x)) = 2$. Thus $\sum_{P \in \mathcal{E}} \nu_P(F_1(x)) > 0$, contradicting the fact that principal divisors have degree 0. It follows that no F exists as stipulated.

These examples set up

Lemma 7.3.2. *Let $\mathcal{E}$ be an elliptic curve over* $\mathbf{k}$.

(a) *For any points* $P, Q \in \mathcal{E}$,

$$\text{the divisor } (P) - (Q) \text{ is principal} \iff P = Q.$$

(b) *For any points* $P, Q \in \mathcal{E}$ *there is a congruence of divisors*

$$(P) - (0_{\mathcal{E}}) + (Q) - (0_{\mathcal{E}}) \equiv (P + Q) - (0_{\mathcal{E}}) \pmod{\mathrm{Div}^{\ell}(\mathcal{E})}.$$

Proof. (a) The implication ($\Longrightarrow$) was shown in the calculation above. The other implication is immediate.

(b) Let the line through P and Q have equation $ax + by + c = 0$ and let $R = -P - Q$ be the third collinear point of $\mathcal{E}$. The formula $\mathrm{div}(F) = (P) + (Q) + (R) - 3(0_{\mathcal{E}})$ for $F(x, y) = ax + by + c$ shows that

$$(P) - (0_{\mathcal{E}}) + (Q) - (0_{\mathcal{E}}) + (R) - (0_{\mathcal{E}}) \in \mathrm{Div}^{\ell}(\mathcal{E}).$$

If $R = 0_\mathcal{E}$ this gives the result. Otherwise let the vertical line through R and $-R$ have equation $x + c = 0$. Changing the example to $F(x, y) = x + c$ shows that

$$(R) - (0_\mathcal{E}) + (-R) - (0_\mathcal{E}) \in \mathrm{Div}^\ell(\mathcal{E}),$$

and since $-R = P + Q$ the result follows from subtracting the two displayed relations.   □

Recall that in the complex analytic environment of Chapter 6 the map of Abel's Theorem followed by the natural identification of the Jacobian of an elliptic curve with the curve itself gave a composite isomorphism from $\mathrm{Pic}^0(\mathcal{E})$ to $\mathcal{E}$. We can now prove an algebraic version of this, no longer requiring $\mathbf{k} = \mathbb{C}$.

Theorem 7.3.3. *Let $\mathcal{E}$ be an elliptic curve. Then the map*

$$\mathrm{Div}(\mathcal{E}) \longrightarrow \mathcal{E}, \qquad \sum n_P(P) \mapsto \sum [n_P]P$$

induces an isomorphism

$$\mathrm{Pic}^0(\mathcal{E}) \xrightarrow{\;\sim\;} \mathcal{E}.$$

Therefore the principal divisors on $\mathcal{E}$ are characterized by the condition

$$\sum n_P(P) \in \mathrm{Div}^\ell(\mathcal{E}) \iff \sum n_P = 0 \text{ and } \sum [n_P]P = 0_\mathcal{E}.$$

Proof. The map $\mathrm{Div}(\mathcal{E}) \longrightarrow \mathcal{E}$ of the theorem is clearly a homomorphism. Its restriction to $\mathrm{Div}^0(\mathcal{E})$ is surjective since $(P) - (0_\mathcal{E}) \mapsto P$ for any $P \in \mathcal{E}$. The restriction has kernel $\mathrm{Div}^\ell(\mathcal{E})$ because for a degree-0 divisor $\sum n_P(P)$,

$$\sum [n_P]P = 0_\mathcal{E} \iff \left(\sum [n_P]P \right) - (0_\mathcal{E}) \in \mathrm{Div}^\ell(\mathcal{E}) \quad \text{by Lemma 7.3.2(a)}$$
$$\iff \sum n_P((P) - (0_\mathcal{E})) \in \mathrm{Div}^\ell(\mathcal{E}) \quad \text{by Lemma 7.3.2(b)}$$
$$\iff \sum n_P(P) \in \mathrm{Div}^\ell(\mathcal{E}) \quad \text{since } \sum n_P = 0.$$

□

If $h : \mathcal{E} \longrightarrow \mathcal{E}'$ is a morphism of elliptic curves over $\mathbf{k}$ such that $h(0_\mathcal{E}) = 0_{\mathcal{E}'}$ then the isomorphisms $\mathrm{Pic}^0(\mathcal{E}) \xrightarrow{\sim} \mathcal{E}$ and $\mathrm{Pic}^0(\mathcal{E}') \xrightarrow{\sim} \mathcal{E}'$ of Theorem 7.3.3 and the fact that $h_* : \mathrm{Pic}^0(\mathcal{E}) \longrightarrow \mathrm{Pic}^0(\mathcal{E}')$ is a homomorphism combine to show that h is a homomorphism as well (Exercise 7.3.5). The characterization of principal divisors on elliptic curves in the theorem generalizes the complex analytic criterion from Exercise 1.4.1(c).

Exercises

7.3.1. (a) Show that the degree of a composition of morphisms over $\mathbf{k}$ is the product of the degrees, and so such a morphism is an isomorphism if and only if its degree is 1.

(b) Let $h : C \longrightarrow C'$ be a nonconstant morphism defined over $\mathbf{k}$. Show that $[\mathbf{k}(C) : h^*(\mathbf{k}(C'))] = [\overline{\mathbf{k}}(C) : h^*(\overline{\mathbf{k}}(C'))]$.

7.3.2. (a) Let $\{v_P : P \in S\}$ be a finite set of nonzero vectors in $\overline{\mathbf{k}}^{n+1}$. Show that there exists some vector $a \in \overline{\mathbf{k}}^{n+1}$ that is not orthogonal to any v_P. (A hint for this exercise is at the end of the book.)

(b) Show that $M_{S,P} \cap M_P^2 = M_{S,P}^2$.

7.3.3. Compute the divisor of the rational function $F(x,y) = y/x$ on the elliptic curve $y^2 = 4x^3 - 4$. Use the results of Exercise 1.4.3 to verify that it corresponds to the divisor of the meromorphic function $F(z) = \wp'_{\mu_3}(z)/\wp_{\mu_3}(z)$ on the complex elliptic curve $\mathbb{C}/\Lambda_{\mu_3}$.

7.3.4. (a) Letting $\mathcal{E}$ be defined by a Weierstrass equation (7.1), show that any rational function $F \in \overline{\mathbf{k}}(\mathcal{E})$ takes the form (7.6). (A hint for this exercise is at the end of the book.)

(b) Show that if $\nu_P(x)$ is even and nonzero, or if $\nu_P(x) = 0$ and $\nu_P(y) \neq 0$, then $\nu_P(F(x))$ is even for any rational function F.

7.3.5. Let $h : \mathcal{E} \longrightarrow \mathcal{E}'$ be a morphism of elliptic curves over $\mathbf{k}$ such that $h(0_{\mathcal{E}}) = 0_{\mathcal{E}'}$. Consider the diagram

$$\begin{array}{ccc} \mathrm{Pic}^0(\mathcal{E}) & \longrightarrow & \mathcal{E} \\ {\scriptstyle h_*}\downarrow & & \downarrow{\scriptstyle h} \\ \mathrm{Pic}^0(\mathcal{E}') & \longrightarrow & \mathcal{E}', \end{array}$$

where the top row is $[(P)-(0_{\mathcal{E}})] \mapsto P$ and similarly for the bottom row. Show that the diagram commutes and therefore h is a homomorphism.

7.4 The Weil pairing algebraically

To define an algebraic version of the Weil pairing using Theorem 7.3.3 let μ_N denote the group of Nth roots of unity in $\overline{\mathbf{k}}$,

$$\mu_N = \{x \in \overline{\mathbf{k}} : x^N = 1\}.$$

Since $\mathrm{char}(\mathbf{k}) = 0$ this can be viewed as the usual cyclic group of order N in $\overline{\mathbb{Q}}$. Recall that $\mathcal{E}[N] \cong (\mathbb{Z}/N\mathbb{Z})^2$. Construct the Weil pairing,

$$e_N : \mathcal{E}[N] \times \mathcal{E}[N] \longrightarrow \mu_N,$$

as follows. Let two N-torsion points $P, Q \in \mathcal{E}[N]$ be given, possibly $P = Q$. By the characterization of principal divisors in Theorem 7.3.3 there exists a function $f = f_Q \in \overline{\mathbf{k}}(\mathcal{E})$ such that

$$\mathrm{div}(f) = N(Q) - N(0_{\mathcal{E}}).$$

(The notation $F = f/g$ is no longer needed, so rational functions are denoted by lower case letters from now on.) Since $[N] : \mathcal{E} \longrightarrow \mathcal{E}$ is unramified, the map $f \circ [N]$ has divisor

$$\operatorname{div}(f \circ [N]) = \sum_{R:[N]R=Q} N(R) - \sum_{S:[N]S=0_{\mathcal{E}}} N(S).$$

If $Q' \in \mathcal{E}[N^2]$ is any point such that $[N]Q' = Q$ then the divisor is

$$\operatorname{div}(f \circ [N]) = N \sum_{S \in \mathcal{E}[N]} (Q' + S) - (S).$$

Again by Theorem 7.3.3, this time applied to the sum on the right, the equality takes the form $\operatorname{div}(f \circ [N]) = N\operatorname{div}(g)$ for some $g = g_Q \in \overline{\mathbf{k}}(\mathcal{E})$. Thus $f \circ [N] = g^N$ after multiplying g by a nonzero constant if necessary. For any point $X \in \mathcal{E}$,

$$g(X + P)^N = f([N]X + [N]P) = f([N]X) = g(X)^N.$$

Consider the rational function $g(X + P)/g(X) \in \overline{\mathbf{k}}(\mathcal{E})$. Since its Nth power is 1 its image is a subset of $\boldsymbol{\mu}_N$, making the map constant since it doesn't surject from $\mathcal{E}$ to $\mathbb{P}^1(\overline{\mathbf{k}})$. The image point is the Weil pairing of P and Q,

$$e_N(P,Q) = \frac{g_Q(X + P)}{g_Q(X)} \in \boldsymbol{\mu}_N.$$

Proposition 7.4.1 (Properties of the Weil pairing).

(a) *The Weil pairing is bilinear,*

$$e_N(aP + bP', cQ + dQ') = e_N(P,Q)^{ac} e_N(P,Q')^{ad} e_N(P',Q)^{bc} e_N(P',Q')^{bd}.$$

(b) *The Weil pairing is alternating and therefore skew symmetric,*

$$e_N(Q,Q) = 1, \qquad e_N(Q,P) = e_N(P,Q)^{-1}.$$

(c) *The Weil pairing is nondegenerate,*

$$\text{if } e_N(P,Q) = 1 \text{ for all } P \in \mathcal{E}[N] \text{ then } Q = 0_{\mathcal{E}}.$$

Consequently the Weil pairing surjects to $\boldsymbol{\mu}_N$.
(d) *The Weil pairing is Galois invariant,*

$$e_N(P,Q)^{\sigma} = e_N(P^{\sigma}, Q^{\sigma}) \quad \text{for all } \sigma \in \operatorname{Gal}(\overline{\mathbf{k}}/\mathbf{k}).$$

(e) *The Weil pairing is invariant under isomorphism, i.e., if $i : \mathcal{E} \longrightarrow \mathcal{E}'$ is an isomorphism and e'_N is the Weil pairing on $\mathcal{E}'$ then*

$$e'_N(i(P), i(Q)) = e_N(P,Q) \quad \text{for all } P, Q \in \mathcal{E}[N].$$

Proof. (a) For linearity in the first argument let $g = g_Q$ and compute

$$e_N(P_1 + P_2, Q) = \frac{g(X + P_1 + P_2)}{g(X)} = \frac{g(X + P_1 + P_2)}{g(X + P_2)} \frac{g(X + P_2)}{g(X)}$$
$$= e_N(P_1, Q)e_N(P_2, Q),$$

using $X + P_2$ for X to evaluate $e_N(P_1, Q)$. For linearity in the second argument, let f_1, f_2, f_3, and g_1, g_2, g_3 be the functions associated to Q_1, Q_2, and $Q_1 + Q_2$. Choose $h \in \overline{\mathbf{k}}(\mathcal{E})$ with divisor

$$\text{div}(h) = (Q_1 + Q_2) - (Q_1) - (Q_2) + (0_\mathcal{E}).$$

Then $\text{div}(f_3/(f_1 f_2)) = N\text{div}(h)$, so $f_3 = cf_1 f_2 h^N$ for some $c \in \overline{\mathbf{k}}^*$. Thus $f_3 \circ [N] = c(f_1 \circ [N])(f_2 \circ [N])(h \circ [N])^N$, or $g_3^N = c(g_1 g_2 (h \circ [N]))^N$, or $g_3 = c' g_1 g_2 (h \circ [N])$. Using this, compute

$$e_N(P, Q_1 + Q_2) = \frac{g_3(X + P)}{g_3(X)} = \frac{g_1(X + P)g_2(X + P)h([N]X + [N]P)}{g_1(X)g_2(X)h([N]X)}$$
$$= e_N(P, Q_1)e_N(P, Q_2).$$

(b) To show $e_N(Q, Q) = 1$ for all $Q \in \mathcal{E}[N]$, let $f = f_Q$ and $g = g_Q$ and compute

$$\text{div}\left(\prod_{n=0}^{N-1} f(X + [n]Q) \right) = \sum_{n=0}^{N-1} N([1 - n]Q) - N([-n]Q) = 0.$$

Hence $\prod_{n=0}^{N-1} f(X + [n]Q)$ is constant, and so if $Q = [N]Q'$ then also $\prod_{n=0}^{N-1} g(X + [n]Q')$ is constant because its Nth power is the previous product at $[N]X$. Equating the second product at X and $X + Q'$ gives $g(X) = g(X + [N]Q') = g(X + Q)$, and $e_N(Q, Q) = g(X + Q)/g(X) = 1$ as desired. Every alternating bilinear form is skew symmetric (Exercise 7.4.1(a)), giving the rest of (b).

(c) For nondegeneracy, if $e_N(P, Q) = 1$ for all $P \in \mathcal{E}[N]$ and $g = g_Q$ then $g(X + P) = g(X)$ for all $P \in \mathcal{E}[N]$. This periodicity implies that $g = h \circ [N]$ for some function h, but we don't yet know that h is rational. Let $\tau^* : \mathcal{E}[N] \longrightarrow \text{Aut}(\overline{\mathbf{k}}(\mathcal{E}))$ map each N-torsion point P to its associated translation automorphism τ_P^* taking $f(X)$ to $f(X + P)$ for all $f \in \overline{\mathbf{k}}(\mathcal{E})$. Then τ^* is a homomorphism. If $P \in \ker(\tau^*)$ then in particular τ_P acts trivially on the function f_P with divisor $N(P) - N(0_\mathcal{E})$ as in the definition of the Weil pairing, so $f_P(0_\mathcal{E}) = \tau_P^* f_P(0_\mathcal{E}) = f_P(P)$ and this forces $P = 0_\mathcal{E}$. Thus τ^* injects, $\overline{\mathbf{k}}(\mathcal{E})$ is a Galois extension of its subfield fixed by $\tau^*(\mathcal{E}[N])$, and the Galois group is isomorphic to $\mathcal{E}[N]$. But the fixed field contains the subfield $[N]^*(\overline{\mathbf{k}}(\mathcal{E})) = \{h \circ [N] : h \in \overline{\mathbf{k}}(\mathcal{E})\}$, while formula (7.5) is $[\overline{\mathbf{k}}(\mathcal{E}) : [N]^*(\overline{\mathbf{k}}(\mathcal{E}))] = N^2$. Since $N^2 = |\mathcal{E}[N]|$ this makes the fixed field $[N]^*(\overline{\mathbf{k}}(\mathcal{E}))$. The condition $g(X + P) = g(X)$ for all $P \in \mathcal{E}[N]$ means that g is in the fixed field, i.e., $g = h \circ [N]$ for some $h \in \overline{\mathbf{k}}(\mathcal{E})$. From this and from the setup of the Weil pairing,

$$\mathrm{div}(h \circ [N]) = \mathrm{div}(g) = \sum_{S \in \mathcal{E}[N]} (Q' + S) - (S) \qquad \text{where } [N]Q' = Q.$$

It follows that $\mathrm{div}(h) = (Q) - (0_{\mathcal{E}})$, and so $Q = 0_{\mathcal{E}}$ by Lemma 7.3.2(a), making the Weil pairing nondegenerate. To see that it surjects, note that the image $e_N(\mathcal{E}[N], \mathcal{E}[N])$ is a subgroup $\boldsymbol{\mu}_M$ of $\boldsymbol{\mu}_N$ for some $M \mid N$. Thus $e_N([M]P, Q) = e_N(P, Q)^M = 1$ for all points $P, Q \in \mathcal{E}[N]$, forcing $[M]P = 0_{\mathcal{E}}$ for all P, hence $M = N$.

(d) For Galois invariance, let $\sigma \in \mathrm{Gal}(\overline{\mathbf{k}}/\mathbf{k})$. Then $f_{Q^\sigma} = f_Q^\sigma$ and $g_{Q^\sigma} = g_Q^\sigma$ (Exercise 7.4.1(b)), and so

$$e_N(P^\sigma, Q^\sigma) = \frac{g^\sigma(X^\sigma + P^\sigma)}{g^\sigma(X^\sigma)} = \left(\frac{g(X + P)}{g(X)} \right)^\sigma = e_N(P, Q)^\sigma.$$

(e) If $i : \mathcal{E} \longrightarrow \mathcal{E}'$ is an isomorphism then running through the definition shows that the function g' for e_N' is $g_{Q'}' = g_{i^{-1}(Q')} \circ i^{-1}$. Thus

$$e_N'(i(P), i(Q)) = \frac{g_{i(Q)}'(X' + i(P))}{g_{i(Q)}'(X')} = \frac{g_Q(X + P)}{g_Q(X)} = e_N(P, Q).$$

$\square$

We will use the following result (Exercise 7.4.2) in the next section. As ever, $\mathrm{char}(\mathbf{k}) = 0$.

Corollary 7.4.2. *Suppose P, Q, P', and Q' are points of $\mathcal{E}[N]$, and P', Q' are linear combinations of P, Q,*

$$\begin{bmatrix} P' \\ Q' \end{bmatrix} = \gamma \begin{bmatrix} P \\ Q \end{bmatrix} \qquad \text{for some } \gamma \in \mathrm{M}_2(\mathbb{Z}/N\mathbb{Z}).$$

Then

$$e_N(P', Q') = e_N(P, Q)^{\det \gamma}.$$

In particular, if (P, Q) is an ordered basis of $\mathcal{E}[N]$ over $\mathbb{Z}/N\mathbb{Z}$ then $e_N(P, Q)$ is a primitive Nth root of unity.

If $e_N' : \mathcal{E}[N] \times \mathcal{E}[N] \longrightarrow \boldsymbol{\mu}_N$ is any function satisfying properties (a), (b), and (c) of the Weil pairing then $e_N' = e_N^n$ for some $n \in (\mathbb{Z}/N\mathbb{Z})^*$ (Exercise 7.4.3). This shows that the Weil pairing as defined on a complex torus $\mathbb{C}/\Lambda$ in Section 1.3 is a power of the algebraic Weil pairing on the elliptic curve $y^2 = 4x^3 - g_2(\Lambda)x - g_3(\Lambda)$. In fact they are equal, though we omit the exercise in complex function theory that shows this.

Exercises

7.4.1. (a) Show that an alternating bilinear form is skew symmetric.

(b) Let $\sigma \in \mathrm{Gal}(\overline{\mathbf{k}}/\mathbf{k})$. Explain why $f_{Q^\sigma} = f_Q^\sigma$ and $g_{Q^\sigma} = g_Q^\sigma$.

7.4.2. Prove Corollary 7.4.2.

7.4.3. Show that any function satisfying properties (a), (b), and (c) of the Weil pairing takes the form e_N^n for some $n \in (\mathbb{Z}/N\mathbb{Z})^*$.

7.5 Function fields over $\mathbb{C}$

From Section 3.2, the function field of the level 1 modular curve $X(1) = \mathrm{SL}_2(\mathbb{Z})\backslash\mathcal{H}^*$ over $\mathbb{C}$ is generated by the modular invariant,

$$\mathbb{C}(X(1)) = \mathcal{A}_0(\mathrm{SL}_2(\mathbb{Z})) = \mathbb{C}(j).$$

This section describes the function fields for the curves $X(N)$, $X_1(N)$, and $X_0(N)$, where $N > 1$. The extension $\mathbb{C}(X(N))/\mathbb{C}(X(1))$ turns out to be Galois with group $\mathrm{SL}_2(\mathbb{Z}/N\mathbb{Z})/\{\pm I\}$, and it reproduces the moduli space description of modular curves from Chapter 1. The end of the section constructs a related function field extension with Galois group $\mathrm{SL}_2(\mathbb{Z}/N\mathbb{Z})$.

For each vector $v \in \mathbb{Z}^2$ such that $\overline{v} \neq \mathbf{0}$, where as before $\overline{v}$ is the reduction of v modulo N, recall from Section 1.5 that the function

$$f_0^{\overline{v}}(\tau) = \frac{g_2(\tau)}{g_3(\tau)} \, \wp_\tau\left(\frac{c_v\tau + d_v}{N}\right), \quad v = (c_v, d_v) \tag{7.7}$$

is weight-0 invariant under $\Gamma(N)$. In fact $f_0^{\overline{v}} \in \mathbb{C}(X(N))$. To see this we must show that $f_0^{\overline{v}}$ is meromorphic on $\mathcal{H}$ and at the cusps. Meromorphy on $\mathcal{H}$ is straightforward (Exercise 7.5.1(a)). For the cusps, first verify that (Exercise 7.5.1(b))

$$f_0^{\overline{v}}(\gamma\tau) = f_0^{\overline{v\gamma}}(\tau), \quad \gamma \in \mathrm{SL}_2(\mathbb{Z}), \ \tau \in \mathcal{H}.$$

Then checking that all $f_0^{\overline{v}}$ are meromorphic at ∞ (Exercise 7.5.1(c)) also checks that they are all meromorphic at the other cusps as well (Exercise 7.5.1(d)). Since the other functions from Section 1.5,

$$f_0^{\overline{d}}(\tau) = f_0^{\overline{(0,d)}}(\tau), \ d \not\equiv 0 \ (\mathrm{mod} \ N) \quad \text{and} \quad f_0(\tau) = \sum_{d=1}^{N-1} f_0^{\overline{d}}(\tau),$$

are sums of special cases of $f_0^{\overline{v}}$ and invariant under $\Gamma_1(N)$ and $\Gamma_0(N)$ respectively, it follows that $f_0^{\overline{d}} \in \mathbb{C}(X_1(N))$ and $f_0 \in \mathbb{C}(X_0(N))$. For another example, let $j_N(\tau) = j(N\tau)$. Then $j_N \in \mathbb{C}(X_0(N))$ by the methods of Exercise 1.2.11 (Exercise 7.5.2). These are all the examples we need. Introduce a little more notation for convenience,

$$f_{1,0} = f_0^{\pm\overline{(1,0)}}, \quad f_{0,1} = f_1 = f_0^{\pm\overline{(0,1)}}.$$

Proposition 7.5.1. *The fields of meromorphic functions on $X(N)$, $X_1(N)$, and $X_0(N)$ are*

$$\mathbb{C}(X(N)) = \mathbb{C}(j, \{f_0^{\pm\overline{v}} : \pm\overline{v} \in ((\mathbb{Z}/N\mathbb{Z})^2 - (0,0))/\pm\})$$
$$= \mathbb{C}(j, f_{1,0}, f_{0,1}),$$
$$\mathbb{C}(X_1(N)) = \mathbb{C}(j, \{f_0^{\pm\overline{d}} : \pm\overline{d} \in (\mathbb{Z}/N\mathbb{Z} - 0)/\pm\})$$
$$= \mathbb{C}(j, f_1),$$
$$\mathbb{C}(X_0(N)) = \mathbb{C}(j, f_0) = \mathbb{C}(j, j_N).$$

Proof. Since $\wp_\tau(z) = \wp_\tau(z')$ if and only if $z \equiv \pm z' \pmod{\Lambda_\tau}$, each pair of functions $f_0^{\bar{v}}$ and $f_0^{-\bar{v}}$ are equal but otherwise the $f_0^{\bar{v}}$ are all distinct. Taking the set $\{f_0^{\pm\bar{v}} : \pm\bar{v} \in ((\mathbb{Z}/N\mathbb{Z})^2 - (0,0))/\pm\}$ gives the field containments

$$\mathbb{C}(X(1)) = \mathbb{C}(j) \subset \mathbb{C}(j, \{f_0^{\pm\bar{v}}\}) \subset \mathbb{C}(X(N)).$$

Consider the map

$$\theta : \mathrm{SL}_2(\mathbb{Z}) \longrightarrow \mathrm{Aut}(\mathbb{C}(X(N))),$$

$$f^{\theta(\gamma)} = f \circ \gamma \quad \text{for } f \in \mathbb{C}(X(N)), \ \gamma \in \mathrm{SL}_2(\mathbb{Z}).$$

This map is a homomorphism (Exercise 7.5.3(a)) and its kernel clearly contains $\{\pm I\}\Gamma(N)$. Since any element of the kernel fixes the subfield $\mathbb{C}(j, \{f_0^{\pm\bar{v}}\})$, the earlier calculation $f_0^{\bar{v}} \circ \gamma = f_0^{\bar{v}\gamma}$ for $\gamma \in \mathrm{SL}_2(\mathbb{Z})$ and the fact that the $f_0^{\pm\bar{v}}$ are distinct show that the kernel is also a subgroup of $\{\pm I\}\Gamma(N)$, making them equal. Thus $\theta(\mathrm{SL}_2(\mathbb{Z})) \cong \mathrm{SL}_2(\mathbb{Z})/\{\pm I\}\Gamma(N)$. Since $\theta(\mathrm{SL}_2(\mathbb{Z}))$ is a group of automorphisms of $\mathbb{C}(X(N))$ and the subfield it fixes is $\mathbb{C}(X(1))$ by definition, the extension $\mathbb{C}(X(N))/\mathbb{C}(X(1))$ is Galois with Galois group $\theta(\mathrm{SL}_2(\mathbb{Z}))$. The formula $f_0^{\bar{v}} \circ \gamma = f_0^{\bar{v}\gamma}$ shows that the intermediate field $\mathbb{C}(j, \{f_0^{\pm\bar{v}}\})$ and its subfield $\mathbb{C}(j, f_{1,0}, f_{0,1})$ both have trivial fixing subgroup and thus both are all of $\mathbb{C}(X(N))$, proving the first part of the proposition. The other two parts are shown by similar arguments (Exercise 7.5.3(b,c)).   □

The function fields in the proposition satisfy the description given before Theorem 7.2.5 of function fields of algebraic curves over $\mathbb{C}$. Thus $X_1(N)$ is birationally equivalent to the plane curve defined by the complex polynomial $\varphi_1 \in \mathbb{C}[x,y]$ such that $\varphi_1(j, f_1) = 0$, and $X_0(N)$ is birationally equivalent to the plane curve defined by the complex polynomial $\varphi_0 \in \mathbb{C}[x,y]$ such that $\varphi_0(j, f_0) = 0$. Section 7.7 will show that in fact the polynomials have rational coefficients. In connection with $X_0(N)$ the proposition also leads to the *modular equation*, the relation between j and j_N, which will also turn out to be defined over $\mathbb{Q}$,

$$\Phi_N(j, j_N) = 0, \qquad \Phi_N(x,y) \in \mathbb{Q}[x,y].$$

This classical topic is discussed in [Cox97], in [Kna93], and briefly in Milne's online notes [Mil], and we will not pursue it except as it bears on the material here.

The proof of Proposition 7.5.1 has shown more than the proposition asserts. Since the field extension $\mathbb{C}(X(N))/\mathbb{C}(X(1))$ is Galois and since $\mathrm{SL}_2(\mathbb{Z})/\Gamma(N) \cong \mathrm{SL}_2(\mathbb{Z}/N\mathbb{Z})$, the map θ in the proof induces an isomorphism

$$\theta^{-1} : \mathrm{Gal}(\mathbb{C}(X(N))/\mathbb{C}(X(1))) \xrightarrow{\sim} \mathrm{SL}_2(\mathbb{Z}/N\mathbb{Z})/\{\pm I\},$$

$$f^\sigma = f \circ \theta^{-1}(\sigma) \quad \text{for } f \in \mathbb{C}(X(N)), \ \sigma \in \mathrm{Gal}(\mathbb{C}(X(N))/\mathbb{C}(X(1))),$$

where $\theta^{-1}(\sigma)$ acts on $\mathcal{H}$ as any of its lifts to $\mathrm{SL}_2(\mathbb{Z})$. In particular, $(f_0^{\pm\bar{v}})^\sigma = f_0^{\pm\bar{v}} \circ \theta^{-1}(\sigma) = f_0^{\pm\bar{v}\theta^{-1}(\sigma)}$ so that θ^{-1} defines an action of the group $\mathrm{Gal}(\mathbb{C}(X(N))/\mathbb{C}(X(1)))$ on the set $((\mathbb{Z}/N\mathbb{Z})^2 - (0,0))/\pm$,

$$\pm\overline{v} \mapsto \pm\overline{v}\theta^{-1}(\sigma). \tag{7.8}$$

Exercise 7.5.3(b,c) shows that the subgroups of $\mathrm{SL}_2(\mathbb{Z}/N\mathbb{Z})/\{\pm I\}$ corresponding to $\Gamma_1(N))$ and $\Gamma_0(N)$ are

$$\theta^{-1}(\mathrm{Gal}(\mathbb{C}(X(N))/\mathbb{C}(X_1(N)))) = \{\pm\left[\begin{smallmatrix}1 & b \\ 0 & 1\end{smallmatrix}\right] \in \mathrm{SL}_2(\mathbb{Z}/N\mathbb{Z})/\{\pm I\}\} \tag{7.9}$$

and

$$\theta^{-1}(\mathrm{Gal}(\mathbb{C}(X(N))/\mathbb{C}(X_0(N)))) = \{\pm\left[\begin{smallmatrix}a & b \\ 0 & d\end{smallmatrix}\right] \in \mathrm{SL}_2(\mathbb{Z}/N\mathbb{Z})/\{\pm I\}\}. \tag{7.10}$$

To understand the results of Proposition 7.5.1 algebraically, recall yet again that the map $(\wp_\tau, \wp'_\tau)$ takes the complex elliptic curve $\mathbb{C}/\Lambda_\tau$ to the algebraic curve $E_\tau : y^2 = 4x^3 - g_2(\tau)x - g_3(\tau)$. (From now on the symbol E interchangeably denotes an elliptic curve, its equation, or its polynomial, and the symbol $\mathcal{E}$ will no longer be used.) Equation (7.7), defining the functions $f_0^{\overline{v}}(\tau)$ in terms of $\wp_\tau((c_v\tau + d_v)/N)$, thus combines with Proposition 7.5.1 to show that the meromorphic functions on $X(N)$ are generated by the modular invariant j and by functions of τ closely related to the x-coordinates of the nonzero N-torsion points on the curve E_τ. We now scale the curve E_τ to a new curve so that indeed $\mathbb{C}(X(N))$ is generated by j and the N-torsion x-coordinates.

Fix any $\tau \in \mathcal{H}$ such that $j(\tau) \notin \{0, 1728\}$. This means that $g_2(\tau)$ and $g_3(\tau)$ are nonzero since $j = 1728g_2^3/(g_2^3 - 27g_3^2)$. Choose either complex square root $(g_2(\tau)/g_3(\tau))^{1/2}$ and consider the map

$$\left(\frac{g_2(\tau)}{g_3(\tau)}\wp_\tau, \left(\frac{g_2(\tau)}{g_3(\tau)}\right)^{3/2}\wp'_\tau\right) : \mathbb{C}/\Lambda_\tau \longrightarrow \mathbb{C}^2 \cup \{\infty\}.$$

This differs from $(\wp_\tau, \wp'_\tau)$ by the admissible change of variable $(x, y) = (u^2x', u^3y')$ where $u = (g_3(\tau)/g_2(\tau))^{1/2}$. Thus $u^{-4} = (g_2(\tau)/g_3(\tau))^2$ and $u^{-6} = (g_2(\tau)/g_3(\tau))^3$, and nonzero points $z + \Lambda_\tau$ map to points (x, y) satisfying the suitable modification of the cubic equation of E_τ according to Exercise 7.1.1,

$$E_{j(\tau)} : y^2 = 4x^3 - \frac{g_2(\tau)^3}{g_3(\tau)^2}x - \frac{g_2(\tau)^3}{g_3(\tau)^2}.$$

Since $g_3^2 = (g_2^3 - \Delta)/27$, so that $g_2^3/g_3^2 = 27g_2^3/(g_2^3 - \Delta) = 27j/(j - 1728)$, the equation is defined in terms of $j(\tau)$, justifying the name of the curve,

$$E_{j(\tau)} : y^2 = 4x^3 - \left(\frac{27j(\tau)}{j(\tau) - 1728}\right)x - \left(\frac{27j(\tau)}{j(\tau) - 1728}\right).$$

(Cf. the method in Section 7.1 reducing the general $y^2 = 4x^3 - g_2x - g_3$ to the form $y^2 = 4x^3 - gx - g$.) The equation is independent of which square root $(g_2(\tau)/g_3(\tau))^{1/2}$ was chosen. The map $\mathbb{C}/\Lambda_\tau \xrightarrow{\sim} E_{j(\tau)}$ restricts to an isomorphism of N-torsion subgroups. In particular it takes the canonical generators $\tau/N + \Lambda_\tau$ and $1/N + \Lambda_\tau$ of $(\mathbb{C}/\Lambda_\tau)[N]$ to the points

$$P_\tau = \left(\frac{g_2(\tau)}{g_3(\tau)} \, \wp_\tau(\tau/N), \, \left(\frac{g_2(\tau)}{g_3(\tau)} \right)^{3/2} \wp'_\tau(\tau/N) \right),$$

$$Q_\tau = \left(\frac{g_2(\tau)}{g_3(\tau)} \, \wp_\tau(1/N), \, \left(\frac{g_2(\tau)}{g_3(\tau)} \right)^{3/2} \wp'_\tau(1/N) \right). \tag{7.11}$$

Negating the square root $(g_2(\tau)/g_3(\tau))^{1/2}$ negates these points, but modulo this (P_τ, Q_τ) is a canonical ordered basis of $E_{j(\tau)}[N]$ over $\mathbb{Z}/N\mathbb{Z}$. The x-coordinates of $\pm P_\tau$ and $\pm Q_\tau$ are $f_{1,0}(\tau)$ and $f_{0,1}(\tau) = f_1(\tau)$ respectively, and more generally the nonzero points of $E_{j(\tau)}[N]$ have x-coordinates $\{f_0^{\pm\bar{v}}(\tau)\}$ as desired. The information $j(\tau)$, $f_{1,0}(\tau)$, $f_{0,1}(\tau)$ thus describes an enhanced elliptic curve for $\Gamma(N)$ modulo negation,

$$(E_{j(\tau)}, \pm(P_\tau, Q_\tau)).$$

This is the sort of element that represents a point $[\mathbb{C}/\Lambda_\tau, (\tau/N + \Lambda_\tau, 1/N + \Lambda_\tau)]$ of the moduli space $S(N)$, excluding the finitely many such points with $j(\tau) \in \{0, 1728\}$. Similarly, the information $j(\tau)$, $f_1(\tau)$ describes $(E_{j(\tau)}, \pm Q_\tau)$, representing a point of $S_1(N)$, and $j(\tau)$, $f_0(\tau)$ describes $(E_{j(\tau)}, \langle Q_\tau \rangle)$, representing a point of $S_0(N)$. The moduli space description of modular curves is emerging from the function field description in Proposition 7.5.1.

Change τ to a variable so that $j = j(\tau)$ varies as well. This gathers the family of elliptic curves $E_{j(\tau)}$ into a single *universal elliptic curve*,

$$E_j : y^2 = 4x^3 - \left(\frac{27j}{j - 1728} \right) x - \left(\frac{27j}{j - 1728} \right), \tag{7.12}$$

whose j-invariant is indeed the variable j (Exercise 7.5.4). The universal elliptic curve specializes to a complex elliptic curve for every complex j except 0 and 1728. To make this idea algebraic view j as a transcendental element over $\mathbb{C}$ and view E_j as an elliptic curve over the field $\mathbb{C}(j)$. The affine points of E_j are ordered pairs from the algebraic closure, elements of $\overline{\mathbb{C}(j)}^2$. The finite N-torsion x-coordinates of E_j are $\{f_0^{\pm\bar{v}}\}$, functions on $X(N)$ rather than $X(1)$, algebraically elements of $\overline{\mathbb{C}(j)}$ rather than of $\mathbb{C}(j)$. The universal enhanced elliptic curves $(E_j, (f_{1,0}, f_{0,1}))$, (E_j, f_1), and (E_j, f_0), each viewed as a family of enhanced elliptic curves, are essentially the same thing as the moduli spaces $S(N)$, $S_1(N)$, and $S_0(N)$, and therefore as the modular curves $X(N)$, $X_1(N)$, and $X_0(N)$.

An argument is required to show that $\{f_0^{\pm\bar{v}}\}$ are the universal N-torsion x-coordinates. Specialize the discussion of division polynomials in Section 7.1 to the curve E_j to show that its N-torsion x-coordinates are characterized by a polynomial condition,

$$\tilde{\psi}_N(g, g, x) = 0, \qquad \tilde{\psi}_N \in \mathbb{Z}[g_2, g_3, x], \qquad g = 27j/(j - 1728).$$

Let $\bar{v}$ be a nonzero element of $(\mathbb{Z}/N\mathbb{Z})^2$. Consider the function field element

$$\tilde{\psi}_N(g, g, f_0^{\pm\bar{v}}) \in \mathbb{C}(X(N)).$$

For any $\tau \in \mathcal{H}$ such that $j(\tau) \notin \{0, 1728\}$, $E_{j(\tau)}$ specializes to an elliptic curve over $\mathbb{C}$, and $g(\tau)$ and $f_0^{\pm \overline{v}}(\tau)$ specialize to complex numbers. Since $f_0^{\pm \overline{v}}(\tau)$ is an N-torsion x-coordinate of $E_{j(\tau)}$,

$$\tilde{\psi}_N(g(\tau), g(\tau), f_0^{\pm \overline{v}}(\tau)) = 0.$$

Thus $\tilde{\psi}_N(g, g, f_0^{\pm \overline{v}})$ is a meromorphic function on the compact Riemann surface $X(N)$ with infinitely many zeros $\Gamma(N)\tau$, forcing it to be the zero function. This makes $f_0^{\pm \overline{v}}$ the x-coordinate of a point R of the universal curve E_j such that $[N]R = 0_{E_j}$. For each $\overline{v} \in (\mathbb{Z}/N\mathbb{Z})^2$, the function $f_0^{\pm \overline{v}}$ determines two N-torsion points of E_j unless $2v = 0$, in which case it determines one (Exercise 7.5.5(a)), and so we have found all N^2 points of $E_j[N]$ (Exercise 7.5.5(b)).

Let $x(E_j[N])$ denote the set of x-coordinates of nonzero points of $E_j[N]$ as an alternative notation to $\{f_0^{\pm \overline{v}}\}$. Thus $x(E_j[N]) \subset \overline{\mathbb{C}(j)}$. Proposition 7.5.1 says that adjoining $f_{1,0}$ and $f_{0,1}$ to $\mathbb{C}(j)$ adjoins all of $x(E_j[N])$. That is, the generators $\tau/N + \Lambda_\tau$ and $1/N + \Lambda_\tau$ of $(\mathbb{C}/\Lambda_\tau)[N]$, where the group law is addition modulo Λ_τ, map to generators of $x(E_j[N])$, where the group law is defined by rational functions. (Making this statement precise requires another specialization argument as in the previous paragraph.) More generally, the first part of Proposition 7.5.1 and of the discussion after its proof rephrase as

Proposition 7.5.2. *Let E_j be the universal elliptic curve* (7.12). *The field of meromorphic functions on $X(N)$ is the field generated by the x-coordinates of nonzero N-torsion points on E_j,*

$$\mathbb{C}(X(N)) = \mathbb{C}(j, x(E_j[N])).$$

Thus the field extension $\mathbb{C}(j, x(E_j[N]))/\mathbb{C}(j)$ is Galois and its group is

$$\mathrm{Gal}(\mathbb{C}(j, x(E_j[N]))/\mathbb{C}(j)) \cong \mathrm{SL}_2(\mathbb{Z}/N\mathbb{Z})/\{\pm I\}.$$

We now adjoin the y-coordinates of $E_j[N]$ to $\mathbb{C}(j)$ as well, thus eliminating the quotient by $\{\pm I\}$ from the Galois group. That is, let $y(E_j[N])$ denote the set of y-coordinates of nonzero N-torsion points of E_j. Analytically these are the functions

$$g_0^{\overline{v}}(\tau) = \left(\frac{g_2(\tau)}{g_3(\tau)}\right)^{3/2} \wp_\tau'\left(\frac{c_v \tau + d_v}{N}\right), \qquad v = (c_v, d_v).$$

Since they are not defined on $X(N)$ because of the square root, but rather on a two-sheeted surface lying over $X(N)$, it is easiest just to think of them as elements of $\overline{\mathbb{C}(j)}$. We will not use the notation $g_0^{\overline{v}}(\tau)$ any further, but special cases of it are implicit in the canonical (up to negation) basis (P_τ, Q_τ) of the universal N-torsion group $E_j[N]$, defined by (7.11) but now viewing τ as a variable. Let $\mathbb{C}(j, E_j[N])$ denote the field generated over $\mathbb{C}(j)$ by $x(E_j[N]) \cup y(E_j[N])$. Consider the field containments

$$\mathbb{C}(j) \subset \mathbb{C}(j, x(E_j[N])) \subset \mathbb{C}(j, E_j[N]) \subset \overline{\mathbb{C}(j)}.$$

Corollary 7.5.3. *The field extension $\mathbb{C}(j, E_j[N])/\mathbb{C}(j)$ is Galois and its group is*

$$\mathrm{Gal}(\mathbb{C}(j, E_j[N])/\mathbb{C}(j)) \cong \mathrm{SL}_2(\mathbb{Z}/N\mathbb{Z}).$$

Proof. Let $\sigma : \mathbb{C}(j, E_j[N]) \longrightarrow \overline{\mathbb{C}(j)}$ be an embedding that fixes $\mathbb{C}(j)$ point-wise. Since the extension $\mathbb{C}(j, x(E_j[N]))/\mathbb{C}(j)$ is Galois, σ restricts to an element of $\mathrm{Gal}(\mathbb{C}(j, x(E_j[N]))/\mathbb{C}(j))$. The set $y(E_j[N])$ consists of pairs of square roots of elements of $\mathbb{C}(j, x(E_j[N]))$ permuted by σ, so σ permutes $y(E_j[N])$ as well, making it an automorphism of $\mathbb{C}(j, E_j[N])$. Thus the extension $\mathbb{C}(j, E_j[N])/\mathbb{C}(j)$ is Galois.

Let H be its Galois group. The ordered basis (P_τ, Q_τ) of $E_j[N]$ over $\mathbb{Z}/N\mathbb{Z}$ specifies an injective representation (Exercise 7.5.6)

$$\rho : H \longrightarrow \mathrm{GL}_2(\mathbb{Z}/N\mathbb{Z}), \qquad \begin{bmatrix} P_\tau^\sigma \\ Q_\tau^\sigma \end{bmatrix} = \rho(\sigma) \begin{bmatrix} P_\tau \\ Q_\tau \end{bmatrix}.$$

We need to prove that $\rho(H) = \mathrm{SL}_2(\mathbb{Z}/N\mathbb{Z})$.

For any $\sigma \in H$ the relation $\begin{bmatrix} P_\tau^\sigma \\ Q_\tau^\sigma \end{bmatrix} = \rho(\sigma) \begin{bmatrix} P_\tau \\ Q_\tau \end{bmatrix}$ combines with Galois invariance of the Weil pairing and Corollary 7.4.2 to show that

$$e_N(P_\tau, Q_\tau)^\sigma = e_N(P_\tau^\sigma, Q_\tau^\sigma) = e_N(P_\tau, Q_\tau)^{\det \rho(\sigma)}.$$

But $e_N(P_\tau, Q_\tau)$ is an Nth root of unity, lying in $\mathbb{C}$ and therefore fixed by σ, so the displayed relation is in fact $e_N(P_\tau, Q_\tau) = e_N(P_\tau, Q_\tau)^{\det \rho(\sigma)}$. Since $e_N(P_\tau, Q_\tau)$ is a primitive Nth root by Corollary 7.4.2, $\det \rho(\sigma) = 1$ in $(\mathbb{Z}/N\mathbb{Z})^*$. This shows that $\rho(H) \subset \mathrm{SL}_2(\mathbb{Z}/N\mathbb{Z})$.

Let $K = \mathrm{Gal}(\mathbb{C}(j, E_j[N])/\mathbb{C}(j, x(E_j[N])))$ be the subgroup of H whose action on $E_j[N]$ preserves x-coordinates. Since points with the same x-coordinate are equal or opposite, if $\sigma \in K$ then $P_\tau^\sigma = \pm P_\tau$ and $Q_\tau^\sigma = \pm Q_\tau$ and $\rho(\sigma) \in \mathrm{SL}_2(\mathbb{Z}/N\mathbb{Z})$, so $\rho(\sigma) \in \{\pm I\}$. Conversely, if $\sigma \in H$ and $\rho(\sigma) \in \{\pm I\}$ then $P^\sigma = \pm P$ for all $P \in E_j[N]$, so $\sigma \in K$. That is, $K = \rho^{-1}(\{\pm I\})$, showing that $|K| \leq 2$. By Proposition 7.5.2 $\mathrm{Gal}(\mathbb{C}(j, x(E_j[N]))/\mathbb{C}(j)) \cong \mathrm{SL}_2(\mathbb{Z}/N\mathbb{Z})/\{\pm I\}$, and consequently $|H| = |\mathrm{SL}_2(\mathbb{Z}/N\mathbb{Z})/\{\pm I\}| \, |K|$. Thus $[\mathrm{SL}_2(\mathbb{Z}/N\mathbb{Z}) : \rho(H)] \leq 2$.

The case $[\mathrm{SL}_2(\mathbb{Z}/N\mathbb{Z}) : \rho(H)] = 2$ can arise only if $|K| = 1$, i.e., $-I \notin \rho(H)$. In this case $\{\pm I\}\rho(H) = \mathrm{SL}_2(\mathbb{Z}/N\mathbb{Z})$ and so one of $\pm \begin{bmatrix} 0 & -1 \\ 1 & 0 \end{bmatrix}$ is in $\rho(H)$. But this gives $-I = (\pm \begin{bmatrix} 0 & -1 \\ 1 & 0 \end{bmatrix})^2 \in \rho(H)$, contradiction, and so $\rho(H) = \mathrm{SL}_2(\mathbb{Z}/N\mathbb{Z})$ as desired. $\square$

The proof has shown that adjoining y-coordinates enlarges the Galois group by adjoining an element that maps under ρ to $-I$, i.e., the element negates points by negating y-coordinates, so it can be denoted -1. Modulo this, the representation ρ is the map θ^{-1} from before (Exercise 7.5.7).

Figure 7.2 summarizes some of the results of this section. The displayed groups are the images under ρ of the actual Galois groups, obtained by lifting (7.9) and (7.10) to $\mathrm{SL}_2(\mathbb{Z}/N\mathbb{Z})$ since ρ and θ^{-1} are compatible. To reiterate, one can think of the transcendental j as parametrizing a family of elliptic curves, of the functions $f_{1,0}$ and $f_{0,1} = f_1$ as N-torsion x-coordinates of the points P_τ and Q_τ on the curves, and of the function f_0 as the sum of x-coordinates of the nonzero points of the N-cyclic subgroup $\langle Q_\tau \rangle$. Thus the algebraic extensions of $\mathbb{C}(j)$ in the figure carry torsion data about the curves to describe moduli spaces, i.e., modular curves. At the top, y-coordinates are adjoined as well to eliminate a quotient by $\{\pm I\}$ from the Galois groups. As the diagram shows, one can think of $X_1(N)$ as being defined by the relation between j and f_1 over $\mathbb{C}$, and of $X_0(N)$ as being defined by the relation between j and f_0. To study the modular curves as algebraic curves over the rational numbers the underlying field needs to be changed from $\mathbb{C}$ to $\mathbb{Q}$. This will be done in the next section.

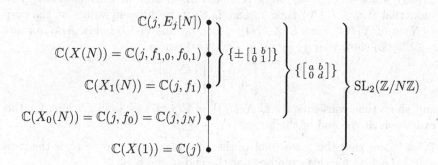

Figure 7.2. Fields and groups over $\mathbb{C}$

Exercises

7.5.1. (a) Show that $f_0^{\bar{v}}$ is meromorphic on $\mathcal{H}$.

(b) Show that $f_0^{\bar{v}} \circ \gamma = f_0^{\bar{v}\gamma}$ for all $\gamma \in \mathrm{SL}_2(\mathbb{Z})$.

(c) Show that $f_0^{\bar{v}}$ is meromorphic at ∞. More specifically, show that for $v = (c_v, d_v)$ as in the section,

$$\lim_{\mathrm{Im}(\tau) \to \infty} \wp_\tau \left(\tfrac{c_v \tau + d_v}{N} \right) = \begin{cases} -2\zeta(2) & \text{if } c_v \not\equiv 0 \pmod{N}, \\ -2\zeta(2) + N^2 \sum_{n \equiv d_v\ (N)} \frac{1}{n^2} & \text{if } c_v \equiv 0 \pmod{N}. \end{cases}$$

(A hint for this exercise is at the end of the book.)

(d) Explain why showing that each $f_0^{\bar{v}}$ is meromorphic at ∞ checks the other cusps as well.

7.5.2. What modifications to Exercise 1.2.11 show that $j_N \in \mathbb{C}(X_0(N))$?

7.5.3. (a) Show that the map θ is a homomorphism.

(b) Consider the set $\{f_0^{\bar{d}} : \bar{d} \in (\mathbb{Z}/N\mathbb{Z} - 0)/\pm\}$ and the field containments

$$\mathbb{C}(X(1)) \subset \mathbb{C}(j, \{f_0^{\vec{d}}\}) \subset \mathbb{C}(X_1(N))) \subset \mathbb{C}(X(N)).$$

Show that since $f_0^{\vec{d}} \circ \gamma = f_0^{\overline{(0,d)\gamma}}$ for $\gamma \in \mathrm{SL}_2(\mathbb{Z})$, the fixing subgroup in $\mathrm{SL}_2(\mathbb{Z})/\{\pm I\}\Gamma(N)$ of the field $\mathbb{C}(j, \{f_0^{\vec{d}}\})$ is $\{\pm I\}\Gamma_1(N)/\{\pm I\}\Gamma(N)$, and thus $\mathbb{C}(X_1(N)) = \mathbb{C}(j, \{f_0^{\vec{d}}\})$ and (7.9) holds.

(c) Similarly, show that

$$f_0 \circ \gamma = \sum_{\delta=1}^{N-1} f_0^{\overline{\delta(c,d)}} \quad \text{for } \gamma = \begin{bmatrix} a & b \\ c & d \end{bmatrix} \in \mathrm{SL}_2(\mathbb{Z}),$$

i.e., $f_0 \circ \gamma$ is the sum of the x-coordinates of the nonzero points of the N-cyclic subgroup of $E_{j(\tau)}$ generated by either point with x-coordinate $f_0^{\overline{(c,d)}}(\tau)$. We already know that $f_0 \in \mathbb{C}(X_0(N))$. Use the result of Exercise 7.5.1(c) to show that if $\gamma \notin \Gamma_0(N)$ then f_0 and $f_0 \circ \gamma$ take different values at the cusp $\Gamma(N)\infty$ of $X(N)$. Thus $\mathbb{C}(X_0(N)) = \mathbb{C}(j, f_0)$ and (7.10) holds. Also, for any $\gamma \in \mathrm{SL}_2(\mathbb{Z})$ show that $j_N(\gamma(\tau)) = j(\gamma'(N\tau))$ where

$$\gamma' = \begin{bmatrix} N & 0 \\ 0 & 1 \end{bmatrix} \gamma \begin{bmatrix} 1/N & 0 \\ 0 & 1 \end{bmatrix},$$

and show that consequently $\mathbb{C}(X_0(N)) = \mathbb{C}(j, j_N)$ as well. (A hint for this exercise is at the end of the book.)

7.5.4. Show that the j-invariant of the universal elliptic curve E_j is the variable j. (A hint for this exercise is at the end of the book.)

7.5.5. (a) Show that for each $\overline{v} \in (\mathbb{Z}/N\mathbb{Z})^2$, the function $f_0^{\pm\overline{v}}$ determines two N-torsion points of E_j unless $2v = 0$, in which case it determines one.

(b) Show that consequently, regardless of whether N is odd or even, we have found all N^2 points of $E_j[N]$.

7.5.6. Show that the map ρ is an injective representation.

7.5.7. Supply details as necessary to the following proof that the representation ρ is the map θ^{-1} from earlier in the section modulo signs. For any nonzero row vector $\overline{v} \in (\mathbb{Z}/N\mathbb{Z})^2$ and any $\sigma \in \mathrm{Gal}(\mathbb{C}(j, E_j[N])/\mathbb{C}(j)) \cong \mathrm{SL}_2(\mathbb{Z}/N\mathbb{Z})$ the definition of ρ gives

$$(\overline{v} \begin{bmatrix} P_\tau \\ Q_\tau \end{bmatrix})^\sigma = \overline{v}\rho(\sigma) \begin{bmatrix} P_\tau \\ Q_\tau \end{bmatrix},$$

and ignoring negation give the relation on x-coordinates,

$$x(\pm\overline{v} \begin{bmatrix} P_\tau \\ Q_\tau \end{bmatrix})^{(\pm\sigma)} = x(\pm\overline{v}\rho(\pm\sigma) \begin{bmatrix} P_\tau \\ Q_\tau \end{bmatrix}).$$

Specifically,

$$\pm\overline{v} \mapsto \pm\overline{v}\rho(\pm\sigma).$$

Since a pair $\pm\sigma$ in the Galois group corresponds to a single element in the quotient group $\mathrm{Gal}(\mathbb{C}(X(N))/\mathbb{C}(X(1))) \cong \mathrm{SL}_2(\mathbb{Z}/N\mathbb{Z})/\{\pm I\}$, this is the same as the earlier action of the quotient group via θ^{-1} as shown in (7.8).

7.6 Function fields over $\mathbb{Q}$

Working over the complex numbers $\mathbb{C}$ we have considered the universal curve E_j and the field containments

$$\mathbb{C}(j) \subset \mathbb{C}(j, E_j[N]) \subset \overline{\mathbb{C}(j)}.$$

Corollary 7.5.3 established that the extension $\mathbb{C}(j, E_j[N])/\mathbb{C}(j)$ is Galois with group $\mathrm{SL}_2(\mathbb{Z}/N\mathbb{Z})$. This section studies the situation when the underlying field is changed to the rational numbers $\mathbb{Q}$. The result will be that the Galois group enlarges to $\mathrm{GL}_2(\mathbb{Z}/N\mathbb{Z})$. Large enough subgroups correspond to intermediate fields that are the function fields of algebraic curves over the rational numbers. The next section will show that the intermediate fields $\mathbb{Q}(j, f_0)$ and $\mathbb{Q}(j, f_1)$ define $X_0(N)$ and $X_1(N)$ over $\mathbb{Q}$. The field $\mathbb{Q}(j, f_{1,0}, f_{0,1})$ defines $X(N)$ over the field $\mathbb{Q}(\boldsymbol{\mu}_N)$ where $\boldsymbol{\mu}_N$ is the group of complex Nth roots of unity.

Since $\mathbb{Q}$ is the prime subfield of $\mathbb{C}$, much of the algebraic structure from the previous section carries over. The equation defining E_j has its coefficients in $\mathbb{Q}(j)$. Viewing the curve as defined over this field means considering points $(x, y) \in \overline{\mathbb{Q}(j)}^2$ satisfying the equation. This includes the nonzero points of $E_j[N]$ over $\mathbb{C}(j)$ from before, and in the field containments

$$\mathbb{Q}(j) \subset \mathbb{Q}(j, E_j[N]) \subset \overline{\mathbb{Q}(j)}$$

the extension $\mathbb{Q}(j, E_j[N])/\mathbb{Q}(j)$ is again Galois. The only difference between the field theory over $\mathbb{Q}$ and over $\mathbb{C}$ will involve $\boldsymbol{\mu}_N$.

Consider the Galois group

$$H_{\mathbb{Q}} = \mathrm{Gal}(\mathbb{Q}(\boldsymbol{\mu}_N, j, E_j[N])/\mathbb{Q}(j))$$

and the representation

$$\rho : H_{\mathbb{Q}} \longrightarrow \mathrm{GL}_2(\mathbb{Z}/N\mathbb{Z})$$

describing how $H_{\mathbb{Q}}$ permutes $E_j[N]$. This is defined as before in terms of the ordered basis (P_τ, Q_τ) of $E_j[N]$ over $\mathbb{Z}/N\mathbb{Z}$ from (7.11), so that

$$\begin{bmatrix} P_\tau^\sigma \\ Q_\tau^\sigma \end{bmatrix} = \rho(\sigma) \begin{bmatrix} P_\tau \\ Q_\tau \end{bmatrix}, \quad \sigma \in H_{\mathbb{Q}}.$$

Lemma 7.6.1. *The function* $\det \rho$ *describes how* $H_{\mathbb{Q}}$ *permutes* $\boldsymbol{\mu}_N$,

$$\mu^\sigma = \mu^{\det \rho(\sigma)}, \quad \mu \in \boldsymbol{\mu}_N, \ \sigma \in H_{\mathbb{Q}}.$$

(Here μ^σ *is* μ *acted on by* σ *while* $\mu^{\det \rho(\sigma)}$ *is* μ *raised to the power* $\det \rho(\sigma)$.)

This is shown with the Weil pairing as in the proof of Corollary 7.5.3 (Exercise 7.6.1). To use the lemma, suppose $\sigma \in H_{\mathbb{Q}}$ fixes $E_j[N]$. This means

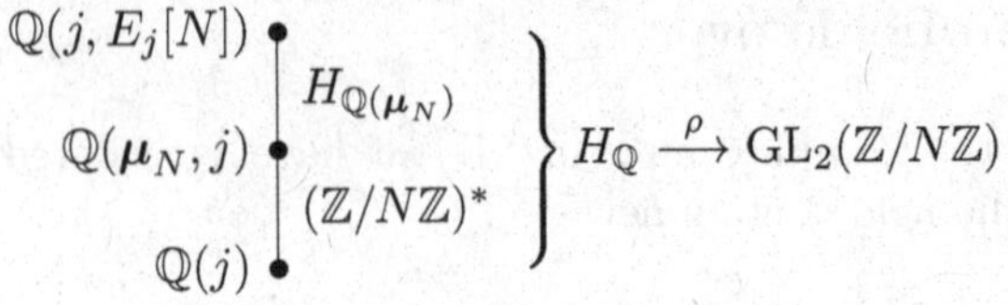

$$\left.\begin{array}{l} \mathbb{Q}(j, E_j[N]) \bullet \\[3ex] \mathbb{Q}(\boldsymbol{\mu}_N, j) \bullet \\[3ex] \mathbb{Q}(j) \bullet \end{array} \right\} \; H_{\mathbb{Q}} \xrightarrow{\rho} \mathrm{GL}_2(\mathbb{Z}/N\mathbb{Z})$$

with $H_{\mathbb{Q}(\boldsymbol{\mu}_N)}$ and $(\mathbb{Z}/N\mathbb{Z})^*$ labels.

Figure 7.3. Fields and groups over $\mathbb{Q}$

that $\sigma \in \ker(\rho)$, so $\sigma \in \ker(\det \rho)$ and the lemma says that σ fixes $\boldsymbol{\mu}_N$. Thus $\boldsymbol{\mu}_N \subset \mathbb{Q}(j, E_j[N])$ by Galois theory, showing that $H_{\mathbb{Q}}$ is in fact the Galois group of $\mathbb{Q}(j, E_j[N])$ over $\mathbb{Q}(j)$, the analog over $\mathbb{Q}$ of the group H in the proof of Corollary 7.5.3. Consider the configuration of fields and groups in Figure 7.3. Since the field extension is generated by $E_j[N]$, now ρ clearly is injective, and by the lemma it restricts to

$$\rho : H_{\mathbb{Q}(\boldsymbol{\mu}_N)} \longrightarrow \mathrm{SL}_2(\mathbb{Z}/N\mathbb{Z}).$$

To analyze the images of $H_{\mathbb{Q}}$ and $H_{\mathbb{Q}(\boldsymbol{\mu}_N)}$ under ρ, recall a result from Galois theory.

Lemma 7.6.2 (Restriction Lemma). *Let* $\mathbf{k}$ *and* $\mathbb{F}$ *be extension fields of* $\mathbf{f}$ *inside* $\mathbb{K}$ *with* $\mathbb{F}/\mathbf{f}$ *Galois. Suppose* $\mathbb{K} = \mathbf{k}\mathbb{F}$. *Then* $\mathbb{K}/\mathbf{k}$ *is Galois, there is a natural injection*

$$\mathrm{Gal}(\mathbb{K}/\mathbf{k}) \longrightarrow \mathrm{Gal}(\mathbb{F}/\mathbf{f}),$$

and the image is $\mathrm{Gal}(\mathbb{F}/(\mathbf{k} \cap \mathbb{F}))$.

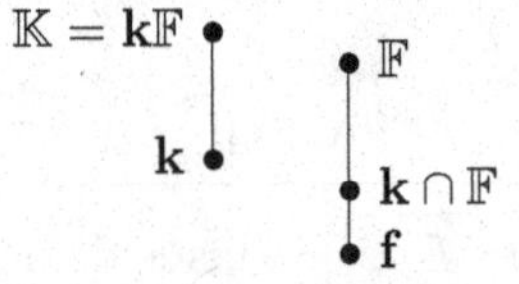

$$\begin{array}{ccc} \mathbb{K} = \mathbf{k}\mathbb{F} \bullet & & \bullet\, \mathbb{F} \\[3ex] \mathbf{k} \bullet & & \bullet\, \mathbf{k} \cap \mathbb{F} \\[1ex] & & \bullet\, \mathbf{f} \end{array}$$

Figure 7.4. Setup for the Restriction Lemma

Proof. The situation is shown in Figure 7.4. Any map $\sigma : \mathbb{K} \longrightarrow \overline{\mathbb{K}}$ fixing $\mathbf{k}$ restricts to a map $\mathbb{F} \longrightarrow \overline{\mathbb{F}}$ fixing $\mathbf{k} \cap \mathbb{F}$. Since the extension $\mathbb{F}/(\mathbf{k} \cap \mathbb{F})$ is Galois, the restriction is an automorphism of $\mathbb{F}$ and therefore σ is an automorphism of $\mathbb{K} = \mathbf{k}\mathbb{F}$. This shows that $\mathbb{K}/\mathbf{k}$ is Galois and that restriction gives a homomorphism

$$\mathrm{Gal}(\mathbb{K}/\mathbf{k}) \longrightarrow \mathrm{Gal}(\mathbb{F}/(\mathbf{k} \cap \mathbb{F})) \subset \mathrm{Gal}(\mathbb{F}/\mathbf{f}).$$

If the restriction of some σ fixes $\mathbb{F}$ along with $\mathbf{k}$ then it fixes $\mathbb{K}$ and is trivial, so the restriction map injects. Since the fixed field of $\mathbb{K}$ under $\mathrm{Gal}(\mathbb{K}/\mathbf{k})$ is $\mathbf{k}$, the fixed field of $\mathbb{F}$ under the restriction is $\mathbf{k} \cap \mathbb{F}$ and so restriction maps to all of $\mathrm{Gal}(\mathbb{F}/\mathbf{k} \cap \mathbb{F})$. □

One application of the lemma is implicit in Figure 7.3, where $(\mathbb{Z}/N\mathbb{Z})^*$ is displayed as $\mathrm{Gal}(\mathbb{Q}(\boldsymbol{\mu}_N, j)/\mathbb{Q}(j))$ (Exercise 7.6.2). For another, consider the situation shown in Figure 7.5.

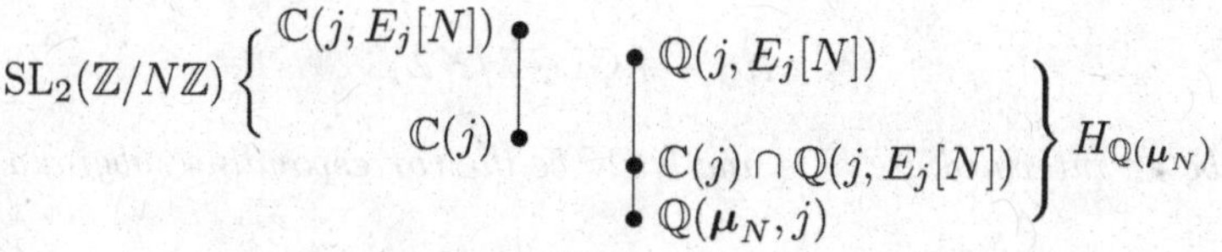

Figure 7.5. Applying the Restriction Lemma

The Restriction Lemma shows that $\mathrm{SL}_2(\mathbb{Z}/N\mathbb{Z})$ injects into $H_{\mathbb{Q}(\boldsymbol{\mu}_N)}$. But also ρ injects in the other direction, making the two groups isomorphic since they are finite,
$$\rho : H_{\mathbb{Q}(\boldsymbol{\mu}_N)} \xrightarrow{\sim} \mathrm{SL}_2(\mathbb{Z}/N\mathbb{Z}).$$
Now the lemma also shows that $\mathbb{C}(j) \cap \mathbb{Q}(j, E_j[N]) = \mathbb{Q}(\boldsymbol{\mu}_N, j)$, and intersecting with $\overline{\mathbb{Q}}$ gives
$$\mathbb{Q}(j, E_j[N]) \cap \overline{\mathbb{Q}} = \mathbb{Q}(\boldsymbol{\mu}_N).$$
Also, Figure 7.3 now shows that
$$|H_{\mathbb{Q}}| = |H_{\mathbb{Q}(\boldsymbol{\mu}_N)}|\,|(\mathbb{Z}/N\mathbb{Z})^*| = |\mathrm{SL}_2(\mathbb{Z}/N\mathbb{Z})|\,|(\mathbb{Z}/N\mathbb{Z})^*|.$$
But $|\mathrm{SL}_2(\mathbb{Z}/N\mathbb{Z})|\,|(\mathbb{Z}/N\mathbb{Z})^*| = |\mathrm{GL}_2(\mathbb{Z}/N\mathbb{Z})|$, so the representation ρ surjects,
$$\rho : H_{\mathbb{Q}} \xrightarrow{\sim} \mathrm{GL}_2(\mathbb{Z}/N\mathbb{Z}).$$

This lets us specify which intermediate fields of $\mathbb{Q}(j, E_j[N])/\mathbb{Q}(j)$ correspond to algebraic curves over $\mathbb{Q}$. Let $\mathbb{K}$ be an intermediate field and let the corresponding subgroup of $H_{\mathbb{Q}}$ be $K = \mathrm{Gal}(\mathbb{Q}(j, E_j[N])/\mathbb{K})$, as in Figure 7.6.

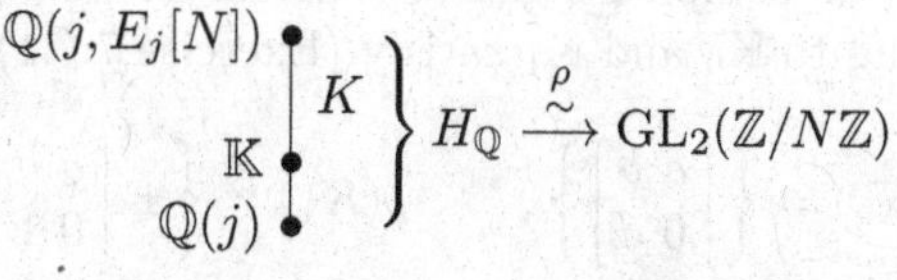

Figure 7.6. Subgroup and fixed field

Recall that $\det \rho$ describes how $H_{\mathbb{Q}}$ permutes $\boldsymbol{\mu}_N$. This gives the equivalences

$$\mathbb{K} \cap \overline{\mathbb{Q}} = \mathbb{Q} \iff \mathbb{K} \cap \mathbb{Q}(\boldsymbol{\mu}_N) = \mathbb{Q}$$
$$\iff \det \rho : K \longrightarrow (\mathbb{Z}/N\mathbb{Z})^* \text{ surjects.}$$

Summing up the results of this section,

Theorem 7.6.3. *Let $H_{\mathbb{Q}}$ denote the Galois group of the field extension $\mathbb{Q}(j, E_j[N])/\mathbb{Q}(j)$. There is an isomorphism*

$$\rho : H_{\mathbb{Q}} \xrightarrow{\sim} GL_2(\mathbb{Z}/N\mathbb{Z}).$$

Let $\mathbb{K}$ be an intermediate field and let K be the corresponding subgroup of $H_{\mathbb{Q}}$. Then

$$\mathbb{K} \cap \overline{\mathbb{Q}} = \mathbb{Q} \iff \det \rho : K \longrightarrow (\mathbb{Z}/N\mathbb{Z})^* \text{ surjects.}$$

Thus $\mathbb{K}$ is the function field of an algebraic curve over $\mathbb{Q}$ if and only if $\det \rho$ surjects.

The last statement in the theorem follows from Theorem 7.2.5.

Exercises

7.6.1. Prove Lemma 7.6.1.

7.6.2. Justify the relation $\mathrm{Gal}(\mathbb{Q}(\boldsymbol{\mu}_N, j)/\mathbb{Q}(j)) \cong (\mathbb{Z}/N\mathbb{Z})^*$ shown in Figure 7.3.

7.7 Modular curves as algebraic curves and Modularity

This section defines the modular curves $X_0(N)$ and $X_1(N)$ as algebraic curves over $\mathbb{Q}$ and then restates the Modularity Theorem algebraically.

Consider two intermediate fields of the extension $\mathbb{Q}(j, E_j[N])/\mathbb{Q}(j)$,

$$\mathbb{K}_0 = \mathbb{Q}(j, f_0), \qquad \mathbb{K}_1 = \mathbb{Q}(j, f_1),$$

analogous to the function fields $\mathbb{C}(j, f_0)$ and $\mathbb{C}(j, f_1)$ of the modular curves $X_0(N)$ and $X_1(N)$ as complex algebraic curves. The subgroups K_0 and K_1 of $H_{\mathbb{Q}}$ corresponding to $\mathbb{K}_0$ and $\mathbb{K}_1$ satisfy (Exercise 7.7.1)

$$\rho(K_0) = \left\{ \begin{bmatrix} a & b \\ 0 & d \end{bmatrix} \right\}, \qquad \rho(K_1) = \left\{ \pm \begin{bmatrix} a & b \\ 0 & 1 \end{bmatrix} \right\}, \tag{7.13}$$

running through all such matrices in $GL_2(\mathbb{Z}/N\mathbb{Z})$. Thus for $j = 0, 1$, the map $\det \rho : K_j \longrightarrow (\mathbb{Z}/N\mathbb{Z})^*$ surjects, and so by Theorem 7.6.3 the fields

$\mathbb{K}_0$ and $\mathbb{K}_1$ are the function fields of nonsingular projective algebraic curves over $\mathbb{Q}$. Denote these curves $X_0(N)_{\text{alg}}$ and $X_1(N)_{\text{alg}}$.

To relate the algebraic curve $X_1(N)_{\text{alg}}$ defined over $\mathbb{Q}$ to the complex modular curve $X_1(N)$ viewed as an algebraic curve over $\mathbb{C}$, we need the following theorem from algebraic geometry.

Theorem 7.7.1. *Let* $\mathbf{k}$ *be a field, let* C *be a nonsingular projective algebraic curve over* $\mathbf{k}$ *defined by polynomials* $\varphi_1, \ldots, \varphi_m \in \mathbf{k}[x_1, \ldots, x_n]$, *and let the function field of* C *be* $\mathbf{k}(C) = \mathbf{k}(t)[u]/\langle p(u) \rangle$. *Let* $\mathbb{K}$ *be a field containing* $\mathbf{k}$. *Then the polynomials* $\varphi_1, \ldots, \varphi_m$, *viewed as elements of* $\mathbb{K}[x_1, \ldots, x_n]$, *define a nonsingular algebraic curve* C' *over* $\mathbb{K}$, *and its function field is* $\mathbb{K}(C') = \mathbb{K}(t)[u]/\langle p(u) \rangle$.

For example, any nonsingular Weierstrass equation over $\mathbb{Q}$ defines an elliptic curve over $\mathbb{C}$, as we earlier observed directly. In the theorem $x_1, \ldots, x_n$ are understood to be indeterminates over $\mathbb{K}$ as well as over $\mathbf{k}$. The theorem says that after enlarging the field of definition from $\mathbf{k}$ to $\mathbb{K}$, the ideal generated by the φ_i remains prime and the nonsingularity condition continues to hold at each point of the curve over the larger field. The form of $\mathbb{K}(C')$ shows that the minimal polynomial p of u over $\mathbf{k}(t)$ is irreducible over $\mathbb{K}(t)$, making it the minimal polynomial of u over $\mathbb{K}(t)$ as well.

To apply the theorem to modular curves, let $\mathbf{k} = \mathbb{Q}$ and $C = X_1(N)_{\text{alg}}$, and let $p_1 \in \mathbb{Q}(j)[x]$ be the minimal polynomial of f_1 over $\mathbb{Q}(j)$, so that

$$\mathbb{Q}(C) = \mathbb{Q}(j, f_1) = \mathbb{Q}(j)[x]/\langle p_1 \rangle.$$

Let $\mathbb{K} = \mathbb{C}$. Since this field is algebraically closed, the theorem says that the $\mathbb{C}$-points (points with coordinates in $\mathbb{C}$) that satisfy the polynomials defining $X_1(N)_{\text{alg}}$ over $\mathbb{Q}$ form a curve $C' = X_1(N)_{\text{alg},\mathbb{C}}$ over $\mathbb{C}$ with function field

$$\mathbb{C}(X_1(N)_{\text{alg},\mathbb{C}}) = \mathbb{C}(j)[x]/\langle p_1 \rangle,$$

and p_1 is also the minimal polynomial of f_1 over $\mathbb{C}(j)$ (see Exercise 7.7.2 for an elementary proof of this last fact). Since the minimal polynomials are the same, the function field $\mathbb{C}(j, f_1)$ of the Riemann surface $X_1(N)$—or of the corresponding nonsingular algebraic curve over $\mathbb{C}$, also denoted $X_1(N)$—is

$$\mathbb{C}(X_1(N)) = \mathbb{C}(j, f_1) = \mathbb{C}(j)[x]/\langle p_1 \rangle,$$

and since the two function fields agree Theorem 7.2.5 says that $X_1(N) = X_1(N)_{\text{alg},\mathbb{C}}$ up to isomorphism over $\mathbb{C}$ as desired. The argument for $X_0(N)$ is virtually identical, using f_0 in place of f_1. For the algebraic model of $X(N)$ see Exercise 7.7.3.

The planar model of $X_1(N)_{\text{alg}}$ is defined by the polynomial p_1. Clear its denominator to get a polynomial $\varphi_1 \in \mathbb{Q}[j, x]$. Then the points of the planar model are

$$X_1(N)_{\text{alg}}^{\text{planar}} = \{(j, x) \in \overline{\mathbb{Q}}^2 : \varphi_1(j, x) = 0\}.$$

As discussed after Theorem 7.2.5 there is a birational equivalence over $\mathbb{Q}$ between the planar model $X_1(N)_{\text{alg}}^{\text{planar}}$ and the actual nonsingular projective curve $X_1(N)_{\text{alg}}$ itself. As a set map this takes all but finitely many points of the planar model to all but finitely many points of the curve. Since the description of $X_1(N)_{\text{alg}}$ is abstract, we will sometimes use its planar model in its stead for computations.

The values $j(\mu_3) = 0$ and $j(i) = 1728$ show that the elliptic points of the Riemann surface $X_1(N)$ correspond to points (j, f_1) with rational first coordinate and thus algebraic coordinates in the planar model of $X_1(N)_{\text{alg},\mathbb{C}}$. That is, no elliptic points are lost in the transition from $X_1(N)$ to $X_1(N)_{\text{alg}}$. The same is true for cusps, viewing $j(\infty) = \infty = [1,0]$ as a rational point of the projective plane. This argument holds for a general modular curve, regardless of whether it is defined over $\mathbb{Q}$ or some larger number field $\mathbb{K} \subset \overline{\mathbb{Q}}$.

Version $X_{\mathbb{C}}$ of the Modularity Theorem, from the end of Chapter 2, has a variant in the language of algebraic geometry.

Theorem 7.7.2 (Modularity Theorem, Version $X_{\mathbb{Q}}$). *Let E be an elliptic curve over $\mathbb{Q}$. Then for some positive integer N there exists a surjective morphism over $\mathbb{Q}$ of curves over $\mathbb{Q}$ from the modular curve $X_0(N)_{\text{alg}}$ to the elliptic curve E,*

$$X_0(N)_{\text{alg}} \longrightarrow E.$$

By contrast, Version $X_{\mathbb{C}}$ assumes a complex elliptic curve E with rational j-invariant and guarantees a holomorphic map $X_0(N) \longrightarrow E$ of Riemann surfaces.

Version $X_{\mathbb{Q}}$ easily implies Version $X_{\mathbb{C}}$. Any complex elliptic curve E with rational j-invariant, viewed as an algebraic curve over $\mathbb{C}$, is isomorphic over $\mathbb{C}$ to the $\mathbb{C}$-points $E'(\mathbb{C})$ of an elliptic curve E' over $\mathbb{Q}$. Version $X_{\mathbb{Q}}$ gives a map $X_0(N)_{\text{alg}} \longrightarrow E'$ over $\mathbb{Q}$ and this extends to $\mathbb{C}$-points, $X_0(N)_{\text{alg}}(\mathbb{C}) \longrightarrow E'(\mathbb{C})$. A map $X_0(N) \longrightarrow E$ follows immediately and can be viewed as a holomorphic map of Riemann surfaces as desired. On the other hand, showing that Version $X_{\mathbb{C}}$ implies Version $X_{\mathbb{Q}}$, i.e., transferring the theorem from complex analysis to algebraic geometry and refining the field of definition from $\mathbb{C}$ to $\mathbb{Q}$, is beyond the scope of this text. The argument is given in the appendix to [Maz91], and it involves replacing the N in Version $X_{\mathbb{C}}$ by a possibly different N in the resulting Version $X_{\mathbb{Q}}$. In any case, the approach to proving Modularity developed by Wiles ultimately gives the *a priori* stronger Version $X_{\mathbb{Q}}$.

Definition 7.7.3. *Let E be an elliptic curve over $\mathbb{Q}$. The smallest N that can occur in Version $X_{\mathbb{Q}}$ of the Modularity Theorem is called the* **analytic conductor** *of E.*

Thus the analytic conductor is well defined on isomorphism classes over $\mathbb{Q}$ of elliptic curves over $\mathbb{Q}$. The next section will show that isogeny over $\mathbb{Q}$ of elliptic curves over $\mathbb{Q}$ is an equivalence relation, so in fact the analytic conductor is well defined on isogeny classes of elliptic curves over $\mathbb{Q}$. We will say more about the analytic conductor in Chapter 8.

Similar algebraic refinements of Versions $J_{\mathbb{C}}$ and $A_{\mathbb{C}}$ of the Modularity Theorem from Chapter 6 require the notion of a *variety*, the higher-dimensional analog of an algebraic curve. As in Section 7.2, let m and n be positive integers, consider a set of m polynomials over $\mathbb{Q}$ in n variables, and suppose that the ideal they generate in the ring of polynomials over $\overline{\mathbb{Q}}$,

$$I = \langle \varphi_1, \ldots, \varphi_m \rangle \subset \overline{\mathbb{Q}}[x_1, \ldots, x_n],$$

is prime. Then the set V of simultaneous solutions of the ideal,

$$V = \{P \in \overline{\mathbb{Q}}^n : \varphi(P) = 0 \text{ for all } \varphi \in I\},$$

is a variety over $\mathbb{Q}$. The coordinate ring of V over $\overline{\mathbb{Q}}$ is the integral domain

$$\overline{\mathbb{Q}}[V] = \overline{\mathbb{Q}}[x_1, \ldots, x_n]/I,$$

and the function field of V over $\overline{\mathbb{Q}}$ is the quotient field of the coordinate ring,

$$\overline{\mathbb{Q}}(V) = \{F = f/g : f, g \in \overline{\mathbb{Q}}[V],\ g \neq 0\}.$$

The *dimension* of V is the transcendence degree of $\overline{\mathbb{Q}}(V)$ over $\overline{\mathbb{Q}}$, i.e., the number of algebraically independent transcendentals in $\overline{\mathbb{Q}}(V)$, the fewest transcendentals one can adjoin to $\overline{\mathbb{Q}}$ so that the remaining extension up to $\overline{\mathbb{Q}}(V)$ is algebraic. In particular a curve is a 1-dimensional variety. If V is a variety over $\mathbb{Q}$ and for each point $P \in V$ the m-by-n derivative matrix $[D_j\varphi_i(P)]$ has rank $n - \dim(V)$ then V is nonsingular. As with curves, homogenizing the defining polynomials leads to a homogeneous version V_{hom} of the variety, and the definition of nonsingularity extends to V_{hom}. Morphisms between nonsingular projective varieties are defined as between curves.

The Jacobians from Chapter 6 have algebraic descriptions as nonsingular projective varieties, defined over $\mathbb{Q}$ since modular curves are defined over $\mathbb{Q}$, and having the same dimensions algebraically as they do as complex tori. We will see in Section 7.9 that the Hecke operators are defined over $\mathbb{Q}$ as well. Consequently the construction of Abelian varieties in Definition 6.6.3 as quotients of Jacobians, defined by the Hecke operators, also gives varieties defined over $\mathbb{Q}$, and again the two notions of dimension agree. Versions $J_{\mathbb{C}}$ and $A_{\mathbb{C}}$ of the Modularity Theorem modify accordingly.

Theorem 7.7.4 (Modularity Theorem, Version $J_{\mathbb{Q}}$). *Let E be an elliptic curve over $\mathbb{Q}$. Then for some positive integer N there exists a surjective morphism over $\mathbb{Q}$ of varieties over $\mathbb{Q}$*

$$J_0(N)_{\text{alg}} \longrightarrow E.$$

Theorem 7.7.5 (Modularity Theorem, Version $A_{\mathbb{Q}}$). *Let E be an elliptic curve over $\mathbb{Q}$. Then for some positive integer N and some newform $f \in \mathcal{S}_2(\Gamma_0(N))$ there exists a surjective morphism over $\mathbb{Q}$ of varieties over $\mathbb{Q}$*

$$A'_{f,\text{alg}} \longrightarrow E.$$

Our policy will be to give statements of Modularity involving varieties but, since we have not studied varieties, to argue using only curves. In particular, we never use the variety structure of the Abelian variety of a modular form. We mention in passing that rational maps between nonsingular projective varieties are defined as for curves, but a rational map between nonsingular projective varieties need not be a morphism; this is particular to curves. The Varieties–Fields Correspondence involves rational maps rather than morphisms, and the rational maps are *dominant* rather than surjective, where dominant means mapping to all of the codomain except possibly a proper subvariety.

Applications of Modularity to number theory typically rely on the algebraic versions of the Modularity Theorem given in this section. A striking example is the construction of rational points on elliptic curves, meaning points whose coordinates are rational. The key idea is that a natural construction of *Heegner points* on modular curves, points with algebraic coordinates, follows from the moduli space point of view. Taking the images of these points on elliptic curves under the map of Version $X_{\mathbb{Q}}$ and then symmetrizing under conjugation gives points with rational coordinates. As mentioned early in this chapter, the Mordell–Weil Theorem states that the group of rational points of an elliptic curve over $\mathbb{Q}$ takes the form $T \oplus \mathbb{Z}^r$ where T is the torsion subgroup, the group of points of finite order, and $r \geq 0$ is the rank. The Birch and Swinnerton-Dyer Conjecture, to be discussed at the end of Chapter 8, provides a formula for r. Gross and Zagier showed that when the conjectured r is 1, the Heegner point construction in fact yields points of infinite order. See Henri Darmon's article in [CSS97] for more on this subject.

Exercises

7.7.1. Show that the images of K_0 and K_1 under ρ are given by (7.13). (A hint for this exercise is at the end of the book.)

7.7.2. Let $p_{1,\mathbb{C}} \in \mathbb{C}(j)[x]$ be the minimal polynomial of f_1 over $\mathbb{C}(j)$ and let $p_{1,\mathbb{Q}} \in \mathbb{Q}(j)[x]$ be the minimal polynomial of f_1 over $\mathbb{Q}(j)$, a multiple of $p_{1,\mathbb{C}}$ in $\mathbb{C}(j)[x]$. Let ϵ_N be 2 if $N \in \{1,2\}$ and 1 if $N > 2$. Show that by Galois theory and Figure 7.2,

$$\deg(p_{1,\mathbb{C}}) = \epsilon_N |\text{SL}_2(\mathbb{Z}/N\mathbb{Z})|/(2N),$$

and show similarly that this is also $\deg(p_{1,\mathbb{Q}})$. Thus the polynomials are equal since both are monic.

7.7.3. The relevant field for the algebraic model of $X(N)$ is

$$\mathbb{K} = \mathbb{Q}(j, f_{1,0}, f_{0,1}),$$

analogous to the function field $\mathbb{C}(j, f_{1,0}, f_{0,1})$ of $X(N)$ as a complex algebraic curve. Show that the corresponding subgroup K of $H_{\mathbb{Q}}$ satisfies $\rho(K) = \{\pm I\}$, so $\mathbb{K} \cap \overline{\mathbb{Q}} = \mathbb{Q}(\mu_N)$. The algebraic model of the modular curve for $\Gamma(N)$ is the curve $X(N)_{\mathrm{alg}}$ over $\mathbb{Q}(\mu_N)$ with function field $\mathbb{Q}(\mu_N)(X(N)_{\mathrm{alg}}) = \mathbb{K}$. Connect this to the complex modular curve $X(N)$.

7.8 Isogenies algebraically

Let N be a positive integer. Recall the moduli space $S_1(N)$ from Section 1.5,

$$S_1(N) = \{\text{equivalence classes } [E, Q] \text{ of enhanced elliptic curves for } \Gamma_1(N)\},$$

and recall the moduli space interpretation of the Hecke operator T_p on the divisor group of $S_1(N)$ from Section 5.2,

$$T_p[E, Q] = \sum_C [E/C, Q + C],$$

where the sum is taken over all order p subgroups C of E with $C \cap \langle Q \rangle = \{0\}$. So far we understand the quotient E/C only when E is a complex torus. This section uses both parts of the Curves–Fields Correspondence to construct the quotient when E is an algebraic elliptic curve instead. Thus T_p can be viewed in algebraic terms. The algebraic formulation of T_p will be elaborated in the next section and used in Chapter 8.

Section 1.3 defined an isogeny of complex tori as a nonzero holomorphic homomorphism, which thus surjects and has finite kernel. Any such isogeny is the quotient map by a finite subgroup followed by an isomorphism. Every isogeny φ of complex tori has a dual isogeny ψ in the other direction such that $\psi \circ \varphi = [N]$ where $[N] = \deg(\varphi) = |\ker(\varphi)|$. Similarly, define an isogeny between algebraic elliptic curves to be a nonzero holomorphic morphism.

For an algebraic construction of the quotient isogeny, let $\mathbf{k}$ be a field of characteristic 0 and let E be an elliptic curve over $\mathbf{k}$. Let C be any finite subgroup of E, so that $C \subset E[N]$ for some N, and let $\mathbf{l} = \mathbf{k}(E[N])$ be the field generated over $\mathbf{k}$ by the coordinates of the nonzero points of $E[N]$. According to Theorem 7.7.1 we may regard E as an elliptic curve over $\mathbf{l}$. When $\mathbf{k}$ is algebraically closed, e.g., $\mathbf{k} = \mathbb{C}$ or $\mathbf{k} = \overline{\mathbb{Q}}$, then $\mathbf{l} = \mathbf{k}$ and there is no need to introduce $\mathbf{l}$ at all, but we cannot always assume this.

For each point $P \in C$ there is a corresponding automorphism τ_P^* of the function field $\mathbf{l}(E)$,

$$\tau_P^* f = f \circ \tau_P \text{ for all } f \in \mathbf{l}(E).$$

Since $\tau_{P+P'}^* = \tau_P^* \circ \tau_{P'}^*$, the group C naturally identifies with a subgroup of $\mathrm{Aut}(\mathbf{l}(E))$,

$$C \cong \{\tau_P^* : P \in C\}.$$

Let $1(E)^C$ denote the corresponding fixed subfield of $1(E)$. This is a function field over 1, so by Theorem 7.2.5 there exists a curve E' over 1 and an 1-isomorphism $i : 1(E') \longrightarrow 1(E)^C$. View this as an 1-injection,

$$i : 1(E') \longrightarrow 1(E).$$

By Theorem 7.2.6 there is a corresponding surjective morphism over 1,

$$\varphi : E \longrightarrow E', \qquad \varphi^* = i.$$

This will turn out to be the desired quotient isogeny.

For any point $X \in E$ let $Y = \varphi(X) \in E'$ and consider the set $\varphi^{-1}(Y) \subset E$. For any $P \in C$ and any $f \in 1(E)^C$ we have $f(X + P) = (\tau_P^* f)(X) = f(X)$, so that the construction of the quotient map given at the end of Section 7.2 shows that $\varphi(X + P) = \varphi(X) = Y$. That is, $X + C \subset \varphi^{-1}(Y)$. But also,

$$|\varphi^{-1}(Y)| \le \sum_{X' \in \varphi^{-1}(Y)} e_{X'}(\varphi) = \deg(\varphi) \qquad \text{by (7.4)}$$
$$= [1(E) : i(1(E'))] \qquad \text{since } i = \varphi^*$$
$$= [1(E) : 1(E)^C] = |C| \quad \text{by Galois theory,}$$

so that in fact $\varphi^{-1}(Y) = X + C$. In the process this has shown that φ is unramified and $\deg(\varphi) = |C|$.

The image curve $E' = \varphi(E)$ is again an elliptic curve with $0_{E'} = \varphi(0_E)$. Indeed, algebraic versions of the genus and the Riemann–Hurwitz formula from Chapter 3 exist, along with a more general notion of an elliptic curve as an algebraic curve of genus 1 with a distinguished point designated as its 0. An argument using the algebraic Riemann–Roch Theorem shows that any such curve is isomorphic to a curve in Weierstrass form. (See Exercise 7.8.1 for an example and then a sketch of the proof. Explaining all of this would lead us too far afield, but the interested reader should see [Sil86].) Since φ is unramified, it follows that $(E', \varphi(0_E))$ is an elliptic curve as claimed. This makes φ a homomorphism as observed after Theorem 7.3.3, and its kernel is

$$\ker(\varphi) = \varphi^{-1}(0'_E) = 0_E + C = C.$$

Thus we can view E' and φ as E/C and the quotient isogeny, transferred from the complex analytic setting of Chapter 1 to morphisms of algebraic elliptic curves over 1,

$$E \xrightarrow{\varphi} E/C, \qquad Q \mapsto Q + C.$$

Exercise 7.8.2 shows that in fact the constructions of E/C and φ are defined over a field that is generally smaller than 1. In the next section Exercise 7.9.2 will construct a rational function r_N over $\mathbb{Q}$ of two variables such that $j(E/C) = r_N(j(E), x(C))$ where E is an elliptic curve over $\overline{\mathbb{Q}}$, C is a cyclic subgroup of order N, and $x(C)$ is the sum of the x-coordinates of the nonzero points of C. This exercise works with a field of definition that is not algebraically closed.

As for the dual isogeny,

Theorem 7.8.1. *Let $\mathbf{k}$ be a field. Isogeny over $\mathbf{k}$ between elliptic curves over $\mathbf{k}$ is an equivalence relation. That is, if $\varphi : E \longrightarrow E'$ is an isogeny over $\mathbf{k}$ of elliptic curves over $\mathbf{k}$ then there exists a dual isogeny $\psi : E' \longrightarrow E$, also defined over $\mathbf{k}$.*

Proof. Given $\varphi : E \longrightarrow E'$, let $C = \ker(\varphi)$, let $N = |C|$, and consider also $[N] : E \longrightarrow E$, again an isogeny over $\mathbf{k}$. Similarly to before, identify each point $P \in E[N]$ with the automorphism τ_P^* of the function field $\mathbf{k}(E)$, and thus identify $E[N]$ with a subgroup of $\mathrm{Aut}(\mathbf{k}(E))$. Since $\ker(\varphi) = C$, the embedded image $\varphi^*(\mathbf{k}(E'))$ in $\mathbf{k}(E)$ is the subfield of functions that don't separate points of C, i.e., it is $\mathbf{k}(E)^C$. Similarly, $[N]^*(\mathbf{k}(E)) = \mathbf{k}(E)^{E[N]}$. Thus in the following configuration the horizontal arrows are $\mathbf{k}$-isomorphisms:

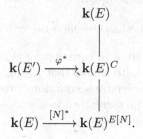

By Theorem 7.2.6 the $\mathbf{k}$-injection

$$\mathbf{k}(E) \xrightarrow{(\varphi^*)^{-1}\circ[N]^*} \mathbf{k}(E')$$

takes the form ψ^* where $\psi : E' \longrightarrow E$ is a morphism over $\mathbf{k}$. And $\varphi^* \circ \psi^* = [N]^* : \mathbf{k}(E) \longrightarrow \mathbf{k}(E)$, so $\psi \circ \varphi = [N]$. That is, ψ is the desired dual isogeny. $\qquad\square$

Exercise 7.8.3 gives an example of an isogeny and its dual. Exercise 7.8.4 is to show that also $\varphi \circ \psi = [\deg(\varphi)]$. Exercise 7.8.5 is to show that $\psi_* = \varphi^*$ as maps from $\mathrm{Pic}^0(E')$ to $\mathrm{Pic}^0(E)$.

The more general notion of elliptic curve that this section has mentioned gives rise to our first examples of the Modularity Theorem. Since the modular curves $X_0(N)_{\mathrm{alg}}$ are defined over $\mathbb{Q}$, those that have genus 1 are modular elliptic curves if they contain a point with rational coordinates. We show inductively that this is the case for all N. Recall that $X_0(N)_{\mathrm{alg}}$ is a subset of $X_0(N)_{\mathrm{alg},\mathbb{C}}$, which is identified with the Riemann surface $X_0(N)$. Similarly to the argument in Section 7.7, $X_0(N)_{\mathrm{alg}}$ retains the cusps of the Riemann surface. Specifically we show that the point of $X_0(N)_{\mathrm{alg}}$ corresponding to the cusp $\Gamma_0(N)(0)$ is rational. Since the map $j : X_0(1) \longrightarrow X_0(1)_{\text{alg},\mathbb{C}} = \mathbb{P}^1(\mathbb{C})$ takes $\text{SL}_2(\mathbb{Z})(0)$ to the rational point $\infty = [1, 0]$, this holds for $N = 1$. Now suppose it holds for some N and let p be prime. Let $P \in X_0(Np)_{\text{alg}}$ be the point corresponding to $\Gamma_0(Np)(0)$. The projection map $\pi : X_0(Np)_{\text{alg}} \longrightarrow X_0(N)_{\text{alg}}$ is a morphism over $\mathbb{Q}$ of curves over $\mathbb{Q}$. Its degree is p or $p + 1$ according to whether $p \mid N$. The point $Q = \pi(P) \in X_0(N)_{\text{alg}}$ corresponds to $\Gamma_0(N)(0)$, and working complex analytically shows that the ramification degree $e_P(\pi)$ is p; cf. Section 3.1. Consider the divisor $D = (Q)$ on $X_0(N)_{\text{alg}}$. By induction Q has rational coordinates, and D is therefore invariant under the automorphisms of $\overline{\mathbb{Q}}$. Its pullback π^*D is of the form $p(P)$ or $p(P) + (P')$ for some $P' \neq P$ according to whether $p \mid N$. For any automorphism σ of $\overline{\mathbb{Q}}$,

$$(\pi^*D)^\sigma = \pi^*(D^\sigma) = \pi^*(D),$$

showing that P has rational coordinates as well. This completes the induction.

Thus if $X_0(N)_{\text{alg}}$ has genus 1 and f is the unique newform of $\mathcal{S}_2(\Gamma_0(N))$ then there are isomorphisms of elliptic curves over $\mathbb{Q}$,

$$X_0(N)_{\text{alg}} \longrightarrow J_0(N)_{\text{alg}} \longrightarrow A'_{f,\text{alg}}.$$

An algorithm in [Cre97] gives an equation for A'_f and therefore for $X_0(N)_{\text{alg}}$. In particular, Exercise 3.1.4(e) shows that $X_0(p)_{\text{alg}}$ is a modular elliptic curve for $p = 11, 17, 19$. The Weierstrass equations produced by the algorithm are in a generalized form to be mentioned in Exercise 7.8.1 and explained at the beginning of the next chapter,

$$X_0(11)_{\text{alg}} : y^2 + y = x^3 - x^2 - 10x - 20,$$
$$X_0(17)_{\text{alg}} : y^2 + xy + y = x^3 - x^2 - x - 14,$$
$$X_0(19)_{\text{alg}} : y^2 + y = x^3 + x^2 - 9x - 15.$$

More generally, for any weight 2 newform $f \in \mathcal{S}_2(\Gamma_0(N))$ with rational coefficients the associated Abelian variety $A'_{f,\text{alg}}$ is a modular elliptic curve and the algorithm in [Cre97] gives an equation for it.

Exercises

7.8.1. (a) Consider the curve $C : x^3 + y^3 = 1$ and its point $P = (0, 1)$. Homogenize to $C : x^3 + y^3 = z^3$ and $P = [0, 1, 1]$. Show that replacing z by $y + z/3$ and then dehomogenizing back to $z = 1$ puts the curve into Weierstrass form E and takes P to 0_E.

(b) Let E' be a nonsingular algebraic curve over $\mathbf{k}$ of genus 1 with a distinguished point P. Let t be a uniformizer at P. Use Corollary 3.4.2(d) of the Riemann–Roch Theorem, now interpreted as a theorem about algebraic curves, to show that $\ell([d]P) = d$ for all $d \in \mathbb{Z}^+$. Use this to show that $L(P) = \mathbf{k}$; that $L([2]P)$ has basis $\{1, X\}$, where X has a double pole at P;

that $L([3]P)$ has basis $\{1, X, Y\}$, where Y has a triple pole at P; that $L([4]P)$ has basis $\{1, X, Y, X^2\}$; that $L([5]P)$ has basis $\{1, X, Y, X^2, XY\}$; and that $L([6]P)$ contains $\{1, X, Y, X^2, XY, X^3, Y^2\}$. Since this last set has seven elements it is linearly dependent. Explain why any linear relation on this set necessarily involves both X^3 and Y^2, so that after rescaling over $\mathbf{k}$ it takes the form

$$Y^2 + a_1 XY + a_3 Y = X^3 + a_2 X^2 + a_4 X + a_6.$$

We will see in the next chapter that easy changes of variable reduce this relation to $(Y')^2 = 4(X')^3 - g_2 X' - g_3$. Thus (E', P) has a planar model in Weierstrass form. (A hint for this exercise is at the end of the book.)

7.8.2. Let E be an elliptic curve over $\mathbf{k}$ and let C be a finite subgroup of E, so $C \subset E[N]$ for some N. Let $\mathbf{l} = \mathbf{k}(E[N])$ and let $G = \mathrm{Gal}(\mathbf{l}/\mathbf{k})$. Let H be the subgroup of G that fixes C as a set,

$$H = \{\sigma \in G : \sigma(C) = C\},$$

and let $\mathbf{k}'$ be the corresponding fixed subfield of $\mathbf{l}$. (See Figure 7.7.) The quotient curve E/C and the quotient isogeny φ have been constructed over $\mathbf{l}$. This exercise shows that in fact they can be defined over $\mathbf{k}'$.

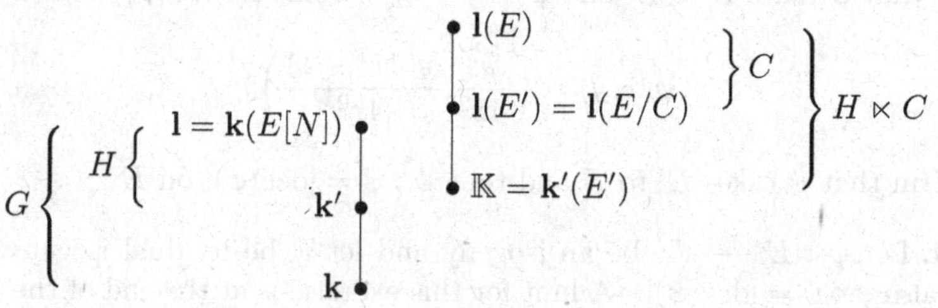

Figure 7.7. Fields and groups for the quotient curve

(a) Regarding E as an elliptic curve over $\mathbf{l}$, consider the corresponding function field $\mathbf{l}(E)$. The group G acts on $\mathbf{l}(E)$ from the left via its action on $\mathbf{l}$. Notate the action of $\sigma \in G$ as $f \mapsto {}^\sigma f$. The group $E[N]$ acts on $\mathbf{l}(E)$ via translation, the action of each point P being $f \mapsto \tau_P^* f$. The semidirect product group $H \ltimes C$ makes sense since H acts on C. Check that this group acts on $\mathbf{l}(E)$ as

$$(\sigma, P) : f \mapsto \tau_P^* {}^\sigma f.$$

(Hints for this exercise are at the end of the book.)

(b) Let $\mathbb{K}$ be the subfield of $\mathsf{l}(E)$ fixed by $H \times C$. Show that $\mathbb{K} \cap \mathsf{l} = \mathbf{k}'$, and explain how this shows that $\mathbb{K}$ is the function field of a curve E' defined over $\mathbf{k}'$. We don't yet know that E' is an elliptic curve.

(c) Use Theorem 7.7.1 to show that $\mathsf{l}(E')$ is the subfield of $\mathsf{l}(E)$ generated by l and $\mathbb{K}$. Show that the field so generated is also the subfield of $\mathsf{l}(E)$ fixed by C, making it the function field $\mathsf{l}(E/C)$ by the construction of quotient isogeny in the section. Therefore E' is isomorphic over l to E/C as desired.

(d) Show that the l-injection $i : \mathsf{l}(E') \longrightarrow \mathsf{l}(E)$ from the section restricts to a $\mathbf{k}'$-injection
$$i : \mathbf{k}'(E') \longrightarrow \mathbf{k}'(E),$$
giving a morphism $\varphi : E \longrightarrow E'$ defined over $\mathbf{k}'$. In particular $\varphi(0_E) \in E'(\mathbf{k}')$, and along with (c) this shows that E' is an elliptic curve over $\mathbf{k}'$.

(e) Over what subfield of $\mathbb{Q}(j, E_j[N])$ does this exercise show that the quotient isogeny $E_j \longrightarrow E_j/\langle Q_\tau \rangle$ is defined, where Q_τ is the N-torsion point as in Section 7.5?

7.8.3. Consider the elliptic curves
$$E : y^2 = 4x^3 - 4x, \qquad E' : \tilde{y}^2 = 4\tilde{x}^3 + \tilde{x}.$$

Let
$$\varphi(x,y) = \left(\frac{y^2}{16x^2}, \frac{y(x^2+1)}{8x^2} \right) \overset{\text{call}}{=} (\tilde{x}, \tilde{y}).$$

Show that φ takes E to E' and $\varphi(0_E) = 0_{E'}$. Compute $\ker(\varphi)$. Let
$$\psi(\tilde{x}, \tilde{y}) = \left(\frac{\tilde{y}^2}{4\tilde{x}^2}, \frac{\tilde{y}(4\tilde{x}^2 - 1)}{4\tilde{x}^2} \right).$$

Confirm that ψ takes E' to E and that $\psi \circ \varphi = [\deg(\varphi)]$ on E.

7.8.4. Let $\varphi : E \longrightarrow E'$ be an isogeny and let ψ be its dual isogeny. Show that also $\varphi \circ \psi = [\deg(\varphi)]$. (A hint for this exercise is at the end of the book.)

7.8.5. Let $\varphi : E \longrightarrow E'$ be an isogeny and let ψ be its dual isogeny. Show that $\psi_* = \varphi^*$ as maps from $\mathrm{Pic}^0(E')$ to $\mathrm{Pic}^0(E)$. (A hint for this exercise is at the end of the book.)

7.9 Hecke operators algebraically

The modular curve for $\Gamma_1(N)$ is now algebraic over $\mathbb{Q}$. This section shows that the Hecke operators $\langle d \rangle$ and T_p are defined over $\mathbb{Q}$ as well, obtaining an algebraic version over $\mathbb{Q}$ of a commutative diagram involving complex analytic objects,

$$\text{Div}(S_1(N)) \xrightarrow{T_p} \text{Div}(S_1(N))$$
$$\psi_1 \downarrow \qquad\qquad\qquad \downarrow \psi_1 \qquad\qquad (7.14)$$
$$\text{Div}(X_1(N)) \xrightarrow{T_p} \text{Div}(X_1(N)).$$

This is diagram (5.8) slightly modified by changing $Y_1(N)$ to $X_1(N)$ in the bottom row. The diagram shows that T_p is defined compatibly on the complex analytic moduli space and modular curve for $\Gamma_1(N)$. The moduli space $S_1(N)$ is defined in Section 1.5. The map ψ_1 down the sides comes from the bijection $\psi_1 : S_1(N) \longrightarrow Y_1(N)$ of Theorem 1.5.1, but it no longer surjects because $X_1(N)$ also includes the cusps. The divisor group interpretations of T_p are given in Section 5.2. This section also gives the similar algebraic diagram over $\mathbb{Q}$ with $\langle d \rangle$ in place of T_p.

To make the top row of diagram (7.14) algebraic, first continue to work over $\mathbb{C}$. An *enhanced complex algebraic elliptic curve for* $\Gamma_1(N)$ is an ordered pair (E, Q) where E is an algebraic elliptic curve over $\mathbb{C}$ and $Q \in E$ is a point of order N, and two such pairs (E, Q) and (E', Q') are equivalent if some isomorphism $E \xrightarrow{\sim} E'$ defined over $\mathbb{C}$ takes Q to Q'. The *complex algebraic moduli space for* $\Gamma_1(N)$ is the set of equivalence classes

$$S_1(N)_{\text{alg},\mathbb{C}} = \{\text{enhanced complex algebraic elliptic curves for } \Gamma_1(N)\}/\sim,$$

and its points are denoted $[E, Q]$. The complex analytic moduli space $S_1(N)$ in (7.14) and the complex algebraic moduli space $S_1(N)_{\text{alg},\mathbb{C}}$ are in bijective correspondence. Specifically, let E_τ be the curve $y^2 = 4x^3 - g_2(\tau)x - g_3(\tau)$ for $\tau \in \mathcal{H}$, as in Section 7.5. Then the bijection from $S_1(N)$ to $S_1(N)_{\text{alg},\mathbb{C}}$ is

$$[\mathbb{C}/\Lambda_\tau, 1/N + \Lambda_\tau] \mapsto [E_\tau, (\wp_\tau(1/N), \wp'_\tau(1/N))]. \qquad (7.15)$$

Since the moduli spaces biject, the formula defining T_p on the top row of (7.14) transfers to $\text{Div}(S_1(N)_{\text{alg},\mathbb{C}})$,

$$T_p[E, Q] = \sum_C [E/C, Q + C]. \qquad (7.16)$$

This makes sense in the algebraic environment because the quotient isogeny construction of the previous section applies to curves over $\mathbb{C}$. As always, the sum is taken over all order p subgroups $C \subset E$ such that $C \cap \langle Q \rangle = \{0\}$, hence over all order p subgroups when $p \nmid N$.

Now work over $\mathbb{Q}$ rather than $\mathbb{C}$. An *enhanced algebraic elliptic curve for* $\Gamma_1(N)$ is an ordered pair (E, Q) where E is an algebraic elliptic curve over $\overline{\mathbb{Q}}$ and $Q \in E$ is a point of order N, and two such pairs (E, Q) and (E', Q') are equivalent if some isomorphism $E \xrightarrow{\sim} E'$ defined over $\overline{\mathbb{Q}}$ takes Q to Q'. The *algebraic moduli space for* $\Gamma_1(N)$ is the set of equivalence classes

$$S_1(N)_{\text{alg}} = \{\text{enhanced algebraic elliptic curves for } \Gamma_1(N)\}/\sim,$$

and its points are denoted $[E, Q]$. The intersection of an equivalence class in $S_1(N)_{\mathrm{alg},\mathbb{C}}$ with $S_1(N)_{\mathrm{alg}}$ is an equivalence class in $S_1(N)_{\mathrm{alg}}$ (Exercise 7.9.1). Thus $S_1(N)_{\mathrm{alg}}$ can be viewed as a subset of $S_1(N)_{\mathrm{alg},\mathbb{C}}$. By Theorem 7.1.3 torsion is retained from $\mathbb{C}$ to $\overline{\mathbb{Q}}$, so if E is defined over $\overline{\mathbb{Q}}$ then its order p subgroups $C \subset E$ such that $C \cap \langle Q \rangle = \{0\}$ are the same as such subgroups of the corresponding complex curve $E_{\mathbb{C}}$. Also, the quotient isogeny construction restricts to curves over $\overline{\mathbb{Q}}$. Thus formula (7.16) restricts to $\mathrm{Div}(S_1(N)_{\mathrm{alg}})$, and this restriction is the desired algebraic version over $\mathbb{Q}$ of the top row of (7.14).

To make the bottom row of diagram (7.14) algebraic, identify the Riemann surface $X_1(N)$ with $X_1(N)_{\mathrm{alg},\mathbb{C}}$, the complex points of the algebraic model over $\mathbb{Q}$. The bottom row becomes

$$T_p : \mathrm{Div}(X_1(N)_{\mathrm{alg},\mathbb{C}}) \longrightarrow \mathrm{Div}(X_1(N)_{\mathrm{alg},\mathbb{C}}).$$

We will see that this map is defined over $\mathbb{Q}$, and so it restricts to $\overline{\mathbb{Q}}$-points,

$$T_p : \mathrm{Div}(X_1(N)_{\mathrm{alg}}) \longrightarrow \mathrm{Div}(X_1(N)_{\mathrm{alg}}).$$

This restriction is the desired algebraic version over $\mathbb{Q}$ of the bottom row of (7.14).

The fact that T_p on $\mathrm{Div}(X_1(N)_{\mathrm{alg},\mathbb{C}})$ is defined over $\mathbb{Q}$ is shown using the Curves–Fields Correspondence and Galois theory. For simplicity first consider the other Hecke operator $\langle d \rangle$. The corresponding field map is its pullback $\langle d \rangle^*$, so to show that $\langle d \rangle$ is defined over $\mathbb{Q}$ it suffices to show that $\langle d \rangle^*$ takes generators of $\mathbb{Q}(X_1(N)_{\mathrm{alg}})$ back to $\mathbb{Q}(X_1(N)_{\mathrm{alg}})$. Since $\mathbb{Q}(X_1(N)_{\mathrm{alg}}) = \mathbb{Q}(j, f_1)$, the definition $\langle d \rangle(\Gamma_1(N)\tau) = \Gamma_1(N)\gamma(\tau)$ where $\gamma = \left[\begin{smallmatrix} a & b \\ c & \delta \end{smallmatrix}\right] \in \Gamma_0(N)$ with $\delta \equiv d \pmod{N}$ shows that this means checking $j(\gamma(\tau))$ and $f_1(\gamma(\tau))$ for such γ. Because $j(\gamma(\tau)) = j(\tau)$, only $f_1(\gamma(\tau))$ is in question. But as in Section 7.5,

$$f_1(\gamma(\tau)) = f_0^{\pm \overline{(0,1)}}(\gamma(\tau)) = f_0^{\pm \overline{(0,1)\gamma}}(\tau) = f_0^{\pm \overline{(0,d)}}(\tau).$$

Thus $f_1(\gamma(\tau))$ is the x-coordinate of $\pm[d]Q_\tau \in E_{j(\tau)}$, making it an element of $\mathbb{Q}(j(\tau), E_{j(\tau)}[N])$. By definition $\mathbb{Q}(X_1(N)_{\mathrm{alg}})$ is the fixed field of the Galois subgroup K_1 such that $\rho(K_1) = \{\pm \left[\begin{smallmatrix} 1 & b \\ 0 & 1 \end{smallmatrix}\right]\} \subset \mathrm{GL}_2(\mathbb{Z}/N\mathbb{Z})$. This group preserves $\pm[d]Q_\tau$ and therefore fixes $f_1(\gamma(\tau))$ as desired, and thus $\langle d \rangle$ is defined over $\mathbb{Q}$.

Recall the planar model $X_1(N)_{\mathrm{alg}}^{\mathrm{planar}}$ from Section 7.7, the ordered pairs (j, x) satisfying the polynomial relation between $j(\tau)$ and $f_1(\tau)$. In terms of this model, the Hecke operator is

$$\langle d \rangle(j, x) = (j, r_d(j, x))$$

where r_d is the rational function over $\mathbb{Q}$ of two variables such that $x([d]Q_\tau) = r_d(j(\tau), x(Q_\tau))$. To provide the reader another opportunity to use the ideas of these last two paragraphs in an easy setting, Exercise 7.9.2 shows that the operator w_N on the modular curve $X_0(N)$ (cf. Exercise 1.5.4) is defined over $\mathbb{Q}$. Then the more elaborate Exercise 7.9.3 shows that T_p is also defined over $\mathbb{Q}$ as claimed.

The sides of diagram (7.14) extend the map of Theorem 1.5.1,

$$\psi_1 : S_1(N) \longrightarrow X_1(N), \qquad [\mathbb{C}/\Lambda_\tau, 1/N + \Lambda_\tau] \mapsto \Gamma_1(N)\tau,$$

to divisor groups. To make ψ_1 algebraic, consider the commutative diagram

$$\begin{array}{ccc} S_1(N) & \longrightarrow & S_1(1) \\ \psi_1 \downarrow & & \downarrow \\ X_1(N) & \longrightarrow & X_1(1) \end{array}$$

given by

$$\begin{array}{ccc} [\mathbb{C}/\Lambda_\tau, 1/N + \Lambda_\tau] & \longmapsto & [\mathbb{C}/\Lambda_\tau] \\ \downarrow & & \downarrow \\ \Gamma_1(N)\tau & \longmapsto & \mathrm{SL}_2(\mathbb{Z})\tau. \end{array}$$

Recall that the complex analytic moduli space $S_1(N)$ identifies with the complex algebraic moduli space $S_1(N)_{\mathrm{alg},\mathbb{C}}$, and the Riemann surface $X_1(N)$ identifies with the complex points $X_1(N)_{\mathrm{alg},\mathbb{C}}$ of the algebraic modular curve. When $N = 1$, the complex algebraic moduli space is the equivalence classes of complex algebraic elliptic curves, and the complex points of the algebraic modular curve are the complex projective line $\mathbb{P}^1(\mathbb{C})$. The diagram becomes

$$\begin{array}{ccc} S_1(N)_{\mathrm{alg},\mathbb{C}} & \longrightarrow & S_1(1)_{\mathrm{alg},\mathbb{C}} \\ \psi_{1,\mathrm{alg}} \downarrow & & \downarrow \\ X_1(N)_{\mathrm{alg},\mathbb{C}} & \longrightarrow & X_1(1)_{\mathrm{alg},\mathbb{C}} \end{array} \qquad (7.17)$$

given by

$$\begin{array}{ccc} [E, Q] & \longmapsto & [E] \\ \downarrow & & \downarrow \\ P & \longmapsto & j(E). \end{array}$$

Since our description of $X_1(N)_{\mathrm{alg}}$ is abstract, we cannot write the maps down the left side and across the bottom row explicitly. Nonetheless, since an element $[E, Q]$ of $S_1(N)_{\mathrm{alg},\mathbb{C}}$ describes an element of the algebraic moduli space $S_1(N)_{\mathrm{alg}}$ if and only if $j(E) \in \overline{\mathbb{Q}}$, as discussed earlier in the section, mapping down and then across in diagram (7.17) takes $S_1(N)_{\mathrm{alg}}$ to the algebraic modular curve $X_1(1)_{\mathrm{alg}} = \mathbb{P}^1(\overline{\mathbb{Q}})$. It follows that the point P in the diagram lies in the algebraic modular curve $X_1(N)_{\text{alg}}$ (Exercise 7.9.4(a)). That is, the left side of diagram (7.17) restricts to the desired algebraic map

$$\psi_{1,\text{alg}} : S_1(N)_{\text{alg}} \longrightarrow X_1(N)_{\text{alg}}.$$

We now have the algebraic counterpart over $\mathbb{Q}$ to diagram (7.14),

$$
\begin{array}{ccc}
\text{Div}(S_1(N)_{\text{alg}}) & \xrightarrow{T_p} & \text{Div}(S_1(N)_{\text{alg}}) \\
\downarrow{\psi_{1,\text{alg}}} & & \downarrow{\psi_{1,\text{alg}}} \\
\text{Div}(X_1(N)_{\text{alg}}) & \xrightarrow{T_p} & \text{Div}(X_1(N)_{\text{alg}}).
\end{array}
$$

The vertical maps restrict to degree-zero divisors and then the bottom row passes to Picard groups, giving a diagram we will need in Chapter 8,

$$
\begin{array}{ccc}
\text{Div}^0(S_1(N)_{\text{alg}}) & \xrightarrow{T_p} & \text{Div}^0(S_1(N)_{\text{alg}}) \\
\downarrow{\psi_{1,\text{alg}}} & & \downarrow{\psi_{1,\text{alg}}} \\
\text{Pic}^0(X_1(N)_{\text{alg}}) & \xrightarrow{T_p} & \text{Pic}^0(X_1(N)_{\text{alg}}).
\end{array}
\tag{7.18}
$$

Similarly, starting from diagram (5.9),

$$
\begin{array}{ccc}
S_1(N) & \xrightarrow{\langle d \rangle} & S_1(N) \\
\downarrow{\psi_1} & & \downarrow{\psi_1} \\
Y_1(N) & \xrightarrow{\langle d \rangle} & Y_1(N),
\end{array}
$$

gives another algebraic diagram we will need in Chapter 8,

$$
\begin{array}{ccc}
\text{Div}^0(S_1(N)_{\text{alg}}) & \xrightarrow{\langle d \rangle} & \text{Div}^0(S_1(N)_{\text{alg}}) \\
\downarrow{\psi_{1,\text{alg}}} & & \downarrow{\psi_{1,\text{alg}}} \\
\text{Pic}^0(X_1(N)_{\text{alg}}) & \xrightarrow{\langle d \rangle_*} & \text{Pic}^0(X_1(N)_{\text{alg}}).
\end{array}
\tag{7.19}
$$

Finally, the map $\psi_{1,\text{alg}}$ can be described more explicitly at the cost of losing some points along the way. Note that $j(\tau) \notin \{0, 1728\}$ for all but finitely many points $[\mathbb{C}/\Lambda_\tau, 1/N + \Lambda_\tau]$ of the complex analytic moduli space $S_1(N)$, so that as in Section 7.5 an admissible change of variable takes E_τ to

$$
E_{j(\tau)} : y^2 = 4x^3 - \left(\frac{27j(\tau)}{j(\tau) - 1728} \right) x - \left(\frac{27j(\tau)}{j(\tau) - 1728} \right)
$$

and takes $(\wp_\tau(1/N), \wp'_\tau(1/N))$ to Q_τ. That is, the bijection (7.15) from $S_1(N)$ to $S_1(N)_{\text{alg},\mathbb{C}}$ restricts to

$$[\mathbb{C}/\Lambda_\tau, 1/N + \Lambda_\tau] \mapsto [E_{j(\tau)}, Q_\tau], \qquad j(\tau) \notin \{0, 1728\}.$$

Consider the diagram

$$
\begin{array}{ccc}
S_1(N)' & \longrightarrow & S_1(N)'_{\text{alg},\mathbb{C}} \\
{\scriptstyle \psi_1}\downarrow & & \\
Y_1(N)' & \longrightarrow & X_1(N)^{\text{planar}'}_{\text{alg},\mathbb{C}}
\end{array}
$$

where the primes mean to remove finitely many points, the map across the top was just given, and the map across the bottom comes from the definition of the planar model,

$$
\begin{array}{ccc}
[C/\Lambda_\tau, 1/N + \Lambda_\tau] & \longmapsto & [E_{j(\tau)}, Q_\tau] \\
\big\uparrow & & \\
\Gamma_1(N)\tau & \longmapsto & (j(\tau), f_1(\tau)).
\end{array}
$$

The complex algebraic map from the moduli space to the planar model of the modular curve comes from completing the diagram,

$$[E_{j(\tau)}, Q_\tau] \mapsto (j(\tau), f_1(\tau)). \tag{7.20}$$

Obtaining $\psi_{1,\text{alg}}$ from this uses one last commutative diagram,

$$
\begin{array}{ccc}
Y_1(N)' & \longrightarrow & X_1(N)^{\text{planar}'}_{\text{alg},\mathbb{C}} \\
\downarrow & & \downarrow \\
X_1(N) & \longrightarrow & X_1(N)_{\text{alg},\mathbb{C}}.
\end{array}
$$

Here the map across the top row is from the previous diagram, the map down the left side is inclusion, the map down the right side is the birational equivalence from the planar model of the modular curve to the actual modular curve as discussed after Theorem 7.2.5, and the map across the bottom is the identification of the Riemann surface with the complex points of the algebraic modular curve. Since the birational map is undefined at finitely many points of the planar model, the primes in the top row now mean to remove finitely more points as necessary for all the maps to make sense. Thus $\psi_{1,\text{alg}}$ on all but finitely many points of $S_1(N)_{\text{alg},\mathbb{C}}$ is described more explicitly than before as (7.20) followed by the birational map. Finally, the map from $S_1(N)'_{\text{alg}}$ to the planar model can be described without reference to the larger complex analytic environment. It is

$$S_1(N)'_{\text{alg}} \longrightarrow X_1(N)^{\text{planar}}_{\text{alg}}, \qquad [E,Q] \mapsto (j(E), x(Q)), \qquad (7.21)$$

where $x(Q)$ is the x-coordinate of Q and again the prime means to preclude $j = 0, 1728$. This follows from (7.20) since $f_1(\tau) = x(Q_\tau)$. We will use this description in Chapter 8.

Exercises

7.9.1. Show that the intersection of an equivalence class in $S_1(N)_{\text{alg},\mathbb{C}}$ with $S_1(N)_{\text{alg}}$ is an equivalence class in $S_1(N)_{\text{alg}}$.

7.9.2. From Exercise 1.5.4 the matrix $\begin{bmatrix} 0 & -1 \\ N & 0 \end{bmatrix}$ normalizes $\Gamma_0(N)$ and so it gives an involution

$$w_N : X_0(N) \longrightarrow X_0(N), \qquad \Gamma_0(N)\tau \mapsto \Gamma_0(N)(-1/(N\tau)).$$

(a) Let $\tau' = -1/(N\tau)$. Explain why showing that w_N is defined over $\mathbb{Q}$ reduces to showing that $j(\tau')$ and $j_N(\tau')$ lie in $\mathbb{Q}(X_0(N)_{\text{alg}})$. Show this. (Hints for this exercise are at the end of the book.)

(b) Now we know that the algebraic version of w_N on the planar model of $X_0(N)$ defined by the minimal polynomial p_0 of f_0 over $\mathbb{Q}(j)$ is

$$w_N : X_0(N)^{\text{planar}}_{\text{alg}} \longrightarrow X_0(N)^{\text{planar}}_{\text{alg}}, \qquad (j,x) \mapsto (j',x'),$$

a map whose first component is a rational function r_N over $\mathbb{Q}$ in two variables. Describe a map

$$S_0(N)'_{\text{alg}} \longrightarrow X_0(N)^{\text{planar}}_{\text{alg}}$$

analogous to (7.21). Explain why $j(E/C) = r_N(j(E), x(C))$ where E is an elliptic curve over $\overline{\mathbb{Q}}$, C is an N-cyclic subgroup of E, and $x(C)$ is the sum of the x-coordinates of its nonzero points.

(c) Describe w_N on the planar model of $X_0(N)$ defined by the modular equation $\Phi(j, j_N) = 0$, cf. Section 7.5. What does this show about the polynomial $\Phi(x, y)$?

7.9.3. This exercise shows that the Hecke operator T_p is defined over $\mathbb{Q}$.

(a) The double coset configuration specialized to T_p (cf. Section 5.2) involves the matrix $\alpha = \begin{bmatrix} 1 & 0 \\ 0 & p \end{bmatrix}$, the group $\Gamma_3 = \Gamma_1^0(N,p) = \Gamma_1(N) \cap \Gamma^0(p)$, and the modular curve $X_1^0(N,p) = X(\Gamma_1^0(N,p))$. Similarly define

$$\Gamma_{1,0}(N,p) = \Gamma_1(N) \cap \Gamma_0(Np), \qquad X_{1,0}(N,p) = X(\Gamma_{1,0}(N,p)).$$

Show that $\Gamma_{1,0}(N,p) = \alpha\Gamma_1^0(N,p)\alpha^{-1}$, i.e., $\Gamma_3' = \Gamma_{1,0}(N,p)$ and $X_3' = X_{1,0}(N,p)$ in the double coset configuration. Show that the paragraph after Theorem 6.3.2 rephrases to say that T_p is the pullback of the map

$$X_{1,0}(N,p) \longrightarrow X_1(N), \qquad \Gamma_{1,0}(N,p)\tau \mapsto \Gamma_1(N)p\tau \qquad (7.22)$$

followed by the pushforward of the map

$$X_{1,0}(N,p) \longrightarrow X_1(N), \qquad \Gamma_{1,0}(N,p)\tau \mapsto \Gamma_1(N)\tau. \tag{7.23}$$

The goal is to show that each of these maps is defined over $\mathbb{Q}$.

(b) Show that $\Gamma_1(Np) \subset \Gamma_{1,0}(N,p) \subset \Gamma_0(Np)$. Zoom in on Figure 7.2 and cite (7.13), both with Np in place of N, to obtain Figure 7.8 here, in which the intermediate field $\mathbb{K}$ on the right corresponds to $\mathbb{C}(X_{1,0}(N,p))$ on the left.

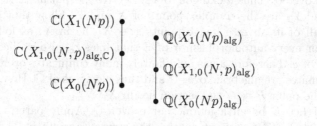

Figure 7.8. Function fields for T_p

Explain why $\mathbb{K}$ is the function field of a curve over $\mathbb{Q}$. Denote this curve $X_{1,0}(N,p)_{\text{alg}}$, so that $\mathbb{Q}(X_{1,0}(N,p)_{\text{alg}}) = \mathbb{K}$. Since $X_1(Np)_{\text{alg},\mathbb{C}}$ is isomorphic over $\mathbb{C}$ to $X_1(Np)$, and similarly for X_0, Theorem 7.7.1 also gives a diagram as in Figure 7.9. Apply the Restriction Lemma twice to this figure to show that $\mathbb{C}(X_{1,0}(N,p)_{\text{alg},\mathbb{C}}) = \mathbb{C}(X_{1,0}(N,p))$ and therefore $X_{1,0}(N,p)_{\text{alg},\mathbb{C}} = X_{1,0}(N,p)$ up to isomorphism over $\mathbb{C}$. Thus $X_{1,0}(N,p)$ is defined over $\mathbb{Q}$.

$$
\begin{array}{ll}
\mathbb{C}(X_1(Np)) \bullet & \\
\qquad\qquad & \bullet\, \mathbb{Q}(X_1(Np)_{\text{alg}}) \\
\mathbb{C}(X_{1,0}(N,p)_{\text{alg},\mathbb{C}}) \bullet & \\
\qquad\qquad & \bullet\, \mathbb{Q}(X_{1,0}(N,p)_{\text{alg}}) \\
\mathbb{C}(X_0(Np)) \bullet & \\
\qquad\qquad & \bullet\, \mathbb{Q}(X_0(Np)_{\text{alg}})
\end{array}
$$

Figure 7.9. Comparing $\mathbb{C}(X_{1,0}(N,p)_{\text{alg},\mathbb{C}})$ and $\mathbb{C}(X_{1,0}(N,p))$

Obtain the tower of fields and ρ-images of subgroups of the Galois group $\text{Gal}(\mathbb{Q}(j, E_j[Np])/\mathbb{Q}(j))$ shown in Figure 7.10. (Hints for this exercise are at the end of the book.)

(c) The two function field injections corresponding to the maps (7.22) and (7.23) are

$$\mathbb{C}(X_1(N)) \longrightarrow \mathbb{C}(X_{1,0}(N,p)), \qquad f(\tau) \mapsto f(p\tau)$$

$$\mathbb{Q}(X_1(Np)_{\text{alg}}) \bullet \{\pm \begin{bmatrix} a & b \\ 0 & 1 \end{bmatrix} \in \mathrm{GL}_2(\mathbb{Z}/Np\mathbb{Z})\}$$

$$\mathbb{Q}(X_{1,0}(N,p)_{\text{alg}}) \bullet \{\pm \begin{bmatrix} a & b \\ 0 & d \end{bmatrix} \in \mathrm{GL}_2(\mathbb{Z}/Np\mathbb{Z}) : d \equiv 1 \ (\mathrm{mod}\ N)\}$$

$$\mathbb{Q}(X_0(Np)_{\text{alg}}) \bullet \{\begin{bmatrix} a & b \\ 0 & d \end{bmatrix} \in \mathrm{GL}_2(\mathbb{Z}/Np\mathbb{Z})\}$$

Figure 7.10. Galois groups

and
$$\mathbb{C}(X_1(N)) \longrightarrow \mathbb{C}(X_{1,0}(N,p)), \qquad f(\tau) \mapsto f(\tau).$$

Since $\mathbb{C}(X_1(N)) = \mathbb{C}(j, f_1)$ the nontrivial map has image $\mathbb{C}(j(p\tau), f_1(p\tau))$. To show that the injections restrict to the function fields over $\mathbb{Q}$ it thus suffices to show

$$j(p\tau) \in \mathbb{Q}(X_{1,0}(N,p)_{\text{alg}}), \qquad f_1(p\tau) \in \mathbb{Q}(X_{1,0}(N,p)_{\text{alg}}).$$

Use the group containment $\Gamma_{1,0}(N,p) \subset \Gamma_0(p)$ to show the first of these.

(d) Let $\tau' = p\tau$ and let $j' = j(\tau')$, both in $\mathbb{Q}(X_{1,0}(N,p)_{\text{alg}})$ and thus in $\mathbb{Q}(j, E_j[Np])$. Show that there is a morphism $\varphi : E_j \longrightarrow E_{j'}$ with kernel $\langle Q_{\tau,N} \rangle$, taking $Q_{\tau,Np}$ to $Q_{\tau',N}$, and defined over some Galois extension $\mathbb{L}$ of $\mathbb{Q}(j, E_j[Np])$. Here the second subscript of each Q denotes its order.

(e) Let $K = \mathrm{Gal}(\mathbb{Q}(j, E_j[Np])/\mathbb{Q}(X_{1,0}(N,p)_{\text{alg}}))$. Any $\sigma \in K$ extends to $\mathrm{Gal}(\mathbb{L}/\mathbb{Q}(X_{1,0}(N,p)_{\text{alg}}))$. Show that then $\varphi^\sigma = \pm\varphi$ and $\varphi(Q_{\tau,Np}) = Q_{\tau',N}$. Deduce that $f_1(p\tau) \in \mathbb{Q}(X_{1,0}(N,p)_{\text{alg}})$ as desired.

7.9.4. (a) Let $h : X \longrightarrow Y$ be a surjective morphism over $\mathbb{Q}$ of algebraic curves over $\mathbb{Q}$. Thus h extends to a surjective morphism $h : X_{\mathbb{C}} \longrightarrow Y_{\mathbb{C}}$ where $X_{\mathbb{C}}$ and $Y_{\mathbb{C}}$ are the complex points of X and Y. Show that if $y \in Y \subset Y_{\mathbb{C}}$ then all of its inverse image points $x \in X_{\mathbb{C}}$ lie in X, as follows. Compute that for every automorphism σ of $\mathbb{C}$ that fixes $\overline{\mathbb{Q}}$, also x^σ lies in the inverse image of y. Therefore x has only finitely many conjugates over $\overline{\mathbb{Q}}$, meaning its coordinates are algebraic over $\overline{\mathbb{Q}}$ and thus elements of $\overline{\mathbb{Q}}$. How does this show that the point P from the section lies in $X_1(N)_{\text{alg}}$?

(b) Let E be an algebraic curve over $\mathbb{Q}$. Apply part (a) to show that $E[N] \cong (\mathbb{Z}/N\mathbb{Z})^2$ without quoting the structure result in Theorem 7.1.3. (A hint for this exercise is at the end of the book.)

8

The Eichler–Shimura Relation and L-functions

This chapter arrives at the first version of the Modularity Theorem stated in the preface to the book: For any elliptic curve E over $\mathbb{Q}$ there exists a weight 2 newform f such that the Fourier coefficients $a_p(f)$ are equal to the solution-counts $a_p(E)$ of a Weierstrass equation for E modulo p. Gathering the $a_p(f)$ and $a_p(E)$ into L-functions, this rephrases as

$$L(s, f) = L(s, E).$$

The techniques that relate this to other versions of Modularity involve working modulo p and expressing both $a_p(E)$ and $a_p(f)$ in terms of the *Frobenius map*,

$$\sigma_p : x \mapsto x^p.$$

The key is the Eichler–Shimura relation, expressing the Hecke operator T_p in characteristic p in terms of σ_p. Since reducing algebraic curves and maps from characteristic 0 to characteristic p is technical, the chapter necessarily quotes many results in quickly sketching the relevant background. The focus here is on the Eichler–Shimura relation itself in Section 8.7 and on its connection to Modularity in Section 8.8.

Chapter 7 transferred the Modularity Theorem from analysis to algebraic geometry and changed its underlying field of definition from $\mathbb{C}$ to $\mathbb{Q}$. Now that the objects are algebraic we drop "alg" from the notation, so

- $X_1(N)$ denotes the modular curve as a nonsingular algebraic curve over $\mathbb{Q}$, with function field $\mathbb{Q}(X_1(N)) = \mathbb{Q}(j, f_1)$.
- $S_1(N)$ denotes the moduli space consisting of equivalence classes $[E, Q]$ where E is an algebraic elliptic curve over $\overline{\mathbb{Q}}$, $Q \in E$ is a point of order N, and the equivalence relation is isomorphism over $\overline{\mathbb{Q}}$.

Also, the Jacobian of a compact Riemann surface X is identified with the Picard group of X, both denoted $\mathrm{Pic}^0(X)$. The notation $\mathrm{Jac}(X)$ will no longer be used.

Related reading: Chapter 7 of [Shi73], Rohrlich's article in [CSS97], [Sil86], [Kna93].

© Springer Science+Business Media New York 2005
F. Diamond, J. Shurman, *A First Course in Modular Forms*,
Graduate Texts in Mathematics 228, DOI 10.1007/978-0-387-27226-9_8

8.1 Elliptic curves in arbitrary characteristic

Much of the material from Section 7.1 on elliptic curves in characteristic 0 is also valid in characteristic p. Let $\mathbf{k}$ be an arbitrary field. A *Weierstrass equation over* $\mathbf{k}$ is any cubic equation of the form

$$E : y^2 + a_1 xy + a_3 y = x^3 + a_2 x^2 + a_4 x + a_6, \quad a_1, \ldots, a_6 \in \mathbf{k}. \tag{8.1}$$

To study this, define

$$b_2 = a_1^2 + 4a_2, \quad b_4 = a_1 a_3 + 2a_4, \quad b_6 = a_3^2 + 4a_6,$$
$$b_8 = a_1^2 a_6 - a_1 a_3 a_4 + a_2 a_3^2 + 4a_2 a_6 - a_4^2,$$

and define the *discriminant* of the equation to be

$$\Delta = -b_2^2 b_8 - 8b_4^3 - 27b_6^2 + 9b_2 b_4 b_6.$$

Further define

$$c_4 = b_2^2 - 24b_4, \quad c_6 = -b_2^3 + 36b_2 b_4 - 216b_6,$$

and if $\Delta \neq 0$ define the *invariant* of the equation to be

$$j = c_4^3/\Delta.$$

Then all $b_i \in \mathbf{k}$, $\Delta \in \mathbf{k}$, all $c_i \in \mathbf{k}$, and $j \in \mathbf{k}$ when it is defined. Also, $4b_8 = b_2 b_6 - b_4^2$ and $1728\Delta = c_4^3 - c_6^2$. (Confirming the calculations in this paragraph is Exercise 8.1.1(a).) If $\mathbf{k}$ does not have characteristic 2 then replacing y by $y - (a_1 x + a_3)/2$ in (8.1) eliminates the xy and y terms from the left side, reducing the Weierstrass equation to

$$E : y^2 = x^3 + (b_2 x^2 + 2b_4 x + b_6)/4, \quad b_2, b_4, b_6 \in \mathbf{k}, \text{char}(\mathbf{k}) \neq 2. \tag{8.2}$$

If $\mathbf{k}$ does not have characteristic 2 or 3 then replacing x by $(x - 3b_2)/36$ and y by $y/216$ in (8.2) eliminates the x^2 term from the right side, further reducing the Weierstrass equation to the form

$$E : y^2 = x^3 - 27c_4 x - 54c_6, \quad c_4, c_6 \in \mathbf{k}, \text{char}(\mathbf{k}) \notin \{2,3\}. \tag{8.3}$$

Since (8.2) and (8.3) are special cases of (8.1), the coefficients of a Weierstrass equation will be referred to as the a_i in all cases.

As before,

Definition 8.1.1. *Let $\overline{\mathbf{k}}$ be an algebraic closure of the field $\mathbf{k}$. When a Weierstrass equation E has nonzero discriminant Δ it is called* **nonsingular** *and the set*

$$E = \{(x,y) \in \overline{\mathbf{k}}^2 \text{ satisfying } E(x,y)\} \cup \{\infty\}$$

is called an **elliptic curve over** $\mathbf{k}$.

Note that an elliptic curve over $\mathbf{k}$ has infinitely many points even when $\mathbf{k}$ is a finite field.

The *general admissible change of variable* is

$$x = u^2 x' + r, \quad y = u^3 y' + su^2 x' + t, \quad u, r, s, t \in \mathbf{k}, \ u \neq 0.$$

These form a group, and they transform Weierstrass equations to Weierstrass equations, taking the discriminant Δ to Δ/u^{12} and preserving the invariant j (Exercise 8.1.1(b)). In particular the changes of variable between Weierstrass equations (8.1), (8.2), and (8.3) are admissible, and thus we may work with (8.3) if $\text{char}(\mathbf{k}) \notin \{2, 3\}$ and with (8.2) if $\text{char}(\mathbf{k}) \neq 2$, needing the general (8.1) only when $\text{char}(\mathbf{k}) = 2$ or when there is no assumption about the characteristic—for example, Section 8.3 will discuss reducing a Weierstrass equation (8.1) over $\mathbb{Q}$ modulo an arbitrary prime p. Since the changes of variable from (8.1) to (8.2) to (8.3) have $u = 1$ and $u = 1/6$ respectively, the discriminant of (8.3) is 6^{12} times the discriminant of (8.1) and (8.2), i.e., $\Delta_{(8.3)} = 2^6 3^9 (c_4^3 - c_6^2)$.

Somewhat awkwardly, the cubic equation $y^2 = 4x^3 - g_2 x - g_3$ from Chapter 7 is no longer a Weierstrass equation by our new definition since the coefficients of y^2 and x^3 are unequal, and replacing y by $2y$ to put it in the form (8.3) is an inadmissible change of variable. Both sorts of cubic equation can be encompassed in more general definitions of Weierstrass equation and admissible change of variable, but since normalizing the coefficients of y^2 and x^3 to 1 simplifies the formulas of this chapter we accept the small inconsistency in terminology instead. Modulo the inadmissible substitution, the earlier definitions of the discriminant and the invariant for equations $y^2 = 4x^3 - g_2 x - g_3$ are the same as their definitions for (8.3) (Exercise 8.1.1(d)).

In particular, replacing y by $2y$ in the universal elliptic curve (7.12) leads to the Weierstrass equation

$$y^2 = x^3 - \frac{1}{4}\left(\frac{27j}{j - 1728}\right) x - \frac{1}{4}\left(\frac{27j}{j - 1728}\right), \tag{8.4}$$

with discriminant $2^6 3^{12} j^2 / (j - 1728)^3$ and invariant j. An admissible change of variable then gives a more general universal curve (Exercise 8.1.1(e))

$$y^2 + xy = x^3 - \left(\frac{36}{j - 1728}\right) x - \left(\frac{1}{j - 1728}\right), \tag{8.5}$$

with discriminant $j^2 / (j - 1728)^3$ and invariant j. The curve (8.5) is well suited for fields of arbitrary characteristic since its discriminant is nonzero even in characteristic 2 or 3. Section 8.6 will use it to define modular curves in prime characteristic as Chapter 7 used the universal elliptic curve to define modular curves over $\mathbb{Q}$.

Most of the results from Section 7.1 hold in arbitrary characteristic. The Weierstrass polynomial associated to (8.1) is

$$E(x,y) = y^2 + a_1xy + a_3y - x^3 - a_2x^2 - a_4x - a_6 \in \mathbf{k}[x,y], \qquad (8.6)$$

and similarly for (8.2) and (8.3). In all cases the Weierstrass equation is non-singular if and only if the corresponding curve E is geometrically nonsingular, i.e., the gradient of the Weierstrass polynomial never vanishes. Verifying this in characteristic 2 requires a different argument from the one given before (Exercise 8.1.2). Again an elliptic curve E lies in $\mathbb{P}^2(\overline{\mathbf{k}})$, it forms an Abelian group with the infinite point $[0,1,0]$ (see Exercise 8.1.3(a)) as its additive identity 0_E, and the addition law is that collinear triples sum to 0_E. Opposite pairs of points P and $-P$ have the same x-coordinate, at most two points with the same x-coordinate satisfy (8.1), and so any two points with the same x-coordinate are equal or opposite, possibly both. Since the y-values satisfying (8.1) for a given x sum to $-a_1x - a_3$ the additive inverse of $P = (x_P, y_P)$ is the natural companion point

$$-P = (x_P, -y_P - a_1x_P - a_3).$$

Given points P and Q of E, let R be their third collinear point. The addition law $P + Q = -R$ says that the sum is the companion point of R,

$$\text{if } P,\, Q,\, R \text{ are collinear then } P + Q = (x_R, -y_R - a_1x_R - a_3). \qquad (8.7)$$

Let $\mathbf{k}_{\text{prime}}$ denote the *prime subfield* of $\mathbf{k}$, meaning the smallest subfield inside $\mathbf{k}$, either the rational numbers $\mathbb{Q}$ if $\text{char}(\mathbf{k}) = 0$ or the finite field $\mathbb{F}_p$ of order p if $\text{char}(\mathbf{k}) = p$. Then the group law is defined by rational functions r and s over the field $\mathbf{k}_{\text{prime}}(\{a_i\})$ where the a_i are the Weierstrass coefficients. If $x_Q = x_P$ and $y_Q = -y_P - a_1x_P - a_3$ then $Q = -P$ and so $P + Q = 0_E$. Otherwise $P + Q$ lies in the affine part of E. Let

$$\lambda = \begin{cases} \dfrac{y_Q - y_P}{x_Q - x_P} & x_P \neq x_Q, \\[3mm] \dfrac{3x_P^2 + 2a_2x_P + a_4 - a_1y_P}{a_1x_P + a_3 + 2y_P} & x_P = x_Q, \end{cases}$$

and

$$\mu = \begin{cases} \dfrac{y_Px_Q - y_Qx_P}{x_Q - x_P} & x_P \neq x_Q, \\[3mm] \dfrac{-x_P^3 + a_4x_P + 2a_6 - a_3y_P}{a_1x_P + a_3 + 2y_P} & x_P = x_Q. \end{cases}$$

The line $y = \lambda x + \mu$ passes through P and Q when $P \neq Q$ and is the tangent line to E at P when $P = Q$ (Exercise 8.1.4(a)). The rational functions giving x_{P+Q} and y_{P+Q} are

$$\begin{aligned} r(x_P, x_Q, y_P, y_Q) &= \lambda^2 + a_1\lambda - a_2 - x_P - x_Q, \\ s(x_P, x_Q, y_P, y_Q) &= -(\lambda + a_1)r(x_P, x_Q, y_P, y_Q) - \mu - a_3. \end{aligned} \qquad (8.8)$$

This algebraic definition of addition corresponds to the geometric description (8.7) (Exercise 8.1.4(b)). As before, the casewise expressions for λ and μ arise from a single rational function (Exercise 8.1.4(c)). For any algebraic extension $\mathbb{K}/\mathbf{k}$ the set of $\mathbb{K}$-points of E is a subgroup of E,

$$E(\mathbb{K}) = \{P \in E - \{0_E\} : (x_P, y_P) \in \mathbb{K}^2\} \cup \{0_E\}.$$

Let N be a positive integer. The structure theorem for the N-torsion subgroup $E[N] = \ker([N])$ of an elliptic curve is

Theorem 8.1.2. *Let E be an elliptic curve over $\mathbf{k}$ and let N be a positive integer. Then*

$$E[N] \cong \prod E[p^{e_p}] \quad \text{where } N = \prod p^{e_p}.$$

Also,

$$E[p^e] \cong (\mathbb{Z}/p^e\mathbb{Z})^2 \quad \text{if } p \neq \text{char}(\mathbf{k}).$$

Thus $E[N] \cong (\mathbb{Z}/N\mathbb{Z})^2$ if $\text{char}(\mathbf{k}) \nmid N$. On the other hand,

$$\left.\begin{array}{c} E[p^e] \cong \mathbb{Z}/p^e\mathbb{Z} \text{ for all } e \geq 1 \\ \text{or} \\ E[p^e] = \{0\} \text{ for all } e \geq 1 \end{array}\right\} \quad \text{if } p = \text{char}(\mathbf{k}).$$

In particular, if $\text{char}(\mathbf{k}) = p$ then either $E[p] \cong \mathbb{Z}/p\mathbb{Z}$, in which case E is called **ordinary**, or $E[p] = \{0\}$ and E is **supersingular**.

In this chapter we will also need to know a bit about singular Weierstrass equations. As already mentioned, the condition $\Delta = 0$ is equivalent to the condition that for some point P satisfying the Weierstrass polynomial (8.6), both partial derivatives vanish at P. The projective point $[0, 1, 0]$ is always nonsingular (this was Exercise 8.1.3(b)), so any such P is affine. The coordinates of P lie in $\mathbf{k}$ (Exercise 8.1.5(a)), so an admissible change of variable over $\mathbf{k}$ takes P to $(0, 0)$. Then the conditions $E(0,0) = D_1 E(0,0) = D_2 E(0,0) = 0$ force the Weierstrass polynomial to be

$$E(x, y) = y^2 + a_1 xy - x^3 - a_2 x^2. \tag{8.9}$$

If $\text{char}(\mathbf{k}) \neq 2$ then letting $\tilde{y} = y + a_1 x/2$ in (8.9) simplifies this to

$$E(x, y) = \tilde{y}^2 - x^3 - a_2' x^2. \tag{8.10}$$

The point $P = (0, 0)$ is the only singular point satisfying E. If $\text{char}(\mathbf{k}) = 2$ then this is easy to verify from (8.9), while if $\text{char}(\mathbf{k}) \neq 2$ then it follows from (8.10) (Exercise 8.1.5(b)).

Rewrite (8.9) as

$$E(x, y) = (y - m_1 x)(y - m_2 x) - x^3, \tag{8.11}$$

where m_1 and m_2 satisfy the quadratic polynomial $f(t) = t^2 + a_1 t - a_2$ over $\mathbb{F}_p$ but need not lie in $\mathbb{F}_p$ themselves. The singular point P is called a *node* if $m_1 \neq m_2$, meaning that two distinct tangent lines pass through the curve at P, and it is called a *cusp* if $m_1 = m_2$, when there is only one tangent line. (See Figure 8.1.) Working from (8.9) and (8.11), it is easy to compute that $c_4 = (m_1 - m_2)^4$ (Exercise 8.1.5(c)), so that the curve has a node if $c_4 \neq 0$ and a cusp if $c_4 = 0$. These conditions apply to the Weierstrass equation in its original form since the admissible change of variable translating the singular point P to $(0,0)$ multiplies c_4 by a nonzero scalar, cf. Exercise 8.1.1(c). In sum,

Proposition 8.1.3. *Let E be a Weierstrass equation over* **k**. *Then*

- E *describes an elliptic curve* $\Longleftrightarrow \Delta \neq 0$,
- E *describes a curve with a node* $\Longleftrightarrow \Delta = 0$ *and* $c_4 \neq 0$,
- E *describes a curve with a cusp* $\Longleftrightarrow \Delta = 0$ *and* $c_4 = 0$.

In the case of a node the set of projective solutions of E other than the singular point forms a multiplicative group isomorphic to $\overline{\mathbf{k}}^*$, and in the case of a cusp the set forms an additive group isomorphic to $\overline{\mathbf{k}}$. (See [Sil86] for the proof of this, and also see Exercise 9.4.2.) The notions of ordinary, supersingular, multiplicative, and additive Weierstrass equations in characteristic p will all figure in this chapter.

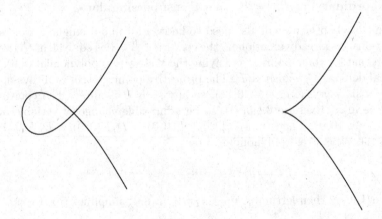

Figure 8.1. Node and cusp

Exercises

8.1.1. (Suggestion: Don't do this problem by hand.)

(a) Confirm that $4b_8 = b_2b_6 - b_4^2$ and that $1728\Delta = c_4^3 - c_6^2$. Confirm that if $\mathrm{char}(\mathbf{k}) \neq 2$ then replacing y by $y - (a_1x + a_3)/2$ in (8.1) gives (8.2). Confirm that if $\mathrm{char}(\mathbf{k}) \notin \{2,3\}$ then replacing (x,y) by $((x - 3b_2)/36, y/216)$ in (8.2) gives (8.3).

(b) Show that the admissible changes of variable $x = u^2x' + r$, $y = u^3y' + su^2x' + t$ where $u, r, s, t \in \mathbf{k}$ and $u \neq 0$ form a group. Show that every admissible change of variable transforms a Weierstrass equation E of the form (8.1) into another such equation E' with

$$ua_1' = a_1 + 2s,$$
$$u^2a_2' = a_2 - sa_1 + 3r - s^2,$$
$$u^3a_3' = a_3 + ra_1 + 2t,$$
$$u^4a_4' = a_4 - sa_3 + 2ra_2 - (rs + t)a_1 + 3r^2 - 2st,$$
$$u^6a_6' = a_6 + ra_4 - ta_3 + r^2a_2 - rta_1 + r^3 - t^2,$$

and

$$u^2b_2' = b_2 + 12r,$$
$$u^4b_4' = b_4 + rb_2 + 6r^2,$$
$$u^6b_6' = b_6 + 2rb_4 + r^2b_2 + 4r^3,$$
$$u^8b_8' = b_8 + 3rb_6 + 3r^2b_4 + r^3b_2 + 3r^4,$$

and

$$u^4c_4' = c_4, \qquad u^6c_6' = c_6,$$

and

$$u^{12}\Delta' = \Delta, \qquad j' = j.$$

(Hints for this exercise are at the end of the book.)

(c) Suppose the seemingly more general change of variable $x = vx' + r$, $y = wy' + vsx' + t$, where $v, w, r, s, t \in \mathbf{k}$ and v, w are nonzero, takes Weierstrass equations to Weierstrass equations. Show that $v = u^2$ and $w = u^3$ for some $u \in \mathbf{k}$.

(d) Replace y by $y/2$ in the third Weierstrass equation (8.3) to get

$$y^2 = 4x^3 - g_2x - g_3, \qquad g_2 = 108c_4, \; g_3 = 216c_6.$$

Show that the previously defined discriminant $\Delta_{\mathrm{old}} = g_2^3 - 27g_3^2$ of this equation is equal to the discriminant $\Delta_{(8.3)} = 2^6 3^9 (c_4^3 - c_6^2)$ of (8.3) in this section. Show that the previously defined invariant $j = 1728g_2^3/\Delta_{\mathrm{old}}$ is equal to the invariant $j = c_4^3/\Delta$ in this section.

(e) Find an admissible change of variable taking the modified universal elliptic curve (8.4) to the more general universal elliptic curve (8.5). Confirm that the discriminants and invariants of the two universal curves are as stated.

8.1.2. Show that algebraic and geometric nonsingularity are equivalent in characteristic 2.

8.1.3. (a) Homogenize the general Weierstrass polynomial by adding in powers of z to make each term cubic,

$$E_{\text{hom}}(x, y, z) = y^2 z + a_1 xyz + a_3 yz^2 - x^3 - a_2 x^2 z - a_4 x z^2 - a_6 z^3.$$

Show that $[0, 1, 0]$ satisfies E_{hom} and no other $[x, y, 0]$ does.
  (b) Dehomogenize E_{hom} by setting $y = 1$ to obtain

$$\tilde{E}(x, z) = z + a_1 xz + a_3 z^2 - x^3 - a_2 x^2 z - a_4 x z^2 - a_6 z^3.$$

In the (x, z) coordinate system, the infinite point 0_E is $(0, 0)$. Working in this affine coordinate system, show that E is geometrically nonsingular at 0_E.

8.1.4. (a) Show that the casewise definitions of λ and μ make the line $y = \lambda x + \mu$ the secant line through P and Q when $P \neq Q$ and the tangent line to E through P when $P = Q$.
  (b) Show that the geometric and algebraic descriptions (8.7) and (8.8) of the group law agree.
  (c) Multiply the numerator and the denominator of the secant case λ by $y_Q + y_P + a_1 x_P + a_3$ and use the Weierstrass equation (8.1) to obtain a new expression for λ when $y_Q + y_P + a_1 x_P + a_3 \neq 0$. Show that this also agrees with the old λ when $x_P = x_Q$, suitably giving ∞ when $P = -Q$. Similarly derive a new expression for μ.

8.1.5. (a) Let P be a singular point of a Weierstrass equation E over $\mathbf{k}$. Show that the coordinates of P lie in $\mathbf{k}$. For $\text{char}(\mathbf{k}) = 2$, assume that every element of $\mathbf{k}$ is a square. (A hint for this exercise is at the end of the book.)
  (b) Show that if $\text{char}(\mathbf{k}) = 2$ then $(0, 0)$ is the only singular point satisfying the Weierstrass polynomial (8.9), and if $\text{char}(\mathbf{k}) \neq 2$ then $(0, 0)$ is the only singular point satisfying the Weierstrass polynomial (8.10).
  (c) Show that $c_4 = (m_1 - m_2)^4$ in the context of equations (8.9) and (8.11).

8.1.6. (a) For what values $a, b \in \mathbb{Q}$ do the Weierstrass equations

$$y^2 = x^3 + ax^2 + bx, \qquad y^2 = x^3 - 2ax^2 + (a^2 - 4b)x$$

both define elliptic curves E and E' over $\mathbb{Q}$?
  (b) For such values a and b show that the map

$$(x, y) \mapsto (y^2/x^2, y(b - x^2)/x^2)$$

defines a map $\varphi : E \longrightarrow E'$ taking 0_E to $0_{E'}$.
  (c) Compute $\ker(\varphi)$ and compute $\varphi^{-1}(0, 0)$.
  (d) Find the dual isogeny $\psi : E' \longrightarrow E$, verifying the compositions $\psi \circ \varphi = [2]_E$ and $\varphi \circ \psi = [2]_{E'}$.

8.2 Algebraic curves in arbitrary characteristic

This section discusses algebraic curves over an arbitrary field $\mathbf{k}$. The definition is as in Section 7.2. Given polynomials $\varphi_1, \ldots, \varphi_m \in \mathbf{k}[x_1, \ldots, x_n]$ such that the ideal

$$I = \langle \varphi_1, \ldots, \varphi_m \rangle \subset \overline{\mathbf{k}}[x_1, \ldots, x_n]$$

is prime, let

$$C = \{P \in \overline{\mathbf{k}}^n : \varphi(P) = 0 \text{ for all } \varphi \in I\}.$$

The function field $\overline{\mathbf{k}}(C)$ of C over $\overline{\mathbf{k}}$ is the quotient field of the coordinate ring $\overline{\mathbf{k}}[C] = \overline{\mathbf{k}}[x_1, \ldots, x_n]/I$. If $\overline{\mathbf{k}}(C)$ is a finite extension of a field $\overline{\mathbf{k}}(t)$ where t is transcendental over $\overline{\mathbf{k}}$ then C is an affine algebraic curve over $\mathbf{k}$. If for each point $P \in C$ the derivative matrix $[D_j\varphi_i(P)]$ has rank $n - 1$ then C is nonsingular. Especially if C is defined by one equation in two variables then it is nonsingular if the gradient never vanishes. The projective version C_{hom} of the curve is defined as before. A nonempty algebraic curve over $\mathbf{k}$ contains infinitely many points even when $\mathbf{k}$ is finite.

The Curves–Fields Correspondence is also unchanged. The map

$$C \mapsto \mathbf{k}(C)$$

induces a bijection from the set of isomorphism classes over $\mathbf{k}$ of nonsingular projective algebraic curves over $\mathbf{k}$ to the set of conjugacy classes over $\mathbf{k}$ of function fields over $\mathbf{k}$. And for any two nonsingular projective algebraic curves C and C' over $\mathbf{k}$, the map

$$(h : C \longrightarrow C') \mapsto (h^* : \mathbf{k}(C') \longrightarrow \mathbf{k}(C))$$

is a bijection from the set of surjective morphisms over $\mathbf{k}$ from C to C' to the set of $\mathbf{k}$-injections of $\mathbf{k}(C')$ in $\mathbf{k}(C)$. Again nonconstant morphisms are surjective.

Let p be prime. Along with the parallels to Section 7.2 just described, algebraic curves and their function fields exhibit new phenomena in characteristic p. Let $\mathbb{F}_p$ denote the field of p elements and let $\overline{\mathbb{F}}_p$ denote its algebraic closure. For any positive integer power $q = p^e$ of p there is a unique field $\mathbb{F}_q$ of order q in $\overline{\mathbb{F}}_p$, the splitting field of the polynomial $x^q - x$ over $\mathbb{F}_p$ (Exercise 8.2.1). The algebraic closure is the union of all such $\mathbb{F}_q$. This is because every element of $\overline{\mathbb{F}}_p$ lies in a finite extension field $\mathbb{K}$ of $\mathbb{F}_p$, a finite-dimensional vector space over $\mathbb{F}_p$ that therefore has order $q = p^e$ for some e, and thus $\mathbb{K} = \mathbb{F}_q$ by uniqueness.

The *Frobenius map on* $\overline{\mathbb{F}}_p$ is

$$\sigma_p : \overline{\mathbb{F}}_p \longrightarrow \overline{\mathbb{F}}_p, \qquad x \mapsto x^p.$$

Because $(x + \tilde{x})^p = x^p + \tilde{x}^p$ in characteristic p, the Frobenius map is a field automorphism. Its inverse is thus again an automorphism of $\overline{\mathbb{F}}_p$ but is not a polynomial function. Its fixed points are $\mathbb{F}_p$, the roots in $\overline{\mathbb{F}}_p$ of the polynomial $x^p = x$, and more generally the fixed points of σ_p^e (e-fold composition where $e \geq 1$) are $\mathbb{F}_q$ where $q = p^e$. For each such q the field extension $\mathbb{F}_q/\mathbb{F}_p$ is Galois and its Galois group is cyclic of order e, generated by σ_p, but the group of automorphisms of $\overline{\mathbb{F}}_p$ is not cyclic.

The Frobenius map on $\overline{\mathbb{F}}_p^n$ is

$$\sigma_p : \overline{\mathbb{F}}_p^n \longrightarrow \overline{\mathbb{F}}_p^n, \qquad (x_1, \ldots, x_n) \mapsto (x_1^p, \ldots, x_n^p).$$

This is a bijection. Its fixed points are $\mathbb{F}_p^n$. It induces a well defined bijection at the level of projective classes,

$$\sigma_p : \mathbb{P}^n(\overline{\mathbb{F}}_p) \longrightarrow \mathbb{P}^n(\overline{\mathbb{F}}_p), \qquad [x_0, x_1, \ldots, x_n] \mapsto [x_0^p, x_1^p, \ldots, x_n^p],$$

with fixed points $\mathbb{P}^n(\mathbb{F}_p)$.

Let $\varphi(x) = \sum_e a_e x^e \in \overline{\mathbb{F}}_p[x_0, x_1, \ldots, x_n]$ be a homogeneous polynomial (where x^e is shorthand for $x_0^{e_0} \cdots x_n^{e_n}$), and let φ^{σ_p} be the polynomial obtained by applying the Frobenius map to its coefficients, $\varphi^{\sigma_p}(x) = \sum_e a_e^{\sigma_p} x^e$. Then (Exercise 8.2.2)

$$\varphi^{\sigma_p}(x^{\sigma_p}) = \varphi(x)^{\sigma_p}. \tag{8.12}$$

This shows that if C is a projective curve defined over $\overline{\mathbb{F}}_p$ by polynomials φ_i and if C^{σ_p} is the corresponding curve defined by the polynomials $\varphi_i^{\sigma_p}$ then σ_p restricts to a morphism from C to C^{σ_p}. That is, if $P \in \mathbb{P}^n(\overline{\mathbb{F}}_p)$ and $\varphi_i(P) = 0$ for all i then $\varphi_i^{\sigma_p}(P^{\sigma_p}) = 0$ for all i as well. Especially, if C is defined over $\mathbb{F}_p$ then $C^{\sigma_p} = C$ and σ_p defines a morphism from C to itself. Summarizing,

Definition 8.2.1. *Let C be a projective curve over $\overline{\mathbb{F}}_p$. The **Frobenius map on** C is*

$$\sigma_p : C \longrightarrow C^{\sigma_p}, \qquad [x_0, x_1, \ldots, x_n] \mapsto [x_0^p, x_1^p, \ldots, x_n^p].$$

For example, the Frobenius map on $\mathbb{P}^1(\overline{\mathbb{F}}_p)$ is $\sigma_p(t) = t^p$ on the affine part, and so the induced map σ_p^* gives the extension of function fields $\mathbb{K}/\mathbf{k}$ where

$$\mathbb{K} = \mathbb{F}_p(t), \qquad \mathbf{k} = \mathbb{F}_p(s), \qquad s = t^p.$$

Thus $\mathbb{K} = \mathbf{k}(t)$. The minimal polynomial of t over $\mathbf{k}$ is $x^p - s$, so the function field extension degree is p even though the Frobenius map bijects and we therefore might expect its degree to be 1. Since $x^p - s = (x - t)^p$ over $\mathbb{K}$, the extension is generated by a pth root that repeats p times as the root of its minimal polynomial.

Similarly the Frobenius map on an elliptic curve over $\mathbb{F}_p$ is $\sigma_p(u, v) = (u^p, v^p)$ on the affine part, so that σ_p^* gives the extension of function fields $\mathbb{K}/\mathbf{k}$ where

$$\mathbb{K} = \mathbb{F}_p(u)[v]/\langle E(u,v) \rangle, \qquad \mathbf{k} = \mathbb{F}_p(s)[t]/\langle E(s,t) \rangle, \qquad s = u^p, \ t = v^p.$$

Thus $\mathbb{K} = \mathbf{k}(u, v)$. The minimal polynomial of u in $\mathbf{k}[x]$ is $x^p - s$, and this factors as $(x - u)^p$ over $\mathbf{k}(u)$. The minimal polynomial of v in $\mathbf{k}(u)[y]$ divides $E(u, y)$, a quadratic polynomial in y. So

$$[\mathbf{k}(u) : \mathbf{k}] = p \quad \text{and} \quad [\mathbf{k}(u, v) : \mathbf{k}(u)] \in \{1, 2\}.$$

A similar argument shows that

$$[\mathbf{k}(v) : \mathbf{k}] = p \quad \text{and} \quad [\mathbf{k}(u, v) : \mathbf{k}(v)] \in \{1, 3\}.$$

Therefore $\mathbb{K} = \mathbf{k}(u) = \mathbf{k}(v)$ and $[\mathbb{K} : \mathbf{k}] = p$. Again the function field extension degree is p even though the Frobenius map is a bijection, and again the extension is generated by a pth root that repeats p times as the root of its minimal polynomial.

Definition 8.2.2. *An algebraic extension of fields $\mathbb{K}/\mathbf{k}$ is **separable** if for every element u of $\mathbb{K}$ the minimal polynomial of u in $\mathbf{k}[x]$ has distinct roots in $\overline{\mathbf{k}}$. Otherwise the extension is **inseparable**. A field extension obtained by adjoining a succession of pth roots that repeat p times as the root of their minimal polynomials is called **purely inseparable**.*

Since a polynomial has a multiple root if and only if it shares a root with its derivative, every algebraic extension of fields of characteristic 0 is separable. Every algebraic extension $\mathbb{K}/\mathbf{k}$ where $\mathbf{k}$ is a finite field is separable as well. The examples given before Definition 8.2.2 are purely inseparable.

Let $h : C \longrightarrow C'$ be a surjective morphism over $\overline{\mathbb{F}}_p$ of nonsingular projective curves over $\overline{\mathbb{F}}_p$. Let $\mathbf{k} = \overline{\mathbb{F}}_p(C')$ and $\mathbb{K} = \overline{\mathbb{F}}_p(C)$, so that the induced $\overline{\mathbb{F}}_p$-injection of function fields is $h^* : \mathbf{k} \longrightarrow \mathbb{K}$. The field extension $\mathbb{K}/h^*(\mathbf{k})$ takes the form

$$h^*(\mathbf{k}) \subset \mathbf{k}_{\text{sep}} \subset \mathbb{K}$$

where $\mathbf{k}_{\text{sep}}/h^*(\mathbf{k})$ is the maximal separable subextension of $\mathbb{K}/h^*(\mathbf{k})$ (this exists since the composite of separable extensions is again separable) and thus $\mathbb{K}/\mathbf{k}_{\text{sep}}$ is purely inseparable. Factoring $h^* : \mathbf{k} \longrightarrow \mathbb{K}$ as

$$\mathbf{k} \xrightarrow{h^*_{\text{sep}}} \mathbf{k}_{\text{sep}} \xrightarrow{h^*_{\text{ins}}} \mathbb{K}$$

gives a corresponding factorization of $h : C \longrightarrow C'$,

$$C \xrightarrow{h_{\text{ins}}} C_{\text{sep}} \xrightarrow{h_{\text{sep}}} C'$$

where the first map is $h_{\text{ins}} = \sigma_p^e$ with $p^e = [\mathbb{K} : \mathbf{k}_{\text{sep}}]$. That is,

$$h = h_{\text{sep}} \circ \sigma_p^e.$$

The morphism h is called *separable*, *inseparable*, or *purely inseparable* according to whether the field extension $\mathbb{K}/h^*(\mathbf{k})$ is separable (i.e., $e = 0$), inseparable ($e > 0$), or purely inseparable ($h_{\text{sep}} = 1$). As in Chapter 7, the degree of h is

$$\deg(h) = [\mathbb{K} : h^*(\mathbf{k})].$$

The *separable* and *inseparable degrees* of h are

$$\deg_{\text{sep}}(h) = \deg(h_{\text{sep}}) = [\mathbf{k}_{\text{sep}} : h^*(\mathbf{k})],$$
$$\deg_{\text{ins}}(h) = \deg(h_{\text{ins}}) = [\mathbb{K} : \mathbf{k}_{\text{sep}}],$$

so that

$$\deg(h) = \deg_{\text{sep}}(h)\,\deg_{\text{ins}}(h).$$

In characteristic p the degree formula from Chapter 7 remains

$$\sum_{P \in h^{-1}(Q)} e_P(h) = \deg(h) \quad \text{for any } Q \in C'.$$

In particular,

$$\sum_{P \in h_{\text{sep}}^{-1}(Q)} e_P(h_{\text{sep}}) = \deg_{\text{sep}}(h) \quad \text{for any } Q \in C',$$

and the ramification index $e_P(h_{\text{sep}}) \in \mathbb{Z}^+$ is 1 at all but finitely many points. Since h_{ins} bijects it follows that $\deg_{\text{sep}}(h) = |h_{\text{sep}}^{-1}(Q)| = |h^{-1}(Q)|$ for all but finitely many $Q \in C'$. This description applies even when h is not surjective (so that h maps to a single point and has degree 0), and this description shows that separable degree is multiplicative, i.e., if $h' : C' \longrightarrow C''$ is another morphism then $\deg_{\text{sep}}(h' \circ h) = \deg_{\text{sep}}(h')\,\deg_{\text{sep}}(h)$. Consequently inseparable degree is multiplicative as well since total degree clearly is.

We have seen that the Frobenius map σ_p is purely inseparable of degree p on $\mathbb{P}^1(\overline{\mathbb{F}}_p)$ and on elliptic curves E over $\mathbb{F}_p$, and in fact this holds on all curves. On elliptic curves, where there is a group law, the map $\sigma_p - 1$ makes sense, and it is separable: for otherwise $\sigma_p - 1 = f \circ \sigma_p$ where $f : E \longrightarrow E$ is a morphism, so $1 = g \circ \sigma_p$ where $g = 1 - f$, giving a contradiction because σ_p is not an isomorphism.

The map $[p]$ is an isogeny of degree p^2 on elliptic curves over $\overline{\mathbb{F}}_p$. (This statement will be justified later by Theorem 8.5.10.) As a special case of the degree formula, if $\varphi : E \longrightarrow E'$ is any isogeny of elliptic curves then

$$\deg_{\text{sep}}(\varphi) = |\ker(\varphi)|. \tag{8.13}$$

By this formula and by the structure of $\ker([p]) = E[p]$ as $\{0_E\}$ or $\mathbb{Z}/p\mathbb{Z}$, $\deg_{\text{sep}}([p])$ is 1 or p. Thus $[p]$ is not separable, and so it takes the form $[p] = f \circ \sigma_p$ for some rational f. This shows that although the inverse σ_p^{-1} of the Frobenius map is not rational, the dual isogeny $\hat{\sigma}_p = f$ is. If $E[p] = \{0_E\}$ then $[p]$ is purely inseparable of degree p^2 and so $\hat{\sigma}_p = i \circ \sigma_p$ where i is an automorphism of E, while if $E[p] \cong \mathbb{Z}/p\mathbb{Z}$ then $[p]$ has separable and inseparable degrees p and so $\hat{\sigma}_p$ is separable of degree p.

This section ends by deriving commutativity properties of the induced forward and reverse maps of the Frobenius map. Let C be a projective curve over $\mathbb{F}_p$. The forward induced map of σ_p on C acts on divisors in the usual way as

$$\sigma_{p,*} : (P) \mapsto (\sigma_p(P)). \qquad (8.14)$$

Since σ_p is bijective and is ramified everywhere with ramification degree p, the reverse induced map acts on divisors as

$$\sigma_p^* : (P) \mapsto p(\sigma_p^{-1}(P)). \qquad (8.15)$$

Let $h : C \longrightarrow C'$ be a map over $\mathbb{F}_p$ of projective curves over $\mathbb{F}_p$. Then the Frobenius map commutes with h. That is, if $\sigma_{p,C}$ denotes the Frobenius map on C and similarly for C' then (8.12) shows that

$$h \circ \sigma_{p,C} = \sigma_{p,C'} \circ h. \qquad (8.16)$$

It follows that the forward induced map of the Frobenius map commutes with the forward induced map of h,

$$h_* \circ (\sigma_{p,C})_* = (\sigma_{p,C'})_* \circ h_*. \qquad (8.17)$$

Since the Frobenius map commutes with h as in (8.16) so does its inverse, for

$$\begin{aligned}
h \circ \sigma_{p,C}^{-1} &= \sigma_{p,C'}^{-1} \circ \sigma_{p,C'} \circ h \circ \sigma_{p,C}^{-1} \\
&= \sigma_{p,C'}^{-1} \circ h \circ \sigma_{p,C} \circ \sigma_{p,C}^{-1} = \sigma_{p,C'}^{-1} \circ h.
\end{aligned} \qquad (8.18)$$

Now compute for any $P \in C$,

$$\begin{aligned}
(h_* \circ \sigma_{p,C}^*)(P) &= p(h \circ \sigma_{p,C}^{-1})(P) \quad \text{by (8.14) and (8.15)} \\
&= p(\sigma_{p,C'}^{-1} \circ h)(P) \quad \text{by (8.18)} \\
&= (\sigma_{p,C'}^* \circ h_*)(P) \quad \text{by (8.14) and (8.15) again.}
\end{aligned}$$

Thus the reverse induced map of the Frobenius map also commutes with the forward induced map of h,

$$h_* \circ \sigma_{p,C}^* = \sigma_{p,C'}^* \circ h_*. \qquad (8.19)$$

We will use formulas (8.17) and (8.19) in Section 8.7 and in Section 8.8.

The only place in Section 7.2 that uses characteristic 0 is the argument that a curve has a planar model. This also holds in characteristic p when the underlying field $\mathbf{k}$ is *perfect*, meaning every element has a pth root in the field. For example, finite fields and their algebraic extensions are perfect. Exercise 8.2.3 sketches the argument for readers equipped with the armamentarium of field theory in characteristic p.

Exercises

8.2.1. Let $q = p^e$. Show that the solutions of $x^q - x$ in $\overline{\mathbb{F}}_p$ form a subfield, and that this is the unique subfield of q elements. (A hint for this exercise is at the end of the book.)

8.2.2. Verify (8.12).

8.2.3. Let $\mathbb{L}/\mathbf{k}(t)$ be a finite extension where t is a variable and $\mathbf{k}$ is perfect. This exercise sketches the argument that the extension is primitive. Supply details as necessary. The Steinitz Criterion states that an extension is primitive if and only if there are only finitely many intermediate fields. Thus to prove that $\mathbb{L}/\mathbf{k}(t)$ is primitive we may replace $\mathbb{L}$ by its normal closure over $\mathbf{k}(t)$. So assume that the finite extension $\mathbb{L}/\mathbf{k}(t)$ is also normal.

The extension decomposes as $\mathbf{k}(t) \subset \mathbb{K}_1 \subset \mathbb{L}$, where $\mathbb{K}_1 = \mathbf{k}(t)_{\text{sep}}$ so that the first extension is the maximal separable subextension of $\mathbb{L}/\mathbf{k}(t)$ while the second is purely inseparable. But also, let $\mathbb{K}_2$ be the subfield of $\mathbb{L}$ fixed by the automorphism group $\text{Aut}(\mathbb{L}/\mathbf{k}(t))$. Then $\mathbf{k}(t) \subset \mathbb{K}_2 \subset \mathbb{L}$, and because $\mathbb{L}/\mathbf{k}(t)$ is normal the first extension is the maximal purely inseparable subextension of $\mathbb{L}/\mathbf{k}(t)$ while the second is separable. These two decompositions of the extension $\mathbb{L}/\mathbf{k}(t)$ show that $\mathbb{L} = \mathbb{K}_1 \mathbb{K}_2$.

Since the extension $\mathbb{K}_1/\mathbf{k}(t)$ is separable, it has a primitive element, i.e., $\mathbb{K}_1 = \mathbf{k}(t)(a)$. Since $\mathbf{k}$ is perfect, $\mathbb{K}_2$ takes the form $\mathbf{k}(t)(b)$ where $b^q = t$, $q = p^e$ for some e. Thus $\mathbb{L} = \mathbf{k}(t)(a, b)$, where a is separable over $\mathbf{k}(t)$. Since $a + b$ has the same number of conjugates under $\text{Aut}(\mathbb{L}/\mathbf{k}(t))$ as a does, and this is the separable degree of $\mathbf{k}(t)(a+b)/\mathbf{k}(t)$, it follows that $\mathbf{k}(t)(a+b)$ contains $\mathbf{k}(t)(a)$. Consequently $\mathbf{k}(t)(a + b) = \mathbf{k}(t)(a, b) = \mathbb{L}$. That is, the extension $\mathbb{L}/\mathbf{k}(t)$ is primitive.

8.3 Elliptic curves over $\mathbb{Q}$ and their reductions

This section discusses reducing elliptic curves defined over $\mathbb{Q}$ modulo a prime p. Consider a general Weierstrass equation E defined over $\mathbb{Q}$,

$$E : y^2 + a_1 xy + a_3 y = x^3 + a_2 x^2 + a_4 x + a_6, \quad a_1, \ldots, a_6 \in \mathbb{Q},$$

and consider admissible changes of variable over $\mathbb{Q}$. In particular the change of variable $(x, y) = (u^2 x', u^3 y')$ gives a Weierstrass equation E' with coefficients $a_i' = a_i/u^i$. A suitable choice of u makes this an integral equation, so we now assume that the original coefficients a_i are integral. View two integral Weierstrass equations as equivalent if they are related by an admissible change of variable over $\mathbb{Q}$.

For any prime p and any nonzero rational number r let $\nu_p(r)$ denote the power of p appearing in r, i.e., $\nu_p(p^e \cdot m/n) = e \in \mathbb{Z}$ where $p \nmid mn$. Also define $\nu_p(0) = +\infty$. This is the *p-adic valuation*, analogous to the valuations from Section 7.2 in that for all $r, r' \in \mathbb{Q}$,

$$\nu_p(rr') = \nu_p(r) + \nu_p(r'),$$

$$\nu_p(r + r') \geq \min\{\nu_p(r), \nu_p(r')\} \text{ with equality if } \nu_p(r) \neq \nu_p(r'),$$

but occurring now in the context of number fields rather than function fields. For each prime p let $\nu_p(E)$ denote the smallest power of p appearing in the discriminant of any integral Weierstrass equation equivalent to E, the minimum of a set of nonnegative integers,

$$\nu_p(E) = \min\{\nu_p(\Delta(E')) : E' \text{ integral, equivalent to } E\}.$$

According to Exercise 8.1.1(b), admissible changes of variable lead to $c_4' = c_4/u^4$, $c_6' = c_6/u^6$, and $\Delta' = \Delta/u^{12}$. Thus if $\nu_p(c_4) < 4$ or $\nu_p(c_6) < 6$ or $\nu_p(\Delta) < 12$ then $\nu_p(E) = \nu_p(\Delta)$. The converse holds as well for $p > 3$ (Exercise 8.3.1). Define the *global minimal discriminant* of E to be

$$\Delta_{\min}(E) = \prod_p p^{\nu_p(E)}.$$

This is a finite product since $\nu_p(E) = 0$ for all $p \nmid \Delta(E)$. A little work (Exercise 8.3.2) shows that the p-adic valuation of the discriminant can be minimized to $\nu_p(E)$ simultaneously for all p under admissible change of variable. That is, E is isomorphic over $\mathbb{Q}$ to an integral model E' with discriminant $\Delta(E') = \Delta_{\min}(E)$. This is the *global minimal Weierstrass equation E'*, the model of E to reduce modulo primes. From now on we freely assume when convenient that elliptic curves over $\mathbb{Q}$ are given in this form.

The field $\mathbb{F}_p$ of p elements can be viewed as $\mathbb{Z}/p\mathbb{Z}$. That is, $\mathbb{F}_p$ is the image of a surjective homomorphism from the ring of rational integers $\mathbb{Z}$, reduction modulo $p\mathbb{Z}$,

$$\tilde{} : \mathbb{Z} \longrightarrow \mathbb{F}_p, \qquad \tilde{n} = n + p\mathbb{Z}. \tag{8.20}$$

This map reduces a global minimal Weierstrass equation E to a Weierstrass equation $\widetilde{E}$ over $\mathbb{F}_p$, and this defines an elliptic curve over $\mathbb{F}_p$ if and only if $p \nmid \Delta_{\min}(E)$. The reduction of E modulo p (also called the reduction of E *at* p) is

1. *good [nonsingular, stable]* if $\widetilde{E}$ is again an elliptic curve,
   a) *ordinary* if $\widetilde{E}[p] \cong \mathbb{Z}/p\mathbb{Z}$,
   b) *supersingular* if $\widetilde{E}[p] = \{0\}$,
2. *bad [singular]* if $\widetilde{E}$ is not an elliptic curve, in which case it has one singular point and can be put over $\mathbb{F}_p$ into the form $(y - m_1 x)(y - m_2 x) = x^3$ (so that $m_1 + m_2, m_1 m_2 \in \mathbb{F}_p$), as explained at the end of section 8.1,
   a) *multiplicative [semistable]* if the singular point is a node ($m_1 \neq m_2$),
      i. *split* if $m_1, m_2 \in \mathbb{F}_p$, i.e., the distinct tangent slopes at the node lie in $\mathbb{F}_p^*$,
      ii. *non-split* if $m_1, m_2 \notin \mathbb{F}_p$, i.e, the distinct tangent slopes at the node do not lie in $\mathbb{F}_p^*$,

b) *additive [unstable]* if the singular point is a cusp ($m_1 = m_2$).

These reduction types are all independent of how E is put into global minimal form. This is because if E has two global minimal Weierstrass equations then the admissible change of variable that takes one to the other has coefficients $r, s, t, u \in \mathbb{Z}$ with $u = \pm 1$ (Exercise 8.3.3), preserving Δ and c_4. Good reduction at p means $\nu_p(\Delta_{\min}(E)) = 0$, and when this holds the two reduced Weierstrass equations differ by the reduced admissible change of variable, making them isomorphic elliptic curves with the same p-torsion and thus preserving ordinary or supersingular reduction as well. Bad reduction occurs when $\nu_p(\Delta_{\min}(E)) > 0$, and either $\nu_p(c_4) = 0$ for multiplicative reduction or $\nu_p(c_4) > 0$ for additive reduction, cf. Proposition 8.1.3. The reason for the terms "multiplicative" and "additive" was also explained at the end of Section 8.1: in these cases the nonsingular points of $\widetilde{E}$ form a multiplicative group isomorphic to $\overline{\mathbb{F}}_p^*$ or an additive group isomorphic to $\overline{\mathbb{F}}_p$. "Semistable" is also used to mean good or multiplicative but not additive, i.e., not unstable. This will be explained in the next section.

Recall from Section 7.7 that the analytic conductor of an elliptic curve E over $\mathbb{Q}$ is the smallest integer N such that $X_0(N)$ maps to E per the Modularity Theorem, i.e., the map is a surjective morphism over $\mathbb{Q}$ of curves over $\mathbb{Q}$. Putting E into global minimal form provides an almost complete description of a related integer N_E called the *algebraic conductor* of E. The global minimal discriminant and the algebraic conductor are divisible by the same primes,

$$p \mid \Delta_{\min}(E) \iff p \mid N_E,$$

so that E has good reduction at all primes p not dividing N_E. More specifically, the algebraic conductor takes the form $N_E = \prod_p p^{f_p}$ where

$$f_p = \begin{cases} 0 & \text{if } E \text{ has good reduction at } p, \\ 1 & \text{if } E \text{ has multiplicative reduction at } p, \\ 2 & \text{if } E \text{ has additive reduction at } p \text{ and } p \notin \{2, 3\}, \\ 2 + \delta_p & \text{if } E \text{ has additive reduction at } p \text{ and } p \in \{2, 3\}. \end{cases}$$

Here $\delta_2 \leq 6$ and $\delta_3 \leq 3$. An algorithm for the algebraic conductor due to Tate and a formula for the algebraic conductor due to Ogg and Saito are presented in [Sil94]. The algebraic conductor, like the analytic conductor, is well defined on isogeny classes of elliptic curves over $\mathbb{Q}$.

Although we have reduced a Weierstrass equation E over $\mathbb{Q}$ to $\widetilde{E}$ over $\mathbb{F}_p$, we have not yet discussed reducing the points of the curve E to points of the curve $\widetilde{E}$. This will be done in the next section, after we extend the reduction map $\mathbb{Z} \longrightarrow \mathbb{F}_p$ suitably.

The preface to this book described the Modularity Theorem in terms of values $a_p(E)$, defined as p minus the solution-count of a Weierstrass equation modulo p. Equivalently,

Definition 8.3.1. *Let E be an elliptic curve over $\mathbb{Q}$. Assume E is in reduced form. Let p be a prime and let $\widetilde{E}$ be the reduction of E modulo p. Then*

$$a_p(E) = p + 1 - |\widetilde{E}(\mathbb{F}_p)|.$$

This definition is the same as before because the reduced curve contains the infinite point $0_{\widetilde{E}}$. That is, subtracting the number of $\mathbb{F}_p$-points on $\widetilde{E}$ from $p+1$ is the same as subtracting the solution-count working in $\mathbb{Z}/p\mathbb{Z}$ from p. The reason to carry out either subtraction is that as x varies through $\mathbb{F}_p$ we expect the Weierstrass equation modulo p to have two solutions (x, y) for roughly half of the values of x modulo p and no solutions for the other x-values, giving a total of p points. Thus the estimated value of $|\widetilde{E}(\mathbb{F}_p)|$ is $p + 1$, and $a_p(E)$ measures how the actual value differs from its estimate. A result at the end of this section and a result outside the scope of this book will combine to show that

$$a_p(E) = 0 \quad \text{if } E \text{ has supersingular reduction at } p \text{ and } p \geq 5.$$

Exercise 9.4.2 will show that the value of $a_p(E)$ for a prime p of bad reduction is

$$a_p(E) = \begin{cases} 1 & \text{if } E \text{ has split reduction at } p, \\ -1 & \text{if } E \text{ has nonsplit reduction at } p, \\ 0 & \text{if } E \text{ has additive reduction at } p. \end{cases}$$

The next proposition expresses $a_p(E)$ in terms of the Frobenius map, cf. the introductory remarks to this chapter.

Proposition 8.3.2. *Let E be an elliptic curve over $\mathbb{Q}$ and let p be a prime such that E has good reduction modulo p. Let $\sigma_{p,*}$ and σ_p^* be the forward and reverse maps of $\mathrm{Pic}^0(\widetilde{E})$ induced by σ_p. Then*

$$a_p(E) = \sigma_{p,*} + \sigma_p^* \quad \text{as endomorphisms of } \mathrm{Pic}^0(\widetilde{E}).$$

(Here the left side means multiplication by $a_p(E)$.)

Proof. An element $x \in \overline{\mathbb{F}}_p$ satisfies $x^p = x$ if and only if $x \in \mathbb{F}_p$. Thus

$$\widetilde{E}(\mathbb{F}_p) = \{P \in \widetilde{E} : P^{\sigma_p} = P\} = \ker(\sigma_p - 1),$$

and so since $\sigma_p - 1$ is separable,

$$|\widetilde{E}(\mathbb{F}_p)| = |\ker(\sigma_p - 1)| = \deg(\sigma_p - 1).$$

The general formula $h_* \circ h^* = \deg(h)$ was given for characteristic 0 in Section 7.3, and we assert without proof that it holds in characteristic p as well. It is clear that the induced forward map of a sum of isogenies is the sum of the induced forward maps. The same fact for reverse maps is not obvious, but it was shown near the end of Section 6.2 in the complex analytic setting and we now assert without proof that it holds in characteristic p as well. Therefore, computing with endomorphisms of $\mathrm{Pic}^0(\widetilde{E})$,

$$
\begin{aligned}
|\widetilde{E}(\mathbb{F}_p)| = \deg(\sigma_p - 1) &= (\sigma_p - 1)_* \circ (\sigma_p - 1)^* \\
&= (\sigma_{p,*} - 1_*) \circ (\sigma_p^* - 1^*) = p + 1 - \sigma_{p,*} - \sigma_p^*.
\end{aligned}
$$

The proposition follows by definition of $a_p(E)$.   □

If E is an elliptic curve over $\mathbb{Q}$ and p is prime then extend Definition 8.3.1 to

$$
t_{p^e}(E) = p^e + 1 - |\widetilde{E}(\mathbb{F}_{p^e})|, \quad e \geq 1.
$$

Thus $t_p(E) = a_p(E)$. The higher-power solution-counts are denoted $t_{p^e}(E)$ rather than $a_{p^e}(E)$ because, as we will see in a moment, they don't quite pattern-match the coefficients $a_{p^e}(f)$ of a normalized eigenform. Proposition 8.3.2 extends to

$$
t_{p^e}(E) = \sigma_{p^e,*} + \sigma_{p^e}^*, \quad e \geq 1.
$$

Further define

$$
t_1(E) = 2,
$$

in contrast to $a_1(f) = 1$ for a normalized eigenform. Then (to be proved in Exercise 9.4.2) these solution-counts of an elliptic curve satisfy the same recurrence as the coefficients $a_{p^e}(f)$ of a normalized eigenform in $\mathcal{S}_2(\Gamma_0(N))$ as in Proposition 5.8.5, despite the different initial value,

$$
t_{p^e}(E) = t_p(E)t_{p^{e-1}}(E) - \mathbf{1}_E(p)pt_{p^{e-2}}(E) \quad \text{for all } e \geq 2. \tag{8.21}
$$

Here $\mathbf{1}_E$ is the trivial character modulo the algebraic conductor N_E of E, so $\mathbf{1}_E(p)$ is 1 for primes of good reduction and 0 for primes of bad reduction. The recurrence further indicates the close relation between elliptic curves and eigenforms. Its consequence that we want here is (Exercise 8.3.5)

Proposition 8.3.3. *Let E be an elliptic curve over $\mathbb{Q}$, and let p be a prime such that E has good reduction at p. Then the reduction is*

$$
\begin{cases}
ordinary & if\ a_p(E) \not\equiv 0 \ (\mathrm{mod}\ p), \\
supersingular & if\ a_p(E) \equiv 0 \ (\mathrm{mod}\ p).
\end{cases}
$$

The estimate $|a_p(E)| \leq 2\sqrt{p}$ due to Hasse (which we do not address in this book) combines with the previous proposition to give the result stated earlier in the section that $a_p(E) = 0$ if E has supersingular reduction at p and $p \geq 5$. Exercise 8.3.6 provides some examples of the ideas in this section.

Exercises

8.3.1. Let $p > 3$ be prime. Let E be a Weierstrass equation over $\mathbb{Q}$ with integral coefficients, and suppose that $\nu_p(\Delta) \geq 12$ and $\nu_p(c_4) \geq 4$. Show that $\nu_p(c_6) \geq 6$ as well. The admissible changes of variable from Section 8.1 put E into the form (8.3) and multiply its discriminant by 6^{12}, so that $\nu_p(\Delta_{(8.3)}) = \nu_p(\Delta)$. Show that from here the admissible change of variable $(x, y) = (p^2 x', p^3 y')$ gives an equation E' in integral form such that $\nu_p(\Delta') = \nu_p(\Delta) - 12$. Thus E is not in global minimal form. (A hint for this exercise is at the end of the book.)

8.3.2. Let E be a Weierstrass equation (8.1) with $a_i \in \mathbb{Z}$ and let Δ be its discriminant. For primes $p \nmid \Delta$ the discriminant is already minimal at p. For each prime $p \mid \Delta$ an admissible change of variable

$$(x, y) = (u_p^2 x' + r_p, u_p^3 y' + s_p u_p^2 x' + t_p), \quad u_p, r_p, s_p, t_p \in \mathbb{Q}$$

transforms E to a new Weierstrass equation E_p' with coefficients $a_{i,p}' \in \mathbb{Z}$ and discriminant $\Delta_p' = \Delta/u_p^{12}$ minimal at p.

(a) Explain why $\nu_p(u_p) \geq 0$ and similarly for r_p, s_p, and t_p. (Hints for this exercise are at the end of the book.)

(b) Let $u = \prod_p p^{\nu_p(u_p)}$. Show that any admissible change of variable

$$(x, y) = (u^2 x' + r, u^3 y' + su^2 x' + t)$$

gives a Weierstrass equation E' with discriminant $\Delta' = \pm\Delta_{\min}$.

(c) Show that r, s, and t can be chosen to make $a_i' \in \mathbb{Z}$ as follows. The results from Exercise 8.1.1(b) give

$$ua_1' = u_p a_{1,p}' + 2(s - s_p),$$
$$u^2 a_2' = u_p^2 a_{2,p}' + 3(r - r_p) - (a_1 + s + s_p)(s - s_p),$$
$$u^3 a_3' = u_p^3 a_{3,p}' + a_1(r - r_p) + 2(t - t_p),$$
$$u^4 a_4' = u_p^4 a_{4,p}' - a_3(s - s_p) + 2a_2(r - r_p) - a_1(rs - r_p s_p + t - t_p)$$
$$\qquad + 3(r^2 - r_p^2) - 2(st - s_p t_p),$$
$$u^6 a_6' = u_p^6 a_{6,p}' + a_4(r - r_p) - a_3(t - t_p) + a_2(r^2 - r_p^2) - a_1(rt - r_p t_p)$$
$$\qquad + (r^3 - r_p^3) - (t^2 - t_p^2).$$

From the first formula,

$$\nu_p(a_1') + \nu_p(u) \geq \min\{\nu_p(a_{1,p}') + \nu_p(u_p), \nu_p(2) + \nu_p(s - s_p)\},$$

and since $\nu_p(u) = \nu_p(u_p)$ and $\nu_p(a_{1,p}') \geq 0$ this shows that $\nu_p(a_1') \geq 0$ if $\nu_p(s - s_p) \geq \nu_p(u)$. Working similarly with the other equations, show that the condition

$$\min\{\nu_p(r - r_p), \nu_p(s - s_p), \nu_p(t - t_p)\} \geq 6\nu_p(u)$$

makes the coefficients a_i' integral. Apply the Chinese Remainder Theorem to show that there are integers r, s, and t satisfying the condition for all primes p.

8.3.3. Let E and E' be two global minimal Weierstrass equations, and suppose an admissible change of variable $x = u^2 x' + r$, $y = u^3 y' + su^2 x' + t$ takes E to E'.

(a) Explain why $u = \pm 1$.

(b) Use the formulas from Exercise 8.1.1(b) relating the coefficients b_i' and b_i to show that $r \in \mathbb{Z}$. (Hints for this exercise are at the end of the book.)

(c) Use the formulas from Exercise 8.1.1(b) relating the coefficients a_i' and a_i to show that $s \in \mathbb{Z}$ and that $t \in \mathbb{Z}$. Thus the change of variable reduces modulo p for all primes p.

8.3.4. Consider an affine algebraic curve over $\mathbb{Q}$,

$$Q : x^2 - dy^2 = 1, \qquad d \in \mathbb{Z}, \ d \neq 0.$$

For any prime p not dividing $2d$ define

$$a_p(Q) = p - |\widetilde{Q}(\mathbb{F}_p)|.$$

Show that the maps

$$t \mapsto \left(\frac{1 + dt^2}{1 - dt^2}, \frac{2t}{1 - dt^2} \right), \qquad \infty \mapsto (-1, 0)$$

and

$$(x, y) \mapsto \frac{y}{x + 1}, \qquad (-1, 0) \mapsto \infty$$

define a bijection between $\mathbb{P}^1(\mathbb{F}_p) - \{t : dt^2 = 1\}$ and $\widetilde{Q}(\mathbb{F}_p)$. Use this to show that $a_p(Q)$ is the Legendre symbol (d/p). As explained in the preface, the values $a_p(Q)$ therefore arise as a system of eigenvalues.

8.3.5. Use the recurrence (8.21) to prove Proposition 8.3.3. (A hint for this exercise is at the end of the book.)

8.3.6. (a) Consider the elliptic curve $E : y^2 = x^3 - 1$ over $\mathbb{Q}$. Show that E is in global minimal form and has good reduction at all $p > 3$. What is the reduction type of E at $p = 2$? $p = 3$? (Hints for this exercise are at the end of the book.)

(b) Let p be prime, $p \equiv 2 \pmod 3$. Show that the map $x \mapsto x^3$ is an automorphism of $\mathbb{F}_p^*$, and therefore the map $x \mapsto x^3 - 1$ bijects $\mathbb{F}_p$ to itself. Use this to count that $|\widetilde{E}(\mathbb{F}_p)| = p + 1$, i.e., $a_p(E) = 0$. Thus E has supersingular reduction at p by Proposition 8.3.3.

(c) More generally, let p be an odd prime and let $E : y^2 = f(x)$ be an elliptic curve over $\mathbb{F}_p$. Recall from elementary number theory that the multiplicative group $\mathbb{F}_p^*$ is cyclic. For each point $x \in \mathbb{F}_p$ show that the number of points $(x, y) \in E(\mathbb{F}_p)$ is $1 + f(x)^{(p-1)/2}$, interpreting $f(x)^{(p-1)/2} \in \mathbb{F}_p$ as one of $-1, 0, 1 \in \mathbb{Z}$. Show that therefore

$$|E(\mathbb{F}_p)| = 1 + p + \sum_{x \in \mathbb{F}_p} f(x)^{(p-1)/2} \equiv 1 + \sum_{x \in \mathbb{F}_p} f(x)^{(p-1)/2} \pmod{p}.$$

Show that for any nonnegative integer i,

$$\sum_{x \in \mathbb{F}_p} x^i = \begin{cases} 0 & \text{if } i = 0 \text{ or } i \not\equiv 0 \pmod{p-1}, \\ -1 & \text{if } i > 0 \text{ and } i \equiv 0 \pmod{p-1}. \end{cases}$$

Therefore the sum needs to be taken only over the terms of $f(x)^{(p-1)/2}$ whose power of x is a positive multiple of $p-1$. Show that the only such is the x^{p-1} term.

(d) Returning to the curve $E : y^2 = x^3 - 1$, let $p > 3$, $p \equiv 1 \pmod{3}$. Use part (c) to show that

$$|E(\mathbb{F}_p)| \equiv 1 - (-1)^{(p-1)/6} \binom{(p-1)/2}{(p-1)/3} \not\equiv 1 \pmod{p},$$

and consequently $a_p(E) \not\equiv 0 \pmod{p}$. Thus E has ordinary reduction at p by Proposition 8.3.3.

(e) Similarly, show that the curve $E : y^2 = x^3 - x$ is in global minimal form with bad reduction only at $p = 2$, and determine whether this reduction is multiplicative or additive. Use part (c) to describe the good reduction type of E at each prime $p > 2$ in terms of $p \pmod 4$.

8.3.7. Show that the Weierstrass equation $E : y^2 + xy = x^3 + x^2 + 44x + 55$ is in global minimal form. What are the primes of good, multiplicative, and additive reduction?

8.4 Elliptic curves over $\overline{\mathbb{Q}}$ and their reductions

This section discusses reducing an elliptic curve over $\overline{\mathbb{Q}}$ modulo a maximal ideal $\mathfrak{p}$ of $\overline{\mathbb{Z}}$, where $\overline{\mathbb{Q}}$ is the field of algebraic numbers and $\overline{\mathbb{Z}}$ is the ring of algebraic integers.

We begin with a brief discussion of the maximal ideal $\mathfrak{p}$ itself. Given such an ideal, the intersection $\mathfrak{p} \cap \mathbb{Z}$ is a maximal ideal of $\mathbb{Z}$, so it takes the form $p\mathbb{Z}$ for some rational prime p and we say that $\mathfrak{p}$ *lies over* p. Conversely, given a rational prime p, the set $p\overline{\mathbb{Z}}$ is an ideal of $\overline{\mathbb{Z}}$, and so by Zorn's Lemma from set theory there exists a maximal ideal $\mathfrak{p}$ containing it. More specifically to this situation, algebraic number theory says that for each number field $\mathbb{K} \subset \overline{\mathbb{Q}}$ with its ring $\mathcal{O}_\mathbb{K}$ of algebraic integers, the ideal $p\mathcal{O}_\mathbb{K}$ of $\mathcal{O}_\mathbb{K}$ factors into a nonempty finite product of positive powers of the maximal ideals of $\mathcal{O}_\mathbb{K}$ containing it,

$$p\mathcal{O}_\mathbb{K} = \prod_{i=1}^{g_\mathbb{K}} \mathfrak{p}_{i,\mathbb{K}}^{e_{i,\mathbb{K}}}. \tag{8.22}$$

(A simple example of this was given in Section 3.7, the factorization of primes p in the number ring $\mathbb{Z}[\mu_6]$.) Since $\overline{\mathbb{Q}} = \bigcup \mathbb{K}$ and $\overline{\mathbb{Z}} = \bigcup_{\mathbb{K}} \mathcal{O}_{\mathbb{K}}$, a maximal ideal $\mathfrak{p}$ of $\overline{\mathbb{Z}}$ containing $p\overline{\mathbb{Z}}$ is a corresponding union of maximal ideals over all number fields,

$$\mathfrak{p} = \bigcup_{\mathbb{K}} \mathfrak{p}_{\mathbb{K}}. \tag{8.23}$$

Conversely, such a union is a maximal ideal of $\overline{\mathbb{Z}}$ if and only if the number field ideals are chosen compatibly, meaning that if $\mathbb{K} \subset \mathbb{K}'$ then $\mathfrak{p}_{\mathbb{K}} \subset \mathfrak{p}_{\mathbb{K}'}$. In sum, every maximal ideal $\mathfrak{p}$ of $\overline{\mathbb{Z}}$ lies over a unique rational prime p, and given a rational prime p, every compatible union $\mathfrak{p}$ of the form (8.23) is a maximal ideal of $\overline{\mathbb{Z}}$ lying over p. From now on, if the symbol $\mathfrak{p}$ is in use then p is understood to denote its underlying rational prime.

The *localization of $\overline{\mathbb{Z}}$ at $\mathfrak{p}$* is

$$\overline{\mathbb{Z}}_{(\mathfrak{p})} = \{x/y : x, y \in \overline{\mathbb{Z}}, \ y \notin \mathfrak{p}\},$$

a subring of $\overline{\mathbb{Q}}$ since maximal ideals are prime. The ideal $\mathfrak{p}\overline{\mathbb{Z}}_{(\mathfrak{p})}$ of $\overline{\mathbb{Z}}_{(\mathfrak{p})}$ is its unique maximal ideal since its complement is the set $\overline{\mathbb{Z}}_{(\mathfrak{p})}^{*}$ of units, making the quotient $\overline{\mathbb{Z}}_{(\mathfrak{p})}/\mathfrak{p}\overline{\mathbb{Z}}_{(\mathfrak{p})}$ a field. There is a natural isomorphism of fields (Exercise 8.4.1(a))

$$\overline{\mathbb{Z}}/\mathfrak{p} \xrightarrow{\sim} \overline{\mathbb{Z}}_{(\mathfrak{p})}/\mathfrak{p}\overline{\mathbb{Z}}_{(\mathfrak{p})}, \qquad \alpha + \mathfrak{p} \mapsto \alpha + \mathfrak{p}\overline{\mathbb{Z}}_{(\mathfrak{p})}.$$

Also, $\overline{\mathbb{Z}}/\mathfrak{p}$ is an algebraic closure of $\mathbb{Z}/p\mathbb{Z}$ (Exercise 8.4.1(b)). So we view $\overline{\mathbb{Z}}_{(\mathfrak{p})}/\mathfrak{p}\overline{\mathbb{Z}}_{(\mathfrak{p})}$ as $\overline{\mathbb{F}}_p$, giving the natural surjection with kernel $\mathfrak{p}\overline{\mathbb{Z}}_{(\mathfrak{p})}$

$$\tilde{} : \overline{\mathbb{Z}}_{(\mathfrak{p})} \longrightarrow \overline{\mathbb{F}}_p, \qquad \tilde{\alpha} = \alpha + \mathfrak{p}\overline{\mathbb{Z}}_{(\mathfrak{p})}. \tag{8.24}$$

Since $\mathfrak{p}\overline{\mathbb{Z}}_{(\mathfrak{p})} \cap \mathbb{Z} = p\mathbb{Z}$ (Exercise 8.4.1(c)), the reduction map (8.24) extends the earlier reduction map $\mathbb{Z} \longrightarrow \mathbb{F}_p$ of (8.20).

Lemma 8.4.1. *Let $\mathfrak{p}$ be a maximal ideal of $\overline{\mathbb{Z}}$. Let α be a nonzero element of $\overline{\mathbb{Q}}$. Then at least one of α or $1/\alpha$ belongs to $\overline{\mathbb{Z}}_{(\mathfrak{p})}$.*

Proof. Consider the number field $\mathbb{K} = \mathbb{Q}(\alpha)$. Let $\mathcal{O}_{\mathbb{K}}$ be its ring of algebraic integers and let $\mathfrak{p}_{\mathbb{K}} = \mathfrak{p} \cap \mathcal{O}_{\mathbb{K}}$. The *localization of $\mathcal{O}_{\mathbb{K}}$ at $\mathfrak{p}_{\mathbb{K}}$* is

$$\mathcal{O}_{\mathbb{K},(\mathfrak{p})} = \{x/y : x, y \in \mathcal{O}_{\mathbb{K}}, \ y \notin \mathfrak{p}_{\mathbb{K}}\},$$

a subring of $\mathbb{K}$ containing $\mathcal{O}_{\mathbb{K}}$ and with unique maximal ideal $\mathfrak{p}_{\mathbb{K}}\mathcal{O}_{\mathbb{K},(\mathfrak{p})}$. There is a valuation

$$\nu_{\mathfrak{p}_{\mathbb{K}}} : \mathbb{K} \longrightarrow \mathbb{Z} \cup \{+\infty\},$$

a surjection taking each nonzero $x \in \mathcal{O}_{\mathbb{K},(\mathfrak{p})}$ to the largest $d \in \mathbb{N}$ such that $x \in \mathfrak{p}_{\mathbb{K}}^d \mathcal{O}_{\mathbb{K},(\mathfrak{p})}$. Constructing this number field valuation uses the facts that the ideal $\mathfrak{p}_{\mathbb{K}}\mathcal{O}_{\mathbb{K},(\mathfrak{p})}$ is principal and that the ring $\mathcal{O}_{\mathbb{K},(\mathfrak{p})}$ is Noetherian, analogously to how we constructed the function field valuation at a nonsingular point of an algebraic curve in Section 7.2. Then $\nu_{\mathfrak{p}_{\mathbb{K}}} = e\nu_p$ on $\mathbb{Q}$, where $e \in \mathbb{Z}^+$ is the power of $\mathfrak{p}_{\mathbb{K}}$ occurring in the factorization (8.22), and ν_p is the p-adic valuation from the previous section. The valuation $\nu = \nu_{\mathfrak{p}_{\mathbb{K}}}$ has the standard properties,

$$\nu(rr') = \nu(r) + \nu(r'),$$
$$\nu(r + r') \geq \min\{\nu(r), \nu(r')\} \text{ with equality if } \nu(r) \neq \nu(r').$$

With the valuation in place the lemma is clear: if $\nu(\alpha) \geq 0$ then $\alpha \in \mathcal{O}_{\mathbb{K},(\mathfrak{p})} \subset \mathbb{Z}_{(\mathfrak{p})}$, and if $\nu(\alpha) \leq 0$ then $\nu(1/\alpha) \geq 0$. $\qquad\square$

The proof of Lemma 8.4.1 used valuation theory in a number field $\mathbb{K}$ because there is no corresponding valuation $\nu = \nu_{\mathfrak{p}}$ on $\overline{\mathbb{Q}}$. But the lemma is set in $\overline{\mathbb{Q}}$, and it enables us to work systematically over $\overline{\mathbb{Q}}$ from now on.

To reduce an elliptic curve over $\overline{\mathbb{Q}}$ modulo $\mathfrak{p}$, consider any Weierstrass equation E over $\overline{\mathbb{Q}}$. By Exercise 6.4.4 every algebraic number takes the form of an algebraic integer divided by a rational integer. It follows that an admissible change of variable $(x, y) = (u^2 x', u^3 y')$ with $1/u \in \mathbb{Z}^+$ transforms E to a Weierstrass equation whose coefficients $a_i' = a_i/u^i$ lie in $\overline{\mathbb{Z}}$ and therefore in $\overline{\mathbb{Z}}_{(\mathfrak{p})}$. Thus we may assume that the Weierstrass equation E is $\mathfrak{p}$-*integral* to begin with, meaning the coefficients are in $\overline{\mathbb{Z}}_{(\mathfrak{p})}$,

$$E : y^2 + a_1 xy + a_3 y = x^3 + a_2 x^2 + a_4 x + a_6, \quad a_1, \ldots, a_6 \in \overline{\mathbb{Z}}_{(\mathfrak{p})}.$$

Any $\mathfrak{p}$-integral Weierstrass equation reduces via (8.24) to a Weierstrass equation over $\overline{\mathbb{F}}_p$,

$$\widetilde{E} : y^2 + \tilde{a}_1 xy + \tilde{a}_3 y = x^3 + \tilde{a}_2 x^2 + \tilde{a}_4 x + \tilde{a}_6, \quad \tilde{a}_1, \ldots, \tilde{a}_6 \in \overline{\mathbb{F}}_p.$$

Proposition 8.1.3 and other results from Section 8.1 combine to show that

- $\widetilde{E}$ defines an elliptic curve over $\overline{\mathbb{F}}_p$ if and only if $\widetilde{\Delta} \neq 0$, in which case the curve either is ordinary with $\widetilde{E}[p] \cong \mathbb{Z}/p\mathbb{Z}$, or supersingular with $\widetilde{E}[p] = \{0\}$,
- $\widetilde{E}$ defines a singular curve over $\overline{\mathbb{F}}_p$ with a node if and only if $\widetilde{\Delta} = 0$ and $\tilde{c}_4 \neq 0$, in which case the rest of $\widetilde{E}$ forms a multiplicative group isomorphic to $\overline{\mathbb{F}}_p^*$,
- $\widetilde{E}$ defines a singular curve over $\overline{\mathbb{F}}_p$ with a cusp if and only if $\widetilde{\Delta} = 0$ and $\tilde{c}_4 = 0$, in which case the rest of $\widetilde{E}$ forms an additive group isomorphic to $\overline{\mathbb{F}}_p$.

Thus we can speak of the ordinary, supersingular, multiplicative, or additive reduction type of any $\mathfrak{p}$-integral Weierstrass equation.

We want the reduction type to make sense for $\overline{\mathbb{Q}}$-isomorphism classes of elliptic curves over $\overline{\mathbb{Q}}$, that is, for $\overline{\mathbb{Q}}$-isomorphism classes of nonsingular Weierstrass equations over $\overline{\mathbb{Q}}$. An admissible change of variable over $\overline{\mathbb{Q}}$ puts any such equation into $\mathfrak{p}$-integral *Legendre form* (Exercise 8.4.2(a,b)),

$$E : y^2 = x(x-1)(x-\lambda), \qquad \lambda \notin \{0,1\}, \qquad \lambda \in \overline{\mathbb{Z}}_{(\mathfrak{p})}.$$

Formulas from Section 8.1 specialize for Legendre form to (Exercise 8.4.2(c))

$$\Delta = 16\lambda^2(1-\lambda)^2, \qquad c_4 = 16(1 - \lambda(1-\lambda)).$$

Assume now that $\mathfrak{p}$ does not lie over the rational prime 2. Then the formulas for Δ and c_4 give the implications (using that maximal ideals are prime for the first one)

$$\Delta \in \mathfrak{p}\overline{\mathbb{Z}}_{(\mathfrak{p})} \implies \lambda(1-\lambda) \in \mathfrak{p}\overline{\mathbb{Z}}_{(\mathfrak{p})} \implies c_4 \notin \mathfrak{p}\overline{\mathbb{Z}}_{(\mathfrak{p})}.$$

That is, the conditions for additive reduction, $\widetilde{\Delta} = \tilde{c}_4 = 0_{\overline{\mathbb{F}}_p}$, are incompatible with $\mathfrak{p}$-integral Legendre form. Therefore every elliptic curve over $\overline{\mathbb{Q}}$ has a $\mathfrak{p}$-integral Weierstrass equation with good or multiplicative reduction when $\mathfrak{p}$ does not lie over 2. This result holds as well when $\mathfrak{p}$ does lie over 2, by an argument using the *Deuring form* of a Weierstrass equation—see Appendix A of [Sil86].

A $\mathfrak{p}$-integral Weierstrass equation with good or multiplicative reduction is called $\mathfrak{p}$-*minimal*. Additive reduction being the worst type, we avoid it, as we now know we can. Consider only $\mathfrak{p}$-minimal Weierstrass equations, and define two such equations to be equivalent when they differ by an admissible change of variable over $\overline{\mathbb{Q}}$.

Proposition 8.4.2. *Ordinary reduction, supersingular reduction, and multiplicative reduction are well defined on equivalence classes of $\mathfrak{p}$-minimal Weierstrass equations. If E and E' are equivalent $\mathfrak{p}$-minimal Weierstrass equations with good reduction at $\mathfrak{p}$ then their reductions define isomorphic elliptic curves over $\overline{\mathbb{F}}_p$.*

Proof. If E and E' are equivalent then by Exercise 8.1.1(b)

$$u^{12}\Delta' = \Delta, \qquad u^4 c_4' = c_4,$$

where u comes from the admissible change of variable taking E to E'. Recall the disjoint union mentioned early in this section,

$$\overline{\mathbb{Z}}_{(\mathfrak{p})} = \overline{\mathbb{Z}}_{(\mathfrak{p})}^* \cup \mathfrak{p}\overline{\mathbb{Z}}_{(\mathfrak{p})}.$$

Assume $\Delta' \in \overline{\mathbb{Z}}_{(\mathfrak{p})}^*$. If also $\Delta \in \mathfrak{p}\overline{\mathbb{Z}}_{(\mathfrak{p})}$ then $u^{12} \in \mathfrak{p}\overline{\mathbb{Z}}_{(\mathfrak{p})}$ and thus $u^4 \in \mathfrak{p}\overline{\mathbb{Z}}_{(\mathfrak{p})}$ so that $c_4 \in \mathfrak{p}\overline{\mathbb{Z}}_{(\mathfrak{p})}$. This means that E has additive reduction, impossible since E is $\mathfrak{p}$-minimal. Thus there is no equivalence between equations of good and multiplicative reduction.

Given a change of variable between two equations of good reduction, we may further assume by Lemma 8.4.1 that $u \in \overline{\mathbb{Z}}_{(\mathfrak{p})}$, and so the relation $u^{12}\Delta' = \Delta$ shows that $u \in \overline{\mathbb{Z}}_{(\mathfrak{p})}^*$. Therefore $\tilde{u} \neq 0$ in $\overline{\mathbb{F}}_p$. Also the other coefficients r, s, t

from the admissible change of variable taking E to E' lie in $\overline{\mathbb{Z}}_{(\mathfrak{p})}$, similarly to Exercise 8.3.3(b,c) (Exercise 8.4.3), so they reduce to $\overline{\mathbb{F}}_p$ under (8.24). The two reduced Weierstrass equations differ by the reduced change of variable, giving the last statement of the proposition. Since isomorphic elliptic curves have the same p-torsion structure, this shows that ordinary and supersingular reduction are preserved under equivalence. $\qquad\square$

The proposition shows that if E is an elliptic curve over $\overline{\mathbb{Q}}$ then its reduction type at $\mathfrak{p}$ is well defined as the ordinary, supersingular, or multiplicative reduction type of any $\mathfrak{p}$-minimal Weierstrass equation for E. Furthermore, the proposition shows that if the reduction is good then it gives a well defined elliptic curve $\widetilde{E}$ over $\overline{\mathbb{F}}_p$ up to isomorphism over $\overline{\mathbb{F}}_p$.

The results of this section explain some earlier terminology. Any elliptic curve E over $\mathbb{Q}$ can be viewed instead as a curve $E_{\overline{\mathbb{Q}}}$ over $\overline{\mathbb{Q}}$. Let $p \in \mathbb{Z}$ be prime and let $\mathfrak{p} \subset \overline{\mathbb{Z}}$ be a maximal ideal lying over p. We have shown that ordinary, supersingular, and multiplicative reduction of E at p do not change upon reducing $E_{\overline{\mathbb{Q}}}$ at $\mathfrak{p}$ instead, but additive reduction of E at p improves to good or multiplicative reduction of $E_{\overline{\mathbb{Q}}}$ at $\mathfrak{p}$. This motivates the words "semistable" for good or multiplicative reduction and "unstable" for additive reduction, as given in the previous section working over $\mathbb{Q}$. Exercise 8.4.4 gives an example of an elliptic curve over $\mathbb{Q}$ whose additive reduction at a rational prime p improves in $\overline{\mathbb{Q}}$.

Good reduction is characterized by the j-invariant of an elliptic curve, something we can check without needing to put the Weierstrass equation into $\mathfrak{p}$-minimal form or even into $\mathfrak{p}$-integral form.

Proposition 8.4.3. *Let E be an elliptic curve over $\overline{\mathbb{Q}}$ and let $\mathfrak{p}$ be a maximal ideal of $\overline{\mathbb{Z}}$. Then E has good reduction at $\mathfrak{p}$ if and only if $j(E) \in \overline{\mathbb{Z}}_{(\mathfrak{p})}$.*

Proof. Suppose E has good reduction at $\mathfrak{p}$. We may assume the Weierstrass equation for E is $\mathfrak{p}$-minimal. Thus $\Delta \in \overline{\mathbb{Z}}_{(\mathfrak{p})}^*$ and $c_4 \in \overline{\mathbb{Z}}_{(\mathfrak{p})}$, so that $j(E) = c_4^3/\Delta \in \overline{\mathbb{Z}}_{(\mathfrak{p})}$ as desired.

For the other direction, assume the Weierstrass equation of E is in $\mathfrak{p}$-integral Legendre form, and again assume that $\mathfrak{p}$ does not lie over 2. Since $\Delta = 16\lambda^2(1-\lambda)^2$, showing that E has good reduction at $\mathfrak{p}$ reduces to showing that $\lambda(1-\lambda) \notin \mathfrak{p}\overline{\mathbb{Z}}_{(\mathfrak{p})}$. But since also $c_4 = 16(1-\lambda(1-\lambda))$, the relation $j = c_4^3/\Delta$ is

$$j\lambda^2(1-\lambda)^2 = 16^2(1 - \lambda(1-\lambda))^3.$$

The assumptions are that $j \in \overline{\mathbb{Z}}_{(\mathfrak{p})}$ and $\mathfrak{p}$ does not lie over 2. Thus if $\lambda(1-\lambda)$ lies in $\mathfrak{p}\overline{\mathbb{Z}}_{(\mathfrak{p})}$ then so does the left side, while the right side does not, contradiction. For the Deuring form proof when $\mathfrak{p}$ does lie over 2, again see Appendix A of [Sil86]. $\qquad\square$

So far we have reduced a Weierstrass equation, but not the points themselves of the elliptic curve. To reduce the points, we first show more generally that for any positive integer n the maximal ideal $\mathfrak{p}$ determines a reduction map

$$\widetilde{} : \mathbb{P}^n(\overline{\mathbb{Q}}) \longrightarrow \mathbb{P}^n(\overline{\mathbb{F}}_p). \tag{8.25}$$

To see this, take any point

$$P = [x_0, \ldots, x_n] \in \mathbb{P}^n(\overline{\mathbb{Q}}),$$

and consider the values $m \in \mathbb{N}$ such that P has a representative with all of $x_0, \ldots, x_m$ in $\overline{\mathbb{Z}}_{(\mathfrak{p})}$ and at least one of them equal to 1. Such values of m exist, e.g., the smallest m such that $x_m \neq 0$. And given such an m that is less than n, if $x_{m+1} \in \overline{\mathbb{Z}}_{(\mathfrak{p})}$ then $m+1$ also works, but otherwise $1/x_{m+1} \in \overline{\mathbb{Z}}_{(\mathfrak{p})}$ by Lemma 8.4.1 and so multiplying P through by $1/x_{m+1}$ shows that again $m+1$ works. Thus P has a representation such that all coordinates lie in $\overline{\mathbb{Z}}_{(\mathfrak{p})}$ and some x_i is 1. This representation reduces to

$$\widetilde{P} = [\tilde{x}_0, \ldots, \tilde{x}_n] \in \mathbb{P}^n(\overline{\mathbb{F}}_p).$$

The scalar quotient of two such representations of P must belong to $\overline{\mathbb{Z}}_{(\mathfrak{p})}^*$ and thus reduce to $\overline{\mathbb{F}}_p^*$, so the two representations reduce to the same element of $\mathbb{P}^n(\overline{\mathbb{F}}_p)$ and the reduction map is well defined. The description of the map shows that an affine point $P = (x_1, \ldots, x_n) = [1, x_1, \ldots, x_n]$ of $\mathbb{P}^n(\overline{\mathbb{Q}})$ reduces to an affine point of $\mathbb{P}^n(\overline{\mathbb{F}}_p)$ if and only if all of its coordinates lie in $\overline{\mathbb{Z}}_{(\mathfrak{p})}$.

In particular, if E is an elliptic curve over $\mathbb{Q}$ or over $\overline{\mathbb{Q}}$ and $\widetilde{E}$ is its reduction at p or at $\mathfrak{p}$ then the points of E, a subset of $\mathbb{P}^2(\overline{\mathbb{Q}})$, reduce to points of $\widetilde{E}$ since

$$\widetilde{E}(\tilde{x}, \tilde{y}) = \widetilde{E(x, y)} = \widetilde{0_{\overline{\mathbb{Q}}}} = 0_{\overline{\mathbb{F}}_p}.$$

Since the only nonaffine point of any elliptic curve is its zero, the end of the previous paragraph shows that reduction from E to $\widetilde{E}$ has for its kernel zero and the affine points whose coordinates are not both $\mathfrak{p}$-integral,

$$\ker(\widetilde{}) = E - \overline{\mathbb{Z}}_{(\mathfrak{p})}^2. \tag{8.26}$$

Let E be an elliptic curve over $\overline{\mathbb{Q}}$ and let N be a positive integer. To study the reduction of N-torsion at a maximal ideal $\mathfrak{p}$, recall the Nth division polynomial ψ_N from Section 7.1, satisfying

$$[N](P) = 0_E \iff \psi_N(P) = 0, \quad \psi_N \in \mathbb{Z}[g_2, g_3, x, y].$$

An elementary argument (see [Sil86]) shows that for Weierstrass equations $y^2 = x^3 - g_2 x - g_3$, i.e., for Weierstrass equations in the form (8.3), ψ_N lies in $\mathbb{Z}[g_2, g_3, x, y^2]$ if N is odd and lies in $y\mathbb{Z}[g_2, g_3, x, y^2]$ if N is even. In either case ψ_N^2 can be viewed as an element of $\mathbb{Z}[g_2, g_3, x]$ after using the Weierstrass relation to eliminate y, and then it takes the form

$$\psi_N^2(x) = N^2 x^{N^2-1} + \cdots \in \mathbb{Z}[g_2, g_3, x].$$

Define the *localization of* $\mathbb{Z}$ *at* p,

$$\mathbb{Z}_{(p)} = \{x/y : x, y \in \mathbb{Z},\ y \notin p\mathbb{Z}\} = \{r \in \mathbb{Q} : \nu_p(r) \geq 0\}.$$

For $p \nmid N$, dividing $\psi_N(x)^2$ through by N^2 gives a monic polynomial that is still p-integral,

$$\psi_N^2(x)/N^2 = x^{N^2-1} + \cdots \in \mathbb{Z}_{(p)}[g_2, g_3, x].$$

Suppose the Weierstrass equation of E is in $\mathfrak{p}$-minimal form, and suppose that $\mathfrak{p}$ does not lie over 2 or 3. The admissible changes of variable at the beginning of Section 8.1 to reduce the general Weierstrass equation (8.1) to the form (8.3) leave E in $\mathfrak{p}$-integral form. Further suppose that $p \nmid N$. Then the monic p-integral polynomial condition on the x-coordinates of $E[N]$ shows that they are $\mathfrak{p}$-integral (Exercise 8.4.5(a)). Consequently the y-coordinates of $E[N]$ are $\mathfrak{p}$-integral as well (Exercise 8.4.5(b)). Thus (8.26) shows that the reduction $E[N] \longrightarrow \widetilde{E}[N]$ is injective, and it is therefore surjective since $E[N]$ and $\widetilde{E}[N]$ are isomorphic finite groups.

Proposition 8.4.4. *Let E be an elliptic curve over $\overline{\mathbb{Q}}$ with good reduction at $\mathfrak{p}$. Then:*

(a) *The reduction map on N-torsion,*

$$E[N] \longrightarrow \widetilde{E}[N],$$

   is surjective for all N.

(b) *Any isogenous image E/C where C is a cyclic subgroup of order p also has good reduction at $\mathfrak{p}$. Furthermore, if E has ordinary reduction at $\mathfrak{p}$ then so does E/C, and if E has supersingular reduction at $\mathfrak{p}$ then so does E/C.*

We have shown part (a) when $p \nmid 6N$, but a complete proof is beyond the scope of this book. The substance of part (b) is that good reduction remains good under the isogeny, also not proved here. Granting this, it follows that ordinary reduction remains ordinary and supersingular reduction remains supersingular. Let $\varphi : E \longrightarrow E'$ be the isogeny from E to $E' = E/C$ and let $\psi : E' \longrightarrow E$ be its dual, so also φ is the dual of ψ. The pending Theorem 8.5.10 says that there exist reduced maps

$$\tilde{\varphi} : \widetilde{E} \longrightarrow \widetilde{E'}, \qquad \tilde{\psi} : \widetilde{E'} \longrightarrow \widetilde{E}$$

such that $\tilde{\psi} \circ \tilde{\varphi} = [p]_{\widetilde{E}}$ and $\tilde{\varphi} \circ \tilde{\psi} = [p]_{\widetilde{E'}}$. Thus the map $\tilde{\varphi} \circ \tilde{\psi} \circ \tilde{\varphi}$ has two descriptions,

$$[p]_{\widetilde{E'}} \circ \tilde{\varphi} = \tilde{\varphi} \circ [p]_{\widetilde{E}} : \widetilde{E} \longrightarrow \widetilde{E'}.$$

The descriptions combine to show that $\deg_{\text{sep}}([p])$ is the same on $\widetilde{E}$ and on $\widetilde{E'}$. That is, $|\widetilde{E}[p]| = |\widetilde{E'}[p]|$ since $\deg_{\text{sep}} = |\ker|$ for isogenies, cf. (8.13).

Exercises

8.4.1. (a) Show that there is a natural isomorphism $\overline{\mathbb{Z}}/\mathfrak{p} \xrightarrow{\sim} \overline{\mathbb{Z}}_{(\mathfrak{p})}/\mathfrak{p}\overline{\mathbb{Z}}_{(\mathfrak{p})}$.

(b) Show that $\overline{\mathbb{Z}}/\mathfrak{p}$ is an algebraic closure of $\mathbb{Z}/p\mathbb{Z}$ as follows. First show that any $\beta \in \overline{\mathbb{Z}}/\mathfrak{p}$ is algebraic over $\mathbb{Z}/p\mathbb{Z}$. Second use the fact that any monic polynomial $g \in (\mathbb{Z}/p\mathbb{Z})[x]$ takes the form $g = \tilde{f}$ where $f \in \mathbb{Z}[x]$, to complete the argument. (A hint for this exercise is at the end of the book.)

(c) Show that $\mathfrak{p}\overline{\mathbb{Z}}_{(\mathfrak{p})} \cap \mathbb{Z} = p\mathbb{Z}$.

8.4.2. (a) Show that the general Weierstrass equation over $\overline{\mathbb{Q}}$ is taken by an admissible change of variable over $\overline{\mathbb{Q}}$ to Legendre form $y^2 = x(x-1)(x-\lambda)$ where $\lambda \in \overline{\mathbb{Q}} - \{0, 1\}$.

(b) Change this to $y^2 = x(x-1)(x-1/\lambda)$ by another admissible change of variable over $\overline{\mathbb{Q}}$. Why may we thus assume that $\lambda \in \overline{\mathbb{Z}}_{(\mathfrak{p})}$?

(c) Confirm the formulas for Δ and c_4 arising from Legendre form.

8.4.3. Show that any admissible change of variable over $\overline{\mathbb{Q}}$ between two $\mathfrak{p}$-integral Weierstrass equations of good reduction entails $r, s, t \in \overline{\mathbb{Z}}_{(\mathfrak{p})}$. (A hint for this exercise is at the end of the book.)

8.4.4. (a) Consider the Weierstrass equation $E : y^2 = x^3 - 9x$. Show that it is in reduced form over $\mathbb{Q}$ and that it defines an elliptic curve over $\mathbb{Q}$ with additive reduction at $p = 3$.

(b) Consider the number field $\mathbb{K} = \mathbb{Q}(\mu_6)$ where $\mu_6 = e^{2\pi i/6}$, satisfying the polynomial $x^2 - x + 1$. Its ring of integers is

$$\mathcal{O}_{\mathbb{K}} = \mathbb{Z}[\mu_6] = \{a + b\mu_6 : a, b \in \mathbb{Z}\},$$

denoted A in Section 3.7. From that section, the factorization of 3 in $\mathcal{O}_{\mathbb{K}}$ is $3\mathcal{O}_{\mathbb{K}} = \mathfrak{p}_{\mathbb{K}}^2$ where $\mathfrak{p}_{\mathbb{K}} = \langle 1 + \mu_6 \rangle$. Let $\pi = 1 + \mu_6$. Show that $3 = \mu_6^5\pi^2$ in $\mathcal{O}_{\mathbb{K}}$.

(c) Now view E as an elliptic curve over $\mathbb{K}$. Apply the admissible change of variable over $\mathbb{K}$

$$(x, y) = (\mu_6^2\pi^2 x', \mu_6^3\pi^3 y').$$

What is the reduction type of E at $\mathfrak{p}_{\mathbb{K}}$? What is the reduction type of E as an elliptic curve over $\overline{\mathbb{Q}}$ at any maximal ideal $\mathfrak{p}$ lying over 3? (A hint for this exercise is at the end of the book.)

8.4.5. (a) Let E be a $\mathfrak{p}$-minimal Weierstrass equation over $\overline{\mathbb{Q}}$ in the form (8.3). Let $\mathfrak{p}$ lie over p where $p \neq 2, 3$. Let N be a positive integer such that $p \nmid N$. Show that the x-coordinates of $E[N]$ are $\mathfrak{p}$-integral.

(b) Show that the y-coordinates of $E[N]$ are $\mathfrak{p}$-integral as well.

8.5 Reduction of algebraic curves and maps

Reducing Weierstrass equations and the points of elliptic curves from characteristic 0 to characteristic p in the previous two sections was fairly straightforward. But already the results of Proposition 8.4.4, that good reduction surjects as a map of N-torsion and is preserved under isogeny, are beyond the scope of this book. Even granting that good reduction is preserved, the argument that consequently so are its ordinary and supersingular subtypes still needed to quote facts about reducing isogenies to characteristic p. The subject of the next section, reducing modular curves from characteristic 0 to characteristic p, is much more substantial than any of this. The present section states the results to be quoted from algebraic geometry, incidentally justifying a few unsupported assertions from earlier. The statements are given in elementary terms, making them cumbersome in places but perhaps accessible to more readers than the scheme-theoretic language of modern algebraic geometry.

Recall the localization of $\mathbb{Z}$ at p,

$$\mathbb{Z}_{(p)} = \{x/y : x, y \in \mathbb{Z}, \ y \notin p\mathbb{Z}\},$$

a subring of $\mathbb{Q}$ containing $\mathbb{Z}$ and having unique maximal ideal $p\mathbb{Z}_{(p)}$. This construction eliminates all primes but p from the ideal structure of $\mathbb{Z}_{(p)}$. There is a natural isomorphism

$$\mathbb{Z}/p\mathbb{Z} \xrightarrow{\sim} \mathbb{Z}_{(p)}/p\mathbb{Z}_{(p)}, \qquad a + p\mathbb{Z} \mapsto a + p\mathbb{Z}_{(p)},$$

so that the reduction map

$$\widetilde{} : \mathbb{Z}_{(p)} \longrightarrow \mathbb{F}_p, \qquad \tilde{\alpha} = \alpha + p\mathbb{Z}_{(p)}$$

is a well defined surjection with kernel $p\mathbb{Z}_{(p)}$. The definition of reducing an algebraic curve over $\mathbb{Q}$ to an algebraic curve over $\mathbb{F}_p$ uses the localization $\mathbb{Z}_{(p)}$.

Definition 8.5.1. *Let C be a nonsingular affine algebraic curve over $\mathbb{Q}$, defined by polynomials $\varphi_1, \dots, \varphi_m \in \mathbb{Z}_{(p)}[x_1, \dots, x_n]$. Then C has **good reduction modulo p** (or **at p**) if*

(1) *The ideal $I = \langle \varphi_1, \dots, \varphi_m \rangle$ of $\mathbb{Z}_{(p)}[x_1, \dots, x_n]$ is prime.*
(2) *The reduced polynomials $\tilde{\varphi}_1, \dots, \tilde{\varphi}_m \in \mathbb{F}_p[x_1, \dots, x_n]$ define a nonsingular affine algebraic curve $\widetilde{C}$ over $\mathbb{F}_p$.*

*In this case $\widetilde{C}$ is the **reduction of C at p.***

Without condition (1) the curve and its reduction need have nothing to do with each other. For instance, consider the nonprime ideal

$$I = \langle p(py - 1), (y - x^2)(py - 1) \rangle \subset \mathbb{Z}_{(p)}[x, y].$$

The ideal $I_{\mathbb{Q}} \subset \mathbb{Q}[x, y]$ with the same generators as I is $\langle py - 1 \rangle$, defining the curve $y = 1/p$ in $\mathbb{Q}^2$, a translate of the x-axis; but the reduction $\tilde{I} \subset \mathbb{F}_p[x, y]$ is $\langle y - x^2 \rangle$, defining a parabola in $\overline{\mathbb{F}}_p^2$. See Exercise 8.5.1 for a generalization of this example.

An observation is required before extending Definition 8.5.1 from affine curves to projective curves. Let $\mathbf{k}$ be any field. Introduce the notations $x = (x_0, \ldots, x_n)$ and $x_{(i)} = (x_0, \ldots, x_{i-1}, 1, x_{i+1}, \ldots, x_n)$ for $i = 0, \ldots, n$, and $\varphi_{(i)} = \varphi(x_{(i)})$ for $\varphi \in \mathbf{k}[x]$. For any homogeneous ideal

$$I = \langle \{\varphi\} \rangle \subset \mathbf{k}[x],$$

its ith dehomogenization for $i = 0, \ldots, n$ is

$$I_{(i)} = \langle \{\varphi_{(i)}\} \rangle \subset \mathbf{k}[x_{(i)}],$$

and its ith rehomogenization is

$$I_{(i),\mathrm{hom}} = \langle \{\varphi_{(i),\mathrm{hom}}\} \rangle \subset \mathbf{k}[x],$$

where $\varphi_{(i),\mathrm{hom}}$ means to multiply each term of $\varphi_{(i)}$ by the smallest power of x_i needed to make all the terms have the same total degree. Thus $\varphi = x_i^e \varphi_{(i),\mathrm{hom}}$ for some $e \geq 0$.

Consider the ideal $I = \langle x_0 x_1 + x_0 x_2 \rangle$. Its 0th dehomogenization is $I_{(0)} = \langle x_1 + x_2 \rangle$, a prime ideal that defines an algebraic curve, but $I_{(0),\mathrm{hom}} = I_{(0)} \neq I$. On the other hand, $I_{(1)} = \langle x_0(1 + x_2) \rangle$ is not a prime ideal, but $I_{(1),\mathrm{hom}} = I$. This example shows that whether a dehomogenization defines a curve and whether it rehomogenizes back to the original ideal are not equivalent. However,

Lemma 8.5.2. *Let $\mathbf{k}$ be a field. Let $I \subset \mathbf{k}[x]$ be the homogenization of a prime ideal $I_{(0)} \subset \mathbf{k}[x_{(0)}]$ that defines an affine algebraic curve. Then I is prime. For $i = 0, \ldots, n$, if $I_{(i)} \neq \mathbf{k}[x_{(i)}]$ then $I_{(i)}$ is prime, $I_{(i),\mathrm{hom}} = I$, and $I_{(i)}$ defines an affine algebraic curve.*

Proof. If $\varphi\psi \in I$, where $\varphi, \psi \in \mathbf{k}[x]$, then $\varphi_{(0)}\psi_{(0)} = (\varphi\psi)_{(0)} \in I_{(0)}$ and so (up to symmetry) $\varphi_{(0)} \in I_{(0)}$, giving $\varphi_{(0),\mathrm{hom}} \in I$. Since $\varphi_{(0),\mathrm{hom}} \mid \varphi$, also $\varphi \in I$. This shows that I is prime.

If $\varphi\psi \in I_{(i)}$, where $\varphi, \psi \in \mathbf{k}[x_{(i)}]$, then $\varphi_{\mathrm{hom}}\psi_{\mathrm{hom}} = (\varphi\psi)_{\mathrm{hom}} \in I$. Thus (up to symmetry) $\varphi_{\mathrm{hom}} \in I$ since I is prime, and consequently $\varphi \in I_{(i)}$ since $\varphi = \varphi_{\mathrm{hom},(i)}$. This shows that $I_{(i)}$ is prime unless it is all of $\mathbf{k}[x_{(i)}]$, not considered a prime ideal since the corresponding trivial quotient is not an integral domain.

As already mentioned, each $\varphi \in I$ takes the form $\varphi = x_i^e \varphi_{(i),\mathrm{hom}}$ for some nonnegative integer e. On the one hand this gives the inclusion $I \subset I_{(i),\mathrm{hom}}$. On the other hand, if $I_{(i)} \neq \mathbf{k}[x_{(i)}]$ then $x_i \notin I$, so $\varphi_{(i),\mathrm{hom}} \in I$ since I is prime. Thus if $I_{(i)} \neq \mathbf{k}[x_{(i)}]$ then $I_{(i),\mathrm{hom}} \subset I$, and this combines with the first inclusion to show that $I_{(i),\mathrm{hom}} = I$.

If $I_{(i)} \neq \mathbf{k}[x_{(i)}]$ then the field of quotients of $\mathbf{k}[x_{(i)}]/I_{(i)}$ can be identified with a subfield of the field of quotients of $\mathbf{k}[x]/I$ that has transcendence degree 1 (Exercise 8.5.2). This shows that $I_{(i)}$ defines a curve. $\qquad \square$

The lemma supports the assertion in Section 7.2 that the function field of any nonempty affine piece of a projective curve is the function field of the entire curve, regardless of which piece is chosen. Examples such as $\langle x_0 x_1 + x_0 x_2 \rangle$ above, with extra powers of a variable, don't arise from homogenizing a prime affine ideal.

Definition 8.5.3. *Let C_{hom} be a nonsingular projective curve over $\mathbb{Q}$ defined by the homogenization $I \subset \mathbb{Z}_{(p)}[x]$ of a prime ideal $I_{(0)} \subset \mathbb{Z}_{(p)}[x_{(0)}]$ as in Definition 8.5.1. Then C_{hom} has **good reduction** at p if for $i = 1, \ldots, n$, either the affine curve C_i defined by $I_{(i)}$ has good reduction at p or $I_{(i)}$ reduces to all of $\mathbb{F}_p[x_{(i)}]$ so that C_i has empty reduction at p. The curve $\widetilde{C}_{\text{hom}}$ defined by the homogenization $(\widetilde{I_{(0)}})_{\text{hom}} \subset \mathbb{F}_p[x]$ is the **reduction of C** at p.*

The reduced curve $\widetilde{C}_{\text{hom}}$ is defined by any nonempty reduction $\widetilde{C}_i$ of an affine piece of C_{hom}, but this is not immediately obvious. Since dehomogenizing and reducing commute,

$$(\widetilde{I_{(i)}})_{\text{hom}} = \widetilde{I}_{(i),\text{hom}}, \quad i = 0, \ldots, n.$$

Let $J = \widetilde{I}_{(0),\text{hom}}$. Then $\widetilde{I} \subset J$ and therefore $\widetilde{I}_{(i)} \subset J_{(i)}$ for each i, with $J_{(0)} = \widetilde{I}_{(0)}$ in particular. Since the reduced curve is defined by the homogenization of $J_{(0)}$, Lemma 8.5.2 says that it is defined by any $J_{(i)}$ that is not all of $\mathbb{F}_p[x_{(i)}]$. But each such $J_{(i)}$ is in fact $\widetilde{I}_{(i)}$ since $J_{(i)}/\widetilde{I}_{(i)}$ is a prime ideal of the coordinate ring $\mathbb{F}_p[\widetilde{C}_i] = \mathbb{F}_p[x_{(i)}]/\widetilde{I}_{(i)}$. As such it is the zero ideal or the maximal ideal m_P of functions vanishing at some point $P \in \widetilde{C}_i$ (see Exercise 8.5.3). But in the latter case the quotient field of the integral domain $\mathbb{F}_p[x_{(i)}]/J_{(i)}$ does not have transcendence degree 1, and this is a contradiction since $J_{(i)}$ defines a curve. Therefore $J_{(i)}/\widetilde{I}_{(i)} = \{0\}$, i.e., $J_{(i)} = \widetilde{I}_{(i)}$ when $J_{(i)} \neq \mathbb{F}_p[x_{(i)}]$. After the pending Theorem 8.5.4 we will be able to show that if on the other hand $J_{(i)} = \mathbb{F}_p[x_{(i)}]$ then also $\widetilde{I}_{(i)} = \mathbb{F}_p[x_{(i)}]$, completing the argument that $\widetilde{C}_{\text{hom}}$ is defined by any nonempty $\widetilde{C}_i$. As in Section 7.2 we now drop the notation C_{hom} to distinguish a projective curve from its affine pieces.

Good reduction at p on one affine piece of a projective curve does not guarantee good reduction of the curve (Exercise 8.5.4). Also, just as a Weierstrass equation over $\mathbb{Q}$ needs to be put into p-minimal form by an isomorphism over $\mathbb{Q}$ in order to optimize its reduction type at p, the same can apply to a nonelliptic curve over $\mathbb{Q}$. For example, the circle $x^2 + y^2 = z^2$ has bad reduction at $p = 2$ since every point is singular, but it is isomorphic over $\mathbb{Q}$ to the line defined by no equations in one variable (Exercise 8.5.5), and this clearly has good reduction at all p.

Curves as defined by equations are now reduced to characteristic p. To reduce the points of a curve, recall the reduction map (8.25) of projective space,

$$\widetilde{} : \mathbb{P}^n(\overline{\mathbb{Q}}) \longrightarrow \mathbb{P}^n(\overline{\mathbb{F}}_p).$$

If C is a nonsingular projective algebraic curve with good reduction at p then this takes C to $\widetilde{C}$.

Theorem 8.5.4. *Let C be a nonsingular projective algebraic curve over $\mathbb{Q}$ with good reduction at p. Then the reduction map $C \longrightarrow \widetilde{C}$ is surjective.*

The idea of the proof is to argue ring theoretically, proceeding from a point in characteristic p to a point in characteristic 0 via ideals. The points of an algebraic curve C over any field $\mathbf{k}$ map to the maximal ideals of its coordinate ring $\mathbf{k}[C]$, each point Q mapping to the maximal ideal m_Q of functions that vanish at Q. The map surjects by Hilbert's Nullstellensatz, mentioned in Exercise 8.5.3. It injects if $\mathbf{k}$ is algebraically closed, and more generally $m_Q = m_{Q'}$ if and only if some automorphism of $\overline{\mathbf{k}}$ over $\mathbf{k}$ takes Q to Q' by acting componentwise. Since C is defined by equations over $\mathbf{k}$ it is closed under this action.

Proof. Let Q' be a point of $\widetilde{C}$. Then Q' lies in some affine piece $\widetilde{C}_i$ of $\widetilde{C}$. Since i is fixed we will simply call the affine piece C throughout the proof. Thus the affine coordinate ring is $\mathbb{F}_p[\widetilde{C}] = \mathbb{F}_p[x_{(i)}]/\widetilde{I}_{(i)}$. Consider the ring $\mathbb{Z}_{(p)}[C] = \mathbb{Z}_{(p)}[x_{(i)}]/I_{(i)}$. The map $\mathbb{Z}_{(p)}[C] \longrightarrow \mathbb{F}_p[\widetilde{C}]$ is a surjection with kernel $\langle p \rangle = p\mathbb{Z}_{(p)}[C]$, so that $\mathbb{Z}_{(p)}[C]/\langle p \rangle$ is naturally identified with $\mathbb{F}_p[\widetilde{C}]$. Let $m \subset \mathbb{Z}_{(p)}[C]$ be the inverse image of the ideal $m_{Q'}$ of $\mathbb{F}_p[\widetilde{C}]$ corresponding to Q'. Then m is a maximal ideal of $\mathbb{Z}_{(p)}[C]$, and $p \in m$. So far we have

$$\mathbb{Z}_{(p)}[C] \longrightarrow \mathbb{F}_p[\widetilde{C}], \qquad m \longrightarrow m_{Q'}.$$

Let $f \in m$ be a lift to $\mathbb{Z}_{(p)}[C]$ of a nonzero function in $m_{Q'}$. Because $f + \langle p \rangle$ is nonzero and hence not a zero-divisor in the integral domain $\mathbb{F}_p[\widetilde{C}]$, it follows that $p + \langle f \rangle$ is not a zero-divisor in $\mathbb{Z}_{(p)}[C]/\langle f \rangle$. Indeed, if $p + \langle f \rangle$ is a zero-divisor then $fg = ph$ for some $g, h \in \mathbb{Z}_{(p)}[C]$ with $h \notin \langle f \rangle$, and thus $g \notin \langle p \rangle$ since $\mathbb{Z}_{(p)}[C]$ is an integral domain, making $f + \langle p \rangle$ a zero-divisor as well, contradiction. Let $P \subset \mathbb{Z}_{(p)}[C]$ be the lift of a minimal prime ideal of $\mathbb{Z}_{(p)}[C]/\langle f \rangle$ (see Exercise 8.5.6(a)) contained in $m/\langle f \rangle$. Thus P is nonzero since $f \in P$. A fact from commutative algebra, that any minimal prime ideal of a ring is a subset of the ring's zero-divisors (Exercise 8.5.6(b–c)), shows that $p + \langle f \rangle \notin P/\langle f \rangle$. Now we have

$$P \subset m \subset \mathbb{Z}_{(p)}[C], \qquad p \notin P.$$

Consider the coordinate ring $\mathbb{Q}[C] = \mathbb{Q}[x_{(i)}]/I_{(i),\mathbb{Q}}$ where $I_{(i),\mathbb{Q}}$ is the ideal of $\mathbb{Q}[x_{(i)}]$ with the same generators as the ideal $I_{(i)}$ of $\mathbb{Z}_{(p)}[x_{(i)}]$. Then $P\mathbb{Q}[C]$ is a nonzero prime ideal of $\mathbb{Q}[C]$ (it is proper because $p \notin P$), and so by Exercise 8.5.3,

$$P\mathbb{Q}[C] = m_Q \quad \text{for some } Q \in C.$$

The quotient $\mathbb{Q}[C]/m_Q$ is a field naturally identified with the values taken by polynomial functions on C at Q, a subfield of the Galois extension $\mathbb{K}$ of $\mathbb{Q}$

generated by the coordinates of Q. Also $P = m_Q \cap \mathbb{Z}_{(p)}[C]$ (the containment "$\subset$" is clear; for the other, if $h \in m_Q$ then $p^N h \in P$ for some N, and if $h \in \mathbb{Z}_{(p)}[C]$ as well then $h \in P$ since $p \notin P$), and so the quotient $\mathbb{Z}_{(p)}[C]/P$ maps to $\mathbb{K}$. Thus we have the map given by evaluating at Q,

$$\epsilon_Q : \mathbb{Z}_{(p)}[C]/P \longrightarrow \mathbb{K}, \qquad g + P \mapsto g(Q).$$

Let $\mathcal{O}_\mathbb{K}$ be the ring of integers of $\mathbb{K}$. Lemma 8.5.5 to follow will show that there is a maximal ideal $\mathfrak{q}_\mathbb{K}$ of $\mathcal{O}_\mathbb{K}$ lying over p such that $\epsilon_Q(\mathbb{Z}_{(p)}[C]/P)$ is contained in the local ring $\mathcal{O}_{\mathbb{K},(\mathfrak{q}_\mathbb{K})}$ and $\epsilon_Q(m/P)$ is contained in its maximal ideal $\mathfrak{q}_\mathbb{K}\mathcal{O}_{\mathbb{K},(\mathfrak{q}_\mathbb{K})}$. This gives a map

$$\epsilon_Q : \mathbb{Z}_{(p)}[C]/m \longrightarrow \mathcal{O}_{\mathbb{K},(\mathfrak{q}_\mathbb{K})}/\mathfrak{q}_\mathbb{K}\mathcal{O}_{\mathbb{K},(\mathfrak{q}_\mathbb{K})}, \qquad g + m \mapsto g(Q) + \mathfrak{q}_\mathbb{K}\mathcal{O}_{\mathbb{K},(\mathfrak{q}_\mathbb{K})}.$$

The reduction map $C \longrightarrow \widetilde{C}$ depends on an initial choice of a maximal ideal $\mathfrak{p}$ of $\overline{\mathbb{Z}}$ lying over p. Let $\mathfrak{p}_\mathbb{K} = \mathfrak{p} \cap \mathbb{K}$. A result from algebraic number theory, to be discussed in Section 9.1, is that some element $\sigma \in \mathrm{Gal}(\mathbb{K}/\mathbb{Q})$ takes $\mathfrak{q}_\mathbb{K}$ to $\mathfrak{p}_\mathbb{K}$. The quotient $\mathcal{O}_{\mathbb{K},(\mathfrak{p}_\mathbb{K})}/\mathfrak{p}_\mathbb{K}\mathcal{O}_{\mathbb{K},(\mathfrak{p}_\mathbb{K})}$ is a finite extension field of $\mathbb{F}_p$, isomorphic to $\mathcal{O}_\mathbb{K}/\mathfrak{p}_\mathbb{K}$. (This generalizes the isomorphism between $\mathbb{Z}_{(p)}/p\mathbb{Z}_{(p)}$ and $\mathbb{Z}/p\mathbb{Z}$ at the beginning of the section.) Another result from algebraic number theory to be discussed in Section 9.1 is that every automorphism of this field is the reduction to characteristic p of an automorphism $\tau \in \mathrm{Gal}(\mathbb{K}/\mathbb{Q})$ such that $\mathfrak{p}_\mathbb{K}^\tau = \mathfrak{p}_\mathbb{K}$. Replacing σ by $\sigma\tau$ for any such τ gives a new σ that still takes $\mathfrak{q}_\mathbb{K}$ to $\mathfrak{p}_\mathbb{K}$. Composing the map ϵ_Q with σ gives

$$\epsilon_Q^\sigma : \mathbb{Z}_{(p)}[C]/m \longrightarrow \mathcal{O}_{\mathbb{K},(\mathfrak{p}_\mathbb{K})}/\mathfrak{p}_\mathbb{K}\mathcal{O}_{\mathbb{K},(\mathfrak{p}_\mathbb{K})}, \qquad g + m \mapsto g(Q)^\sigma + \mathfrak{p}_\mathbb{K}\mathcal{O}_{\mathbb{K},(\mathfrak{p}_\mathbb{K})}.$$

Note that $g(Q)^\sigma = g(Q^\sigma)$, so this last map can also be denoted ϵ_{Q^σ}. In particular, specializing g to the coordinate functions x_j shows that Q^σ has $\mathfrak{p}$-integral coordinates.

We now have a commutative diagram

$$\begin{array}{ccc} \mathbb{Z}_{(p)}[C] & \longrightarrow & \mathcal{O}_{\mathbb{K},(\mathfrak{p}_\mathbb{K})} \\ \downarrow & & \downarrow \\ \mathbb{F}_p[\widetilde{C}] & \longrightarrow & \overline{\mathbb{F}}_p, \end{array}$$

where the top row is $g \mapsto g(Q^\sigma)$, the vertical arrows are reduction to characteristic p, and the bottom row is $\tilde{g} \mapsto \tilde{g}(\widetilde{Q^\sigma})$ for $\tilde{g} \in \mathbb{F}_p[\widetilde{C}]$. The kernel of the bottom row is thus $m_{\widetilde{Q^\sigma}}$, but since the diagram commutes the kernel of the bottom row is also the reduction of m, i.e., $m_{Q'}$. As explained before the proof this means that $Q' = (\widetilde{Q^\sigma})^\tau$ for some automorphism τ of $\overline{\mathbb{F}}_p$. But this τ restricts to an automorphism of $\mathcal{O}_{\mathbb{K},(\mathfrak{p}_\mathbb{K})}/\mathfrak{p}_\mathbb{K}\mathcal{O}_{\mathbb{K},(\mathfrak{p}_\mathbb{K})}$, and so as explained in the previous paragraph we may assume that it is the reduction of an automorphism $\tau \in \mathrm{Gal}(\mathbb{K}/\mathbb{Q})$ that fixes $\mathfrak{p}_\mathbb{K}$. That is, we may assume that τ has already been incorporated into σ, giving the desired result $Q' = \widetilde{Q^\sigma}$. $\square$

The following lemma was used in the proof of Theorem 8.5.4.

Lemma 8.5.5. *Let p be prime. Suppose S is a $\mathbb{Z}_{(p)}$-subalgebra of a number field $\mathbb{K}$ and M is a maximal ideal of S containing p. Then there is a maximal ideal $\mathfrak{q}$ of $\mathcal{O}_{\mathbb{K}}$ lying over p such that $S \subset \mathcal{O}_{\mathbb{K},(\mathfrak{q})}$ and $M \subset \mathfrak{q}\mathcal{O}_{\mathbb{K},(\mathfrak{q})}$.*

Specifically, the lemma was applied with $S = \mathbb{Z}_{(p)}[C]/P$ and $M = m/P$. The more experienced reader should supply details for the following proof, while the reader with less background can skim it.

Proof. (Sketch.) Let S' be the localization of S at M. Then $S \subset S'$ and S' is local with maximal ideal M' containing M. Let $S'' = S'\mathcal{O}_{\mathbb{K}}$. Then S'' is finitely generated as an S'-module, so $S''/M'S''$ is nonzero by Nakayama's Lemma and thus M' is contained in some maximal ideal M'' of S''. It suffices to prove the lemma with S'' and M'' in place of S and M. Now, S'' contains the set

$$\mathcal{O}_{\mathbb{K},(p)} = \{s \in \mathbb{K} : \nu_{\mathfrak{q}}(s) \geq 0 \text{ for all } \mathfrak{q} \text{ over } p\}.$$

This is the localization of $\mathcal{O}_{\mathbb{K}}$ at the multiplicative set $\mathbb{Z} - \langle p \rangle$. It is a principal ideal domain with only finitely many primes, these being $\mathfrak{q}\mathcal{O}_{\mathbb{K},(p)}$ for $\mathfrak{q}$ over p. Showing this relies on the fact that for each $\mathfrak{q}$ there exists a uniformizer that is a unit at the finitely many other $\mathfrak{q}'$ over p. (We proved a similar result for function fields in Chapter 7.) Using these uniformizers we can also show that there is a subset $\mathcal{Q}$ of the set of $\mathfrak{q}$ over p such that

$$S'' = \{s \in \mathbb{K} : \nu_{\mathfrak{q}}(s) \geq 0 \text{ for all } \mathfrak{q} \in \mathcal{Q}\}. \tag{8.27}$$

Indeed, for any $\mathfrak{q}$ such that there exists an $s \in S''$ with $\nu_{\mathfrak{q}}(s) < 0$ we can multiply through by suitable powers of uniformizers to get an $s' \in S''$ with $\nu_{\mathfrak{q}}(s') = -1$ and $\nu_{\mathfrak{q}'}(s') \geq 0$ for other $\mathfrak{q}'$, and then add the product of the uniformizers at $\mathfrak{q}'$ such that $\nu_{\mathfrak{q}'}(s') > 0$ to get an $s'' \in S''$ such that $\nu_{\mathfrak{q}'}(s'') = 0$ for $\mathfrak{q}' \neq \mathfrak{q}$. Doing this for each such $\mathfrak{q}$ we get that S'' takes the form (8.27). Since S'' has a nonzero maximal ideal and since the maximal ideals of S'' are $M''_{\mathfrak{q}} = \{s \in S'' : \nu_{\mathfrak{q}}(s) > 0\}$ for each $\mathfrak{q} \in \mathcal{Q}$, the set $\mathcal{Q}$ is nonempty. Thus S'' and M'' satisfy the desired conclusions of the lemma. □

We can now complete the argument that the reduced curve $\widetilde{C}_{\mathrm{hom}}$ is defined by any nonempty reduction $\widetilde{C}_i$ of an affine piece of C_{hom}. Let $\mathfrak{p}$ be a maximal ideal of $\overline{\mathbb{Q}}$ lying over p, and let $C_{i,1}$ denote the points of C whose ith coordinate can be normalized to 1 with the others $\mathfrak{p}$-integral. The proof of Theorem 8.5.4 shows that the map $C_{i,1} \longrightarrow \widetilde{C}_i$ is surjective. The case remaining to be proved is when $J_{(i)}$ is all of $\mathbb{F}_p[x_{(i)}]$. But then $x_i^m \in J$ for some m, and so $x_i^m \in \widetilde{I}_{(0)} = J_{(0)}$, i.e., $x_i \in \widetilde{I}_{(0)}$ since $\widetilde{I}_{(0)}$ is prime. Writing $x_i = f + pg$ for some $f \in I_{(0)}$ shows that for any $Q = (a_1, \ldots, a_n) = [1, a_1, \ldots, a_n]$ in C_0 we have $a_i \in p\mathbb{Z}_{(p)}$, so $Q \notin C_{i,1}$. Therefore $C_{i,1}$ is contained in $C - C_0$, which is finite (Exercise 8.5.7(a)), therefore $\widetilde{C}_i$ is finite and thus empty (Exercise 8.5.7(b)), i.e., $\widetilde{I}_{(i)} = \overline{\mathbb{F}}_p[x_{(i)}] = J_{(i)}$.

With curves and their points reduced, the next object is to reduce morphisms. Doing so requires the following result.

Lemma 8.5.6 (Krull Intersection Theorem, special case). *Let R be a Noetherian domain and let $\alpha \in R$ be a nonunit. Then $\bigcap_{e=0}^{\infty} \alpha^e R = \{0\}$.*

Proof. Let $J = \bigcap_{e=0}^{\infty} \alpha^e R$, let $m_1, \ldots, m_k$ be a set of generators for J, and let m be the column vector that they form. Since $\alpha J = J$ (Exercise 8.5.8), also αm is a column vector of generators, and so there exists a k-by-k matrix A with entries in R such that

$$m = \alpha A m.$$

That is, $(1 - \alpha A)m = 0$ where 1 is the k-by-k identity matrix. But $\det(1 - \alpha A)$ is 1 modulo αR and hence nonzero, making $1 - \alpha A$ invertible over the quotient field of the integral domain R. This shows that $m = 0$, giving $J = 0$ as desired. $\qquad \square$

Let

$$h : C \longrightarrow C'$$

be a morphism over $\mathbb{Q}$ of nonsingular projective algebraic curves over $\mathbb{Q}$ with good reduction at p. To reduce h to a morphism $\tilde{h}$ of the reduced curves, take a representation $h = [h_0, \ldots, h_r]$. Let $I_{(0)} \subset \mathbb{Z}_{(p)}[x_{(0)}]$ be the ideal whose homogenization defines C. We may assume that each h_i lies in the subring $R = \mathbb{Z}_{(p)}[x_{(0)}]/I_{(0)}$ of the coordinate ring $\mathbb{Q}[C_0] = \mathbb{Q}[x_{(0)}]/I_{\mathbb{Q},(0)}$. Let the p-adic valuation of each h_i be

$$\nu_p(h_i) = \max\{e : h_i \in p^e R\} \in \mathbb{N} \cup \{+\infty\},$$

and note that at least one $\nu_p(h_i)$ is finite by the lemma. Then the p-adic valuation of h is

$$\nu_p(h) = \min\{\nu_p(h_i) : i = 0, \ldots, r\},$$

and h rewrites as

$$h = [h'_0, \ldots, h'_r], \qquad h'_i = p^{-\nu_p(h_{(0)})} h_i \text{ for } i = 0, \ldots, r.$$

Each entry h'_i still lies in R but now some entry lies in $R - pR$, giving it nonzero reduction in $\mathbb{F}_p[\tilde{C}_0] = R/pR$. The reduction $\tilde{h} = [\tilde{h}'_0, \ldots, \tilde{h}'_r]$ is a rational map from $\tilde{C}_0$ to $\tilde{C}'$. To see this, note that each point of $\tilde{C}_0$ takes the form $\tilde{P}$ where $P \in C_0$, and $\tilde{h}(\tilde{P}) \neq 0$ for all but finitely many such points. If, say, $\tilde{h}'_i(\tilde{P}) \neq 0$ then we want to show that $\tilde{h}(\tilde{P})$ lies in $\tilde{C}'_i$. But any element of the defining ideal $\tilde{I}'_{(i)}$ takes the form $\tilde{g}$ where $g \in I'_{(i)}$. Thus for $\tilde{P}$, i, and $\tilde{g}$ as in this discussion, $\tilde{g}(\tilde{h}(\tilde{P}))$ is the reduction of $g(h(P))$, and this is 0 because $h(P)$ lies in C'_i. The argument has shown that $\tilde{h}(\tilde{P}) = \widetilde{h(P)}$ for all $P \in C$ that reduce to the complement of a finite subset of $\tilde{C}$. This statement is independent of which affine piece of C and which representatives h_i were used to define h, although the finite set can change if these do. In sum, $\tilde{h}$ is well defined as a rational map from $\tilde{C}$ to $\widetilde{C'}$. Finally, by the general theory of algebraic curves the rational map extends to the desired reduced morphism,

$$\tilde{h} : \tilde{C} \longrightarrow \widetilde{C'}.$$

The genus of an algebraic curve was mentioned in Section 7.8. An algebraic curve C over $\mathbb{Q}$ has genus 0 if and only if it is isomorphic over $\overline{\mathbb{Q}}$ to the projective line, or equivalently if and only if its function field $\overline{\mathbb{Q}}(C)$ is $\overline{\mathbb{Q}}$-isomorphic to a field $\overline{\mathbb{Q}}(t)$. Genus 0 curves are exceptional, but as long as the target curve C' has positive genus, the morphism h and its reduction $\tilde{h}$ act compatibly on points and have the same degree.

The proofs of the following theorem and of the pending Theorem 8.5.9 are beyond our scope. The theorems follow from Theorem 9.5.1 of [BLR90].

Theorem 8.5.7. *Let C and C' be nonsingular projective algebraic curves over $\mathbb{Q}$ with good reduction at p, and let C' have positive genus. For any morphism $h : C \longrightarrow C'$ the following diagram commutes:*

$$
\begin{array}{ccc}
C & \xrightarrow{\ h\ } & C' \\
\downarrow & & \downarrow \\
\tilde{C} & \xrightarrow{\ \tilde{h}\ } & \widetilde{C'}.
\end{array}
$$

Also, $\deg(\tilde{h}) = \deg(h)$.

Since $C \longrightarrow \tilde{C}$ surjects, $\tilde{h}$ is the unique morphism from $\tilde{C}$ to $\widetilde{C'}$ that makes the diagram commute. Such a diagram can fail to commute when C' has genus 0, for example the map $h : \mathbb{P}^1 \longrightarrow \mathbb{P}^1$ where $h[x, y] = [px, y]$ surjects over $\mathbb{Q}$ but it reduces at p to the zero map. Theorem 8.5.7 has the following consequences (Exercise 8.5.9):

Corollary 8.5.8. *Let C' have positive genus. Then*

(a) *If $h : C \longrightarrow C'$ surjects then so does $\tilde{h} : \tilde{C} \longrightarrow \widetilde{C'}$.*
(b) *If also $k : C' \longrightarrow C''$ and C'' has positive genus then*

$$\widetilde{k \circ h} = \tilde{k} \circ \tilde{h}.$$

(c) *If h is an isomorphism then so is $\tilde{h}$.*

Part (c) of the corollary shows that if C has positive genus then the isomorphism class of $\tilde{C}$ over $\mathbb{F}_p$ depends only on the isomorphism class of C over $\mathbb{Q}$.

With curves and maps reduced compatibly, we also need the following functorial result.

Theorem 8.5.9. *Let C be a nonsingular projective algebraic curve over $\mathbb{Q}$ with good reduction at p. The map on degree-0 divisors induced by reduction,*

$$\mathrm{Div}^0(C) \longrightarrow \mathrm{Div}^0(\widetilde{C}), \qquad \sum n_P(P) \mapsto \sum n_P(\widetilde{P}).$$

sends principal divisors to principal divisors and therefore further induces a surjection of Picard groups,

$$\mathrm{Pic}^0(C) \longrightarrow \mathrm{Pic}^0(\widetilde{C}), \qquad [\sum n_P(P)] \mapsto [\sum n_P(\widetilde{P})].$$

Let C' also be a nonsingular projective algebraic curve over $\mathbb{Q}$ with good reduction at p, and let C' have positive genus. Let $h : C \longrightarrow C'$ be a morphism over $\mathbb{Q}$, and let $h_* : \mathrm{Pic}^0(C) \longrightarrow \mathrm{Pic}^0(C')$ and $\tilde{h}_* : \mathrm{Pic}^0(\widetilde{C}) \longrightarrow \mathrm{Pic}^0(\widetilde{C'})$ be the induced forward maps of h and $\tilde{h}$. Then the following diagram commutes:

$$
\begin{array}{ccc}
\mathrm{Pic}^0(C) & \xrightarrow{\;h_*\;} & \mathrm{Pic}^0(C') \\
\downarrow & & \downarrow \\
\mathrm{Pic}^0(\widetilde{C}) & \xrightarrow{\;\tilde{h}_*\;} & \mathrm{Pic}^0(\widetilde{C'}).
\end{array}
$$

The first statement of Theorem 8.5.9 is not obvious because the induced map on divisors does not send the principal divisor $\mathrm{div}(f)$ to $\mathrm{div}(\tilde{f})$. The diagram follows from the first statement.

These results on reducing curves over $\mathbb{Q}$ at a prime p don't yet support the assertions in the previous section about elliptic curves over $\overline{\mathbb{Q}}$ and their reductions at a maximal ideal $\mathfrak{p}$. However, the ideas here extend to reducing curves over any number field $\mathbb{K}$ at a maximal ideal $\mathfrak{p}_{\mathbb{K}}$, using the local ring $\mathcal{O}_{\mathbb{K},(\mathfrak{p})}$ and the valuation $\nu_{\mathfrak{p}_{\mathbb{K}}}$ as in the proof of Lemma 8.4.1. The ideas thus extend to reducing curves over $\overline{\mathbb{Q}}$ at a maximal ideal $\mathfrak{p}$.

Theorem 8.5.10. *Let*

$$\varphi : E \longrightarrow E'$$

be an isogeny over $\overline{\mathbb{Q}}$ of elliptic curves over $\overline{\mathbb{Q}}$. Then there is a reduction

$$\tilde{\varphi} : \widetilde{E} \longrightarrow \widetilde{E'}$$

with the properties

(a) *$\tilde{\varphi}$ is an isogeny.*
(b) *If $\psi : E' \longrightarrow E''$ is also an isogeny then $\widetilde{\psi \circ \varphi} = \tilde{\psi} \circ \tilde{\varphi}$.*
(c) *The following diagram commutes:*

$$
\begin{array}{ccc}
E & \xrightarrow{\;\varphi\;} & E' \\
\downarrow & & \downarrow \\
\widetilde{E} & \xrightarrow{\;\tilde{\varphi}\;} & \widetilde{E'}.
\end{array}
$$

(d) $\deg(\tilde{\varphi}) = \deg(\varphi)$.

Part (a) of Theorem 8.5.10 follows from the general results of this section and from the fact that $\tilde{\varphi}$ is a nonconstant map taking 0 to 0. The statement in Section 8.2 that the map $[p]$ is an isogeny of degree p^2 on elliptic curves over $\mathbb{F}_p$ follows from Theorem 8.5.10 since any such elliptic curve is the reduction $\tilde{E}$ of an elliptic curve E over $\overline{\mathbb{Q}}$, and $[p]_{\tilde{E}}$ is the reduction of $[p]_E$. The facts quoted after Proposition 8.4.4 about reducing isogenies also follow from Theorem 8.5.10.

Exercises

8.5.1. Given polynomials $\varphi_1, \ldots, \varphi_m, \phi, \psi_1, \ldots, \psi_l$ in $\mathbb{Z}_{(p)}[x_1, \ldots, x_n]$, find an ideal $I \subset \mathbb{Z}_{(p)}[x_1, \ldots, x_n]$ such that $I = \langle \varphi_1, \ldots, \varphi_m, p\phi - 1 \rangle$ as an ideal of $\mathbb{Q}[x_1, \ldots, x_n]$, but $\tilde{I} = \langle \tilde{\psi}_1, \ldots, \tilde{\psi}_l \rangle$ in $\mathbb{F}_p[x_1, \ldots, x_n]$. (A hint for this exercise is at the end of the book.)

8.5.2. (a) In the context of Lemma 8.5.2, show that the elements $(f+I)/(g+I)$ of the field of fractions of $\mathbf{k}[x]/I$ where f and g are homogeneous of the same degree form a subfield $\mathbb{K}$.

(b) Assume that $I_{(i)}$ is a proper ideal of $\mathbf{k}[x_{(i)}]$, so that $x_i \notin I$. Define a map $\mathbf{k}[x_{(i)}]/I_{(i)} \longrightarrow \mathbb{K}$ by $f \mapsto (f_{\mathrm{hom}} + I)/(x_i^d + I)$ where $d = \deg f$. Show that this map is well defined and injective, and show that it extends to an isomorphism from the field of quotients of $\mathbf{k}[x_{(i)}]/I_{(i)}$ to $\mathbb{K}$.

(c) Show that $\mathbb{K}$ has transcendence degree 1 and therefore so does the field of quotients of $\mathbf{k}[x_{(i)}]/I_{(i)}$. (A hint for this exercise is at the end of the book.)

8.5.3. Let C be an algebraic curve over a field $\mathbf{k}$ and let $\mathbf{k}[C]$ be its coordinate ring. Let P be a point of C, let $m = m_P$ be the maximal ideal of $\mathbf{k}[C]$ associated to P, and let $M = M_P$ be the corresponding maximal ideal of the local ring $\mathbf{k}[C]_P$, cf. Section 7.2. Show that there is a bijection from the prime ideals of m and the prime ideals of M, given by $I \mapsto I\mathbf{k}[C]_P$ and having inverse $J \mapsto J \cap m$. Since the local ring is a discrete valuation ring, its only primes are $\{0\}$ and M. Consequently, every nonzero prime ideal of $\mathbf{k}[C]$ is maximal since it lies in some maximal ideal and by a result called Hilbert's Nullstellensatz any such ideal takes the form m_P.

8.5.4. Find an ideal $I_{(0)}$ of $\mathbb{Z}_{(p)}[x_1, x_2]$ defining a nonsingular curve with good reduction at p, but such that the homogenized ideal I of $\mathbb{Z}_{(p)}[x_0, x_1, x_2]$ also dehomogenizes to an ideal $I_{(2)}$ defining a nonsingular curve with bad reduction at p. (A hint for this exercise is at the end of the book.)

8.5.5. Let $\mathbf{k}$ be any field with $\mathrm{char}(\mathbf{k}) \neq 2$. Show that the map $[t, u] \mapsto [t^2 - u^2, 2tu, t^2 + u^2]$ is an isomorphism from $\mathbb{P}^1(\mathbf{k})$ to the curve $x^2 + y^2 = z^2$ in $\mathbb{P}^2(\mathbf{k})$ with inverse $[x, y, z] \mapsto [x + z, y]$ by showing that both compositions give the identity. Why does this fail if $\mathrm{char}(\mathbf{k}) = 2$?

8.5.6. Let R be a commutative ring with 1. This exercise shows that R has minimal prime ideals and that any such ideal is a subset of the zero-divisors of R.

(a) Let $I_0 \supset I_1 \supset \cdots$ be a descending chain of prime ideals of R. (The subscripts do not denote dehomogenizations.) Show that $\bigcap_{n=0}^{\infty} I_n$ is again a prime ideal. Thus R has minimal prime ideals.

(b) Let P be a minimal prime ideal of R. The localization of R at P is the ring $R_P = R(R - P)^{-1}$ whose definition should be clear from Exercise 6.5.5 and the various examples of localizing that we have seen since then. This ring is Noetherian and PR_P is its only prime ideal. Part (c) will show that PR_P contains only zero-divisors of R_P. Show that consequently P contains only zero-divisors of R.

(c) To complete the argument, let S be a Noetherian ring with only one prime ideal I. To show that I contains only zero-divisors of S, consider ideals of the form

$$\mathrm{Ann}_S(x) = \{s \in S : sx = 0\}, \quad x \in S, \, x \neq 0.$$

Note that any such ideal contains only zero-divisors of S. Since S is Noetherian any ascending chain of such ideals eventually stabilizes, and so the set of such ideals contains maximal elements. Let $J = \mathrm{Ann}_S(x)$ be such a maximal element. Show that J is prime, so that $J = I$ and the result follows. (A hint for this exercise is at the end of the book.)

8.5.7. (a) Show that any nonempty affine piece of a nonsingular projective algebraic curve is all but finitely many points of the curve. (Hints for this exercise are at the end of the book.)

(b) Show that any nonempty affine piece of a nonsingular projective algebraic curve is infinite.

8.5.8. In the proof of Lemma 8.5.6, show that $\alpha J = J$.

8.5.9. Prove Corollary 8.5.8.

8.6 Modular curves in characteristic p: Igusa's Theorem

Let N be a positive integer and let p be a prime, $p \nmid N$. The modular curves $X_1(N)$ and $X_0(N)$ have good reduction at p. This section describes the reductions.

First we reduce the moduli space at p. Let $\mathfrak{p}$ be a maximal ideal of $\overline{\mathbb{Q}}$ lying over p. Recall from Proposition 8.4.3 that an elliptic curve E over $\overline{\mathbb{Q}}$ with good reduction at $\mathfrak{p}$ has $j(E) \in \overline{\mathbb{Z}}_{(\mathfrak{p})}$, so this reduces at $\mathfrak{p}$ to $\widetilde{j(E)}$ in $\overline{\mathbb{F}}_p$. We want to avoid $j = 0, 1728$ in the image. Denote this with a prime in the notation, i.e., the suitable restriction of the moduli space $S_1(N)$ over $\mathbb{Q}$ is

$$S_1(N)'_{\mathrm{gd}} = \{[E, Q] \in S_1(N) : E \text{ has good reduction at } \mathfrak{p}, \ \widetilde{j(E)} \notin \{0, 1728\}\}.$$

In characteristic p, let $\widetilde{S}_1(N)$ denote the moduli space over $\overline{\mathbb{F}}_p$, i.e., it consists of equivalence classes $[E, Q]$ where E is an elliptic curve over $\overline{\mathbb{F}}_p$ and $Q \in E$ is a point of order N and the equivalence relation is isomorphism over $\overline{\mathbb{F}}_p$. Again avoid $j = 0, 1728$ by defining

$$\widetilde{S}_1(N)' = \{[E, Q] \in \widetilde{S}_1(N) : j(E) \notin \{0, 1728\}\}.$$

The resulting reduction map is

$$S_1(N)'_{\mathrm{gd}} \longrightarrow \widetilde{S}_1(N)', \qquad [E_j, Q] \mapsto [\widetilde{E}_j, \widetilde{Q}].$$

This surjects. Indeed, any Weierstrass equation over $\overline{\mathbb{F}}_p$ lifts to $\overline{\mathbb{Z}}$, and the discriminant of a lift is a lift of the discriminant, so nonzero discriminants lift to nonzero discriminants, i.e., elliptic curves lift to elliptic curves. Proposition 8.4.4(a) shows that each N-torsion point has a lift as well. Thus reduction surjects from $S_1(N)_{\mathrm{gd}}$ to $\widetilde{S}_1(N)$, and since $S_1(N)'_{\mathrm{gd}}$ is the inverse image of $\widetilde{S}_1(N)'$ the displayed reduction map surjects as well. Priming the moduli space in characteristic p removes only finitely many points. Even though priming in characteristic 0 removes infinitely many points, all that matters for our purposes is that the reduction surjects.

To define the reduction $\widetilde{X}_1(N)$ of $X_1(N)$ at p, consider the universal elliptic curve $\widetilde{E}_j$ defined over $\mathbb{F}_p(j)$,

$$\widetilde{E}_j : y^2 + xy = x^3 - \left(\frac{36}{j - 1728}\right) x - \left(\frac{1}{j - 1728}\right).$$

This is the curve of (8.5), with discriminant $j^2/(j - 1728)^3$ and invariant j. The points (x, y) of $\widetilde{E}_j$ are a subset of $\mathbb{P}^2(\overline{\mathbb{F}_p(j)})$. Let Q be any point of $\widetilde{E}_j$ with order N, i.e., $[N]Q = 0_{\widetilde{E}}$ but $[n]Q \neq 0_{\widetilde{E}}$ for $0 < n < N$. Let $\varphi_{1,N} \in \mathbb{F}_p(j)[x]$ be the minimal polynomial of its x-coordinate $x(Q)$. Define a field

$$\mathbb{K}_1(N) = \mathbb{F}_p(j)[x]/\langle \varphi_{1,N}(x)\rangle.$$

We assert without proof that $\mathbb{K}_1(N) \cap \overline{\mathbb{F}}_p = \mathbb{F}_p$, so $\mathbb{K}_1(N)$ is a function field over $\mathbb{F}_p$. The polynomial $\varphi_{1,N}$ defines a planar curve $\widetilde{X}_1(N)^{\mathrm{planar}}$, possibly singular, whose points are ordered pairs $(j, x) \in \overline{\mathbb{F}}_p^2$. There is a birational equivalence over $\mathbb{F}_p$ from the planar curve to any nonsingular projective algebraic curve whose function field is isomorphic to $\mathbb{K}_1(N)$, as discussed after Theorem 7.2.5.

Theorem 8.6.1 (Igusa). *Let N be a positive integer and let p be a prime with $p \nmid N$. The modular curve $X_1(N)$ has good reduction at p. There is an isomorphism of function fields*

$$\mathbb{F}_p(\widetilde{X}_1(N)) \xrightarrow{\sim} \mathbb{K}_1(N).$$

Moreover, reducing the modular curve is compatible with reducing the moduli space in that the following diagram commutes:

$$
\begin{array}{ccc}
S_1(N)'_{\mathrm{gd}} & \xrightarrow{\ \psi_1\ } & X_1(N) \\
\downarrow & & \downarrow \\
\widetilde{S}_1(N)' & \xrightarrow{\ \tilde{\psi}_1\ } & \widetilde{X}_1(N).
\end{array}
$$

Here the top row is the map $[E_j, Q] \mapsto (j, x(Q))$ to the planar model followed by the birational equivalence to $X_1(N)$, and similarly for the bottom row but in characteristic p.

The primes in the diagram thus now mean to remove finitely many additional points as necessary for all the maps to make sense. A similar commutative diagram but with the planar models is easy to establish (Exercise 8.6.1), but this is not enough to show that the diagram in the theorem commutes as well. The diagram extends to divisor groups, restricts to degree 0 divisors, and takes principal divisors to principal divisors, giving a modified diagram that we will use in the next section,

$$
\begin{array}{ccc}
\mathrm{Div}^0(S_1(N)'_{\mathrm{gd}}) & \longrightarrow & \mathrm{Pic}^0(X_1(N)) \\
\downarrow & & \downarrow \\
\mathrm{Div}^0(\widetilde{S}_1(N)') & \longrightarrow & \mathrm{Pic}^0(\widetilde{X}_1(N)).
\end{array}
\tag{8.28}
$$

In this diagram the maps down the left side and across the bottom surject. The map across the bottom surjects by Proposition 7.3.1 since $\tilde{\psi}_1(\widetilde{S}_1(N)')$ is all but finitely many points of $\widetilde{X}_1(N)$. Although Chapter 7 was set in characteristic 0, the proof of the proposition holds in any characteristic.

Starting instead from a point Q as above and from the minimal polynomial $\varphi_{0,N}$ of $\sum_{d=1}^{N-1} x([d]Q)$, a similar discussion applies to reducing $X_0(N)$. As usual, we will discuss $X_1(N)$ but then use $X_0(N)$ in connection with the Modularity Theorem. For more on Igusa's Theorem see [Igu59], [DR73], [KM85].

Exercise

8.6.1. Show that a diagram as in Igusa's Theorem but with the planar models of $X_1(N)$ and $\widetilde{X}_1(N)$ commutes. (A hint for this exercise is at the end of the book.)

8.7 The Eichler–Shimura Relation

Let N be a positive integer and let $p \nmid N$ be prime. This section gives a description of the Hecke operator T_p at the level of Picard groups of reduced modular curves,

$$\widetilde{T}_p : \mathrm{Pic}^0(\widetilde{X}_1(N)) \longrightarrow \mathrm{Pic}^0(\widetilde{X}_1(N)).$$

By Theorem 8.5.9 the Hecke operator $\langle d \rangle$ on $X_1(N)$ reduces modulo p and passes to Picard groups to give a commutative diagram

$$
\begin{array}{ccc}
\mathrm{Pic}^0(X_1(N)) & \xrightarrow{\langle d \rangle_*} & \mathrm{Pic}^0(X_1(N)) \\
\downarrow & & \downarrow \\
\mathrm{Pic}^0(\widetilde{X}_1(N)) & \xrightarrow{\widetilde{\langle d \rangle}_*} & \mathrm{Pic}^0(\widetilde{X}_1(N)).
\end{array}
\tag{8.29}
$$

We want a similar diagram for the Hecke operator T_p on $\mathrm{Pic}^0(X_1(N))$,

$$
\begin{array}{ccc}
\mathrm{Pic}^0(X_1(N)) & \xrightarrow{T_p} & \mathrm{Pic}^0(X_1(N)) \\
\downarrow & & \downarrow \\
\mathrm{Pic}^0(\widetilde{X}_1(N)) & \xrightarrow{\widetilde{T}_p} & \mathrm{Pic}^0(\widetilde{X}_1(N)).
\end{array}
\tag{8.30}
$$

However, since the top row here is not the pushforward of a morphism from $X_1(N)$ to $X_1(N)$, Theorem 8.5.9 doesn't give this diagram as it gives (8.29). Recall from Section 6.3 and from Exercise 7.9.3 that in fact the top row is a pullback followed by a pushforward via an intermediate curve, either $X_1^0(N, p)$ or $X_{1,0}(N, p)$. These curves do not have good reduction at p, but each is defined by a system of equations over $\mathbb{Z}$ whose reduction at p is isomorphic to two copies of $\widetilde{X}_1(N)$ crossing at the values of j where E_j is supersingular. This structure makes it possible to show that T_p reduces and even to compute its reduction. Alternatively, results on Abelian varieties (e.g., Theorem 9.5.1 of [BLR90]) also show that T_p reduces. Both of these approaches are beyond the scope of this book, so instead we assert without proof that T_p reduces, i.e., that a commutative diagram (8.30) exists, and then granting that there is a reduction of T_p, we will compute it. The resulting description of $\widetilde{T}_p$ is called the Eichler–Shimura Relation.

The way to compute the reduction $\widetilde{T}_p$ on $\mathrm{Pic}^0(\widetilde{X}_1(N))$ is to compute it first in the moduli space environment. We will establish a moduli space diagram analogous to (8.30) and then use Igusa's Theorem to translate it to modular curves. Recall the moduli space interpretation of the Hecke operator T_p on the divisor group of $S_1(N)$,

$$T_p[E, Q] = \sum_C [E/C, Q + C]. \tag{8.31}$$

The sum is taken over all order p subgroups $C \subset E$ such that $C \cap \langle Q \rangle = \{0\}$, in this case all order p subgroups since $p \nmid N$. Let $\mathfrak{p} \subset \overline{\mathbb{Z}}$ be a maximal ideal lying over p. By Proposition 8.4.4(b), if the curve E on the left side of (8.31) has ordinary reduction at $\mathfrak{p}$ then so do all the curves E/C on the right side. Restricting to such curves, this section begins by describing the right side of (8.31) over $\overline{\mathbb{F}}_p$ rather than over $\overline{\mathbb{Q}}$, the description being in terms of the Frobenius map. The same description will hold for curves with supersingular reduction at $\mathfrak{p}$.

Let E be an elliptic curve over $\overline{\mathbb{Q}}$ with ordinary reduction at $\mathfrak{p}$ and let $Q \in E$ be a point of order N. Let C_0 be the kernel of the reduction map $E[p] \longrightarrow \tilde{E}[p]$, an order p subgroup of E since the map surjects and the reduction is ordinary.

Lemma 8.7.1. *For any order p subgroup C of E,*

$$[\widetilde{E/C}, \widetilde{Q+C}] = \begin{cases} [\tilde{E}^{\sigma_p}, \tilde{Q}^{\sigma_p}] & \text{if } C = C_0, \\ [\tilde{E}^{\sigma_p^{-1}}, [p]\tilde{Q}^{\sigma_p^{-1}}] & \text{if } C \neq C_0. \end{cases}$$

Proof. First suppose $C = C_0$. Let $E' = E/C$ and let $Q' = Q + C = \varphi(Q)$ where $\varphi : E \longrightarrow E'$ is the quotient isogeny. Let $\psi : E' \longrightarrow E$ be the dual isogeny of φ and consider the commutative diagram

$$
\begin{array}{ccc}
E'[p] & \xrightarrow{\psi} & E[p] \\
\downarrow & & \downarrow \\
\tilde{E}'[p] & \xrightarrow{\tilde{\psi}} & \tilde{E}[p].
\end{array}
$$

Since E has ordinary reduction at $\mathfrak{p}$ so does its isogenous image E', and thus $|\tilde{E}'[p]| = p$. The subgroup $\psi(E'[p])$ of $E[p]$ has order p since $|\ker(\psi)| = \deg(\psi) = p$, it is a subgroup of $C = \ker(\varphi)$ since $\varphi(\psi(E'[p])) = [p](E'[p]) = \{0\}$, and so it is all of C since both have order p. Thus $\psi(E'[p]) = C = C_0$ is the kernel of the right vertical arrow, and since the left vertical arrow surjects, the diagram shows that the bottom arrow is the zero map, i.e., $\tilde{E}'[p] \subset \ker(\tilde{\psi})$. On the other hand, the relation $\tilde{\varphi} \circ \tilde{\psi} = [p]_{\tilde{E}'}$ shows that $\ker(\tilde{\psi}) \subset \ker([p]) = \tilde{E}'[p]$. In sum, $\ker(\tilde{\psi})$ is $\tilde{E}'[p]$, a group of order p.

By Theorem 8.5.10(d), reduction at $\mathfrak{p}$ preserves the degrees of the three maps in the relation $[p] = \psi \circ \varphi$,

$$\deg([p]_{\tilde{E}'}) = p^2, \qquad \deg(\tilde{\varphi}) = p, \qquad \deg(\tilde{\psi}) = p.$$

Since $\ker([p]_{\tilde{E}'}) = \tilde{E}'[p] = \ker(\tilde{\psi})$ has order p, since $\deg_{\text{sep}} = |\ker|$ for isogenies as discussed in Section 8.2, and since the total degree is the product of the separable and inseparable degrees, this implies

$$\deg_{\text{sep}}([p]_{\tilde{E}'}) = p, \qquad \deg_{\text{ins}}([p]_{\tilde{E}'}) = p$$

and

$$\deg_{\text{sep}}(\tilde{\psi}) = p, \qquad \deg_{\text{ins}}(\tilde{\psi}) = 1.$$

Also, $\tilde{\varphi} \circ \tilde{\psi} = [p]_{\widetilde{E'}}$ by Theorem 8.5.10(b), and so since separable and inseparable degrees are multiplicative,

$$\deg_{\mathrm{sep}}(\tilde{\varphi}) = 1, \qquad \deg_{\mathrm{ins}}(\tilde{\varphi}) = p.$$

The general factorization $\tilde{\varphi} = \tilde{\varphi}_{\mathrm{sep}} \circ \sigma_p^e$ where $\deg(\tilde{\varphi}_{\mathrm{sep}}) = \deg_{\mathrm{sep}}(\tilde{\varphi})$ and
$p^e = \deg_{\mathrm{ins}}(\tilde{\varphi})$ specializes in this case to

$$\tilde{\varphi} = i \circ \sigma_p$$

where $i : \tilde{E}^{\sigma_p} \longrightarrow \widetilde{E'}$ is an isomorphism taking $\tilde{Q}^{\sigma_p}$ to $\widetilde{Q'}$. The isomorphism
gives an equivalence of enhanced elliptic curves so that

$$[\widetilde{E'}, \widetilde{Q'}] = [\tilde{E}^{\sigma_p}, \tilde{Q}^{\sigma_p}],$$

proving the first case of the lemma.

Now suppose $C \neq C_0$. Again let $E' = E/C$, let $Q' = Q + C = \varphi(Q)$ where
φ is the quotient isogeny, and let ψ be the dual isogeny of φ. Analogously
to $C = \ker(\varphi)$ and C_0, let $C' = \ker(\psi)$ and let C_0' be the kernel of the reduction
map $E'[p] \longrightarrow \widetilde{E'}[p]$, an order p subgroup of $E'[p]$. This time consider the
commutative diagram

$$\begin{array}{ccc} E[p] & \xrightarrow{\ \varphi\ } & E'[p] \\ \downarrow & & \downarrow \\ \tilde{E}[p] & \xrightarrow{\ \tilde{\varphi}\ } & \widetilde{E'}[p] \end{array}$$

and consider the subgroup $\varphi(C_0)$ of $E'[p]$. This is an order p subgroup since
$C_0 \neq C = \ker(\varphi)$. It is a subgroup of $C' = \ker(\psi)$ since $\psi(\varphi(C_0)) = [p](C_0) =$
$\{0\}$, making $\varphi(C_0)$ all of C' since both have order p. Also $\varphi(C_0)$ is a subgroup
of C_0' since the diagram commutes and C_0 is the kernel of the left arrow,
making $\varphi(C_0)$ all of C_0'. Thus $C' = C_0'$. Now the argument for the first case
applies to E' and Q' and ψ (which takes Q' to $[p]Q$) in place of E and Q
and φ, showing that $\tilde{\psi} = i \circ \sigma_p$ where $i : \widetilde{E'}^{\sigma_p} \longrightarrow \tilde{E}$ is an isomorphism
taking $\widetilde{Q'}^{\sigma_p}$ to $[p]\tilde{Q}$. Applying σ_p^{-1} to the coefficients of i gives an isomorphism
$i^{\sigma_p^{-1}} : \widetilde{E'} \longrightarrow \tilde{E}^{\sigma_p^{-1}}$ taking $\widetilde{Q'}$ to $[p]\tilde{Q}^{\sigma_p^{-1}}$. That is, it gives an equivalence of
enhanced elliptic curves so that

$$[\widetilde{E'}, \widetilde{Q'}] = [\tilde{E}^{\sigma_p^{-1}}, [p]\tilde{Q}^{\sigma_p^{-1}}],$$

proving the second case of the lemma. $\qquad\qquad\qquad\qquad\qquad\qquad \square$

There are $p + 1$ order p subgroups C of E, one of which is C_0. Define the
moduli space diamond operator in characteristic p to be

$$\langle d \rangle : \tilde{S}_1(N) \longrightarrow \tilde{S}_1(N), \qquad [E, Q] \mapsto [E, [d]Q], \quad (d, N) = 1.$$

Then summing the formula of Lemma 8.7.1 over all order p subgroups $C \subset E$ gives for a curve E with ordinary reduction at $\mathfrak{p}$,

$$\sum_C [\widetilde{E/C}, \widetilde{Q+C}] = (\sigma_p + p\langle\widetilde{p}\rangle\sigma_p^{-1})[\widetilde{E}, \widetilde{Q}]. \tag{8.32}$$

Lemma 8.7.1 extends to supersingular curves, as follows. If E is an elliptic curve over $\overline{\mathbb{Q}}$ with supersingular reduction at $\mathfrak{p}$ and $Q \in E$ is a point of order N, then for any order p subgroup C of E (Exercise 8.7.1),

$$[\widetilde{E/C}, \widetilde{Q+C}] = [\widetilde{E}^{\sigma_p}, \widetilde{Q}^{\sigma_p}] = [\widetilde{E}^{\sigma_p^{-1}}, [p]\widetilde{Q}^{\sigma_p^{-1}}].$$

Summing this over the $p+1$ such subgroups C of E, using the first expression for $[\widetilde{E/C}, \widetilde{Q+C}]$ once and the second expression p times, shows that formula (8.32) applies to curves with supersingular reduction at $\mathfrak{p}$ as well. Therefore it applies to all curves with good reduction at $\mathfrak{p}$.

If an elliptic curve $\widetilde{E}$ over $\overline{\mathbb{F}}_p$ has invariant $j \notin \{0, 1728\}$ then the same holds for $\widetilde{E}^{\sigma_p}$ and $\widetilde{E}^{\sigma_p^{-1}}$. Formulas (8.31) and (8.32) combine with this observation to give a commutative diagram, where as before the primes mean to avoid some points, only finitely many in characteristic p,

$$
\begin{array}{ccc}
S_1(N)'_{\mathrm{gd}} & \xrightarrow{\;\;T_p\;\;} & \mathrm{Div}(S_1(N)'_{\mathrm{gd}}) \\
\downarrow & & \downarrow \\
\widetilde{S}_1(N)' & \xrightarrow{\;\sigma_p + p\langle\widetilde{p}\rangle\sigma_p^{-1}\;} & \mathrm{Div}(\widetilde{S}_1(N)').
\end{array}
$$

That is, T_p on $S_1(N)'_{\mathrm{gd}}$ reduces at $\mathfrak{p}$ to $\sigma_p + p\langle\widetilde{p}\rangle\sigma_p^{-1}$. This extends to divisor groups and then restricts to degree-0 divisors,

$$
\begin{array}{ccc}
\mathrm{Div}^0(S_1(N)'_{\mathrm{gd}}) & \xrightarrow{\;\;T_p\;\;} & \mathrm{Div}^0(S_1(N)'_{\mathrm{gd}}) \\
\downarrow & & \downarrow \\
\mathrm{Div}^0(\widetilde{S}_1(N)') & \xrightarrow{\;\sigma_p + p\langle\widetilde{p}\rangle\sigma_p^{-1}\;} & \mathrm{Div}^0(\widetilde{S}_1(N)').
\end{array}
\tag{8.33}
$$

We also have a commutative diagram (Exercise 8.7.2)

$$
\begin{array}{ccc}
\mathrm{Div}^0(\widetilde{S}_1(N)') & \xrightarrow{\;\sigma_p + p\langle\widetilde{p}\rangle\sigma_p^{-1}\;} & \mathrm{Div}^0(\widetilde{S}_1(N)') \\
\downarrow & & \downarrow \\
\mathrm{Pic}^0(\widetilde{X}_1(N)) & \xrightarrow{\;\sigma_{p,*} + \langle\widetilde{p}\rangle_*\sigma_p^*\;} & \mathrm{Pic}^0(\widetilde{X}_1(N)).
\end{array}
\tag{8.34}
$$

A cube-shaped diagram now exists as follows, where all squares except possibly the back one commute.

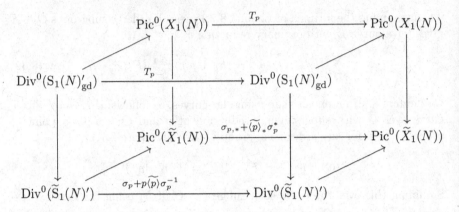

To establish this, note that

- The top square, a diagram in characteristic 0 relating T_p on the moduli space and on the modular curve, is a restriction of commutative diagram (7.18) (page 308).
- The side squares, relating reduction at p to the map from the moduli space to the modular curve, are commutative diagram (8.28) from Igusa's Theorem.
- The front square, relating T_p on the moduli space to reduction at p, is commutative diagram (8.33) resulting from Lemma 8.7.1.
- The bottom square, relating maps in characteristic p on the moduli space and on the modular curve, is commutative diagram (8.34).

Let σ temporarily denote $\sigma_{p,*} + \widetilde{\langle p \rangle}_* \sigma_p^*$ on $\mathrm{Pic}^0(\widetilde{X}_1(N))$. Consider the map from the top front left corner of the cube down, then into the page, and then across,

$$\mathrm{Div}^0(S_1(N)'_{\mathrm{gd}}) \longrightarrow \mathrm{Div}^0(\widetilde{S}_1(N)') \longrightarrow \mathrm{Pic}^0(\widetilde{X}_1(N)) \xrightarrow{\ \sigma\ } \mathrm{Pic}^0(\widetilde{X}_1(N)).$$

$$(8.35)$$

The composite of the first two stages surjects as explained after diagram (8.28). Since the left square of the cube commutes, (8.35) is

$$\mathrm{Div}^0(S_1(N)'_{\mathrm{gd}}) \longrightarrow \mathrm{Pic}^0(X_1(N)) \longrightarrow \mathrm{Pic}^0(\widetilde{X}_1(N)) \xrightarrow{\ \sigma\ } \mathrm{Pic}^0(\widetilde{X}_1(N)),$$

$$(8.36)$$

where again the composite of the first two stages surjects. But since the bottom, front, right, and top squares of the cube commute, (8.35) also becomes

$$\mathrm{Div}^0(S_1(N)'_{\mathrm{gd}}) \longrightarrow \mathrm{Pic}^0(X_1(N)) \xrightarrow{\ T_p\ } \mathrm{Pic}^0(X_1(N)) \longrightarrow \mathrm{Pic}^0(\widetilde{X}_1(N)).$$

Diagram (8.30) shows that a map $\widetilde{T}_p : \mathrm{Pic}^0(\widetilde{X}_1(N)) \longrightarrow \mathrm{Pic}^0(\widetilde{X}_1(N))$ makes the back square of the cube commute. This changes the previous composite to

$$\text{Div}^0(\text{S}_1(N)'_{\text{gd}}) \longrightarrow \text{Pic}^0(X_1(N)) \longrightarrow \text{Pic}^0(\widetilde{X}_1(N)) \xrightarrow{\widetilde{T}_p} \text{Pic}^0(\widetilde{X}_1(N)).$$

$$(8.37)$$

Since (8.36) and (8.37) are equal, and their shared composite of the first two stages surjects, their respective maps $\sigma = \sigma_{p,*} + \widetilde{\langle p \rangle}_* \sigma_p^*$ and $\widetilde{T}_p$ at the last stage are equal. That is, $\sigma_{p,*} + \widetilde{\langle p \rangle}_* \sigma_p^*$ is the map that makes the back square of the cube commute. This gives

Theorem 8.7.2 (Eichler–Shimura Relation). *Let $p \nmid N$. The following diagram commutes:*

$$
\begin{array}{ccc}
\text{Pic}^0(X_1(N)) & \xrightarrow{\quad T_p \quad} & \text{Pic}^0(X_1(N)) \\
\downarrow & & \downarrow \\
\text{Pic}^0(\widetilde{X}_1(N)) & \xrightarrow{\sigma_{p,*} + \widetilde{\langle p \rangle}_* \sigma_p^*} & \text{Pic}^0(\widetilde{X}_1(N)).
\end{array}
$$

In particular since $\widetilde{\langle p \rangle}$ acts trivially on $\widetilde{X}_0(N)$, the following diagram commutes as well:

$$
\begin{array}{ccc}
\text{Pic}^0(X_0(N)) & \xrightarrow{\quad T_p \quad} & \text{Pic}^0(X_0(N)) \\
\downarrow & & \downarrow \\
\text{Pic}^0(\widetilde{X}_0(N)) & \xrightarrow{\sigma_{p,*} + \sigma_p^*} & \text{Pic}^0(\widetilde{X}_0(N)).
\end{array}
$$

$$(8.38)$$

The Hecke operator $\widetilde{T}_p$ in characteristic p is now described in terms of the Frobenius map σ_p, as desired.

Exercises

8.7.1. Let E be an elliptic curve over $\overline{\mathbb{Q}}$ with supersingular reduction at $\mathfrak{p}$ and let $Q \in E$ be a point of order N. Extend Lemma 8.7.1 to show that for any order p subgroup C of E,

$$[\widetilde{E/C}, \widetilde{Q + C}] = [\widetilde{E}^{\sigma_p}, \widetilde{Q}^{\sigma_p}] = [\widetilde{E}^{\sigma_p^{-1}}, [p]\widetilde{Q}^{\sigma_p^{-1}}].$$

(A hint for this exercise is at the end of the book.)

8.7.2. This exercise shows that diagram (8.34), the bottom square of the cube in the section, commutes.

(a) Show that the following diagrams commute, where the map down the sides is described in Theorem 8.6.1:

$$
\begin{array}{ccc}
\text{Div}^0(\widetilde{S}_1(N)') & \xrightarrow{\quad \sigma_p \quad} & \text{Div}^0(\widetilde{S}_1(N)') \\
\downarrow & & \downarrow \\
\text{Div}^0(\widetilde{X}_1(N)^{\text{planar}}) & \xrightarrow{\sigma_{p,*}} & \text{Div}^0(\widetilde{X}_1(N)^{\text{planar}}),
\end{array}
$$

$$\mathrm{Div}^0(\widetilde{S}_1(N)') \xrightarrow{p\sigma_p^{-1}} \mathrm{Div}^0(\widetilde{S}_1(N)')$$
$$\downarrow \qquad\qquad\qquad \downarrow$$
$$\mathrm{Div}^0(\widetilde{X}_1(N)^{\mathrm{planar}}) \xrightarrow{\sigma_p^*} \mathrm{Div}^0(\widetilde{X}_1(N)^{\mathrm{planar}}).$$

(Hints for this exercise are at the end of the book.)

(b) Use Theorem 8.6.1 to show that the diagrams as in part (a) but with $\mathrm{Pic}^0(\widetilde{X}_1(N))$ in place of $\mathrm{Div}^0(\widetilde{X}_1(N)^{\mathrm{planar}})$ on the bottom row commute.

(c) Recall as a special case of Exercise 1.5.3 and Section 5.2 that the moduli space interpretation of the diamond operator $\langle d \rangle$ is

$$\langle d \rangle : S_1(N) \longrightarrow S_1(N), \qquad [E, Q] \mapsto [E, [d]Q].$$

The section defined the moduli space diamond operator in characteristic p by the same formula,

$$\widetilde{\langle d \rangle} : \widetilde{S}_1(N) \longrightarrow \widetilde{S}_1(N), \qquad [E, Q] \mapsto [E, [d]Q].$$

This immediately gives a commutative diagram relating $\langle d \rangle$ on the moduli space to reduction, and the diagram extends to degree 0 divisors,

$$\mathrm{Div}^0(S_1(N)'_{\mathrm{gd}}) \xrightarrow{\langle d \rangle} \mathrm{Div}^0(S_1(N)'_{\mathrm{gd}})$$
$$\downarrow \qquad\qquad\qquad \downarrow$$
$$\mathrm{Div}^0(\widetilde{S}_1(N)') \xrightarrow{\widetilde{\langle d \rangle}} \mathrm{Div}^0(\widetilde{S}_1(N)').$$

Show that the diamond operator in characteristic p on the moduli space and on the modular curve (cf. diagram (8.29)) are compatible in that the following diagram commutes:

$$\mathrm{Div}^0(\widetilde{S}_1(N)') \xrightarrow{\widetilde{\langle d \rangle}} \mathrm{Div}^0(\widetilde{S}_1(N)')$$
$$\downarrow \qquad\qquad\qquad \downarrow \qquad\qquad (8.39)$$
$$\mathrm{Pic}^0(\widetilde{X}_1(N)) \xrightarrow{\widetilde{\langle d \rangle}_*} \mathrm{Pic}^0(\widetilde{X}_1(N)).$$

(d) Use part (b) and diagram (8.39) to complete the proof that diagram (8.34) commutes.

8.8 Fourier coefficients, L-functions, and Modularity

Recall that if E is an elliptic curve over $\mathbb{Q}$ then its analytic conductor was defined in Section 7.7 as the smallest N such that $X_0(N)$ surjects to E, and its algebraic conductor N_E was described in Section 8.3. Both conductors depend only on the isogeny class over $\mathbb{Q}$ of E. In this section we simply call the algebraic conductor the conductor en route to explaining why the two conductors are in fact equal and no longer need to be distinguished.

Theorem 8.8.1 (Modularity Theorem, Version a_p). *Let E be an elliptic curve over $\mathbb{Q}$ with conductor N_E. Then for some newform $f \in S_2(\Gamma_0(N_E))$,*

$$a_p(f) = a_p(E) \quad \text{for all primes } p.$$

This version of Modularity is most obviously related to Version $A_{\mathbb{Q}}$, the version that provides a map $A'_f \longrightarrow E$, since each version involves a new-form f, and unsurprisingly the two f's are the same. But since A'_f is a variety rather than a curve, and our policy is to argue using only curves, we give instead a partial proof that Version $X_{\mathbb{Q}}$, providing a map $X_0(N) \longrightarrow E$, im-plies this version. The argument necessarily requires a little effort to extract f from $X_0(N)$ in consequence of avoiding varieties. Specifically, we prove

Theorem 8.8.2. *Let E be an elliptic curve over $\mathbb{Q}$ with conductor N_E, let N be a positive integer, and let*

$$\alpha : X_0(N) \longrightarrow E$$

be a nonzero morphism over $\mathbb{Q}$ of curves over $\mathbb{Q}$. Then for some newform $f \in S_2(\Gamma_0(M_f))$ where $M_f \mid N$,

$$a_p(f) = a_p(E) \quad \text{for all primes } p \nmid N_E N.$$

After proving this we will touch on the rest of the argument that Ver-sion $X_{\mathbb{Q}}$ of Modularity implies Version a_p, and on the more natural-seeming argument starting from Version $A_{\mathbb{Q}}$.

Proof. For any $p \nmid N_E N$ the route from $a_p(f)$ for some f to $a_p(E)$ is that

- $a_p(f)$ on A'_f is T_p for each f, by a variant of diagram (6.15), and a sum of factors A'_f over all f is isogenous to $\mathrm{Pic}^0(X_0(N))$, by a variant of The-orem 6.6.6,
- T_p on $\mathrm{Pic}^0(X_0(N))$ reduces to $\sigma_{p,*} + \sigma_p^*$ on $\mathrm{Pic}^0(\widetilde{X}_0(N))$, by the Eichler–Shimura Relation ($X_0(N)$ has good reduction at p because $p \nmid N$),
- $\sigma_{p,*} + \sigma_p^*$ on $\mathrm{Pic}^0(\widetilde{X}_0(N))$ commutes with $\tilde{\alpha}_*$ to become $\sigma_{p,*} + \sigma_p^*$ on $\mathrm{Pic}^0(\widetilde{E})$, by formulas (8.17) and (8.19) (E has good reduction at p because $p \nmid N_E$),
- and finally $\sigma_{p,*} + \sigma_p^*$ on $\mathrm{Pic}^0(\widetilde{E})$ is $a_p(E)$, by Proposition 8.3.2.

Recall some ideas from Section 6.6, given there for the group $\Gamma_1(N)$ and now modified for $\Gamma_0(N)$. The complex vector space $S_2(\Gamma_0(N))$ has basis

$$\mathcal{B}'_2(N) = \bigcup_f \bigcup_n \bigcup_\sigma f^\sigma(n\tau)$$

where the first union is taken over equivalence class representatives of new-forms $f \in \mathcal{S}_2(\Gamma_0(M_f))$ with M_f dividing N, the second over divisors n of N/M_f, and the third over embeddings $\sigma : \mathbb{K}_f \hookrightarrow \mathbb{C}$. Work over $\mathbb{C}$ now, identifying complex algebraic curves and Riemann surfaces, and identifying Picard groups and Jacobians. Then the Picard group associated to $\Gamma_0(N)$ is isogenous to a direct sum of Abelian varieties, both sides being viewed as complex tori,

$$\mathrm{Pic}^0(X_0(N)_{\mathbb{C}}) \longrightarrow \bigoplus_{f,n} A'_{f,\mathbb{C}},$$

and there exists a dual isogeny

$$\bigoplus_{f,n} A'_{f,\mathbb{C}} \longrightarrow \mathrm{Pic}^0(X_0(N)_{\mathbb{C}}).$$

The given map $\alpha : X_0(N) \longrightarrow E$ extends to

$$\alpha_{\mathbb{C}} : X_0(N)_{\mathbb{C}} \longrightarrow E_{\mathbb{C}},$$

viewed as a morphism of complex algebraic curves or as a holomorphic map of compact Riemann surfaces, a surjection in either case.

For any $p \nmid N_E N$ the diagram

$$
\begin{array}{ccc}
\displaystyle\bigoplus_{f,n} A'_{f,\mathbb{C}} & \xrightarrow{\;\Pi_{f,n}(a_p(f)-a_p(E))\;} & \displaystyle\bigoplus_{f,n} A'_{f,\mathbb{C}} \\
\big\downarrow & & \big\downarrow \\
\mathrm{Pic}^0(X_0(N)_{\mathbb{C}}) & \xrightarrow{\;T_p-a_p(E)\;} \mathrm{Pic}^0(X_0(N)_{\mathbb{C}}) & \xrightarrow{\;\alpha_{\mathbb{C},*}\;} \mathrm{Pic}^0(E_{\mathbb{C}})
\end{array}
\tag{8.40}
$$

has the following properties, to be proved in a moment:

(a) On the top row, if $a_p(f) \neq a_p(E)$ for some f then the corresponding restriction $\bigoplus_n A'_{f,\mathbb{C}} \longrightarrow \bigoplus_n A'_{f,\mathbb{C}}$ surjects.
(b) The square commutes.
(c) The composite map across the bottom row is zero.

Now, if $a_p(f) \neq a_p(E)$ for some f and some $p \nmid N_E N$ then in the diagram for that p, mapping $\bigoplus_n A'_{f,\mathbb{C}}$ across the top row takes it to all of $\bigoplus_n A'_{f,\mathbb{C}}$ by property (a), and then mapping this down gives its isogenous image in $\mathrm{Pic}^0(X_0(N)_{\mathbb{C}})$. By property (b) the isogenous image also comes from mapping $\bigoplus_n A'_{f,\mathbb{C}}$ down the left side of the diagram and then halfway across the bottom row. Property (c) now shows that the isogenous image of $\bigoplus_n A'_{f,\mathbb{C}}$ lies in $\ker(\alpha_{\mathbb{C},*})$. Therefore, if for each f there exists a $p \nmid N_E N$ such that $a_p(f) \neq a_p(E)$ then all of $\mathrm{Pic}^0(X_0(N)_{\mathbb{C}})$ lies in $\ker(\alpha_{\mathbb{C},*})$. But $\alpha_{\mathbb{C},*}$ surjects, so this is impossible. That is, there is a newform f such that $a_p(f) = a_p(E)$ for all $p \nmid N_E N$, as we needed to prove.

Returning to diagram (8.40), to prove its property (a) let $a_p(f) \neq a_p(E)$. Recall that $a_p(f)$ is an algebraic integer, and $a_p(E)$ is a rational integer. Thus their difference δ satisfies a minimal monic polynomial with rational integer coefficients,

$$\delta^e + a_1\delta^{e-1} + \cdots + a_{e-1}\delta + a_e = 0, \quad a_1, \ldots, a_e \in \mathbb{Z}, \ a_e \neq 0.$$

The resulting relation $\delta(\delta^{e-1} + a_1\delta^{e-2} + \cdots + a_{e-1}) = -a_e$ shows that δ is surjective on $A'_{f,\mathbb{C}}$ as desired, since $-a_e$ is.

Property (b) follows quickly from diagram (6.20) suitably modified from $\Gamma_1(N)$ to $\Gamma_0(N)$.

To prove property (c) switch back to working over $\mathbb{Q}$ and consider the following diagram:

$$
\begin{array}{ccccc}
\mathrm{Pic}^0(X_0(N)) & \xrightarrow{T_p-a_p(E)} & \mathrm{Pic}^0(X_0(N)) & \xrightarrow{\ \alpha_*\ } & \mathrm{Pic}^0(E) \\
\downarrow & & \downarrow & & \downarrow \\
\mathrm{Pic}^0(\widetilde{X}_0(N)) & \xrightarrow{\sigma_{p,*}+\sigma_p^*-a_p(E)} & \mathrm{Pic}^0(\widetilde{X}_0(N)) & \xrightarrow{\ \tilde\alpha_*\ } & \mathrm{Pic}^0(\widetilde{E}) \\
1\downarrow & & & & \downarrow 1 \\
\mathrm{Pic}^0(\widetilde{X}_0(N)) & \xrightarrow{\ \tilde\alpha_*\ } & \mathrm{Pic}^0(\widetilde{E}) & \xrightarrow{\sigma_{p,*}+\sigma_p^*-a_p(E)} & \mathrm{Pic}^0(\widetilde{E}).
\end{array}
\qquad (8.41)
$$

The top row here is a restriction of the bottom row of diagram (8.40), and it suffices to prove that this restriction doesn't surject (Exercise 8.8.1). The top left square commutes since it is essentially (8.38) from the Eichler–Shimura Relation, and the top right square commutes by Corollary 8.5.9. The bottom rectangle commutes because formulas (8.17) and (8.19) from earlier in the chapter show that $\sigma_{p,*}$ and σ_p^* commute with $\tilde\alpha_*$, and multiplication by $a_p(E)$ certainly commutes with the linear map $\tilde\alpha_*$ as well. The second map on the bottom row is zero by Proposition 8.3.2, making the bottom row zero and thus making the middle row zero. So the top row followed by the right vertical arrow to the second row is zero. But the vertical arrow surjects by Theorem 8.5.9. This shows that the top row can't surject, as desired. With properties (a), (b), and (c) of diagram (8.40) established, the proof of Theorem 8.8.2 is complete. $\qquad \square$

For Version $X_{\mathbb{Q}}$ of Modularity to imply Version a_p, the newform f of Theorem 8.8.2 needs to have level $M_f = N_E$ and to have Fourier coefficients $a_p(f) = a_p(E)$ for all p.

Strong Multiplicity One shows that all the Fourier coefficients of f are rational integers. To see this, recall that if σ is any automorphism of $\overline{\mathbb{Q}}$ then the conjugate of f under σ is defined as

$$f^\sigma(\tau) = \sum_{n=1}^{\infty} a_n(f)^\sigma q^n.$$

According to Theorem 6.5.4 all conjugates of f are again newforms. In this case they are all equal to f itself because for all $p \nmid N$ and all σ,

$$a_p(f) = a_p(E) \implies a_p(f) \in \mathbb{Z} \implies a_p(f)^\sigma = a_p(f),$$

so that $f^\sigma = f$ by Strong Multiplicity One. This means that the Fourier coefficients of f are algebraic integers invariant under $\mathrm{Aut}(\overline{\mathbb{Q}})$, i.e., the Fourier coefficients are rational integers as claimed. Thus f has number field $\mathbb{K}_f = \mathbb{Q}$.

Obtaining Version a_p of Modularity from here requires results beyond the scope of this book, but we sketch the ideas. As quoted in Section 7.7, the Abelian variety A_f', viewed as a variety over $\mathbb{Q}$, has dimension $[\mathbb{K}_f : \mathbb{Q}] = 1$, i.e., A_f' is an elliptic curve over $\mathbb{Q}$. The map $X_0(N) \longrightarrow \mathrm{Pic}^0(X_0(N)) \longrightarrow A_f'$ is defined over $\mathbb{Q}$, so we can run the proof of Theorem 8.8.2 again to show that there exists a newform g such that $a_p(g) = a_p(A_f')$ for all but finitely many p. The proof in this case shows that $g = f$, so

$$a_p(f) = a_p(A_f') \quad \text{for all but finitely many } p. \tag{8.42}$$

Carayol [Car86], building on the work of Eichler–Shimura, Langlands, and Deligne, showed that in fact $a_p(f) = a_p(A_f')$ for all p, and the level of f is the conductor of A_f', notated $M_f = N_f$, and this is also the analytic conductor of A_f. But also, the work so far gives an isogeny over $\mathbb{Q}$ from A_f' to E, so that $a_p(A_f') = a_p(E)$ for all p and the conductor of A_f' is the conductor of E, that is, $N_f = N_E$. This gives $a_p(f) = a_p(E)$ for all p and $M_f = N_E$, i.e., Version a_p of Modularity. It also shows that the analytic and algebraic conductors of E are the same. That is, the newform f associated to E has the same level as the smallest modular curve $X_0(N)$ that maps to E.

Starting instead from Version $A_{\mathbb{Q}}$ of Modularity, that is, starting from a map $\alpha : A_f' \longrightarrow E$, a more natural-looking proof that $a_p(f) = a_p(E)$ for all $p \nmid N_E M_f$ (where N_E is the conductor of E and M_f is the level of f) would set up a diagram analogous to (8.41),

$$
\begin{array}{ccccc}
A_f' & \xrightarrow{a_p(f)-a_p(E)} & A_f' & \xrightarrow{\ \alpha\ } & E \\
\downarrow & & \downarrow & & \downarrow \\
\widetilde{A_f'} & \xrightarrow{\sigma_{p,*}+\sigma_p^*-a_p(E)} & \widetilde{A_f'} & \xrightarrow{\ \widetilde{\alpha}\ } & \widetilde{E} \\
{\scriptstyle 1}\downarrow & & & & \downarrow{\scriptstyle 1} \\
\widetilde{A_f'} & \xrightarrow{\ \widetilde{\alpha}\ } & \widetilde{E} & \xrightarrow{\sigma_{p,*}+\sigma_p^*-a_p(E)} & \widetilde{E}.
\end{array}
$$

If $a_p(f) \neq a_p(E)$ then the top row is surjective, as is $E \longrightarrow \widetilde{E}$, but the bottom row is zero and hence the middle row is zero, giving a contradiction because all the rectangles commute. However, this argument makes use of the structure of A'_f as a variety over $\mathbb{Q}$, reduces the variety modulo p, and then makes further use of the structure of the reduction as a variety over $\mathbb{F}_p$, invoking algebraic geometry well beyond the scope of this book. By contrast our proof of Theorem 8.8.2 makes no reference to the variety structure of $\mathrm{Pic}^0(X_0(N))$, and it uses $\mathrm{Pic}^0(\widetilde{X}_0(N))$, the Picard group of the reduced curve, rather than reducing an Abelian variety.

Section 1.3 mentioned that some complex tori have complex multiplication, endomorphisms other than $\{[N] : N \in \mathbb{Z}\}$. This notion extends to algebraic elliptic curves, providing more examples of Modularity. For example the elliptic curve

$$E : y^2 = x^3 - d, \quad d \in \mathbb{Z}, \ d \neq 0$$

has the order 3 automorphism $(x, y) \mapsto (\mu_3 x, y)$ over $\mathbb{Q}(\mu_3)$, and its ring of endomorphisms is isomorphic to $A = \mathbb{Z}[\mu_3]$. As in Exercise 8.3.6(b), $a_p(E) = 0$ for all $p \equiv 2 \pmod 3$ for which the displayed Weierstrass equation is minimal. The theory of complex multiplication in fact describes $a_p(E)$ for all p: there exist an integer $N \mid 12 \prod_{p \mid d} p$ and a character $\chi : (A/NA)^* \longrightarrow \mathbb{C}^*$ of order 6 such that $\chi(u) = u^{-1}$ for $u \in A^*$, $\chi(a) = (a/3)$ (Legendre symbol) for $a \in (\mathbb{Z}/N\mathbb{Z})^*$, and

$$a_p(E) = \tfrac{1}{6} \sum_{\substack{n \in A \\ |n|^2 = p}} \chi(n) n.$$

(See Chapter 2 of [Sil94] for proofs of these results, and see Exercise II.5.7 of [Kob93] for the case $d = -16$, giving $N = 3$.) The solution-counts of E are the Fourier coefficients of the function

$$\theta_{2,\chi}(\tau) = \tfrac{1}{6} \sum_{n \in A} \chi(n) n e^{2\pi i |n|^2 \tau}, \quad \tau \in \mathcal{H}.$$

This function is a modular form of level $3N^2$ by the weight 2 version of Hecke's construction in Section 4.11. It is a normalized eigenform by an argument similar to the one in Section 5.9 for the weight 1 theta function. For the minimal choice of N, it is in fact a newform of level $3N^2 = N_E$, thus illustrating Version a_p of Modularity for these elliptic curves E. The elliptic curves E over $\mathbb{Q}$ with complex multiplication form a small class of examples—for instance they have only a finite set of j-invariants—but they are important in number theory, and Shimura's proof of Version $X_\mathbb{Q}$ of Modularity for such curves [Shi71] provided evidence for the algebraic formulations of Modularity in general.

Version a_p of the Modularity Theorem rephrases in terms of L-functions. Recall from Chapter 5 that if $f \in \mathcal{S}_2(\Gamma_0(N))$ is a newform then its L-function is

$$L(s, f) = \sum_{n=1}^{\infty} a_n(f) n^{-s} = \prod_p (1 - a_p(f) p^{-s} + \mathbf{1}_N(p) p^{1-2s})^{-1}, \qquad (8.43)$$

with convergence in a right half plane. For any elliptic curve E over $\mathbb{Q}$ let $\mathbf{1}_E$ be the trivial character modulo the conductor N of E as in Section 8.3. For any prime p the *local counting zeta-function of E*, encoding the normalized solution-counts $t_{p^e}(E) = p^e + 1 - |\widetilde{E}(\mathbb{F}_{p^e})|$, is

$$Z_p(X, E) = \exp\left(\sum_{e=1}^{\infty} \frac{t_{p^e}(E)}{e} X^e\right).$$

Taking logarithmic derivatives shows (exercise 8.8.2) that in fact for $X = p^{-s}$ the zeta-function takes the form of an Euler factor,

$$Z_p(p^{-s}, E) = (1 - a_p(E)p^{-s} + \mathbf{1}_E(p)p^{1-2s})^{-1}.$$

The fact that $\mathbf{1}_E(p)$ equals 1 at a prime of good reduction and 0 at a prime of bad reduction, and the value of $a_p(E)$ for a prime of bad reduction as given in section 8.3, show that in fact

$$Z_p(p^{-s}, E) = \begin{cases} (1 - a_p(E)p^{-s} + p^{1-2s})^{-1} & \text{if } E \text{ has good reduction at } p, \\ (1 - p^{-s})^{-1} & \text{if } E \text{ has split reduction at } p, \\ (1 + p^{-s})^{-1} & \text{if } E \text{ has nonsplit reduction at } p, \\ 1 & \text{if } E \text{ has additive reduction at } p. \end{cases}$$

The *Hasse-Weil L-function of E* is the product of these Euler factors,

$$L(s, E) = \prod_p (1 - a_p(E)p^{-s} + \mathbf{1}_E(p)p^{1-2s})^{-1}.$$

By the methods of the proof of Theorem 5.9.2, the Dirichlet series form of the L-function is

$$L(s, E) = \sum_{n=1}^{\infty} a_n(E)n^{-s}$$

where similarly to the Fourier coefficients of a newform, the $a_n(E)$ satisfy

$$a_1(E) = 1,$$
$$a_p(E) = p + 1 - |\widetilde{E}(\mathbb{F}_p)|,$$
$$a_{p^e}(E) = a_p(E)a_{p^{e-1}}(E) - \mathbf{1}_E(p)p a_{p^{e-2}}(E), \quad e \geq 2,$$
$$a_{mn}(E) = a_m(E)a_n(E), \quad (m, n) = 1,$$

(The relation between these a-values and the solution-counts t is

$$t_{p^e}(E) = a_{p^e}(E) - \mathbf{1}_E(p)p a_{p^{e-2}}(E), \quad e \geq 2,$$

and it is not hard to invert the relation and express $a_{p^e}(E)$ in terms of t-values as well. A more conceptual connection between the a-values and the t-values will be given in exercise 9.4.2(d).) In sum, as in Chapter 5 the L-function of an elliptic curve is now

$$L(s, E) = \sum_{n=1}^{\infty} a_n(E) n^{-s} = \prod_p (1 - a_p(E) p^{-s} + 1_E(p) p^{1-2s})^{-1}. \qquad (8.44)$$

Half plane convergence of $L(s, E)$ can be established as well by estimating $a_p(E)$. But this is unnecessary since comparing the products in (8.43) and (8.44) shows that Version a_p of Modularity is equivalent to

Theorem 8.8.3 (Modularity Theorem, Version L). *Let E be an elliptic curve over $\mathbb{Q}$ with conductor N_E. Then for some newform $f \in \mathcal{S}_2(\Gamma_0(N_E))$,*

$$L(s, f) = L(s, E).$$

Faltings's Isogeny Theorem (Corollary 5.2 in Chapter 2 of [CS86]), a deep result, now shows that this Version L of Modularity implies

Theorem 8.8.4 (Modularity Theorem, strong Version $A_{\mathbb{Q}}$). *Let E be an elliptic curve over $\mathbb{Q}$ with conductor N_E. Then for some newform $f \in \mathcal{S}_2(\Gamma_0(N_E))$ the Abelian variety A'_f is also an elliptic curve over $\mathbb{Q}$ and there exists an isogeny over $\mathbb{Q}$*

$$A'_f \longrightarrow E.$$

To see this, suppose that by Version L we have $L(s, f) = L(s, E)$. Then f has rational coefficients, making A'_f an elliptic curve. Equation (8.42) shows that $L(s, A'_f)$ and $L(s, f)$ have the same Euler product factors for all but finitely many primes p. So the same is true of $L(s, A'_f)$ and $L(s, E)$, and now Faltings's Theorem gives an isogeny $A'_f \longrightarrow E$. Faltings's Isogeny Theorem will be explained a bit further in Exercise 9.4.3.

Version L of the Modularity Theorem shows that the half plane convergence, analytic continuation, and functional equation of $L(s, f)$ from Theorem 5.10.2 now apply to $L(s, E)$. This is important because the continued $L(s, E)$ is conjectured to contain sophisticated information about the group structure of E. Specifically, since $E(\mathbb{Q})$ is a finitely generated Abelian group it takes the form

$$E(\mathbb{Q}) \cong T \oplus \mathbb{Z}^r,$$

where T is the torsion subgroup and r is the rank. The rank is much harder to compute than the torsion. However,

Conjecture 8.8.5 (Weak Birch and Swinnerton-Dyer Conjecture). *Let E be an elliptic curve defined over $\mathbb{Q}$. Then the order of vanishing of $L(s, E)$ at $s = 1$ is the rank of $E(\mathbb{Q})$. That is, if $E(\mathbb{Q})$ has rank r then*

$$L(s, E) = (s - 1)^r g(s), \quad g(1) \neq 0, \infty.$$

The original half plane of convergence of $L(s, E)$ is $\{\text{Re}(s) > 2\}$, and the functional equation then determines $L(s, E)$ for $\text{Re}(s) < 0$, but the behavior of $L(s, E)$ at the center of the remaining strip $\{0 \leq \text{Re}(s) \leq 2\}$ is what conjecturally determines the rank of $E(\mathbb{Q})$. The Birch and Swinnerton-Dyer Conjecture would give an algorithm for finding all rational points on elliptic curves, and it would give an effective method for finding imaginary quadratic fields with a given class number. For more on the Birch and Swinnerton-Dyer Conjecture see [Tat02], Appendix C.16 of [Sil86], or Chapter 17 of [Hus04].

Exercises

8.8.1. Let $\beta_{\mathbb{C}}$ denote the map of the bottom row of diagram (8.40),

$$\beta_{\mathbb{C}} : \text{Pic}^0(X_0(N)_{\mathbb{C}}) \xrightarrow{\alpha_* \circ (T_p - a_p(E))} \text{Pic}^0(E_{\mathbb{C}}).$$

Augment $\beta_{\mathbb{C}}$ to get a map of complex algebraic curves,

$$\gamma_{\mathbb{C}} : X_0(N) \longrightarrow \text{Pic}^0(X_0(N)_{\mathbb{C}}) \xrightarrow{\beta_{\mathbb{C}}} \text{Pic}^0(E_{\mathbb{C}}) \longrightarrow E_{\mathbb{C}}.$$

Here the first stage is $P \mapsto [(P) - (P_0)]$ where $P_0 \in X_0(N)_{\mathbb{C}}$ is a base point, and the third stage is $[\sum (Q_i)] \mapsto \sum Q_i$. Since $\gamma_{\mathbb{C}}$ is a composite of holomorphic maps, it is holomorphic as a map of Riemann surfaces and therefore it is a morphism as a map of complex algebraic curves.

(a) Show that if $\gamma_{\mathbb{C}}$ is zero then $\beta_{\mathbb{C}}$ is zero. (A hint for this exercise is at the end of the book.)

(b) Assume that the base point P_0 in part (a) has algebraic coordinates. Show that $\gamma_{\mathbb{C}}$ is defined over $\overline{\mathbb{Q}}$ as follows. It suffices to show that $\gamma_{\mathbb{C}}^\sigma = \gamma_{\mathbb{C}}$ for all $\sigma \in \text{Aut}(\mathbb{C}/\overline{\mathbb{Q}})$. Compute that for any $P \in X_0(N)_{\mathbb{C}}$ and any σ,

$$\gamma_{\mathbb{C}}(P^\sigma) = \gamma_{\mathbb{C}}(P)^\sigma = \gamma_{\mathbb{C}}^\sigma(P^\sigma).$$

Since P^σ can be any point of $X_0(N)_{\mathbb{C}}$, this gives the result.

(c) Let β denote the map of the top row of diagram (8.41),

$$\beta : \text{Pic}^0(X_0(N)) \xrightarrow{\alpha_* \circ (T_p - a_p(E))} \text{Pic}^0(E).$$

This is the restriction of $\beta_{\mathbb{C}}$ to $\overline{\mathbb{Q}}$-points. Consider the corresponding restriction of $\gamma_{\mathbb{C}}$, viewed as a morphism over $\overline{\mathbb{Q}}$ of algebraic curves over $\overline{\mathbb{Q}}$ according to part (b),

$$\gamma : X_0(N) \longrightarrow \text{Pic}^0(X_0(N)) \xrightarrow{\beta} \text{Pic}^0(E) \longrightarrow E.$$

Use the maps of curves γ and $\gamma_{\mathbb{C}}$ to show that if β does not surject then $\beta_{\mathbb{C}}$ is zero. This was used in proving property (c) of diagram (8.40). The result is immediate from quoting that as morphisms from varieties to curves, β and $\beta_{\mathbb{C}}$ are both constant or surjective, but the argument in this exercise uses only the algebraic geometry of curves.

8.8.2. This exercise establishes the formal equality

$$\sum_{e=1}^{\infty} \frac{t_{p^e}(E)}{e} X^e = -\log(1 - a_p(E)X + \mathbf{1}_E(p)pX^2).$$

Since both sides are 0 when $X = 0$, it suffices to show that their derivatives are equal,

$$\sum_{e=1}^{\infty} t_{p^e}(E)X^{e-1} = \frac{a_p(E) - \mathbf{1}_E(p)2pX}{1 - a_p(E)X + \mathbf{1}_E(p)pX^2}.$$

Show that this relation is equivalent to the initial value and the recurrence characterizing the $t_{p^e}(E)$-values at the end of Section 8.3.

9

Galois Representations

This book has explained the idea that all elliptic curves over $\mathbb{Q}$ arise from modular forms. Chapters 1 and 2 introduced elliptic curves and modular curves as Riemann surfaces, and Chapter 1 described elliptic curves as algebraic curves over $\mathbb{C}$. As a general principle, information about mathematical objects can be obtained from related algebraic structures. Elliptic curves already form Abelian groups. Modular curves do not, but Chapter 3 showed that the complex vector space of weight 2 cusp forms associated to a modular curve has dimension equal to the genus of the curve, then Chapter 5 defined the Hecke operators, linear operators that act on the vector space, and Chapter 6 showed that integral homology is a lattice in the dual space and is stable under the Hecke action.

As number theorists we are interested in polynomial equations over number fields, in particular elliptic curves over $\mathbb{Q}$. Chapter 7 showed that modular curves are defined over $\mathbb{Q}$ as well. As another general principle, information about equations can be obtained by reducing them modulo primes p. Chapter 8 reduced the equations of elliptic curves and modular curves to obtain similar relations for the two kinds of curve: for an elliptic curve E over $\mathbb{Q}$,

$$a_p(E) = \sigma_{p,*} + \sigma_p^* \qquad \text{as an endomorphism of } \operatorname{Pic}^0(\widetilde{E}),$$

while for the modular curve $X_0(N)$ the Eichler–Shimura Relation is

$$T_p = \sigma_{p,*} + \sigma_p^* \qquad \text{as an endomorphism of } \operatorname{Pic}^0(\widetilde{X}_0(N)).$$

These relations hold for all but finitely many p, and each involves different geometric objects as p varies.

Using more algebraic structure, this last chapter lifts the two relations from characteristic p back to characteristic 0 but in a different setting. For any prime ℓ the ℓ-power torsion groups of an elliptic curve give rise to a vector space $V_\ell(E)$ over the ℓ-adic number field $\mathbb{Q}_\ell$. Similarly, the ℓ-power torsion groups of the Picard group of a modular curve give an ℓ-adic vector

© Springer Science+Business Media New York 2005
F. Diamond, J. Shurman, *A First Course in Modular Forms*,
Graduate Texts in Mathematics 228, DOI 10.1007/978-0-387-27226-9_9

space $V_\ell(X)$. The vector spaces $V_\ell(E)$ and $V_\ell(X)$ are acted on by the *absolute Galois group* of $\mathbb{Q}$, the group $G_\mathbb{Q}$ of automorphisms of the algebraic closure $\overline{\mathbb{Q}}$. This group subsumes the Galois groups of all Galois number fields, and it contains *absolute Frobenius elements* $\mathrm{Frob}_\mathfrak{p}$ for maximal ideals $\mathfrak{p}$ of $\overline{\mathbb{Z}}$ lying over rational primes p. The vector spaces $V_\ell(X)$ are also acted on by the Hecke algebra. The two relations in the previous paragraph lead to the relations

$$\mathrm{Frob}_\mathfrak{p}^2 - a_p(E)\mathrm{Frob}_\mathfrak{p} + p = 0 \qquad \text{as an endomorphism of } V_\ell(E)$$

and

$$\mathrm{Frob}_\mathfrak{p}^2 - T_p\,\mathrm{Frob}_\mathfrak{p} + p = 0 \qquad \text{as an endomorphism of } V_\ell(X_0(N)).$$

These hold for a dense set of elements $\mathrm{Frob}_\mathfrak{p}$ in $G_\mathbb{Q}$, but now each involves a single vector space as $\mathrm{Frob}_\mathfrak{p}$ varies. The second relation connects the Hecke action and the Galois action on the vector spaces associated to modular curves. Remarkably, both relations are independent of the prime ℓ used to construct them.

The vector spaces V_ℓ are *Galois representations*, representation spaces of the group $G_\mathbb{Q}$. The Galois representation $V_\ell(X_0(N))$ associated to a modular curve decomposes into pieces $V_\ell(A_f)$ for the Abelian varieties of eigenforms $f \in \mathcal{S}_2(\Gamma_0(N))$, where $a_p(f)$ acts as T_p, and these decompose further into pieces $V_\lambda(f)$ for primes λ dividing ℓ in the number field generated by the coefficients of f. If the number field is $\mathbb{Q}$ then the only such λ is ℓ, and the second relation in the previous paragraph gives

$$\mathrm{Frob}_\mathfrak{p}^2 - a_p(f)\mathrm{Frob}_\mathfrak{p} + p = 0 \qquad \text{as an endomorphism of } V_\ell(f).$$

The Modularity Theorem in this context is that any Galois representation $V_\ell(E)$ where E is an elliptic curve over $\mathbb{Q}$ arises from some $V_\ell(f)$. This fits into a larger problem, to show that Galois representations from algebraic geometry arise from modular forms.

This chapter provides less background than the rest of the book. It quotes results from algebraic number theory and it uses techniques from algebra without comment. Related reading: Chapter 15 of [Hus04], Chapter III.7 of [Sil86], Section 7.4 of [Shi73]. The volumes [Mur95] and [CSS97] contain lectures on the proof of Fermat's Last Theorem, and [DDT94] is a survey article. Henri Darmon's article in [Mur95] discusses the conjecture of Serre—since proved by Khare and Wintenberger—mentioned at the end of the chapter.

9.1 Galois number fields

Recall from Section 6.4 that a number field is a field $\mathbb{F} \subset \overline{\mathbb{Q}}$ such that the degree $[\mathbb{F} : \mathbb{Q}]$ is finite, and each number field has its ring of algebraic integers $\mathcal{O}_\mathbb{F}$. This chapter will work with number fields $\mathbb{F}$ such that the extension

$\mathbb{F}/\mathbb{Q}$ is Galois. These fields are denoted $\mathbb{F}$ rather than $\mathbb{K}$ to emphasize that they play a different role from other number fields earlier in the book, but the reader is cautioned that $\mathbb{F}_q$ continues to denote a finite field. The purpose of this section is to illustrate some results from algebraic number theory in the Galois case by giving specific examples, without proof, to convey a concrete sense of the ideas before we start using them. The reader without background in algebraic number theory is encouraged to refer to a text on the subject.

Let $\mathbb{F}$ be a Galois number field and let $p \in \mathbb{Z}$ be prime. There are positive integers e, f, and g that describe the ideal $p\mathcal{O}_\mathbb{F}$ as a product of maximal ideals of $\mathcal{O}_\mathbb{F}$,

$$p\mathcal{O}_\mathbb{F} = (\mathfrak{p}_1 \cdots \mathfrak{p}_g)^e, \qquad \mathcal{O}_\mathbb{F}/\mathfrak{p}_i \cong \mathbb{F}_{p^f} \text{ for } i = 1, \ldots, g, \qquad efg = [\mathbb{F} : \mathbb{Q}].$$

The first formula is (8.22) specialized to the Galois case. The *ramification degree* e says how many times each maximal ideal $\mathfrak{p}$ of $\mathcal{O}_\mathbb{F}$ that lies over p repeats as a factor of $p\mathcal{O}_\mathbb{F}$. There are only finitely many p such that $e > 1$, the primes that *ramify in* $\mathbb{F}$. The *residue degree* f is the dimension of the *residue field* $\mathbf{f}_\mathfrak{p} = \mathcal{O}_\mathbb{F}/\mathfrak{p}$ (a finite field) as a vector space over $\mathbb{F}_p = \mathbb{Z}/p\mathbb{Z}$ for any $\mathfrak{p}$ over p. The *decomposition index* g is the number of distinct $\mathfrak{p}$ over p. The condition $efg = [\mathbb{F} : \mathbb{Q}]$ says that the product of the ramification degree, residue degree, and decomposition index associated to each rational prime p is the field extension degree. Equivalently, $efg = |\mathrm{Gal}(\mathbb{F}/\mathbb{Q})|$.

The simplest nontrivial Galois number fields are the quadratic fields. Let

$$\mathbb{F} = \mathbb{Q}(\sqrt{d}), \quad d \in \mathbb{Z} - \{0, 1\} \text{ squarefree.}$$

Then $[\mathbb{F} : \mathbb{Q}] = 2$ and the extension $\mathbb{F}/\mathbb{Q}$ is Galois with its group generated by the automorphism taking $\sqrt{d}$ to $-\sqrt{d}$. The *discriminant* of the field is

$$\Delta_\mathbb{F} = \begin{cases} d & \text{if } d \equiv 1 \pmod 4, \\ 4d & \text{if } d \not\equiv 1 \pmod 4, \end{cases}$$

and the ring of algebraic integers in this field is

$$\mathcal{O}_\mathbb{F} = \mathbb{Z}\left[\frac{\Delta_\mathbb{F} + \sqrt{\Delta_\mathbb{F}}}{2}\right] = \left\{a + b\frac{\Delta_\mathbb{F} + \sqrt{\Delta_\mathbb{F}}}{2} : a, b \in \mathbb{Z}\right\}.$$

This says that $\mathcal{O}_\mathbb{F} = \mathbb{Z}[\frac{1+\sqrt{d}}{2}]$ if $d \equiv 1 \pmod 4$ while $\mathcal{O}_\mathbb{F} = \mathbb{Z}[\sqrt{d}]$ if $d \not\equiv 1 \pmod 4$, but phrasing results in terms of $\Delta_\mathbb{F}$ and thus making no further direct reference to cases is tidier. The Legendre symbol from elementary number theory, (a/p) where $a \in \mathbb{Z}$ and p is an odd prime, extends to incorporate the *Kronecker symbol*, defined only for $a \equiv 0, 1 \pmod 4$,

$$\left(\frac{a}{2}\right) = \begin{cases} 1 & \text{if } a \equiv 1 \pmod 8, \\ -1 & \text{if } a \equiv 5 \pmod 8, \\ 0 & \text{if } a \equiv 0 \pmod 4. \end{cases}$$

This makes the behavior of a rational prime p in $\mathcal{O}_\mathbb{F}$ easy to notate,

$$pO_\mathbb{F} = \begin{cases} \mathfrak{p}\mathfrak{q} & \text{if } (\Delta_\mathbb{F}/p) = 1, \\ \mathfrak{p} & \text{if } (\Delta_\mathbb{F}/p) = -1, \\ \mathfrak{p}^2 & \text{if } (\Delta_\mathbb{F}/p) = 0. \end{cases}$$

The content of this formula is that the odd primes p such that d is a quadratic residue modulo p decompose in $\mathcal{O}_\mathbb{F}$, the odd primes such that d is a nonresidue remain inert, and the odd primes dividing d ramify, while 2 decomposes if $d \equiv 1 \pmod 8$, remains inert if $d \equiv 5 \pmod 8$, and ramifies if $d \not\equiv 1 \pmod 4$. The quadratic reciprocity law from elementary number theory says that the Legendre symbol $(\Delta_\mathbb{F}/p)$ for odd primes p depends only on p modulo $\Delta_\mathbb{F}\mathbb{Z}$; thus so does the decomposition of $pO_\mathbb{F}$.

Another family of simple Galois number fields is the cyclotomic fields. Let N be a positive integer and let

$$\mathbb{F} = \mathbb{Q}(\mu_N), \quad \mu_N = e^{2\pi i/N}.$$

Then $[\mathbb{F} : \mathbb{Q}] = \phi(N)$ (Euler totient) and the extension $\mathbb{F}/\mathbb{Q}$ is Galois with group isomorphic to $(\mathbb{Z}/N\mathbb{Z})^*$,

$$\text{Gal}(\mathbb{F}/\mathbb{Q}) \xrightarrow{\sim} (\mathbb{Z}/N\mathbb{Z})^*, \quad (\mu_N \mapsto \mu_N^a) \longmapsto a \pmod N. \tag{9.1}$$

The cyclotomic integers are

$$\mathcal{O}_\mathbb{F} = \mathbb{Z}[\mu_N] = \{a_0 + a_1\mu_N + \cdots + a_{N-1}\mu_N^{N-1} : a_0, \ldots, a_{N-1} \in \mathbb{Z}\}.$$

(No claim is being made here that the powers of μ_N are independent over $\mathbb{Z}$.) Each rational prime not dividing N is unramified in $\mathbb{F}$,

$$pO_\mathbb{F} = \mathfrak{p}_1 \cdots \mathfrak{p}_g, \quad p \nmid N,$$

and its residue degree f is the order of $p \pmod N$ in $(\mathbb{Z}/N\mathbb{Z})^*$. The primes dividing N ramify in $\mathbb{Q}(\mu_N)$ (except that 2 does not ramify if $N \equiv 2 \pmod 4$, but then $\mathbb{Q}(\mu_N) = \mathbb{Q}(\mu_{N/2})$ and $N/2$ is odd). We do not need a precise description of their behavior since we will focus on the unramified primes.

For the simplest non-Abelian Galois group, let $d > 1$ be a cubefree integer, let $d^{1/3}$ denote the real cube root of d, and let

$$\mathbb{F} = \mathbb{Q}(d^{1/3}, \mu_3).$$

In this case $[\mathbb{F} : \mathbb{Q}] = 6$ and $\text{Gal}(\mathbb{F}/\mathbb{Q})$ is isomorphic to S_3, the symmetric group on three letters. The Galois group is generated by

$$\sigma : \begin{pmatrix} d^{1/3} \mapsto \mu_3 d^{1/3} \\ \mu_3 \mapsto \mu_3 \end{pmatrix}, \qquad \tau : \begin{pmatrix} d^{1/3} \mapsto d^{1/3} \\ \mu_3 \mapsto \mu_3^2 \end{pmatrix},$$

and the isomorphism (noncanonical) is

$$\mathrm{Gal}(\mathbb{F}/\mathbb{Q}) \xrightarrow{\sim} S_3, \qquad \sigma \mapsto (1\,2\,3), \ \tau \mapsto (2\,3).$$

The rational primes not dividing $3d$ are unramified in $\mathbb{F}$, and their behavior is (Exercise 9.1.1)

$$p\mathcal{O}_{\mathbb{F}} = \begin{cases} \mathfrak{p}_1 \cdots \mathfrak{p}_6 & \text{if } p \equiv 1 \ (\mathrm{mod}\ 3) \text{ and } d \text{ is a cube modulo } p, \\ \mathfrak{p}_1\mathfrak{p}_2 & \text{if } p \equiv 1 \ (\mathrm{mod}\ 3) \text{ and } d \text{ is not a cube modulo } p, \quad (9.2) \\ \mathfrak{p}_1\mathfrak{p}_2\mathfrak{p}_3 & \text{if } p \equiv 2 \ (\mathrm{mod}\ 3). \end{cases}$$

Returning to the general situation, again let $\mathbb{F}$ be a Galois number field and let p be a rational prime. For each maximal ideal $\mathfrak{p}$ of $\mathcal{O}_{\mathbb{F}}$ lying over p the *decomposition group of* $\mathfrak{p}$ is the subgroup of the Galois group that fixes $\mathfrak{p}$ as a set,

$$D_{\mathfrak{p}} = \{\sigma \in \mathrm{Gal}(\mathbb{F}/\mathbb{Q}) : \mathfrak{p}^{\sigma} = \mathfrak{p}\}.$$

The decomposition group has order ef, so its index in $\mathrm{Gal}(\mathbb{F}/\mathbb{Q})$ is indeed the decomposition index g. By its definition it acts on the residue field $\mathbf{f}_{\mathfrak{p}} = \mathcal{O}_{\mathbb{F}}/\mathfrak{p}$,

$$(x + \mathfrak{p})^{\sigma} = x^{\sigma} + \mathfrak{p}, \quad x \in \mathcal{O}_{\mathbb{F}}, \ \sigma \in D_{\mathfrak{p}}.$$

The *inertia group of* $\mathfrak{p}$ is the kernel of the action,

$$I_{\mathfrak{p}} = \{\sigma \in D_{\mathfrak{p}} : x^{\sigma} \equiv x \ (\mathrm{mod}\ \mathfrak{p}) \text{ for all } x \in \mathcal{O}_{\mathbb{F}}\}.$$

The inertia group has order e, so it is trivial for all $\mathfrak{p}$ lying over any unramified p. Recall the Frobenius automorphism $\sigma_p : x \mapsto x^p$ in characteristic p from Chapter 8. If we view $\mathbb{F}_p = \mathbb{Z}/p\mathbb{Z}$ as a subfield of $\mathbf{f}_{\mathfrak{p}} = \mathcal{O}_{\mathbb{F}}/\mathfrak{p} \cong \mathbb{F}_{p^f}$ then there is an injection

$$D_{\mathfrak{p}}/I_{\mathfrak{p}} \longrightarrow \mathrm{Gal}(\mathbf{f}_{\mathfrak{p}}/\mathbb{F}_p) = \langle \sigma_p \rangle.$$

Since both groups have order f, the injection is an isomorphism and the quotient $D_{\mathfrak{p}}/I_{\mathfrak{p}}$ has a generator that maps to σ_p. Any representative of this generator in $D_{\mathfrak{p}}$ is called a *Frobenius element* of $\mathrm{Gal}(\mathbb{F}/\mathbb{Q})$ and denoted $\mathrm{Frob}_{\mathfrak{p}}$. That is, $\mathrm{Frob}_{\mathfrak{p}}$ is any element of a particular coset $\sigma I_{\mathfrak{p}}$ in the subgroup $D_{\mathfrak{p}}$ of $\mathrm{Gal}(\mathbb{F}/\mathbb{Q})$. Its action on $\mathbb{F}$, restricted to $\mathcal{O}_{\mathbb{F}}$, descends to the residue field $\mathbf{f}_{\mathfrak{p}} = \mathcal{O}_{\mathbb{F}}/\mathfrak{p}$, where it is the action of σ_p. When p is unramified, making the inertia group $I_{\mathfrak{p}}$ trivial, $\mathrm{Frob}_{\mathfrak{p}}$ is unique. To summarize,

Definition 9.1.1. *Let* $\mathbb{F}/\mathbb{Q}$ *be a Galois extension. Let p be a rational prime and let $\mathfrak{p}$ be a maximal ideal of $\mathcal{O}_{\mathbb{F}}$ lying over p. A* **Frobenius element** *of* $\mathrm{Gal}(\mathbb{F}/\mathbb{Q})$ *is any element* $\mathrm{Frob}_{\mathfrak{p}}$ *satisfying the condition*

$$x^{\mathrm{Frob}_{\mathfrak{p}}} \equiv x^p \ (\mathrm{mod}\ \mathfrak{p}) \quad \textit{for all } x \in \mathcal{O}_{\mathbb{F}}.$$

Thus $\mathrm{Frob}_{\mathfrak{p}}$ *acts on the residue field* $\mathbf{f}_{\mathfrak{p}}$ *as the Frobenius automorphism* σ_p.

When $\mathbb{F}/\mathbb{Q}$ is Galois the Galois group acts transitively on the maximal ideals lying over p, i.e., given any two such ideals $\mathfrak{p}$ and $\mathfrak{p}'$ there is an automorphism $\sigma \in \mathrm{Gal}(\mathbb{F}/\mathbb{Q})$ such that

$$\mathfrak{p}^\sigma = \mathfrak{p}'.$$

(We made reference to this fact, and to the earlier-mentioned fact that $D_\mathfrak{p}$ surjects to $\mathrm{Gal}(\mathbf{f}_\mathfrak{p}/\mathbb{F}_p)$, in the proof of Theorem 8.5.4.) The associated decomposition and inertia groups satisfy

$$D_{\mathfrak{p}^\sigma} = \sigma^{-1} D_\mathfrak{p} \sigma, \qquad I_{\mathfrak{p}^\sigma} = \sigma^{-1} I_\mathfrak{p} \sigma,$$

and the relation between corresponding Frobenius elements is

$$\mathrm{Frob}_{\mathfrak{p}^\sigma} = \sigma^{-1} \mathrm{Frob}_\mathfrak{p} \, \sigma.$$

If p is ramified then this means that the conjugate of a Frobenius is a Frobenius of the conjugate. The relation shows that if the Galois group is Abelian then $\mathrm{Frob}_\mathfrak{p}$ for any $\mathfrak{p}$ lying over p depends only on p and thus can be denoted Frob_p.

To compute the Frobenius element in the quadratic field case, again let $\mathbb{F} = \mathbb{Q}(\sqrt{d})$ where $d \neq 0, 1$ is squarefree. The Galois group $\mathrm{Gal}(\mathbb{F}/\mathbb{Q})$ consists of the identity and the map taking $\sqrt{\Delta_\mathbb{F}}$ to $-\sqrt{\Delta_\mathbb{F}}$. Let $\mathfrak{p}$ be a maximal ideal of $\mathcal{O}_\mathbb{F}$ lying over an odd prime p. Each $a + b(\Delta_\mathbb{F} + \sqrt{\Delta_\mathbb{F}})/2 \in \mathcal{O}_\mathbb{F}$ reduces to an element of the residue field $\mathbf{f}_\mathfrak{p}$, with a, b, and $\Delta_\mathbb{F}$ reducing to its subfield $\mathbb{F}_p$ and with 2 reducing to $\mathbb{F}_p^*$. Using the same symbols for the reductions, compute in the residue field that

$$\left(a + b \frac{\Delta_\mathbb{F} + \sqrt{\Delta_\mathbb{F}}}{2}\right)^p = a + b \frac{\Delta_\mathbb{F} + \Delta_\mathbb{F}^{(p-1)/2}\sqrt{\Delta_\mathbb{F}}}{2}.$$

This shows that the Frobenius element is the Legendre symbol in that Frob_p multiplies $\sqrt{\Delta_\mathbb{F}}$ by $\Delta_\mathbb{F}^{(p-1)/2} = (\Delta_\mathbb{F}/p)$ for odd primes p. There are infinitely many such p such that $(\Delta_\mathbb{F}/p) = 1$, and similarly for $(\Delta_\mathbb{F}/p) = -1$. Therefore every element of the Galois group of $\mathbb{F}$ takes the form Frob_p for infinitely many p, and there is an isomorphism

$$\mathrm{Gal}(\mathbb{F}/\mathbb{Q}) \xrightarrow{\sim} \{\pm 1\}, \qquad \mathrm{Frob}_p \mapsto (\Delta_\mathbb{F}/p) \text{ for odd } p \nmid \Delta_\mathbb{F}. \qquad (9.3)$$

The Frobenius element has a natural description in the cyclotomic case $\mathbb{F} = \mathbb{Q}(\mu_N)$ as well. For any prime $p \nmid N$ let $\mathfrak{p}$ lie over p and note that in the residue field $\mathbf{f}_\mathfrak{p} = \mathbb{F}_p[\mu_N]$, again using the same symbols to denote reductions,

$$\left(\sum_m a_m \mu_N^m\right)^p = \sum_m a_m \mu_N^{pm}.$$

This shows that Frob_p is the element of $\mathrm{Gal}(\mathbb{Q}(\mu_N)/\mathbb{Q})$ that takes μ_N to μ_N^p, and thus the isomorphism (9.1) takes Frob_p to $p \pmod{N}$. By Dirichlet's

Theorem on Arithmetic Progressions, for each $a \in (\mathbb{Z}/N\mathbb{Z})^*$ there are infinitely many p such that $p \equiv a \pmod{N}$. Therefore every element of the Galois group of $\mathbb{F}$ again takes the form Frob_p for infinitely many p, and the isomorphism (9.1) is

$$\mathrm{Gal}(\mathbb{F}/\mathbb{Q}) \xrightarrow{\sim} (\mathbb{Z}/N\mathbb{Z})^*, \qquad \mathrm{Frob}_p \mapsto p \pmod{N} \text{ for } p \nmid N. \qquad (9.4)$$

The isomorphisms (9.3) and (9.4) give a clear proof of Quadratic Reciprocity (Exercise 9.1.2).

In the non-Abelian example $\mathbb{F} = \mathbb{Q}(d^{1/3}, \mu_3)$ since the conjugacy classes in any symmetric group S_n are specified by the cycle structure of their elements, in this case of S_3 they are

$$\{1\}, \quad \{(1\,2),(2\,3),(3\,1)\}, \quad \{(1\,2\,3),(1\,3\,2)\}.$$

So the conjugacy class of an element of S_3 is determined by the element's order, and therefore this holds in $\mathrm{Gal}(\mathbb{F}/\mathbb{Q})$ as well. To determine the conjugacy class of $\mathrm{Frob}_{\mathfrak{p}}$ it thus suffices to determine its order, the residue degree f of the prime p lying under $\mathfrak{p}$. Formula (9.2) and the formula $efg = 6$ combine to show that for any unramified rational prime p, i.e., $p \nmid 3d$, the associated conjugacy class

$$\{\mathrm{Frob}_{\mathfrak{p}} : \mathfrak{p} \text{ lies over } p\} \qquad (9.5)$$

is

$$\text{the elements of order } \begin{cases} 1 & \text{if } p \equiv 1 \pmod{3} \text{ and } d \text{ is a cube modulo } p, \\ 3 & \text{if } p \equiv 1 \pmod{3} \text{ and } d \text{ is not a cube modulo } p, \\ 2 & \text{if } p \equiv 2 \pmod{3}. \end{cases}$$

Each conjugacy class takes the form (9.5) for infinitely many p, as with the previous two fields. This fact, as well as Dirichlet's Theorem, is a special case of

Theorem 9.1.2 (Tchebotarov Density Theorem, weak version). *Let $\mathbb{F}$ be a Galois number field. Then every element of $\mathrm{Gal}(\mathbb{F}/\mathbb{Q})$ takes the form $\mathrm{Frob}_{\mathfrak{p}}$ for infinitely many maximal ideals $\mathfrak{p}$ of $\mathcal{O}_{\mathbb{F}}$.*

We end this section with a motivating example. There is an embedding of S_3 in $\mathrm{GL}_2(\mathbb{Z})$ (Exercise 9.1.3) such that

$$(1\,2\,3) \mapsto \begin{bmatrix} 0 & 1 \\ -1 & -1 \end{bmatrix}, \qquad (2\,3) \mapsto \begin{bmatrix} 0 & 1 \\ 1 & 0 \end{bmatrix}. \qquad (9.6)$$

Again letting $\mathbb{F} = \mathbb{Q}(d^{1/3}, \mu_3)$ this gives a representation

$$\rho : \mathrm{Gal}(\mathbb{F}/\mathbb{Q}) \longrightarrow \mathrm{GL}_2(\mathbb{Z}).$$

The trace of ρ is a well defined function on conjugacy classes (9.5) and therefore depends only on the underlying unramified rational primes p,

$$\operatorname{tr} \rho(\operatorname{Frob}_{\mathfrak{p}}) = \begin{cases} 2 & \text{if } p \equiv 1 \pmod 3 \text{ and } d \text{ is a cube modulo } p, \\ -1 & \text{if } p \equiv 1 \pmod 3 \text{ and } d \text{ is not a cube modulo } p, \\ 0 & \text{if } p \equiv 2 \pmod 3. \end{cases} \quad (9.7)$$

Recall from Section 4.11 the theta function $\theta_\chi(\tau) \in \mathcal{M}_1(3N^2, \psi)$ where $N = 3\prod_{p|d} p$ and ψ is the quadratic character with conductor 3. Formula (9.7) for $\operatorname{tr} \rho(\operatorname{Frob}_{\mathfrak{p}})$ matches formula (4.51) for the Fourier coefficient $a_p(\theta_\chi)$ when $p \nmid 3d$. Similarly the determinant of ρ is defined on conjugacy classes over unramified primes,

$$\det \rho(\operatorname{Frob}_{\mathfrak{p}}) = \begin{cases} 1 & \text{if } p \equiv 1 \pmod 3, \\ -1 & \text{if } p \equiv 2 \pmod 3. \end{cases}$$

This is $\psi(p)$. So the Galois group representation ρ, as described by its trace and determinant on Frobenius elements, arises from the modular form θ_χ. This modular form is a normalized eigenform by Section 5.9 and a cusp form by Exercise 5.11.3. The idea of this chapter is that 2-dimensional representations of Galois groups arise from such modular forms in great generality.

Exercises

9.1.1. Let $\mathbb{F} = \mathbb{Q}(d^{1/3}, \mu_3)$ be the non-Abelian number field from the section. Granting that the factorization of $p \nmid 3d$ in $\mathbb{Q}(d^{1/3})$ comes from reducing $x^3 - d$ modulo p, obtain the factorization of p in $\mathbb{F}$. (A hint for this exercise is at the end of the book.)

9.1.2. Let q be an odd prime, and let $q^* = (-1)^{(q-1)/2}q$ (this is whichever of $\pm q$ is congruent to 1 modulo 4). Consider a quadratic field and a cyclotomic field,

$$\mathbb{F}_0 = \mathbb{Q}(\sqrt{q^*}), \qquad \mathbb{F} = \mathbb{Q}(\mu_q).$$

(a) Show that $\mathbb{F}_0 \subset \mathbb{F}$ as follows. Substitute $X = 1$ in the relation

$$\sum_{i=0}^{q-1} X^i = \prod_{i=1}^{q-1} (X - \mu_q^i)$$

to obtain the first equality in

$$q = \prod_{i=1}^{q-1} (1 - \mu_q^i) = \prod_{i=1}^{(q-1)/2} (1 - \mu_q^i)(1 - \mu_q^{-i}) = \prod_{i=1}^{(q-1)/2} (-\mu_q^{-i})(1 - \mu_q^i)^2,$$

Note that $\mu_q = (\mu_q^{(q+1)/2})^2$ is a square in $\mathbb{Q}(\mu_q)$. Explain why consequently the display shows that q^* is a square as well, giving the result.

(b) There is a commutative diagram as follows, in which the maps across the bottom and the top are (9.3) and (9.4), in which the map down the left side is restriction, and in which the map down the right side is therefore determined by the other three maps:

$$
\begin{array}{ccc}
\mathrm{Gal}(\mathbb{F}/\mathbb{Q}) & \longrightarrow & (\mathbb{Z}/q\mathbb{Z})^* \\
\downarrow & & \downarrow \\
\mathrm{Gal}(\mathbb{F}_0/\mathbb{Q}) & \longrightarrow & \{\pm 1\}.
\end{array}
$$

Show that for every odd prime $p \neq q$ the determined map down the right side is $p \pmod{q} \mapsto (q^*/p)$. The fact that this map is well defined says that the Legendre symbol (q^*/p) depends only on $p \pmod{q}$. This is the main part of Quadratic Reciprocity, essentially as stated at the beginning of the preface to this book.

9.1.3. Show that (9.6) defines an embedding of S_3 in $\mathrm{GL}_2(\mathbb{Z})$.

9.2 The ℓ-adic integers and the ℓ-adic numbers

For the remainder of this chapter let ℓ denote a prime number.

Consider an affine algebraic curve over $\mathbb{Q}$,

$$ C : xy = 1. $$

The points of C are identified with the Abelian group $\overline{\mathbb{Q}}^*$ via $(x,y) \mapsto x$, so C has an Abelian group structure as well. For each $n \in \mathbb{Z}^+$ the points of order ℓ^n form a subgroup $C[\ell^n]$. Since this is identified with the cyclic group of ℓ^nth roots of unity there is an isomorphism

$$ C[\ell^n] \longrightarrow \mathbb{Z}/\ell^n\mathbb{Z}, \qquad \mu_{\ell^n}^a \mapsto a. \tag{9.8} $$

Consequently there is also an isomorphism

$$ \mathrm{Aut}(C[\ell^n]) \longrightarrow (\mathbb{Z}/\ell^n\mathbb{Z})^*, \qquad (\mu_{\ell^n} \mapsto \mu_{\ell^n}^m) \mapsto m. $$

This extends to the isomorphism (9.1) with $N = \ell^n$,

$$ \mathrm{Gal}(\mathbb{Q}(\mu_{\ell^n})/\mathbb{Q}) \longrightarrow (\mathbb{Z}/\ell^n\mathbb{Z})^*, \qquad (\mu_{\ell^n} \mapsto \mu_{\ell^n}^m) \mapsto m. \tag{9.9} $$

Amalgamate the number fields $\mathbb{Q}(\mu_{\ell^n})$ for all n by defining

$$ \mathbb{Q}(\boldsymbol{\mu}_{\ell^\infty}) = \bigcup_{n=1}^{\infty} \mathbb{Q}(\mu_{\ell^n}). $$

This is again a field, a subfield of $\overline{\mathbb{Q}}$ but not a number field since it has infinite degree over $\mathbb{Q}$, placing it outside classical Galois theory. Still, since $\mathbb{Q}(\boldsymbol{\mu}_{\ell^\infty})$ is a union of Galois number fields whose Galois groups act on finite groups, its automorphism group should be a limit of Galois groups and act on a limit of finite groups. To make this idea precise, let

$$G_{\mathbb{Q},\ell} = \mathrm{Aut}(\mathbb{Q}(\boldsymbol{\mu}_{\ell^\infty})).$$

Every $\sigma \in G_{\mathbb{Q},\ell}$ restricts to $\sigma_n \in \mathrm{Gal}(\mathbb{Q}(\mu_{\ell^n})/\mathbb{Q})$ for each $n \in \mathbb{Z}^+$. The restrictions form a compatible sequence,

$$(\sigma_1, \sigma_2, \sigma_3, \dots), \qquad \sigma_{n+1}|_{\mathbb{Q}(\mu_{\ell^n})} = \sigma_n \text{ for all } n.$$

Conversely, every such compatible sequence of automorphisms determines an automorphism $\sigma \in G_{\mathbb{Q},\ell}$. Thus $G_{\mathbb{Q},\ell}$ can be viewed as a group of compatible sequences, where the group operation is componentwise composition. Each sequence acts componentwise on the group of compatible sequences of ℓ-power roots of unity, the ℓ-adic Tate module of C,

$$\mathrm{Ta}_\ell(C) = \{(\mu_\ell^{a_1}, \mu_{\ell^2}^{a_2}, \mu_{\ell^3}^{a_3}, \dots) : \mu_{\ell^n}^{a_{n+1}} = \mu_{\ell^n}^{a_n} \text{ for all } n\},$$

where the group operation is componentwise multiplication. Under the isomorphisms (9.8) and (9.9) at each component, $\mathrm{Ta}_\ell(C)$ is isomorphic to the Abelian group of sequences

$$\{(a_1, a_2, a_3, \dots) : a_n \in \mathbb{Z}/\ell^n\mathbb{Z} \text{ and } a_{n+1} \equiv a_n \pmod{\ell^n} \text{ for all } n\},$$

where the group operation is componentwise addition, and $G_{\mathbb{Q},\ell}$ is isomorphic to the Abelian group of sequences

$$\{(m_1, m_2, m_3, \dots) : m_n \in (\mathbb{Z}/\ell^n\mathbb{Z})^* \text{ and } m_{n+1} \equiv m_n \pmod{\ell^n} \text{ for all } n\},$$

where the group operation is componentwise multiplication. The action of $G_{\mathbb{Q},\ell}$ on $\mathrm{Ta}_\ell(C)$ transfers to an action of this group on the previous one by componentwise multiplication as well. These constructions motivate

Definition 9.2.1. Let ℓ be prime. An ℓ-**adic integer** is a sequence

$$\alpha = (a_1, a_2, a_3, \dots)$$

with $a_n \in \mathbb{Z}/\ell^n\mathbb{Z}$ and $a_{n+1} \equiv a_n \pmod{\ell^n}$ for each $n \in \mathbb{Z}^+$. The ring of ℓ-adic integers, where the operations are componentwise addition and multiplication, is denoted $\mathbb{Z}_\ell$.

Thus each entry a_n in an ℓ-adic integer determines the preceding entries a_{n-1} down to a_1, while the entry a_{n+1} to its right is one of its ℓ lifts from $\mathbb{Z}/\ell^n\mathbb{Z}$ to $\mathbb{Z}/\ell^{n+1}\mathbb{Z}$. This makes the ℓ-adic integers a special case of an algebraic construct called the *inverse limit*, in this case the inverse limit of the rings $\mathbb{Z}/\ell^n\mathbb{Z}$ for $n \in \mathbb{Z}^+$,

$$\mathbb{Z}_\ell = \varprojlim_n \{\mathbb{Z}/\ell^n\mathbb{Z}\}.$$

The ring $\mathbb{Z}_\ell$ is an integral domain. The natural map

$$\mathbb{Z} \longrightarrow \mathbb{Z}_\ell, \qquad a \mapsto (a + \ell\mathbb{Z}, a + \ell^2\mathbb{Z}, a + \ell^3\mathbb{Z}, \dots)$$

is a ring injection, so we view $\mathbb{Z}$ as a subring of $\mathbb{Z}_\ell$. The map induces a natural isomorphism

$$\mathbb{Z}/\ell\mathbb{Z} \xrightarrow{\sim} \mathbb{Z}_\ell/\ell\mathbb{Z}_\ell, \qquad a + \ell\mathbb{Z} \mapsto a + \ell\mathbb{Z}_\ell,$$

so we identify $\mathbb{Z}/\ell\mathbb{Z}$ and $\mathbb{Z}_\ell/\ell\mathbb{Z}_\ell$. As the inverse limit of a system of finite rings, $\mathbb{Z}_\ell$ is *profinite*.

The multiplicative group of units in $\mathbb{Z}_\ell$ is

$$\mathbb{Z}_\ell^* = \{(a_1, a_2, a_3, \dots) \in \mathbb{Z}_\ell : a_n \in (\mathbb{Z}/\ell^n\mathbb{Z})^* \text{ for all } n\},$$

for given such a compatible sequence, the sequence of inverses modulo ℓ^n is again compatible. (Thus the examples leading up to Definition 9.2.1 are now seen to show that $\mathrm{Ta}_\ell(C) \cong \mathbb{Z}_\ell$ and $G_{\mathbb{Q},\ell} \cong \mathbb{Z}_\ell^*$.) Every ℓ-adic integer α with $a_1 \neq 0$ in $\mathbb{Z}/\ell\mathbb{Z}$ is invertible. The ideal $\ell\mathbb{Z}_\ell$ is the unique maximal ideal of $\mathbb{Z}_\ell$, and $\mathbb{Z}_\ell^* = \mathbb{Z}_\ell - \ell\mathbb{Z}_\ell$. The ideal structure of $\mathbb{Z}_\ell$ is $\mathbb{Z}_\ell \supset \ell\mathbb{Z}_\ell \supset \ell^2\mathbb{Z}_\ell \supset \cdots$.

To put a topology on $\mathbb{Z}_\ell$ let

$$U_x(n) = x + \ell^n\mathbb{Z}_\ell, \qquad x \in \mathbb{Z}_\ell, \ n \in \mathbb{Z}^+.$$

If $x \in \mathbb{Z}_\ell^*$ then also $U_x(n) = x(1 + \ell^n\mathbb{Z}_\ell)$, or

$$U_x(n) = x \cdot \ker\left(\mathbb{Z}_\ell^* \longrightarrow (\mathbb{Z}/\ell^n\mathbb{Z})^*\right), \qquad x \in \mathbb{Z}_\ell^*, \ n \in \mathbb{Z}^+.$$

Then a basis of the topology of $\mathbb{Z}_\ell$ is $\{U_x(n) : x \in \mathbb{Z}_\ell, n \in \mathbb{Z}^+\}$, meaning that the topology is the smallest topology containing all $U_x(n)$. The neighborhoods $U_1(n)$ will figure frequently in the discussion, so they are abbreviated to $U(n)$. The ring operations of $\mathbb{Z}_\ell$ are continuous (Exercise 9.2.1(a)). As the inverse limit of finite rings, the topological ring $\mathbb{Z}_\ell$ is compact (Exercise 9.2.1(b)). For any positive integer d the $\mathbb{Z}_\ell$-module $\mathbb{Z}_\ell^d$ has the product topology, a basis being $\{U_v(n) : v \in \mathbb{Z}_\ell^d, n \in \mathbb{Z}^+\}$ where now $U_v(n) = v + \ell^n\mathbb{Z}_\ell^d$. We remark that the topology of $\mathbb{Z}_\ell$ arises from extending the ℓ-adic valuation of Chapter 8 from $\mathbb{Z}$ to $\mathbb{Z}_\ell$ and then defining a corresponding metric, but in this chapter only the topology matters.

The field $\mathbb{Q}_\ell$ of *ℓ-adic numbers* is the field of quotients of $\mathbb{Z}_\ell$. A basis of the topology of $\mathbb{Q}_\ell$ is $\{U_x(n) : x \in \mathbb{Q}_\ell, n \in \mathbb{Z}^+\}$ where as before $U_x(n) = x + \ell^n\mathbb{Z}_\ell$. The field operations of $\mathbb{Q}_\ell$ are continuous (Exercise 9.2.1(c)). For any positive integer d the $\mathbb{Q}_\ell$-vector space $\mathbb{Q}_\ell^d$ has the product topology, a basis being $\{U_v(n) : v \in \mathbb{Q}_\ell^d, n \in \mathbb{Z}^+\}$. The matrix group $\mathrm{GL}_d(\mathbb{Q}_\ell)$ acquires a topology as a subset of $\mathbb{Q}_\ell^{d^2}$. If $m \in \mathrm{GL}_d(\mathbb{Z}_\ell)$ then $U_m(n) = m(I + \ell^n\mathrm{M}_d(\mathbb{Z}_\ell))$, or

$$U_m(n) = m \cdot \ker\left(\mathrm{GL}_d(\mathbb{Z}_\ell) \longrightarrow \mathrm{GL}_d(\mathbb{Z}/\ell^n\mathbb{Z})\right), \qquad m \in \mathrm{GL}_d(\mathbb{Z}_\ell),$$

and as before, $U(n)$ means $U_I(n)$. Since matrix multiplication and matrix inversion are defined by rational functions of the matrix entries, they are continuous under this topology. Vector-by-matrix multiplication,

$$\mathbb{Q}_\ell^d \times \mathrm{GL}_d(\mathbb{Q}_\ell) \longrightarrow \mathbb{Q}_\ell^d,$$

is continuous as well. Here we view vectors as rows so that matrices multiply from the right; this is not the usual convention but in this chapter such multiplication will represent the action of Galois groups, and these have been acting from the right since their appearance in Chapter 7.

Let V be a finite-dimensional vector space over $\mathbb{Q}_\ell$. Any basis $B = (\beta_1, \ldots, \beta_d)$ of V determines a topology $\mathcal{T}$ on V to make the coordinate map $c_B : V \longrightarrow \mathbb{Q}_\ell^d$ given by $\sum a_j \beta_j \mapsto [a_1, \ldots, a_d]$ a homeomorphism. We show that this topology is independent of the basis. If B' is another basis determining another topology $\mathcal{T}'$ then the identity map $(V, \mathcal{T}) \longrightarrow (V, \mathcal{T}')$ is $c_{B'}^{-1} \circ m \circ c_B$ where $m : \mathbb{Q}_\ell^d \longrightarrow \mathbb{Q}_\ell^d$ is multiplication by a transition matrix between the bases. This is a homeomorphism, so $\mathcal{T}' = \mathcal{T}$.

Let $\mathbb{K}$ be any number field, not necessarily Galois over $\mathbb{Q}$, and let $\mathcal{O}_\mathbb{K}$ be its ring of integers. The factorization (8.22) of $\ell\mathcal{O}_\mathbb{K}$ into maximal ideals is written

$$\ell\mathcal{O}_\mathbb{K} = \prod_{\lambda | \ell} \lambda^{e_\lambda},$$

where the notation $\lambda \mid \ell$ means that λ lies over the rational prime ℓ. Similarly to $\mathbb{Z}_\ell$ and $\mathbb{Q}_\ell$, for each λ the ring of λ-adic integers is the inverse limit

$$\mathcal{O}_{\mathbb{K},\lambda} = \varprojlim_n \{\mathcal{O}_\mathbb{K}/\lambda^n\}$$

and then the field of λ-adic numbers is the field of quotients $\mathbb{K}_\lambda$ of $\mathcal{O}_{\mathbb{K},\lambda}$. We may view $\mathbb{Z}_\ell$ as a subring of $\mathcal{O}_{\mathbb{K},\lambda}$ and $\mathbb{Q}_\ell$ as a subfield of $\mathbb{K}_\lambda$ (Exercise 9.2.3(a)). Define the residue degree f_λ to be $[\mathbf{k}_\lambda : \mathbb{F}_\ell]$ where $\mathbf{k}_\lambda = \mathcal{O}_\mathbb{K}/\lambda$. Then the containments $\mathbb{Z}_\ell \subset \mathcal{O}_{\mathbb{K},\lambda}$ and $\mathbb{Q}_\ell \subset \mathbb{K}_\lambda$ are equalities when $e_\lambda f_\lambda = 1$ (Exercise 9.2.3(b)), and in fact $[\mathbb{K}_\lambda : \mathbb{Q}_\ell] = e_\lambda f_\lambda$. There is a ring isomorphism

$$\mathbb{K} \otimes_\mathbb{Q} \mathbb{Q}_\ell \cong \prod_{\lambda | \ell} \mathbb{K}_\lambda. \tag{9.10}$$

We provide the proof to illustrate properties of tensors and inverse limits for the less experienced reader. Compute that

$$\mathcal{O}_\mathbb{K} \otimes \mathbb{Z}_\ell \cong \varprojlim_n \{\mathcal{O}_\mathbb{K} \otimes \mathbb{Z}/\ell^n\mathbb{Z}\} \cong \varprojlim_n \{\mathcal{O}_\mathbb{K}/\ell^n\mathcal{O}_\mathbb{K}\}.$$

But $\mathcal{O}_\mathbb{K}/\ell^n\mathcal{O}_\mathbb{K} = \mathcal{O}_\mathbb{K}/\prod_\lambda \lambda^{ne_\lambda} \cong \prod_\lambda \mathcal{O}_\mathbb{K}/\lambda^{ne_\lambda}$. Thus

$$\mathcal{O}_\mathbb{K} \otimes \mathbb{Z}_\ell \cong \varprojlim_n \{\prod_\lambda \mathcal{O}_\mathbb{K}/\lambda^{ne_\lambda}\} \cong \prod_\lambda \varprojlim_n \{\mathcal{O}_\mathbb{K}/\lambda^{ne_\lambda}\} \cong \prod_\lambda \mathcal{O}_{\mathbb{K},\lambda},$$

and this gives

$$\mathbb{K} \otimes_\mathbb{Q} \mathbb{Q}_\ell \cong \mathcal{O}_\mathbb{K} \otimes \mathbb{Q}_\ell \cong \mathcal{O}_\mathbb{K} \otimes \mathbb{Z}_\ell \otimes_{\mathbb{Z}_\ell} \mathbb{Q}_\ell \cong \prod_\lambda (\mathcal{O}_{\mathbb{K},\lambda} \otimes_{\mathbb{Z}_\ell} \mathbb{Q}_\ell) \cong \prod_\lambda \mathbb{K}_\lambda.$$

Each $\mathbb{K}_\lambda$ acquires a topology as a finite-dimensional vector space over $\mathbb{Q}_\ell$. Let V be a finite-dimensional vector space over some $\mathbb{K}_\lambda$. Then V is also a finite-dimensional vector space over $\mathbb{Q}_\ell$. Thus V acquires two topologies, from $\mathbb{K}_\lambda$ and from $\mathbb{Q}_\ell$, but since $\mathbb{K}_\lambda$ acquires its topology in turn from $\mathbb{Q}_\ell$ the two topologies are the same. Any finitely generated $\mathbb{Z}_\ell$-submodule of V is compact since $\mathbb{Z}_\ell$ is compact. The results in these last two paragraphs will be used in Sections 9.3 and 9.5.

Exercises

9.2.1. (a) Show that the ring operations of $\mathbb{Z}_\ell$ are continuous under the ℓ-adic topology.

(b) Show that the ℓ-adic topology makes $\mathbb{Z}_\ell$ a topological subgroup of the product $P = \prod_{n=1}^{\infty} \mathbb{Z}/\ell^n \mathbb{Z}$. The product is compact by the Tychonov Theorem (see for example [Mun00]). For each n consider the finite product $P_n = \prod_{m=1}^{n} \mathbb{Z}/\ell^m \mathbb{Z}$ and the projection $\pi_n : P \longrightarrow P_n$. The product topology makes the projections continuous. Show that each P_n contains a closed subgroup C_n of compatible elements such that $\mathbb{Z}_\ell = \bigcap_{n=1}^{\infty} \pi^{-1}(C_n)$. This shows that $\mathbb{Z}_\ell$ is a closed subgroup of P, making it compact. (A hint for this exercise is at the end of the book.)

(c) Show that the field operations of $\mathbb{Q}_\ell$ are continuous under the ℓ-adic topology.

9.2.2. Since $U(n)$ is a subgroup of $\mathbb{Z}_\ell^*$ for all $n \in \mathbb{Z}^+$, every neighborhood U of 1 in $\mathbb{Q}_\ell^*$ contains a nontrivial subgroup of $\mathbb{Q}_\ell^*$. Show that the analogous result fails with $\mathbb{C}$ in place of $\mathbb{Q}_\ell$, and similarly for $\mathrm{GL}_d(\mathbb{Q}_\ell)$ versus $\mathrm{GL}_d(\mathbb{C})$. (A hint for this exercise is at the end of the book.)

9.2.3. (a) Show that $\lambda^{n e_\lambda} \cap \mathbb{Z} = \ell^n \mathbb{Z}$ for all $n \in \mathbb{Z}^+$, and that therefore we may view $\mathbb{Z}_\ell$ as a subring of $\mathcal{O}_{\mathbb{K},\lambda}$ and $\mathbb{Q}_\ell$ as a subfield of $\mathbb{K}_\lambda$. (Hints for this exercise are at the end of the book.)

(b) Show that the containments are equalities when $e_\lambda f_\lambda = 1$.

(c) What does (9.10) say for the three Galois number fields of Section 9.1?

9.3 Galois representations

As always, let $\overline{\mathbb{Q}}$ denote the algebraic closure of $\mathbb{Q}$. The automorphisms of $\overline{\mathbb{Q}}$ form the *absolute Galois group of* $\mathbb{Q}$,

$$G_\mathbb{Q} = \mathrm{Aut}(\overline{\mathbb{Q}}).$$

Recall that ℓ is prime. This section defines ℓ-adic Galois representations as continuous homomorphisms from $G_\mathbb{Q}$ into ℓ-adic matrix groups, or alternatively as ℓ-adic vector spaces that are $G_\mathbb{Q}$-modules.

The algebraic closure $\overline{\mathbb{Q}}$ is the union of all Galois number fields. Any automorphism $\sigma \in G_{\mathbb{Q}}$ fixes $\mathbb{Q}$ pointwise and restricts to an automorphism $\sigma|_{\mathbb{F}} \in \mathrm{Gal}(\mathbb{F}/\mathbb{Q})$ for every Galois number field $\mathbb{F}$. Restriction from $G_{\mathbb{Q}}$ to $\mathrm{Gal}(\mathbb{F}/\mathbb{Q})$ surjects. The restrictions are compatible in that

$$\sigma_{\mathbb{F}} = \sigma_{\mathbb{F}'}|_{\mathbb{F}} \quad \text{if} \quad \mathbb{F} \subset \mathbb{F}'.$$

Conversely, every compatible system of automorphisms $\{\sigma_{\mathbb{F}}\}$ over all Galois number fields $\mathbb{F}$ defines an automorphism of $\overline{\mathbb{Q}}$, i.e., an element of $G_{\mathbb{Q}}$. As in the previous section this describes $G_{\mathbb{Q}}$ as an inverse limit,

$$G_{\mathbb{Q}} = \varprojlim_{\mathbb{F}} \{\mathrm{Gal}(\mathbb{F}/\mathbb{Q})\}.$$

Since the Galois groups are finite, $G_{\mathbb{Q}}$ is a profinite group. A good theory arises from giving it a suitable topology, the *Krull topology*, and then considering only its continuous actions. For each automorphism $\sigma \in G_{\mathbb{Q}}$ and each Galois number field $\mathbb{F}$ there is an associated coset of a normal subgroup that gives rise to a finite quotient, $U_{\sigma}(\mathbb{F}) = \{\sigma\tau : \tau|_{\mathbb{F}} = 1\}$, or

$$U_{\sigma}(\mathbb{F}) = \sigma \cdot \ker\left(G_{\mathbb{Q}} \longrightarrow \mathrm{Gal}(\mathbb{F}/\mathbb{Q})\right).$$

Again $U(\mathbb{F})$ denotes $U_1(\mathbb{F})$ where now $1 = 1_{\overline{\mathbb{Q}}}$. A basis of the Krull topology on $G_{\mathbb{Q}}$ is

$$\{U_{\sigma}(\mathbb{F}) : \sigma \in G_{\mathbb{Q}}, \mathbb{F} \text{ is a Galois number field}\}.$$

Thus every $U(\mathbb{F})$ is an open normal subgroup of $G_{\mathbb{Q}}$, and conversely every open normal subgroup of $G_{\mathbb{Q}}$ takes the form $U(\mathbb{F})$ for a Galois number field $\mathbb{F}$ (Exercise 9.3.1). As the inverse limit of finite groups, the topological group $G_{\mathbb{Q}}$ is compact.

One element of $G_{\mathbb{Q}}$ is complex conjugation. To introduce an important family of elements, let $p \in \mathbb{Z}$ be any prime and let $\mathfrak{p} \subset \overline{\mathbb{Z}}$ be any maximal ideal over p. Use $\mathfrak{p}$ as the kernel in defining the reduction map $\overline{\mathbb{Z}} \longrightarrow \overline{\mathbb{F}}_p$. The *decomposition group of* $\mathfrak{p}$ is

$$D_{\mathfrak{p}} = \{\sigma \in G_{\mathbb{Q}} : \mathfrak{p}^{\sigma} = \mathfrak{p}\}.$$

Thus each $\sigma \in D_{\mathfrak{p}}$ acts on $\overline{\mathbb{Z}}/\mathfrak{p}$ as $(x + \mathfrak{p})^{\sigma} = x^{\sigma} + \mathfrak{p}$, and this can be viewed as an action on $\overline{\mathbb{F}}_p$. Let $G_{\overline{\mathbb{F}}_p}$ denote the absolute Galois group $\mathrm{Aut}(\overline{\mathbb{F}}_p)$ of $\mathbb{F}_p$. The reduction map

$$D_{\mathfrak{p}} \longrightarrow G_{\overline{\mathbb{F}}_p}$$

is surjective. An *absolute Frobenius element over* p is any preimage $\mathrm{Frob}_{\mathfrak{p}} \in D_{\mathfrak{p}}$ of the Frobenius automorphism $\sigma_p \in G_{\overline{\mathbb{F}}_p}$. Thus $\mathrm{Frob}_{\mathfrak{p}}$ is defined only up to the kernel of the reduction map, the *inertia group of* $\mathfrak{p}$,

$$I_{\mathfrak{p}} = \{\sigma \in D_{\mathfrak{p}} : x^{\sigma} \equiv x \pmod{\mathfrak{p}} \text{ for all } x \in \overline{\mathbb{Z}}\}.$$

For each Galois number field $\mathbb{F}$ the restriction map $G_{\mathbb{Q}} \longrightarrow \mathrm{Gal}(\mathbb{F}/\mathbb{Q})$ takes an absolute Frobenius element to a corresponding Frobenius element for $\mathbb{F}$,

$$\text{Frob}_\mathfrak{p}|_\mathbb{F} = \text{Frob}_{\mathfrak{p}_\mathbb{F}} \quad \text{where } \mathfrak{p}_\mathbb{F} = \mathfrak{p} \cap \mathbb{F}. \tag{9.11}$$

All maximal ideals of $\overline{\mathbb{Z}}$ over p are conjugate to $\mathfrak{p}$, and the definition of $\text{Frob}_\mathfrak{p}$ shows that analogously to before,

$$\text{Frob}_{\mathfrak{p}^\sigma} = \sigma^{-1}\text{Frob}_\mathfrak{p}\,\sigma, \quad \sigma \in G_\mathbb{Q}.$$

Again this means that the conjugate of a Frobenius is a Frobenius of the conjugate. Theorem 9.1.2 and the definition of the Krull topology combine to show that we now have many examples (Exercise 9.3.2):

Theorem 9.3.1. *For each maximal ideal $\mathfrak{p}$ of $\overline{\mathbb{Z}}$ lying over any but a finite set of rational primes p, choose an absolute Frobenius element $\text{Frob}_\mathfrak{p}$. The set of such elements forms a dense subset of $G_\mathbb{Q}$.*

Let $\chi : (\mathbb{Z}/N\mathbb{Z})^* \longrightarrow \mathbb{C}^*$ be a primitive Dirichlet character. Consider the following diagram, in which π_N is restriction to $\text{Gal}(\mathbb{Q}(\mu_N)/\mathbb{Q})$, the horizontal arrow is the isomorphism (9.1), and $\rho_{\chi,N}$ is defined to make the triangle commute.

$$\begin{array}{c} G_\mathbb{Q} \\ \downarrow{\scriptstyle \pi_N} \\ \text{Gal}(\mathbb{Q}(\mu_N)/\mathbb{Q}) \xrightarrow{\ \sim\ } (\mathbb{Z}/N\mathbb{Z})^* \\ {\scriptstyle \rho_{\chi,N}}\searrow \quad \swarrow{\scriptstyle \chi} \\ \mathbb{C}^* \end{array} \tag{9.12}$$

The diagram shows that χ determines a homomorphism

$$\rho_\chi = \rho_{\chi,N} \circ \pi_N : G_\mathbb{Q} \longrightarrow \mathbb{C}^*.$$

In particular, since complex conjugation in $G_\mathbb{Q}$ restricts to $\mu_N \mapsto \mu_N^{-1}$ in the finite Galois group it follows that $\rho_\chi(\text{conj}) = \chi(-1)$. Similarly, since (9.11) and then (9.4) show that an absolute Frobenius element $\text{Frob}_\mathfrak{p}$ in $G_\mathbb{Q}$ lying over a prime $p \nmid N$ maps to $p \pmod N$ in $(\mathbb{Z}/N\mathbb{Z})^*$ it follows that $\rho_\chi(\text{Frob}_\mathfrak{p}) = \chi(p)$ for such p. To show that the homomorphism ρ_χ is continuous it suffices to check that $\rho_\chi^{-1}(1)$ is open (see Exercise 9.3.3), and this holds because the inverse image is $U(\mathbb{F})$ for some Galois number field $\mathbb{F} \subset \mathbb{Q}(\mu_N)$.

Conversely, any continuous homomorphism $\rho : G_\mathbb{Q} \longrightarrow \mathbb{C}^*$ has finite image (Exercise 9.3.4) and thus factors through some Abelian Galois group $\text{Gal}(\mathbb{F}/\mathbb{Q})$. The Kronecker–Weber Theorem states that we may take $\mathbb{F} = \mathbb{Q}(\mu_N)$ for some N, and now diagram (9.12) shows that $\rho = \rho_\chi$ for some Dirichlet character χ_ρ. Taking N as small as possible (see Exercise 9.3.5) makes χ_ρ primitive. That is, Dirichlet characters lead to homomorphisms of a certain type, and all such homomorphisms arise from Dirichlet characters. Section 9.6 will discuss a conjectured 2-dimensional analog of this 1-dimensional correspondence, encompassing the Modularity Theorem.

Exercise 9.3.4 shows more generally that any continuous homomorphism $\rho : G_{\mathbb{Q}} \longrightarrow \mathrm{GL}_d(\mathbb{C})$ has finite image and therefore captures only a finite amount of the structure of $G_{\mathbb{Q}}$. However, the image of a Dirichlet character χ lies in a number field $\mathbb{K}$ and therefore in a field $\mathbb{K}_\lambda$ where λ lies over any given rational prime ℓ. So $\mathbb{C}^*$ can be replaced by $\mathbb{K}_\lambda^*$ in diagram (9.12), and the argument that the associated representation $\rho_\chi : G_{\mathbb{Q}} \longrightarrow \mathbb{K}_\lambda^*$ is continuous still applies. This suggests the following definition.

Definition 9.3.2. *Let d be a positive integer. A d-**dimensional ℓ-adic Galois representation** is a continuous homomorphism*

$$\rho : G_{\mathbb{Q}} \longrightarrow \mathrm{GL}_d(\mathbb{L})$$

*where $\mathbb{L}$ is a finite extension field of $\mathbb{Q}_\ell$. If $\rho' : G_{\mathbb{Q}} \longrightarrow \mathrm{GL}_d(\mathbb{L})$ is another such representation and there is a matrix $m \in \mathrm{GL}_d(\mathbb{L})$ such that $\rho'(\sigma) = m^{-1}\rho(\sigma)m$ for all $\sigma \in G_{\mathbb{Q}}$ then ρ and ρ' are **equivalent**. Equivalence is notated $\rho \sim \rho'$.*

We state without proof that every finite extension field $\mathbb{L}$ of $\mathbb{Q}_\ell$ takes the form $\mathbb{K}_\lambda$ for some number field $\mathbb{K}$ and maximal ideal $\lambda \mid \ell$ of $\mathcal{O}_{\mathbb{K}}$, and that for such $\mathbb{L}$ the ring $\mathcal{O}_{\mathbb{L}} = \mathcal{O}_{\mathbb{K},\lambda}$ is well defined independently of $\mathbb{K}$ and λ. The ring $\mathcal{O}_{\mathbb{L}}$ is a lattice in $\mathbb{L}$, i.e., there is $\mathbb{Z}_\ell$-basis of $\mathcal{O}_{\mathbb{L}}$ that is also a $\mathbb{Q}_\ell$-basis of $\mathbb{L}$. The ℓ-adic Galois representations in this chapter will often use $\mathbb{L} = \mathbb{Q}_\ell$, but for example Section 9.5 will construct representations using fields $\mathbb{L} = \mathbb{K}_{f,\lambda}$ where f is a modular form and $\mathbb{K}_f$ is the number field generated by its Fourier coefficients. Complex Galois representations are defined similarly to Definition 9.3.2, but because their images are finite they will not play a role here until the end of the chapter. Galois representations are assumed from now on to be ℓ-adic rather than complex unless otherwise specified.

For an example involving infinite structure, first note that the field containment $\mathbb{Q}(\boldsymbol{\mu}_{\ell^\infty}) \subset \overline{\mathbb{Q}}$ gives the group surjection $G_{\mathbb{Q}} \longrightarrow G_{\mathbb{Q},\ell}$ restricting each automorphism of $\overline{\mathbb{Q}}$ to $\mathbb{Q}(\boldsymbol{\mu}_{\ell^\infty})$. Follow this by the group isomorphism $G_{\mathbb{Q},\ell} \longrightarrow \mathbb{Z}_\ell^*$ described before Definition 9.2.1 to get the *ℓ-adic cyclotomic character of $G_{\mathbb{Q}}$,*

$$\chi_\ell : G_{\mathbb{Q}} \longrightarrow \mathbb{Q}_\ell^*, \qquad \sigma \mapsto (m_1, m_2, m_3, \dots) \text{ where } \mu_{\ell^n}^\sigma = \mu_{\ell^n}^{m_n} \text{ for all } n,$$

and in fact the image of χ_ℓ lies in $\mathbb{Z}_\ell^*$. To show that the cyclotomic character is continuous it suffices to check that $\chi_\ell^{-1}(U(n))$ is open for any n (again see Exercise 9.3.3), and this holds because the inverse image is $U(\mathbb{Q}(\mu_{\ell^n}))$. Thus χ_ℓ is a 1-dimensional Galois representation. Similarly to the Dirichlet character example, $\chi_\ell(\mathrm{conj}) = -1$, and we will frequently use the formula (Exercise 9.3.6)

$$\chi_\ell(\mathrm{Frob}_{\mathfrak{p}}) = p \quad \text{for } p \neq \ell. \tag{9.13}$$

This is independent of how $\mathfrak{p}$ and $\mathrm{Frob}_{\mathfrak{p}}$ are chosen. The formula shows that despite being 1-dimensional χ_ℓ has infinite image, already indicating that ℓ-adic Galois representations have a richer theory than complex ones. Profinite groups $\mathrm{GL}_d(\mathbb{L})$ give better partial views of the profinite group $G_{\mathbb{Q}}$.

Given a Galois representation ρ we naturally want to know values $\rho(\sigma)$ for $\sigma \in G_\mathbb{Q}$. In particular we want to evaluate ρ at absolute Frobenius elements. But each $\mathrm{Frob}_\mathfrak{p}$ is defined only up to the absolute inertia group $I_\mathfrak{p}$, so the notion of $\rho(\mathrm{Frob}_\mathfrak{p})$ is well defined if and only if $I_\mathfrak{p} \subset \ker \rho$. Furthermore, if $\mathfrak{p}$ and $\mathfrak{p}'$ lie over the same rational prime p then the inertia groups $I_\mathfrak{p}$ and $I_{\mathfrak{p}'}$ are conjugate in $G_\mathbb{Q}$, so the condition $I_\mathfrak{p} \subset \ker \rho$ depends only on the underlying prime p since $\ker \rho$ is normal in $G_\mathbb{Q}$. The condition makes sense at the level of the similarity class of ρ, again since $\ker \rho$ is normal in $G_\mathbb{Q}$. Although $\rho(\mathrm{Frob}_\mathfrak{p})$ does depend on the choice of $\mathfrak{p}$ over p when it is defined, its characteristic polynomial depends only on the conjugacy class of $\rho(\mathrm{Frob}_\mathfrak{p})$ and therefore only on p.

Definition 9.3.3. *Let ρ be a Galois representation and let p be prime. Then ρ is **unramified at** p if $I_\mathfrak{p} \subset \ker \rho$ for any maximal ideal $\mathfrak{p} \subset \overline{\mathbb{Z}}$ lying over p.*

For example, let $\chi : (\mathbb{Z}/N\mathbb{Z})^* \longrightarrow \mathbb{C}^*$ be a primitive Dirichlet character, let p be any prime not dividing N, and let $\mathfrak{p}$ lie over p. Then in diagram (9.12) the map π_N takes $I_\mathfrak{p}$ into $I_{\mathfrak{p}_N}$ where $\mathfrak{p}_N = \mathfrak{p} \cap \mathbb{Q}(\mu_N)$. But $I_{\mathfrak{p}_N} = \{1\}$ since p is unramified in $\mathbb{Q}(\mu_N)$, and so $I_\mathfrak{p} \subset \ker \rho_\chi$, i.e., ρ_χ is unramified at p. Similarly for the cyclotomic character χ_ℓ, let $p \neq \ell$ and let $\mathfrak{p}$ lie over p. Any $\sigma \in I_\mathfrak{p}$ acts trivially on $\mathbb{F}_n = \mathbb{Q}(\mu_{\ell^n})$ for all positive n since p is unramified in $\mathbb{F}_n$. This shows that $I_\mathfrak{p} \subset \ker \chi_\ell$, i.e., χ_ℓ is unramified at p. In general, the Galois representations that we will consider are unramified at all but finitely many primes p. For such a representation the values $\rho(\mathrm{Frob}_\mathfrak{p})$ for $\mathfrak{p}$ over unramified p determine ρ everywhere by continuity.

To think about χ_ℓ in another way, recall that it arises from the $G_\mathbb{Q}$-module structure of the Tate module of the curve C from Section 9.2. Let

$$V_\ell(C) = \mathrm{Ta}_\ell(C) \otimes \mathbb{Q} \cong \mathbb{Q}_\ell.$$

Then the following diagram commutes, where the map from $V_\ell(C)$ to $\mathbb{Q}_\ell$ is the $\mathbb{Q}_\ell$-linear extension of the isomorphism $\mathrm{Ta}_\ell(C) \longrightarrow \mathbb{Z}_\ell$ described before Definition 9.2.1, and the map from $G_\mathbb{Q}$ to $\mathbb{Q}_\ell^*$ is χ_ℓ.

$$
\begin{array}{ccc}
V_\ell(C) \times G_\mathbb{Q} & \longrightarrow & V_\ell(C) \\
\downarrow & & \downarrow \\
\mathbb{Q}_\ell \times \mathbb{Q}_\ell^* & \longrightarrow & \mathbb{Q}_\ell
\end{array}
$$

Once $V_\ell(C)$ is topologized compatibly with $\mathbb{Q}_\ell$ the action of $G_\mathbb{Q}$ on $V_\ell(C)$ is continuous by the continuity of all the other maps in the diagram. This example motivates a second definition of an ℓ-adic Galois representation as a vector space over $\mathbb{Q}_\ell$ with $G_\mathbb{Q}$-module structure.

Definition 9.3.4. *Let d be a positive integer. A d-**dimensional ℓ-adic Galois representation** is a d-dimensional topological vector space V over $\mathbb{L}$,* *where* $\mathbb{L}$ *is a finite extension field of* $\mathbb{Q}_\ell$, *that is also a* $G_\mathbb{Q}$-*module such that the action*

$$V \times G_\mathbb{Q} \longrightarrow V, \qquad (v, \sigma) \mapsto v^\sigma$$

is continuous. If V' *is another such representation and there is a continuous* $G_\mathbb{Q}$-*module isomorphism of* $\mathbb{L}$-*vector spaces* $V \xrightarrow{\sim} V'$ *then* V *and* V' *are* **equivalent**.

The two definitions of an ℓ-adic Galois representation are compatible (Exercise 9.3.7). Similarly to how the 1-dimensional representation $V_\ell(C)$ arises from the $G_\mathbb{Q}$-module structure of $\mathrm{Ta}_\ell(C)$, the next two sections will consider higher-dimensional Galois representations arising from the Tate modules of elliptic curves, Jacobians, and Abelian varieties associated to modular forms. Any 1-dimensional Galois representation has an Abelian image but the higher-dimensional representations capture non-Abelian structure of $G_\mathbb{Q}$ as well.

The values taken by a Galois representation $\rho_\chi : G_\mathbb{Q} \longrightarrow \mathbb{K}_\lambda$ arising from a Dirichlet character are roots of unity and therefore they lie in the subring $\mathcal{O}_{\mathbb{K},\lambda}$ of $\mathbb{K}_\lambda$. The values taken by the cyclotomic character χ_ℓ lie in the subring $\mathbb{Z}_\ell$ of $\mathbb{Q}_\ell$. The second definition of Galois representation helps to show that this is a general phenomenon.

Proposition 9.3.5. *Let* $\rho : G_\mathbb{Q} \longrightarrow \mathrm{GL}_d(\mathbb{L})$ *be a Galois representation. Then* ρ *is similar to a Galois representation* $\rho' : G_\mathbb{Q} \longrightarrow \mathrm{GL}_d(\mathcal{O}_\mathbb{L})$.

Proof. Let $V = \mathbb{L}^d$ and let $\Lambda = \mathcal{O}_\mathbb{L}^d$. Then Λ is a lattice of V, hence a finitely generated $\mathbb{Z}_\ell$-module, hence compact as noted at the end of Section 9.2. Since $G_\mathbb{Q}$ is compact as well, so is the image Λ' of $\Lambda \times G_\mathbb{Q}$ under the map $V \times G_\mathbb{Q} \longrightarrow V$. Thus the image lies in $\lambda^{-r}\Lambda$ for some $r \in \mathbb{Z}^+$. The image is finitely generated, it contains Λ so its rank is at least d, it is free since $\mathcal{O}_\mathbb{L}$ is a principal ideal domain, and so its rank is exactly d. It is taken to itself by $G_\mathbb{Q}$. All of this combines to show that any $\mathcal{O}_\mathbb{L}$-basis of Λ' gives the desired ρ'. $\square$

This result will be used in Section 9.6.

We end this section with an example. Let ℓ be prime and let $d \geq 2$ be an integer. For each positive integer n introduce the number field $\mathbb{F}_n = \mathbb{Q}(d^{1/\ell^n}, \mu_{\ell^n})$ (the splitting field over $\mathbb{Q}$ of the polynomial $X^{\ell^n} - d$), and let $G_n = \mathrm{Gal}(\mathbb{F}_n/\mathbb{Q})$. The nth component $\chi_{\ell,n}$ of the ℓ-adic cyclotomic character of $G_\mathbb{Q}$, defined by the condition $\mu_{\ell^n}^\sigma = \mu_{\ell^n}^{\chi_{\ell,n}(\sigma)}$ for $\sigma \in G_\mathbb{Q}$, factors through $\mathrm{Gal}(\mathbb{Q}(\mu_{\ell^n})/\mathbb{Q})$ and hence through G_n. Define a function $b_n = b_{\ell,d,n} : G_n \longrightarrow \mathbb{Z}/\ell^n\mathbb{Z}$ by the condition $(d^{1/\ell^n})^\sigma/d^{1/\ell^n} = \mu_{\ell^n}^{b_n(\sigma)}$ for all $\sigma \in G_n$. Consider the map

$$\rho_n = \rho_{\ell,d,n} : G_n \longrightarrow \mathrm{GL}_2(\mathbb{Z}/\ell^n\mathbb{Z}), \qquad \rho_n(\sigma) = \begin{bmatrix} 1 & b_n(\sigma) \\ 0 & \chi_{\ell,n}(\sigma) \end{bmatrix}.$$

For any $\sigma, \tau \in G_n$,

$$\rho_n(\sigma)\rho_n(\tau) = \begin{bmatrix} 1 & b_n(\tau) + b_n(\sigma)\chi_{\ell,n}(\tau) \\ 0 & \chi_{\ell,n}(\sigma\tau) \end{bmatrix}.$$

and so ρ_n is a representation because

$$\mu_{\ell^n}^{b_n(\tau)+b_n(\sigma)\chi_{\ell,n}(\tau)} = (d^{1/\ell^n})^\tau/d^{1/\ell^n} \cdot ((d^{1/\ell^n})^\sigma/d^{1/\ell^n})^\tau = (d^{1/\ell^n})^{\sigma\tau}/d^{1/\ell^n}.$$

The diagram

$$
\begin{array}{ccc}
G_{n+1} & \longrightarrow & G_n \\
\downarrow{\scriptstyle \rho_{n+1}} & & \downarrow{\scriptstyle \rho_n} \\
\mathrm{GL}_2(\mathbb{Z}/\ell^{n+1}\mathbb{Z}) & \longrightarrow & \mathrm{GL}_2(\mathbb{Z}/\ell^n\mathbb{Z})
\end{array}
$$

commutes if $b_{n+1}(\sigma) + \ell^n\mathbb{Z} = b_n(\sigma|_{\mathbb{F}_n})$ and $\chi_{\ell,n+1}(\sigma) + \ell^n\mathbb{Z} = \chi_{\ell,n}(\sigma|_{\mathbb{F}_n})$ for all $\sigma \in G_{n+1}$, and these equalities hold because

$$\mu_{\ell^n}^{b_{n+1}(\sigma)} = (\mu_{\ell^{n+1}}^{b_{n+1}(\sigma)})^\ell = ((d^{1/\ell^{n+1}})^\sigma/d^{1/\ell^{n+1}})^\ell = (d^{1/\ell^n})^\sigma/d^{1/\ell^n}$$

and

$$\mu_{\ell^n}^{\chi_{\ell,n+1}(\sigma)} = (\mu_{\ell^{n+1}}^{\chi_{\ell,n+1}(\sigma)})^\ell = (\mu_{\ell^{n+1}}^\sigma)^\ell = \mu_{\ell^n}^\sigma.$$

Now prepending the projection $\pi_n : G_{\mathbb{Q}} \longrightarrow G_n$ to each ρ_n, the maps $\rho_{n+1} : G_{\mathbb{Q}} \longrightarrow \mathrm{GL}_2(\mathbb{Z}/\ell^{n+1}\mathbb{Z})$ and $\rho_n : G_{\mathbb{Q}} \longrightarrow \mathrm{GL}_2(\mathbb{Z}/\ell^n\mathbb{Z})$ are compatible with the transition map from $\mathrm{GL}_2(\mathbb{Z}/\ell^{n+1}\mathbb{Z})$ to $\mathrm{GL}_2(\mathbb{Z}/\ell^n\mathbb{Z})$, and so altogether we have a representation

$$\rho_{\ell,d} : G_{\mathbb{Q}} \longrightarrow \mathrm{GL}_2(\mathbb{Z}_\ell), \qquad \rho_{\ell,d}(\sigma) = \begin{bmatrix} 1 & b_{\ell,d}(\sigma) \\ 0 & \chi_\ell(\sigma) \end{bmatrix},$$

with χ_ℓ the ℓ-adic cyclotomic character of $G_{\mathbb{Q}}$. The triangular structure of the matrices $\rho_{\ell,d}(G_{\mathbb{Q}})$ shows that this representation is reducible but not semisimple: the subspace $0 \times \mathbb{Z}_\ell$ (row vectors) of $\mathbb{Z}_\ell^2$ is fixed under the representation but it has no fixed complement.

Exercises

9.3.1. Show that every open normal subgroup of $G_{\mathbb{Q}}$ takes the form $U(\mathbb{F})$ for a Galois number field $\mathbb{F}$. (A hint for this exercise is at the end of the book.)

9.3.2. Show that Theorem 9.3.1 follows from Theorem 9.1.2. (A hint for this exercise is at the end of the book.)

9.3.3. Let $\rho : G \longrightarrow H$ be a homomorphism of topological groups. Show that ρ is continuous if and only if $\rho^{-1}(V)$ is open for each V in a basis of neighborhoods of the identity 1_H.

9.3.4. Show that any continuous homomorphism $\rho : G_{\mathbb{Q}} \longrightarrow \mathrm{GL}_d(\mathbb{C})$ factors through $\mathrm{Gal}(\mathbb{F}/\mathbb{Q}) \longrightarrow \mathrm{GL}_d(\mathbb{C})$ for some Galois number field $\mathbb{F}$. Hence its image is finite. (A hint for this exercise is at the end of the book.)

9.3.5. Suppose that $\rho : G_{\mathbb{Q}} \longrightarrow \mathbb{C}^*$ factors through $\mathrm{Gal}(\mathbb{Q}(\mu_N)/\mathbb{Q})$ and also factors through $\mathrm{Gal}(\mathbb{Q}(\mu_{N'})/\mathbb{Q})$. Show that ρ factors through $\mathrm{Gal}(\mathbb{Q}(\mu_n)/\mathbb{Q})$ where $n = \gcd(N, N')$. (A hint for this exercise is at the end of the book.)

9.3.6. Establish formula (9.13).

9.3.7. (a) Let $\rho : G_{\mathbb{Q}} \longrightarrow \mathrm{GL}_d(\mathbb{L})$ be a Galois representation as in Definition 9.3.2. Show that this makes $\mathbb{L}^d$ a Galois representation as in Definition 9.3.4. On the other hand, let V be a d-dimensional ℓ-adic Galois representation as in Definition 9.3.4. Show that any choice of ordered basis of V gives a representation $\rho : G_{\mathbb{Q}} \longrightarrow \mathrm{GL}_d(\mathbb{L})$ as in Definition 9.3.2. Thus the two definitions of Galois representation are compatible. Remember to check continuity. (Hints for this exercise are at the end of the book.)

(b) Show that the notions of equivalence in the two definitions are also compatible.

9.4 Galois representations and elliptic curves

This section associates 2-dimensional Galois representations to elliptic curves.

Let E be an elliptic curve over $\mathbb{Q}$ and let ℓ be prime. Multiplication by ℓ on ℓ-power torsion of E gives maps

$$E[\ell] \longleftarrow E[\ell^2] \longleftarrow E[\ell^3] \longleftarrow \cdots$$

The ℓ-adic Tate module of E is the resulting inverse limit,

$$\mathrm{Ta}_\ell(E) = \varprojlim_n \{E[\ell^n]\}.$$

Choose a basis (P_n, Q_n) of $E[\ell^n]$ for each $n \in \mathbb{Z}^+$ with the choices compatible in that each basis is a lift of its predecessor,

$$[\ell]P_{n+1} = P_n \quad \text{and} \quad [\ell]Q_{n+1} = Q_n, \quad n \in \mathbb{Z}^+.$$

Each basis determines an isomorphism $E[\ell^n] \overset{\sim}{\longrightarrow} (\mathbb{Z}/\ell^n\mathbb{Z})^2$, and so

$$\mathrm{Ta}_\ell(E) \cong \mathbb{Z}_\ell^2.$$

Any automorphism $\sigma \in G_{\mathbb{Q}}$ acts on $\overline{\mathbb{Q}}^3$ coordinatewise, and since $(\overline{\mathbb{Q}}^*)^\sigma = \overline{\mathbb{Q}}^*$ the action descends to $\mathbb{P}^2(\overline{\mathbb{Q}})$,

$$P = [x_0, x_1, x_2] \mapsto [x_0^\sigma, x_1^\sigma, x_2^\sigma] = P^\sigma.$$

Hence each $\sigma \in G_{\mathbb{Q}}$ defines an automorphism of $\mathrm{Div}^0(E)$,

$$\left(\sum n_P(P) \right)^\sigma = \sum n_P(P^\sigma).$$

Since $\mathrm{div}(f)^\sigma = \mathrm{div}(f^\sigma)$ for any $f \in \overline{\mathbb{Q}}(E)$, the automorphism descends to $\mathrm{Pic}^0(E)$. Under the isomorphic identification of E with $\mathrm{Pic}^0(E)$, multiplication by ℓ^n on E for any $n \in \mathbb{Z}^+$ becomes purely formal on $\mathrm{Pic}^0(E)$, and so it clearly commutes with the $G_{\mathbb{Q}}$-action. Thus the actions on E commute as well, and so the Galois action restricts to ℓ^n-torsion,

$$G_{\mathbb{Q}} \longrightarrow \mathrm{Aut}(E[\ell^n]).$$

These maps are compatible in that for each n the following diagram commutes.

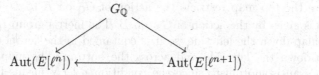

That is, the Tate module is indeed a $G_{\mathbb{Q}}$-module. Each basis (P_n, Q_n) determines an isomorphism $\mathrm{Aut}(E[\ell^n]) \overset{\sim}{\longrightarrow} \mathrm{GL}_2(\mathbb{Z}/\ell^n\mathbb{Z})$, and these combine to give $\mathrm{Aut}(\mathrm{Ta}_\ell(E)) \overset{\sim}{\longrightarrow} \mathrm{GL}_2(\mathbb{Z}_\ell)$. Since $G_{\mathbb{Q}}$ acts on $\mathrm{Ta}_\ell(E)$ the net result is a homomorphism

$$\rho_{E,\ell} : G_{\mathbb{Q}} \longrightarrow \mathrm{GL}_2(\mathbb{Z}_\ell) \subset \mathrm{GL}_2(\mathbb{Q}_\ell).$$

The homomorphism is continuous by an argument similar to the one for χ_ℓ in the previous section (Exercise 9.4.1). Thus $\rho_{E,\ell}$ is a Galois representation, the 2-dimensional Galois representation associated to E.

Theorem 9.4.1. *Let ℓ be prime and let E be an elliptic curve over $\mathbb{Q}$ with conductor N. The Galois representation $\rho_{E,\ell}$ is unramified at every prime $p \nmid \ell N$. For any such p let $\mathfrak{p} \subset \overline{\mathbb{Z}}$ be any maximal ideal over p. Then the characteristic equation of $\rho_{E,\ell}(\mathrm{Frob}_\mathfrak{p})$ is*

$$x^2 - a_p(E)x + p = 0.$$

The Galois representation $\rho_{E,\ell}$ is irreducible.

Alternatively, the vector space

$$V_\ell(E) = \mathrm{Ta}_\ell(E) \otimes \mathbb{Q} \cong \mathbb{Q}_\ell^2$$

is the 2-dimensional Galois representation as in Definition 9.3.4 associated to E, and a diagram as follows commutes.

$$
\begin{array}{ccc}
V_\ell(E) \times G_{\mathbb{Q}} & \longrightarrow & V_\ell(E) \\
\downarrow & & \downarrow \\
\mathbb{Q}_\ell^2 \times \mathrm{GL}_2(\mathbb{Q}_\ell) & \longrightarrow & \mathbb{Q}_\ell^2
\end{array}
$$

The theorem can be rephrased appropriately, e.g., its displayed equation is the characteristic equation of Frob_p itself as an automorphism of $V_\ell(E)$. As noted in the introduction to this chapter, the characteristic relation of $\mathrm{Frob}_\mathfrak{p}$ in $\mathrm{End}(V_\ell(E))$ is independent of ℓ.

Proof. Let $p \nmid \ell N$ and let $\mathfrak{p}$ lie over p. Recall the absolute Galois group of $\mathbb{F}_p$, $G_{\mathbb{F}_p} = \mathrm{Aut}(\overline{\mathbb{F}}_p)$. For each $n \in \mathbb{Z}^+$ there is a commutative diagram

$$
\begin{array}{ccc}
D_{\mathfrak{p}} & \longrightarrow & \mathrm{Aut}(E[\ell^n]) \\
\downarrow & & \downarrow \\
G_{\mathbb{F}_p} & \longrightarrow & \mathrm{Aut}(\widetilde{E}[\ell^n])
\end{array}
$$

where the top map restricts the action of $G_{\mathbb{Q}}$ on E to $D_{\mathfrak{p}}$ and the bottom map is given by the action of $G_{\mathbb{F}_p}$ on $\widetilde{E}$. The inertia group $I_{\mathfrak{p}}$ is the kernel of the map down the left side, so it is contained in the kernel of the composite map down the left side and across the bottom. The map down the right side is an isomorphism since the condition $p \nmid \ell N$ means that E has good reduction at p and the reduction preserves ℓ^n-torsion structure, cf. Section 8.3, Theorem 8.1.2, and Proposition 8.4.4. Consequently, $I_{\mathfrak{p}}$ is contained in the kernel of the map across the top. Since n is arbitrary this means that $I_{\mathfrak{p}} \subset \ker \rho_{E,\ell}$ as desired.

The relation $a_p(E) = \sigma_{p,*} + \sigma_p^*$ as endomorphisms of $\mathrm{Pic}^0(\widetilde{E})$ (Proposition 8.3.2) and the preservation of ℓ^n-torsion under reduction combine to show that $\mathrm{Frob}_{\mathfrak{p}}$ satisfies its asserted characteristic equation. Consider the diagram

$$
\begin{array}{ccc}
E[\ell^n] & \xrightarrow{\ a_p(E)\ } & E[\ell^n] \\
\downarrow & & \downarrow \\
\widetilde{E}[\ell^n] & \xrightarrow{\ \sigma_p + p\,\sigma_p^{-1}\ } & \widetilde{E}[\ell^n].
\end{array}
$$

Identifying elliptic curves with their degree-0 Picard groups as earlier, and recalling from equations (8.14) and (8.15) that $\sigma_p = \sigma_{p,*}$ and $p\sigma_p^{-1} = \sigma_p^*$ under the identification, we see that the diagram commutes. The same diagram but instead with $\mathrm{Frob}_{\mathfrak{p}} + p\,\mathrm{Frob}_{\mathfrak{p}}^{-1}$ across the top row also commutes. Since the vertical arrows are isomorphisms, $a_p(E) = \mathrm{Frob}_{\mathfrak{p}} + p\,\mathrm{Frob}_{\mathfrak{p}}^{-1}$ on $E[\ell^n]$, and since n is arbitrary, the equality extends to $\mathrm{Ta}_\ell(E)$. Multiply the equality through by $\mathrm{Frob}_{\mathfrak{p}}$ to get $\mathrm{Frob}_{\mathfrak{p}}^2 - a_p(E)\mathrm{Frob}_{\mathfrak{p}} + p = 0$.

The previous paragraph shows that the minimal polynomial of $\mathrm{Frob}_{\mathfrak{p}}$ divides $x^2 - a_p(E)x + p$ but not yet that this is the characteristic polynomial. (For example, the identity operator on a 2-dimensional vector space satisfies any quadratic polynomial $(x - 1)(x - a)$, not only its characteristic polynomial $(x-1)^2$.) To finish establishing the characteristic polynomial of $\mathrm{Frob}_{\mathfrak{p}}$ for $p \nmid \ell N$, we show that $\det \rho_{E,\ell}(\mathrm{Frob}_{\mathfrak{p}}) = p$. Let let $\rho_n : G_{\mathbb{Q}} \longrightarrow \mathrm{GL}_2(\mathbb{Z}/\ell^n\mathbb{Z})$ be the nth entry of $\rho_{E,\ell}$ for $n \in \mathbb{Z}^+$. As in Lemma 7.6.1, the Weil pairing shows that the action of $\sigma \in G_{\mathbb{Q}}$ on the root of unity μ_{ℓ^n} is given by the determinant, but by definition the action is also to raise μ_{ℓ^n} to the nth entry of the cyclotomic character $\chi_\ell(\sigma)$,

$$\mu_{\ell^n}^\sigma = \mu_{\ell^n}^{\det \rho_n(\sigma)} = \mu_{\ell^n}^{\chi_{\ell,n}(\sigma)}.$$

That is, $\det \rho_n(\sigma) = \chi_{\ell,n}(\sigma)$ in $(\mathbb{Z}/\ell^n\mathbb{Z})^*$ for all n, so $\det \rho_{E,\ell}(\sigma) = \chi_\ell(\sigma)$ in $\mathbb{Z}_\ell^*$. In particular (9.13) gives $\det \rho_{E,\ell}(\mathrm{Frob}_p) = p$, as desired.

Proving the last statement of the theorem is beyond the scope of this book. $\qquad\square$

Exercises

9.4.1. Show that $\rho_{E,\ell}$ is continuous.

9.4.2. Recall from Section 8.3 that we have defined

$$t_{p^e}(E) = p^e + 1 - |\widetilde{E}(\mathbb{F}_{p^e})| \quad \text{for } e \geq 1, \qquad t_1(E) = 2$$

(so that especially $t_p(E) = a_p(E)$), shown that

$$t_{p^e}(E) = \sigma_{p^e,*} + \sigma_{p^e}^*, \quad e \geq 1,$$

and stated that there is a recurrence

$$t_{p^e}(E) = t_p(E)t_{p^{e-1}}(E) - \mathbf{1}_E(p)pt_{p^{e-2}}(E) \quad \text{for } e \geq 2,$$

where $\mathbf{1}_E$ is the trivial character modulo the conductor N of E. This exercise proves the recurrence.

(a) Show that the end of the proof of Theorem 9.4.1 extends to show that for $p \nmid \ell N$,

$$\mathrm{tr}\,(\mathrm{Frob}_\mathfrak{p}^e) = t_{p^e}(E), \quad e \geq 1.$$

Note that the equality holds for $e = 0$ as well. Let λ_1 and λ_2 be the eigenvalues of $\mathrm{Frob}_\mathfrak{p}$. Show that for $e \geq 2$,

$$\mathrm{tr}\,(\mathrm{Frob}_\mathfrak{p}^e) = \lambda_1^e + \lambda_2^e = (\lambda_1 + \lambda_2)(\lambda_1^{e-1} + \lambda_2^{e-1}) - \lambda_1\lambda_2(\lambda_1^{e-2} + \lambda_2^{e-2}).$$

Use this and the identities $\det \mathrm{Frob}_\mathfrak{p} = p$, $\mathrm{tr}\,\mathrm{Frob}_\mathfrak{p} = t_p(E)$ to prove the recurrence for $p \nmid \ell N$, so that letting ℓ vary proves the recurrence for $p \nmid N$.

(b) For $p \mid N$, (8.11) says that we may take $\widetilde{E} : (y - m_1 x)(y - m_2 x) = x^3$ with $m_1 + m_2, m_1 m_2 \in \mathbb{F}_p$. Show that the formula

$$t \mapsto ((t - m_1)(t - m_2), t(t - m_1)(t - m_2))$$

describes a map from $\mathbb{P}^1(\mathbb{F}_q)$ to $\widetilde{E}(\mathbb{F}_q)$. By considering the map $(x, y) \mapsto y/x$ from $\widetilde{E}(\mathbb{F}_q) - \{(0,0)\}$ to $\mathbb{P}^1(\mathbb{F}_q)$ also, show that the displayed map injects except for possibly hitting $(0,0)$ more than once (when m_1, m_2 are distinct and lie in $\mathbb{F}_q$) and that the map surjects except for possibly missing $(0,0)$ (when m_1, m_2 do not lie in $\mathbb{F}_q$), and so the map bijects when $m_1 = m_2$ lies in $\mathbb{F}_q$.

The reduction $\widetilde{E}$ is multiplicative if $m_1 \neq m_2$. Show that if the reduction is split, i.e., $m_1, m_2 \in \mathbb{F}_p$, then $t_{p^e}(E) = 1$ for all $e \geq 1$. Show that if the reduction is nonsplit, i.e., $m_1, m_2 \notin \mathbb{F}_p$, then $t_{p^e}(E) = (-1)^e$ for all $e \geq 1$. Show that the recurrence is satisfied in both cases.

The reduction is additive if $m_1 = m_2$. Show that the common value m lies in $\mathbb{F}_p$ (the argument will be different for $p = 2$). Show that $t_{p^e}(E) = 0$ for all $e \geq 1$, and show that the recurrence is satisfied in this case as well.

(c) Again assume that $p \nmid N$. Show that

$$(1 - a_p(E)x + px^2)^{-1} = (1 - \lambda_1 x)^{-1}(1 - \lambda_2 x)^{-1} = \sum_{e=0}^{\infty} \left(\sum_{c+d=e} \lambda_1^c \lambda_2^d \right) x^e.$$

Explain why it follows that whereas the normalized prime-power solution-counts of the elliptic curve are $t_{p^e}(E) = \lambda_1^e + \lambda_2^e$, the corresponding prime-power Dirichlet coefficients of $L(s, E)$ are $a_{p^e}(E) = \sum_{c+d=e} \lambda_1^c \lambda_2^d$.

9.4.3. Let E and E' be isogenous elliptic curves over $\mathbb{Q}$. Show that their Galois representations $V_\ell(E)$ and $V_\ell(E')$ are equivalent. Faltings's Isogeny Theorem, mentioned in Section 8.8, states the converse. (A hint for this exercise is at the end of the book.)

9.4.4. This exercise gives an example of a *mod ℓ representation*. Such representations will play a role in Section 9.6.

(a) Let E be an elliptic curve over $\mathbb{Q}$ with conductor N and let ℓ be prime. Thus $E[\ell] \cong \mathbb{F}_\ell^2$. The mod ℓ representation of E is

$$\overline{\rho}_{E,\ell} : G_\mathbb{Q} \longrightarrow \mathrm{Gal}(\mathbb{Q}(E[\ell])/\mathbb{Q}) \longrightarrow \mathrm{GL}_2(\mathbb{F}_\ell).$$

Explain why for any maximal ideal $\mathfrak{p} \subset \overline{\mathbb{Z}}$ over any $p \nmid \ell N$, the characteristic equation of $\overline{\rho}_{E,\ell}(\mathrm{Frob}_\mathfrak{p})$ is $x^2 - a_p(E)x + p = 0$. (Here the coefficients are in $\mathbb{F}_\ell$.)

(b) Consider the elliptic curve $E : y^2 = x^3 - d$ where $d \in \mathbb{Z}$ is cubefree, and let $\ell = 2$. Show that $\mathbb{Q}(E[2])$ is the non-Abelian Galois number field $\mathbb{F} = \mathbb{Q}(d^{1/3}, \mu_3)$ from Section 9.1. Therefore the embedding (9.6) followed by reduction modulo 2 gives an isomorphism $\mathrm{Gal}(\mathbb{Q}(E[2])/\mathbb{Q}) \longrightarrow \mathrm{GL}_2(\mathbb{F}_2)$. Use this to describe $a_p(E)$ (mod 2).

(c) Let E be any elliptic curve over $\mathbb{Q}$ with conductor N. Suppose that $E[\ell](\mathbb{Q}) \neq \{0\}$, i.e., E has an ℓ-torsion point with rational coordinates. Show that for any maximal ideal $\mathfrak{p} \subset \overline{\mathbb{Z}}$ over any $p \nmid \ell N$, the characteristic equation of $\overline{\rho}_{E,\ell}(\mathrm{Frob}_\mathfrak{p})$ is $(x - 1)(x - p) = 0$. Explain how this shows that $E[\ell](\mathbb{Q})$ cannot be all of $E[\ell]$ for $\ell > 2$, cf. the function field result from Section 7.6 that $\mathbb{Q}(j, E_j[N])$ contains $\boldsymbol{\mu}_N$. (A hint for this exercise is at the end of the book.)

9.5 Galois representations and modular forms

This section associates Galois representations to normalized Hecke eigenforms. The process involves three stages:

- Given a level N, construct an ℓ-adic representation $V_\ell(X_1(N))$ associated to the modular curve $X_1(N)$. The representation has dimension $2g$ over $\mathbb{Q}_\ell$, where g is the genus of $X_1(N)$. The representation is a $\mathbb{T}_\mathbb{Z}$-module along with being a $G_\mathbb{Q}$-module.
- Given a normalized Hecke eigenform $f \in \mathcal{S}_2(N, \chi)$, construct an ℓ-adic representation $V_\ell(A_f)$ associated to the Abelian variety A_f. Since the Abelian variety A_f is a quotient of the Jacobian $J_1(N)$, its associated representation $V_\ell(A_f)$ is correspondingly a quotient of $V_\ell(X_1(N))$. The representation has dimension $2d$ over $\mathbb{Q}_\ell$, where $d = [\mathbb{K}_f : \mathbb{Q}]$ is the degree of the number field generated by the Fourier coefficients of f. The representation $V_\ell(A_f)$ is an $\mathcal{O}_f$-module along with being a $G_\mathbb{Q}$-module.
- In fact $V_\ell(A_f)$ is a free module of rank 2 over $\mathbb{K}_f \otimes_\mathbb{Q} \mathbb{Q}_\ell$. Thus the decomposition $\mathbb{K}_f \otimes_\mathbb{Q} \mathbb{Q}_\ell \cong \prod_{\lambda|\ell} \mathbb{K}_{f,\lambda}$ correspondingly decomposes $V_\ell(A_f)$ into representations $V_\lambda(f)$, each such representation having dimension 2 over $\mathbb{K}_{f,\lambda}$. These representations $V_\lambda(f)$, or the corresponding homomorphisms $\rho_{f,\lambda} : G_\mathbb{Q} \longrightarrow \mathrm{GL}_2(\mathbb{K}_{f,\lambda})$, are the end products of this section.

Let N be a positive integer. The modular curve $X_1(N)$ is a projective nonsingular algebraic curve over $\mathbb{Q}$, and its Picard group is the Abelian group of divisor classes on the points of $X_1(N)$,

$$\mathrm{Pic}^0(X_1(N)) = \mathrm{Div}^0(X_1(N))/\mathrm{Div}^\ell(X_1(N)).$$

Let $X_1(N)_\mathbb{C}$ denote the curve over $\mathbb{C}$ defined by the same equations as $X_1(N)$. The commutative diagram of inclusions

$$
\begin{array}{ccc}
\mathrm{Div}^0(X_1(N)) & \longrightarrow & \mathrm{Div}^0(X_1(N)_\mathbb{C}) \\
\uparrow & & \uparrow \\
\mathrm{Div}^\ell(X_1(N)) & \longrightarrow & \mathrm{Div}^\ell(X_1(N)_\mathbb{C})
\end{array}
$$

gives rise to a map of quotients $\mathrm{Pic}^0(X_1(N)) \longrightarrow \mathrm{Pic}^0(X_1(N)_\mathbb{C})$. The map is injective if $\mathrm{Div}^\ell(X_1(N))$ is all of $\mathrm{Div}^0(X_1(N)) \cap \mathrm{Div}^\ell(X_1(N)_\mathbb{C})$; that is, the map of Picard groups is injective if any function in $f \in \mathbb{C}(X_1(N))$ whose zeros and poles are algebraic points is a scalar multiple of some $\tilde{f} \in \overline{\mathbb{Q}}(X_1(N))$. We show that this is the case. Given f as just above, consider its divisor $D = \mathrm{div}(f)$. Let $L(D)$ denote the linear space of the divisor as in Section 3.4, a nonzero space because it contains f^{-1}, and let $L_{\overline{\mathbb{Q}}}(D)$ denote $L(D) \cap \overline{\mathbb{Q}}(X_1(N))$. A basic fact of algebraic geometry is

$$L(D) = L_{\overline{\mathbb{Q}}}(D) \otimes_{\overline{\mathbb{Q}}} \mathbb{C},$$

in consequence of the analogous fact for affine curves. Thus $L_{\overline{\mathbb{Q}}}(D) \neq 0$, meaning that there exists a function $\tilde{f} \in \overline{\mathbb{Q}}(X_1(N))$ such that $\operatorname{div}(\tilde{f}^{-1}) \geq -D$, and hence $\operatorname{div}(\tilde{f}^{-1}) = -D$ because both divisors have degree 0. That is, $\operatorname{div}(\tilde{f}) = D$, which is to say that f is a scalar multiple of $\tilde{f}$ as desired. So $\operatorname{Pic}^0(X_1(N))$ can be identified with a subgroup of $\operatorname{Pic}^0(X_1(N)_{\mathbb{C}})$.

Let g denote the genus of $X_1(N)$. The curve $X_1(N)_{\mathbb{C}}$ can also be viewed as a compact Riemann surface, and as in Chapter 6 its Jacobian is a g-dimensional complex torus obtained from integration modulo homology,

$$ J_1(N) = \operatorname{Jac}(X_1(N)_{\mathbb{C}}) = \mathcal{S}_2(\Gamma_1(N))^{\wedge}/H_1(X_1(N)_{\mathbb{C}}, \mathbb{Z}) \cong \mathbb{C}^g/\Lambda_g. $$

The complex Picard group is naturally isomorphic to the Jacobian by Abel's Theorem as in Section 6.1. Thus altogether there is an inclusion of ℓ^n-torsion for any prime ℓ,

$$ i_n : \operatorname{Pic}^0(X_1(N))[\ell^n] \longrightarrow \operatorname{Pic}^0(X_1(N)_{\mathbb{C}})[\ell^n] \cong (\mathbb{Z}/\ell^n\mathbb{Z})^{2g}. $$

Recall that Igusa's Theorem (Theorem 8.6.1) states that $X_1(N)$ has good reduction at primes $p \nmid N$, so also there is a natural surjective reduction map $\operatorname{Pic}^0(X_1(N)) \longrightarrow \operatorname{Pic}^0(\tilde{X}_1(N))$ restricting to

$$ \pi_n : \operatorname{Pic}^0(X_1(N))[\ell^n] \longrightarrow \operatorname{Pic}^0(\tilde{X}_1(N))[\ell^n]. $$

We state without proof two generalizations of facts we have used about elliptic curves:

- The inclusion i_n is in fact an isomorphism.
- So is the surjection π_n for $p \nmid \ell N$.

These follow from results of algebraic geometry. Specifically, if a curve X over a field $\mathbf{k}$ has genus g and M is coprime to $\operatorname{char}(\mathbf{k})$ then $\operatorname{Pic}^0(X)[M] \cong (\mathbb{Z}/M\mathbb{Z})^{2g}$, and if a curve X over $\mathbb{Q}$ has good reduction at a prime $p \nmid M$ then the reduction map is injective on $\operatorname{Pic}^0(X)[M]$. The upshot is that the prime power torsion group structure of $\operatorname{Pic}^0(X_1(N))$ is the same over $\mathbb{C}$ (analytically and algebraically), over $\mathbb{Q}$, and over $\mathbb{F}_p$ for good primes p away from ℓ.

The ℓ-adic Tate module of $X_1(N)$ is

$$ \operatorname{Ta}_\ell(\operatorname{Pic}^0(X_1(N))) = \varprojlim_n \{\operatorname{Pic}^0(X_1(N))[\ell^n]\}. $$

Similarly to the previous section but now using the first of the two quoted facts, choosing bases of $\operatorname{Pic}^0(X_1(N))[\ell^n]$ compatibly for all n shows that

$$ \operatorname{Ta}_\ell(\operatorname{Pic}^0(X_1(N))) \cong \mathbb{Z}_\ell^{2g}. $$

Any automorphism $\sigma \in G_{\mathbb{Q}}$ defines an automorphism of $\operatorname{Div}^0(X_1(N))$,

$$ \left(\sum n_P(P)\right)^{\sigma} = \sum n_P(P^{\sigma}). $$

Since $\mathrm{div}(f)^\sigma = \mathrm{div}(f^\sigma)$ for any $f \in \overline{\mathbb{Q}}(X_1(N))$, the automorphism descends to $\mathrm{Pic}^0(X_1(N))$,

$$\mathrm{Pic}^0(X_1(N)) \times G_{\mathbb{Q}} \longrightarrow \mathrm{Pic}^0(X_1(N)). \qquad (9.14)$$

The Galois action commutes with the purely formal action of multiplication by ℓ^n for any $n \in \mathbb{Z}^+$, so the action restricts to $\mathrm{Pic}^0(X_1(N))[\ell^n]$. For each n there is a commutative diagram

$$
\begin{array}{ccc}
 & G_{\mathbb{Q}} & \\
 \swarrow & & \searrow \\
\mathrm{Aut}(\mathrm{Pic}^0(X_1(N))[\ell^n]) & \longleftarrow & \mathrm{Aut}(\mathrm{Pic}^0(X_1(N))[\ell^{n+1}]).
\end{array}
$$

Again as in the previous section this leads to a continuous homomorphism

$$\rho_{X_1(N),\ell} : G_{\mathbb{Q}} \longrightarrow \mathrm{GL}_{2g}(\mathbb{Z}_\ell) \subset \mathrm{GL}_{2g}(\mathbb{Q}_\ell).$$

This is the $2g$-dimensional Galois representation associated to $X_1(N)$.

Recall from Chapter 6 that the Hecke algebra over $\mathbb{Z}$ is the algebra of endomorphisms of $\mathcal{S}_2(\Gamma_1(N))$ generated over $\mathbb{Z}$ by the Hecke operators,

$$\mathbb{T}_{\mathbb{Z}} = \mathbb{Z}[\{T_n, \langle n \rangle : n \in \mathbb{Z}^+\}].$$

The Hecke algebra acts on $\mathrm{Pic}^0(X_1(N))$, cf. the bottom rows of diagrams (7.18) and (7.19),

$$\mathbb{T}_{\mathbb{Z}} \times \mathrm{Pic}^0(X_1(N)) \longrightarrow \mathrm{Pic}^0(X_1(N)). \qquad (9.15)$$

Since the action is linear it restricts to ℓ-power torsion, and so it extends to $\mathrm{Ta}_\ell(\mathrm{Pic}^0(X_1(N)))$. From Section 7.9 the Hecke action is defined over $\mathbb{Q}$. So the Galois action (9.14) and the Hecke action (9.15) on $\mathrm{Pic}^0(X_1(N))$ commute, and therefore so do the two actions on $\mathrm{Ta}_\ell(\mathrm{Pic}^0(X_1(N)))$.

Theorem 9.5.1. *Let ℓ be prime and let N be a positive integer. The Galois representation $\rho_{X_1(N),\ell}$ is unramified at every prime $p \nmid \ell N$. For any such p let $\mathfrak{p} \subset \overline{\mathbb{Z}}$ be any maximal ideal over p. Then $\rho_{X_1(N),\ell}(\mathrm{Frob}_{\mathfrak{p}})$ satisfies the polynomial equation*

$$x^2 - T_p\, x + \langle p \rangle p = 0.$$

Similarly to how we have abbreviated $T_{p,*}$ to T_p on $\mathrm{Pic}^0(X_1(N))$ all along, the last formula in the theorem omits the asterisk from the subscript of both Hecke operators on $\mathrm{Ta}_\ell(\mathrm{Pic}^0(X_1(N)))$. As in the previous section, the vector space

$$V_\ell(X_1(N)) = \mathrm{Ta}_\ell(\mathrm{Pic}^0(X_1(N))) \otimes \mathbb{Q} \cong \mathbb{Q}_\ell^{2g}$$

can be taken as the Galois representation rather than $\rho_{X_1(N),\ell}$, giving a corresponding commutative diagram, and the theorem can be rephrased appropriately.

Proof. Let $p \nmid \ell N$ and let $\mathfrak{p}$ lie over p. As in the proof of Theorem 9.4.1, for each $n \in \mathbb{Z}^+$ there is a commutative diagram

$$
\begin{array}{ccc}
D_{\mathfrak{p}} & \longrightarrow & \operatorname{Aut}(\operatorname{Pic}^0(X_1(N))[\ell^n]) \\
\downarrow & & \downarrow \\
G_{\mathbb{F}_p} & \longrightarrow & \operatorname{Aut}(\operatorname{Pic}^0(\widetilde{X}_1(N))[\ell^n]).
\end{array}
$$

The map down the right side is an isomorphism by the second quoted fact at the beginning of the section. Similarly to before, $I_{\mathfrak{p}} \subset \ker \rho_{X_1(N),\ell}$.

The second part of the theorem follows from the Eichler–Shimura Relation (Theorem 8.7.2) and the preservation of ℓ^n-torsion under reduction. The Eichler–Shimura Relation restricts to ℓ^n-torsion,

$$
\begin{array}{ccc}
\operatorname{Pic}^0(X_1(N))[\ell^n] & \xrightarrow{\ T_p\ } & \operatorname{Pic}^0(X_1(N))[\ell^n] \\
\downarrow & & \downarrow \\
\operatorname{Pic}^0(\widetilde{X}_1(N))[\ell^n] & \xrightarrow{\sigma_{p,*}+\widetilde{\langle p \rangle}_*\sigma_p^*} & \operatorname{Pic}^0(\widetilde{X}_1(N))[\ell^n].
\end{array}
$$

The same diagram but instead with $\operatorname{Frob}_{\mathfrak{p}} + \langle p \rangle p \operatorname{Frob}_{\mathfrak{p}}^{-1}$ across the top row also commutes, cf. (8.15). Since the vertical arrows are isomorphisms, $T_p = \operatorname{Frob}_{\mathfrak{p}} + \langle p \rangle p \operatorname{Frob}_{\mathfrak{p}}^{-1}$ on $\operatorname{Pic}^0(X_1(N))[\ell^n]$, and since n is arbitrary, the equality extends to $\operatorname{Ta}_\ell(\operatorname{Pic}^0(X_1(N)))$. The result follows.   □

Now we move to the second bullet at the beginning of the section. To proceed from modular curves to modular forms, consider a normalized eigenform

$$
f \in \mathcal{S}_2(N, \chi).
$$

Recall from Chapter 6 that the Hecke algebra contains an ideal associated to f, the kernel of the eigenvalue map,

$$
I_f = \{T \in \mathbb{T}_{\mathbb{Z}} : Tf = 0\},
$$

and the Abelian variety of f is defined as

$$
A_f = \mathrm{J}_1(N)/I_f\mathrm{J}_1(N).
$$

This is a complex analytic object. We do not define an algebraic version of it because its role here is auxiliary. By (6.12) and Exercise 6.5.2 there is an isomorphism

$$
\mathbb{T}_{\mathbb{Z}}/I_f \xrightarrow{\ \sim\ } \mathcal{O}_f \quad \text{where} \quad \mathcal{O}_f = \mathbb{Z}[\{a_n(f) : n \in \mathbb{Z}^+\}].
$$

Under this isomorphism each Fourier coefficient $a_p(f)$ acts on A_f as $T_p + I_f$. Also, $\mathcal{O}_f$ contains the values $\chi(n)$ for $n \in \mathbb{Z}^+$ (this was shown in Exercise 6.5.2) and $\chi(p)$ acts on A_f as $\langle p \rangle + I_f$. The ring $\mathcal{O}_f$ generates the number field of f, denoted $\mathbb{K}_f$. The extension degree $d = [\mathbb{K}_f : \mathbb{Q}]$ is also the dimension of A_f as a complex torus. As with elliptic curves and modular curves, the Abelian variety has an ℓ-adic Tate module,

$$\mathrm{Ta}_\ell(A_f) = \varprojlim_n \{A_f[\ell^n]\} \cong \mathbb{Z}_\ell^{2d}.$$

The action of $\mathcal{O}_f$ on A_f is defined on ℓ-power torsion and thus extends to an action on $\mathrm{Ta}_\ell(A_f)$. The following lemma shows that $G_\mathbb{Q}$ acts on $\mathrm{Ta}_\ell(A_f)$ as well.

Lemma 9.5.2. *The map* $\mathrm{Pic}^0(X_1(N))[\ell^n] \longrightarrow A_f[\ell^n]$ *is a surjection. Its kernel is stable under* $G_\mathbb{Q}$.

Proof. Multiplication by ℓ^n is surjective on $I_f J_1(N)$. Indeed, it is surjective on the complex torus $J_1(N)$, and the commutative Hecke algebra $\mathbb{T}_\mathbb{Z}$ contains both I_f and ℓ^n, so that $\ell^n I_f J_1(N) = I_f \ell^n J_1(N) = I_f J_1(N)$.

To show the first statement of the lemma, take any $y \in A_f[\ell^n]$. Then $y = x + I_f J_1(N)$ for some $x \in J_1(N)$ such that $\ell^n x \in I_f J_1(N)$. Thus $\ell^n x = \ell^n x'$ for some $x' \in I_f J_1(N)$ by the previous paragraph. The difference $x - x'$ lies in $J_1(N)[\ell^n] = \mathrm{Pic}^0(X_1(N))[\ell^n]$ and maps to y as desired.

The kernel is $\mathrm{Pic}^0(X_1(N))[\ell^n] \cap I_f J_1(N) = (I_f J_1(N))[\ell^n]$. We claim that the containment

$$(I_f \mathrm{Pic}^0(X_1(N)))[\ell^n] \subset (I_f J_1(N))[\ell^n],$$

is in fact equality. Granting the equality, the second statement of the lemma follows quickly: the kernel is now $(I_f \mathrm{Pic}^0(X_1(N)))[\ell^n]$. That is, the kernel is $\mathrm{Pic}^0(X_1(N))[\ell^n] \cap I_f \mathrm{Pic}^0(X_1(N))$, which is stable under the Galois action— the first intersectand is stable because the Galois action on $\mathrm{Pic}^0(X_1(N))$ preserves ℓ^n-torsion, and the second is stable because the Galois and Hecke actions on $\mathrm{Pic}^0(X_1(N))$ commute.

To prove that the containment is equality, note that it is a containment of torsion of I_f-images, while if instead we were considering I_f-images of torsion then there would be nothing to show, i.e., $\mathrm{Pic}^0(X_1(N))[\ell^n] = J_1(N)[\ell^n]$ and thus $I_f(\mathrm{Pic}^0(X_1(N))[\ell^n]) = I_f(J_1(N)[\ell^n])$. So the argument will relate the given containment of torsion of I_f-images to an equality of I_f-images of torsion, though not quite ℓ^n-torsion. To do so, let $\mathcal{S}_2 = \mathcal{S}_2(\Gamma_1(N))$ and $H_1 = \mathrm{H}_1(X_1(N)_\mathbb{C}, \mathbb{Z}) \subset \mathcal{S}_2^\wedge$. Thus $J_1(N) = \mathcal{S}_2^\wedge / H_1$ and

$$I_f J_1(N) = (I_f \mathcal{S}_2^\wedge + H_1)/H_1 \cong I_f \mathcal{S}_2^\wedge / (H_1 \cap I_f \mathcal{S}_2^\wedge).$$

Now suppose that $y \in (I_f J_1(N))[\ell^n]$. Then $y = x + H_1$ where by the previous display we may take

$$x \in I_f \mathcal{S}_2^\wedge \quad \text{and} \quad \ell^n x \in H_1 \cap I_f \mathcal{S}_2^\wedge.$$

Proposition 6.2.4 shows that $H_1 \cap I_f \mathcal{S}_2^\wedge$ contains $I_f H_1$ as a subgroup of some finite index M. Consequently $H_1 \cap I_f \mathcal{S}_2^\wedge \subset I_f M^{-1} H_1$. From the previous display and the containment, $\ell^n x \in I_f M^{-1} H_1$, and so

$$x \in I_f M^{-1} \ell^{-n} H_1.$$

That is, $x = T x_0$ where $T \in I_f$ and $x_0 \in \mathcal{S}_2^\wedge$ and $M\ell^n x_0 \in H_1$, and so $y = T(x_0 + H_1)$ where $x_0 + H_1 \in \mathrm{J}_1(N)$ and $M\ell^n(x_0 + H_1) = 0$. In sum, our y from $(I_f \mathrm{J}_1(N))[\ell^n]$ lies in $I_f\big(\mathrm{J}_1(N)[M\ell^n]\big)$, and we are set up to use the equality of I_f-images of torsion,

$$y \in I_f\big(\mathrm{J}_1(N)[M\ell^n]\big) = I_f\big(\mathrm{Pic}^0(X_1(N))[M\ell^n]\big) \subset I_f \mathrm{Pic}^0(X_1(N)).$$

And since $\ell^n y = 0$ in fact $y \in (I_f \mathrm{Pic}^0(X_1(N)))[\ell^n]$. Thus the opposite containment $(I_f \mathrm{J}_1(N))[\ell^n] \subset (I_f \mathrm{Pic}^0(X_1(N)))[\ell^n]$ is proved, establishing the desired equality. As explained above, the proof of the lemma is complete. $\qquad\square$

So $G_\mathbb{Q}$ acts on $A_f[\ell^n]$ and therefore on $\mathrm{Ta}_\ell(A_f)$. The action commutes with the action of $\mathcal{O}_f$ since the $G_\mathbb{Q}$-action and the $\mathbb{T}_\mathbb{Z}$-action commute on $\mathrm{Ta}_\ell(\mathrm{Pic}^0(X_1(N)))$. Choosing coordinates appropriately gives a Galois representation

$$\rho_{A_f,\ell} : G_\mathbb{Q} \longrightarrow \mathrm{GL}_{2d}(\mathbb{Q}_\ell).$$

This is continuous because $\rho_{X_1(N),\ell}$ is continuous and (Exercise 9.5.1)

$$\rho_{X_1(N),\ell}^{-1}(U(n,g)) \subset \rho_{A_f,\ell}^{-1}(U(n,d)), \tag{9.16}$$

where $U(n,g) = \ker\big(\mathrm{GL}_{2g}(\mathbb{Z}_\ell) \longrightarrow \mathrm{GL}_{2g}(\mathbb{Z}/\ell^n\mathbb{Z})\big)$ and similarly for $U(n,d)$. The representation is unramified at all primes $p \nmid \ell N$ since its kernel contains $\ker \rho_{X_1(N),\ell}$. For any such p let $\mathfrak{p} \subset \overline{\mathbb{Z}}$ be any maximal ideal over p. At the level of Abelian varieties, since T_p acts as $a_p(f)$ and $\langle p \rangle$ acts as $\chi(p)$, $\rho_{A_f,\ell}(\mathrm{Frob}_\mathfrak{p})$ satisfies the polynomial equation

$$x^2 - a_p(f)x + \chi(p)p = 0.$$

Since the Tate module $\mathrm{Ta}_\ell(A_f) \cong \mathbb{Z}_\ell^{2d}$ is a module over $\mathcal{O}_f$, the tensor product

$$V_\ell(A_f) = \mathrm{Ta}_\ell(A_f) \otimes \mathbb{Q} \cong \mathbb{Q}_\ell^{2d}$$

is a module over $\mathcal{O}_f \otimes \mathbb{Q} = \mathbb{K}_f$. Also, it is a module over $\mathbb{Q}_\ell$, with the two actions commuting and with the restrictions of the two actions to $\mathbb{Q}$ agreeing. Thus $V_\ell(A_f)$ is a module over $\mathbb{K}_f \otimes_\mathbb{Q} \mathbb{Q}_\ell$. The goal of this section is to decompose it into 2-dimensional representations $V_\lambda(f)$ where $\lambda \mid \ell$. We need the following lemma.

Lemma 9.5.3. $V_\ell(A_f)$ *is a free module of rank 2 over* $\mathbb{K}_f \otimes_\mathbb{Q} \mathbb{Q}_\ell$.

Proof. Since $\mathrm{Ta}_\ell(A_f)$ is the inverse limit of the torsion groups $A_f[\ell^n]$, we need to describe $A_f[\ell^n]$ in a fashion that will help establish the freeness.

As above, let $\mathcal{S}_2 = S_2(\Gamma_1(N))$ and let $H_1 = \mathrm{H}_1(X_1(N)_{\mathbb{C}}, \mathbb{Z}) \subset \mathcal{S}_2^\wedge$. Consider the quotients $\overline{\mathcal{S}_2^\wedge} = \mathcal{S}_2^\wedge / I_f \mathcal{S}_2^\wedge$ and $\overline{H}_1 = (H_1 + I_f \mathcal{S}_2^\wedge)/I_f \mathcal{S}_2^\wedge$, both $\mathcal{O}_f$-modules. Compute that

$$
\begin{aligned}
A_f = \mathrm{J}_1(N)/I_f \mathrm{J}_1(N) &= \left(\mathcal{S}_2^\wedge/H_1\right) / \left((I_f \mathcal{S}_2^\wedge + H_1)/H_1\right) \\
&\cong \mathcal{S}_2^\wedge/(I_f \mathcal{S}_2^\wedge + H_1) \\
&\cong \left(\mathcal{S}_2^\wedge/I_f \mathcal{S}_2^\wedge\right) / \left((H_1 + I_f \mathcal{S}_2^\wedge)/I_f \mathcal{S}_2^\wedge\right) = \overline{\mathcal{S}_2^\wedge}/\overline{H}_1.
\end{aligned}
$$

Thus $A_f[\ell^n] \cong \ell^{-n}\overline{H}_1/\overline{H}_1$ for any $n \in \mathbb{Z}^+$. The $\mathcal{O}_f$-linear isomorphisms $\ell^{-n}\overline{H}_1/\overline{H}_1 \longrightarrow \overline{H}_1/\ell^n \overline{H}_1$ induced by multiplication by ℓ^n on $\ell^{-n}\overline{H}_1$ assemble to give an isomorphism of $\mathcal{O}_f \otimes \mathbb{Z}_\ell$-modules,

$$
\mathrm{Ta}_\ell(A_f) = \varprojlim_n \{A_f[\ell^n]\} = \varprojlim_n \{\ell^{-n}\overline{H}_1/\overline{H}_1\} \cong \varprojlim_n \{\overline{H}_1/\ell^n \overline{H}_1\} \cong \overline{H}_1 \otimes \mathbb{Z}_\ell,
$$

where the transition maps in the last inverse limit are the natural projection maps.

The fact that A_f is a complex torus of dimension d and the calculation a moment ago that $A_f \cong \overline{\mathcal{S}_2^\wedge}/\overline{H}_1$ combine to show that the $\mathcal{O}_f$-module $\overline{H}_1 \cong H_1/(H_1 \cap I_f \mathcal{S}_2^\wedge)$ has $\mathbb{Z}$-rank $2d$. Since $\mathbb{K}_f$ is a field, $\overline{H}_1 \otimes \mathbb{Q}$ is a free $\mathbb{K}_f$-module whose $\mathbb{Q}$-rank is $2d$ and whose $\mathbb{K}_f$-rank is therefore 2. Consequently, $\overline{H}_1 \otimes \mathbb{Q}_\ell = \overline{H}_1 \otimes \mathbb{Q} \otimes_{\mathbb{Q}} \mathbb{Q}_\ell$ is free of rank 2 over $\mathbb{K}_f \otimes_{\mathbb{Q}} \mathbb{Q}_\ell$. So finally,

$$
V_\ell(A_f) = \mathrm{Ta}_\ell(A_f) \otimes \mathbb{Q} \cong \overline{H}_1 \otimes \mathbb{Z}_\ell \otimes \mathbb{Q} \cong \overline{H}_1 \otimes \mathbb{Q}_\ell
$$

is an isomorphism of $\mathbb{K}_f \otimes_{\mathbb{Q}} \mathbb{Q}_\ell$-modules, and the proof is complete. $\qquad \square$

With the lemma in hand, decomposing $V_\ell(A_f)$ into representations $V_\lambda(f)$ where $\lambda \mid \ell$ is straightforward. The absolute Galois group $G_{\mathbb{Q}}$ acts $(\mathbb{K}_f \otimes_{\mathbb{Q}} \mathbb{Q}_\ell)$-linearly on $V_\ell(A_f)$, and $V_\ell(A_f) \cong (\mathbb{K}_f \otimes_{\mathbb{Q}} \mathbb{Q}_\ell)^2$ by the lemma. Choose a basis B of $V_\ell(A_f)$ to get a homomorphism $G_{\mathbb{Q}} \longrightarrow \mathrm{GL}_2(\mathbb{K}_f \otimes_{\mathbb{Q}} \mathbb{Q}_\ell)$. Also, (9.10) specializes to give $\mathbb{K}_f \otimes_{\mathbb{Q}} \mathbb{Q}_\ell \cong \prod_{\lambda \mid \ell} \mathbb{K}_{f,\lambda}$, so for each λ we can compose the homomorphism with a projection to get

$$
\rho_{f,\lambda} : G_{\mathbb{Q}} \longrightarrow \mathrm{GL}_2(\mathbb{K}_{f,\lambda}).
$$

This is continuous (Exercise 9.5.2(b)), making it a Galois representation. And $\ker(\rho_{A_f,\ell}) \subset \ker(\rho_{f,\lambda})$ (Exercise 9.5.2(c)). We have proved

Theorem 9.5.4. *Let $f \in S_2(N, \chi)$ be a normalized eigenform with number field $\mathbb{K}_f$. Let ℓ be prime. For each maximal ideal λ of $\mathcal{O}_{\mathbb{K}_f}$ lying over ℓ there is a 2-dimensional Galois representation*

$$
\rho_{f,\lambda} : G_{\mathbb{Q}} \longrightarrow \mathrm{GL}_2(\mathbb{K}_{f,\lambda}).
$$

This representation is unramified at every prime $p \nmid \ell N$. For any such p let $\mathfrak{p} \subset \overline{\mathbb{Z}}$ be any maximal ideal lying over p. Then $\rho_{f,\lambda}(\mathrm{Frob}_{\mathfrak{p}})$ satisfies the polynomial equation

$$x^2 - a_p(f)x + \chi(p)p = 0.$$

In particular, if $f \in \mathcal{S}_2(\Gamma_0(N))$ then the relation is $x^2 - a_p(f)x + p = 0$.

Exercises

9.5.1. Establish (9.16). (A hint for this exercise is at the end of the book.)

9.5.2. (a) Let $i : \mathbb{K}_f \otimes_{\mathbb{Q}} \mathbb{Q}_\ell \longrightarrow \prod_{\lambda | \ell} \mathbb{K}_{f,\lambda}$ be the isomorphism of (9.10). For each λ, let e_λ be the element of $\mathbb{K}_f \otimes_{\mathbb{Q}} \mathbb{Q}_\ell$ that is taken by i to $(0, \ldots, 0, 1_{\mathbb{K}_{f,\lambda}}, 0, \ldots, 0)$ and let

$$V_\lambda(f) = e_\lambda(V_\ell(A_f)).$$

Show that each $V_\lambda(f)$ is a 2-dimensional vector space over $\mathbb{K}_{f,\lambda}$ and that

$$V_\ell(A_f) = \bigoplus_{\lambda | \ell} V_\lambda(f).$$

Show that each $V_\lambda(f)$ is invariant under the $G_{\mathbb{Q}}$-action on $V_\ell(A_f)$. Show that if each $V_\lambda(f)$ is given the basis $e_\lambda B$ over $\mathbb{K}_{f,\lambda}$ where B is the basis of $V_\ell(A_f)$ over $\mathbb{K}_f \otimes_{\mathbb{Q}} \mathbb{Q}_\ell$ in the section then each $\rho_{f,\lambda}$ is defined by the action of $G_{\mathbb{Q}}$ on $V_\lambda(f)$. (Hints for this exercise are at the end of the book.)

(b) To show that $\rho_{f,\lambda}$ is continuous it suffices to show that the action

$$V_\lambda(f) \times G_{\mathbb{Q}} \longrightarrow V_\lambda(f)$$

is continuous. Explain why this statement is independent of whether $V_\lambda(f)$ is viewed as a vector space over $\mathbb{K}_{f,\lambda}$ or over $\mathbb{Q}_\ell$. Explain why the action is continuous in the latter case.

(c) Use the decomposition from (a) to show that $\ker(\rho_{A_f,\ell}) \subset \ker(\rho_{f,\lambda})$ for each λ.

9.6 Galois representations and Modularity

This last section states the Modularity Theorem in terms of Galois representations, connects it to the arithmetic versions in Chapter 8, and describes how the modularity of elliptic curves is part of a broader conjecture. Finally we discuss how the modularity of Galois representations and of mod ℓ representations are related.

Definition 9.6.1. *An irreducible Galois representation*

$$\rho : G_{\mathbb{Q}} \longrightarrow \mathrm{GL}_2(\mathbb{Q}_\ell)$$

such that $\det \rho = \chi_\ell$ *is* **modular** *if there exists a newform* $f \in \mathcal{S}_2(\Gamma_0(M_f))$ *such that* $\mathbb{K}_{f,\lambda} = \mathbb{Q}_\ell$ *for some maximal ideal* λ *of* $\mathcal{O}_{\mathbb{K}_f}$ *lying over* ℓ *and such that* $\rho_{f,\lambda} \sim \rho$.

In particular, if E is an elliptic curve over $\mathbb{Q}$ and ℓ is prime then the Galois representation $\rho_{E,\ell}$ is a candidate to be modular since it is irreducible and its determinant is χ_ℓ by the proof of Theorem 9.4.1.

Theorem 9.6.2 (Modularity Theorem, Version R). *Let E be an elliptic curve over $\mathbb{Q}$. Then $\rho_{E,\ell}$ is modular for some ℓ.*

This is the version that was proved, for semistable curves in [Wil95] and [TW95] and then for all curves in [BCDT01]. We will explain how Version R of Modularity leads to Version a_p from Chapter 8, which in turn implies a stronger Version R,

Theorem 9.6.3 (Modularity Theorem, strong Version R). *Let E be an elliptic curve over $\mathbb{Q}$ with conductor N. Then for some newform $f \in \mathcal{S}_2(\Gamma_0(N))$ with number field $\mathbb{K}_f = \mathbb{Q}$,*

$$\rho_{f,\ell} \sim \rho_{E,\ell} \quad \text{for all } \ell.$$

Given Version R of Modularity, let E be an elliptic curve over $\mathbb{Q}$ with conductor N. Then there exists a newform $f \in \mathcal{S}_2(\Gamma_0(M_f))$ as in Definition 9.6.1, so $\rho_{f,\lambda} \sim \rho_{E,\ell}$ for some suitable maximal ideal λ of its number field $\mathbb{K}_f$. Thus $\rho_{E,\ell}(\mathrm{Frob}_\mathfrak{p})$ satisfies the polynomial $x^2 - a_p(f)x + p$ for any absolute Frobenius element $\mathrm{Frob}_\mathfrak{p}$ where $\mathfrak{p}$ lies over any prime $p \nmid \ell M_f$, since $\rho_{f,\lambda}$ does. But the characteristic polynomial of $\rho_{E,\ell}(\mathrm{Frob}_\mathfrak{p})$ for any $\mathrm{Frob}_\mathfrak{p}$ where $p \nmid \ell N$ is $x^2 - a_p(E)x + p$. Therefore $a_p(f) = a_p(E)$ for all but finitely many p. The work of Carayol mentioned at the end of Chapter 8 shows equality for all p and shows that $M_f = N$. This is Version a_p.

On the other hand, given Version a_p of Modularity, again let E be an elliptic curve over $\mathbb{Q}$ with conductor N. There exists a newform $f \in \mathcal{S}_2(\Gamma_0(N))$ such that $a_p(f) = a_p(E)$ for all p. Since $a_p(f) \in \mathbb{Z}$ for all p, the number field of f is $\mathbb{Q}$ and the Abelian variety A_f is an elliptic curve. Consider the representations $\rho_{f,\ell} = \rho_{A_f,\ell}$ and $\rho_{E,\ell}$ for any ℓ. The respective characteristic polynomials of $\rho_{f,\ell}(\mathrm{Frob}_\mathfrak{p})$ and $\rho_{E,\ell}(\mathrm{Frob}_\mathfrak{p})$ are $x^2 - a_p(f)x + p$ and $x^2 - a_p(E)X + p$ for all but finitely many p. Thus the characteristic polynomials are equal on a dense subset of $G_{\mathbb{Q}}$, and since trace and determinant are continuous this makes the characteristic polynomials equal. Consequently the representations are equivalent (Exercise 9.6.1). Since ℓ is arbitrary, this is the strong Version R.

The reasoning from Version R of Modularity to Version a_p and then back to the strong Version R (or see Exercise 9.6.2) proves

Proposition 9.6.4. *Let E be an elliptic curve over $\mathbb{Q}$. Then*

if $\rho_{E,\ell}$ is modular for some ℓ then $\rho_{E,\ell}$ is modular for all ℓ.

Theorem 9.5.4 generalizes to a result that Galois representations are associated to modular forms of weights other than 2.

Theorem 9.6.5. *Let $f \in \mathcal{S}_k(N, \chi)$ be a normalized eigenform with number field $\mathbb{K}_f$. Let ℓ be prime. For each maximal ideal λ of $\mathcal{O}_{\mathbb{K}_f}$ lying over ℓ there is an irreducible 2-dimensional Galois representation*

$$\rho_{f,\lambda} : G_\mathbb{Q} \longrightarrow \mathrm{GL}_2(\mathbb{K}_{f,\lambda}).$$

This representation is unramified at all primes $p \nmid \ell N$. For any $\mathfrak{p} \subset \overline{\mathbb{Z}}$ lying over such p, the characteristic equation of $\rho_{f,\lambda}(\mathrm{Frob}_\mathfrak{p})$ is

$$x^2 - a_p(f)x + \chi(p)p^{k-1} = 0.$$

Similarly to the remark after Theorem 9.4.1, note that the characteristic equation is independent of λ. The set $\{\rho_{f,\lambda}\}$ (for fixed f and varying λ) form what is called a *compatible system* of ℓ-adic Galois representations. The characteristic equation shows that

$$\det \rho_{f,\lambda} = \chi \chi_\ell^{k-1}$$

where the Dirichlet character χ is being identified with the Galois representation ρ_χ from Section 9.3 (Exercise 9.6.3). It follows that the representation $\rho_{f,\lambda}$ satisfies

$$\det \rho_{f,\lambda}(\mathrm{conj}) = -1$$

where as before, *conj* denotes complex conjugation. Indeed, $\chi(\mathrm{conj}) = \chi(-1)$ as noted in Section 9.3, and necessarily $\chi(-1) = (-1)^k$ for $\mathcal{S}_k(N, \chi)$ to be nontrivial. Since $\chi_\ell(\mathrm{conj}) = -1$ from Section 9.3 as well, the result follows. In general, a Galois representation ρ such that $\det \rho(\mathrm{conj}) = -1$ is called *odd*.

Section 9.5 constructed $\rho_{f,\lambda}$ when $k = 2$ but didn't prove that $\rho_{f,\lambda}$ is irreducible or that the equation is the characteristic equation. The construction for $k > 2$, due to Deligne [Del71], is similar but uses more sophisticated machinery. The dual of $\mathcal{S}_k(\Gamma_1(N))$ contains a lattice L that is stable under the action of $\mathbb{T}_\mathbb{Z}$ and such that $L \otimes_\mathbb{Q} \mathbb{Q}_\ell$ admits a compatible action of $G_\mathbb{Q}$. To define the Galois action and generalize the Eichler–Shimura Relation, Deligne used *étale cohomology*. The construction for $k = 1$, due to Deligne and Serre [DS74], is different. It uses congruences between f and modular forms of higher weight (Exercise 9.6.4 will illustrate how to produce such congruences), and it produces a single representation with finite image,

$$\rho_f : G_\mathbb{Q} \longrightarrow \mathrm{GL}_2(\mathbb{K}_f),$$

that gives rise to all the $\rho_{f,\lambda}$. By embedding $\mathbb{K}_f$ in $\mathbb{C}$ we can view ρ_f as a complex representation $G_{\mathbb{Q}} \longrightarrow \mathrm{GL}_2(\mathbb{C})$, a phenomenon unique to $k = 1$. For example the representation ρ from the end of Section 9.1 is ρ_{θ_χ}.

Theorem 9.5.4 also generalizes to Eisenstein series and reducible representations. Recall the Eisenstein series $E_k^{\psi,\varphi} \in \mathcal{E}_k(N,\chi)$ from Chapter 4, where ψ and φ are primitive Dirichlet characters modulo u and v where $uv \mid N$ and $\psi\varphi = \chi$ at level N. The Fourier coefficients given in Theorem 4.5.1 for $n \geq 1$ are

$$a_n(E_k^{\psi,\varphi}) = 2\sigma_{k-1}^{\psi,\varphi}(n), \quad \text{where } \sigma_{k-1}^{\psi,\varphi}(n) = \sum_{\substack{m|n \\ m>0}} \psi(n/m)\varphi(m)m^{k-1}.$$

Recall also the Eisenstein series

$$E_k^{\psi,\varphi,t}(\tau) = \begin{cases} E_k^{\psi,\varphi}(t\tau) & \text{unless } k = 2, \ \psi = \varphi = \mathbf{1}, \\ E_2^{1,1}(\tau) - tE_2^{1,1}(t\tau) & \text{if } k = 2, \ \psi = \varphi = \mathbf{1}. \end{cases}$$

Here t is a positive integer and $tuv \mid N$. These form a basis of $\mathcal{E}_k(N,\chi)$ as the triples (ψ,φ,t) run through the elements of a set $A_{N,k}$ such that $\psi\varphi = \chi$, cf. Theorem 4.5.2, Theorem 4.6.2, and Theorem 4.8.1. By Proposition 5.2.3, $E_k^{\psi,\varphi,t}$ for such a triple is an eigenform if $uv = N$ or if $k = 2$ and $\psi = \varphi = \mathbf{1}$ and t is prime and N is a power of t.

Theorem 9.6.6. *Let $f = \frac{1}{2}E_k^{\psi,\varphi,t} \in \mathcal{E}_k(N,\chi)$ where $E_k^{\psi,\varphi,t}$ is an eigenform as just described. Let ℓ be prime. For each maximal ideal λ of $\mathcal{O}_{\mathbb{K}_f}$ lying over ℓ, the reducible 2-dimensional Galois representation*

$$\rho_{f,\lambda} = \begin{bmatrix} \psi & 0 \\ 0 & \varphi\chi_\ell^{k-1} \end{bmatrix}$$

is unramified at all primes $p \nmid \ell N$. For any $\mathfrak{p} \subset \overline{\mathbb{Z}}$ lying over such p, the characteristic equation of $\rho_{f,\lambda}(\mathrm{Frob}_\mathfrak{p})$ is

$$x^2 - a_p(f)x + \chi(p)p^{k-1} = 0.$$

To prove this, again recall from Section 9.3 that any primitive Dirichlet character ϕ modulo N acts as a Galois representation unramified at the primes p not dividing its conductor, and it takes $\mathrm{Frob}_\mathfrak{p}$ to $\phi(p)$ for such p. Thus if also $p \nmid \ell$ then $\det \rho_{f,\lambda}(\mathrm{Frob}_\mathfrak{p}) = \psi(p)\varphi(p)p^{k-1} = \chi(p)p^{k-1}$ as desired. Also $\mathrm{tr}\, \rho_{f,\lambda}(\mathrm{Frob}_\mathfrak{p}) = \psi(p) + \varphi(p)p^{k-1} = \sigma_{k-1}^{\psi,\varphi}(p)$. If $t = 1$ then this is $a_p(f)$. In the exceptional case when $k = 2$ and $\psi = \varphi = \mathbf{1}$ and $t > 1$ it is $\sigma_1(p)$, which is $\sigma_1(p) - t\sigma_1(p/t)$ since $p \nmid N = t^e$. Again this is $a_p(f)$.

All reducible representations that plausibly arise from Theorem 9.6.6 indeed do so.

Theorem 9.6.7. *Let the Galois representation $\rho : G_{\mathbb{Q}} \longrightarrow \mathrm{GL}_2(\mathbb{L})$ be odd, reducible, and semisimple, i.e. $\rho \sim \begin{bmatrix} \psi & 0 \\ 0 & \phi \end{bmatrix}$. If ψ has finite image and $\det \rho =$*

$\chi\chi_\ell^{k-1}$ *where χ has finite image then for some eigenform $f = \frac{1}{2}E_k^{\psi,\varphi,t}$ with $\psi\varphi = \chi$ as described before Theorem 9.6.6,*

$$\rho_{f,\lambda} \sim \rho.$$

Here we are using the result from Section 9.3 that 1-dimensional Galois representations with finite image can be viewed as Dirichlet characters.

Note that the conclusion of Theorem 9.6.6 holds also for any representation of the form $\rho = \begin{bmatrix} \psi & * \\ 0 & \varphi\chi_\ell^{k-1} \end{bmatrix}$ that is unramified at all $p \nmid \ell N$, even if ρ does not decompose as the direct sum of two 1-dimensional representations. Such representations arise naturally, and they have a claim to be considered as associated to the eigenforms $f = \frac{1}{2}E_k^{\psi,\varphi,t}$ appearing in Theorem 9.6.6. For example, with p prime and d a power of p the non-semisimple representation $\rho_{\ell,d} = \begin{bmatrix} 1 & * \\ 0 & \chi_\ell \end{bmatrix}$ constructed at the end of Section 9.3 satisfies the conclusion of the theorem for $f = \frac{1}{2}E_2^{1,1,p}$ and $N = p$. A broadened statement of Theorem 9.6.6 that includes normalized eigenforms in the spaces $\mathcal{E}_k(N, \psi, \varphi)$ from Section 5.9 covers the class of examples $\rho_{\ell,d}$ where d is not a prime power. Assuming this broadening and given d, let N be the product of the prime factors of d once each. The Galois representation $\rho_{\ell,d}$ from the end of Section 9.3 can then be viewed as associated to the eigenform f of level N from Exercise 5.2.6(d), whose L-function is $\zeta(s)\zeta(s-1)$ but with Euler factors $(1 - p^{-s})^{-1}$ rather than $(1 - p^{-s})^{-1}(1 - p^{1-s})^{-1}$ at primes $p \mid N$ (this was Exercise 5.9.6(d)).

A modularity result is conjectured for irreducible representations as well. This requires slightly broadening the definition of a modular representation.

Definition 9.6.8. *Consider a Galois representation*

$$\rho : G_\mathbb{Q} \longrightarrow \mathrm{GL}_2(\mathbb{L}).$$

*Suppose that ρ is irreducible and odd and that $\det \rho = \chi\chi_\ell^{k-1}$ where χ has finite image. Then ρ is **modular** if there exists a newform $f \in \mathcal{S}_k(M_f, \chi)$ such that for some maximal ideal λ of $\mathcal{O}_{\mathbb{K}_f}$ lying over ℓ, $\mathbb{K}_{f,\lambda}$ embeds in $\mathbb{L}$ and $\rho_{f,\lambda} \sim \rho$.*

The Fontaine–Mazur–Langlands Conjecture [FM93] is

Conjecture 9.6.9. *Let $\rho : G_\mathbb{Q} \longrightarrow \mathrm{GL}_2(\mathbb{L})$ be irreducible, odd, and geometric. Then $\chi_\ell^r \rho$ is modular for some integer r.*

The exact definition of a *geometric* Galois representation is beyond our scope. Modular Galois representations are geometric, and Conjecture 9.6.9 states a converse. Besides representations $\rho_{E,\ell}$ and the representations in Theorem 9.6.7, the geometric Galois representations also include the representations of $G_\mathbb{Q}$ with finite image. The statement that any irreducible, odd, 2-dimensional Galois representation with finite image is modular is a version of *Artin's Conjecture*. In this case $k = 1$ and $\mathbb{L}$ can be replaced by a number field in Definition 9.6.8. The result was proved for solvable finite image by Langlands [Lan80] and Tunnell [Tun81]. In the case of projectively dihedral representations the relevant weight 1 eigenform giving rise to ρ is a theta function already considered by Hecke [Hec26], generalizing the theta function θ_χ from Section 4.11 that gives rise to the representation with image D_3 at the end of Section 9.1. The only possible nonsolvable finite image is projectively the alternating group A_5. Modularity in this case was proved by Buzzard, Dickinson, Shepherd-Barron, and Taylor [BDSBT01] [Tay03] under technical hypotheses, building on ideas in Wiles's proof. Modularity in this case was proved in its entirety by Khare and Wintenberger [KW09a, KW09b] as a consequence of their proof of Serre's Conjecture (Theorem 9.6.11 just below).

Mod ℓ representations as in Exercise 9.4.4 play a prominent role in the proofs of Fermat's Last Theorem and Modularity. Let $f \in \mathcal{S}_2(\Gamma_1(M_f))$ be a newform and let $\lambda \subset \mathcal{O}_{\mathbb{K}_f}$ lie over ℓ. By Proposition 9.3.5 we may assume up to similarity that the representation $\rho_{f,\lambda}$ maps to $\mathrm{GL}_2(\mathcal{O}_{\mathbb{K}_f,\lambda})$. So it has a mod ℓ reduction

$$\overline{\rho}_{f,\lambda} : G_\mathbb{Q} \longrightarrow \mathrm{GL}_2(\mathcal{O}_{\mathbb{K}_f,\lambda}/\lambda\mathcal{O}_{\mathbb{K}_f,\lambda}).$$

More generally we consider mod ℓ representations $\overline{\rho} : G_\mathbb{Q} \longrightarrow \mathrm{GL}_2(\overline{\mathbb{F}}_\ell)$ where $\overline{\mathbb{F}}_\ell$ has the discrete topology and $\overline{\rho}$ is continuous. Since $G_\mathbb{Q}$ is compact this means that the image is finite and therefore lies in $\mathrm{GL}_2(\mathbb{F}_{\ell^r})$ for some r. The notion of modularity applies to mod ℓ representations as well.

Definition 9.6.10. *An irreducible representation* $\overline{\rho} : G_\mathbb{Q} \longrightarrow \mathrm{GL}_2(\overline{\mathbb{F}}_\ell)$ *is* **modular of level** M *if there exists a newform* $f \in \mathcal{S}_2(\Gamma_1(M))$ *and a maximal ideal* $\lambda \subset \mathcal{O}_{\mathbb{K}_f}$ *lying over* ℓ *such that* $\overline{\rho}_{f,\lambda} \sim \overline{\rho}$.

A modularity conjecture for mod ℓ representations is due to Serre [Ser87]. Its formulation uses a recipe giving a minimal level $M(\overline{\rho})$ in terms of the ramification of $\overline{\rho}$. As at the end of the previous paragraph, Serre's conjecture was proved by Khare and Wintenberger.

Theorem 9.6.11. *Let* $\overline{\rho} : G_\mathbb{Q} \longrightarrow \mathrm{GL}_2(\overline{\mathbb{F}}_\ell)$ *be irreducible and odd. Then* $\overline{\rho}$ *is modular of level* $M(\overline{\rho})$.

Excepting some cases with $\ell = 2$ it was already known, building on work of Ribet [Rib90], that

if $\overline{\rho}$ *is modular then* $\overline{\rho}$ *is modular of level* $M(\overline{\rho})$.

The relation of this to Fermat's Last Theorem is that a nontrivial solution to the Fermat equation

$$a^\ell + b^\ell + c^\ell = 0, \quad \ell \geq 3 \text{ prime}$$

would give rise to the corresponding Frey curve

$$E_{\mathrm{F}} : y^2 = x(x - a^\ell)(x + b^\ell).$$

The Fermat equation reduces to the case $\gcd(a, b, c) = 1$, $a \equiv -1$ (mod 4), and b even. The Frey curve is then semistable. Also, the number field $\mathbb{Q}(E_{\mathrm{F}}[\ell])$ is then unramified outside 2 and ℓ, and even the ramification there is unusually small, making $M(\overline{\rho}_{E_{\mathrm{F}}, \ell}) = 2$. But $S_2(\Gamma_0(2)) = \{0\}$ since the corresponding modular curve has genus 0, cf. Figure 3.4. Thus $\overline{\rho}_{E_{\mathrm{F}}, \ell}$ is not modular and so $\rho_{E_{\mathrm{F}}, \ell}$ is not modular.

On the other hand, [Wil95] begins with any elliptic curve E over $\mathbb{Q}$ and considers the mod 3 representation

$$\overline{\rho}_{E,3} : G_{\mathbb{Q}} \longrightarrow \mathrm{GL}_2(\mathbb{F}_3).$$

There is an embedding $i : \mathrm{GL}_2(\mathbb{F}_3) \longrightarrow \mathrm{GL}_2(\mathcal{O}_{\mathbb{K}})$ where $\mathbb{K}$ is a number field (Exercise 9.6.4(a)), and $\mathrm{GL}_2(\mathbb{F}_3)$ is solvable. So if $\overline{\rho}_{E,3}$ is irreducible then the Langlands–Tunnell result shows that the 3-adic representation $i \circ \overline{\rho}_{E,3}$ arises from a weight 1 newform. One can show that therefore $\overline{\rho}_{E,3}$ arises from a weight 2 newform, making it modular in the sense of Definition 9.6.10 (Exercise 9.6.4(b–h)). After this the core of the argument in [Wil95] and [TW95] shows that under technical hypotheses on a representation $\rho : G_{\mathbb{Q}} \longrightarrow \mathrm{GL}_2(\mathbb{Z}_\ell)$ with mod ℓ reduction $\overline{\rho}$,

$$\text{if } \overline{\rho} \text{ is modular then } \rho \text{ is modular.} \tag{9.17}$$

The technical hypotheses apply when E is semistable and $\rho = \rho_{E,3}$. This all assumes that $\overline{\rho}_{E,3}$ is irreducible, but if it isn't then the proof argues that $\overline{\rho}_{E,5}$ can be used as a starting point instead. Thus Wiles and Taylor–Wiles show that for any semistable E one of $\rho_{E,3}$ or $\rho_{E,5}$ is modular. By Proposition 9.6.4, $\rho_{E,\ell}$ is modular for all ℓ. In particular, because the Frey curve has the non-modular representation $\rho_{E_{\mathrm{F}}, \ell}$ it can not exist, and so there is no nontrivial solution to the Fermat equation.

The contribution of [BCDT01] to Modularity is to relax the technical hypotheses so that $\rho_{E,3}$ or $\rho_{E,5}$ meets them even when E is not semistable. Kisin [Kis09c, Kis09a, Kis09b] has further relaxed the hypotheses so that (9.17) applies to a much larger class of geometric Galois representations. Especially, with Serre's Conjecture now a theorem, so that $\overline{\rho}$ is known to be modular, (9.17) gives many more cases where Conjecture 9.6.9 is proved.

We briefly touch on the proof of Serre's Conjecture by Khare and Wintenberger [KW09a, KW09b], which incorporates ideas of Dieulefait [Die04]. First note that the conjecture implies that every odd, irreducible $\overline{\rho}$ lifts to a geometric ℓ-adic representation; moreover, this lift should be part of a compatible system, as described after Theorem 9.6.5. Ramakrishna [Ram02] proved the existence of geometric ℓ-adic lifts without assuming the conjecture, and a key point in the Khare–Wintenberger argument is to show that the lifts can indeed be taken to be part of a compatible system. Now using results of the form (9.17) one can try to argue by induction on ℓ. First lift $\overline{\rho}$ to a suitable family $\{\rho_{\ell'}\}$. If $\overline{\rho}_{\ell'}$ is modular for all $\ell' < \ell$ then (9.17) implies that the corresponding $\rho_{\ell'}$ are modular, and then the definitions of modularity and compatibility imply that so is ρ_ℓ, and hence $\overline{\rho} = \overline{\rho}_\ell$. This is an elaboration of how Wiles switches between $\ell = 3$ and $\ell = 5$. To start the induction, one knows by results of Serre and Tate that for $\ell = 2$ and $\ell = 3$, an odd $\overline{\rho}$ with $M(\overline{\rho}) = 1$ is reducible and hence modular in a sense afforded by Theorem 9.6.6. Khare [Kha06] proved the conjecture using such an induction argument under the assumption $M(\overline{\rho}) = 1$, and Khare and Wintenberger proved the conjecture in general by combining this strategy with an induction on $M(\overline{\rho})$.

Exercises

9.6.1. Consider two representations $\rho, \rho' : G_\mathbb{Q} \longrightarrow \mathrm{GL}_2(\overline{\mathbb{Q}}_\ell)$ with ρ irreducible and $\det \rho(\mathrm{conj}) = -1$. Assume that ρ and ρ' have the same characteristic polynomial. Prove that ρ and ρ' are equivalent as follows.

(a) Show that $\rho(\mathrm{conj})$ has distinct eigenvalues in $\overline{\mathbb{Q}}_\ell$. After conjugating if necessary we may assume that $\rho(\mathrm{conj}) = \rho'(\mathrm{conj}) = \left[\begin{smallmatrix} 1 & 0 \\ 0 & -1 \end{smallmatrix}\right]$. (Hints for this exercise are at the end of the book.)

(b) Show that there is some $\sigma \in G_\mathbb{Q}$ such that $\rho(\sigma) = \left[\begin{smallmatrix} a & b \\ c & d \end{smallmatrix}\right]$ with b and c nonzero. Write $\rho'(\sigma) = \left[\begin{smallmatrix} a' & b' \\ c' & d' \end{smallmatrix}\right]$. Use the relations $\mathrm{tr}\,\rho(\sigma) = \mathrm{tr}\,\rho'(\sigma)$, $\mathrm{tr}\,\rho(\mathrm{conj}\,\sigma) = \mathrm{tr}\,\rho'(\mathrm{conj}\,\sigma)$, and $\det \rho(\sigma) = \det \rho'(\sigma)$ to show that $a = a'$, $d = d'$, and $bc = b'c' \neq 0$. Show that after a further conjugation if necessary, $\rho(\sigma) = \rho'(\sigma)$.

(c) For arbitrary $\tau \in G_\mathbb{Q}$ show that $\rho(\tau) = \rho'(\tau)$ by applying the equality $\mathrm{tr}\,\rho = \mathrm{tr}\,\rho'$ to τ, $\mathrm{conj}\,\tau$, $\sigma\tau$, and $\mathrm{conj}\,\sigma\tau$.

9.6.2. The proof of Proposition 9.6.4 does not require proceeding via Version a_p of Modularity and Carayol's work. If $\rho_{E,\ell}$ is modular for some ℓ then there exists a newform $f \in S_2(\Gamma_0(M_f))$ such that $\mathbb{K}_{f,\lambda} = \mathbb{Q}_\ell$ for some λ over ℓ and such that $\rho_{f,\lambda} \sim \rho_{E,\ell}$. Explain why therefore $a_p(f) = a_p(E)$ for almost all p. As explained in Section 8.8, Strong Multiplicity One now shows that $\mathbb{K}_f = \mathbb{Q}$, so the Galois representations associated to f take the form $\rho_{f,\ell} : G_\mathbb{Q} \longrightarrow \mathrm{GL}_2(\mathbb{Q}_\ell)$ for every ℓ. Use Exercise 9.6.1 to show that therefore $\rho_{f,\ell} \sim \rho_{E,\ell}$, i.e., $\rho_{E,\ell}$ is modular for all ℓ.

9.6.3. (a) Let $\rho_{f,\lambda}$ be the representation in Theorem 9.6.5. Use the characteristic equation of $\rho_{f,\lambda}(\mathrm{Frob}_p)$ for $p \nmid \ell N$ to show that $\det \rho_{f,\lambda} = \chi\chi_\ell^{k-1}$.

(b) Conversely, suppose that the theorem stated instead that $\det \rho_{f,\lambda} = \chi\chi_\ell^{k-1}$ and that $\rho_{f,\lambda}(\mathrm{Frob}_p)$ satisfies the displayed equation for all primes $p \nmid \ell N$. Show that the equation is the characteristic equation of $\rho_{f,\lambda}(\mathrm{Frob}_p)$ for such p.

9.6.4. This last exercise carries out the first part of the argument in [Wil95].

(a) It is known that the projective group $\mathrm{PGL}_2(\mathbb{F}_3)$ is isomorphic to the symmetric group S_4. Show that the map

$$\begin{bmatrix} -1 & 1 \\ -1 & 0 \end{bmatrix} \mapsto \begin{bmatrix} -1 & 1 \\ -1 & 0 \end{bmatrix}, \qquad \begin{bmatrix} 1 & -1 \\ 1 & 1 \end{bmatrix} \mapsto \begin{bmatrix} 1 & -1 \\ -\sqrt{-2} & -1 + \sqrt{-2} \end{bmatrix}$$

defines an embedding $i : \mathrm{GL}_2(\mathbb{F}_3) \longrightarrow \mathrm{GL}_2(\mathbb{Z}[\sqrt{-2}])$. Note that $\mathbb{Z}[\sqrt{-2}]$ is the number ring $\mathcal{O}_{\mathbb{K}}$ where $\mathbb{K} = \mathbb{Q}(\sqrt{-2})$. Show that the ideal $\lambda = \langle 1 + \sqrt{-2} \rangle$ is a maximal ideal of $\mathcal{O}_{\mathbb{K}}$ lying over 3 such that the residue field $\mathbf{k}_\lambda$ is isomorphic to $\mathbb{F}_3$. Show that following i by reduction modulo λ gives the identity. (Hints for this exercise are at the end of the book.)

(b) Let E be an elliptic curve over $\mathbb{Q}$ and let $f \in \mathcal{S}_1(M_f, \psi)$ be the newform provided by the Langlands–Tunnell result such that $\rho_{f,\lambda} \sim i \circ \overline{\rho}_{E,3}$. By Proposition 9.3.5 we may assume that $\rho_{f,\lambda}$ maps $G_{\mathbb{Q}}$ to $\mathrm{GL}_2(\mathcal{O}_{\mathbb{K}})$. Show that the reduction $\overline{\rho}_{f,\lambda}$ of $\rho_{f,\lambda}$ modulo λ is $\overline{\rho}_{E,3}$. Show that $a_p(f) \equiv a_p(E) \pmod{\lambda}$ for all but finitely many p. Show that ψ is the quadratic character with conductor 3, so that it makes sense to write $\psi(p) \equiv p \pmod{\lambda}$ for all p.

(c) Recall from Section 4.8 the weight 1 Eisenstein series $E_1^{\psi,1} \in \mathcal{M}_1(3, \psi)$. Show that

$$3E_1^{\psi,1} = 1 + \sum_{n=1}^{\infty} a_n q^n$$

where $a_n \in 3\mathbb{Z}$ for all $n \geq 1$. Let $g = 3E_1^{\psi,1} f$. Show that g lies in $\mathcal{S}_2(\Gamma_0(3M_f))$ and that $f \equiv g \pmod{\lambda}$, meaning $a_n(f) \equiv a_n(g) \pmod{\lambda}$ for all n.

(d) Although f and $3E_1^{\psi,1}$ are eigenforms their product g need not be. Nonetheless, show that $T_p g \equiv a_p(E) g \pmod{\lambda}$ for all but finitely many p and $T g \equiv a_1(Tg) g \pmod{\lambda}$ for all $T \in \mathbb{T}_{\mathbb{Z}}$. Deduce that the map

$$\phi : \mathbb{T}_{\mathbb{Z}} \longrightarrow \mathbb{F}_3, \qquad T \mapsto a_1(Tg) \pmod{\lambda}$$

is a homomorphism taking T_p to $a_p(E) \pmod{3}$ and $\langle p \rangle$ to 1 for all but finitely many primes p.

(e) We want to obtain an eigenform $g' \in \mathcal{S}_2(\Gamma_0(3M_f))$ that gives rise to $\overline{\rho}_{E,3}$. This requires using ideas from Chapter 6 and commutative algebra from Chapter 8. Let $m = \ker \phi$ and let P be a minimal prime ideal of $\mathbb{T}_{\mathbb{Z}}$ contained in m, cf. Exercise 8.5.6. Since $\mathbb{T}_{\mathbb{Z}}$ is a finitely generated $\mathbb{Z}$-module, $\mathbb{T}_{\mathbb{Z}}/P$ is an integral domain that is finitely generated as a $\mathbb{Z}$-module. And since $\mathbb{T}_{\mathbb{Z}}$ is free over $\mathbb{Z}$ (cf. Exercise 6.5.1) no rational prime p is a zero-divisor in $\mathbb{T}_{\mathbb{Z}}$, so no rational prime is contained in P. Therefore $\mathbb{T}_{\mathbb{Z}}/P$ contains $\mathbb{Z}$, i.e., it is of characteristic 0. Its field of quotients is thus a number field $\mathbb{K}'$ and $\mathbb{T}_{\mathbb{Z}}/P$ is contained in its ring of integers $\mathcal{O}_{\mathbb{K}'}$. This argument shows that there is a homomorphism

$$\phi' : \mathbb{T}_{\mathbb{Z}} \longrightarrow \mathcal{O}_{\mathbb{K}'}.$$

A maximal ideal λ' of $\mathcal{O}_{\mathbb{K}'}$ containing $\phi'(m)$ exists by Lemma 8.5.5 applied to the localization of $\mathbb{T}_{\mathbb{Z}}/P$ at m/P or by the lying-over theorem in commutative algebra. Show that $\phi'(T_p) \equiv a_p(E) \pmod{\lambda'}$ and $\phi'(\langle p \rangle) \equiv 1 \pmod{\lambda'}$ for all but finitely many p.

(f) Before continuing we need to show that the natural surjection

$$\pi : \mathbb{T}_{\mathbb{Z}} \otimes \mathbb{C} \longrightarrow \mathbb{T}_{\mathbb{C}}, \qquad \sum_i T_i \otimes z_i \mapsto \sum_i z_i T_i$$

is an isomorphism. Reread the proof of Theorem 6.5.4. Note that the vector space V in the proof is $H_1(X_1(N), \mathbb{Z}) \otimes \mathbb{C}$. Since the action of $\mathbb{T}_{\mathbb{Z}}$ on $H_1(X_1(N), \mathbb{Z})$ is faithful, the action of $\mathbb{T}_{\mathbb{Z}} \otimes \mathbb{C}$ on V is faithful as well. Explain how this shows that (using the notation in the proof) $\mathbb{T}_{\mathbb{Z}} \otimes \mathbb{C}$ acts faithfully on $\mathcal{S}_2 \oplus \mathcal{S}_2^{\wedge}$,

$$\left(\sum_i T_i \otimes z_i \right)(f, \varphi) = \left(\sum_i z_i T_i f, \varphi \circ \sum_i z_i T_i \right).$$

Use this to show that π is injective.

(g) Using the identification $\mathbb{T}_{\mathbb{Z}} \otimes \mathbb{C} = \mathbb{T}_{\mathbb{C}}$ and the containment $\mathcal{O}_{\mathbb{K}'}$ we may extend ϕ' to a homomorphism $\phi' : \mathbb{T}_{\mathbb{C}} \longrightarrow \mathbb{C}$ given by $\phi'(\sum_i T_i \otimes z_i) = \sum_i z_i \phi'(T_i)$. Convince yourself that the pairing in the proof of Proposition 6.6.4 gives rise to a corresponding eigenform $g' \in \mathcal{S}_2(\Gamma_1(3M_f))$ with coefficients in $\mathcal{O}_{\mathbb{K}'}$ such that

$$\phi'(T) = a_1(Tg'), \quad T \in \mathbb{T}_{\mathbb{Z}}.$$

Show that $a_p(g') \equiv a_p(E) \pmod{\lambda'}$ and $\chi_{g'}(p) \equiv 1 \pmod{\lambda'}$ (where $\chi_{g'}$ is the character of g', cf. Exercise 5.8.1) for all but finitely many primes p. Parts (e) and (g) of this exercise are the *Deligne–Serre Lifting Lemma*.

(h) Let g'' be the newform associated to g' by Proposition 5.8.4. Show that $\mathbb{K}_{g''} \subset \mathbb{K}_{g'}$ and that $a_p(g'') \equiv a_p(E) \pmod{\lambda''}$ for all but finitely many primes p, where $\lambda'' = \mathbb{K}_{g''} \cap \lambda'$. Use the argument in Exercise 9.6.1 to conclude that $\overline{\rho}_{g'', \lambda''} \sim \overline{\rho}_{E,3}$. Hence $\overline{\rho}_{E,3}$ is modular in the sense of Definition 9.6.10.

A

Hints and Answers to the Exercises

Chapter 1

1.1.4. (b) For $|\delta| < 1$, $|\tau + \delta| \geq B = B \cdot \sup\{1, |\delta|\}$ for all $\tau \in \Omega$. For $1 \leq |\delta| \leq 3A$ and $\mathrm{Im}(\tau) > A$, $|\tau + \delta| > A \geq |\delta|/3 = \sup\{1, |\delta|\}/3$. For $1 \leq |\delta| \leq 3A$ and $B \leq \mathrm{Im}(\tau) \leq A$, the quantity $|\tau + \delta|/|\delta|$ takes a nonzero minimum m, so $|\tau + \delta| \geq m|\delta| = m \cdot \sup\{1, |\delta|\}$. For $|\delta| > 3A$, $|\tau + \delta| \geq |\delta| - A \geq 2|\delta|/3 = 2\sup\{1, |\delta|\}/3$. So any positive C less than $\inf\{B, 1/3, m\}$ works. For (c), break the sum into $2\zeta(k) + \sum_{c \neq 0, d} |c\tau + d|^{-k}$. For $c \neq 0$, $|c\tau + d| = |c||\tau + \delta|$ where $\delta = d/c$. Apply (b) and then (a). Since any compact subset of $\mathcal{H}$ sits inside a suitable Ω, holomorphy follows from a theorem of complex analysis.

1.1.5. For the first formula, take the logarithmic derivative of the product expansion $\sin \pi\tau = \pi\tau \prod_{n=1}^{\infty}(1 - \tau^2/n^2)$. The second formula follows from the definitions of sin and cos in terms of the complex exponential.

1.1.6. (d) To show $\hat{h}(0) = 0$, simply antidifferentiate. To show $\hat{h}(m) = 0$ for $m < 0$ replace τ by $-\tau$ and reason as in part (c) except now the rectangle no longer goes around the singularity at the origin.

1.2.2. (a) Let $g = \gcd(c, d)$ and note that g is relatively prime to N. If $c \neq 0$, set $s = 0$, and use the Chinese Remainder Theorem to find t congruent to 1 modulo primes $p \mid g$ and congruent to 0 modulo primes $p \nmid g$, $p \mid c$. If $c = 0$ then $d \neq 0$ (unless $N = 1$, in which case the whole problem is trivial) and a similar argument works.

1.2.3. (a) One way is by induction on e. For $e = 1$, $|\mathrm{GL}_2(\mathbb{Z}/p\mathbb{Z})|$ is the number of bases of $(\mathbb{Z}/p\mathbb{Z})^2$, and $\mathrm{SL}_2(\mathbb{Z}/p\mathbb{Z})$ is the kernel of the surjective determinant map to $(\mathbb{Z}/p\mathbb{Z})^*$. For the induction, count $\ker(\mathrm{SL}_2(\mathbb{Z}/p^{e+1}\mathbb{Z}) \longrightarrow \mathrm{SL}_2(\mathbb{Z}/p^e\mathbb{Z}))$. The map surjects since $\mathrm{SL}_2(\mathbb{Z}) \longrightarrow \mathrm{SL}_2(\mathbb{Z}/p^e\mathbb{Z})$ does. For (b), $\mathrm{SL}_2(\mathbb{Z}/N\mathbb{Z}) \cong \prod_i \mathrm{SL}_2(\mathbb{Z}/p_i^{e_i}\mathbb{Z})$ where $N = \prod_i p_i^{e_i}$, by the Chinese Remainder Theorem.

© Springer Science+Business Media New York 2005
F. Diamond, J. Shurman, *A First Course in Modular Forms*,
Graduate Texts in Mathematics 228, DOI 10.1007/978-0-387-27226-9

1.2.6. (b) Since $(f[\alpha]_k)(\tau+N) = f[\alpha]_k(\tau)$ we may estimate $|f[\alpha]_k|$ as $q_N \to 0$ letting $y \to \infty$ and assuming $0 \le x < N$ where $\tau = x + iy$. So $|c\tau + d|$ grows as y and $|(f[\alpha]_k)(\tau)| = |f(\alpha(\tau))||c\tau+d|^{-k}$ is bounded by $(C_0 + C/\mathrm{Im}(\alpha(\tau))^r) \cdot |c\tau + d|^{-k}$, or $(C_0 + C|c\tau + d|^{2r}/y^r)|c\tau + d|^{-k}$. We may take $r \ge 1$ and thus absorb the first term into the second. The result follows.

1.2.8. (b) For the inverse, note $E_2(\tau) = E_2(\gamma(\gamma^{-1}(\tau)))$. For (e), note that for $\gamma \in \Gamma_0(N)$, $N\gamma(\tau) = \gamma'(N\tau)$ for a certain $\gamma' \in \mathrm{SL}_2(\mathbb{Z})$. Use the Fourier series and Proposition 1.2.4 to check the holomorphy condition.

1.2.9. See Exercise 1.1.7(b) for the relevant ζ-values.

1.2.10. (a) The sum is $-\sum_{n=1}^{\infty}\sum_{m=1}^{\infty} q^{nm}/m$. The sum of absolute values is $\sum_m \sum_n |q|^{nm}/m = \sum_m |q|^m/(m(1 - |q|^m))$, and this converges since its terms are bounded by $e^{-2\pi m \mathrm{Im}(\tau)}$ as $m \to \infty$. The convergence is uniform on compact sets because any compact subset of $\mathcal{H}$ has a point with minimal imaginary part.

1.2.11. (b) If $c \neq 0$ then write $a/c = a'/c'$ with $a', c' \in \mathbb{Z}$, $\gcd(a', c') = 1$. Now left multiplying by an $\mathrm{SL}_2(\mathbb{Z})$ matrix with top row $(c', -a')$ clears out the lower left entry. For the second part, if $f[\alpha]_k$ has period h then $f[\gamma]_k$ has period dh.

1.3.2. Show (1) $\implies$ (3) $\implies$ (2) $\implies$ (1).

1.3.3. (c) The entire exercise is set in one E, so multiplying by d takes points $P + \Lambda$ to $dP + \Lambda$, not to $dP + d\Lambda$.

1.4.1. (c) The integrals along opposing pairs of boundary edges cancel down to $1/(2\pi i)(\omega_2 \int_0^{\omega_1} f'(z)dz/f(z) - \omega_1 \int_0^{\omega_2} f'(z)dz/f(z))$. Each of these integrals is $\int d\log f(z)$ along a path with equal f-values at the endpoints, so $\log f$ changes by an integer multiple of $2\pi i$ along each path. For the second part, if $f(z) = (z - x)^{\nu_x(f)} g(z)$ where $g(x) \neq 0$ then $zf'(z)/f(z) = \nu_x(f)z/(z - x) + zg'(z)/g(z)$ has residue $\nu_x(f)x$ at x, so the Residue Theorem gives the result.

1.4.3. For Λ_{μ_3} show that $\wp(\mu_3 z) = \mu_3 \wp(z)$ and $\wp(\bar{z}) = \overline{\wp(z)}$. If $\wp(z) = 0$ then also $\wp(\mu_3 z) = 0$, so $\mu_3 z \equiv \pm z \pmod{\Lambda}$ since $\wp$ is even and has only two zeros. The minus sign can not arise because it would imply $z \equiv -z \pmod{\Lambda}$ but $\wp$ is nonzero at the order 2 points. The zeros are $z = (1 + 2\mu_3)/3$ and $z = (2 + \mu_3)/3$.

1.5.1. For $\Gamma(N)$, take any point $[E, (P, Q)]$ of $S(N)$. After applying an isomorphism (which preserves the Weil pairing by Exercise 1.3.3(d)) we may assume $E = \mathbb{C}/\Lambda_{\tau'}$, $P = (a\tau' + b)/N + \Lambda_{\tau'}$, $Q = (c\tau' + d)/N + \Lambda_{\tau'}$ with $a, b, c, d \in \mathbb{Z}/N\mathbb{Z}$ and $\tau' \in \mathcal{H}$. Since P and Q generate $E[N]$ and have Weil pairing $e_N(P, Q) = e^{2\pi i/N}$, $\left[\begin{smallmatrix} a & b \\ c & d \end{smallmatrix}\right] \in \mathrm{SL}_2(\mathbb{Z}/N\mathbb{Z})$. Lift this matrix back to some $\gamma \in \mathrm{SL}_2(\mathbb{Z})$, let $\tau = \gamma(\tau')$ and $m = c\tau' + d$, so that $m\tau = a\tau' + b$, and show that $[E, (P, Q)] = [\mathbb{C}/\Lambda_\tau, (\tau/N + \Lambda_\tau, 1/N + \Lambda_\tau)]$.

1.5.2. The map $S(N) \longrightarrow S_1(N)$ takes $\psi^{-1}(\Gamma(N)\tau)$ to $\psi_1^{-1}(\Gamma_1(N)\tau)$. Compute this first for points in the form $[\mathbb{C}/\Lambda_\tau, (\tau/N + \Lambda_\tau, 1/N + \Lambda_\tau)]$, and then show that in general the map works out to $[E, (P, Q)] \mapsto [E, Q]$.

1.5.3. Each $b \pmod{N} \in \mathbb{Z}/N\mathbb{Z}$ acts on the modular curve $Y(N)$ as $b \pmod{N} : \Gamma(N)\tau \mapsto \Gamma(N)\gamma(\tau)$ where $\gamma = \left[\begin{smallmatrix} 1 & b \\ 0 & 1 \end{smallmatrix}\right] \in \Gamma_1(N)$. The corresponding action on the moduli space $S(N)$ under ψ^{-1} is therefore

$$b \pmod{N} : [E_\tau, (\tau/N + \Lambda_\tau, 1/N + \Lambda_\tau)]$$
$$\mapsto [E_{\gamma(\tau)}, (\gamma(\tau)/N + \Lambda_{\gamma(\tau)}, 1/N + \Lambda_{\gamma(\tau)})].$$

By the methods of the section, letting $m = j(\gamma, \tau)$ shows that the image is $[E_\tau, (m\gamma(\tau)/N + \Lambda_\tau, m/N + \Lambda_\tau)]$. Since $\gamma = \left[\begin{smallmatrix} 1 & b \\ 0 & 1 \end{smallmatrix}\right]$, $m = 1$ and this is $[E_\tau, ((\tau + b)/N + \Lambda_\tau, 1/N + \Lambda_\tau)]$, showing that the map is

$$b \pmod{N} : [E, (P, Q)] \mapsto [E, (P + bQ, Q)].$$

Argue similarly for $d \bmod N \in (\mathbb{Z}/N\mathbb{Z})^*$ acting on $S_1(N)$, this time letting $\gamma = \left[\begin{smallmatrix} a & k \\ N & d \end{smallmatrix}\right] \in \Gamma_0(N)$ for suitable a, k, N.

1.5.4. Note that multiplying $\Lambda_{-1/(N\tau)}$ by τ gives the lattice $\tau\mathbb{Z} \oplus (1/N)\mathbb{Z}$. The map is $[E, C] \mapsto [E/C, E[N]/C]$.

1.5.6. (a) Recall the condition $C \cap \langle Q \rangle = \{0_E\}$. If $p \nmid N$ then the bottom row of the matrix is not always $(0, 1)$.

Chapter 2

2.1.3. (a) Let $y_1 = \inf\{\mathrm{Im}(\tau) : \tau \in U_1'\}$, $Y_1 = \sup\{\mathrm{Im}(\tau) : \tau \in U_1'\}$, $y_2 = \inf\{\mathrm{Im}(\tau) : \tau \in U_2'\}$, all positive. Then for $\gamma = \left[\begin{smallmatrix} a & b \\ c & d \end{smallmatrix}\right] \in \mathrm{SL}_2(\mathbb{Z})$ and $\tau \in U_1'$, the formula $\mathrm{Im}(\gamma(\tau)) = \mathrm{Im}(\tau)/|c\tau + d|^2$ shows that $\mathrm{Im}(\gamma(\tau)) \leq \min\{1/(c^2 y_1), Y_1/(c\mathrm{Re}(\tau) + d)^2\}$. The first of these is less than y_2 for all but finitely many values of c; for each exceptional c, the second is less than y_2 uniformly in τ for all but finitely many values of d.

2.2.5. Both π and ψ are open and continuous.

2.3.4. (b) For part (b), $\mathbb{Z}[\gamma]$ is ring isomorphic to $\mathbb{Z}[i]$. For part (a), if γ has order 3 then $-\gamma$ has order 6.

2.3.5. (a) The matrices fixing μ_3 are $\left[\begin{smallmatrix} a & b \\ -b & a-b \end{smallmatrix}\right]$, $(2a - b)^2 + 3b^2 = 4$.

2.3.7. (c) It suffices to show the result for $\Gamma_0(p)$, $p \equiv -1 \pmod{12}$. If $\left[\begin{smallmatrix} a & b \\ c & d \end{smallmatrix}\right]$ fixes a point then $a + d \in \{0, \pm 1\}$ and $ad \equiv 1 \pmod{p}$. If $a + d = 0$ then this means $a^2 \equiv -1 \pmod{p}$, impossible by the nature of $p \pmod 4$. The other two cases are similar with $p \pmod 3$.

2.3.8. The proof of Lemma 2.3.1 shows that given any $\tau \in \mathcal{H}$, some $\gamma \in$ $SL_2(\mathbb{Z})$ transforms some $\tau_0 \in \mathcal{D}$ to τ.

2.4.4. (a) $\Gamma(N) \subset \delta\Gamma\delta^{-1}$ for some $N \in \mathbb{Z}^+$. (b) Note that $SL_2(\mathbb{Z}) = \delta SL_2(\mathbb{Z})\delta^{-1}$.

2.4.6. See Corollary 2.3.4, recall that $\text{Im}(\gamma(\tau)) = \text{Im}(\tau)/|c\tau + d|^2$ for $\gamma = \begin{bmatrix} a & b \\ c & d \end{bmatrix} \in SL_2(\mathbb{Z})$, and note that if $\tau \in \{i, \mu_3\}$ then $|c\tau + d| \geq 1$ for all nonzero integer pairs (c, d).

Chapter 3

3.1.1. Let $\{\pm I\}\Gamma_2 = \bigcup_j \{\pm I\}\Gamma_1\gamma_j$. Then $f^{-1}(\Gamma_2\tau) = \{\Gamma_1\gamma_j\tau\}$ and these are distinct when τ is not an elliptic point for Γ_2.

3.1.2. Show that for $\gamma \in \Gamma_2$, the indices $[\{\pm I\}\Gamma_{2,\gamma(\tau)} : \{\pm I\}\Gamma_{1,\gamma(\tau)}]$ and $[\{\pm I\}\Gamma_{2,\tau} : \{\pm I\}\Gamma_{1,\tau}]$ are equal.

3.1.4. (b) The cusps are 0 and ∞, cf. Figure 3.1. For (c), $\gamma\alpha_j(i) = \alpha_j(i)$ if and only if $\alpha_j^{-1}\gamma\alpha_j(i) = i$, and by Corollary 2.3.4 the nontrivial transformations in $SL_2(\mathbb{Z})_i$ are $\pm \begin{bmatrix} 0 & -1 \\ 1 & 0 \end{bmatrix}$. Similarly for (d). The data to compute g in this and the next two problems are tabulated in Figure H.1.

Γ	d	ε_2	ε_3	ε_∞
$\Gamma_0(p)$	$p+1$	1 if $p = 2$ 2 if $p \equiv 1$ (4) 0 if $p \equiv 3$ (4)	1 if $p = 3$ 2 if $p \equiv 1$ (3) 0 if $p \equiv 2$ (3)	2
$\Gamma_1(2) = \Gamma_0(2)$	3	1	0	2
$\Gamma_1(3)$	4	0	1	2
$\Gamma_1(p)$, $p > 3$	$\frac{p^2-1}{2}$	0	0	$p - 1$
$\Gamma(2)$	6	0	0	3
$\Gamma(p)$, $p > 2$	$\frac{p(p^2-1)}{2}$	0	0	$\frac{p^2-1}{2}$

Figure H.1. Data to compute the genus

3.1.5. (b) For $p > 3$ the answer is $g = 1 + (p-1)(p-11)/24$.

3.1.6. The cusp ∞ has ramification degree p under $\pi : X(p) \longrightarrow SL_2(\mathbb{Z})\backslash\mathcal{H}^*$, and since $\Gamma(p)$ is normal in $SL_2(\mathbb{Z})$ so do all the other cusps, cf. Exercise 2.4.4(c). The number of cusps follows from the degree of the map and the number of cusps of $SL_2(\mathbb{Z})\backslash\mathcal{H}^*$. For $p > 2$ the answer is $g = 1 + (p^2 - 1)(p - 6)/24$.

3.2.2. Let $A = 240 \sum_{n=1}^{\infty} \sigma_3(n)q^n$ and $B = -504 \sum_{n=1}^{\infty} \sigma_5(n)q^n$. Cite Exercise 1.1.5 to show that $(60E_4(\tau))^3 = (64\pi^{12}/27)(1+A)^3$ and $\Delta(\tau) = (64\pi^{12}/27)((1+A)^3 - (1+B)^2)$. Consequently $j(\tau) = 1728(1+A)^3/(3A+3A^2+A^3-2B-B^2)$. The numerator lies in $1728(1+q\mathbb{Z}[[q]])$, where the double brackets denote power series. In the denominator $3A^2$ and A^3 and B^2 lie in $1728q^2\mathbb{Z}[[q]]$, and since $d^3 \equiv d^5 \pmod{12}$ for all integers d, also $3A - 2B$ lies in $1728q(1 + q\mathbb{Z}[[q]])$.

3.2.3. We may assume $\Gamma = \mathrm{SL}_2(\mathbb{Z})$. For weight-2 $\mathrm{SL}_2(\mathbb{Z})$-invariance use the chain rule relation $(j \circ \gamma)'(\tau) = j'(\gamma(\tau))\gamma'(\tau)$. Also remember to check meromorphy at ∞. Show that $(j')^{k/2} \in \mathcal{A}_k(\mathrm{SL}_2(\mathbb{Z}))$ for positive even k.

3.2.5. The group is the matrices $\pm\left[\begin{smallmatrix} 1 & h \\ 0 & 1 \end{smallmatrix}\right]$ such that $\pm\alpha\left[\begin{smallmatrix} 1 & h \\ 0 & 1 \end{smallmatrix}\right]\alpha^{-1} \in \Gamma_1(4)$.

3.5.3. For the last part show that for all $k \in \mathbb{Z}$, multiplying by the cusp form Δ defines an isomorphism $\mathcal{M}_k(\mathrm{SL}_2(\mathbb{Z})) \longrightarrow \mathcal{S}_{k+12}(\mathrm{SL}_2(\mathbb{Z}))$.

3.5.5. Recall Exercise 3.5.4(b) with $N = 11$.

3.7.1. (a) Suppose $\gamma \in \mathrm{SL}_2(\mathbb{Z})$ has order 4 or 6 and fixes the point $\tau \in \mathcal{H}$, and suppose $\gamma^{-1} = \alpha\gamma\alpha^{-1}$ with $\alpha \in \mathrm{GL}_2^+(\mathbb{Q})$. Then γ^{-1} fixes $\alpha(\tau)$, so α fixes τ since α takes $\mathcal{H}$ to $\mathcal{H}$ and the other fixed point of γ^{-1} is $\bar{\tau}$. Thus $\alpha \in \langle\gamma\rangle$ and $\gamma^{-1} = \gamma$, contradiction. For (c), recall Proposition 2.3.3.

3.7.4. The bottom entry of $u \odot_\gamma l = (aI + c\gamma)\left[\begin{smallmatrix} x \\ y \end{smallmatrix}\right]$ works out to $cc_\gamma x + (a + cd_\gamma)y$. Use information about c_γ and d_γ to show that this is 0 (mod N).

3.7.5. (b) For period 2, the ring $A = \mathbb{Z}[i]$ is a principal ideal domain and its maximal ideals are

- for each prime $p \equiv 1 \pmod 4$, two ideals $J_p = \langle a + bi \rangle$ and $\bar{J}_p = \langle a - bi \rangle$ such that $\langle p \rangle = J_p\bar{J}_p$ and the quotients A/J_p^e and $A/\bar{J}_p^e$ are isomorphic to $\mathbb{Z}/p^e\mathbb{Z}$ for all $e \in \mathbb{N}$,
- for each prime $p \equiv -1 \pmod 4$, the ideal $J_p = \langle p \rangle$ such that the quotient A/J_p^e is isomorphic to $(\mathbb{Z}/p^e\mathbb{Z})^2$ for all $e \in \mathbb{N}$,
- for $p = 2$, the ideal $J_2 = \langle 1 + i \rangle$ such that $\langle 2 \rangle = J_2^2$ and the quotient A/J_2^e is isomorphic to $(\mathbb{Z}/2^{e/2}\mathbb{Z})^2$ for even $e \in \mathbb{N}$ and is isomorphic to $\mathbb{Z}/2^{(e+1)/2}\mathbb{Z} \oplus \mathbb{Z}/2^{(e-1)/2}\mathbb{Z}$ for odd $e \in \mathbb{N}$.

3.7.6. (a) This is another elementary number theory problem. Use the Chinese Remainder Theorem to reduce to prime power N, and then see what happens when you try to lift solutions modulo primes p to solutions modulo powers p^e.

3.8.1. (a) The numerator and denominator linearly combine back to a and c under γ^{-1}.

3.8.2. (b) Since the summand $(N/d)\phi(d)\phi(N/d)$ is multiplicative in d, the sum is multiplicative in N, so it suffices to take $N = p^e$.

3.8.7. Since $\Gamma(4)$ is normal in $\mathrm{SL}_2(\mathbb{Z})$, and since the cusp ∞ is regular, all cusps are regular. See Exercise 5 of Section 3.2 for the argument that $s = 1/2$ is an irregular cusp of $\Gamma_1(4)$. Checking the cusp $s = 0$ is similar.

3.9.1. Since the calculations involve only one value of N such that $-I$ belongs to the groups, only one irregular cusp, and only one nonzero value of ε_2, the formulas for even k and odd k usually agree.

Chapter 4

4.2.4. (b) Since $\bar{v}$ is a point of order N, $\gcd(\gcd(c_v, d_v), N) = 1$.

4.3.1. If χ is nontrivial then replace n by $n_0 n$ in the sum for some value n_0 such that $\chi(n_0) \neq 0$. The second relation is proved similarly since if $n \neq 1$ in $(\mathbb{Z}/N\mathbb{Z})^*$ then the proof of duality shows that $\chi(n) \neq 1$ for some character χ.

4.3.4. (a) This is a standard result from representation theory. Alternatively, for each $d \in (\mathbb{Z}/N\mathbb{Z})^*$ define the operator $\langle d \rangle$ on $\mathcal{M}_k(\Gamma_1(N))$ to be $\langle d \rangle = [[\begin{smallmatrix} a & b \\ c & \delta \end{smallmatrix}]]_k$ for any $[\begin{smallmatrix} a & b \\ c & \delta \end{smallmatrix}] \in \Gamma_0(N)$ with $\delta \equiv d \pmod N$. Chapter 5 will show that this operator is well defined and multiplicative. For each character $\chi :$ $(\mathbb{Z}/N\mathbb{Z})^* \longrightarrow \mathbb{C}^*$, define the operator

$$\pi_\chi = \frac{1}{\phi(N)} \sum_{d \in (\mathbb{Z}/N\mathbb{Z})^*} \chi(d)^{-1} \langle d \rangle$$

on $\mathcal{M}_k(\Gamma_1(N))$. Show that $\pi_\chi^2 = \pi_\chi$, so π_χ is a projection; show that $\pi_\chi(\mathcal{M}_k(\Gamma_1(N))) \subset \mathcal{M}_k(N, \chi)$ and that $\pi_\chi = 1$ on $\mathcal{M}_k(N, \chi)$, so the projection is on $\mathcal{M}_k(N, \chi)$. Show that $\sum_\chi \pi_\chi = 1$ and that $\pi_\chi \circ \pi_{\chi'} = 0$ when $\chi \neq \chi'$, so the subspaces $\mathcal{M}_k(N, \chi)$ span and are linearly disjoint.

4.4.1. For (b), change variables to get a Gaussian integral. For (c), integrate by parts.

4.4.2. Compute that

$$\frac{(-2\pi i)^k}{2\Gamma(k)} = \frac{2^{\frac{k}{2}} \pi^k (-1)^{k/2}}{(1 \cdot 3 \cdots (k-1))(1 \cdot 2 \cdots (\frac{k}{2} - 1))}$$

$$= \frac{\pi^{\frac{k-1}{2}} \Gamma\left(\frac{1}{2}\right)}{(-\frac{1}{2})(-\frac{3}{2}) \cdots \frac{1-k}{2} \pi^{-\frac{k}{2}} \Gamma\left(\frac{k}{2}\right)} = \frac{\pi^{-\frac{1-k}{2}} \Gamma\left(\frac{1-k}{2}\right)}{\pi^{-\frac{k}{2}} \Gamma\left(\frac{k}{2}\right)}.$$

4.4.5. For (c), Exercise 4.4.4 and the fact that $\zeta(s)$ has a simple pole at $s = 1$ with residue 1 show that $L(1_N, s) \sim \phi(N)/(N(s-1))$ as $s \to 1$. And $(1 + (-1)^{-s})/(s-1) = (e^{-\pi i s} - e^{-\pi i})/(s-1)$ is a difference quotient for the derivative of $e^{-\pi i s}$ at $s = 1$. For (d), compute that

$$\sum_{\chi \neq 1_N} \chi(n^{-1}) L(1, \chi) = \sum_{\chi \neq 1_N} \chi(n^{-1}) \sum_{\substack{m=1 \\ (m,N)=1}}^{N-1} \chi(m) \sum_{d=0}^{\infty} \frac{1}{m + dN}$$

$$= \frac{1}{N} \sum_{d} \sum_{m} \frac{1}{m/N + d} \sum_{\chi \neq 1_N} \chi(mn^{-1}(N)).$$

The inner sum is $\phi(N) - 1$ if $m \equiv n \pmod{N}$ and -1 otherwise. The middle sum is taken over $\phi(N)$ values of m, so now

$$\sum_{\chi \neq 1_N} \chi(n^{-1}) L(1, \chi) = \frac{1}{N} \sum_{d=0}^{\infty} \sum_{\substack{m=1 \\ (m,N)=1}}^{N-1} \left(\frac{1}{n/N + d} - \frac{1}{m/N + d} \right),$$

where in this calculation the equivalence class $n \pmod{N}$ and its representative in $\{1, \ldots, N-1\}$ are identified. Replacing n by $-n$ and its representative by $N - n$ gives

$$- \sum_{\chi \neq 1_N} \chi((-n)^{-1}) L(1, \chi) = \frac{1}{N} \sum_{d=0}^{\infty} \sum_{\substack{m=1 \\ (m,N)=1}}^{N-1} \left(\frac{1}{n/N - d - 1} + \frac{1}{m/N + d} \right).$$

Thus the sum $\dfrac{1}{\phi(N)} \displaystyle\sum_{\chi \neq 1_N} (\chi(n^{-1}) - \chi((-n)^{-1})) L(1, \chi)$ in $\zeta^{\overline{n}}(1)$ is

$$\frac{1}{N} \sum_{d=0}^{\infty} \left(\frac{1}{n/N + d} + \frac{1}{n/N - d - 1} \right).$$

This is

$$\frac{1}{N} \sum_{d=0}^{\infty} \left(\frac{1}{n/N + d + 1} + \frac{1}{n/N - d - 1} + \frac{1}{n/N + d} - \frac{1}{n/N + d + 1} \right)$$

$$= \frac{1}{N} \left[\sum_{d=0}^{\infty} \left(\frac{1}{n/N + d + 1} + \frac{1}{n/N - d - 1} \right) + \frac{1}{n/N} \right] = \frac{1}{N} \cot \left(\frac{\pi n}{N} \right).$$

4.4.6. Use the expression for Γ in the exercise and switch to polar coordinates to get

$$\Gamma(a) \Gamma(b) = 4 \int_{r=0}^{\infty} e^{-r^2} r^{2(a+b-1)} \, r \, dr \int_{\theta=0}^{\pi/2} \cos^{2a-1} \theta \sin^{2b-1} \theta \, d\theta.$$

Let $x = \cos^2 \theta$.

4.5.3. (c) For each character ψ modulo d where $d \mid N$, Lemma 4.3.2 shows that half of the characters φ modulo N/d combine with ψ to meet the parity condition unless $N/d = 1$ or $N/d = 2$. For the exceptional values $d = N$ and $d = N/2$ (if N is even), for each character φ modulo N/d, Lemma 4.3.2 shows that half of the characters ψ modulo d combine with φ to meet the parity condition unless $d = 1$ or $d = 2$. Thus $|B_{N,k}| = (1/2) \sum \phi(d)\phi(N/d)$ for $N \nmid 4$, as required by formula (4.3). For the exceptional cases, count that $|B_{N,k}|$ for even k is 1 when $N = 1$, 2 when $N = 2$, and 3 when $N = 4$, and that $|B_{N,k}|$ for odd k is 0 when $N = 1$ or $N = 2$, and 2 when $N = 4$. Again these match formula (4.3).

4.6.4. To show that $\theta(\tau, 4)$ and $E_2^{1_1, 1_1, 4}(\tau)$ are proportional, it suffices to show that in both cases the first Fourier coefficient is 8 times the zeroth. The difference $E_2^{1_1, 1_1, 4}(\tau) - 3E_2^{1_1, 1_1, 2}(\tau)$ leads to the expression $\sum_{\substack{m|n \\ 4 \nmid m}} m - 3 \sum_{\substack{m|n \\ 2 \nmid m}} m$, and this is $-2\sigma_1(n)$ for odd n and is 0 for even n.

4.7.3. Cite the Monotone Convergence Theorem from real analysis, the fact that absolute integrability implies integrability, and the Dominated Convergence Theorem from real analysis.

4.7.5. (c) Let $z = \varepsilon e^{i\theta}$ and bound the absolute value of the integral by a quantity of the order $|\varepsilon^{s-1}|$.

4.8.2. This is very similar to Exercise 1.4.1, but integrate Z_Λ over $t + \partial P$ instead.

4.8.7. For (b) See Figure 3.4 and Theorem 3.6.1. For (c), $\mathcal{M}_1(\Gamma_1(4)) = \mathbb{C} E_1^{\chi, 1, 1}$ where χ is the nontrivial character of $(\mathbb{Z}/4\mathbb{Z})^*$. A basis of $\mathcal{M}_3(\Gamma_1(4))$ is $\{E_3^{\chi, 1, 1}, E_3^{1, \chi, 1}\}$. A basis of $\mathcal{M}_4(\Gamma_1(4))$ is $\{E_4^{1, 1, 1}, E_4^{1, 1, 2}, E_4^{1, 1, 4}\}$. Here all three basis elements contribute to $\theta(\tau, 8)$.

4.9.1. $\hat{f}(x)$ is uniformly approximated within ε by an integral over a compact subset K of $\mathbb{R}^l$. In the integral $\hat{f}(x + \delta x) - \hat{f}(x)$, note that $e^{-2\pi i \langle y, x + \delta x \rangle} - e^{-2\pi i \langle y, x \rangle} = e^{-2\pi i \langle y, x \rangle}(e^{-2\pi i \langle y, \delta x \rangle} - 1)$, and if δx is small enough then the quantity in parentheses is uniformly small as y runs through K.

4.9.3. (b) The square of the integral is $\iint_{x,y} e^{-\pi(x^2 + y^2)} dx\, dy$. Change to polar coordinates.

4.9.4. The Fourier transform is $\hat{f}(x) = \int_{y=0}^{\infty} y^{s-1} e^{2\pi i y(\tau - x)} dy$. Replace y by $z/(-2\pi i(\tau - x))$, going from 0 to ∞ along a ray in the right half plane. The resulting integral equals a gamma function integral by complex contour integration.

4.9.5. (b) $\Gamma(s)$ is the Mellin transform of e^{-t}. Use part (a).

4.10.2. Compute that

$$\int_{y\in\mathbb{R}^2} f(y\gamma r)e^{-2\pi i\langle y,x\rangle}\,dy = \int_{y\in\mathbb{R}^2} f(y)e^{-2\pi i\langle y\gamma^{-1}/r,x\rangle}\,d(y\gamma^{-1}/r)$$

$$= r^{-2}\int_{y\in\mathbb{R}^2} f(y)e^{-2\pi i\langle y,x\gamma^{-T}/r\rangle}\,dy.$$

4.10.3. For nonzero $\overline{x}\in G$, note that $\mu_N^{\langle x,e_j S\rangle}\neq 1$ for $j=1$ or $j=2$ where the e_j are the standard basis vectors, and that $\sum_v \mu_N^{\langle x,vS\rangle}=\mu_N^{\langle x,e_j S\rangle}\sum_v \mu_N^{\langle x,vS\rangle}$, so the sum is 0.

4.10.5. (a) For the first part, show that $\sum_u a(u)b(u)=\sum_v \hat{a}(-v)\hat{b}(v)$ in general. For the second part, replace b by $\hat{b}$.

4.10.6. (b) For $k<0$, since $f_k(x)=\overline{f}_{-k}(x)$ and in general $\hat{\overline{\varphi}}(x)=\overline{\hat{\varphi}}(-x)$ (show this), it follows from $\hat{f}_{-k}=(-i)^{-k}f_{-k}$ that $\hat{f}_k=(-i)^k f_k$ as before. Similarly since $f_{-k}(xS)=(-i)^{-k}f_{-k}(x)$, also $f_k(xS)=(-i)^k f_k(x)$ as before. Note that $h_k(xt^{1/2})=h_k(x)t^{|k|/2}$, that $|n\gamma|^{-|k|-2s}=y^{-k/2+s}/|c\tau+d|^{-k+2s}$, and that $h_k(n\gamma)=\overline{h}_{-k}(n\gamma)=y^{k/2}/(c\tau+d)^k$.

4.10.8. For (a), recall formula (4.11). For (c), show that $g(\psi)g(\overline{\psi})=\psi(-1)u$ similarly to the calculation after (4.11) and recall that $(\psi\varphi)(-1)=(-1)^k$.

Chapter 5

5.1.2. Take integers N_1, N_2 such that $\Gamma(N_i)\subset G_i$ for $i=1,2$, and let $N_3=\mathrm{lcm}(N_1,N_2)$; use the fact that each $[G_i:\Gamma(N_3)]$ is finite.

5.1.3. Suppose that β and β' represent the same orbit in $\Gamma_1\backslash\Gamma_1\alpha\Gamma_2$, i.e., $\Gamma_1\beta=\Gamma_1\beta'$. Letting $\beta=\gamma_1\alpha\gamma_2$ and $\beta'=\gamma_1'\alpha\gamma_2'$ translates this condition into $\alpha\gamma_2\in\Gamma_1\alpha\gamma_2'$. Since f is weight-k invariant under Γ_1, it quickly follows that $f[\beta]_k=f[\beta']_k$.

5.1.4. Set $h=\mathrm{lcm}(\{h_j\})$ where each g_j has period h_j.

5.2.5. (c) If $uv=N$ then $t=1$ and part (b) holds for all p. Suppose $p\nmid N$; if $t\mid n$ then use part (b), if $t\nmid n$ but $t\mid np$ then $p\mid t\mid N$, contradiction and hence this case can't arise, and if $t\nmid np$ then there is nothing to check. For $n=0$ there is nothing to check unless $\psi=1_1$. For (d), $a_n(E)=\sigma_1(n)-t\sigma_1(n/t)$ and $a_n(T_pE)=\sigma_1(np)-1_N(p)p\sigma_1(n/p)-t(\sigma_1(np/t)-1_N(p)p\sigma_1(n/(tp)))$ when $n\geq 1$. If $p\nmid N$ then use parts (b) and (c). If $p\mid N$ then the assumption is $t=p$, $N=p^f$. Verify the result for $p\nmid n$ and for $p\mid n$. Also verify the result for $n=0$.

5.2.6. (a) For $n\geq 1$, let $n=p^e m$ with $p\nmid m$ and then compute both $a_n(T_pf_t)$ and $a_n(\text{right side})$ in each of the three cases. One can also check $n=0$ or note that the difference of the two sides is constant and therefore zero.

5.2.8. (a) In the desired equality $((\tau+j)/p)\mathbb{Z}\oplus\mathbb{Z}=((\tau+j)/p)\mathbb{Z}+\tau\mathbb{Z}\oplus\mathbb{Z}$ one containment is obvious, and for the other note that $\tau=p(\tau+j)/p-j$. The groups $(Np\tau+p)\Lambda_{[\begin{smallmatrix}m&n\\N&p\end{smallmatrix}](p\tau)}$ and $\Lambda_{p\tau}$ are equal by Lemma 1.3.1, and multiplying through by $1/p$ gives the second desired equality of groups.

5.2.9. Recall the proof of Lemma 2.3.1 and Lemma 2.3.2.

5.3.1. Checking $n=p$ is straightforward. For $e>1$ show that $M_{p^{e-1}}M_p=M_{p^e}$ if $p\mid N$ and $M_{p^{e-1}}M_p=M_{p^e}\cup\bigcup_{j=0}^{p-1}\left[\begin{smallmatrix}p&0\\0&p\end{smallmatrix}\right]M_{p^{e-2}}\left[\begin{smallmatrix}1&j\\0&1\end{smallmatrix}\right]$ if $p\nmid N$, and the result follows for $n=p^e$. When $\gcd(m,n)=1$, M_{mn} isn't quite M_mM_n but since $\left[\begin{smallmatrix}1&1\\0&1\end{smallmatrix}\right]\in\Gamma_0(N)$ the difference doesn't affect the weight-k operator.

5.4.1. (a) Start by writing $x=(\tau+\overline{\tau})/2$, $y=(\tau-\overline{\tau})/2i$, and recall that in the algebra of differential forms, $d\tau\,d\tau=d\overline{\tau}\,d\overline{\tau}=0$ and $d\overline{\tau}\,d\tau=-d\tau\,d\overline{\tau}$.

5.4.3. If the unions $\mathrm{SL}_2(\mathbb{Z})=\bigcup(\{\pm I\}\Gamma)\alpha_i$, $\{\pm I\}\Gamma=\bigcup(\{\pm I\}\Gamma')\beta_j$ are disjoint then so is $\mathrm{SL}_2(\mathbb{Z})=\bigcup(\{\pm I\}\Gamma')\beta_j\alpha_i$.

5.4.4. Let $\mathrm{SL}_2(\mathbb{Z})=\bigcup_j\Gamma\beta_j$ so that also $\mathrm{SL}_2(\mathbb{Z})=\bigcup_{i,j}\Gamma'\alpha_i\beta_j$.

5.5.1. (c) Use Proposition 5.5.2 for the first part.

5.6.3. (a) Add a parenthesized subscript to the Hecke operators to denote level. Checking the diagram reduces to showing that for $f,g\in\mathcal{S}_k(\Gamma_1(Np^{-1}))$, (1) $T_{(Np^{-1})}f=T_{(N)}f$ and (2) $(T_{(Np^{-1})}g)[\alpha_p]_k=T_{(N)}(g[\alpha_p]_k)$.
For $T=\langle d\rangle$, show that if $\gamma\in\Gamma_0(N)$ has bottom right entry d then (1) since also $\gamma\in\Gamma_0(Np^{-1})$ it follows that $\langle d\rangle_{(Np^{-1})}$ is $\langle d\rangle_{(N)}$ restricted to level Np^{-1}; and (2) since $\alpha_p\gamma\alpha_p^{-1}\in\Gamma_0(Np^{-1})$ has bottom right entry d as well, it follows that $\langle d\rangle_{(Np^{-1})}$ is $[\alpha_p]_k\cdot\langle d\rangle_{(N)}\cdot[\alpha_p^{-1}]_k$ (composing left to right).
For $T=T_{p'}$, show that (1) $T_{p',(Np^{-1})}$ is $T_{p',(N)}$ restricted to level Np^{-1}; and (2) if $g\in\mathcal{S}_k(Np^{-1},\chi)$ for some character $\chi:(\mathbb{Z}/Np^{-1}\mathbb{Z})^*\longrightarrow\mathbb{C}^*$, then $g[\alpha_p]_k\in\mathcal{S}_k(N,\chi)$ where χ is now lifted to $(\mathbb{Z}/N\mathbb{Z})^*$. Use Proposition 5.2.2(b) to show $(T_{p',(Np^{-1})}g)[\alpha_p]_k=T_{p',(N)}(g[\alpha_p]_k)$.
(b) Going down and across takes (f,g) to $\sum_j f[\left[\begin{smallmatrix}1&j\\0&p\end{smallmatrix}\right]]_k+\sum_j g[\left[\begin{smallmatrix}p&0\\0&1\end{smallmatrix}\right]]\left[\begin{smallmatrix}1&j\\0&p\end{smallmatrix}\right]]_k$ (this relies on $p\mid N$); at level Np^{-1} the first sum is $T_pf-(\langle p\rangle f)[\alpha_p]_k$ regardless of whether $p\mid Np^{-1}$; the second sum is $\sum_j g[\left[\begin{smallmatrix}p&0\\0&p\end{smallmatrix}\right]\left[\begin{smallmatrix}1&j\\0&1\end{smallmatrix}\right]]_k$ which is $p^{k-1}g$. Going across and then down gives the same expression.
(e) Show that $w_{(N)}f=(w_{(Np^{-1})}f)[\alpha_p]_k$ and $w_{(N)}(g[\alpha_p]_k)=p^{k-2}w_{(Np^{-1})}g$, and now checking the diagram is straightforward.

5.7.2. If $\gamma_2=\left[\begin{smallmatrix}\alpha&\beta\\\gamma&\delta\end{smallmatrix}\right]\in\Gamma_d$ then $\det\gamma_2=1$ and $\gamma_2\equiv\left[\begin{smallmatrix}1&\beta\\0&1\end{smallmatrix}\right]\pmod N$ and $\beta=kN/d$ for some $k\in\mathbb{Z}$. Write $k=qd+b$, $0\le b<d$, and compute

$$\gamma_2\begin{bmatrix}1&-bN/d\\0&1\end{bmatrix}=\begin{bmatrix}\alpha&\beta-\alpha bN/d\\\gamma&\delta-\gamma bN/d\end{bmatrix}\overset{\text{call}}{=}\gamma_1.$$

Then $\det \gamma_1 = 1$, and $\beta - \alpha bN/d \equiv (1 - \alpha)bN/d \pmod{N} \equiv 0 \pmod{N}$, and $\delta - \gamma bN/d \equiv 1 \pmod{N}$, so $\gamma_1 \equiv I \pmod{N}$. Thus $\gamma_1 \in \Gamma(N)$ and $\gamma_2 \in \Gamma(N) \left[\begin{smallmatrix} 1 & bN/d \\ 0 & 1 \end{smallmatrix}\right]$. For uniqueness, the coset $\Gamma(N) \left[\begin{smallmatrix} 1 & bN/d \\ 0 & 1 \end{smallmatrix}\right]$ has its upper right entry (mod N) determined by $b \pmod{d}$.

5.7.4. If π is a projection then so is $1 - \pi$. If π_1 and π_2 are commuting projections then $\ker(\pi_1\pi_2) = \ker(\pi_1) + \ker(\pi_2)$ (one containment is clear; for the other, write $x = y + z$ where $y = \pi_2(x)$ and $z = x - y$ and show $y \in \ker(\pi_1)$, $z \in \ker(\pi_2)$). If π is a projection then $x \in \text{im}(\pi)$ if and only if $\pi(x) = x$ (if $x = \pi(y)$ then $\pi(x) = \pi^2(y) = \pi(y) = x$), and it follows that $\ker(1 - \pi) = \text{im}(\pi)$.

5.7.5. For example, the right side summand in Theorem 5.7.5 is

$$\mathcal{S}_k(\Gamma^1(N/p)) = \mathcal{S}_k(\Gamma(N))^{\Gamma^1(N/p)/\Gamma(N)}$$

$$= \mathcal{S}_k(\Gamma(N))^{\prod_{j \neq i} \Gamma^1(p_j^{e_j})/\Gamma(p_j^{e_j}) \times \Gamma^1(p_i^{e_i-1})/\Gamma(p_i^{e_i})}$$

$$= \mathcal{S}_k(\Gamma(N))^{\prod_{j \neq i} H_j \times \langle H_i, K_i \rangle} \quad \text{by Lemma 5.7.6}$$

$$= \mathcal{S}_k(\Gamma(N))^{\langle H, K_i \rangle}.$$

5.7.6. Setting $V = \mathcal{S}_k(\Gamma(N))$ isn't enough since the proposition requires V to be irreducible.

5.8.3. (c) Formula (5.4) with $\chi = 1_{11}$ describes the T_p-action on $\mathcal{S}_2(\Gamma_0(11))$, and the same formula with $\chi = 1_{88}$ describes the T_p-action on $\mathcal{S}_2(\Gamma_0(88))$.

5.8.4. By the methods early in the section, if $a_1(f) = 0$ then $f = 0$ and if $a_1(f) \neq 0$ then normalizing to $a_1(f) = 1$ gives $T_n f = a_n(f)f$ for all $n \in \mathbb{Z}^+$. Let $f = g + h$ with g old and h new. Applying T_n gives $a_n(f)f = T_n g + T_n h$, and since T_n preserves the decomposition of $\mathcal{S}_k(\Gamma_1(N))$ as a direct sum of old and new subspaces necessarily $T_n g = a_n(f)g$ and $T_n h = a_n(f)h$. Similarly g and h are $\langle n \rangle$-eigenforms for all $n \in \mathbb{Z}^+$. Thus g and h are eigenforms with T_n-eigenvalues $a_n(f)$. If $h = 0$ then $f = g$ is old. If $h \neq 0$ then again by the methods early in the section $a_1(h) \neq 0$ and $T_n h = (a_n(h)/a_1(h))h$, showing that $a_n(f) = a_n(h)/a_1(h)$ and thus $f = h/a_1(h)$ is new.

5.8.6. (a) The condition $p_i \nmid N/M_i$ gives $T_{p_i}(f_i(n\tau)) = (T_{p_i}f_i)(n\tau)$ by Proposition 5.6.2, and also the condition makes $T_{p_i}f_i$ at level N match $T_{p_i}f_i = a_{p_i}(f_i)f_i$ at level M_i, so that altogether $T_{p_i}f_{i,n} = a_{p_i}(f_i)f_{i,n}$. For the last statement of the proposition, the diamond operator $\langle n \rangle$ is the same at levels M and N for n coprime to N.

5.9.1. (a) See the more general argument in Section 5.4.

5.9.2. First consider the product over a finite set of primes and the sum over n divisible only by these primes.

5.9.6. (b) Consider characters ψ, ψ', φ, φ' modulo N, not necessarily primitive, such that $\psi(p) + \varphi(p)p^{k-1} = \psi'(p) + \varphi'(p)p^{k-1}$ for all $p \nmid N$. If $\psi \neq \psi'$ then $\psi(a) \neq \psi'(a)$ for some a. By the Dirichlet theorem on primes in an arithmetic progression there exist arbitrarily large primes p congruent to a modulo N, but the condition $0 < p^{k-1}|\varphi'(a) - \varphi(a)| = |\psi(a) - \psi'(a)| \leq 2$ gives a contradiction for large enough p.

5.11.2. For (a), let $T_p^* f = \tilde{a}_p f$ and compute that

$$\tilde{a}_p \langle f, f \rangle = \langle T_p^* f, f \rangle = \langle f, T_p f \rangle = \bar{a}_p \langle f, f \rangle.$$

For (b), if $b_p = a_p$ for all p then Proposition 5.8.5 shows that $E_k^{\psi,\varphi}/2 - f$ is constant, making it the zero function, but this violates the linear disjointness of $\mathcal{S}_k(\Gamma_1(N))$ and $\mathcal{E}_k(\Gamma_1(N))$. For (c), compute similarly to (a) that

$$a_p \langle E_k^{\psi,\varphi}, f \rangle = \langle E_k^{\psi,\varphi}, T_p^* f \rangle = \langle T_p E_k^{\psi,\varphi}, f \rangle = b_p \langle E_k^{\psi,\varphi}, f \rangle.$$

For (d), use Proposition 5.5.2(a) and then Exercise 5.4.3.

Chapter 6

6.1.3. If $\deg(f) = 1$ then there is no ramification.

6.1.4. By Proposition 6.1.4 $\varphi(z + \Lambda_g) = mz + \Lambda$ for some row vector $m \in \mathbb{C}^g$ such that $m\Lambda_g \subset \Lambda$, and $m \neq \mathbf{0}$ since φ surjects. Thus $\ker(\varphi)$ takes the form $V + \Lambda_g$ where $V \subset \mathbb{C}^g$ is a vector subspace of dimension $g - 1$. On the other hand $\operatorname{im}(f)$ contains $\int_{x_0}^x$ for all $x \in X_0(N)$. This includes $0_{J_0(N)}$ so $0_E \in \operatorname{im}(\varphi \circ f)$, and $\operatorname{span}(\operatorname{im}(f)) = J_0(N)$ by Abel's Theorem so $\operatorname{im}(f)$ can't be a subset of $\ker(\varphi)$. This makes $\varphi \circ f$ a nonconstant holomorphic map of compact Riemann surfaces, therefore a surjection.

6.2.2. (a) The result is clear on the subset Y' of Y defined later in the section. Since $\operatorname{norm}_h f$ extends continuously to Y as a function to $\widehat{\mathbb{C}}$ it is meromorphic.

6.3.1. (a) Regardless of whether the weight-k operator in general is defined by $(f[\alpha]_k)(\tau) = (\det \alpha)^e j(\alpha, \tau)^{-k} f(\alpha(\tau))$ with $e = k - 1$ or with $e = k/2$, the exponent is $e = 1$ for $k = 2$. The diagram says that $g(\alpha(\tau))d(\alpha(\tau)) = (g[\alpha]_2)(\tau)d\tau$ for $g \in \mathcal{S}_2(\Gamma_Y)$, and this is easy to check for any $g : \mathcal{H} \longrightarrow \mathbb{C}$.

6.4.1. By definition of r as a resultant, $r(u) = 0$ if and only if there exists some t such that $\tilde{q}(t, u) = 0$ and $q(t) = 0$, and by definition of $\tilde{q}$ as a resultant, there exists some t such that $\tilde{q}(t, u) = 0$ if and only if there exists some s such that $p(s) = 0$ and $u = s + t$. Thus $r(u) = 0$ if and only if there exist s and t such that $p(s) = 0$, $q(t) = 0$, and $u = s + t$. In particular, $r(\alpha + \beta) = 0$.

6.4.3. If $\alpha \in \overline{\mathbb{Z}} \cap \mathbb{Q}$ then $\alpha = s/t$ with $\gcd(s,t) = 1$ and $t^n p(\alpha) = 0$ for some monic polynomial p with integer coefficients. This last relation implies $t \mid s$, so $t = 1$. The other containment is clear.

6.4.4. Each algebraic number α satisfies a polynomial $x^n + (c_1/d)x^{n-1} + \cdots + (c_n/d)$ with $c_1, \ldots, c_n, d \in \mathbb{Z}$. Consider $d^n \alpha$.

6.5.2. For each $d \in (\mathbb{Z}/N\mathbb{Z})^*$ take two primes p and p' both congruent to d modulo N. Use formula (5.10) with $r = 2$ to express $\chi(d)$ in terms of $a_p(f)$, $a_{p^2}(f)$, $a_{p'}(f)$, and $a_{p'^2}(f)$.

6.5.3. For the first isomorphism, if $\varphi \in (M/JM)^{\wedge}$ then the map $\tilde{\varphi}$: $m \mapsto \varphi(m + JM)$ is an element of $M^{\wedge}[J]$, and if $\psi \in M^{\wedge}[J]$ then the map $\check{\psi} : m + JM \mapsto \psi(m)$ is a well defined element of $(M/JM)^{\wedge}$. Show that $\varphi \mapsto \tilde{\varphi}$ and $\psi \mapsto \check{\psi}$ invert each other and that either of them is an A-module homomorphism. The second isomorphism follows from the first since a finite-dimensional vector space is naturally isomorphic to its double dual. In particular, if $\varphi + JM^{\wedge} \in M^{\wedge}/JM^{\wedge}$ then the restriction $\tilde{\varphi} = \varphi|_{M[J]}$ is the corresponding element of $M[J]^{\wedge}$.

6.6.5. (a) Taking $\varphi \in J_1(N)$ and omitting cosets from the notation, $((\Psi_{f,n} \circ T_p)\varphi)(f^\sigma) = n\varphi((T_p f^\sigma) \circ n)$ while $((a_p(f) \circ \Psi_{f,n})\varphi)(f^\sigma) = n\varphi(T_p(f^\sigma \circ n))$. These are the same by Section 5.6 since $p \nmid N$.

(b) Stack (6.19) on (6.20), show that the outer rectangle commutes, combine this with the top square commuting and isogenies surjecting to show that the bottom square commutes.

Chapter 7

7.1.4. (b) Substitute the relation $y = \lambda x + \mu$ into (7.1) to get a cubic equation $4x^3 - \lambda^2 x^2 + \cdots = 0$. Show that the roots are x_P, x_Q, and r as in (7.3). Also letting s be as in (7.3), the sum $P + Q = (r, s)$ agrees with (7.2).

7.2.3. (a) Since φ is a combination $\sum_i f_i \varphi_i$ where the f_i are polynomials, use the product rule and then the fact that $\varphi_i(P) = 0$ for all i. (c) If $D_2 E(P) \neq 0$ then $(0, 1) \notin T_p(\mathcal{E})$, so $T_p(\mathcal{E})$ is spanned by some (a, b) with $a \neq 0$. Thus $x - x_P + m_P^2$ can serve as the dual basis under the pairing. (d) One containment is clear. For the other, suppose $s \in m_P \cap M_P^2$. Then $s = r/t$ where $r \in m_P^2$ and $t \in \overline{\mathbf{k}}[C] - m_P$. Since m_P is maximal, $1 - tv \in m_P$ for some $v \in \overline{\mathbf{k}}[C]$. But $s = s(1 - tv) + rv$, showing that $s \in m_P^2$.

7.3.2. Renotate $\{v_P\}$ as $\{v_1, \ldots, v_N\}$ and use induction on N. The case $N = 1$ is clear since $v_1^{\perp}$ is a subspace of codimension 1. For the induction step suppose some a_{N-1} satisfies $a_{N-1} \cdot v_i \neq 0$ for $i = 1, \ldots, N-1$. If $a_{N-1} \cdot v_N \neq 0$ then there is nothing to show. Otherwise take some u such that $u \cdot v_N \neq 0$ and consider vectors $a_N = u - ka_{N-1}$.

7.3.4. (a) Let $F_1(x,y) = (F(x,y) - F(x,-y))/(2y)$ and $F_2(x,y) = (F(x,y) + F(x,-y))/2$. These are both invariant under $y \mapsto -y$, making them functions of x.

7.5.1. (c) First evaluate the limit termwise and then convince yourself that doing so is justified.

7.5.3. (c) For the first part, $f_0(\Gamma(N)\infty) = \frac{3N(N-1)}{2}$ while $(f_0 \circ \gamma)(\Gamma(N)\infty)$ is strictly smaller if $\gamma \notin \Gamma_0(N)$. Geometrically, as $\mathrm{Im}(\tau) \to +\infty$ so that $j \to \infty$, the universal elliptic curve is degenerating to the singular curve $y^2 = 4x^3 - 27x - 27$, whose singular point $(-3/2, 0)$ is an isolated point of the curve's real points. All N-torsion points coming from complex torus points $(c\tau + d)/N$ with $c \neq 0 \pmod{N}$ go to the isolated point as $\mathrm{Im}(\tau) \to +\infty$, but the N-torsion points coming from complex torus points d/N stay on the other real piece. Section 8.1 will discuss singular curves such as this one. For the second part, the formula before the display shows that j_N is γ-invariant if and only if j is γ'-invariant. Clearly γ' has rational entries and determinant 1. If $\gamma' \in \mathrm{SL}_2(\mathbb{Z})$ then j is γ'-invariant. If $\gamma' \notin \mathrm{SL}_2(\mathbb{Z})$ then it identifies points that are incongruent under $\mathrm{SL}_2(\mathbb{Z})$, because any $\tau \in \mathcal{H}$ such that $\gamma'(\tau) = \delta(\tau)$ for some $\delta \in \mathrm{SL}_2(\mathbb{Z})$ satisfies a quadratic equation over $\mathbb{Q}$. But j takes a different value at each point of $\mathrm{SL}_2(\mathbb{Z}) \backslash \mathcal{H}$, so it is not γ'-invariant. Thus j_N is γ-invariant if and only if $\gamma' \in \mathrm{SL}_2(\mathbb{Z})$, and it follows quickly from the display that this condition is $\gamma \in \Gamma_0(N)$.

7.5.4. First compute j for the curve $y^2 = 4x^3 - g_2(\tau)x - g_3(\tau)$ obtained from the map $(\wp, \wp')$. Since the curve E_j differs from this curve by an admissible change of variable it has the same invariant.

7.7.1. Since $f_0 = x(\langle Q_\tau \rangle)$ is the sum of the finite x-coordinates of $\langle Q_\tau \rangle$, it follows that $f_0^\sigma = x(\langle Q_\tau^\sigma \rangle)$ for $\sigma \in H_\mathbb{Q}$. The results obtained over $\mathbb{C}$ show that the fixing subgroup is the subgroup that preserves $\langle Q_\tau \rangle$, i.e., $Q_\tau^\sigma = dQ_\tau$ for some d.

7.8.1. (b) Since $\ell(P) = 1$ and $L(P)$ contains $\mathbf{k}$, $L(P)$ is no more than $\mathbf{k}$. Thus there is no function with a simple pole at P. Any set of elements with distinct valuations at P is linearly independent, by properties of the valuation. Thus the linear relation at the end must involve at least one of X^3 and Y^2 since the other five elements are linearly independent. Take the coefficient of X^3 to be 1 and the coefficient of Y^2 to be a. Substitute aX and aY for X and Y and then divide by a^3.

7.8.2. (a) The group law is $(\sigma, P)(\sigma', P') = (\sigma\sigma', \sigma'(P) + P')$. Check that $(\sigma, P)((\sigma', P')f) = ((\sigma, P)(\sigma', P'))f$.

(b) The semidirect product acts on constant functions in $\mathbb{K}$ as H since translating the variable has no effect. Showing that $\mathbb{K} \cap \mathbf{1} = \mathbf{k}'$ in fact shows $\mathbb{K} \cap \overline{\mathbf{k}'} = \mathbf{k}'$ by the nature of the rest of the configuration.

(d) Write the restriction of i as $\mathbf{1}(E)^{H \ltimes C} \longrightarrow \mathbf{1}(E)^H$.

7.8.4. $\varphi \circ \psi \circ \varphi = \varphi \circ [\deg(\varphi)] = [\deg(\varphi)] \circ \varphi$ since φ is a homomorphism. Since φ surjects it cancels on the right, giving the result.

7.8.5. Take a point $Q \in E'$ and any point $P \in E$ such that $\varphi(P) = Q$. Then at the level of divisors, $\hat{\psi}_*((Q) - (0_{E'})) = ([\deg(\varphi)]P) - (0_E)$ while since φ is unramified,

$$\varphi^*((Q) - (0_{E'})) = \sum_{R \in \ker(\varphi)} (P + R) - \sum_{R \in \ker(\varphi)} (R).$$

These are not equal, but their difference satisfies the characterization of principal divisors in Theorem 7.3.3 since $\deg(\varphi) = |\ker(\varphi)|$, so their classes are equal in $\mathrm{Pic}^0(E)$.

7.9.2. (a) $w_N : X_0(N) \longrightarrow X_0(N)$ given by $\Gamma_0(N)\tau \mapsto \Gamma_0(N)\tau'$ is a holomorphic map of compact Riemann surfaces. By Section 7.3 it can be viewed instead as a morphism over $\mathbb{C}$ of algebraic curves over $\mathbb{C}$. Its pullback is $w_N^* : \mathbb{C}(X_0(N)) \longrightarrow \mathbb{C}(X_0(N))$, a $\mathbb{C}$-injection of function fields over $\mathbb{C}$. By Section 7.5 this is $w_N^* : \mathbb{C}(j(\tau), j_N(\tau)) \longrightarrow \mathbb{C}(j(\tau), j_N(\tau))$, taking $j(\tau)$ to $j(\tau')$ and $j_N(\tau)$ to $j_N(\tau')$. If $j(\tau')$ and $j_N(\tau')$ are in $\mathbb{Q}(j(\tau), j_N(\tau))$ then it restricts to $w_N^* : \mathbb{Q}(j(\tau), j_N(\tau)) \longrightarrow \mathbb{Q}(j(\tau), j_N(\tau))$, a $\mathbb{Q}$-injection of function fields over $\mathbb{Q}$. This is $w_N^* : \mathbb{Q}(X_0(N)_{\mathrm{alg}}) \longrightarrow \mathbb{Q}(X_0(N)_{\mathrm{alg}})$ by Section 7.7, and now Theorem 7.2.6 gives $w_N : X_0(N)_{\mathrm{alg}} \longrightarrow X_0(N)_{\mathrm{alg}}$, a morphism over $\mathbb{Q}$ of algebraic curves over $\mathbb{Q}$. For the last part, compute that $j(\tau') = j_N(\tau)$ and $j_N(\tau') = j(\tau)$.

(b) A point of $X_0(N)_{\mathrm{alg}}^{\mathrm{planar}}$ is $(j, x) \in \overline{\mathbb{Q}}^2$ such that $p_0(j, x) = 0$. The map is $[E, C] \mapsto (j(E), x(C))$ where $x(C)$ is the sum of the x-coordinates of the nonzero points of C. For the second part, recall Exercise 1.5.4.

7.9.3. (b) The full extensions have the same Abelian Galois group. The upper and lower Galois extensions on the left inject into the ones on the right, so the extension degrees must match. Now use the fact that the Galois group is cyclic.

(e) Compute

$$f_1(p\tau)^\sigma = x(Q_{\tau',N})^\sigma = x(Q_{\tau',N}^\sigma) = x(\varphi(Q_{\tau,Np})^\sigma) = x(\varphi^\sigma(Q_{\tau,Np}^\sigma))$$
$$= x(\pm\varphi(Q_{\tau,Np})) = f_1(p\tau),$$

the second-to-last step using parts (b) and (d) to show that $\varphi(Q_{\tau,Np}^\sigma) = \varphi(Q_{\tau,Np})$.

7.9.4. Consider the map $[N] : E \longrightarrow E$ and recall that the structure of $E_{\mathbb{C}}[N]$ was established in Chapter 1.

Chapter 8

8.1.1. (b) Consider the matrices

$$\begin{bmatrix} u^2 & 0 & r \\ su^2 & u^3 & t \\ 0 & 0 & 1 \end{bmatrix}.$$

For (c), the condition implies $v^3 = w^2$. Take $u = w/v$. For (e), take $r = -3/2$, $s = i3\sqrt{2}/2$, $t = 0$, and $u = i3\sqrt{2}$.

8.1.5. (a) If $\text{char}(\mathbf{k}) \neq 2$ then we may take $a_1 = a_3 = 0$ and so $P = (x, 0)$ where x is a repeated root of a cubic polynomial over $\mathbf{k}$. If $\text{char}(\mathbf{k}) = 2$ and $a_1 \neq 0$ then $x = a_1^{-1} a_3$ and $y = a_1^{-1}(x^2 + a_4)$. If $\text{char}(\mathbf{k}) = 2$ and $a_1 = 0$ then use the fact that every element of $\mathbf{k}$ is a square.

8.2.1. For uniqueness, the $q - 1$ nonzero elements of such a subfield satisfy $x^{q-1} = 1$.

8.3.1. Recall that $1728\Delta = c_4^3 - c_6^2$.

8.3.2. (a) Since $\Delta_p' = \Delta/u_p^{12}$ and $\nu_p(\Delta_p') \leq \nu_p(\Delta)$ it follows that $\nu_p(u_p) \geq 0$. If $\nu_p(u_p) = 0$ then we may take $u_p = 1$ and $r_p = s_p = t_p = 0$, so now assume $\nu_p(u_p) > 0$. Since $\nu_p(a_i) \geq 0$ and $\nu_p(a_{i,p}') \geq 0$ the formula $a_{1,p}' = (a_1 + 2s_p)/u_p$ shows that $\nu_p(s_p) \geq 0$. (Note that for $p = 2$ this requires $\nu_2(u_2) > 0$.) Similarly the formula $a_{2,p}' = (a_2 - s_p a_1 + 3r_p - s_p^2)/u_p^2$ shows that $\nu_p(r_p) \geq 0$, and $a_{3,p}' = (a_3 + r_p a_1 + 2t_p)/u_p^3$ shows that $\nu_p(t_p) \geq 0$.
   (c) For each $p \mid \Delta$ let $r_p = p^{e_p} m_p/n_p$. Then $\nu_p(r - r_p) = \nu_p(n_p r - p^{e_p} m_p)$ for $r \in \mathbb{Z}$. The congruence $n_p r \equiv p^{e_p} m_p \pmod{p^{6\nu_p(u)}}$ has a solution $r \pmod{p^{6\nu_p(u)}}$, and thus $\nu_p(r - r_p) \geq 6\nu_p(u)$. The Chinese Remainder Theorem gives an integer r simultaneously satisfying the condition for all primes $p \mid \Delta$.

8.3.3. (b) If the a_i and the a_i' are integral then so are the b_i and the b_i'. The relation $b_2' - b_2 = 12r$ shows that $\nu_p(r) \geq 0$ for all primes p except possibly 2 and 3. If $\nu_3(r) < 0$ then the relation $\nu_3(b_6' - b_6) \geq 0$ is impossible, and similarly for 2 and $b_8' - b_8$.
   (c) The relation $\pm a_1' - a_1 = 2s$ shows that $\nu_p(s) \geq 0$ for all primes p except possibly 2, and the relations $\nu_2(s) < 0$ and $a_2' - a_2 = -sa_1 + 3r - s^2$ are incompatible. Similarly for t, using the relations between $\pm a_3'$ and a_3, and a_6' and a_6.

8.3.5. Show that $a_p(E) \equiv 0 \pmod p$ if and only if $|\widetilde{E}(\mathbb{F}_{p^e})| \not\equiv 0 \pmod p$ for all $e \geq 1$. By Theorem 8.1.2 this holds if and only if $\widetilde{E}(\mathbb{F}_{p^e}) \cap \widetilde{E}[p] = \{0_E\}$ for all $e \geq 1$, and this in turn holds if and only if $\widetilde{E}[p] = \{0\}$.

8.3.6. (a) The discriminant is $-2^4 3^3$ and $c_4 = 0$. Recall Proposition 8.1.3.

(b) If $x^3 = 1$ and $x \neq 1$ then $x^2 + x + 1 = 0$, impossible in $\mathbb{F}_p$ when $p \equiv 2 \pmod 3$ by Quadratic Reciprocity. So as x runs through $\mathbb{F}_p$ so does $x^3 - 1$, giving one point (x, y) on the curve for each $y \in \mathbb{F}_p$. Remember 0_E as well.

(c) For the first displayed equality, remember 0_E. For the last part, remember that f is cubic.

(e) The reduction is ordinary for $p \equiv 1 \pmod 4$, supersingular for $p \equiv 3 \pmod 4$.

8.4.1. (b) For the first part, $\beta = \tilde{\alpha}$ for some $\alpha \in \mathbb{Z}$, this α satisfies some monic polynomial $f \in \mathbb{Z}[x]$, and so β satisfies the monic polynomial $g = \tilde{f} \in (\mathbb{Z}/p\mathbb{Z})[x]$ obtained by reducing the coefficients of f modulo p,

$$g(\beta) = \tilde{f}(\tilde{\alpha}) = \widetilde{f(\alpha)} = \widetilde{0_{\overline{\mathbb{Z}}}} = 0_{\overline{\mathbb{Z}}/\mathfrak{p}}.$$

For the second part, since $f(x) = \prod(x - \alpha_i)$ in $\overline{\mathbb{Z}}[x]$ it follows that $g(x) = \prod(x - \tilde{\alpha}_i)$ in $(\overline{\mathbb{Z}}/\mathfrak{p})[x]$.

8.4.3. Since we don't have a valuation on $\mathbb{Q}$, one method is to work in the number field $\mathbb{K}$ generated by the Weierstrass coefficients and the change of variable parameters, arguing as in the proof of Lemma 8.4.1 that this field has a valuation and then continuing as in Exercise 8.3.3(b,c). Another method is to use the lemma itself as follows. The conditions $u \in \overline{\mathbb{Z}}_{(\mathfrak{p})}^*$ and $u^2 b_2' - b_2 = 12r$ show that $r \in \overline{\mathbb{Z}}_{(\mathfrak{p})}$ unless $\mathfrak{p}$ lies over 2 or 3. If $r \notin \overline{\mathbb{Z}}_{(\mathfrak{p})}$ then $1/r \in \mathfrak{p}\overline{\mathbb{Z}}_{(\mathfrak{p})}$ by the lemma and the union $\overline{\mathbb{Z}}_{(\mathfrak{p})} = \overline{\mathbb{Z}}_{(\mathfrak{p})}^* \cup \mathfrak{p}\overline{\mathbb{Z}}_{(\mathfrak{p})}$. If $\mathfrak{p}$ lies over 3 then the relation

$$u^6 b_6' - b_6 = (2b_4/r^2 + b_2/r + 4)r^3$$

gives a contradiction since the left side lies in $\overline{\mathbb{Z}}_{(\mathfrak{p})}$ and the first factor on the right side lies in $\overline{\mathbb{Z}}_{(\mathfrak{p})}^*$, but $r^3 \notin \overline{\mathbb{Z}}_{(\mathfrak{p})}$. If $\mathfrak{p}$ lies over 2 then use the relation

$$u^8 b_8' - b_8 = (3b_6/r^3 + 3b_4/r^2 + b_2/r + 3)r^4.$$

So $r \in \overline{\mathbb{Z}}_{(\mathfrak{p})}$. Argue similarly for s and t.

8.4.4. (c) Recall Exercise 8.3.6(e).

8.5.1. $I = \langle \{p\varphi_i\}, p(p\phi - 1), \{(p\phi - 1)\psi_j\} \rangle$.

8.5.2. (c) Part (b) applies with $i = 0$.

8.5.4. $I_{(0)} = \langle x_1 + px_2^2 \rangle$ works.

8.5.6. (c) If $yz \in J$ and $y \notin J$ then $z \in \text{Ann}_S(xy)$, so $\langle J, z \rangle \subset \text{Ann}_S(xy)$, contradicting maximality unless $z \in J$.

8.5.7. (a) Let C be the curve and C_i the nonempty affine piece. Then $P \in C - C_i$ if and only if $(x_i/x_j)(P) = 0$ for some $j \neq i$, but each x_i/x_j (where x_j is not identically 0 on C) has a finite set of zeros.

(b) This is immediate from (a) since the projective curve is infinite.

8.6.1. Taking $[E_j, Q]$ across and then down gives $(j, x(Q))$ and then $(\tilde{j}, \widetilde{x(Q)})$. Taking it down and then across gives $[\tilde{E}_j, \tilde{Q}]$ and then $(j(\tilde{E}_j), x(\tilde{Q}))$. These are the same.

8.7.1. Let $E' = E/C$ and $Q' = Q + C$. Let $\varphi : E \longrightarrow E'$ be the quotient isogeny, so that $Q' = \varphi(Q)$. Properties of degree and the calculation $\ker(\tilde{\varphi}) \subset \ker([p]_{\tilde{E}}) = \tilde{E}[p] = \{0\}$ combine to show that $\tilde{\varphi} = i \circ \sigma_p$ where $i : \tilde{E}^{\sigma_p} \longrightarrow \tilde{E}'$ is an isomorphism taking $\tilde{Q}^{\sigma_p}$ to $\tilde{Q}'$. The first equality follows. The second is shown similarly using the dual isogeny ψ, citing Proposition 8.4.4(b), and applying σ_p^{-1} to the coefficients of the resulting isomorphism i in this case.

8.7.2. (a) For the first diagram, going across and then down takes $[\tilde{E}_j, Q]$ to $[\tilde{E}_j^{\sigma_p}, Q^{\sigma_p}]$ and then to $(j(\tilde{E}_j^{\sigma_p}), x(Q^{\sigma_p}))$, while going down and then across takes it to $(j, x(Q))$ and then to $(j^{\sigma_p}, x(Q)^{\sigma_p})$. For the second diagram, going across and then down takes $[\tilde{E}_j, Q]$ to $p[\tilde{E}_j^{\sigma_p^{-1}}, Q^{\sigma_p^{-1}}]$ and then to $p(j(\tilde{E}_j^{\sigma_p^{-1}}), x(Q^{\sigma_p^{-1}}))$, while going down and then across takes it to $(j, x(Q))$ and then to $p(j^{\sigma_p^{-1}}, x(Q)^{\sigma_p^{-1}})$, cf. (8.15). In both cases the results are the same.

(b) There is a birational equivalence h from $\tilde{X}_1(N)^{\text{planar}}$ to $\tilde{X}_1(N)$ as described in Theorem 8.6.1. Consider the diagram

$$
\begin{array}{ccc}
\mathrm{Div}^0(\tilde{X}_1(N)^{\text{planar}}) & \xrightarrow{\sigma_{p,*}} & \mathrm{Div}^0(\tilde{X}_1(N)^{\text{planar}}) \\
\downarrow{\scriptstyle h_*} & & \downarrow{\scriptstyle h_*} \\
\mathrm{Div}^0(\tilde{X}_1(N)) & \xrightarrow{\sigma_{p,*}} & \mathrm{Div}^0(\tilde{X}_1(N))
\end{array}
$$

and recall formula (8.17). The map across the bottom row descends to Picard groups. The other diagram is similar by formula (8.19).

(c) Set up a cube diagram as in the section, but with $\langle d \rangle$ and $\langle d \rangle_*$ across the top rows and with the reductions $\widetilde{\langle d \rangle}$ and $\widetilde{\langle d \rangle}_*$ across the bottom rows. Thus the bottom square is diagram (8.39). Explain why all the other squares of the cube commute and why the map from the top front left to the bottom front left surjects. Complete the argument.

(d) Combine the second diagram from part (b) with diagram (8.39) to get a commutative diagram

$$
\begin{array}{ccccc}
\mathrm{Div}^0(\tilde{S}_1(N)') & \xrightarrow{p\sigma_p^{-1}} & \mathrm{Div}^0(\tilde{S}_1(N)') & \xrightarrow{\widetilde{\langle d \rangle}} & \mathrm{Div}^0(\tilde{S}_1(N)') \\
\downarrow & & \downarrow & & \downarrow \\
\mathrm{Pic}^0(\tilde{X}_1(N)) & \xrightarrow{\sigma_p^*} & \mathrm{Pic}^0(\tilde{X}_1(N)) & \xrightarrow{\widetilde{\langle d \rangle}_*} & \mathrm{Pic}^0(\tilde{X}_1(N)).
\end{array}
$$

Along with the first diagram from part (b) this gives (8.34).

8.8.1. (a) $\mathrm{Pic}^0(X_0(N)_{\mathbb{C}})$ is generated by the image of $X_0(N)_{\mathbb{C}}$, and the third stage of $\beta_{\mathbb{C}}$ is an isomorphism.

Chapter 9

9.1.1. The extension $\mathbb{Q}(d^{1/3})/\mathbb{Q}$ is not Galois but the extensions $\mathbb{F}/\mathbb{Q}(d^{1/3})$, $\mathbb{F}/\mathbb{Q}(\mu_3)$, and $\mathbb{Q}(\mu_3)/\mathbb{Q}$ are. Recall from Exercise 8.3.6(b) that if $p \equiv 2 \pmod 3$ then every a is a cube modulo p.

9.2.1. (b) A basis of the product topology is the subsets $S = \prod_n S_n$ where $S_n = \mathbb{Z}/\ell^n\mathbb{Z}$ for all but finitely many n. Each C_n is the subgroup of compatible elements, naturally isomorphic to $\mathbb{Z}/\ell^n\mathbb{Z}$.

9.2.2. Any nonidentity $m \in \mathrm{GL}_d(\mathbb{C})$ that is close enough to I takes the form $m = \exp(a)$ where $a \in \mathrm{M}_d(\mathbb{C})$ is nonzero, and $m^n = \exp(na)$ for all $n \in \mathbb{Z}$.

9.2.3. (a) Since λ lies over ℓ it follows that $\lambda^{ne_\lambda} \cap \mathbb{Z} = \ell^{n'}\mathbb{Z}$ for some n'. The condition $\ell^m \in \lambda^{ne_\lambda}$ is $\prod_\lambda \lambda^{me_\lambda} \subset \lambda^{ne_\lambda}$, and the unique factorization of ideals in $\mathcal{O}_{\mathbb{K}}$ shows that this holds if and only if $m \geq n$. Thus $n' = n$. So the map $\mathbb{Z} \longrightarrow \mathcal{O}_{\mathbb{K}}/\lambda^{ne_\lambda}$ has kernel $\ell^n\mathbb{Z}$, making $\mathbb{Z}/\ell^n\mathbb{Z} \longrightarrow \mathcal{O}_{\mathbb{K}}/\lambda^{ne_\lambda}$ an injection for all n and λ. This gives an injection $\mathbb{Z}_\ell \longrightarrow \mathcal{O}_{\mathbb{K},\lambda}$ for all λ.
(b) The injection surjects if $|\mathcal{O}_{\mathbb{K}}/\lambda^{ne_\lambda}| = \ell^n$ for all n, i.e., $|\mathcal{O}_{\mathbb{K}}/\lambda|^{e_\lambda} = \ell$, i.e., $e_\lambda f_\lambda = 1$.

9.3.1. Let U be any open normal subgroup. Then $U(\mathbb{F}) \subset U$ for some Galois number field $\mathbb{F}$, giving a surjection $\mathrm{Gal}(\mathbb{F}/\mathbb{Q}) = G_{\mathbb{Q}}/U(\mathbb{F}) \longrightarrow G_{\mathbb{Q}}/U$. This shows that $G_{\mathbb{Q}}/U = \mathrm{Gal}(\mathbb{F}'/\mathbb{Q})$ for some $\mathbb{F}' \subset \mathbb{F}$, and so $U = U(\mathbb{F}')$.

9.3.2. Consider any neighborhood $U = U_\sigma(\mathbb{F})$ in $G_{\mathbb{Q}}$. We want to find some $\mathrm{Frob}_{\mathfrak{p}} \in U$. This holds if $\mathrm{Frob}_{\mathfrak{p}}|_{\mathbb{F}} = \sigma|_{\mathbb{F}}$. But $\sigma|_{\mathbb{F}}$ takes the form $\mathrm{Frob}_{\mathfrak{p}_{\mathbb{F}}}$ for some maximal ideal of $\mathcal{O}_{\mathbb{F}}$ by Theorem 9.1.2. Lift $\mathfrak{p}_{\mathbb{F}}$ to a maximal ideal $\mathfrak{p}$ of $\overline{\mathbb{Z}}$.

9.3.4. Take a neighborhood V of I in $\mathrm{GL}_d(\mathbb{C})$ containing no nontrivial subgroup, cf. Exercise 9.2.2. Let $U = \rho^{-1}(V)$. As a neighborhood of 1 in $G_{\mathbb{Q}}$, U contains $U(\mathbb{F})$ for some Galois number field $\mathbb{F}$. So ρ is defined on $\mathrm{Gal}(\mathbb{F}/\mathbb{Q})$.

9.3.5. Let $M = \mathrm{lcm}(N, N')$, so $\rho|_{\mathbb{Q}(\mu_M)}$ can be viewed via the isomorphism (9.1) as a character χ of $(\mathbb{Z}/M\mathbb{Z})^*$ that factors through $(\mathbb{Z}/N\mathbb{Z})^*$ and $(\mathbb{Z}/N'\mathbb{Z})^*$. Since $\chi(n \bmod M)$ is trivial if $n \equiv 1 \pmod N$ and if $n \equiv 1 \pmod{N'}$ it follows that χ is defined modulo $\gcd(N, N')$.

9.3.7. (a) $\mathbb{L}^d$ is a d-dimensional topological vector space over $\mathbb{L}$. The groups $\mathrm{GL}_d(\mathbb{L})$ and $\mathrm{Aut}(\mathbb{L}^d)$ are naturally identified. Since ρ is continuous so is its composition with vector-by-matrix multiplication,

$$\mathbb{L}^d \times G_{\mathbb{Q}} \longrightarrow \mathbb{L}^d, \qquad (v, \sigma) \mapsto v\rho(\sigma).$$

Thus ρ makes $\mathbb{L}^d$ a $G_{\mathbb{Q}}$-module satisfying the continuity condition of Definition 9.3.4. On the other hand, any choice of ordered basis of V identifies $\mathrm{Aut}(V)$ with $\mathrm{GL}_d(\mathbb{L})$, and then the map $G_{\mathbb{Q}} \longrightarrow \mathrm{Aut}(V)$ gives a map $\rho : G_{\mathbb{Q}} \longrightarrow \mathrm{GL}_d(\mathbb{L})$. Specifically, the matrix entries are $\rho(\sigma)_{ij} = x_j(e_i\sigma)$ where the ordered basis is $(e_1, \ldots, e_d)$ and $x_i : V \longrightarrow \mathbb{L}$ is $\sum_\iota a_\iota e_\iota \mapsto a_i$. Each ρ_{ij} is a continuous function, making ρ a Galois representation as in Definition 9.3.2.

(b) Equivalent representations ρ and ρ' as in Definition 9.3.2 determine $G_{\mathbb{Q}}$-linear isomorphic $G_{\mathbb{Q}}$-module structures of $\mathbb{L}^d$, and if V and V' are equivalent as in Definition 9.3.4 then any choice of ordered bases B and B' determines equivalent representations as in Definition 9.3.2. Also, going from ρ to $\mathbb{L}^d$ and then choosing a basis B gives a representation ρ' equivalent to ρ, while on the other hand starting with V and then choosing a basis to define ρ gives $\mathbb{L}^d$ a $G_{\mathbb{Q}}$-module structure equivalent to V.

9.4.3. The isogeny $E \longrightarrow E'$ induces a map $V_\ell(E) \longrightarrow V_\ell(E')$. The map is an isomorphism since the dual isogeny induces a similar map in the other direction and the composite is multiplication by the degree of the isogeny, an automorphism because $\mathbb{Q}_\ell$ has characteristic 0.

9.4.4. (c) Since ℓ-torsion contains an ℓ-cyclic subgroup of points with rational coordinates, $\overline{\rho}_{E,\ell} \sim \left[\begin{smallmatrix} 1 & 0 \\ 0 & \overline{\chi_\ell} \end{smallmatrix}\right]$ where $\overline{\chi_\ell} : G_{\mathbb{Q}} \longrightarrow \mathbb{F}_2$ is the mod 2 reduction of the cyclotomic character. Alternatively, the ℓ-cyclic subgroup of points with rational coordinates implies that $\ell \mid |\tilde{E}(\mathbb{F}_\ell)|$, i.e., $a_p(E) \equiv p + 1 \pmod{\ell}$.

9.5.1. This follows from Lemma 9.5.2.

9.5.2. (a) Each $\mathbb{K}_{f,\lambda}$ acts on $V_\lambda(f)$ via i^{-1}. The dimension is 2 because $V_\lambda(f) = e_\lambda V_\ell(A_f) \cong e_\lambda(\mathbb{K}_f \otimes_{\mathbb{Q}} \mathbb{Q}_\ell)^2 \cong \mathbb{K}_{f,\lambda}^2$. Apply each e_λ to any linear dependence $\sum_{\lambda'} e_{\lambda'} v_{\lambda'} = 0$ to show that the sum $\sum_\lambda V_\lambda(f)$ is direct. The relation $v = \sum_\lambda e_\lambda v$ for any $v \in V_\ell(A_f)$ shows that the sum spans $V_\ell(A_f)$. The $G_{\mathbb{Q}}$-action restricts to $V_\lambda(f)$ because it commutes with e_λ.

(b) The first part is explained at the end of Section 9.2. For the second part, $\rho_{A_f,\ell}$ is continuous, making $V_\ell(A_f) \times G_{\mathbb{Q}} \longrightarrow V_\ell(A_f)$ continuous, and $V_\lambda(f)$ is a $\mathbb{Q}_\ell$-vector subspace of $V_\ell(A_f)$.

9.6.1. (a) Since $conj$ has order 2 the only possible eigenvalues are ± 1. Recall that $\det \rho(\mathrm{conj}) = -1$.

(b) Irreducibility immediately gives σ with $b \neq 0$ and σ' with $c \neq 0$. If neither σ nor σ' works then $\sigma\sigma'$ does. For the last part, conjugate by a matrix of the form $\left[\begin{smallmatrix} m & 0 \\ 0 & 1 \end{smallmatrix}\right]$.

9.6.4. (a) To verify that the map is an embedding check the orders of the elements by using characteristic polynomials to compute eigenvalues. This also shows that the two elements of $\mathrm{GL}_2(\mathbb{F}_3)$ generate a subgroup H of order divisible by 24. Since A_4 is the only order 12 subgroup of S_4, $\mathrm{SL}_2(\mathbb{F}_3)$ is the only order 24 subgroup of $GL_2(\mathbb{F}_3)$, but the second generator of H has determinant -1.

(b) The first statement follows from the end of part (a). Working modulo λ, compute for $p \nmid 3M_f N_E$ that

$$a_p(f) \equiv \operatorname{tr} \overline{\rho}_{f,\lambda}(\operatorname{Frob}_\mathfrak{p}) = \operatorname{tr} \overline{\rho}_{E,3}(\operatorname{Frob}_\mathfrak{p}) \equiv a_p(E).$$

Since $\psi = \det \rho_{f,\lambda}$ is a lift of $\det \overline{\rho}_{E,3} = \chi_3 \pmod 3$ from $\mathbb{F}_3^*$ to $\mathcal{O}_K^*$ it suffices to consider the latter character, $\det \overline{\rho}_{E,3} = \chi_3 \pmod 3$. This surjects, making ψ quadratic, and since as a Galois representation it is defined on $\operatorname{Gal}(\mathbb{Q}(\mu_3)/\mathbb{Q})$, as a Dirichlet character ψ has conductor 3.

(c) Use results in Chapter 4 to evaluate the leading coefficient of the Eisenstein series.

(d) Since $\psi(p) \equiv p \pmod \lambda$ the operators T_p on $\mathcal{S}_1(M_f, \psi)$ and T_p on $\mathcal{S}_2(\Gamma_0(M_f))$ are congruent modulo λ. This observation and part (c) show that $T_p g \equiv T_p f = a_p(f) f \pmod \lambda$. On the one hand the right side is congruent modulo λ to $a_p(E)g$ for all but finitely many p by parts (b) and (c), giving the first statement. On the other hand the right is also congruent modulo λ to $a_p(g)g = a_1(T_p g)g$ for all p by part (c). The second statement follows.

(e) For all but finitely many primes $\phi(T_p) \equiv a_p(E) \pmod \lambda$, and so $T_p - a_p(E) \in \ker \phi = m$, implying $\phi'(T_p) - a_p(E) \in \phi'(m) \subset \lambda'$.

(h) The argument in Exercise 9.6.1 applies over any field whose characteristic is not 2. Let $\mathbb{L}$ be a finite Galois extension of $\mathbb{Q}$ containing $\mathbb{K}_{g''}$ and $\mathbb{K}_{g'}$. Use Strong Multiplicity One to show that if $\sigma \in \operatorname{Gal}(\mathbb{L}/\mathbb{Q})$ then $(g'')^\sigma$ is the newform associated to $(g')^\sigma$; in particular if σ fixes $\mathbb{K}_{g'}$ then it fixes $\mathbb{K}_{g''}$.

List of Symbols

© Springer Science+Business Media New York 2005

F. Diamond, J. Shurman, *A First Course in Modular Forms*, Graduate Texts in Mathematics 228, DOI 10.1007/978-0-387-27226-9

Index

Abel's Theorem, 84, 218
Abelian variety associated to an
    eigenform, 245
absolute Galois group of $\mathbb{Q}$, 383
addition law of elliptic curves
  algebraic, 258, 316
  geometric, 257, 316
adjoint, 184
admissible change of variable, 255, 315
affine plane, 256
algebraic closure of a field, 254
algebraic curve
  affine, 262
  nonsingular, 262
algebraic curves
  isomorphic, 267
  isomorphic over $\mathbf{k}$, 268
algebraic element over a field, 254
algebraic field extension, 254
algebraic integer, 235
algebraic number, 234
algebraically closed, 235
algebraically closed field, 254
Artin's Conjecture, 407
automorphic form of weight k, 72

base point, 218
Bernoulli number, 9, 134
  of ψ, 135
Bernoulli polynomial, 134
Bézout's Theorem, 256
birational equivalence of algebraic
    curves, 268

Birch and Swinnerton-Dyer Conjecture,
  367
canonical divisor on a Riemann surface,
  84
commensurable congruence subgroups,
  164
compatibility condition for meromorphic differentials on a Riemann
  surface, 78
complex multiplication, 30
complex torus, 25
conductor
  of a Dirichlet character, 117
  of a normalized eigenform, 199
  of an elliptic curve
    algebraic, 328
    analytic, 296
congruence subgroup, 13
conjugate function fields, 268
coordinate ring of an algebraic curve,
  261
Cubic Reciprocity Theorem, 160
Curves–Fields Correspondence
  Part 1, 268
  Part 2, 269
cusp form of weight k, 6
  with respect to a congruence
    subgroup, 17
cusp of a modular curve, 58
  irregular, 89
  regular, 89
cusp of an elliptic curve, 318
cyclic quotient isogeny, 27

© Springer Science+Business Media New York 2005
F. Diamond, J. Shurman, *A First Course in Modular Forms*,
Graduate Texts in Mathematics 228, DOI 10.1007/978-0-387-27226-9

References

[AL70]   A. O. L. Atkin and J. Lehner. Hecke operators on $\Gamma_0(m)$. *Mathematische Annalen*, 185:134–160, 1970.

[BBB00]   Lennert Berggren, Jonathan Borwein, and Peter Borwein. *Pi: A Source Book*. Springer-Verlag, second edition, 2000.

[BCDT01]   C. Breuil, B. Conrad, F. Diamond, and R. Taylor. On the modularity of ellipic curves over $\mathbb{Q}$: wild 3-adic exercises. *J. Amer. Math. Soc.*, 14(4):843–939, 2001.

[BDSBT01]   K. Buzzard, M. Dickinson, N. Shepherd-Barron, and R. Taylor. On icosahedral Artin representations. *Duke Math. J.*, 109(2):283–318, 2001.

[BLR90]   Siegfried Bosch, Werner Lütkebohmert, and Michel Raynaud. *Néron Models*, volume 21 of *Ergebnisse der Mathematik und ihrer Grenzgebiete (3)*. Springer-Verlag, 1990.

[Bum97]   Daniel Bump. *Automorphic Forms and Representations*. Studies in Advanced Mathematics **55**. Cambridge University Press, 1997.

[Car86]   H. Carayol. Sur les représentations ℓ-adiques associées aux formes modulaires de Hilbert. *Ann. Sci. E. N. S.*, 19:409–468, 1986.

[Car99]   D. Carlton. On a result of Atkin and Lehner. http://math.stanford.edu/~carlton/math/, 1999.

[Car01]   D. Carlton. Moduli for pairs of elliptic curves with isomorphic N-torsion. *Manuscripta Mathematica*, 105(2):201–234, 2001.

[Cox84]   D. Cox. The arithmetic-geometric mean of Gauss. *L'Enseignement Mathématique*, 30:275–330, 1984.

[Cox97]   David Cox. *Primes of the Form $x^2 + ny^2$: Fermat, Class Field Theory, and Complex Multiplication*. Wiley, second edition, 1997.

[Cre97]   J. E. Cremona. *Algorithms for Modular Elliptic Curves*. Cambridge University Press, second edition, 1997.

[CS86]   Gary Cornell and Joseph H. Silverman, editors. *Arithmetic Geometry*. Springer-Verlag, 1986.

[CS05]   D. Cox and J. Shurman. Geometry and number theory on clovers. *Amer. Math. Monthly*, 112, October 2005.

[CSS97]   Gary Cornell, Joseph H. Silverman, and Glenn Stevens, editors. *Modular Forms and Fermat's Last Theorem*. Springer-Verlag, 1997.

[DDT94]   H. Darmon, F. Diamond, and R. Taylor. Fermat's last theorem. *Current developments in mathematics, 1995*, pages 1–154, 1994.

© Springer Science+Business Media New York 2005
F. Diamond, J. Shurman, *A First Course in Modular Forms*,
Graduate Texts in Mathematics 228, DOI 10.1007/978-0-387-27226-9

[Del71]   P. Deligne. Forms modulaires et représentations ℓ-adiques. In *Lecture Notes in Math.*, volume 179, pages 139–172. Springer-Verlag, 1971.

[Die04]   L. Dieulefait. Existence of families of Galois representations and new cases of the Fontaine–Mazur conjecture. *J. Reine Angew. Math.*, 577:147–151, 2004.

[DR73]   P. Deligne and M. Rapoport. Les schémas de modules de courbes elliptiques. In *Modular Functions of One Variable, II*, volume 349 of *Lecture Notes in Math.*, pages 143–316. Springer-Verlag, 1973.

[DS74]   P. Deligne and J.-P. Serre. Forms modulaires de poids 1. *Ann. Sci. Ec. Norm. Sup.*, 7:507–530, 1974.

[FH91]   William Fulton and Joe Harris. *Representation Theory: A First Course*. Graduate Texts in Mathematics **129**. Springer-Verlag, 1991.

[FK80]   Hershel M. Farkas and Irwin Kra. *Riemann Surfaces*. Graduate Texts in Mathematics **71**. Springer-Verlag, 1980.

[FM93]   J.-M. Fontaine and B. Mazur. Geometric Galois representations. *Elliptic Curves, Modular Forms and Fermat's Last Theorem*, 17:41–78, 1993.

[Ful69]   William Fulton. *Algebraic Curves*. Benjamin, 1969.

[God66]   R. Godement. The decomposition of $L^2(G/\Gamma)$ for $\Gamma = \mathrm{SL}_2(\mathbb{Z})$. *Proc. Symp. Pure Math IX*, pages 211–224, 1966.

[Gun62]   R. P. Gunning. *Lectures on Modular Forms*. Annals of Mathematics studies. Princeton University Press, 1962.

[Hec26]   E. Hecke. Zur Theorie der elliptischen Modulfunktionen. *Math. Annalen*, 97:210–242, 1926.

[Hec27]   E. Hecke. Theorie der Eisensteinschen Reihen höherer Stufe und ihre Anwendung auf Funktionentheorie un Arithmetik. *Abh. Math. Sem. Hamburg*, 5:199–224, 1927.

[Hid86]   H. Hida. Galois representations into $\mathrm{GL}_2(\mathbf{Z}_p[[X]])$ attached to ordinary cusp forms. *Inv. Math.*, 85:545–613, 1986.

[Hid93]   Haruzo Hida. *Elementary theory of L-functions and Eisenstein series*. London Math. Soc. Student Texts **26**. Cambridge University Press, second edition, 1993.

[Hus04]   Dale Husemöller. *Elliptic Curves*. Graduate Texts in Mathematics **111**. Springer-Verlag, second edition, 2004.

[Igu59]   J. Igusa. Kroneckerian model of fields of elliptic modular functions. *Amer. J. Math.*, 81:561–577, 1959.

[IR92]   Kenneth Ireland and Michael Rosen. *A Classical Introduction to Modern Number Theory*. Graduate Texts in Mathematics **84**. Springer-Verlag, second edition, 1992.

[JS87]   Gareth A. Jones and David Singerman. *Complex Functions, an Algebraic and Geometric Viewpoint*. Cambridge University Press, 1987.

[Kha06]   C. Khare. Serre's modularity conjecture: the level one case. *Duke Math. J.*, 134(3):557–589, 2006.

[Kis09a]   M. Kisin. The Fontaine–Mazur conjecture for GL_2. *J. Amer. Math. Soc.*, 22(3):641–690, 2009.

[Kis09b]   M. Kisin. Modularity of 2-adic Barsotti-Tate representations. *Invent. Math.*, 178(3):587–634, 2009.

[Kis09c]   M. Kisin. Moduli of finite flat group schemes, and modularity. *Ann. of Math. (2)*, 170(3):1085–1180, 2009.

[KM85]   Nicholas M. Katz and Barry Mazur. *Arithmetic Moduli of Elliptic Curves*. Annals of Math. Studies **108**. Princeton University Press, 1985.

[Kna93]   Anthony W. Knapp. *Elliptic Curves*. Princeton Math. Notes **40**. Princeton University Press, 1993.

[Kob93]   Neal Koblitz. *Introduction to Elliptic Curves and Modular Forms*. Graduate Texts in Mathematics **97**. Springer-Verlag, second edition, 1993.

[KW09a]   C. Khare and J.-P. Wintenberger. Serre's modularity conjecture. I. *Invent. Math.*, 178(3):485–504, 2009.

[KW09b]   C. Khare and J.-P. Wintenberger. Serre's modularity conjecture. II. *Invent. Math.*, 178(3):505–586, 2009.

[Lan73]   Serge Lang. *Elliptic Functions*. Advanced book program. Addison-Wesley, 1973.

[Lan76]   Serge Lang. *Introduction to Modular Forms*. Grundl. Math. Wiss. **222**. Springer-Verlag, 1976.

[Lan80]   R. P. Langlands. *Base change for GL(2)*, volume 96 of *Annals of Math. Studies*. Princeton Univ. Press, 1980.

[Mar]   G. Martin. Dimensions of spaces of cusp forms and newforms on $\Gamma_0(N)$ and $\Gamma_1(N)$. http://arXiv.org/abs/math/0306128.

[Mar89]   Daniel Marcus. *Algebraic Number Theory*. Springer-Verlag, 1989.

[Maz91]   B. Mazur. Number theory as gadfly. *Amer. Math. Monthly*, 7:593–610, 1991.

[Mil]   J. S. Milne. Modular functions and modular forms. http://www.jmilne.org/math.

[Miy89]   Toshitsune Miyake. *Modular Forms*. Springer-Verlag, 1989.

[Mun00]   James R. Munkres. *Topology*. Prentice-Hall, second edition, 2000.

[Mur95]   V. Kumar Murty, editor. *Seminar on Fermat's Last Theorem*, volume 17 of *CMF Conf. Proc.* Amer. Math.Soc., 1995.

[Ogg69]   Andrew Ogg. *Modular Forms and Dirichlet Series*. Benjamin, 1969.

[Ram02]   R. Ramakrishna. Deforming Galois representations and the conjectures of Serre and Fontaine–Mazur. *Ann. of Math. (2)*, 156(1):115–154, 2002.

[Ran39]   R. Rankin. Contributions to the theory of Ramanujan's function $\tau(n)$ and similar arithmetical functions, i and ii. *Proc. Cambridge Phil. Soc.*, 35:351–356, 1939.

[Rib90]   K. Ribet. On modular representations of $\mathrm{Gal}(\overline{\mathbb{Q}}/\mathbb{Q})$ arising from modular forms. *Invent. Math.*, 100:431–476, 1990.

[Ros81]   M. Rosen. Abel's theorem on the lemniscate. *Amer. Math. Monthly*, 88:387–395, 1981.

[Rud74]   Walter Rudin. *Real and Complex Analysis*. McGraw-Hill, second edition, 1974.

[Sam72]   Pierre Samuel. *Algebraic Theory of Numbers*. Kershaw, 1972.

[Sch74]   Bruno Schoeneberg. *Elliptic Modular Functions*. Springer-Verlag, 1974.

[Sel40]   A. Selberg. Bemerkungen über eine Dirichletsche Reihe, die mit der Theorie der Modulformen nahe verbunden ist. *Arch. Math. Naturvid.*, 43:47–50, 1940.

[Ser73]   Jean-Pierre Serre. *A Course in Arithmetic*. Graduate Texts in Mathematics **7**. Springer-Verlag, 1973.

[Ser87]   J.-P. Serre. Sur les représentations modulaires de degré 2 de $\mathrm{Gal}(\overline{\mathbb{Q}}/\mathbb{Q})$. *Duke Math. J.*, 54:179–230, 1987.

[Shi71]   G. Shimura. On elliptic curves with complex multiplication as factors of the Jacobians of modular function fields. *Nagoya Math. J.*, 43:199–208, 1971.

[Shi73]   Goro Shimura. *Introduction to the Arithmetic Theory of Automorphic Functions*. Iwanami Shoten and Princeton University Press, 1973.

[Sil86]   Joseph H. Silverman. *The Arithmetic of Elliptic Curves*. Graduate Texts in Mathematics **106**. Springer-Verlag, 1986.

[Sil94]   Joseph H. Silverman. *Advanced Topics in the Arithmetic of Elliptic Curves*. Graduate Texts in Mathematics **151**. Springer-Verlag, 1994.

[ST92]   Joseph H. Silverman and John Tate. *Rational Points on Elliptic Curves*. Undergraduate Texts in Mathematics **7**. Springer-Verlag, 1992.

[Swi74]   H. P. F. Swinnerton-Dyer. *Analytic Theory of Abelian Varieties*. Cambridge, 1974.

[Tat02]   J. Tate. The millennium prize problems. A lecture by John Tate. Springer VideoMATH, 2002.

[Tay03]   R. Taylor. On icosahedral Artin representations, II. *Amer. J. Math.*, 125(3):549–566, 2003.

[Tun81]   J. Tunnell. Artin's conjecture for representations of the octahedral type. *Bull. Amer. Math. Soc.*, 5:173–175, 1981.

[TW95]   R. Taylor and A. Wiles. Ring-theoretic properties of certain Hecke algebras. *Ann. of Math.*, 141(3):553–572, 1995.

[Ver]   H. A. Verrill. Fundamental domains.
http://www.math.lsu.edu/~verrill/.

[Was03]   Lawrence C. Washington. *Elliptic Curves: Number Theory and Cryptography*. Discrete Mathematics and Its Applications. Chapman and Hall/CRC, 2003.

[Wei67]   A. Weil. Über die Bestimmung Dirichletscher Reihen durch Funktionalgleichungen. *Math. Annalen*, 168:149–156, 1967.

[Wil95]   A. Wiles. Modular elliptic curves and Fermat's last theorem. *Ann. of Math.*, 141(3):443–551, 1995.

(continued from page ii)

Printed in the United States
By Bookmasters